CAMBRIDGE STUDIES IN ADVANCED MATHEMATICS 118

An Introduction to Random Matrices

The theory of random matrices plays an important role in many areas of pure mathematics and employs a variety of sophisticated mathematical tools (analytical, probabilistic and combinatorial). This diverse array of tools, while attesting to the vitality of the field, presents several formidable obstacles to the newcomer, and even the expert probabilist.

This rigorous introduction to the basic theory is sufficiently self-contained to be accessible to graduate students in mathematics or related sciences, who have mastered probability theory at the graduate level, but have not necessarily been exposed to advanced notions of functional analysis, algebra or geometry. Useful background material is collected in the appendices and exercises are also included throughout to test the reader's understanding. Enumerative techniques, stochastic analysis, large deviations, concentration inequalities, disintegration and Lie algebras all are introduced in the text, which will enable readers to approach the research literature with confidence.

GREG W. ANDERSON is Professor of Mathematics at the University of Minnesota.

ALICE GUIONNET is CNRS Research Director at the Ecole Normale Supérieure in Lyon (ENS-Lyon).

OFER ZEITOUNI is Professor of Mathematics at both the University of Minnesota and the Weizmann Institute of Science in Rehovot, Israel.

CAMBRIDGE STUDIES IN ADVANCED MATHEMATICS

All the titles listed below can be obtained from good booksellers or from Cambridge University Press. For a complete series listing visit: www.cambridge.org/series/sSeries.asp?code=CSAM

Already published

65 A. J. Berrick & M. E. Keating *An introduction to rings and modules with K-theory in view*
66 S. Morosawa *et al. Holomorphic dynamics*
67 A. J. Berrick & M. E. Keating *Categories and modules with K-theory in view*
68 K. Sato *Lévy processes and infinitely divisible distributions*
69 H. Hida *Modular forms and Galois cohomology*
70 R. Iorio & V. Iorio *Fourier analysis and partial differential equations*
71 R. Blei *Analysis in integer and fractional dimensions*
72 F. Borceux & G. Janelidze *Galois theories*
73 B. Bollobás *Random graphs (2nd Edition)*
74 R. M. Dudley *Real analysis and probability (2nd Edition)*
75 T. Sheil-Small *Complex polynomials*
76 C. Voisin *Hodge theory and complex algebraic geometry, I*
77 C. Voisin *Hodge theory and complex algebraic geometry, II*
78 V. Paulsen *Completely bounded maps and operator algebras*
79 F. Gesztesy & H. Holden *Soliton equations and their algebro-geometric solutions, I*
81 S. Mukai *An introduction to invariants and moduli*
82 G. Tourlakis *Lectures in logic and set theory, I*
83 G. Tourlakis *Lectures in logic and set theory, II*
84 R. A. Bailey *Association schemes*
85 J. Carlson, S. Müller-Stach & C. Peters *Period mappings and period domains*
86 J. J. Duistermaat & J. A. C. Kolk *Multidimensional real analysis, I*
87 J. J. Duistermaat & J. A. C. Kolk *Multidimensional real analysis, II*
89 M. C. Golumbic & A. N. Trenk *Tolerance graphs*
90 L. H. Harper *Global methods for combinatorial isoperimetric problems*
91 I. Moerdijk & J. Mrčun *Introduction to foliations and Lie groupoids*
92 J. Kollár, K. E. Smith & A. Corti *Rational and nearly rational varieties*
93 D. Applebaum *Lévy processes and stochastic calculus (1st Edition)*
94 B. Conrad *Modular forms and the Ramanujan conjecture*
95 M. Schechter *An introduction to nonlinear analysis*
96 R. Carter *Lie algebras of finite and affine type*
97 H. L. Montgomery & R. C. Vaughan *Multiplicative number theory, I*
98 I. Chavel *Riemannian geometry (2nd Edition)*
99 D. Goldfeld *Automorphic forms and L-functions for the group GL(n,R)*
100 M. B. Marcus & J. Rosen *Markov processes, Gaussian processes, and local times*
101 P. Gille & T. Szamuely *Central simple algebras and Galois cohomology*
102 J. Bertoin *Random fragmentation and coagulation processes*
103 E. Frenkel *Langlands correspondence for loop groups*
104 A. Ambrosetti & A. Malchiodi *Nonlinear analysis and semilinear elliptic problems*
105 T. Tao & V. H. Vu *Additive combinatorics*
106 E. B. Davies *Linear operators and their spectra*
107 K. Kodaira *Complex analysis*
108 T. Ceccherini-Silberstein, F. Scarabotti & F. Tolli *Harmonic analysis on finite groups*
109 H. Geiges *An introduction to contact topology*
110 J. Faraut *Analysis on Lie groups: An Introduction*
111 E. Park *Complex topological K-theory*
112 D. W. Stroock *Partial differential equations for probabilists*
113 A. Kirillov, Jr *An introduction to Lie groups and Lie algebras*
114 F. Gesztesy *et al. Soliton equations and their algebro-geometric solutions, II*
115 E. de Faria & W. de Melo *Mathematical tools for one-dimensional dynamics*
116 D. Applebaum *Lévy processes and stochastic calculus (2nd Edition)*
117 T. Szamuely *Galois groups and fundamental groups*

An Introduction to Random Matrices

GREG W. ANDERSON
University of Minnesota

ALICE GUIONNET
Ecole Normale Supérieure de Lyon

OFER ZEITOUNI
University of Minnesota and
Weizmann Institute of Science

CAMBRIDGE
UNIVERSITY PRESS

CAMBRIDGE UNIVERSITY PRESS
Cambridge, New York, Melbourne, Madrid, Cape Town, Singapore, São Paulo, Delhi, Dubai, Tokyo

Cambridge University Press
The Edinburgh Building, Cambridge CB2 8RU, UK

Published in the United States of America by Cambridge University Press, New York

www.cambridge.org
Information on this title: www.cambridge.org/9780521194525

First published 2010

Digitally reprinted ICM edition 2010
Digitally printed in India at Images & Impressions, New Delhi-110002

A catalogue record for this publication is available from the British Library

ISBN 978-0-521-16886-1 paperback

To Meredith, Benoit and Naomi

Contents

Preface

The study of random matrices, and in particular the properties of their eigenvalues, has emerged from the applications, first in data analysis and later as statistical models for heavy-nuclei atoms. Thus, the field of random matrices owes its existence to applications. Over the years, however, it became clear that models related to random matrices play an important role in areas of pure mathematics. Moreover, the tools used in the study of random matrices came themselves from different and seemingly unrelated branches of mathematics.

At this point in time, the topic has evolved enough that the newcomer, especially if coming from the field of probability theory, faces a formidable and somewhat confusing task in trying to access the research literature. Furthermore, the background expected of such a newcomer is diverse, and often has to be supplemented before a serious study of random matrices can begin.

We believe that many parts of the field of random matrices are now developed enough to enable one to expose the basic ideas in a systematic and coherent way. Indeed, such a treatise, geared toward theoretical physicists, has existed for some time, in the form of Mehta's superb book [Meh91]. Our goal in writing this book has been to present a rigorous introduction to the basic theory of random matrices, including free probability, that is sufficiently self-contained to be accessible to graduate students in mathematics or related sciences who have mastered probability theory at the graduate level, but have not necessarily been exposed to advanced notions of functional analysis, algebra or geometry. Along the way, enough techniques are introduced that we hope will allow readers to continue their journey into the current research literature.

This project started as notes for a class on random matrices that two of us (G. A. and O. Z.) taught in the University of Minnesota in the fall of 2003, and notes for a course in the probability summer school in St. Flour taught by A. G. in the summer of 2006. The comments of participants in these courses, and in particular A. Bandyopadhyay, H. Dong, K. Hoffman-Credner, A. Klenke, D. Stanton and P.M. Zamfir, were extremely useful. As these notes evolved, we taught from them again at the University of Minnesota, the University of California at Berkeley, the Technion and the Weizmann Institute, and received more much appreciated feedback from the participants in those courses. Finally, when expanding and refining these course notes, we have profited from the comments and questions of many colleagues. We would like in particular to thank G. Ben Arous, F. Benaych-Georges, P. Biane, P. Deift, A. Dembo, P. Diaconis, U. Haagerup, V. Jones, M. Krishnapur, Y. Peres, R. Pinsky, G. Pisier, B. Rider, D. Shlyakhtenko, B. Solel, A. Soshnikov, R. Speicher, T. Suidan, C. Tracy, B. Virag and D. Voiculescu for their suggestions, corrections and patience in answering our questions or explaining their work to us. Of course, any remaining mistakes and unclear passages are fully our responsibility.

GREG ANDERSON — MINNEAPOLIS, MINNESOTA
ALICE GUIONNET — LYON, FRANCE
OFER ZEITOUNI — REHOVOT, ISRAEL

1
Introduction

This book is concerned with random matrices. Given the ubiquitous role that matrices play in mathematics and its application in the sciences and engineering, it seems natural that the evolution of probability theory would eventually pass through random matrices. The reality, however, has been more complicated (and interesting). Indeed, the study of random matrices, and in particular the properties of their eigenvalues, has emerged from the applications, first in data analysis (in the early days of statistical sciences, going back to Wishart [Wis28]), and later as statistical models for heavy-nuclei atoms, beginning with the seminal work of Wigner [Wig55]. Still motivated by physical applications, at the able hands of Wigner, Dyson, Mehta and co-workers, a mathematical theory of the spectrum of random matrices began to emerge in the early 1960s, and links with various branches of mathematics, including classical analysis and number theory, were established. While much progress was initially achieved using enumerative combinatorics, gradually, sophisticated and varied mathematical tools were introduced: Fredholm determinants (in the 1960s), diffusion processes (in the 1960s), integrable systems (in the 1980s and early 1990s), and the Riemann–Hilbert problem (in the 1990s) all made their appearance, as well as new tools such as the theory of free probability (in the 1990s). This wide array of tools, while attesting to the vitality of the field, presents, however, several formidable obstacles to the newcomer, and even to the expert probabilist. Indeed, while much of the recent research uses sophisticated probabilistic tools, it builds on layers of common knowledge that, in the aggregate, few people possess.

Our goal in this book is to present a rigorous introduction to the basic theory of random matrices that would be sufficiently self-contained to be accessible to graduate students in mathematics or related sciences who have mastered probability theory at the graduate level, but have not necessarily been exposed to advanced notions of functional analysis, algebra or geometry. With such readers in mind, we present some background material in the appendices, that novice and expert alike can consult; most material in the appendices is stated without proof, although the details of some specialized computations are provided.

Keeping in mind our stated emphasis on accessibility over generality, the book is essentially divided into two parts. In Chapters 2 and 3, we present a self-contained analysis of random matrices, quickly focusing on the Gaussian ensembles and culminating in the derivation of the gap probabilities at 0 and the Tracy–Widom law. These chapters can be read with very little background knowledge, and are particularly suitable for an introductory study. In the second part of the book, Chapters 4 and 5, we use more advanced techniques, requiring more extensive background, to emphasize and generalize certain aspects of the theory, and to introduce the theory of *free probability*.

So what is a random matrix, and what questions are we about to study? Throughout, let $\mathbb{F}=\mathbb{R}$ or $\mathbb{F}=\mathbb{C}$, and set $\beta=1$ in the former case and $\beta=2$ in the latter. (In Section 4.1, we will also consider the case $\mathbb{F}=\mathbb{H}$, the skew-field of quaternions, see Appendix E for definitions and details.) Let $\mathrm{Mat}_N(\mathbb{F})$ denote the space of N-by-N matrices with entries in $\mathbb{F}$, and let $\mathscr{H}_N^{(\beta)}$ denote the subset of self-adjoint matrices (i.e., real symmetric if $\beta=1$ and Hermitian if $\beta=2$). One can always consider the sets $\mathrm{Mat}_N(\mathbb{F})$ and $\mathscr{H}_N^{(\beta)}$, $\beta=1,2$, as submanifolds of an appropriate Euclidean space, and equip it with the induced topology and (Borel) sigma-field.

Recall that a probability space is a triple $(\Omega,\mathscr{F},P)$ so that $\mathscr{F}$ is a sigma-algebra of subsets of Ω and P is a probability measure on $(\Omega,\mathscr{F})$. In that setting, a *random matrix* X_N is a measurable map from $(\Omega,\mathscr{F})$ to $\mathrm{Mat}_N(\mathbb{F})$.

Our main interest is in the *eigenvalues* of random matrices. Recall that the eigenvalues of a matrix $H\in\mathrm{Mat}_N(\mathbb{F})$ are the roots of the characteristic polynomial $P_N(z)=\det(zI_N-H)$, with I_N the identity matrix. Therefore, on the (open) set where the eigenvalues are all simple, they are smooth functions of the entries of X_N (a more complete discussion can be found in Section 4.1).

We will be mostly concerned in this book with self-adjoint matrices $H\in\mathscr{H}_N^{(\beta)}$, $\beta=1,2$, in which case the eigenvalues are all real and can be ordered. Thus, for $H\in\mathscr{H}_N^{(\beta)}$, we let $\lambda_1(H)\leq\cdots\leq\lambda_N(H)$ be the eigenvalues of H. A consequence of the perturbation theory of normal matrices (see Lemma A.4) is that the eigenvalues $\{\lambda_i(H)\}$ are continuous functions of H (this also follows from the Hoffman–Wielandt theorem, Theorem 2.1.19). In particular, if X_N is a random matrix then the eigenvalues $\{\lambda_i(X_N)\}$ are random variables.

We present now a guided tour of the book. We begin by considering *Wigner matrices* in Chapter 2. These are symmetric (or Hermitian) matrices X_N whose entries are independent and identically distributed, except for the symmetry constraints. For $x \in \mathbb{R}$, let δ_x denote the *Dirac* measure at x, that is, the unique probability measure satisfying $\int f d\delta_x = f(x)$ for all continuous functions on $\mathbb{R}$. Let $L_N = N^{-1}\sum_{i=1}^{N} \delta_{\lambda_i(X_N)}$ denote the *empirical measure* of the eigenvalues of X_N. Wigner's Theorem (Theorem 2.1.1) asserts that, under appropriate assumptions on the law of the entries, L_N converges (with respect to the weak convergence of measures) towards a deterministic probability measure, the *semicircle law*. We present in Chapter 2 several proofs of Wigner's Theorem. The first, in Section 2.1, involves a combinatorial machinery that is also exploited to yield central limit theorems and estimates on the spectral radius of X_N. After first introducing in Section 2.3 some useful estimates on the deviation between the empirical measure and its mean, we define in Section 2.4 the *Stieltjes transform* of measures and use it to give another quick proof of Wigner's Theorem.

Having discussed techniques valid for entries distributed according to general laws, we turn attention to special situations involving additional symmetry. The simplest of these concerns the *Gaussian ensembles*, the GOE and GUE, so named because their law is invariant under conjugation by orthogonal (resp., unitary) matrices. The latter extra symmetry is crucial in deriving in Section 2.5 an explicit joint distribution for the eigenvalues (thus effectively reducing consideration from a problem involving order of N^2 random variables, namely the matrix entries, to one involving only N variables). (The GSE, or Gaussian symplectic ensemble, also shares this property and is discussed briefly.) A large deviations principle for the empirical distribution, which leads to yet another proof of Wigner's Theorem, follows in Section 2.6.

The expression for the joint density of the eigenvalues in the Gaussian ensembles is the starting point for obtaining *local* information on the eigenvalues. This is the topic of Chapter 3. The bulk of the chapter deals with the GUE, because in that situation the eigenvalues form a *determinantal process*. This allows one to effectively represent the probability that no eigenvalues are present in a set as a *Fredholm determinant*, a notion that is particularly amenable to asymptotic analysis. Thus, after representing in Section 3.2 the joint density for the GUE in terms of a determinant involving appropriate orthogonal polynomials, the *Hermite* polynomials, we develop in Section 3.4 in an elementary way some aspects of the theory of Fredholm determinants. We then present in Section 3.5 the asymptotic analysis required in order to study the *gap probability at 0*, that is the probability that no eigenvalue is present in an interval around the origin. Relevant tools, such as the Laplace method, are developed along the way. Section 3.7 repeats this analysis for the edge of the spectrum, introducing along the way the method of steepest descent. The link with integrable systems and the *Painlevé equations* is established in Sections 3.6 and 3.8.

As mentioned before, the eigenvalues of the GUE are an example of a determinantal process. The other Gaussian ensembles (GOE and GSE) do not fall into this class, but they do enjoy a structure where certain Pfaffians replace determinants. This leads to a considerably more involved analysis, the details of which are provided in Section 3.9.

Chapter 4 is a hodge-podge of results whose common feature is that they all require new tools. We begin in Section 4.1 with a re-derivation of the joint law of the eigenvalues of the Gaussian ensemble, in a geometric framework based on Lie theory. We use this framework to derive the expressions for the joint distribution of eigenvalues of Wishart matrices, of random matrices from the various unitary groups and of matrices related to random projectors. Section 4.2 studies in some depth determinantal processes, including their construction, associated central limit theorems, convergence and ergodic properties. Section 4.3 studies what happens when in the GUE (or GOE), the Gaussian entries are replaced by Brownian motions. The powerful tools of stochastic analysis can then be brought to bear and lead to functional laws of large numbers, central limit theorems and large deviations. Section 4.4 consists of an in-depth treatment of concentration techniques and their application to random matrices; it is a generalization of the discussion in the short Section 2.3. Finally, in Section 4.5, we study a family of tri-diagonal matrices, parametrized by a parameter β, whose distribution of eigenvalues coincides with that of members of the Gaussian ensembles for $\beta = 1,2,4$. The study of the maximal eigenvalue for this family is linked to the spectrum of an appropriate random Schrödinger operator.

Chapter 5 is devoted to *free probability theory*, a probability theory for certain noncommutative variables, equipped with a notion of independence called free independence. Invented in the early 1990s, free probability theory has become a versatile tool for analyzing the laws of noncommutative polynomials in several random matrices, and of the limits of the empirical measure of eigenvalues of such polynomials. We develop the necessary preliminaries and definitions in Section 5.2, introduce free independence in Section 5.3, and discuss the link with random matrices in Section 5.4. We conclude the chapter with Section 5.5, in which we study the convergence of the spectral radius of noncommutative polynomials of random matrices.

Each chapter ends with bibliographical notes. These are not meant to be comprehensive, but rather guide the reader through the enormous literature and give some hint of recent developments. Although we have tried to represent accurately the historical development of the subject, we have necessarily omitted important references, misrepresented facts, or plainly erred. Our apologies to those authors whose work we have thus unintentionally slighted.

Of course, we have barely scratched the surface of the subject of random matrices. We mention now the most glaring omissions, together with references to some recent books that cover these topics. We have not discussed the theory of the Riemann–Hilbert problem and its relation to integrable systems, Painlevé equations, asymptotics of orthogonal polynomials and random matrices. The interested reader is referred to the books [FoIKN06], [Dei99] and [DeG09] for an in-depth treatment. We do not discuss the relation between asymptotics of random matrices and combinatorial problems – a good summary of these appears in [BaDS09]. We barely discuss applications of random matrices, and in particular do not review the recent increase in applications to statistics or communication theory – for a nice introduction to the latter we refer to [TuV04]. We have presented only a partial discussion of ensembles of matrices that possess explicit joint distribution of eigenvalues. For a more complete discussion, including also the case of non-Hermitian matrices that are not unitary, we refer the reader to [For05]. Finally, we have not discussed the link between random matrices and number theory; the interested reader should consult [KaS99] for a taste of that link. We further refer to the bibliographical notes for additional reading, less glaring omissions and references.

2

Real and complex Wigner matrices

2.1 Real Wigner matrices: traces, moments and combinatorics

We introduce in this section a basic model of random matrices. Nowhere do we attempt to provide the weakest assumptions or sharpest results available. We point out in the bibliographical notes (Section 2.7) some places where the interested reader can find finer results.

Start with two independent families of independent and identically distributed (i.i.d.) zero mean, real-valued random variables $\{Z_{i,j}\}_{1\le i<j}$ and $\{Y_i\}_{1\le i}$, such that $EZ_{1,2}^2 = 1$ and, for all integers $k \ge 1$,

$$r_k := \max\left(E|Z_{1,2}|^k, E|Y_1|^k\right) < \infty. \tag{2.1.1}$$

Consider the (symmetric) $N \times N$ matrix X_N with entries

$$X_N(j,i) = X_N(i,j) = \begin{cases} Z_{i,j}/\sqrt{N}, & \text{if } i < j, \\ Y_i/\sqrt{N}, & \text{if } i = j. \end{cases} \tag{2.1.2}$$

We call such a matrix a *Wigner matrix*, and if the random variables $Z_{i,j}$ and Y_i are Gaussian, we use the term *Gaussian Wigner matrix*. The case of Gaussian Wigner matrices in which $EY_1^2 = 2$ is of particular importance, and for reasons that will become clearer in Chapter 3, such matrices (rescaled by $\sqrt{N}$) are referred to as Gaussian orthogonal ensemble (GOE) matrices.

Let λ_i^N denote the (real) eigenvalues of X_N, with $\lambda_1^N \le \lambda_2^N \le \cdots \le \lambda_N^N$, and define the *empirical distribution* of the eigenvalues as the (random) probability measure on $\mathbb{R}$ defined by

$$L_N = \frac{1}{N}\sum_{i=1}^{N} \delta_{\lambda_i^N}.$$

Define the *semicircle distribution* (or *law*) as the probability distribution $\sigma(x)dx$

on $\mathbb{R}$ with density

$$\sigma(x) = \frac{1}{2\pi}\sqrt{4-x^2}\mathbf{1}_{|x|\le 2}\,. \tag{2.1.3}$$

The following theorem, contained in [Wig55], can be considered the starting point of random matrix theory (RMT).

Theorem 2.1.1 (Wigner) *For a Wigner matrix, the empirical measure L_N converges weakly, in probability, to the semicircle distribution.*

In greater detail, Theorem 2.1.1 asserts that for any $f \in C_b(\mathbb{R})$, and any $\varepsilon > 0$,

$$\lim_{N\to\infty} P(|\langle L_N, f\rangle - \langle \sigma, f\rangle| > \varepsilon) = 0\,.$$

Remark 2.1.2 The assumption (2.1.1) that $r_k < \infty$ for all k is not really needed. See Theorem 2.1.21 in Section 2.1.5.

We will see many proofs of Wigner's Theorem 2.1.1. In this section, we give a direct combinatorics-based proof, mimicking the original argument of Wigner. Before doing so, however, we need to discuss some properties of the semicircle distribution.

2.1.1 The semicircle distribution, Catalan numbers and Dyck paths

Define the moments $m_k := \langle \sigma, x^k\rangle$. Recall the Catalan numbers

$$C_k = \frac{\binom{2k}{k}}{k+1} = \frac{(2k)!}{(k+1)!k!}\,.$$

We now check that, for all integers $k \geq 1$,

$$m_{2k} = C_k\,, \quad m_{2k+1} = 0\,. \tag{2.1.4}$$

Indeed, $m_{2k+1} = 0$ by symmetry, while

$$\begin{aligned} m_{2k} &= \int_{-2}^{2} x^{2k}\sigma(x)dx = \frac{2\cdot 2^{2k}}{\pi}\int_{-\pi/2}^{\pi/2}\sin^{2k}(\theta)\cos^2(\theta)d\theta \\ &= \frac{2\cdot 2^{2k}}{\pi}\int_{-\pi/2}^{\pi/2}\sin^{2k}(\theta)d\theta - (2k+1)m_{2k}\,. \end{aligned}$$

Hence,

$$m_{2k} = \frac{2\cdot 2^{2k}}{\pi(2k+2)}\int_{-\pi/2}^{\pi/2}\sin^{2k}(\theta)d\theta = \frac{4(2k-1)}{2k+2}m_{2k-2}\,, \tag{2.1.5}$$

from which, together with $m_0 = 1$, one concludes (2.1.4).

The Catalan numbers possess many combinatorial interpretations. To introduce a first one, say that an integer-valued sequence $\{S_n\}_{0\le n\le \ell}$ is a *Bernoulli walk* of length ℓ if $S_0 = 0$ and $|S_{t+1} - S_t| = 1$ for $t \le \ell - 1$. Of particular relevance here is the fact that C_k counts the number of *Dyck paths* of length $2k$, that is, the number of nonnegative Bernoulli walks of length $2k$ that terminate at 0. Indeed, let β_k denote the number of such paths. A classical exercise in combinatorics is

Lemma 2.1.3 $\beta_k = C_k < 4^k$. *Further, the generating function* $\hat\beta(z) := 1 + \sum_{k=1}^\infty z^k \beta_k$ *satisfies, for* $|z| < 1/4$,

$$\hat\beta(z) = \frac{1 - \sqrt{1-4z}}{2z}. \tag{2.1.6}$$

Proof of Lemma 2.1.3 Let B_k denote the number of Bernoulli walks $\{S_n\}$ of length $2k$ that satisfy $S_{2k} = 0$, and let $\bar B_k$ denote the number of Bernoulli walks $\{S_n\}$ of length $2k$ that satisfy $S_{2k} = 0$ and $S_t < 0$ for some $t < 2k$. Then, $\beta_k = B_k - \bar B_k$. By reflection at the first hitting of -1, one sees that $\bar B_k$ equals the number of Bernoulli walks $\{S_n\}$ of length $2k$ that satisfy $S_{2k} = -2$. Hence,

$$\beta_k = B_k - \bar B_k = \binom{2k}{k} - \binom{2k}{k-1} = C_k\,.$$

Turning to the evaluation of $\hat\beta(z)$, considering the first return time to 0 of the Bernoulli walk $\{S_n\}$ gives the relation

$$\beta_k = \sum_{j=1}^{k} \beta_{k-j}\beta_{j-1}\,,\ k \ge 1\,, \tag{2.1.7}$$

with the convention that $\beta_0 = 1$. Because the number of Bernoulli walks of length $2k$ is bounded by 4^k, one has that $\beta_k \le 4^k$, and hence the function $\hat\beta(z)$ is well defined and analytic for $|z| < 1/4$. But, substituting (2.1.7),

$$\hat\beta(z) - 1 = \sum_{k=1}^{\infty} z^k \sum_{j=1}^{k} \beta_{k-j}\beta_{j-1} = z\sum_{k=0}^{\infty} z^k \sum_{j=0}^{k} \beta_{k-j}\beta_j\,,$$

while

$$\hat\beta(z)^2 = \sum_{k,k'=0}^{\infty} z^{k+k'}\beta_k\beta_{k'} = \sum_{q=0}^{\infty}\sum_{\ell=0}^{q} z^q \beta_{q-\ell}\beta_\ell\,.$$

Combining the last two equations, one sees that

$$\hat\beta(z) = 1 + z\hat\beta(z)^2\,,$$

from which (2.1.6) follows (using that $\hat{\beta}(0) = 1$ to choose the correct branch of the square-root). □

We note in passing that, expanding (2.1.6) in power series in z in a neighborhood of zero, one gets (for $|z| < 1/4$)

$$\hat{\beta}(z) = \frac{2\sum_{k=1}^{\infty} \frac{z^k(2k-2)!}{k!(k-1)!}}{2z} = \sum_{k=0}^{\infty} \frac{(2k)!}{k!(k+1)!} z^k = \sum_{k=0}^{\infty} z^k C_k \,,$$

which provides an alternative proof of the fact that $\beta_k = C_k$.

Another useful interpretation of the Catalan numbers is that C_k counts the number of rooted planar trees with k edges. (A *rooted planar tree* is a planar graph with no cycles, with one distinguished vertex, and with a choice of ordering at each vertex; the ordering defines a way to "explore" the tree, starting at the root.) It is not hard to check that the Dyck paths of length $2k$ are in bijection with such rooted planar trees. See the proof of Lemma 2.1.6 in Section 2.1.3 for a formal construction of this bijection.

We note in closing that a third interpretation of the Catalan numbers, particularly useful in the context of Chapter 5, is that they count the *non-crossing partitions* of the ordered set $\mathscr{K}_k := \{1, 2, \ldots, k\}$.

Definition 2.1.4 A partition of the set $\mathscr{K}_k := \{1, 2, \ldots, k\}$ is called *crossing* if there exists a quadruple (a, b, c, d) with $1 \le a < b < c < d \le k$ such that a, c belong to one part while b, d belong to another part. A partition which is not crossing is a *non-crossing partition.*

Non-crossing partitions form a lattice with respect to refinement. A look at Figure 2.1.1 should explain the terminology "non-crossing": one puts the points $1, \ldots, k$ on the circle, and connects each point with the next member of its part (in cyclic order) by an internal path. Then, the partition is non-crossing if this can be achieved without arcs crossing each other.

It is not hard to check that C_k is indeed the number γ_k of non-crossing partitions of $\mathscr{K}_k$. To see that, let π be a non-crossing partition of $\mathscr{K}_k$ and let j denote the largest element connected to 1 (with $j = 1$ if the part containing 1 is the set $\{1\}$). Then, because π is non-crossing, it induces non-crossing partitions on the sets $\{1, \ldots, j-1\}$ and $\{j+1, \ldots, k\}$. Therefore, $\gamma_k = \sum_{j=1}^{k} \gamma_{k-j}\gamma_{j-1}$. With $\gamma_1 = 1$, and comparing with (2.1.7), one sees that $\beta_k = \gamma_k$.

Exercise 2.1.5 Prove that for $z \in \mathbb{C}$ such that $z \notin [-2, 2]$, the Stieltjes transform

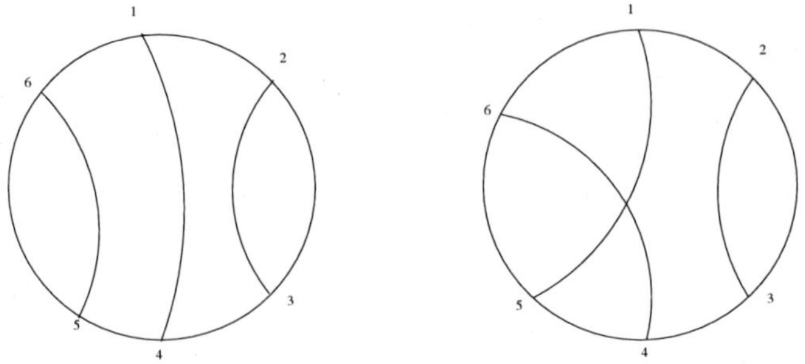

Fig. 2.1.1. Non-crossing (left, $(1,4),(2,3),(5,6)$) and crossing (right, $(1,5),(2,3),(4,6)$) partitions of the set $\mathscr{K}_6$.

$S(z)$ of the semicircle law (see Definition 2.4.1) equals

$$S(z) = \int \frac{1}{\lambda - z}\sigma(d\lambda) = \frac{-z+\sqrt{z^2-4}}{2z}.$$

Hint: Either use the residue theorem, or relate $S(z)$ to the generating function $\hat{\beta}(z)$, see Remark 2.4.2.

2.1.2 Proof #1 of Wigner's Theorem 2.1.1

Define the probability distribution $\bar{L}_N = EL_N$ by the relation $\langle \bar{L}_N, f\rangle = E\langle L_N, f\rangle$ for all $f \in C_b$, and set $m_k^N := \langle \bar{L}_N, x^k\rangle$. Theorem 2.1.1 follows from the following two lemmas.

Lemma 2.1.6 *For every* $k \in \mathbb{N}$,

$$\lim_{N\to\infty} m_k^N = m_k .$$

(See (2.1.4) for the definition of m_k.)

Lemma 2.1.7 *For every* $k \in \mathbb{N}$ *and* $\varepsilon > 0$,

$$\lim_{N\to\infty} P\left(\left|\langle L_N, x^k\rangle - \langle \bar{L}_N, x^k\rangle\right| > \varepsilon\right) = 0 .$$

Indeed, assume that Lemmas 2.1.6 and 2.1.7 have been proved. To conclude the proof of Theorem 2.1.1, one needs to check that for any bounded continuous function f,

$$\lim_{N\to\infty} \langle L_N, f\rangle = \langle \sigma, f\rangle, \quad \text{in probability.} \tag{2.1.8}$$

Toward this end, note first that an application of the Chebyshev inequality yields

$$P\left(\langle L_N, |x|^k \mathbf{1}_{|x|>B}\rangle > \varepsilon\right) \le \frac{1}{\varepsilon} E\langle L_N, |x|^k \mathbf{1}_{|x|>B}\rangle \le \frac{\langle \bar{L}_N, x^{2k}\rangle}{\varepsilon B^k}.$$

Hence, by Lemma 2.1.6,

$$\limsup_{N\to\infty} P\left(\langle L_N, |x|^k \mathbf{1}_{|x|>B}\rangle > \varepsilon\right) \le \frac{\langle \sigma, x^{2k}\rangle}{\varepsilon B^k} \le \frac{4^k}{\varepsilon B^k},$$

where we used that $C_k \le 4^k$. Thus, with $B=5$, noting that the left side above is increasing in k, it follows that

$$\limsup_{N\to\infty} P\left(\langle L_N, |x|^k \mathbf{1}_{|x|>B}\rangle > \varepsilon\right) = 0. \tag{2.1.9}$$

In particular, when proving (2.1.8), we may and will assume that f is supported on the interval $[-5,5]$.

Fix next such an f and $\delta > 0$. By the Weierstrass approximation theorem, one can find a polynomial $Q_\delta(x) = \sum_{i=0}^{L} c_i x^i$ such that

$$\sup_{x:|x|\le B} |Q_\delta(x) - f(x)| \le \frac{\delta}{8}.$$

Then,

$$\begin{aligned}
P(|\langle L_N, f\rangle - \langle \sigma, f\rangle| > \delta) &\le P\left(|\langle L_N, Q_\delta\rangle - \langle \bar{L}_N, Q_\delta\rangle| > \frac{\delta}{4}\right) \\
&+P\left(|\langle \bar{L}_N, Q_\delta\rangle - \langle \sigma, Q_\delta\rangle| > \frac{\delta}{4}\right) + P\left(|\langle L_N, Q_\delta 1_{|x|>B}\rangle| > \frac{\delta}{4}\right) \\
&=: P_1 + P_2 + P_3.
\end{aligned}$$

By an application of Lemma 2.1.7, $P_1 \to_{N\to\infty} 0$. Lemma 2.1.6 implies that $P_2 = 0$ for N large, while (2.1.9) implies that $P_3 \to_{N\to\infty} 0$. This completes the proof of Theorem 2.1.1 (modulo Lemmas 2.1.6 and 2.1.7). □

2.1.3 Proof of Lemma 2.1.6: words and graphs

The starting point of the proof of Lemma 2.1.6 is the following identity:

$$\begin{aligned}
\langle \bar{L}_N, x^k\rangle &= \frac{1}{N} E \operatorname{tr} X_N^k \\
&= \frac{1}{N} \sum_{i_1,\ldots,i_k=1}^{N} E X_N(i_1,i_2) X_N(i_2,i_3) \cdots X_N(i_{k-1},i_k) X_N(i_k,i_1) \\
&=: \frac{1}{N} \sum_{i_1,\ldots,i_k=1}^{N} E T_{\mathbf{i}}^N =: \frac{1}{N} \sum_{i_1,\ldots,i_k=1}^{N} \bar{T}_{\mathbf{i}}^N,
\end{aligned} \tag{2.1.10}$$

where we use the notation $\mathbf{i} = (i_1, \ldots, i_k)$.

The proof of Lemma 2.1.6 now proceeds by considering which terms contribute to (2.1.10). Let us provide first an informal sketch that explains the emergence of the Catalan numbers, followed by a formal proof. For the purpose of this sketch, assume that the variables Y_i vanish, and that the law of $Z_{1,2}$ is symmetric, so that all odd moments vanish (and in particular, $\langle \bar{L}_N, x^k \rangle = 0$ for k odd).

A first step in the sketch (that is fully justified in the actual proof below) is to check that the only terms in (2.1.10) that survive the passage to the limit involve only second moments of $Z_{i,j}$, because there are order $N^{k/2+1}$ nonzero terms but only at most order $N^{k/2}$ terms that involve moments higher than or equal to 4. One then sees that

$$\langle \bar{L}_N, x^{2k} \rangle = (1+O(N^{-1}))\frac{1}{N} \sum_{\substack{\forall p, \exists! j \neq p: \\ (i_p, i_{p+1}) = (i_j, i_{j+1}) \text{ or } (i_{j+1}, i_j)}} \bar{T}^N_{i_1,\ldots,i_{2k}}, \tag{2.1.11}$$

where the notation $\exists!$ means "there exists a unique". Considering the index $j > 1$ such that either $(i_j, i_{j+1}) = (i_2, i_1)$ or $(i_j, i_{j+1}) = (i_1, i_2)$, and recalling that $i_2 \neq i_1$ since $Y_{i_1} = 0$, one obtains

$$\langle \bar{L}_N, x^{2k} \rangle = (1+O(N^{-1}))\frac{1}{N} \sum_{j=2}^{2k} \sum_{i_1 \neq i_2 = 1}^{N} \sum_{\substack{i_3,\ldots,i_{j-1}, \\ i_{j+2},\ldots,i_{2k}=1}}^{N} \tag{2.1.12}$$
$$\Big(EX_N(i_2,i_3)\cdots X_N(i_{j-1},i_2)X_N(i_1,i_{j+2})\cdots X_N(i_{2k},i_1)$$
$$+ EX_N(i_2,i_3)\cdots X_N(i_{j-1},i_1)X_N(i_2,i_{j+2})\cdots X_N(i_{2k},i_1) \Big).$$

Hence, *if we could prove* that $E[\langle L_N - \bar{L}_N, x^k \rangle]^2 = O(N^{-2})$ and hence

$$E[\langle L_N, x^j \rangle \langle L_N, x^{2k-j-2} \rangle] = \langle \bar{L}_N, x^j \rangle \langle \bar{L}_N, x^{2k-j-2} \rangle (1+O(N^{-1})),$$

we would obtain

$$\begin{aligned}
\langle \bar{L}_N, x^{2k} \rangle &= (1+O(N^{-1})) \sum_{j=0}^{2(k-1)} \Big(\langle \bar{L}_N, x^j \rangle \langle \bar{L}_N, x^{2k-j-2} \rangle \\
&\qquad + \frac{1}{N} \langle \bar{L}_N, x^{2k-2} \rangle \Big) \\
&= (1+O(N^{-1})) \sum_{j=0}^{2k-2} \langle \bar{L}_N, x^j \rangle \langle \bar{L}_N, x^{2k-j-2} \rangle \\
&= (1+O(N^{-1})) \sum_{j=0}^{k-1} \langle \bar{L}_N, x^{2j} \rangle \langle \bar{L}_N, x^{2(k-j-1)} \rangle,
\end{aligned} \tag{2.1.13}$$

where we have used the fact that by induction $\langle \bar{L}_N, x^{2k-2}\rangle$ is uniformly bounded and also the fact that odd moments vanish. Further,

$$\langle \bar{L}_N, x^2\rangle = \frac{1}{N}\sum_{i,j=1}^{N} EX_N(i,j)^2 \to_{N\to\infty} 1 = C_1 \,. \tag{2.1.14}$$

Thus, we conclude from (2.1.13) by induction that $\langle \bar{L}_N, x^{2k}\rangle$ converges to a limit a_k with $a_0 = a_1 = 1$, and further that the family $\{a_k\}$ satisfies the recursions $a_k = \sum_{j=1}^{k} a_{k-j}a_{j-1}$. Comparing with (2.1.7), we deduce that $a_k = C_k$, as claimed.

We turn next to the actual proof. To handle the summation in expressions like (2.1.10), it is convenient to introduce some combinatorial machinery that will serve us also in the sequel. We thus first digress and discuss the combinatorics intervening in the evaluation of the sum in (2.1.10). This is then followed by the actual proof of Lemma 2.1.6.

In the following definition, the reader may think of $\mathscr{S}$ as a subset of the integers.

Definition 2.1.8 ($\mathscr{S}$-words) Given a set $\mathscr{S}$, an *$\mathscr{S}$-letter s* is simply an element of $\mathscr{S}$. An *$\mathscr{S}$-word w* is a finite sequence of letters $s_1 \cdots s_n$, at least one letter long. An $\mathscr{S}$-word w is *closed* if its first and last letters are the same. Two $\mathscr{S}$-words w_1, w_2 are called *equivalent*, denoted $w_1 \sim w_2$, if there is a bijection on $\mathscr{S}$ that maps one into the other.

When $\mathscr{S} = \{1,\dots,N\}$ for some finite N, we use the term N-word. Otherwise, if the set $\mathscr{S}$ is clear from the context, we refer to an $\mathscr{S}$-word simply as a word.

For any $\mathscr{S}$-word $w = s_1 \cdots s_k$, we use $\ell(w) = k$ to denote the *length* of w, define the *weight* $\mathrm{wt}(w)$ as the number of distinct elements of the set $\{s_1,\dots,s_k\}$ and the *support* of w, denoted $\mathrm{supp}\, w$, as the set of letters appearing in w. With any word w we may associate an undirected graph, with $\mathrm{wt}(w)$ vertices and $\ell(w) - 1$ edges, as follows.

Definition 2.1.9 (Graph associated with an $\mathscr{S}$-word) Given a word $w = s_1 \cdots s_k$, we let $G_w = (V_w, E_w)$ be the graph with set of vertices $V_w = \mathrm{supp}\, w$ and (undirected) edges $E_w = \{\{s_i, s_{i+1}\}, i = 1,\dots,k-1\}$. We define the set of *self edges* as $E_w^{\mathrm{s}} = \{e \in E_w : e = \{u,u\}, u \in V_w\}$ and the set of *connecting edges* as $E_w^{\mathrm{c}} = E_w \setminus E_w^{\mathrm{s}}$.

The graph G_w is connected since the word w defines a path connecting all the vertices of G_w, which further starts and terminates at the same vertex if the word is closed. For $e \in E_w$, we use N_e^w to denote the number of times this path traverses the edge e (in any direction). We note that equivalent words generate the same graphs G_w (up to graph isomorphism) and the same passage-counts N_e^w.

Coming back to the evaluation of $\bar{T}_{\mathbf{i}}^N$, see (2.1.10), note that any k-tuple of integers $\mathbf{i}$ defines a closed word $w_{\mathbf{i}} = i_1 i_2 \cdots i_k i_1$ of length $k+1$. We write $\mathrm{wt}_{\mathbf{i}} = \mathrm{wt}(w_{\mathbf{i}})$, which is nothing but the number of distinct integers in $\mathbf{i}$. Then,

$$\bar{T}_{\mathbf{i}}^N = \frac{1}{N^{k/2}} \prod_{e \in E_{w_{\mathbf{i}}}^c} E(Z_{1,2}^{N_e^{w_{\mathbf{i}}}}) \prod_{e \in E_{w_{\mathbf{i}}}^s} E(Y_1^{N_e^{w_{\mathbf{i}}}}) . \tag{2.1.15}$$

In particular, $\bar{T}_{\mathbf{i}}^N = 0$ unless $N_e^{w_{\mathbf{i}}} \geq 2$ for all $e \in E_{w_{\mathbf{i}}}$, which implies that $\mathrm{wt}_{\mathbf{i}} \leq k/2+1$. Also, (2.1.15) shows that if $w_{\mathbf{i}} \sim w_{\mathbf{i}'}$ then $\bar{T}_{\mathbf{i}}^N = \bar{T}_{\mathbf{i}'}^N$. Further, if $N \geq t$ then there are exactly

$$C_{N,t} := N(N-1)(N-2)\cdots(N-t+1)$$

N-words that are equivalent to a given N-word of weight t. We make the following definition:

> $\mathscr{W}_{k,t}$ denotes a set of representatives for equivalence classes of closed t-words w of length $k+1$ and weight t with $N_e^w \geq 2$ for each $e \in E_w$. (2.1.16)

One deduces from (2.1.10) and (2.1.15) that

$$\langle \bar{L}_N, x^k \rangle = \sum_{t=1}^{\lfloor k/2 \rfloor + 1} \frac{C_{N,t}}{N^{k/2+1}} \sum_{w \in \mathscr{W}_{k,t}} \prod_{e \in E_w^c} E(Z_{1,2}^{N_e^w}) \prod_{e \in E_w^s} E(Y_1^{N_e^w}) . \tag{2.1.17}$$

Note that the cardinality of $\mathscr{W}_{k,t}$ is bounded by the number of closed $\mathscr{S}$-words of length $k+1$ when the cardinality of $\mathscr{S}$ is $t \leq k$, that is, $|\mathscr{W}_{k,t}| \leq t^k \leq k^k$. Thus, (2.1.17) and the finiteness of r_k, see (2.1.1), imply that

$$\lim_{N\to\infty} \langle \bar{L}_N, x^k \rangle = 0 \text{, if } k \text{ is odd} ,$$

while, for k even,

$$\lim_{N\to\infty} \langle \bar{L}_N, x^k \rangle = \sum_{w \in \mathscr{W}_{k,k/2+1}} \prod_{e \in E_w^c} E(Z_{1,2}^{N_e^w}) \prod_{e \in E_w^s} E(Y_1^{N_e^w}) . \tag{2.1.18}$$

We have now motivated the following definition. Note that for the purpose of this section, the case $k=0$ in Definition 2.1.10 is not really needed. It is introduced in this way here in anticipation of the analysis in Section 2.1.6.

Definition 2.1.10 A closed word w of length $k+1 \geq 1$ is called a *Wigner word* if either $k=0$ or k is even and w is equivalent to an element of $\mathscr{W}_{k,k/2+1}$.

We next note that if $w \in \mathscr{W}_{k,k/2+1}$ then G_w is a tree: indeed, G_w is a connected graph with $|V_w| = k/2+1$, hence $|E_w| \geq k/2$, while the condition $N_e^w \geq 2$ for each $e \in E_w$ implies that $|E_w| \leq k/2$. Thus, $|E_w| = |V_w| - 1$, implying that G_w is a tree, that is a connected graph with no loops. Further, the above implies that E_w^s is empty for $w \in \mathscr{W}_{k,k/2+1}$, and thus, for k even,

$$\lim_{N\to\infty} \langle \bar{L}_N, x^k \rangle = |\mathscr{W}_{k,k/2+1}| \,. \tag{2.1.19}$$

We may now complete the

Proof of Lemma 2.1.6 Let k be even. It is convenient to choose the set of representatives $\mathscr{W}_{k,k/2+1}$ such that each word $w = v_1 \cdots v_{k+1}$ in that set satisfies, for $i = 1, \ldots, k+1$, the condition that $\{v_1, \ldots, v_i\}$ is an interval in $\mathbb{Z}$ beginning at 1. (There is a unique choice of such representatives.) Each element $w \in \mathscr{W}_{k,k/2+1}$ determines a path $v_1, v_2, \ldots, v_k, v_{k+1} = v_1$ of length k on the tree G_w. We refer to this path as the *exploration process* associated with w. Let $d(v, v')$ denote the distance between vertices v, v' on the tree G_w, i.e. the length of the shortest path on the tree beginning at v and terminating at v'. Setting $x_i = d(v_{i+1}, v_1)$, one sees that each word $w \in \mathscr{W}_{k,k/2+1}$ defines a Dyck path $D(w) = (x_1, x_2, \ldots, x_k)$ of length k. See Figure 2.1.2 for an example of such coding. Conversely, given a Dyck path $\mathbf{x} = (x_1, \ldots, x_k)$, one may construct a word $w = T(\mathbf{x}) \in \mathscr{W}_{k,k/2+1}$ by recursively constructing an increasing sequence $w_2, \ldots,\ w_k = w$ of words, as follows. Put $w_2 = (1,2)$. For $i > 2$, if $x_{i-1} = x_{i-2} + 1$, then w_i is obtained by adjoining on the right of w_{i-1} the smallest positive integer not appearing in w_{i-1}. Otherwise, w_i is obtained by adjoining on the right of w_{i-1} the next-to-last letter of w_{i-1}. Note that for all i, G_{w_i} is a tree (because G_{w_2} is a tree and, inductively, at stage i, either a backtrack is added to the exploration process on $G_{w_{i-1}}$ or a leaf is added to $G_{w_{i-1}}$). Furthermore, the distance in G_{w_i} between first and last letters of w_i equals x_{i-1}, and therefore, $D(w) = (x_1, \ldots, x_k)$. With our choice of representatives, $T(D(w)) = w$, because each uptick in the Dyck path $D(w)$ starting at location $i-2$ corresponds to adjoinment on the right of w_{i-1} of a new letter, which is uniquely determined by $\operatorname{supp} w_{i-1}$, whereas each downtick at location $i-2$ corresponds to the adjoinment of the next-to-last letter in w_{i-1}. This establishes a bijection between Dyck paths of length k and $\mathscr{W}_{k,k/2+1}$. Lemma 2.1.3 then establishes that

$$|\mathscr{W}_{k,k/2+1}| = C_{k/2} \,. \tag{2.1.20}$$

This completes the proof of Lemma 2.1.6. $\square$

From the proof of Lemma 2.1.6 we extract as a further benefit a proof of a fact needed in Chapter 5. Let k be an even positive integer and let $\mathscr{K}_k = \{1, \ldots, k\}$. Recall the notion of non-crossing partition of $\mathscr{K}_k$, see Definition 2.1.4. We define

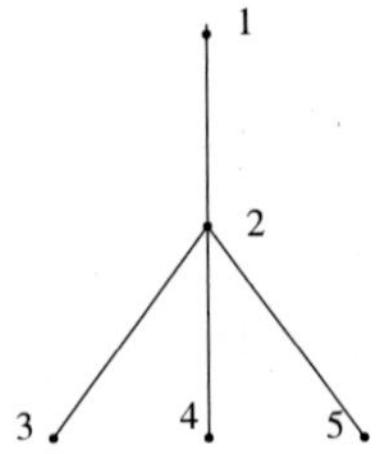

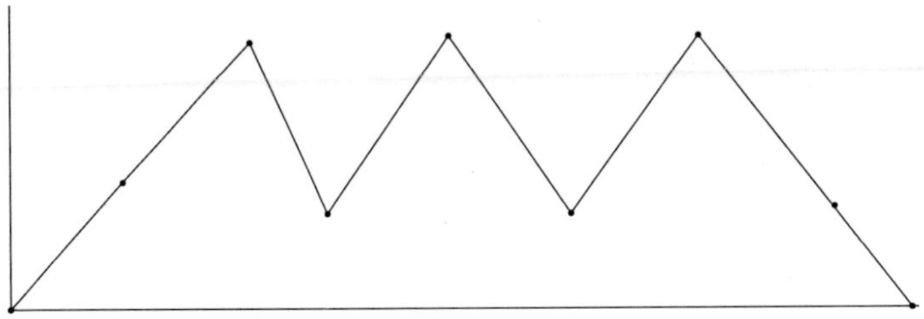

Fig. 2.1.2. Coding of the word $w = 123242521$ into a tree and a Dyck path of length 8. Note that $\ell(w) = 9$ and $\mathrm{wt}(w) = 5$.

a *pair partition* of $\mathscr{K}_k$ to be a partition all of whose parts are two-element sets. The fact we need is that the equivalence classes of Wigner words of length $k+1$ and the non-crossing pair partitions of $\mathscr{K}_k$ are in canonical bijective correspondence. More precisely, we have the following result which describes the bijection in detail.

Proposition 2.1.11 *Given a Wigner word $w = i_1 \cdots i_{k+1}$ of length $k+1$, let Π_w be the partition generated by the function $j \mapsto \{i_j, i_{j+1}\} : \{1,\dots,k\} \to E_w$. (Here, recall, E_w is the set of edges of the graph G_w associated with w.) Then the following hold:*
(i) Π_w is a non-crossing pair partition;
(ii) every non-crossing pair partition of $\mathscr{K}_k$ is of the form Π_w for some Wigner word w of length $k+1$;
(iii) if two Wigner words w and w' of length $k+1$ satisfy $\Pi_w = \Pi_{w'}$, then w and w' are equivalent.

Proof (i) Because a Wigner word w viewed as a walk on its graph G_w crosses every edge exactly twice, Π_w is a pair partition. Because the graph G_w is a tree, the pair partition Π_w is non-crossing.
(ii) The non-crossing pair partitions of $\mathscr{K}_k$ correspond bijectively to Dyck paths. More precisely, given a non-crossing pair partition Π of $\mathscr{K}_k$, associate with it a path $f_\Pi = (f_\Pi(1),\dots,f_\Pi(k))$ by the rules that $f_\Pi(1) = 1$ and, for $i = 2,\dots,k$,

$f_\Pi(i) = f_\Pi(i-1)+1$ (resp., $f_\Pi(i) = f_\Pi(i-1)-1$) if i is the first (resp., second) member of the part of Π to which i belongs. It is easy to check that f_Π is a Dyck path, and furthermore that the map $\Pi \mapsto f_\Pi$ puts non-crossing pair partitions of $\mathcal{K}_k$ into bijective correspondence with Dyck paths of length k. Now choose a Wigner word w whose associated Dyck path $D(w)$, see the proof of Lemma 2.1.6, equals f_Π. One can verify that $\Pi_w = \Pi$.
(iii) Given $\Pi_w = \Pi_{w'}$, one can verify that $D(w) = D(w')$, from which the equivalence of w and w' follows. □

2.1.4 Proof of Lemma 2.1.7: sentences and graphs

By Chebyshev's inequality, it is enough to prove that

$$\lim_{N\to\infty} |E\left(\langle L_N, x^k\rangle^2\right) - \langle \bar{L}_N, x^k\rangle^2| = 0\,.$$

Proceeding as in (2.1.10), one has

$$E(\langle L_N, x^k\rangle^2) - \langle \bar{L}_N, x^k\rangle^2 = \frac{1}{N^2} \sum_{\substack{i_1,\ldots,i_k=1\\ i'_1,\ldots,i'_k=1}}^{N} \bar{T}^N_{\mathbf{i},\mathbf{i}'}\,, \tag{2.1.21}$$

where

$$\bar{T}^N_{\mathbf{i},\mathbf{i}'} = \left[ET^N_{\mathbf{i}}T^N_{\mathbf{i}'} - ET^N_{\mathbf{i}}ET^N_{\mathbf{i}'}\right]. \tag{2.1.22}$$

The role of words in the proof of Lemma 2.1.6 is now played by pairs of words, which is a particular case of a *sentence*.

Definition 2.1.12 ($\mathcal{S}$-sentences) Given a set $\mathcal{S}$, an *$\mathcal{S}$-sentence a* is a finite sequence of $\mathcal{S}$-words $w_1,\ldots,w_n$, at least one word long. Two $\mathcal{S}$-sentences a_1, a_2 are called *equivalent*, denoted $a_1 \sim a_2$, if there is a bijection on $\mathcal{S}$ that maps one into the other.

As with words, for a sentence $a = (w_1, w_2, \ldots, w_n)$, we define the *support* as $\operatorname{supp}(a) = \bigcup_{i=1}^n \operatorname{supp}(w_i)$, and the *weight* $\operatorname{wt}(a)$ as the cardinality of $\operatorname{supp}(a)$.

Definition 2.1.13 (Graph associated with an $\mathcal{S}$-sentence) Given a sentence $a = (w_1,\ldots,w_k)$, with $w_i = s^i_1 s^i_2 \cdots s^i_{\ell(w_i)}$, we set $G_a = (V_a, E_a)$ to be the graph with set of vertices $V_a = \operatorname{supp}(a)$ and (undirected) edges

$$E_a = \{\{s^i_j, s^i_{j+1}\}, j = 1,\ldots,\ell(w_i)-1, i = 1,\ldots,k\}\,.$$

We define the set of *self edges* as $E^{\mathrm{s}}_a = \{e \in E_a : e = \{u,u\}, u \in V_a\}$ and the set of *connecting edges* as $E^{\mathrm{c}}_a = E_a \setminus E^{\mathrm{s}}_a$.

In words, the graph associated with a sentence $a=(w_1,\ldots,w_k)$ is obtained by piecing together the graphs of the individual words w_i (and in general, it differs from the graph associated with the word obtained by concatenating the words w_i). Unlike the graph of a word, the graph associated with a sentence may be disconnected. Note that the sentence a defines k paths in the graph G_a. For $e\in E_a$, we use N_e^a to denote the number of times the union of these paths traverses the edge e (in any direction). We note that equivalent sentences generate the same graphs G_a and the same passage-counts N_e^a.

Coming back to the evaluation of $\bar{T}_{\mathbf{i},\mathbf{i}'}$, see (2.1.21), recall the closed words $w_{\mathbf{i}},w_{\mathbf{i}'}$ of length $k+1$, and define the two-word sentence $a_{\mathbf{i},\mathbf{i}'}=(w_{\mathbf{i}},w_{\mathbf{i}'})$. Then,

$$\begin{aligned}\bar{T}^N_{\mathbf{i},\mathbf{i}'} &= \frac{1}{N^k}\Big[\prod_{e\in E^c_{a_{\mathbf{i},\mathbf{i}'}}} E(Z_{1,2}^{N_e^a})\prod_{e\in E^s_{a_{\mathbf{i},\mathbf{i}'}}} E(Y_1^{N_e^a}) \\ &\quad -\prod_{e\in E^c_{w_{\mathbf{i}}}} E(Z_{1,2}^{N_e^{w_{\mathbf{i}}}})\prod_{e\in E^s_{w_{\mathbf{i}}}} E(Y_1^{N_e^{w_{\mathbf{i}}}})\prod_{e\in E^c_{w_{\mathbf{i}'}}} E(Z_{1,2}^{N_e^{w_{\mathbf{i}'}}})\prod_{e\in E^s_{w_{\mathbf{i}'}}} E(Y_1^{N_e^{w_{\mathbf{i}'}}})\Big].\end{aligned} \tag{2.1.23}$$

In particular, $\bar{T}^N_{\mathbf{i},\mathbf{i}'}=0$ unless $N_e^{a_{\mathbf{i},\mathbf{i}'}}\geq 2$ for all $e\in E_{a_{\mathbf{i},\mathbf{i}'}}$. Also, $\bar{T}^N_{\mathbf{i},\mathbf{i}'}=0$ unless $E_{w_{\mathbf{i}}}\cap E_{w_{\mathbf{i}'}}\neq\emptyset$. Further, (2.1.23) shows that if $a_{\mathbf{i},\mathbf{i}'}\sim a_{\mathbf{j},\mathbf{j}'}$ then $\bar{T}^N_{\mathbf{i},\mathbf{i}'}=\bar{T}^N_{\mathbf{j},\mathbf{j}'}$. Finally, if $N\geq t$ then there are exactly $C_{N,t}$ N-sentences that are equivalent to a given N-sentence of weight t. We make the following definition:

> $\mathscr{W}^{(2)}_{k,t}$ denotes a set of representatives for equivalence classes of sentences a of weight t consisting of two closed t-words (w_1,w_2), each of length $k+1$, with $N_e^a\geq 2$ for each $e\in E_a$, and $E_{w_1}\cap E_{w_2}\neq\emptyset$. (2.1.24)

One deduces from (2.1.21) and (2.1.23) that

$$\begin{aligned}&E(\langle L_N,x^k\rangle^2)-\langle\bar{L}_N,x^k\rangle^2 \\ &= \sum_{t=1}^{2k}\frac{C_{N,t}}{N^{k+2}}\sum_{a=(w_1,w_2)\in\mathscr{W}^{(2)}_{k,t}}\Big(\prod_{e\in E^c_a}E(Z_{1,2}^{N_e^a})\prod_{e\in E^s_a}E(Y_1^{N_e^a}) \\ &\quad -\prod_{e\in E^c_{w_1}}E(Z_{1,2}^{N_e^{w_1}})\prod_{e\in E^s_{w_1}}E(Y_1^{N_e^{w_1}})\prod_{e\in E^c_{w_2}}E(Z_{1,2}^{N_e^{w_2}})\prod_{e\in E^s_{w_2}}E(Y_1^{N_e^{w_2}})\Big).\end{aligned} \tag{2.1.25}$$

We have completed the preliminaries to

Proof of Lemma 2.1.7 In view of (2.1.25), it suffices to check that $\mathscr{W}^{(2)}_{k,t}$ is empty for $t\geq k+2$. Since we need it later, we prove a slightly stronger claim, namely that $\mathscr{W}^{(2)}_{k,t}$ is empty for $t\geq k+1$.

Toward this end, note that if $a\in\mathscr{W}^{(2)}_{k,t}$ then G_a is a connected graph, with t vertices and at most k edges (since $N_e^a\geq 2$ for $e\in E_a$), which is impossible when

$t > k+1$. Considering the case $t = k+1$, it follows that G_a is a tree, and each edge must be visited by the paths generated by a exactly twice. Because the path generated by w_1 in the tree G_a starts and end at the same vertex, it must visit each edge an even number of times. Thus, the set of edges visited by w_1 is disjoint from the set of edges visited by w_2, contradicting the definition of $\mathscr{W}^{(2)}_{k,t}$. □

Remark 2.1.14 Note that in the course of the proof of Lemma 2.1.7, we actually showed that for $N > 2k$,

$$E(\langle L_N, x^k\rangle^2) - \langle \bar{L}_N, x^k\rangle^2 \tag{2.1.26}$$
$$= \sum_{t=1}^{k} \frac{C_{N,t}}{N^{k+2}} \sum_{a=(w_1,w_2)\in\mathscr{W}^{(2)}_{k,t}} \Big[\prod_{e\in E_a^c} E(Z_{1,2}^{N_e^a}) \prod_{e\in E_a^s} E(Y_1^{N_e^a})$$
$$- \prod_{e\in E_{w_1}^c} E(Z_{1,2}^{N_e^{w_1}}) \prod_{e\in E_{w_1}^s} E(Y_1^{N_e^{w_1}}) \prod_{e\in E_{w_2}^c} E(Z_{1,2}^{N_e^{w_2}}) \prod_{e\in E_{w_2}^s} E(Y_1^{N_e^{w_2}})\Big],$$

that is, that the summation in (2.1.25) can be restricted to $t \le k$.

Exercise 2.1.15 Consider symmetric random matrices X_N, with the zero mean independent random variables $\{X_N(i,j)\}_{1\le i\le j\le N}$ no longer assumed identically distributed nor all of variance $1/N$. Check that Theorem 2.1.1 still holds if one assumes that for all $\varepsilon > 0$,

$$\lim_{N\to\infty} \frac{\#\{(i,j) : |1 - NEX_N(i,j)^2| < \varepsilon\}}{N^2} = 1,$$

and for all $k \ge 1$, there exists a finite r_k independent of N such that

$$\sup_{1\le i\le j\le N} E\left|\sqrt{N}X_N(i,j)\right|^k \le r_k.$$

Exercise 2.1.16 Check that the conclusion of Theorem 2.1.1 remains true when convergence in probability is replaced by almost sure convergence.
Hint: Using Chebyshev's inequality and the Borel–Cantelli Lemma, it is enough to verify that for all positive integers k, there exists a constant $C = C(k)$ such that

$$|E\left(\langle L_N, x^k\rangle^2\right) - \langle \bar{L}_N, x^k\rangle^2| \le \frac{C}{N^2}.$$

Exercise 2.1.17 In the setup of Theorem 2.1.1, assume that $r_k < \infty$ for all k but not necessarily that $E[Z_{1,2}^2] = 1$. Show that, for any positive integer k,

$$\sup_{N\in\mathbb{N}} E[\langle L_N, x^k\rangle] =: C(r_\ell, \ell \le k) < \infty.$$

Exercise 2.1.18 We develop in this exercise the limit theory for *Wishart* matrices. Let $M = M(N)$ be a sequence of positive integers such that

$$\lim_{N\to\infty} M(N)/N = \alpha \in [1,\infty) .$$

Consider an $N \times M(N)$ matrix Y_N with i.i.d. entries of mean zero and variance $1/N$, and such that $E\left(N^{k/2}|Y_N(1,1)|^k\right) \le r_k < \infty$. Define the $N \times N$ Wishart matrix as $W_N = Y_N Y_N^{\mathrm{T}}$, and let L_N denote the empirical measure of the eigenvalues of W_N. Set $\bar{L}_N = EL_N$.
(a) Write $N\langle \bar{L}_N, x^k\rangle$ as

$$\sum_{\substack{i_1,\ldots,i_k\\ j_1,\ldots,j_k}} EY_N(i_1,j_1)Y_N(i_2,j_1)Y_N(i_2,j_2)Y_N(i_3,j_2)\cdots Y_N(i_k,j_k)Y_N(i_1,j_k)$$

and show that the only contributions to the sum (divided by N) that survive the passage to the limit are those in which each term appears exactly twice.
Hint: use the words $i_1 j_1 i_2 j_2 \ldots j_k i_1$ and a bi-partite graph to replace the Wigner analysis.
(b) Code the contributions as Dyck paths, where the even heights correspond to i indices and the odd heights correspond to j indices. Let $\ell = \ell(\mathbf{i},\mathbf{j})$ denote the number of times the excursion makes a descent from an odd height to an even height (this is the number of distinct j indices in the tuple!), and show that the combinatorial weight of such a path is asymptotic to $N^{k+1}\alpha^{\ell}$.
(c) Let $\bar{\ell}$ denote the number of times the excursion makes a descent from an even height to an odd height, and set

$$\beta_k = \sum_{\text{Dyck paths of length } 2k} \alpha^{\ell}, \qquad \gamma_k = \sum_{\text{Dyck paths of length } 2k} \alpha^{\bar{\ell}} .$$

(The β_k are the kth moments of any weak limit of $\bar{L}_N$.) Prove that

$$\beta_k = \alpha \sum_{j=1}^{k} \gamma_{k-j}\beta_{j-1}, \quad \gamma_k = \sum_{j=1}^{k} \beta_{k-j}\gamma_{j-1}, \quad k \ge 1 .$$

(d) Setting $\hat{\beta}_\alpha(z) = \sum_{k=0}^{\infty} z^k \beta_k$, prove that $\hat{\beta}_\alpha(z) = 1 + z\hat{\beta}_\alpha(z)^2 + (\alpha-1)z\hat{\beta}_\alpha(z)$, and thus the limit F_α of $\bar{L}_N$ possesses the Stieltjes transform (see Definition 2.4.1) $-z^{-1}\hat{\beta}_\alpha(1/z)$, where

$$\hat{\beta}_\alpha(z) = \frac{1-(\alpha-1)z-\sqrt{1-4z\left[\frac{\alpha+1}{2}-\frac{(\alpha-1)^2 z}{4}\right]}}{2z} .$$

(e) Conclude that F_α possesses a density f_α supported on $[b_-,b_+]$, with $b_- = (1-\sqrt{\alpha})^2$, $b_+ = (1+\sqrt{\alpha})^2$, satisfying

$$f_\alpha(x) = \frac{\sqrt{(x-b_-)(b_+-x)}}{2\pi x}, \quad x \in [b_-,b_+]. \tag{2.1.27}$$

(This is the famous *Marčenko–Pastur* law, due to [MaP67].)
(f) Prove the analog of Lemma 2.1.7 for Wishart matrices, and deduce that $L_N \to F_\alpha$ weakly, in probability.
(g) Note that F_1 is the image of the semicircle distribution under the transformation $x \mapsto x^2$.

2.1.5 Some useful approximations

This section is devoted to the following simple observation that often allows one to considerably simplify arguments concerning the convergence of empirical measures.

Lemma 2.1.19 (Hoffman–Wielandt) *Let A, B be $N \times N$ symmetric matrices, with eigenvalues $\lambda_1^A \le \lambda_2^A \le \ldots \le \lambda_N^A$ and $\lambda_1^B \le \lambda_2^B \le \ldots \le \lambda_N^B$. Then*

$$\sum_{i=1}^N |\lambda_i^A - \lambda_i^B|^2 \le \mathrm{tr}(A-B)^2 .$$

Proof Note that $\mathrm{tr}A^2 = \sum_i (\lambda_i^A)^2$ and $\mathrm{tr}B^2 = \sum_i (\lambda_i^B)^2$. Let U denote the matrix diagonalizing B written in the basis determined by A, and let D_A, D_B denote the diagonal matrices with diagonal elements λ_i^A, λ_i^B respectively. Then,

$$\mathrm{tr}AB = \mathrm{tr}D_A U D_B U^{\mathrm{T}} = \sum_{i,j} \lambda_i^A \lambda_j^B u_{ij}^2 .$$

The last sum is linear in the coefficients $v_{ij} = u_{ij}^2$, and the orthogonality of U implies that $\sum_j v_{ij} = 1, \sum_i v_{ij} = 1$. Thus

$$\mathrm{tr}AB \le \sup_{v_{ij}\ge 0: \sum_j v_{ij}=1, \sum_i v_{ij}=1} \sum_{i,j} \lambda_i^A \lambda_j^B v_{ij} . \tag{2.1.28}$$

But this is a maximization of a linear functional over the convex set of doubly stochastic matrices, and the maximum is obtained at the extreme points, which are well known to correspond to permutations The maximum among permutations is then easily checked to be $\sum_i \lambda_i^A \lambda_i^B$. Collecting these facts together implies Lemma 2.1.19. Alternatively, one sees directly that a maximizing $V = \{v_{ij}\}$ in (2.1.28) is the identity matrix. Indeed, assume w.l.o.g. that $v_{11} < 1$. We then construct a matrix $\bar{V} = \{\bar{v}_{ij}\}$ with $\bar{v}_{11} = 1$ and $\bar{v}_{ii} = v_{ii}$ for $i > 1$ such that $\bar{V}$ is also a maximizing matrix. Indeed, because $v_{11} < 1$, there exist a j and a k with $v_{1j} > 0$ and $v_{k1} > 0$. Set $v = \min(v_{1j}, v_{k1}) > 0$ and define $\bar{v}_{11} = v_{11} + v$, $\bar{v}_{kj} = v_{kj} + v$ and $\bar{v}_{1j} = v_{1j} - v$, $\bar{v}_{k1} = v_{k1} - v$, and $\bar{v}_{ab} = v_{ab}$ for all other pairs ab. Then,

$$\begin{aligned}\sum_{i,j} \lambda_i^A \lambda_j^B (\bar{v}_{ij} - v_{ij}) &= v(\lambda_1^A \lambda_1^B + \lambda_k^A \lambda_j^B - \lambda_k^A \lambda_1^B - \lambda_1^A \lambda_j^B) \\ &= v(\lambda_1^A - \lambda_k^A)(\lambda_1^B - \lambda_j^B) \geq 0 .\end{aligned}$$

Thus, $\bar{V} = \{\bar{v}_{ij}\}$ satisfies the constraints, is also a maximum, and the number of zero elements in the first row and column of $\bar{V}$ is larger by 1 at least from the corresponding one for V. If $\bar{v}_{11} = 1$, the claims follows, while if $\bar{v}_{11} < 1$, one repeats this (at most $2N-2$ times) to conclude. Proceeding in this manner with all diagonal elements of V, one sees that indeed the maximum of the right side of (2.1.28) is $\sum_i \lambda_i^A \lambda_i^B$, as claimed. □

Remark 2.1.20 The statement and proof of Lemma 2.1.19 carry over to the case where A and B are both Hermitian matrices.

Lemma 2.1.19 allows one to perform all sorts of truncations when proving convergence of empirical measures. For example, let us prove the following variant of Wigner's Theorem 2.1.1.

Theorem 2.1.21 *Assume X_N is as in (2.1.2), except that instead of (2.1.1), only $r_2 < \infty$ is assumed. Then, the conclusion of Theorem 2.1.1 still holds.*

Proof Fix a constant C and consider the symmetric matrix $\hat{X}_N$ whose elements satisfy, for $1 \leq i \leq j \leq N$,

$$\hat{X}_N(i,j) = X_N(i,j) \mathbf{1}_{\sqrt{N}|X_N(i,j)| \leq C} - E(X_N(i,j) \mathbf{1}_{\sqrt{N}|X_N(i,j)| \leq C}).$$

Then, with $\hat{\lambda}_i^N$ denoting the eigenvalues of $\hat{X}_N$, ordered, it follows from Lemma 2.1.19 that

$$\frac{1}{N} \sum_{i=1}^N |\lambda_i^N - \hat{\lambda}_i^N|^2 \leq \frac{1}{N} \mathrm{tr}(X_N - \hat{X}_N)^2 .$$

But

$$\begin{aligned} W_N &:= \frac{1}{N} \mathrm{tr}(X_N - \hat{X}_N)^2 \\ &\leq \frac{1}{N^2} \sum_{i,j} \left[\sqrt{N} X_N(i,j) \mathbf{1}_{|\sqrt{N} X_N(i,j)| \geq C} - E(\sqrt{N} X_N(i,j) \mathbf{1}_{|\sqrt{N} X_N(i,j)| \geq C}) \right]^2 .\end{aligned}$$

Since $r_2 < \infty$, and the involved random variables are identical in law to either $Z_{1,2}$ or Y_1, it follows that $E[(\sqrt{N} X_N(i,j))^2 \mathbf{1}_{|\sqrt{N} X_N(i,j)| \geq C}]$ converges to 0 uniformly in

N, i, j, when C converges to infinity. Hence, one may chose for each ε a large enough C such that $P(|W_N| > \varepsilon) < \varepsilon$. Further, let

$$\mathrm{Lip}(\mathbb{R}) = \{f \in C_b(\mathbb{R}) : \sup_x |f(x)| \le 1, \sup_{x \neq y} \frac{|f(x) - f(y)|}{|x-y|} \le 1\}.$$

Then, on the event $\{|W_N| < \varepsilon\}$, it holds that for $f \in \mathrm{Lip}(\mathbb{R})$,

$$|\langle L_N, f\rangle - \langle \hat{L}_N, f\rangle| \le \frac{1}{N}\sum_i |\lambda_i^N - \hat{\lambda}_i^N| \le \sqrt{\varepsilon},$$

where $\hat{L}_N$ denotes the empirical measure of the eigenvalues of $\hat{X}_N$, and Jensen's inequality was used in the second inequality. This, together with the weak convergence in probability of $\hat{L}_N$ toward the semicircle law assured by Theorem 2.1.1, and the fact that weak convergence is equivalent to convergence with respect to the Lipschitz bounded metric, see Theorem C.8, complete the proof of Theorem 2.1.21. □

2.1.6 Maximal eigenvalues and Füredi–Komlós enumeration

Wigner's theorem asserts the weak convergence of the empirical measure of eigenvalues to the compactly supported semicircle law. One immediately is led to suspect that the maximal eigenvalue of X_N should converge to the value 2, the largest element of the support of the semicircle distribution. This fact, however, does not follow from Wigner's Theorem. Nonetheless, the combinatorial techniques we have already seen allow one to prove the following, where we use the notation introduced in (2.1.1) and (2.1.2).

Theorem 2.1.22 (Maximal eigenvalue) *Consider a Wigner matrix X_N satisfying $r_k \le k^{Ck}$ for some constant C and all positive integers k. Then, λ_N^N converges to* 2 *in probability.*

Remark The assumption of Theorem 2.1.22 holds if the random variables $|Z_{1,2}|$ and $|Y_1|$ possess a finite exponential moment.

Proof of Theorem 2.1.22 Fix $\delta > 0$ and let $g : \mathbb{R} \mapsto \mathbb{R}_+$ be a continuous function supported on $[2-\delta, 2]$, with $\langle \sigma, g\rangle = 1$. Then, applying Wigner's Theorem 2.1.1,

$$P(\lambda_N^N < 2-\delta) \le P(\langle L_N, g\rangle = 0) \le P(|\langle L_N, g\rangle - \langle \sigma, g\rangle| > \frac{1}{2}) \to_{N\to\infty} 0. \quad (2.1.29)$$

We thus need to provide a complementary estimate on the probability that λ_N^N is large. We do that by estimating $\langle \bar{L}_N, x^{2k}\rangle$ for k growing with N, using the bounds on r_k provided in the assumptions. The key step is contained in the following combinatorial lemma that gives information on the sets $\mathscr{W}_{k,t}$, see (2.1.16).

Lemma 2.1.23 *For all integers $k > 2t - 2$ one has the estimate*

$$|\mathscr{W}_{k,t}| \leq 2^k k^{3(k-2t+2)} . \tag{2.1.30}$$

The proof of Lemma 2.1.23 is deferred to the end of this section.

Equipped with Lemma 2.1.23, we have for $2k < N$, using (2.1.17),

$$\begin{aligned} \langle \bar{L}_N, x^{2k} \rangle &\leq \sum_{t=1}^{k+1} N^{t-(k+1)} |\mathscr{W}_{2k,t}| \sup_{w \in \mathscr{W}_{2k,t}} \prod_{e \in E_w^c} E(Z_{1,2}^{N_e^w}) \prod_{e \in E_w^s} E(Y_1^{N_e^w}) \quad (2.1.31) \\ &\leq 4^k \sum_{t=1}^{k+1} \left(\frac{(2k)^6}{N} \right)^{k+1-t} \sup_{w \in \mathscr{W}_{2k,t}} \prod_{e \in E_w^c} E(Z_{1,2}^{N_e^w}) \prod_{e \in E_w^s} E(Y_1^{N_e^w}) . \end{aligned}$$

To evaluate the last expectation, fix $w \in \mathscr{W}_{2k,t}$, and let l denote the number of edges in E_w^c with $N_e^w = 2$. Hölder's inequality then gives

$$\prod_{e \in E_w^c} E(Z_{1,2}^{N_e^w}) \prod_{e \in E_w^s} E(Y_1^{N_e^w}) \leq r_{2k-2l} ,$$

with the convention that $r_0 = 1$. Since G_w is connected, $|E_w^c| \geq |V_w| - 1 = t - 1$. On the other hand, by noting that $N_e^w \geq 3$ for $|E_w^c| - l$ edges, one has $2k \geq 3(|E_w^c| - l) + 2l + 2|E_w^s|$. Hence, $2k - 2l \leq 6(k+1-t)$. Since r_{2q} is a nondecreasing function of q bounded below by 1, we get, substituting back in (2.1.31), that for some constant $c_1 = c_1(C) > 0$ and all $k < N$,

$$\begin{aligned} \langle \bar{L}_N, x^{2k} \rangle &\leq 4^k \sum_{t=1}^{k+1} \left(\frac{(2k)^6}{N} \right)^{k+1-t} r_{6(k+1-t)} \quad (2.1.32) \\ &\leq 4^k \sum_{t=1}^{k+1} \left(\frac{(2k)^6 (6(k+1-t))^{6C}}{N} \right)^{k+1-t} \leq 4^k \sum_{i=0}^{k} \left(\frac{k^{c_1}}{N} \right)^i . \end{aligned}$$

Choose next a sequence $k(N) \to_{N \to \infty} \infty$ such that

$$k(N)^{c_1}/N \to_{N \to \infty} 0 \quad \text{but} \quad k(N)/\log N \to_{N \to \infty} \infty .$$

Then, for any $\delta > 0$, and all N large,

$$\begin{aligned} P(\lambda_N^N > (2+\delta)) &\leq P(N \langle L_N, x^{2k(N)} \rangle > (2+\delta)^{2k(N)}) \\ &\leq \frac{N \langle \bar{L}_N, x^{2k(N)} \rangle}{(2+\delta)^{2k(N)}} \leq \frac{2N 4^{k(N)}}{(2+\delta)^{2k(N)}} \to_{N \to \infty} 0 , \end{aligned}$$

completing the proof of Theorem 2.1.22, modulo Lemma 2.1.23. □

Proof of Lemma 2.1.23 The idea of the proof is to keep track of the number of possibilities to prevent words in $\mathscr{W}_{k,t}$ from having weight $\lfloor k/2 \rfloor + 1$. Toward this end, let $w \in \mathscr{W}_{k,t}$ be given. A *parsing* of the word w is a sentence $a_w = (w_1, \ldots, w_n)$ such that the word obtained by concatenating the words w_i is w. One can imagine creating a parsing of w by introducing commas between parts of w.

We say that a parsing $a = a_w$ of w is an *FK parsing* (after Füredi and Komlós), and call the sentence a an *FK sentence*, if the graph associated with a is a tree, if $N_e^a \leq 2$ for all $e \in E_a$, and if for any $i = 1, \ldots, n-1$, the first letter of w_{i+1} belongs to $\bigcup_{j=1}^{i} \operatorname{supp} w_j$. If the one-word sentence $a = w$ is an FK parsing, we say that w is an *FK word*. Note that the constituent words in an FK parsing are FK words.

As will become clear next, the graph of an FK word consists of trees whose edges have been visited twice by w, glued together by edges that have been visited only once. Recalling that a Wigner word is either a one-letter word or a closed word of odd length and maximal weight (subject to the constraint that edges are visited at least twice), this leads to the following lemma.

Lemma 2.1.24 *Each FK word can be written in a unique way as a concatenation of pairwise disjoint Wigner words. Further, there are at most 2^{n-1} equivalence classes of FK words of length n.*

Proof of Lemma 2.1.24 Let $w = s_1 \cdots s_n$ be an FK word of length n. By definition, G_w is a tree. Let $\{s_{i_j}, s_{i_j+1}\}_{j=1}^r$ denote those edges of G_w visited only once by the walk induced by w. Defining $i_0 = 1$, one sees that the words $\bar{w}_j = s_{i_{j-1}+1} \cdots s_{i_j}$, $j \geq 1$, are closed, disjoint, and visit each edge in the tree $G_{\bar{w}_j}$ exactly twice. In particular, with $l_j := i_j - i_{j-1} - 1$, it holds that l_j is even (possibly, $l_j = 0$ if $\bar{w}_j$ is a one-letter word), and further if $l_j > 0$ then $\bar{w}_j \in \mathscr{W}_{l_j, l_j/2+1}$. This decomposition being unique, one concludes that for any z, with N_n denoting the number of equivalence classes of FK words of length n, and with $|\mathscr{W}_{0,1}| := 1$,

$$\begin{aligned} \sum_{n=1}^{\infty} N_n z^n &= \sum_{r=1}^{\infty} \sum_{\substack{\{l_j\}_{j=1}^r \\ l_j \text{ even}}} \prod_{j=1}^{r} z^{l_j+1} |\mathscr{W}_{l_j, l_j/2+1}| \\ &= \sum_{r=1}^{\infty} \left(z + \sum_{l=1}^{\infty} z^{2l+1} |\mathscr{W}_{2l,l+1}| \right)^r, \end{aligned} \tag{2.1.33}$$

in the sense of formal power series. By the proof of Lemma 2.1.6, $|\mathscr{W}_{2l,l+1}| = C_l = \beta_l$. Hence, by Lemma 2.1.3, for $|z| < 1/4$,

$$z + \sum_{l=1}^{\infty} z^{2l+1} |\mathscr{W}_{2l,l+1}| = z\hat{\beta}(z^2) = \frac{1 - \sqrt{1 - 4z^2}}{2z}.$$

Substituting in (2.1.33), one sees that (again, in the sense of power series)

$$\sum_{n=1}^{\infty} N_n z^n = \frac{z\hat{\beta}(z^2)}{1-z\hat{\beta}(z^2)} = \frac{1-\sqrt{1-4z^2}}{2z-1+\sqrt{1-4z^2}} = -\frac{1}{2} + \frac{z+\frac{1}{2}}{\sqrt{1-4z^2}}.$$

Using the fact that

$$\sqrt{\frac{1}{1-t}} = \sum_{k=0}^{\infty} \frac{t^k}{4^k} \binom{2k}{k},$$

one concludes that

$$\sum_{n=1}^{\infty} N_n z^n = z + \frac{1}{2}(1+2z) \sum_{n=1}^{\infty} z^{2n} \binom{2n}{n},$$

from which Lemma 2.1.24 follows. □

Our interest in FK parsings is the following FK parsing w' of a word $w = s_1 \cdots s_n$. Declare an edge e of G_w to be *new* (relative to w) if for some index $1 \le i < n$ we have $e = \{s_i, s_{i+1}\}$ and $s_{i+1} \notin \{s_1, \ldots, s_i\}$. If the edge e is not new, then it is *old*. Define w' to be the sentence obtained by breaking w (that is, "inserting a comma") at all visits to old edges of G_w and at third and subsequent visits to new edges of G_w.

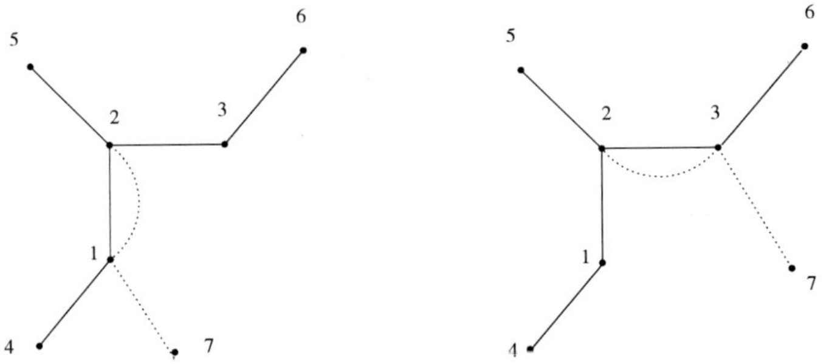

Fig. 2.1.3. Two inequivalent FK sentences $[x_1, x_2]$ corresponding to (solid line) $b = 141252363$ and (dashed line) $c = 1712$ (in left) ~ 3732 (in right).

Since a word w can be recovered from its FK parsing by omitting the extra commas, and since the number of equivalence classes of FK words is estimated by Lemma 2.1.24, one could hope to complete the proof of Lemma 2.1.23 by controlling the number of possible parsed FK sequences. A key step toward this end is the following lemma, which explains how FK words are fitted together to form FK sentences. Recall that any FK word w can be written in a unique way as a concatenation of disjoint Wigner words w_i, $i = 1, \ldots, r$. With s_i denoting the first (and last) letter of w_i, define the *skeleton* of w as the word $s_1 \cdots s_r$. Finally, for a sentence a with graph G_a, let $G_a^1 = (V_a^1, E_a^1)$ be the graph with vertex set $V_a = V_a^1$ and edge set $E_a^1 = \{e \in E_a : N_e^a = 1\}$. Clearly, when a is an FK sentence, G_a^1 is always a *forest*, that is a disjoint union of trees.

Lemma 2.1.25 *Suppose b is an FK sentence with $n-1$ words and c is an FK word with skeleton $s_1 \cdots s_r$ such that $s_1 \in \operatorname{supp}(b)$. Let ℓ be the largest index such that $s_\ell \in \operatorname{supp} b$, and set $d = s_1 \cdots s_\ell$. Then $a = (b,c)$ is an FK sentence only if $\operatorname{supp} b \cap \operatorname{supp} c = \operatorname{supp} d$ and d is a geodesic in G_b^1.*

(A *geodesic* connecting $x, y \in G_b^1$ is a path of minimal length starting at x and terminating at y.) A consequence of Lemma 2.1.25 is that there exist at most $(\mathrm{wt}(b))^2$ equivalence classes of FK sentences $x_1, \ldots, x_n$ such that $b \sim x_1, \ldots, x_{n-1}$ and $c \sim x_n$. See Figure 2.1.3 for an example of two such equivalence classes and their pictorial description.

Before providing the proof of Lemma 2.1.25, we explain how it leads to

Completion of proof of Lemma 2.1.23 Let $\Gamma(t,\ell,m)$ denote the set of equivalence classes of FK sentences $a = (w_1, \ldots, w_m)$ consisting of m words, with total length $\sum_{i=1}^m \ell(w_i) = \ell$ and $\mathrm{wt}(a) = t$. An immediate corollary of Lemma 2.1.25 is that

$$|\Gamma(t,\ell,m)| \le 2^{\ell-m} t^{2(m-1)} \binom{\ell-1}{m-1}. \tag{2.1.34}$$

Indeed, there are $c_{\ell,m} := \binom{\ell-1}{m-1}$ m-tuples of positive integers summing to ℓ, and thus at most $2^{\ell-m} c_{\ell,m}$ equivalence classes of sentences consisting of m pairwise disjoint FK words with sum of lengths equal to ℓ. Lemma 2.1.25 then shows that there are at most $t^{2(m-1)}$ ways to "glue these words into an FK sentence", whence (2.1.34) follows.

For any FK sentence a consisting of m words with total length ℓ, we have that

$$m = |E_a^1| - 2\mathrm{wt}(a) + 2 + \ell. \tag{2.1.35}$$

Indeed, the word obtained by concatenating the words of a generates a list of $\ell - 1$ (not necessarily distinct) unordered pairs of adjoining letters, out of which $m-1$ correspond to commas in the FK sentence a and $2|E_a| - |E_a^1|$ correspond to edges of G_a. Using that $|E_a| = |V_a| - 1$, (2.1.35) follows.

Consider a word $w \in \mathscr{W}_{k,t}$ that is parsed into an FK sentence w' consisting of m words. Note that if an edge e is retained in $G_{w'}$, then no comma is inserted at e at the first and second passage on e (but is introduced if there are further passages on e). Therefore, $E_{w'}^1 = \emptyset$. By (2.1.35), this implies that for such words,

$m-1=k+2-2t$. Inequality (2.1.34) then allows one to conclude the proof of Lemma 2.1.23. □

Proof of Lemma 2.1.25 Assume a is an FK sentence. Then G_a is a tree, and since the Wigner words composing c are disjoint, d is the unique geodesic in $G_c \subset G_a$ connecting s_1 to s_ℓ. Hence, it is also the unique geodesic in $G_b \subset G_a$ connecting s_1 to s_ℓ. But d visits only edges of G_b that have been visited exactly once by the constituent words of b, for otherwise (b,c) would not be an FK sentence (that is, a comma would need to be inserted to split c). Thus, $E_d \subset E_b^1$. Since c is an FK word, $E_c^1 = E_{s_1 \cdots s_r}$. Since a is an FK sentence, $E_b \cap E_c = E_b^1 \cap E_c^1$. Thus, $E_b \cap E_c = E_d$. But, recall that G_a, G_b, G_c, G_d are trees, and hence

$$\begin{aligned} |V_a| &= 1+|E_a| = 1+|E_b|+|E_c|-|E_b\cap E_c| = 1+|E_b|+|E_c|-|E_d| \\ &= 1+|E_b|+1+|E_c|-1-|E_d| = |V_b|+|V_c|-|V_d|. \end{aligned}$$

Since $|V_b|+|V_c|-|V_b\cap V_c| = |V_a|$, it follows that $|V_d| = |V_b \cap V_c|$. Since $V_d \subset V_b \cap V_c$, one concludes that $V_d = V_b \cap V_c$, as claimed. □

Remark 2.1.26 The result described in Theorem 2.1.22 is not optimal, in the sense that even with uniform bounds on the (rescaled) entries, i.e. r_k uniformly bounded, the estimate one gets on the displacement of the maximal eigenvalue to the right of 2 is $O(n^{-1/6}\log n)$, whereas the true displacement is known to be of order $n^{-2/3}$ (see Section 2.7 for more details, and, in the context of complex Gaussian Wigner matrices, see Theorems 3.1.4 and 3.1.5).

Exercise 2.1.27 Prove that the conclusion of Theorem 2.1.22 holds with convergence in probability replaced by either almost sure convergence or L^p convergence.

Exercise 2.1.28 Prove that the statement of Theorem 2.1.22 can be strengthened to yield that for some constant $\delta = \delta(C) > 0$, $N^\delta(\lambda_N^N - 2)$ converges to 0, almost surely.

Exercise 2.1.29 Assume that for some constants $\lambda > 0$, C, the independent (but not necessarily identically distributed) entries $\{X_N(i,j)\}_{1\le i\le j\le N}$ of the symmetric matrices X_N satisfy

$$\sup_{i,j,N} E(e^{\lambda\sqrt{N}|X_N(i,j)|}) \le C.$$

Prove that there exists a constant $c_1 = c_1(C)$ such that $\limsup_{N\to\infty} \lambda_N^N \le c_1$, almost surely, and $\limsup_{N\to\infty} E\lambda_N^N \le c_1$.

Exercise 2.1.30 We develop in this exercise an alternative proof, that avoids moment computations, to the conclusion of Exercise 2.1.29, under the stronger assumption that for some $\lambda > 0$,

$$\sup_{i,j,N} E(e^{\lambda(\sqrt{N}|X_N(i,j)|)^2}) \leq C.$$

(a) Prove (using Chebyshev's inequality and the assumption) that there exists a constant c_0 independent of N such that for any fixed $z \in \mathbb{R}^N$, and all C large enough,

$$P(\|z^T X_N\|_2 > C) \leq e^{-c_0 C^2 N}. \tag{2.1.36}$$

(b) Let $\mathcal{N}_\delta = \{z_i\}_{i=1}^{N_\delta}$ be a minimal deterministic net in the unit ball of $\mathbb{R}^N$, that is $\|z_i\|_2 = 1$, $\sup_{z:\|z\|_2=1} \inf_i \|z - z_i\|_2 \leq \delta$, and N_δ is the minimal integer with the property that such a net can be found. Check that

$$(1-\delta^2) \sup_{z:\|z\|_2=1} z^T X_N z \leq \sup_{z_i \in \mathcal{N}_\delta} z_i^T X_N z_i + 2 \sup_i \sup_{z:\|z-z_i\|_2 \leq \delta} z^T X_N z_i. \tag{2.1.37}$$

(c) Combine steps (a) and (b) and the estimate $N_\delta \leq c_\delta^N$, valid for some $c_\delta > 0$, to conclude that there exists a constant c_2 independent of N such that for all C large enough, independently of N,

$$P(\lambda_N^N > C) = P(\sup_{z:\|z\|_2=1} z^T X_N z > C) \leq e^{-c_2 C^2 N}.$$

2.1.7 Central limit theorems for moments

Our goal here is to derive a simple version of a central limit theorem (CLT) for linear statistics of the eigenvalues of Wigner matrices. With X_N a Wigner matrix and L_N the associated empirical measure of its eigenvalues, set $W_{N,k} := N[\langle L_N, x^k \rangle - \langle \bar{L}_N, x^k \rangle]$. Let

$$\Phi(x) = \frac{1}{\sqrt{2\pi}} \int_{-\infty}^{x} e^{-u^2/2} du$$

denote the Gaussian distribution. We set σ_k^2 as in (2.1.44) below, and prove the following.

Theorem 2.1.31 *The law of the sequence of random variables $W_{N,k}/\sigma_k$ converges weakly to the standard Gaussian distribution. More precisely,*

$$\lim_{N\to\infty} P\left(\frac{W_{N,k}}{\sigma_k} \leq x\right) = \Phi(x). \tag{2.1.38}$$

Proof of Theorem 2.1.31 Most of the proof consists of a variance computation. The reader interested only in a proof of convergence to a Gaussian distribution (without worrying about the actual variance) can skip to the text following equation (2.1.45).

Recall the notation $\mathscr{W}_{k,t}^{(2)}$, see (2.1.24). Using (2.1.26), we have

$$\begin{aligned}\lim_{N\to\infty} E(W_{N,k}^2) &= \lim_{N\to\infty} N^2 \left[E(\langle L_N, x^k\rangle^2) - \langle \bar{L}_N, x^k\rangle^2\right] \\ &= \sum_{a=(w_1,w_2)\in\mathscr{W}_{k,k}^{(2)}} \Big[\prod_{e\in E_a^c} E(Z_{1,2}^{N_e^a}) \prod_{e\in E_a^s} E(Y_1^{N_e^a}) \\ &\quad - \prod_{e\in E_{w_1}^c} E(Z_{1,2}^{N_e^{w_1}}) \prod_{e\in E_{w_1}^s} E(Y_1^{N_e^{w_1}}) \prod_{e\in E_{w_2}^c} E(Z_{1,2}^{N_e^{w_2}}) \prod_{e\in E_{w_2}^s} E(Y_1^{N_e^{w_2}})\Big].\end{aligned} \tag{2.1.39}$$

Note that if $a=(w_1,w_2)\in\mathscr{W}_{k,k}^{(2)}$ then G_a is connected and possesses k vertices and at most k edges, each visited at least twice by the paths generated by a. Hence, with k vertices, G_a possesses either $k-1$ or k edges. Let $\mathscr{W}_{k,k,+}^{(2)}$ denote the subset of $\mathscr{W}_{k,k}^{(2)}$ such that $|E_a|=k$ (that is, G_a is *unicyclic*, i.e. "possesses one edge too many to be a tree") and let $\mathscr{W}_{k,k,-}^{(2)}$ denote the subset of $\mathscr{W}_{k,k}^{(2)}$ such that $|E_a|=k-1$.

Suppose first $a\in\mathscr{W}_{k,k,-}^{(2)}$. Then, G_a is a tree, $E_a^s=\emptyset$, and necessarily G_{w_i} is a subtree of G_a. This implies that k is even and that $|E_{w_i}|\le k/2$. In this case, for $E_{w_1}\cap E_{w_2}\neq\emptyset$ one must have $|E_{w_i}|=k/2$, which implies that all edges of G_{w_i} are visited twice by the walk generated by w_i, and exactly one edge is visited twice by both w_1 and w_2. In particular, w_i are both closed Wigner words of length $k+1$. The emerging picture is of two trees with $k/2$ edges each "glued together" at one edge. Since there are $C_{k/2}$ ways to chose each of the trees, $k/2$ ways of choosing (in each tree) the edge to be glued together, and 2 possible orientations for the gluing, we deduce that

$$|\mathscr{W}_{k,k,-}^{(2)}| = 2\left(\frac{k}{2}\right)^2 C_{k/2}^2. \tag{2.1.40}$$

Further, for each $a\in\mathscr{W}_{k,k,-}^{(2)}$,

$$\begin{aligned}&\Big[\prod_{e\in E_a^c} E(Z_{1,2}^{N_e^a}) \prod_{e\in E_a^s} E(Y_1^{N_e^a}) \\ &\quad - \prod_{e\in E_{w_1}^c} E(Z_{1,2}^{N_e^{w_1}}) \prod_{e\in E_{w_1}^s} E(Y_1^{N_e^{w_1}}) \prod_{e\in E_{w_2}^c} E(Z_{1,2}^{N_e^{w_2}}) \prod_{e\in E_{w_2}^s} E(Y_1^{N_e^{w_2}})\Big] \\ &= E(Z_{1,2}^4)[E(Z_{1,2}^2)]^{k-2} - [E(Z_{1,2}^2)]^k = E(Z_{1,2}^4)-1.\end{aligned} \tag{2.1.41}$$

We next turn to consider $\mathscr{W}^{(2)}_{k,k,+}$. In order to do so, we need to understand the structure of unicyclic graphs.

Definition 2.1.32 A graph $G = (V,E)$ is called a *bracelet* if there exists an enumeration $\alpha_1, \alpha_2, \ldots, \alpha_r$ of V such that

$$E = \begin{cases} \{\{\alpha_1,\alpha_1\}\} & \text{if } r=1, \\ \{\{\alpha_1,\alpha_2\}\} & \text{if } r=2, \\ \{\{\alpha_1,\alpha_2\},\{\alpha_2,\alpha_3\},\{\alpha_3,\alpha_1\}\} & \text{if } r=3, \\ \{\{\alpha_1,\alpha_2\},\{\alpha_2,\alpha_3\},\{\alpha_3,\alpha_4\},\{\alpha_4,\alpha_1\}\} & \text{if } r=4, \end{cases}$$

and so on. We call r the *circuit length* of the bracelet G.

We need the following elementary lemma, allowing one to decompose a unicyclic graph as a bracelet and its associated pendant trees. Recall that a graph $G = (V,E)$ is unicyclic if it is connected and $|E| = |V|$.

Lemma 2.1.33 *Let $G = (V,E)$ be a unicyclic graph. Let Z be the subgraph of G consisting of all $e \in E$ such that $G \setminus e$ is connected, along with all attached vertices. Let r be the number of edges of Z. Let F be the graph obtained from G by deleting all edges of Z. Then, Z is a bracelet of circuit length r, F is a forest with exactly r connected components, and Z meets each connected component of F in exactly one vertex. Further, $r = 1$ if $E^s \neq \emptyset$ while $r \geq 3$ otherwise.*

We call Z the *bracelet* of G. We call r the *circuit length* of G, and each of the components of F we call a *pendant tree*. (The case $r = 2$ is excluded from Lemma 2.1.33 because a bracelet of circuit length 2 is a tree and thus never unicyclic.) See Figure 2.1.4.

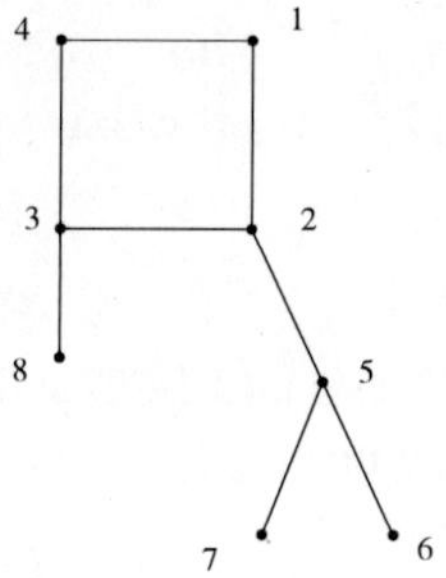

Fig. 2.1.4. The bracelet 1234 of circuit length 4, and the pendant trees, associated with the unicyclic graph corresponding to $[12565752341, 2383412]$

Coming back to $a \in \mathscr{W}^{(2)}_{k,k,+}$, let Z_a be the associated bracelet (with circuit length $r=1$ or $r\geq 3$). Note that for any $e\in E_a$ one has $N_e^a=2$. We claim next that $e\in Z_a$ if and only if $N_e^{w_1}=N_e^{w_2}=1$. On the one hand, if $e\in Z_a$ then $(V_a,E_a\setminus e)$ is a tree. If one of the paths determined by w_1 and w_2 fail to visit e then all edges visited by this path determine a walk on a tree and therefore the path visits each edge exactly twice. This then implies that the set of edges visited by the walks are disjoint, a contradiction. On the other hand, if $e=(x,y)$ and $N_e^{w_i}=1$, then all vertices in V_{w_i} are connected to x and to y by a path using only edges from $E_{w_i}\setminus e$. Hence, $(V_a,E_a\setminus e)$ is connected, and thus $e\in Z_a$.

Thus, any $a=(w_1,w_2)\in\mathscr{W}^{(2)}_{k,k,+}$ with bracelet length r can be constructed from the following data: the pendant trees $\{T_j^i\}_{j=1}^r$ (possibly empty) associated with each word w_i and each vertex j of the bracelet Z_a, the starting point for each word w_i on the graph consisting of the bracelet Z_a and trees $\{T_j^i\}$, and whether Z_a is traversed by the words w_i in the same or in opposing directions (in the case $r\geq 3$). In view of the above, counting the number of ways to attach trees to a bracelet of length r, and then the distinct number of non-equivalent ways to choose starting points for the paths on the resulting graph, there are exactly

$$\frac{2^{\mathbf{1}_{r\geq 3}}k^2}{r}\left(\sum_{\substack{k_i\geq 0:\\ 2\sum_{i=1}^r k_i=k-r}}\prod_{i=1}^r C_{k_i}\right)^2 \tag{2.1.42}$$

elements of $\mathscr{W}^{(2)}_{k,k,+}$ with bracelet of length r. Further, for $a\in\mathscr{W}^{(2)}_{k,k,+}$ we have

$$\begin{aligned}
&\Big[\prod_{e\in E_a^c}E(Z_{1,2}^{N_e^a})\prod_{e\in E_a^s}E(Y_1^{N_e^a})\\
&\qquad-\prod_{e\in E_{w_1}^c}E(Z_{1,2}^{N_e^{w_1}})\prod_{e\in E_{w_1}^s}E(Y_1^{N_e^{w_1}})\prod_{e\in E_{w_2}^c}E(Z_{1,2}^{N_e^{w_2}})\prod_{e\in E_{w_2}^s}E(Y_1^{N_e^{w_2}})\Big]\\
&=\begin{cases}(E(Z_{1,2}^2))^k-0 & \text{if } r\geq 3,\\ (E(Z_{1,2}^2))^{k-1}EY_1^2-0 & \text{if } r=1\end{cases}\\
&=\begin{cases}1 & \text{if } r\geq 3,\\ EY_1^2 & \text{if } r=1.\end{cases}
\end{aligned}\tag{2.1.43}$$

Combining (2.1.39), (2.1.40), (2.1.41), (2.1.42) and (2.1.43), and setting $C_x=0$ if x is not an integer, one obtains, with

$$\sigma_k^2=k^2C^2_{\frac{k-1}{2}}EY_1^2+\frac{k^2}{2}C^2_{\frac{k}{2}}[EZ_{1,2}^4-1]+\sum_{r=3}^{\infty}\frac{2k^2}{r}\left(\sum_{\substack{k_i\geq 0:\\ 2\sum_{i=1}^r k_i=k-r}}\prod_{i=1}^r C_{k_i}\right)^2, \tag{2.1.44}$$

that

$$\sigma_k^2 = \lim_{N\to\infty} EW_{N,k}^2 \,. \tag{2.1.45}$$

The rest of the proof consists in verifying that, for $j \geq 3$,

$$\lim_{N\to\infty} E\left(\frac{W_{N,k}}{\sigma_k}\right)^j = \begin{cases} 0 & \text{if } j \text{ is odd}, \\ (j-1)!! & \text{if } j \text{ is even}, \end{cases} \tag{2.1.46}$$

where $(j-1)!! = (j-1)(j-3)\cdots 1$. Indeed, this completes the proof of the theorem since the right hand side of (2.1.46) coincides with the moments of the Gaussian distribution Φ, and the latter moments determine the Gaussian distribution by an application of Carleman's theorem (see, e.g., [Dur96]), since

$$\sum_{n=1}^{\infty} [(2j-1)!!]^{(-1/2j)} = \infty \,.$$

To see (2.1.46), recall, for a multi-index $\mathbf{i} = (i_1, \ldots, i_k)$, the terms $\bar{T}_{\mathbf{i}}^N$ of (2.1.15), and the associated closed word $w_{\mathbf{i}}$. Then, as in (2.1.21), one has

$$E(W_{N,k}^j) = \sum_{\substack{i_1^n,\ldots,i_k^n=1 \\ n=1,2,\ldots j}}^{N} \bar{T}_{\mathbf{i}^1,\mathbf{i}^2,\ldots,\mathbf{i}^j}^N \,, \tag{2.1.47}$$

where

$$\bar{T}_{\mathbf{i}^1,\mathbf{i}^2,\ldots,\mathbf{i}^j}^N = E\left[\prod_{n=1}^{j} (T_{\mathbf{i}^n}^N - ET_{\mathbf{i}^n}^N)\right] \,. \tag{2.1.48}$$

Note that $\bar{T}_{\mathbf{i}^1,\mathbf{i}^2,\ldots,\mathbf{i}^j}^N = 0$ if the graph generated by any word $w_n := w_{\mathbf{i}^n}$ does not have an edge in common with any graph generated by the other words $w_{n'}$, $n' \neq n$. Motivated by that and our variance computation, let

> $\mathscr{W}_{k,t}^{(j)}$ denote a set of representatives for equivalence classes of sentences a of weight t consisting of j closed words $(w_1, w_2, \ldots, w_j)$, each of length $k+1$, with $N_e^a \geq 2$ for each $e \in E_a$, and such that for each n there is an $n' = n'(n) \neq n$ such that $E_{w_n} \cap E_{w_{n'}} \neq \emptyset$. (2.1.49)

As in (2.1.25), one obtains

$$E(W_{N,k}^j) = \sum_{t=1}^{jk} C_{N,t} \sum_{a=(w_1,w_2,\ldots,w_j)\in\mathscr{W}_{k,t}^{(j)}} \bar{T}_{w_1,w_2,\ldots,w_j}^N := \sum_{t=1}^{jk} \frac{C_{N,t}}{N^{jk/2}} \sum_{a\in\mathscr{W}_{k,t}^{(j)}} \bar{T}_a \,. \tag{2.1.50}$$

The next lemma, whose proof is deferred to the end of the section, is concerned with the study of $\mathscr{W}_{k,t}^{(j)}$.

Lemma 2.1.34 *Let c denote the number of connected components of G_a for $a \in \bigcup_t \mathscr{W}_{k,t}^{(j)}$. Then, $c \leq \lfloor j/2 \rfloor$ and $\mathrm{wt}(a) \leq c - j + \lfloor (k+1)j/2 \rfloor$.*

In particular, Lemma 2.1.34 and (2.1.50) imply that

$$\lim_{N\to\infty} E(W_{N,k}^j) = \begin{cases} 0 & \text{if } j \text{ is odd}, \\ \sum_{a \in \mathscr{W}_{k,kj/2}^{(j)}} \bar{T}_a & \text{if } j \text{ is even}. \end{cases} \tag{2.1.51}$$

By Lemma 2.1.34, if $a \in \mathscr{W}_{k,kj/2}^{(j)}$ for j even then G_a possesses exactly $j/2$ connected components. This is possible only if there exists a permutation

$$\pi : \{1,\dots,j\} \to \{1,\dots,j\},$$

all of whose cycles have length 2 (that is, a *matching*), such that the connected components of G_a are the graphs $\{G_{(w_i,w_{\pi(i)})}\}$. Letting Σ_j^m denote the collection of all possible matchings, one thus obtains that for j even,

$$\begin{aligned} \sum_{a \in \mathscr{W}_{k,kj/2}^{(j)}} \bar{T}_a &= \sum_{\pi \in \Sigma_j^m} \prod_{i=1}^{j/2} \sum_{(w_i,w_{\pi(i)}) \in \mathscr{W}_{k,k}^{(2)}} \bar{T}_{w_i,w_{\pi(i)}} \\ &= \sum_{\pi \in \Sigma_j^m} \sigma_k^j = |\Sigma_j^m| \sigma_k^j = \sigma_k^j (j-1)!!, \end{aligned} \tag{2.1.52}$$

which, together with (2.1.51), completes the proof of Theorem 2.1.31. □

Proof of Lemma 2.1.34 That $c \leq \lfloor j/2 \rfloor$ is immediate from the fact that the subgraph corresponding to any word in a must have at least one edge in common with at least one subgraph corresponding to another word in a.

Next, put

$$a = [[\alpha_{i,n}]_{n=1}^k]_{i=1}^j, \; I = \bigcup_{i=1}^j \{i\} \times \{1,\dots,k\}, \; A = [\{\alpha_{i,n}, \alpha_{i,n+1}\}]_{(i,n)\in I}.$$

We visualize A as a left-justified table of j rows. Let $G' = (V', E')$ be any spanning forest in G_a, with c connected components. Since every connected component of G' is a tree, we have

$$\mathrm{wt}(a) = c + |E'|. \tag{2.1.53}$$

Now let $X = \{X_{in}\}_{(i,n)\in I}$ be a table of the same "shape" as A, but with all entries equal either to 0 or 1. We call X an *edge-bounding table* under the following conditions.

- For all $(i,n) \in I$, if $X_{i,n} = 1$, then $A_{i,n} \in E'$.

We next note that if $w \in \mathcal{W}_{k,k/2+1}$ then G_w is a tree: indeed, G_w is a connected graph with $|V_w| = k/2+1$, hence $|E_w| \geq k/2$, while the condition $N_e^w \geq 2$ for each $e \in E_w$ implies that $|E_w| \leq k/2$. Thus, $|E_w| = |V_w| - 1$, implying that G_w is a tree, that is a connected graph with no loops. Further, the above implies that E_w^s is empty for $w \in \mathcal{W}_{k,k/2+1}$, and thus, for k even,

$$\lim_{N\to\infty} \langle \bar{L}_N, x^k \rangle = |\mathcal{W}_{k,k/2+1}|. \tag{2.1.19}$$

We may now complete the

Proof of Lemma 2.1.6 Let k be even. It is convenient to choose the set of representatives $\mathcal{W}_{k,k/2+1}$ such that each word $w = v_1 \cdots v_{k+1}$ in that set satisfies, for $i = 1, \ldots, k+1$, the condition that $\{v_1, \ldots, v_i\}$ is an interval in $\mathbb{Z}$ beginning at 1. (There is a unique choice of such representatives.) Each element $w \in \mathcal{W}_{k,k/2+1}$ determines a path $v_1, v_2, \ldots, v_k, v_{k+1} = v_1$ of length k on the tree G_w. We refer to this path as the *exploration process* associated with w. Let $d(v, v')$ denote the distance between vertices v, v' on the tree G_w, i.e. the length of the shortest path on the tree beginning at v and terminating at v'. Setting $x_i = d(v_{i+1}, v_1)$, one sees that each word $w \in \mathcal{W}_{k,k/2+1}$ defines a Dyck path $D(w) = (x_1, x_2, \ldots, x_k)$ of length k. See Figure 2.1.2 for an example of such coding. Conversely, given a Dyck path $\mathbf{x} = (x_1, \ldots, x_k)$, one may construct a word $w = T(\mathbf{x}) \in \mathcal{W}_{k,k/2+1}$ by recursively constructing an increasing sequence $w_2, \ldots, w_k = w$ of words, as follows. Put $w_2 = (1,2)$. For $i > 2$, if $x_{i-1} = x_{i-2} + 1$, then w_i is obtained by adjoining on the right of w_{i-1} the smallest positive integer not appearing in w_{i-1}. Otherwise, w_i is obtained by adjoining on the right of w_{i-1} the next-to-last letter of w_{i-1}. Note that for all i, G_{w_i} is a tree (because G_{w_2} is a tree and, inductively, at stage i, either a backtrack is added to the exploration process on $G_{w_{i-1}}$ or a leaf is added to $G_{w_{i-1}}$). Furthermore, the distance in G_{w_i} between first and last letters of w_i equals x_{i-1}, and therefore, $D(w) = (x_1, \ldots, x_k)$. With our choice of representatives, $T(D(w)) = w$, because each uptick in the Dyck path $D(w)$ starting at location $i-2$ corresponds to adjoinment on the right of w_{i-1} of a new letter, which is uniquely determined by $\operatorname{supp} w_{i-1}$, whereas each downtick at location $i-2$ corresponds to the adjoinment of the next-to-last letter in w_{i-1}. This establishes a bijection between Dyck paths of length k and $\mathcal{W}_{k,k/2+1}$. Lemma 2.1.3 then establishes that

$$|\mathcal{W}_{k,k/2+1}| = C_{k/2}. \tag{2.1.20}$$

This completes the proof of Lemma 2.1.6. □

From the proof of Lemma 2.1.6 we extract as a further benefit a proof of a fact needed in Chapter 5. Let k be an even positive integer and let $\mathcal{K}_k = \{1, \ldots, k\}$. Recall the notion of non-crossing partition of $\mathcal{K}_k$, see Definition 2.1.4. We define on r_k provided in the assumptions. The key step is contained in the following combinatorial lemma that gives information on the sets $\mathscr{W}_{k,t}$, see (2.1.16).

Lemma 2.1.23 *For all integers $k > 2t-2$ one has the estimate*

$$|\mathscr{W}_{k,t}| \le 2^k k^{3(k-2t+2)}. \tag{2.1.30}$$

The proof of Lemma 2.1.23 is deferred to the end of this section.

Equipped with Lemma 2.1.23, we have for $2k < N$, using (2.1.17),

$$\begin{aligned}\langle \bar{L}_N, x^{2k}\rangle &\le \sum_{t=1}^{k+1} N^{t-(k+1)}|\mathscr{W}_{2k,t}| \sup_{w\in\mathscr{W}_{2k,t}} \prod_{e\in E_w^c} E(Z_{1,2}^{N_e^w}) \prod_{e\in E_w^s} E(Y_1^{N_e^w}) \qquad (2.1.31)\\ &\le 4^k \sum_{t=1}^{k+1}\left(\frac{(2k)^6}{N}\right)^{k+1-t} \sup_{w\in\mathscr{W}_{2k,t}} \prod_{e\in E_w^c} E(Z_{1,2}^{N_e^w}) \prod_{e\in E_w^s} E(Y_1^{N_e^w}).\end{aligned}$$

To evaluate the last expectation, fix $w \in \mathscr{W}_{2k,t}$, and let l denote the number of edges in E_w^c with $N_e^w = 2$. Hölder's inequality then gives

$$\prod_{e\in E_w^c} E(Z_{1,2}^{N_e^w}) \prod_{e\in E_w^s} E(Y_1^{N_e^w}) \le r_{2k-2l},$$

with the convention that $r_0 = 1$. Since G_w is connected, $|E_w^c| \ge |V_w| - 1 = t-1$. On the other hand, by noting that $N_e^w \ge 3$ for $|E_w^c| - l$ edges, one has $2k \ge 3(|E_w^c| - l) + 2l + 2|E_w^s|$. Hence, $2k - 2l \le 6(k+1-t)$. Since r_{2q} is a nondecreasing function of q bounded below by 1, we get, substituting back in (2.1.31), that for some constant $c_1 = c_1(C) > 0$ and all $k < N$,

$$\begin{aligned}\langle \bar{L}_N, x^{2k}\rangle &\le 4^k \sum_{t=1}^{k+1}\left(\frac{(2k)^6}{N}\right)^{k+1-t} r_{6(k+1-t)} \qquad (2.1.32)\\ &\le 4^k \sum_{t=1}^{k+1}\left(\frac{(2k)^6(6(k+1-t))^{6C}}{N}\right)^{k+1-t} \le 4^k \sum_{i=0}^{k}\left(\frac{k^{c_1}}{N}\right)^i.\end{aligned}$$

Choose next a sequence $k(N) \to_{N\to\infty} \infty$ such that

$$k(N)^{c_1}/N \to_{N\to\infty} 0 \quad \text{but} \quad k(N)/\log N \to_{N\to\infty} \infty.$$

Then, for any $\delta > 0$, and all N large,

$$\begin{aligned}P(\lambda_N^N > (2+\delta)) &\le P(N\langle L_N, x^{2k(N)}\rangle > (2+\delta)^{2k(N)})\\ &\le \frac{N\langle \bar{L}_N, x^{2k(N)}\rangle}{(2+\delta)^{2k(N)}} \le \frac{2N4^{k(N)}}{(2+\delta)^{2k(N)}} \to_{N\to\infty} 0,\end{aligned}$$

completing the proof of Theorem 2.1.22, modulo Lemma 2.1.23. □

- For each $e \in E'$ there exist distinct $(i_1, n_1), (i_2, n_2) \in I$ such that $X_{i_1,n_1} = X_{i_2,n_2} = 1$ and $A_{i_1,n_1} = A_{i_2,n_2} = e$.
- For each $e \in E'$ and index $i \in \{1, \ldots, j\}$, if e appears in the ith row of A then there exists $(i, n) \in I$ such that $A_{i,n} = e$ and $X_{i,n} = 1$.

For any edge-bounding table X the corresponding quantity $\frac{1}{2}\sum_{(i,n)\in I} X_{i,n}$ bounds $|E'|$. At least one edge-bounding table exists, namely the table with a 1 in position (i, n) for each $(i, n) \in I$ such that $A_{i,n} \in E'$ and 0 elsewhere. Now let X be an edge-bounding table such that for some index i_0 all the entries of X in the i_0th row are equal to 1. Then the closed word w_{i_0} is a walk in G', and hence every entry in the i_0th row of A appears there an even number of times and *a fortiori* at least twice. Now choose $(i_0, n_0) \in I$ such that $A_{i_0,n_0} \in E'$ appears in more than one row of A. Let Y be the table obtained by replacing the entry 1 of X in position (i_0, n_0) by the entry 0. Then Y is again an edge-bounding table. Proceeding in this way we can find an edge-bounding table with 0 appearing at least once in every row, and hence we have $|E'| \le \lfloor \frac{|I|-j}{2} \rfloor$. Together with (2.1.53) and the definition of I, this completes the proof. □

Exercise 2.1.35 (from [AnZ05]) Prove that the random vector $\{W_{N,i}\}_{i=1}^k$ satisfies a multidimensional CLT (as $N \to \infty$). (See Exercise 2.3.7 for an extension of this result.)

2.2 Complex Wigner matrices

In this section we describe the (minor) modifications needed when one considers the analog of Wigner's theorem for Hermitian matrices. Compared with (2.1.2), we will have complex-valued random variables $Z_{i,j}$. That is, start with two independent families of i.i.d. random variables $\{Z_{i,j}\}_{1\le i<j}$ (complex-valued) and $\{Y_i\}_{1\le i}$ (real-valued), zero mean, such that $EZ_{1,2}^2 = 0$, $E|Z_{1,2}|^2 = 1$ and, for all integers $k \ge 1$,

$$r_k := \max\left(E|Z_{1,2}|^k, E|Y_1|^k\right) < \infty. \tag{2.2.1}$$

Consider the (Hermitian) $N \times N$ matrix X_N with entries

$$X_N^*(j,i) = X_N(i,j) = \begin{cases} Z_{i,j}/\sqrt{N} & \text{if } i < j, \\ Y_i/\sqrt{N} & \text{if } i = j. \end{cases} \tag{2.2.2}$$

We call such a matrix a *Hermitian Wigner matrix*, and if the random variables $Z_{i,j}$ and Y_i are Gaussian, we use the term *Gaussian Hermitian Wigner matrix*. The case of Gaussian Hermitian Wigner matrices in which $EY_1^2 = 1$ is of particular importance, and for reasons that will become clearer in Chapter 3, such matrices (rescaled by $\sqrt{N}$) are referred to as Gaussian unitary ensemble (GUE) matrices.

As before, let λ_i^N denote the (real) eigenvalues of X_N, with $\lambda_1^N \leq \lambda_2^N \leq \cdots \leq \lambda_N^N$, and recall that the empirical distribution of the eigenvalues is the probability measure on $\mathbb{R}$ defined by

$$L_N = \frac{1}{N}\sum_{i=1}^{N} \delta_{\lambda_i^N}.$$

The following is the analog of Theorem 2.1.1.

Theorem 2.2.1 (Wigner) *For a Hermitian Wigner matrix, the empirical measure L_N converges weakly, in probability, to the semicircle distribution.*

As in Section 2.1.2, the proof of Theorem 2.2.1 is a direct consequence of the following two lemmas.

Lemma 2.2.2 *For any $k \in \mathbb{N}$,*

$$\lim_{N\to\infty} m_k^N = m_k.$$

Lemma 2.2.3 *For any $k \in \mathbb{N}$ and $\varepsilon > 0$,*

$$\lim_{N\to\infty} P\left(\left|\langle L_N, x^k\rangle - \langle \bar{L}_N, x^k\rangle\right| > \varepsilon\right) = 0.$$

Proof of Lemma 2.2.2 We recall the machinery introduced in Section 2.1.3. Thus, an N-word $w = (s_1, \ldots, s_k)$ defines a graph $G_w = (V_w, E_w)$ and a path on the graph. For our purpose, it is convenient to keep track of the direction in which edges are traversed by the path. Thus, given an edge $e = \{s, s'\}$, with $s < s'$, we define $N_e^{w,+}$ as the number of times the edge is traversed from s to s', and we set $N_e^{w,-} = N_e^w - N_e^{w,+}$ as the number of times it is traversed in the reverse direction.

Recalling the equality (2.1.10), we now have instead of (2.1.15) the equation

$$\bar{T}_{\mathbf{i}}^N = \frac{1}{N^{k/2}} \prod_{e\in E_{w_{\mathbf{i}}}^c} E(Z_{1,2}^{N_e^{w_{\mathbf{i}},+}} (Z_{1,2}^*)^{N_e^{w_{\mathbf{i}},-}}) \prod_{e\in E_{w_{\mathbf{i}}}^s} E(Y_1^{N_e^{w_{\mathbf{i}}}}). \qquad (2.2.3)$$

In particular, $\bar{T}_{\mathbf{i}}^N = 0$ unless $N_e^{w_{\mathbf{i}}} \geq 2$ for all $e \in E_{w_{\mathbf{i}}}$. Furthermore, since $EZ_{1,2}^2 = 0$, one has $T_{\mathbf{i}}^N = 0$ if $N_e^{w_{\mathbf{i}}} = 2$ and $N_e^{w_{\mathbf{i}},+} \neq 1$ for some $e \in E_{w_{\mathbf{i}}}$.

A slight complication occurs since the function

$$g_w(N_e^{w,+}, N_e^{w,-}) := E(Z_{1,2}^{N_e^{w,+}} (Z_{1,2}^*)^{N_e^{w,-}})$$

is not constant over equivalence classes of words (since changing the letters determining w may switch the role of $N_e^{w,+}$ and $N_e^{w,-}$ in the above expression). Note however that, for any $w \in \mathscr{W}_{k,t}$, one has

$$|g_w(N_e^{w,+}, N_e^{w,-})| \leq E(|Z_{1,2}|^{N_e^w}).$$

On the other hand, any $w \in \mathscr{W}_{k,k/2+1}$ satisfies that G_w is a tree, with each edge visited exactly twice by the path determined by w. Since the latter path starts and ends at the same vertex, one has $N_e^{w,+} = N_e^{w,-} = 1$ for each $e \in E_w$. Thus, repeating the argument in Section 2.1.3, the finiteness of r_k implies that

$$\lim_{N\to\infty} \langle \bar{L}_N, x^k \rangle = 0, \text{ if } k \text{ is odd},$$

while, for k even,

$$\lim_{N\to\infty} \langle \bar{L}_N, x^k \rangle = |\mathscr{W}_{k,k/2+1}| g_w(1,1). \tag{2.2.4}$$

Since $g_w(1,1) = 1$, the proof is completed by applying (2.1.20). □

Proof of Lemma 2.2.3 The proof is a rerun of the proof of Lemma 2.1.7, using the functions $g_w(N_e^{w,+}, N_e^{w,-})$, defined in the course of proving Lemma 2.2.2. The proof boils down to showing that $\mathscr{W}_{k,k+2}^{(2)}$ is empty, a fact that was established in the course of proving Lemma 2.1.7. □

Exercise 2.2.4 We consider in this exercise *Hermitian self-dual* matrices, which in the Gaussian case reduce to matrices from the Gaussian symplectic ensemble discussed in greater detail in Section 4.1. For any $a, b \in \mathbb{C}$, set

$$m_{a,b} = \begin{pmatrix} a & b \\ -b^* & a^* \end{pmatrix} \in \text{Mat}_2(\mathbb{C}).$$

Let $\{Z_{i,j}^{(k)}\}_{1\leq i<j, 1\leq k\leq 4}$ and $\{Y_i\}_{1\leq i\leq N}$ be independent zero mean real-valued random variables of unit variance satisfying the condition (2.1.1). For $1 \leq i < j \leq N$, set $a_{i,j} = (Z_{i,j}^{(1)} + \mathbf{i}Z_{i,j}^{(2)})/(2\sqrt{N})$, $b_{i,j} = (Z_{i,j}^{(3)} + \mathbf{i}Z_{i,j}^{(4)})/(2\sqrt{N})$, $a_{i,i} = Y_i/\sqrt{N}$, $b_{i,i} = 0$, and write $m_{i,j} = m_{a_{i,j}, b_{i,j}}$ for $1 \leq i \leq j \leq N$. Finally, construct a Hermitian matrix $X_N \in \mathscr{H}_{2N}^{(2)}$ from the 2-by-2 matrices $m_{i,j}$ by setting $X_N(i,j) = m_{i,j}$, $1 \leq i \leq j \leq N$.
(a) Let

$$J_1 = \begin{pmatrix} 0 & 1 \\ -1 & 0 \end{pmatrix} \in \text{Mat}_2(\mathbb{R}),$$

and let $J_N = \text{diag}(J_1, \ldots, J_1) \in \text{Mat}_{2N}(\mathbb{R})$ be the block diagonal matrix with blocks J_1 on the diagonal. Check that $X_N = J_N X_N J_N^{-1}$. This justifies the name "self-dual".
(b) Verify that the eigenvalues of X_N occur in pairs, and that Wigner's Theorem continues to hold.

2.3 Concentration for functionals of random matrices and logarithmic Sobolev inequalities

In this short section we digress slightly and prove that certain functionals of random matrices have the concentration property, namely, with high probability these functionals are close to their mean value. A more complete treatment of concentration inequalities and their application to random matrices is postponed to Section 4.4. The results of this section will be useful in Section 2.4, where they will play an important role in the proof of Wigner's Theorem via the Stieltjes transform.

2.3.1 Smoothness properties of linear functions of the empirical measure

Let us recall that if X is a symmetric (Hermitian) matrix and f is a bounded measurable function, $f(X)$ is defined as the matrix with the same eigenvectors as X but with eigenvalues that are the image by f of those of X; namely, if e is an eigenvector of X with eigenvalue λ, $Xe = \lambda e$, $f(X)e := f(\lambda)e$. In terms of the spectral decomposition $X = UDU^*$ with U orthogonal (unitary) and D diagonal real, one has $f(X) = Uf(D)U^*$ with $f(D)_{ii} = f(D_{ii})$. For $M \in \mathbb{N}$, we denote by $\langle \cdot, \cdot \rangle$ the Euclidean scalar product on $\mathbb{R}^M$ (or $\mathbb{C}^M$), $\langle x, y \rangle = \sum_{i=1}^M x_i y_i$ ($\langle x, y \rangle = \sum_{i=1}^M x_i y_i^*$), and by $||\cdot||_2$ the associated norm $||x||_2^2 = \langle x, x \rangle$.

General functions of independent random variables need not, in general, satisfy a concentration property. Things are different when the functions involved satisfy certain regularity conditions. It is thus reassuring to see that linear functionals of the empirical measure, viewed as functions of the matrix entries, do possess some regularity properties.

Throughout this section, we denote the Lipschitz constant of a function $G : \mathbb{R}^M \to \mathbb{R}$ by

$$|G|_{\mathscr{L}} := \sup_{x \neq y \in \mathbb{R}^M} \frac{|G(x) - G(y)|}{||x - y||_2},$$

and call G a *Lipschitz function* if $|G|_{\mathscr{L}} < \infty$. The following lemma is an immediate application of Lemma 2.1.19. In its statement, we identify $\mathbb{C}$ with $\mathbb{R}^2$.

Lemma 2.3.1 *Let $g : \mathbb{R}^N \to \mathbb{R}$ be Lipschitz with Lipschitz constant $|g|_{\mathscr{L}}$. Then, with X denoting the Hermitian matrix with entries $X(i,j)$, the map*

$$\{X(i,j)\}_{1 \leq i \leq j \leq N} \mapsto g(\lambda_1(X), \ldots, \lambda_N(X))$$

is a Lipschitz function on $\mathbb{R}^{N^2}$ with Lipschitz constant bounded by $\sqrt{2}|g|_{\mathscr{L}}$. In *particular, if f is a Lipschitz function on $\mathbb{R}$,*

$$\{X(i,j)\}_{1\le i\le j\le N} \mapsto \mathrm{tr}(f(X))$$

is a Lipschitz function on $\mathbb{R}^{N(N+1)}$ with Lipschitz constant bounded by $\sqrt{2N}|f|_{\mathscr{L}}$.

2.3.2 Concentration inequalities for independent variables satisfying logarithmic Sobolev inequalities

We derive in this section concentration inequalities based on the logarithmic Sobolev inequality.

To begin with, recall that a probability measure P on $\mathbb{R}$ is said to satisfy the *logarithmic Sobolev inequality* (LSI) with constant c if, for any differentiable function f in $L^2(P)$,

$$\int f^2 \log \frac{f^2}{\int f^2 dP} dP \le 2c \int |f'|^2 dP.$$

It is not hard to check, by induction, that if P_i satisfy the LSI with constant c and if $P^{(M)} = \otimes_{i=1}^M P_i$ denotes the product measure on $\mathbb{R}^M$, then $P^{(M)}$ satisfies the LSI with constant c in the sense that, for every differentiable function F on $\mathbb{R}^M$,

$$\int F^2 \log \frac{F^2}{\int F^2 dP^{(M)}} dP^{(M)} \le 2c \int ||\nabla F||_2^2 dP^{(M)}, \tag{2.3.1}$$

where ∇F denotes the gradient of F. (See Exercise 2.3.4 for hints.) We note that if the law of a random variable X satisfies the LSI with constant c, then for any fixed $\alpha \neq 0$, the law of αX satisfies the LSI with constant $\alpha^2 c$.

Before discussing consequences of the logarithmic Sobolev inequality, we quote from [BoL00] a general sufficient condition for it to hold.

Lemma 2.3.2 *Let $V : \mathbb{R}^M \to \mathbb{R} \cup \infty$ satisfy that for some positive constant C, $V(x) - \|x\|_2^2/2C$ is convex. Then, the probability measure $\nu(dx) = Z^{-1}e^{-V(x)}\, dx$, where $Z = \int e^{-V(x)}dx$, satisfies the logarithmic Sobolev inequality with constant C. In particular, the standard Gaussian law on $\mathbb{R}^M$ satisfies the logarithmic Sobolev inequality with constant* 1.

The lemma is also a consequence of the Bakry–Emery criterion, see Theorem 4.4.18 in Section 4.4 for details.

The interest in the logarithmic Sobolev inequality, in the context of concentration inequalities, lies in the following argument, that among other things, shows that LSI implies sub-Gaussian tails.

Lemma 2.3.3 (Herbst) *Assume that P satisfies the LSI on $\mathbb{R}^M$ with constant c. Let G be a Lipschitz function on $\mathbb{R}^M$, with Lipschitz constant $|G|_{\mathscr{L}}$. Then for all $\lambda \in \mathbb{R}$,*

$$E_P[e^{\lambda(G-E_P(G))}] \leq e^{c\lambda^2|G|^2_{\mathscr{L}}/2}, \tag{2.3.2}$$

and so for all $\delta > 0$

$$P(|G - E_P(G)| \geq \delta) \leq 2e^{-\delta^2/2c|G|^2_{\mathscr{L}}}. \tag{2.3.3}$$

Note that part of the statement in Lemma 2.3.3 is that $E_P G$ is finite.

Proof of Lemma 2.3.3 Note first that (2.3.3) follows from (2.3.2). Indeed, by Chebyshev's inequality, for any $\lambda > 0$,

$$\begin{aligned} P(|G - E_P G| > \delta) &\leq e^{-\lambda\delta} E_P[e^{\lambda|G-E_P G|}] \\ &\leq e^{-\lambda\delta}(E_P[e^{\lambda(G-E_P G)}] + E_P[e^{-\lambda(G-E_P G)}]) \\ &\leq 2e^{-\lambda\delta} e^{c|G|^2_{\mathscr{L}}\lambda^2/2}. \end{aligned}$$

Optimizing with respect to λ (by taking $\lambda = \delta/c|G|^2_{\mathscr{L}}$) yields the bound (2.3.3).

Turning to the proof of (2.3.2), let us first assume that G is a bounded differentiable function such that

$$|| \, ||\nabla G||_2^2 ||_\infty := \sup_{x \in \mathbb{R}^M} \sum_{i=1}^{M} (\partial_{x_i} G(x))^2 < \infty.$$

Define

$$A_\lambda = \log E_P e^{2\lambda(G-E_P G)}.$$

Then, taking $F = e^{\lambda(G-E_P G)}$ in (2.3.1), some algebra reveals that for $\lambda > 0$,

$$\frac{d}{d\lambda}\left(\frac{A_\lambda}{\lambda}\right) \leq 2c|| \, ||\nabla G||_2^2||_\infty.$$

Now, because $G - E_P(G)$ is centered,

$$\lim_{\lambda \to 0+} \frac{A_\lambda}{\lambda} = 0$$

and hence integrating with respect to λ yields

$$A_\lambda \leq 2c|| \, ||\nabla G||_2^2||_\infty \lambda^2,$$

first for $\lambda \geq 0$ and then for any $\lambda \in \mathbb{R}$ by considering the function $-G$ instead of G. This completes the proof of (2.3.2) in the case that G is bounded and differentiable.

Let us now assume only that G is Lipschitz with $|G|_{\mathscr{L}} < \infty$. For $\varepsilon > 0$, define $\bar{G}_\varepsilon = G \wedge (-1/\varepsilon) \vee (1/\varepsilon)$, and note that $|\bar{G}_\varepsilon|_{\mathscr{L}} \leq |G|_{\mathscr{L}} < \infty$. Consider the regularization $G_\varepsilon(x) = p_\varepsilon * \bar{G}_\varepsilon(x) = \int \bar{G}_\varepsilon(y) p_\varepsilon(x-y) dy$ with the Gaussian density

$p_\varepsilon(x) = e^{-|x|^2/2\varepsilon}dx/\sqrt{(2\pi\varepsilon)^M}$ such that $p_\varepsilon(x)dx$ converges weakly towards the atomic measure δ_0 as ε converges to 0. Since, for any $x \in \mathbb{R}^M$,

$$|G_\varepsilon(x) - \bar{G}_\varepsilon(x)| \le |G|_{\mathscr{L}} \int ||y||_2 p_\varepsilon(y)dy = M|G|_{\mathscr{L}}\sqrt{\varepsilon},$$

G_ε converges pointwise towards G. Moreover, G_ε is Lipschitz, with Lipschitz constant bounded by $|G|_{\mathscr{L}}$ independently of ε. G_ε is also continuously differentiable and

$$\begin{aligned} \| \|\nabla G_\varepsilon\|_2^2\|_\infty &= \sup_{x\in\mathbb{R}^M} \sup_{u\in\mathbb{R}^M} \{2\langle\nabla G_\varepsilon(x), u\rangle - \|u\|_2^2\} \\ &\le \sup_{u,x\in\mathbb{R}^M} \sup_{\delta>0} \{2\delta^{-1}(G_\varepsilon(x+\delta u) - G_\varepsilon(x)) - \|u\|_2^2\} \\ &\le \sup_{u\in\mathbb{R}^M} \{2|G|_{\mathscr{L}}\|u\|_2 - \|u\|_2^2\} = |G|_{\mathscr{L}}^2. \end{aligned} \tag{2.3.4}$$

Thus, we can apply (2.3.2) in the bounded differentiable case to find that for any $\varepsilon > 0$ and all $\lambda \in \mathbb{R}$,

$$E_P[e^{\lambda G_\varepsilon}] \le e^{\lambda E_P G_\varepsilon} e^{c\lambda^2|G|_{\mathscr{L}}^2/2}. \tag{2.3.5}$$

Therefore, by Fatou's Lemma,

$$E_P[e^{\lambda G}] \le e^{\liminf_{\varepsilon\to0} \lambda E_P G_\varepsilon} e^{c\lambda^2|G|_{\mathscr{L}}^2/2}. \tag{2.3.6}$$

We next show that $\lim_{\varepsilon\to0} E_P G_\varepsilon = E_P G$, which, in conjunction with (2.3.6), will conclude the proof. Indeed, (2.3.5) implies that

$$P(|G_\varepsilon - E_P G_\varepsilon| > \delta) \le 2e^{-\delta^2/2c|G|_{\mathscr{L}}^2}. \tag{2.3.7}$$

Consequently,

$$\begin{aligned} E[(G_\varepsilon - E_P G_\varepsilon)^2] &= 2\int_0^\infty xP(|G_\varepsilon - E_P G_\varepsilon| > x)\,dx \\ &\le 4\int_0^\infty xe^{-\frac{x^2}{2c|G|_{\mathscr{L}}^2}}dx = 4c|G|_{\mathscr{L}}^2, \end{aligned} \tag{2.3.8}$$

so that the sequence $(G_\varepsilon - E_P G_\varepsilon)_{\varepsilon\ge0}$ is uniformly integrable. Now, G_ε converges pointwise towards G and therefore there exists a constant K, independent of ε, such that for $\varepsilon < \varepsilon_0$, $P(|G_\varepsilon| \le K) \ge \frac{3}{4}$. On the other hand, (2.3.7) implies that $P(|G_\varepsilon - E_P G_\varepsilon| \le r) \ge \frac{3}{4}$ for some r independent of ε. Thus,

$$\{|G_\varepsilon - E_P G_\varepsilon| \le r\} \cap \{|G_\varepsilon| \le K\} \subset \{|E_P G_\varepsilon| \le K + r\}$$

is not empty, providing a uniform bound on $(E_P G_\varepsilon)_{\varepsilon<\varepsilon_0}$. We thus deduce from (2.3.8) that $\sup_{\varepsilon<\varepsilon_0} E_P G_\varepsilon^2$ is finite, and hence $(G_\varepsilon)_{\varepsilon<\varepsilon_0}$ is uniformly integrable. In particular,

$$\lim_{\varepsilon\to0} E_P G_\varepsilon = E_P G < \infty,$$

which finishes the proof. □

Exercise 2.3.4 (From [Led01], page 98)
(a) Let $f \geq 0$ be a measurable function and set $\text{Ent}_P(f) = \int f \log(f/E_P f) dP$. Prove that

$$\text{Ent}_P(f) = \sup\{E_P f g : E_P e^g \leq 1\}.$$

(b) Use induction and the above representation to prove (2.3.1).

2.3.3 *Concentration for Wigner-type matrices*

We consider in this section (symmetric) matrices X_N with independent (and not necessarily identically distributed) entries $\{X_N(i,j)\}_{1 \leq i \leq j \leq N}$. The following is an immediate corollary of Lemmas 2.3.1 and 2.3.3.

Theorem 2.3.5 *Suppose that the laws of the independent entries $\{X_N(i,j)\}_{1 \leq i \leq j \leq N}$ all satisfy the LSI with constant c/N. Then, for any Lipschitz function f on $\mathbb{R}$, for any $\delta > 0$,*

$$P\left(|\text{tr}(f(X_N)) - E[\text{tr}(f(X_N))]| \geq \delta N\right) \leq 2e^{-\frac{1}{4c|f|_{\mathscr{L}}^2} N^2 \delta^2}. \tag{2.3.9}$$

Further, for any $k \in \{1, \ldots, N\}$,

$$P\left(|f(\lambda_k(X_N)) - Ef(\lambda_k(X_N))| \geq \delta\right) \leq 2e^{-\frac{1}{4c|f|_{\mathscr{L}}^2} N \delta^2}. \tag{2.3.10}$$

We note that under the assumptions of Theorem 2.3.5, $E\lambda_N(X_N)$ is uniformly bounded, see Exercise 2.1.29 or Exercise 2.1.30. In the Gaussian case, more information is available, see the bibliographical notes (Section 2.7).

Proof of Theorem 2.3.5 To see (2.3.9), take

$$G(X_N(i,j), 1 \leq i \leq j \leq N) = \text{tr}(f(X_N)).$$

By Lemma 2.3.1, we see that if f is Lipschitz, G is also Lipschitz with constant bounded by $\sqrt{2N}|f|_{\mathscr{L}}$ and hence Lemma 2.3.3 with $M = N(N+1)/2$ yields the result. To see (2.3.10), apply the same argument to the function

$$\bar{G}(X_N(i,j), 1 \leq i \leq j \leq N) = f(\lambda_k(X_N)).$$

□

Remark 2.3.6 The assumption of Theorem 2.3.5 is satisfied for Gaussian matrices whose entries on or above the diagonal are independent, with variance bounded by c/N. In particular, the assumptions hold for Gaussian Wigner matrices. We emphasize that Theorem 2.3.5 applies also when the variance of $X_N(i,j)$ depends on i,j, e.g. when $X_N(i,j) = a_N(i,j)Y_N(i,j)$ with $Y_N(i,j)$ i.i.d. with law P satisfying the log-Sobolev inequality and $a(i,j)$ uniformly bounded (since if P satisfies the log-Sobolev inequality with constant c, the law of ax under P satisfies it also with a constant bounded by a^2c).

Exercise 2.3.7 (From [AnZ05]) Using Exercise 2.1.35, prove that if X_N is a Gaussian Wigner matrix and $f : \mathbb{R} \to \mathbb{R}$ is a C_b^1 function, then $N[\langle f, L_N\rangle - \langle f, \bar{L}_N\rangle]$ satisfies a central limit theorem.

2.4 Stieltjes transforms and recursions

We begin by recalling some classical results concerning the Stieltjes transform of a probability measure.

Definition 2.4.1 Let μ be a positive, finite measure on the real line. The *Stieltjes transform* of μ is the function

$$S_\mu(z) := \int_{\mathbb{R}} \frac{\mu(dx)}{x-z}, \; z \in \mathbb{C}\setminus\mathbb{R}.$$

Note that for $z \in \mathbb{C}\setminus\mathbb{R}$, both the real and imaginary parts of $1/(x-z)$ are continuous bounded functions of $x \in \mathbb{R}$ and, further, $|S_\mu(z)| \le \mu(\mathbb{R})/|\Im z|$. These crucial observations are used repeatedly in what follows.

Remark 2.4.2 The generating function $\hat{\beta}(z)$, see (2.1.6), is closely related to the Stieltjes transform of the semicircle distribution σ: for $|z| < 1/4$,

$$\begin{aligned}
\hat{\beta}(z) &= \sum_{k=0}^{\infty} z^k \int x^{2k}\sigma(x)dx = \int \left(\sum_{k=0}^{\infty} (zx^2)^k\right)\sigma(x)dx \\
&= \int \frac{1}{1-zx^2}\sigma(x)dx \\
&= \int \frac{1}{1-\sqrt{z}x}\sigma(x)dx = \frac{-1}{\sqrt{z}} S_\sigma(1/\sqrt{z}),
\end{aligned}$$

where the third equality uses the fact that the support of σ is the interval $[-2,2]$, and the fourth uses the symmetry of σ.

Stieltjes transforms can be inverted. In particular, one has

Theorem 2.4.3 *For any open interval I with neither endpoint on an atom of μ,*

$$\begin{aligned}\mu(I) &= \lim_{\varepsilon\to 0}\frac{1}{\pi}\int_I \frac{S_\mu(\lambda+\mathrm{i}\varepsilon)-S_\mu(\lambda-\mathrm{i}\varepsilon)}{2\mathrm{i}}d\lambda \\ &= \lim_{\varepsilon\to 0}\frac{1}{\pi}\int_I \Im S_\mu(\lambda+\mathrm{i}\varepsilon)d\lambda\,. \end{aligned}\tag{2.4.1}$$

Proof Note first that because

$$\Im S_\mu(\mathrm{i}) = \int \frac{1}{1+x^2}\mu(dx)\,,$$

we have that $S_\mu \equiv 0$ implies $\mu = 0$. So assume next that S_μ does not vanish identically. Then, since

$$\lim_{y\uparrow+\infty} y\Im S_\mu(\mathrm{i}y) = \lim_{y\uparrow+\infty}\int \frac{y^2}{x^2+y^2}\mu(dx) = \mu(\mathbb{R})$$

by bounded convergence, we may and will assume that $\mu(\mathbb{R}) = 1$, i.e. that μ is a probability measure.

Let X be distributed according to μ, and denote by C_ε a random variable, independent of X, Cauchy distributed with parameter ε, i.e. the law of C_ε has density

$$\frac{\varepsilon dx}{\pi(x^2+\varepsilon^2)}\,.\tag{2.4.2}$$

Then, $\Im S_\mu(\lambda+\mathrm{i}\varepsilon)/\pi$ is nothing but the density (with respect to Lebesgue measure) of the law of $X+C_\varepsilon$ evaluated at $\lambda\in\mathbb{R}$. The convergence in (2.4.1) is then just a rewriting of the weak convergence of the law of $X+C_\varepsilon$ to that of X, as $\varepsilon\to 0$. □

Theorem 2.4.3 allows for the reconstruction of a measure from its Stieltjes transform. Further, one has the following.

Theorem 2.4.4 *Let $\mu_n \in M_1(\mathbb{R})$ be a sequence of probability measures.*
(a) If μ_n converges weakly to a probability measure μ then $S_{\mu_n}(z)$ converges to $S_\mu(z)$ for each $z\in\mathbb{C}\setminus\mathbb{R}$.
(b) If $S_{\mu_n}(z)$ converges for each $z\in\mathbb{C}\setminus\mathbb{R}$ to a limit $S(z)$, then $S(z)$ is the Stieltjes transform of a sub-probability measure μ, and μ_n converges vaguely to μ.
(c) If the probability measures μ_n are random and, for each $z\in\mathbb{C}\setminus\mathbb{R}$, $S_{\mu_n}(z)$ converges in probability to a deterministic limit $S(z)$ that is the Stieltjes transform of a probability measure μ, then μ_n converges weakly in probability to μ.

(We recall that μ_n *converges vaguely* to μ if, for any continuous function f on $\mathbb{R}$ that decays to 0 at infinity, $\int f d\mu_n \to \int f d\mu$. Recall also that a positive measure μ on $\mathbb{R}$ is a *sub-probability measure* if it satisfies $\mu(\mathbb{R}) \le 1$.)

Proof Part (a) is a restatement of the notion of weak convergence. To see part (b), let n_k be a subsequence on which μ_{n_k} converges vaguely (to a sub-probability measure μ). (Such a subsequence always exists by Helly's selection theorem.) Because $x \mapsto 1/(z-x)$, for $z \in \mathbb{C} \setminus \mathbb{R}$, is continuous and decays to zero at infinity, one obtains the convergence $S_{\mu_{n_k}}(z) \to S_\mu(z)$ pointwise for such z. From the hypothesis, it follows that $S(z) = S_\mu(z)$. Applying Theorem 2.4.3, we conclude that all vaguely convergent subsequences converge to the same μ, and hence $\mu_n \to \mu$ vaguely.

To see part (c), fix a sequence $z_i \to z_0$ in $\mathbb{C} \setminus \mathbb{R}$ with $z_i \neq z_0$, and define, for $\nu_1, \nu_2 \in M_1(\mathbb{R})$, $\rho(\nu_1, \nu_2) = \sum_i 2^{-i} |S_{\nu_1}(z_i) - S_{\nu_2}(z_i)|$. Note that $\rho(\nu_n, \nu) \to 0$ implies that ν_n converges weakly to ν. Indeed, moving to a subsequence if necessary, ν_n converges vaguely to some sub-probability measure θ, and thus $S_{\nu_n}(z_i) \to S_\theta(z_i)$ for each i. On the other hand, the uniform (in i, n) boundedness of $S_{\nu_n}(z_i)$ and $\rho(\nu_n, \nu) \to 0$ imply that $S_{\nu_n}(z_i) \to S_\nu(z_i)$. Thus, $S_\nu(z) = S_\theta(z)$ for all $z = z_i$ and hence, for all $z \in \mathbb{C} \setminus \mathbb{R}$ since the set $\{z_i\}$ possesses an accumulation point and S_ν, S_θ are analytic. By the inversion formula (2.4.1), it follows that $\nu = \theta$ and in particular θ is a probability measure and ν_n converges weakly to $\theta = \nu$. From the assumption of part (c) we have that $\rho(\mu_n, \mu) \to 0$, in probability, and thus μ_n converges weakly to μ in probability, as claimed. □

For a matrix X, define $\mathbf{S}_X(z) := (X - zI)^{-1}$. Taking $A = X$ in the matrix inversion lemma (Lemma A.1), one gets

$$\mathbf{S}_X(z) = z^{-1}(X\mathbf{S}_X(z) - I), \quad z \in \mathbb{C} \setminus \mathbb{R}. \tag{2.4.3}$$

Note that with L_N denoting the empirical measure of the eigenvalues of X_N,

$$S_{L_N}(z) = \frac{1}{N} \mathrm{tr} \mathbf{S}_{X_N}(z), \quad S_{\bar{L}_N}(z) = \frac{1}{N} E \mathrm{tr} \mathbf{S}_{X_N}(z).$$

2.4.1 Gaussian Wigner matrices

We consider in this section the case when X_N is a Gaussian Wigner matrix, providing

Proof #2 of Theorem 2.1.1 (X_N a Gaussian Wigner matrix).
Recall first the following identity, characterizing the Gaussian distribution, which is proved by integration by parts.

Lemma 2.4.5 *If ζ is a zero mean Gaussian random variable, then for f differentiable, with polynomial growth of f and f',*

$$E(\zeta f(\zeta)) = E(f'(\zeta))E(\zeta^2).$$

Define next the matrix $\Delta_N^{i,k}$ as the symmetric $N \times N$ matrix satisfying

$$\Delta_N^{i,k}(j,l) = \begin{cases} 1, & (i,k) = (j,l) \text{ or } (i,k) = (l,j), \\ 0, & \text{otherwise}. \end{cases}$$

Then, with X an $N \times N$ symmetric matrix,

$$\frac{\partial}{\partial X(i,k)} \mathbf{S}_X(z) = -\mathbf{S}_X(z) \Delta_N^{i,k} \mathbf{S}_X(z). \tag{2.4.4}$$

Using now (2.4.3) in the first equality and Lemma 2.4.5 and (2.4.4) (conditioning on all entries of X_N but one) in the second, one concludes that

$$\begin{aligned} \frac{1}{N} E \mathrm{tr} \mathbf{S}_{X_N}(z) &= -\frac{1}{z} + \frac{1}{z}\frac{1}{N} E\left(\mathrm{tr} X_N \mathbf{S}_{X_N}(z)\right) \\ &= -\frac{1}{z} - \frac{1}{zN^2} E\left(\sum_{i,k} [\mathbf{S}_{X_N}(z)(i,i)\mathbf{S}_{X_N}(z)(k,k) + \mathbf{S}_{X_N}(z)(i,k)^2]\right) \\ &\quad - \frac{1}{zN^2} \sum_i \left((EY_i^2 - 2)\, E\mathbf{S}_{X_N}(z)(i,i)^2\right) \\ &= -\frac{1}{z} - \frac{1}{z} E[\langle L_N, (x-z)^{-1}\rangle^2] - \frac{1}{zN} \langle \bar{L}_N, (x-z)^{-2}\rangle \\ &\quad - \frac{1}{zN^2} \sum_i \left((EY_i^2 - 2)\, E\mathbf{S}_{X_N}(z)(i,i)^2\right). \end{aligned} \tag{2.4.5}$$

Since $(x-z)^{-1}$ is a Lipschitz function for any fixed $z \in \mathbb{C} \setminus \mathbb{R}$, it follows from Theorem 2.3.5 and Remark 2.3.6 that

$$|E[\langle L_N, (x-z)^{-1}\rangle^2] - \langle \bar{L}_N, (x-z)^{-1}\rangle^2| \to_{N\to\infty} 0.$$

This, and the boundedness of $1/(z-x)^2$ for a fixed z as above, imply the existence of a sequence $\varepsilon_N(z) \to_{N\to\infty} 0$ such that, letting $\bar{S}_N(z) := N^{-1} E \mathrm{tr} \mathbf{S}_{X_N}(z)$, one has

$$\bar{S}_N(z) = -\frac{1}{z} - \frac{1}{z}\bar{S}_N(z)^2 + \varepsilon_N(z).$$

Thus any limit point $s(z)$ of $\bar{S}_N(z)$ satisfies

$$s(z)(z + s(z)) + 1 = 0. \tag{2.4.6}$$

Further, let $\mathbb{C}_+ = \{z \in \mathbb{C} : \Im z > 0\}$. Then, for $z \in \mathbb{C}_+$, by its definition, $s(z)$ must have a nonnegative imaginary part, while for $z \in \mathbb{C} \setminus (\mathbb{R} \cup \mathbb{C}_+)$, $s(z)$ must have a nonpositive imaginary part. Hence, for all $z \in \mathbb{C}$, with the choice of the branch of the square-root dictated by the last remark,

$$s(z) = -\frac{1}{2}\left[z - \sqrt{z^2 - 4}\right]. \tag{2.4.7}$$

Comparing with (2.1.6) and using Remark 2.4.2, one deduces that $s(z)$ is the Stieltjes transform of the semicircle law σ, since $s(z)$ coincides with the latter for $|z| > 2$ and hence for all $z \in \mathbb{C} \setminus \mathbb{R}$ by analyticity. Applying again Theorem 2.3.5 and Remark 2.3.6, it follows that $S_{L_N}(z)$ converges in probability to $s(z)$, solution of (2.4.7), for all $z \in \mathbb{C} \setminus \mathbb{R}$. The proof is completed by using part (c) of Theorem 2.4.4. □

2.4.2 General Wigner matrices

We consider in this section the case when X_N is a Wigner matrix. We give now:

Proof #3 of Theorem 2.1.1 (X_N a Wigner matrix).
We begin again with a general fact valid for arbitrary symmetric matrices.

Lemma 2.4.6 *Let $W \in \mathcal{H}_N^{(1)}$ be a symmetric matrix, and let w_i denote the ith column of W with the entry $W(i,i)$ removed (i.e., w_i is an $N-1$-dimensional vector). Let $W^{(i)} \in \mathcal{H}_{N-1}^{(1)}$ denote the matrix obtained by erasing the ith column and row from W. Then, for every $z \in \mathbb{C} \setminus \mathbb{R}$,*

$$(W - zI)^{-1}(i,i) = \frac{1}{W(i,i) - z - w_i^{\mathrm{T}}(W^{(i)} - zI_{N-1})^{-1} w_i}. \tag{2.4.8}$$

Proof of Lemma 2.4.6 Note first that from Cramer's rule,

$$(W - zI_N)^{-1}(i,i) = \frac{\det(W^{(i)} - zI_{N-1})}{\det(W - zI)}. \tag{2.4.9}$$

Write next

$$W - zI_N = \begin{pmatrix} W^{(N)} - zI_{N-1} & w_N \\ w_N^{\mathrm{T}} & W(N,N) - z \end{pmatrix},$$

and use the matrix identity (A.1) with $A = W^{(N)} - zI_{N-1}$, $B = w_N$, $C = w_N^{\mathrm{T}}$ and $D = W(N,N) - z$ to conclude that

$$\det(W - zI_N) = \det(W^{(N)} - zI_{N-1}) \det\left[W(N,N) - z - w_N^{\mathrm{T}}(W^{(N)} - zI_{N-1})^{-1} w_N\right].$$

The last formula holds in the same manner with $W^{(i)}$, w_i and $W(i,i)$ replacing $W^{(N)}$, w_N and $W(N,N)$ respectively. Substituting in (2.4.9) completes the proof of Lemma 2.4.6. □

We are now ready to return to the proof of Theorem 2.1.1. Repeating the truncation argument used in the proof of Theorem 2.1.21, we may and will assume in the sequel that $X_N(i,i)=0$ for all i and that for some constant C independent of N, it holds that $|\sqrt{N}X_N(i,j)|\le C$ for all i,j. Define $\bar{\alpha}_k(i)=X_N(i,k)$, i.e. $\bar{\alpha}_k$ is the kth column of the matrix X_N. Let α_k denote the $N-1$ dimensional vector obtained from $\bar{\alpha}_k$ by erasing the entry $\alpha_k(k)=0$. Denote by $X_N^{(k)}\in\mathscr{H}_N^{(1)}$ the matrix consisting of X_N with the kth row and column removed. By Lemma 2.4.6, one gets that

$$\begin{aligned}\frac{1}{N}\mathrm{tr}\mathbf{S}_{X_N}(z) &= \frac{1}{N}\sum_{i=1}^N \frac{1}{-z-\alpha_i^{\mathrm{T}}(X_N^{(i)}-zI_{N-1})^{-1}\alpha_i}\\ &= -\frac{1}{z+N^{-1}\mathrm{tr}\mathbf{S}_{X_N}(z)}-\delta_N(z)\,,\end{aligned}\qquad(2.4.10)$$

where

$$\delta_N(z)=\frac{1}{N}\sum_{i=1}^N\frac{\varepsilon_{i,N}}{(-z-N^{-1}\mathrm{tr}\mathbf{S}_{X_N}(z)+\varepsilon_{i,N})(-z-N^{-1}\mathrm{tr}\mathbf{S}_{X_N}(z))}\,,\qquad(2.4.11)$$

and

$$\varepsilon_{i,N}=N^{-1}\mathrm{tr}\mathbf{S}_{X_N}(z)-\alpha_i^{\mathrm{T}}(X_N^{(i)}-zI_{N-1})^{-1}\alpha_i\,.\qquad(2.4.12)$$

Our next goal is to prove the convergence in probability of $\delta_N(z)$ to zero for each fixed $z\in\mathbb{C}\setminus\mathbb{R}$ with $|\Im z|=\delta_0>0$. Toward this end, note that the term $-z-N^{-1}\mathrm{tr}\mathbf{S}_{X_N}(z))$ in the right side of (2.4.11) has modulus at least δ_0, since $|\Im z|=\delta_0$ and all eigenvalues of X_N are real. Thus, if we prove the convergence of $\sup_{i\le N}|\varepsilon_{i,N}|$ to zero in probability, it will follow that $\delta_N(z)$ converges to 0 in probability. Toward this end, let $\bar{X}_N^{(i)}$ denote the matrix X_N with the ith column and row set to zero. Then, the eigenvalues of $\bar{X}_N^{(i)}$ and $X_N^{(i)}$ coincide except that $\bar{X}_N^{(i)}$ has one more zero eigenvalue. Hence,

$$\frac{1}{N}|\mathrm{tr}\mathbf{S}_{\bar{X}_N^{(i)}}(z)-\mathrm{tr}\mathbf{S}_{X_N^{(i)}}(z)|\le\frac{1}{\delta_0 N}\,,$$

whereas, with the eigenvalues of $\bar{X}_N^{(i)}$ denoted $\lambda_1^{(i)}\le\lambda_2^{(i)}\le\cdots\le\lambda_N^{(i)}$, and those of X_N denoted $\lambda_1^N\le\lambda_2^N\le\cdots\le\lambda_N^N$, one has

$$\begin{aligned}\frac{1}{N}|\mathrm{tr}\mathbf{S}_{\bar{X}_N^{(i)}}(z)-\mathrm{tr}\mathbf{S}_{X_N}(z)| &\le \frac{1}{\delta_0^2 N}\sum_{k=1}^N|\lambda_k^{(i)}-\lambda_k^N|\le\frac{1}{\delta_0^2}\left(\frac{1}{N}\sum_{k=1}^N|\lambda_k^{(i)}-\lambda_k^N|^2\right)^{1/2}\\ &\le \frac{1}{\delta_0^2}\left(\frac{2}{N}\sum_{k=1}^N X_N(i,k)^2\right)^{1/2},\end{aligned}$$

where Lemma 2.1.19 was used in the last inequality. Since $|\sqrt{N}X_N(i,j)|\le C$, we get that $\sup_i N^{-1}|\mathrm{tr}\mathbf{S}_{\bar{X}_N^{(i)}}(z)-\mathrm{tr}\mathbf{S}_{X_N}(z)|$ converges to zero (deterministically).

Combining the above, it follows that to prove the convergence of $\sup_{i\le N}|\varepsilon_{i,N}|$ to zero in probability, it is enough to prove the convergence to 0 in probability of $\sup_{i\le N}|\bar{\varepsilon}_{i,N}|$, where

$$\begin{aligned}\bar{\varepsilon}_{i,N} &= \alpha_i^{\mathrm{T}} B_N^{(i)}(z)\alpha_i - \frac{1}{N}\mathrm{tr} B_N^{(i)}(z)\\ &= \frac{1}{N}\sum_{k=1}^{N-1}\left(\left[\sqrt{N}\alpha_i(k)\right]^2 - 1\right) B_N^{(i)}(z)(k,k) + \sum_{k,k'=1,k\neq k'}^{N-1} \alpha_i(k)\alpha_i(k')B_N^{(i)}(z)(k,k')\\ &=: \bar{\varepsilon}_{i,N}(1) + \bar{\varepsilon}_{i,N}(2),\end{aligned} \tag{2.4.13}$$

where $B_N^{(i)}(z) = (X_N^{(i)} - zI_{N-1})^{-1}$. Noting that α_i is independent of $B_N^{(i)}(z)$, and possesses zero mean independent entries of variance $1/N$, one observes by conditioning on the sigma-field $\mathscr{F}_{i,N}$ generated by $X_N^{(i)}$ that $E\bar{\varepsilon}_{i,N} = 0$. Further, since

$$N^{-1}\mathrm{tr}\left(B_N^{(i)}(z)^2\right) \le \frac{1}{\delta_0^2},$$

and the random variables $|\sqrt{N}\alpha_i(k)|$ are uniformly bounded, it follows that

$$E|\bar{\varepsilon}_{i,N}(1)|^4 \le \frac{c_1}{N^2}.$$

for some constant c_1 that depends only on δ_0 and C. Similarly, one checks that

$$E|\bar{\varepsilon}_{i,N}(2)|^4 \le \frac{c_2}{N^2},$$

for some constant c_2 depending only on C, δ_0. One obtains then, by Chebyshev's inequality, the claimed convergence of $\sup_{i\le N}|\varepsilon_{i,N}(z)|$ to 0 in probability.

The rest of the argument is similar to what has already been done in Section 2.4.1, and is omitted. □

Remark 2.4.7 We note that reconstruction and continuity results that are stronger than those contained in Theorems 2.4.3 and 2.4.4 are available. An accessible introduction to these and their use in RMT can be found in [Bai99]. For example, in Theorem 2.4.3, if μ possesses a Hölder continuous density m then, for $\lambda \in \mathbb{R}$,

$$S_\mu(\lambda + \mathrm{i}0) := \lim_{\varepsilon\downarrow 0} S_\mu(\lambda+\varepsilon) = \mathrm{i}\pi m(\lambda) + \mathrm{P.V.}\int_{\mathbb{R}} \frac{\mu(dx)}{x-\lambda} \tag{2.4.14}$$

exists, where the notation P.V. stands for "principal value". Also, in the context of Theorem 2.4.4, if the μ and ν are probability measures supported on $[-B,B]$, a,γ are constants satisfying

$$\gamma := \frac{1}{\pi}\int_{|u|\le a}\frac{1}{u^2+1}du > \frac{1}{2},$$

and A is a constant satisfying

$$\kappa := \frac{4B}{\pi(A-B)(2\gamma-1)} \in (0,1),$$

then for any $v > 0$,

$$\begin{aligned}\pi(1-\kappa)(2\gamma-1)\sup_{|x|\le B}|\mu([-B,x]) - \nu([-B,x]) \le \\ \Big[\int_{-A}^{A}|S_\mu(u+iv) - S_\nu(u+iv)|du \\ +\frac{1}{v}\sup_x \int_{|y|\le 2va}|\mu([-B,x+y]) - \mu([-B,x])|dy\Big].\end{aligned} \tag{2.4.15}$$

In the context of random matrices, equation (2.4.15) is useful in obtaining the rate of convergence of L_N to its limit, but we will not discuss this issue here at all.

Exercise 2.4.8 Let $Y(N)$ be a sequence of matrices as in Exercise 2.1.18. By writing $W_N = Y_N Y_N^T = \sum_{i=1}^{M(N)} y_i y_i^T$ for appropriate vectors y_i, and again using Lemma A.1, provide a proof of points (d) and (e) of Exercise 2.1.18 based on Stieltjes transforms, showing that $N^{-1}\mathrm{tr}\mathbf{S}_{W_N}(z)$ converges to the solution of the equation $m(z) = -1/(z - \alpha/(1+m(z))$.
Hint: use the equality

$$I_N + (z-x)(W_N - zI_N)^{-1} = (W_N - xI_N)(W_N - zI_N)^{-1}, \tag{2.4.16}$$

and then use the equality

$$y_i^T(B + y_i y_i^T)^{-1} = \frac{1}{1 + y_i^T B^{-1} y_i} y_i^T B^{-1},$$

with the matrices $B_i = W_N - zI - y_i y_i^T$, to show that the normalized trace of the right side of (2.4.16) converges to 0.

2.5 Joint distribution of eigenvalues in the GOE and the GUE

We are going to calculate the joint distribution of eigenvalues of a random symmetric or Hermitian matrix under a special type of probability law which displays a high degree of symmetry but still makes on-or-above-diagonal entries independent so that the theory of Wigner matrices applies.

2.5.1 Definition and preliminary discussion of the GOE and the GUE

Let $\{\xi_{i,j}, \eta_{i,j}\}_{i,j=1}^{\infty}$ be an i.i.d. family of real mean 0 variance 1 Gaussian random variables. We define

$$P_2^{(1)}, P_3^{(1)}, \dots$$

to be the laws of the random matrices

$$\begin{bmatrix} \sqrt{2}\xi_{1,1} & \xi_{1,2} \\ \xi_{1,2} & \sqrt{2}\xi_{2,2} \end{bmatrix} \in \mathscr{H}_2^{(1)}, \begin{bmatrix} \sqrt{2}\xi_{1,1} & \xi_{1,2} & \xi_{1,3} \\ \xi_{1,2} & \sqrt{2}\xi_{2,2} & \xi_{2,3} \\ \xi_{1,3} & \xi_{2,3} & \sqrt{2}\xi_{3,3} \end{bmatrix} \in \mathscr{H}_3^{(1)}, \dots,$$

respectively. We define

$$P_2^{(2)}, P_3^{(2)}, \dots$$

to be the laws of the random matrices

$$\begin{bmatrix} \xi_{1,1} & \frac{\xi_{1,2}+i\eta_{1,2}}{\sqrt{2}} \\ \frac{\xi_{1,2}-i\eta_{1,2}}{\sqrt{2}} & \xi_{2,2} \end{bmatrix} \in \mathscr{H}_2^{(2)}, \begin{bmatrix} \xi_{11} & \frac{\xi_{1,2}+i\eta_{1,2}}{\sqrt{2}} & \frac{\xi_{1,3}+i\eta_{1,3}}{\sqrt{2}} \\ \frac{\xi_{1,2}-i\eta_{1,2}}{\sqrt{2}} & \xi_{2,2} & \frac{\xi_{2,3}+i\eta_{2,3}}{\sqrt{2}} \\ \frac{\xi_{1,3}-i\eta_{1,3}}{\sqrt{2}} & \frac{\xi_{2,3}-i\eta_{2,3}}{\sqrt{2}} & \xi_{3,3} \end{bmatrix} \in \mathscr{H}_3^{(2)}, \dots,$$

respectively. A random matrix $X \in \mathscr{H}_N^{(\beta)}$ with law $P_N^{(\beta)}$ is said to belong to the *Gaussian orthogonal ensemble (GOE)* or the *Gaussian unitary ensemble (GUE)* according as $\beta = 1$ or $\beta = 2$, respectively. (We often write GOE(N) and GUE(N) when an emphasis on the dimension is needed.) The theory of Wigner matrices developed in previous sections of this book applies here. In particular, for fixed β, given for each N a random matrix $X(N) \in \mathscr{H}_N^{(\beta)}$ with law $P_N^{(\beta)}$, the empirical distribution of the eigenvalues of $X_N := X(N)/\sqrt{N}$ tends to the semicircle law of mean 0 and variance 1.

So what's special about the law $P_N^{(\beta)}$ within the class of laws of Wigner matrices? The law $P_N^{(\beta)}$ is highly symmetrical. To explain the symmetry, as well as to explain the presence of the terms "orthogonal" and "unitary" in our terminology, let us calculate the density of $P_N^{(\beta)}$ with respect to Lebesgue measure $\ell_N^{(\beta)}$ on $\mathscr{H}_N^{(\beta)}$. To fix $\ell_N^{(\beta)}$ unambiguously (rather than just up to a positive constant factor) we use the following procedure. In the case $\beta = 1$, consider the one-to-one onto mapping $\mathscr{H}_N^{(1)} \to \mathbb{R}^{N(N+1)/2}$ defined by taking on-or-above-diagonal entries as coordinates, and normalize $\ell_N^{(1)}$ by requiring it to push forward to Lebesgue measure on $\mathbb{R}^{N(N+1)/2}$. Similarly, in the case $\beta = 2$, consider the one-to-one onto mapping $\mathscr{H}_N^{(2)} \to \mathbb{R}^N \times \mathbb{C}^{N(N-1)/2} = \mathbb{R}^{N^2}$ defined by taking on-or-above-diagonal entries as coordinates, and normalize $\ell_N^{(2)}$ by requiring it to push forward to Lebesgue measure on $\mathbb{R}^{N^2}$. Let $H_{i,j}$ denote the entry of $H \in \mathscr{H}_N^{(\beta)}$ in row i and column j. Note that

$$\mathrm{tr}H^2 = \mathrm{tr}HH^* = \sum_{i=1}^{N} H_{i,i}^2 + 2 \sum_{1 \le i < j \le N} |H_{i,j}|^2 .$$

It is a straightforward matter now to verify that

$$\frac{dP_N^{(\beta)}}{d\ell_N^{(\beta)}}(H) = \begin{cases} 2^{-N/2}(2\pi)^{-N(N+1)/4}\exp(-\mathrm{tr}H^2/4) & \text{if } \beta = 1, \\ \\ 2^{-N/2}\pi^{-N^2/2}\exp(-\mathrm{tr}H^2/2) & \text{if } \beta = 2. \end{cases} \tag{2.5.1}$$

The latter formula clarifies the symmetry of $P_N^{(\beta)}$. The main thing to notice is that the density at H depends only on the eigenvalues of H. It follows that if X is a random element of $\mathscr{H}_N^{(1)}$ with law $P_N^{(1)}$, then for any $N \times N$ orthogonal matrix U, again UXU^* has law $P_N^{(1)}$; and similarly, if X is a random element of $\mathscr{H}_N^{(2)}$ with law $P_N^{(2)}$, then for any $N \times N$ unitary matrix U, again UXU^* has law $P_N^{(2)}$. As we already observed, for random $X \in \mathscr{H}_N^{(\beta)}$ it makes sense to talk about the joint distribution of the eigenvalues $\lambda_1(X) \le \cdots \le \lambda_N(X)$.

Definition 2.5.1 Let $x = (x_1, \ldots, x_N) \in \mathbb{C}^N$. The *Vandermonde determinant* associated with x is

$$\Delta(x) = \det(\{x_i^{j-1}\}_{i,j=1}^n) = \prod_{i<j}(x_j - x_i) . \tag{2.5.2}$$

(For an easy verification of the second equality in (2.5.2), note that the determinant is a polynomial that must vanish when $x_i = x_j$ for any pair $i \neq j$.)

The main result in this section is the following.

Theorem 2.5.2 (Joint distribution of eigenvalues: GOE and GUE) *Let $X \in \mathscr{H}_N^{(\beta)}$ be random with law $P_N^{(\beta)}$, $\beta = 1, 2$. The joint distribution of the eigenvalues $\lambda_1(X) \le \cdots \le \lambda_N(X)$ has density with respect to Lebesgue measure which equals*

$$N!\bar{C}_N^{(\beta)} 1_{x_1 \le \cdots \le x_N} |\Delta(x)|^\beta \prod_{i=1}^{N} e^{-\beta x_i^2/4} , \tag{2.5.3}$$

where

$$\begin{aligned} N!\bar{C}_N^{(\beta)} &= N! \left(\int_{-\infty}^{\infty} \cdots \int_{-\infty}^{\infty} |\Delta(x)|^\beta \prod_{i=1}^{N} e^{-\beta x_i^2/4} dx_i \right)^{-1} \\ &= (2\pi)^{-N/2} \left(\frac{\beta}{2}\right)^{\beta N(N-1)/4 + N/2} \prod_{j=1}^{N} \frac{\Gamma(\beta/2)}{\Gamma(j\beta/2)} . \end{aligned} \tag{2.5.4}$$

Here, for any positive real s,

$$\Gamma(s) = \int_0^\infty x^{s-1} e^{-x} dx \qquad (2.5.5)$$

is Euler's *Gamma function*.

Remark 2.5.3 We refer to the probability measure $\mathscr{P}_N^{(\beta)}$ on $\mathbb{R}^N$ with density

$$\frac{d\mathscr{P}_N^{(\beta)}}{d\mathrm{Leb}_N} = \bar{C}_N^{(\beta)} |\Delta(x)|^\beta \prod_{i=1}^N e^{-\beta x_i^2/4}, \qquad (2.5.6)$$

where Leb_N is the Lebesgue measure on $\mathbb{R}^N$ and $\bar{C}_N^\beta$ is given in (2.5.4), as the *law of the unordered eigenvalues* of the GOE(N) (when $\beta = 1$) or GUE(N) (when $\beta = 2$). The special case $\beta = 4$ corresponds to the GSE(N) (see Section 4.1 for details on the explicit construction of random matrices whose eigenvalues are distributed according to $\mathscr{P}_N^{(4)}$).

The distributions $\mathscr{P}_N^{(\beta)}$ for $\beta \geq 1$, $\beta \neq 1,2,4$ also appear as the law of the unordered eigenvalues of certain random matrices, although with a very different structure, see Section 4.5.

A consequence of Theorem 2.5.2 is that a.s., the eigenvalues of the GOE and GUE are all distinct. Let $v_1, \ldots, v_N$ denote the eigenvectors corresponding to the eigenvalues $(\lambda_1^N, \ldots, \lambda_N^N)$ of a matrix X from GOE(N) or GUE(N), with their first nonzero entry positive real. Recall that $O(N)$ (the group of orthogonal matrices) and $U(N)$ (the group of unitary matrices) admit a unique Haar probability measure (see Theorem F.13). The invariance of the law of X under arbitrary orthogonal (unitary) transformations implies then the following.

Corollary 2.5.4 *The collection $(v_1, \ldots, v_N)$ is independent of the eigenvalues $(\lambda_1^N, \ldots, \lambda_N^N)$. Each of the eigenvectors $v_1, \ldots, v_N$ is distributed uniformly on*

$$S_+^{N-1} = \{\mathbf{x} = (x_1, \ldots, x_N) : x_i \in \mathbb{R}, \|\mathbf{x}\|_2 = 1, x_1 > 0\}$$

(for the GOE), or on

$$S_{\mathbb{C},+}^{N-1} = \{\mathbf{x} = (x_1, \ldots, x_N) : x_1 \in \mathbb{R}, x_i \in \mathbb{C} \text{ for } i \geq 2, \|\mathbf{x}\|_2 = 1, x_1 > 0\}$$

(for the GUE). Further, $(v_1, \ldots, v_N)$ is distributed like a sample of Haar measure on $O(N)$ (for the GOE) or $U(N)$ (for the GUE), with each column multiplied by a norm one scalar so that the columns all belong to S_+^{N-1} (for the GOE) and $S_{\mathbb{C},+}^{N-1}$ (for the GUE).

Proof Write $X = UDU^*$. Since TXT^* possesses the same eigenvalues as X and is distributed like X for any orthogonal (in the GOE case) or unitary (in the GUE case) T independent of X, and since choosing T uniformly according to Haar measure and independent of U makes TU Haar distributed and hence of law independent of that of U, the independence of the eigenvectors and the eigenvalues follows. All other statements are immediate consequences of this and the fact that each column of a Haar distributed orthogonal (resp., unitary) matrix is distributed, after multiplication by a scalar that makes its first entry real and nonnegative, uniformly on S_+^{N-1} (resp. $S_{\mathbb{C},+}^{N-1}$). $\square$

2.5.2 *Proof of the joint distribution of eigenvalues*

We present in this section a proof of Theorem 2.5.2 that has the advantage of being direct, elementary, and not requiring much in terms of computations. On the other hand, this proof is not enough to provide one with the evaluation of the normalization constant $\bar{C}_N^\beta$ in (2.5.4). The evaluation of the latter is postponed to subsection 2.5.3, where the *Selberg integral formula* is derived. Another approach to evaluating the normalization constants, in the case of the GUE, is provided in Section 3.2.1.

The idea behind the proof of Theorem 2.5.2 is as follows. Since $X \in \mathscr{H}_N^{(\beta)}$, there exists a decomposition $X = UDU^*$, with eigenvalue matrix $D \in \mathscr{D}_N$, where $\mathscr{D}_N$ denotes diagonal matrices with real entries, and with eigenvector matrix $U \in \mathscr{U}_N^{(\beta)}$, where $\mathscr{U}_N^{(\beta)}$ denotes the collection of orthogonal matrices (when $\beta = 1$) or unitary matrices (when $\beta = 2$). Suppose this map were a bijection (which it is not, at least at the matrices X without distinct eigenvalues) and that one could parametrize $\mathscr{U}_N^{(\beta)}$ using $\beta N(N-1)/2$ parameters in a smooth way (which one cannot). An easy computation shows that the Jacobian of the transformation would then be a polynomial in the eigenvalues with coefficients that are functions of the parametrization of $\mathscr{U}_N^{(\beta)}$, of degree $\beta N(N-1)/2$. Since the bijection must break down when $D_{ii} = D_{jj}$ for some $i \neq j$, the Jacobian must vanish on that set; symmetry and degree considerations then show that the Jacobian must be proportional to the factor $\Delta(x)^\beta$. Integrating over the parametrization of $\mathscr{U}_N^{(\beta)}$ then yields (2.5.3).

In order to make the above construction work, we need to throw away subsets of $\mathscr{H}_N^{(\beta)}$ that fortunately turn out to have zero Lebesgue measure. Toward this end, we say that $U \in \mathscr{U}_N^{(\beta)}$ is *normalized* if every diagonal entry of U is strictly positive real. We say that $U \in \mathscr{U}_N^{(\beta)}$ is *good* if it is normalized and every entry of U is nonzero. The collection of good matrices is denoted $\mathscr{U}_N^{(\beta),\mathrm{g}}$. We also say that $D \in \mathscr{D}_N$ is *distinct* if its entries are all distinct, denoting by $\mathscr{D}_N^{\mathrm{d}}$ the collection of distinct matrices, and by $\mathscr{D}_N^{\mathrm{do}}$ the subset of matrices with decreasing entries, that is $\mathscr{D}_N^{\mathrm{do}} = \{D \in \mathscr{D}_N^{\mathrm{d}} : D_{i,i} > D_{i+1,i+1}\}$.

Let $\mathscr{H}_N^{(\beta),\mathrm{dg}}$ denote the subset of $\mathscr{H}^{(\beta)}$ consisting of those matrices that possess a decomposition $X = UDU^*$ where $D \in \mathscr{D}_N^{\mathrm{d}}$ and $U \in \mathscr{U}_N^{(\beta),\mathrm{g}}$. The first step is contained in the following lemma.

Lemma 2.5.5 *$\mathscr{H}_N^{(\beta)} \setminus \mathscr{H}_N^{(\beta),\mathrm{dg}}$ has null Lebesgue measure. Further, the map $(\mathscr{D}_N^{\mathrm{do}}, \mathscr{U}_N^{(\beta),\mathrm{g}}) \to \mathscr{H}_N^{(\beta),\mathrm{dg}}$ given by $(D,U) \mapsto UDU^*$ is one-to-one and onto, while $(\mathscr{D}_N^{\mathrm{d}}, \mathscr{U}_N^{(\beta),\mathrm{g}}) \to \mathscr{H}_N^{(\beta),\mathrm{dg}}$ given by the same map is $N!$-to-one.*

Proof of Lemma 2.5.5 In order to prove the first part of the lemma, we note that for any nonvanishing polynomial function p of the entries of X, the set $\{X : p(X) = 0\}$ is closed and has zero Lebesgue measure (this fact can be checked by applying Fubini's Theorem). So it is enough to exhibit a nonvanishing polynomial p with $p(X) = 0$ if $X \in \mathscr{H}_N^{(\beta)} \setminus \mathscr{H}_N^{(\beta),\mathrm{dg}}$. Toward this end, we will show that for such X, either X has some multiple eigenvalue, or, for some k, X and the matrix $X^{(k)}$ obtained by erasing the kth row and column of X possess a common eigenvalue.

Given any n by n matrix H, for $i, j = 1, \dots, n$ let $H^{(i,j)}$ be the $n-1$ by $n-1$ matrix obtained by deleting the ith column and jth row of H, and write $H^{(k)}$ for $H^{(k,k)}$. We begin by proving that if $X = UDU^*$ with $D \in \mathscr{D}_N^{\mathrm{d}}$, and X and $X^{(k)}$ do not have eigenvalues in common for any $k = 1, 2, \dots, N$, then all entries of U are nonzero. Indeed, let λ be an eigenvalue of X, set $A = X - \lambda I$, and define A^{adj} as the N by N matrix with $A^{\mathrm{adj}}_{i,j} = (-1)^{i+j} \det(A^{(i,j)})$. Using the identity $AA^{\mathrm{adj}} = \det(A)I$, one concludes that $AA^{\mathrm{adj}} = 0$. Since the eigenvalues of X are assumed distinct, the null space of A has dimension 1, and hence all columns of A^{adj} are scalar multiple of some vector v_λ, which is then an eigenvector of X corresponding to the eigenvalue λ. Since $v_\lambda(i) = A^{\mathrm{adj}}_{i,i} = \det(X^{(i)} - \lambda I) \neq 0$ by assumption, it follows that all entries of v_λ are nonzero. But each column of U is a nonzero scalar multiple of some v_λ, leading to the conclusion that all entries of U do not vanish.

We recall, see Appendix A.4, that the resultant of the characteristic polynomials of X and $X^{(k)}$, which can be written as a polynomial in the entries of X and $X^{(k)}$, and hence as a polynomial P_1 in the entries of X, vanishes if and only if X and $X^{(k)}$ have a common eigenvalue. Further, the discriminant of X, which is a polynomial P_2 in the entries of X, vanishes if and only if not all eigenvalues of X are distinct. Taking $p(X) = P_1(X)P_2(X)$, one obtains a nonzero polynomial p with $p(X) = 0$ if $X \in \mathscr{H}_N^{(\beta)} \setminus \mathscr{H}_N^{(\beta),\mathrm{dg}}$. This completes the proof of the first part of Lemma 2.5.5.

The second part of the lemma is immediate since the eigenspace corresponding to each eigenvalue is of dimension 1, the eigenvectors are fixed by the normalization condition, and the multiplicity arises from the possible permutations of the order of the eigenvalues. □

Next, we say that $U \in \mathscr{U}_N^{(\beta),\mathrm{g}}$ is *very good* if all minors of U have nonvanishing determinant. Let $\mathscr{U}_N^{(\beta),\mathrm{vg}}$ denote the collection of very good matrices. The interest in such matrices is that they possess a particularly nice parametrization.

Lemma 2.5.6 *The map* $T : \mathscr{U}_N^{(\beta),\mathrm{vg}} \to \mathbb{R}^{\beta N(N-1)/2}$ *defined by*

$$T(U) = \left(\frac{U_{1,2}}{U_{1,1}}, \ldots, \frac{U_{1,N}}{U_{1,1}}, \frac{U_{2,3}}{U_{2,2}}, \ldots, \frac{U_{2,N}}{U_{2,2}}, \ldots, \frac{U_{N-1,N}}{U_{N-1,N-1}} \right) \tag{2.5.7}$$

(where $\mathbb{C}$ is identified with $\mathbb{R}^2$ in the case $\beta = 2$) is one-to-one with smooth inverse. Further, the set $\left(T(\mathscr{U}_N^{(\beta),\mathrm{vg}})\right)^c$ is closed and has zero Lebesgue measure.

Proof of Lemma 2.5.6 We begin with the first part. The proof is by an inductive construction. Clearly, $U_{1,1}^{-2} = 1 + \sum_{j=2}^N |U_{1,j}|^2/|U_{1,1}|^2$. So suppose that $U_{i,j}$ are given for $1 \le i \le i_0$ and $1 \le j \le N$. Let $v_i = (U_{i,1}, \ldots, U_{i,i_0})$, $i = 1, \ldots, i_0$. One then solves the equation

$$\begin{pmatrix} v_1 \\ v_2 \\ \vdots \\ v_{i_0} \end{pmatrix} Z = - \begin{pmatrix} U_{1,i_0+1} + \sum_{i=i_0+2}^N U_{1,i} \left(\frac{U_{i_0+1,i}}{U_{i_0+1,i_0+1}} \right)^* \\ U_{2,i_0+1} + \sum_{i=i_0+2}^N U_{2,i} \left(\frac{U_{i_0+1,i}}{U_{i_0+1,i_0+1}} \right)^* \\ \vdots \\ U_{i_0,i_0+1} + \sum_{i=i_0+2}^N U_{i_0,i} \left(\frac{U_{i_0+1,i}}{U_{i_0+1,i_0+1}} \right)^* \end{pmatrix}.$$

The very good condition on U ensures that the vector Z is uniquely determined by this equation, and one then sets

$$U_{i_0+1,i_0+1}^{-2} = 1 + \sum_{k=1}^{i_0} |Z_k|^2 + \sum_{i=i_0+2}^{N} \left| \frac{U_{i_0+1,i}}{U_{i_0+1,i_0+1}} \right|^2$$

and

$$U_{i_0+1,j} = Z_j^* U_{i_0+1,i_0+1}, \qquad \text{for } 1 \le j \le i_0 .$$

(All entries $U_{i_0+1,j}$ with $j > i_0 + 1$ are then determined by $T(U)$.) This completes the proof of the first part.

To see the second part, let $\mathscr{Z}_N^{(\beta)}$ be the space of matrices whose columns are orthogonal, whose diagonal entries all equal to 1, and all of whose minors have nonvanishing determinants. Define the action of T on $\mathscr{Z}_N^{(\beta)}$ using (2.5.7). Then, $T(\mathscr{U}_N^{(\beta),\mathrm{vg}}) = T(\mathscr{Z}_N^{(\beta)})$. Applying the previous constructions, one immediately obtains a polynomial type condition for a point in $\mathbb{R}^{\beta N(N-1)/2}$ to not belong to the set $T(\mathscr{Z}_N^{(\beta)})$. □

Let $\mathscr{H}_N^{(\beta),\mathrm{vg}}$ denote the subset of $\mathscr{H}_N^{(\beta),\mathrm{dg}}$ consisting of those matrices X that can be written as $X = UDU^*$ with $D \in \mathscr{D}_N^{\mathrm{d}}$ and $U \in \mathscr{U}_N^{(\beta),\mathrm{vg}}$.

Lemma 2.5.7 *The Lebesgue measure of $\mathscr{H}_N^{(\beta)} \setminus \mathscr{H}_N^{(\beta),\mathrm{vg}}$ is zero.*

Proof of Lemma 2.5.7 We identify a subset of $\mathscr{H}_N^{(\beta),\mathrm{vg}}$ which we will prove to be of full Lebesgue measure. We say that a matrix $D \in \mathscr{D}_N^{\mathrm{d}}$ is *strongly distinct* if for any integer $r = 1,2,\ldots,N-1$ and subsets I,J of $\{1,2,\ldots,N\}$,

$$I = \{i_1 < \cdots < i_r\}, \quad J = \{j_1 < \cdots < j_r\}$$

with $I \neq J$, it holds that $\prod_{i\in I} D_{i,i} \neq \prod_{i\in J} D_{i,i}$. We consider the subset $\mathscr{H}_N^{(\beta),\mathrm{sdg}}$ of $\mathscr{H}_N^{(\beta),\mathrm{vg}}$ consisting of those matrices $X = UDU^*$ with D strongly distinct and $U \in \mathscr{U}_N^{(\beta),\mathrm{vg}}$.

Given a positive integer r and subsets I,J as above, put

$$(\bigwedge^r X)_{IJ} := \det_{\mu,\nu=1}^{r} X_{i_\mu,j_\nu},$$

thus defining a square matrix $\bigwedge^r X$ with rows and columns indexed by r-element subsets of $\{1,\ldots,n\}$. If we replace each entry of X by its complex conjugate, we replace each entry of $\bigwedge^r X$ by its complex conjugate. If we replace X by its transpose, we replace $\bigwedge^r X$ by its transpose. Given another N by N matrix Y with complex entries, by the Cauchy–Binet Theorem A.2 we have $\bigwedge^r(XY) = (\bigwedge^r X)(\bigwedge^r Y)$. Thus, if $U \in \mathscr{U}_N^{(\beta)}$ then $\bigwedge^r U \in \mathscr{U}_{c_N^r}^{(\beta)}$ where $c_N^r = N!/(N-r)!r!$. We thus obtain that if $X = UDU^*$ then $\bigwedge^r X$ can be decomposed as $\bigwedge^r X = (\bigwedge^r U)(\bigwedge^r D)(\bigwedge^r U^*)$. In particular, if D is not strongly distinct then, for some r, $\bigwedge^r X$ does not possess all eigenvalues distinct. Similarly, if D is strongly distinct but $U \notin \mathscr{U}_N^{(\beta),\mathrm{vg}}$, then some entry of $\bigwedge^r U$ vanishes. Repeating the argument presented in the proof of the first part of Lemma 2.5.5, we conclude that the Lebesgue measure of $\mathscr{H}_N^{(\beta)} \setminus \mathscr{H}_N^{(\beta),\mathrm{sdg}}$ vanishes. This completes the proof of the lemma. □

We are now ready to provide the

Proof of (2.5.3) Recall the map T introduced in Lemma 2.5.6, and define the map $\hat{T} : T(\mathscr{U}_N^{(\beta),\mathrm{vg}}) \times \mathbb{R}^N \to \mathscr{H}_N^{(\beta)}$ by setting, for $\lambda \in \mathbb{R}^N$ and $z \in T(\mathscr{U}_N^{(\beta),\mathrm{vg}})$, $D \in \mathscr{D}_N$ with $D_{i,i} = \lambda_i$ and $\hat{T}(z,\lambda) = T^{-1}(z)DT^{-1}(z)^*$. By Lemma 2.5.6, $\hat{T}$ is smooth, whereas by Lemma 2.5.5, it is $N!$-to-1 on a set of full Lebesgue measure and is locally one-to-one on a set of full Lebesgue measure. Letting $J\hat{T}$ denote the

Jacobian of $\hat{T}$, we note that $J\hat{T}(z,\lambda)$ is a homogeneous polynomial in λ of degree (at most) $\beta N(N-1)/2$, with coefficients that are functions of z (since derivatives of $\hat{T}(z,\lambda)$ with respect to the λ-variables do not depend on λ, while derivatives with respect to the z variables are linear in λ). Note next that $\hat{T}$ fails to be locally one-to-one when $\lambda_i = \lambda_j$ for some $i \neq j$. In particular, it follows by the implicit function theorem that $J\hat{T}$ vanishes at such points. Hence, $\Delta(\lambda) = \prod_{i<j}(\lambda_j - \lambda_i)$ is a factor of $J\hat{T}$. In fact, we have that

$$\Delta(\lambda)^\beta \text{ is a factor of } J\hat{T}\,. \tag{2.5.8}$$

We postpone the proof of (2.5.8) in the case $\beta = 2$. Since $\Delta(\lambda)$ is a polynomial of degree $N(N-1)/2$, it follows from (2.5.8) that $J\hat{T}(z,\lambda) = g(z)\Delta(\lambda)^\beta$ for some (continuous, hence measurable) function g. By Lemma 2.5.7, we conclude that for any function f that depends only on the eigenvalues of X, it holds that

$$N! \int f(H)dP_N^{(\beta)} = \int |g(z)|dz \int f(\lambda)|\Delta(\lambda)|^\beta \prod_{i=1}^{N} e^{-\beta\lambda_i^2/4} d\lambda_i\,.$$

Up to the normalization constant $(\int |g(z)|dz)/N!$, this is (2.5.3).

It only remains to complete the proof of (2.5.8) in the case $\beta = 2$. Writing for brevity $W = T^{-1}(z)$, we have $\hat{T} = WDW^*$, and $W^*W = I$. Using the notation $d\hat{T}$ for the matrix of differentials of $\hat{T}$, we have $d\hat{T} = (dW)DW^* + W(dD)W^* + WD(dW^*)$. Using the relation $d(W^*W) = (dW^*)W + W^*(dW) = 0$, we deduce that

$$W^*(d\hat{T})W = W^*(dW)D - DW^*(dW) + (dD)\,.$$

Therefore, when $\lambda_i = \lambda_j$ for some $i \neq j$, a *complex* entry (above the diagonal) of $W^*(d\hat{T})W$ vanishes. This implies that, when $\lambda_i = \lambda_j$, there exist two linear (real) relations between the on-and-above diagonal entries of $d\hat{T}$, which implies in turn that $(\lambda_i - \lambda_j)^2$ must divide $J\hat{T}$. □

2.5.3 Selberg's integral formula and proof of (2.5.4)

To complete the description of the joint distribution of eigenvalues of the GOE, GUE and GSE, we derive in this section an expression for the normalization constant in (2.5.4). The value of the normalization constant does not play a role in the rest of this book, except for Section 2.6.2.

We begin by stating Selberg's integral formula. We then describe in Corollary 2.5.9 a couple of limiting cases of Selberg's formula. The evaluation of the normalization constant in (2.5.4) is immediate from Corollary 2.5.9. Recall, see Definition 2.5.1, that $\Delta(x)$ denotes the Vandermonde determinant of x.

Theorem 2.5.8 (Selberg's integral formula) *For all positive numbers a, b and c we have*

$$\frac{1}{n!}\int_0^1 \cdots \int_0^1 |\Delta(x)|^{2c} \prod_{i=1}^{n} x_i^{a-1}(1-x_i)^{b-1} dx_i = \prod_{j=0}^{n-1} \frac{\Gamma(a+jc)\Gamma(b+jc)\Gamma((j+1)c)}{\Gamma(a+b+(n+j-1)c)\Gamma(c)}. \tag{2.5.9}$$

Corollary 2.5.9 *For all positive numbers a and c we have*

$$\frac{1}{n!}\int_0^\infty \cdots \int_0^\infty |\Delta(x)|^{2c} \prod_{i=1}^{n} x_i^{a-1} e^{-x_i} dx_i = \prod_{j=0}^{n-1} \frac{\Gamma(a+jc)\Gamma((j+1)c)}{\Gamma(c)}, \tag{2.5.10}$$

and

$$\frac{1}{n!}\int_{-\infty}^\infty \cdots \int_{-\infty}^\infty |\Delta(x)|^{2c} \prod_{i=1}^{n} e^{-x_i^2/2} dx_i = (2\pi)^{n/2} \prod_{j=0}^{n-1} \frac{\Gamma((j+1)c)}{\Gamma(c)}. \tag{2.5.11}$$

Remark 2.5.10 The identities in Theorem 2.5.8 and Corollary 2.5.9 hold under rather less stringent conditions on the parameters a, b and c. For example, one can allow a, b and c to be complex with positive real parts. We refer to the bibliographical notes for references. We note also that only (2.5.11) is directly relevant to the study of the normalization constants for the GOE and GUE. The usefulness of the other more complicated formulas will become apparent in Section 4.1.

We will prove Theorem 2.5.8 following Anderson's method [And91], after first explaining how to deduce Corollary 2.5.9 from (2.5.9) by means of the *Stirling approximation*, which we recall is the statement

$$\Gamma(s) = \sqrt{\frac{2\pi}{s}}\left(\frac{s}{e}\right)^s (1+o_{s\to+\infty}(1)), \tag{2.5.12}$$

where s tends to $+\infty$ along the positive real axis. (For a proof of (2.5.12) by an application of Laplace's method, see Exercise 3.5.5.)

Proof of Corollary 2.5.9 We denote the left side of (2.5.9) by $S_n(a,b,c)$. Consider first the integral

$$I_s = \frac{1}{n!}\int_0^s \cdots \int_0^s \Delta(x)^{2c} \prod_{i=1}^{n} x_i^{a-1}(1-x_i/s)^s dx_i,$$

where s is a large positive number. By monotone convergence, the left side of (2.5.10) equals $\lim_{s\to\infty} I_s$. By rescaling the variables of integration, we find that

$$I_s = s^{n(a+(n-1)c)} S_n(a,s+1,c).$$

From (2.5.12) we deduce the formula

$$\frac{\Gamma(s+1+A)}{\Gamma(s+1+B)} = s^{A-B}(1+o_{s\to+\infty}(1)), \tag{2.5.13}$$

in which A and B are any real constants. Finally, assuming the validity of (2.5.9), we can evaluate $\lim_{s\to\infty} I_s$ with the help of (2.5.13), thus verifying (2.5.10).

Turning to the proof of (2.5.11), consider the integral

$$J_s = \frac{1}{n!}\int_{-\sqrt{2s}}^{\sqrt{2s}} \cdots \int_{-\sqrt{2s}}^{\sqrt{2s}} |\Delta(x)|^{2c} \prod_{i=1}^{n}\left(1-\frac{x_i^2}{2s}\right)^{s} dx_i,$$

where s is a large positive number. By monotone convergence the left side of (2.5.11) equals $\lim_{s\to\infty} J_s$. By shifting and rescaling the variables of integration, we find that

$$J_s = 2^{3n(n-1)/2+3n/2+2ns} s^{n(n-1)c/2+n/2} S_n(s+1,s+1,c).$$

From (2.5.12) we deduce the formula

$$\frac{\Gamma(2s+2+A)}{\Gamma(s+1+B)^2} = \frac{2^{A+3/2+2s} s^{A-2B+1/2}}{\sqrt{2\pi}}(1+o_{s\to+\infty}(1)), \tag{2.5.14}$$

where A and B are any real constants. Assuming the validity of (2.5.9), we can evaluate $\lim_{s\to\infty} J_s$ with the help of (2.5.14), thus verifying (2.5.11). □

Before providing the proof of Theorem 2.5.8, we note the following identity involving the *beta integral* in the left side:

$$\int_{\{x\in\mathbb{R}^n:\min_{i=1}^n x_i>0,\sum_{i=1}^n x_i<1\}} \left(1-\sum_{i=1}^{n} x_i\right)^{s_{n+1}-1} \prod_{i=1}^{n} x_i^{s_i-1} dx_i = \frac{\Gamma(s_1)\cdots\Gamma(s_{n+1})}{\Gamma(s_1+\cdots+s_{n+1})}. \tag{2.5.15}$$

The identity (2.5.15) is proved by substituting $u_1 = tx_1, \ldots, u_n = tx_n$, $u_{n+1} = t(1-x_1-\cdots-x_n)$ in the integral

$$\int_0^\infty \cdots \int_0^\infty \prod_{i=1}^{n+1} u_i^{s_i-1} e^{-u_i} du_i,$$

and applying Fubini's Theorem both before and after the substitution.

Proof of Theorem 2.5.8 We aim now to rewrite the left side of (2.5.9) in an intuitive way, see Lemma 2.5.12 below. Toward this end, we introduce some notation.

Let $\mathscr{D}_n$ be the space consisting of monic polynomials $P(t)$ of degree n in a variable t with real coefficients such that $P(t)$ has n distinct real roots. More generally, given an open interval $I \subset \mathbb{R}$, let $\mathscr{D}_n I \subset \mathscr{D}_n$ be the subspace consisting of polynomials with n distinct roots in I. Given $x \in \mathbb{R}^n$, let $P_x(t) = t^n + \sum_{i=1}^n (-1)^i x_{n-i} t^{n-i}$. For any open interval $I \subset \mathbb{R}$, the set $\{x \in \mathbb{R}^n \mid P_x \in \mathscr{D}_n I\}$ is open, since the perturbation of a degree n polynomial by the addition of a degree $n-1$ polynomial with small real coefficients does not destroy the property of having n distinct real roots, nor does it move the roots very much. By definition a set $A \subset \mathscr{D}_n$ is measurable if and only if $\{x \in \mathbb{R}^n \mid P_x \in A\}$ is Lebesgue measurable. Let ℓ_n be the measure on $\mathscr{D}_n$ obtained by pushing Lebesgue measure on the open set $\{x \in \mathbb{R}^n \mid P_x \in \mathscr{D}_n\}$ forward to $\mathscr{D}_n$ via $x \mapsto P_x$ (that is, under ℓ_n, monic polynomials of degree n have coefficients that are jointly Lebesgue distributed). Given $P \in \mathscr{D}_n$, we define $\sigma_k(P) \in \mathbb{R}$ for $k = 0, \ldots, n$ by the rule $P(t) = \sum_{k=0}^n (-1)^k \sigma_k(P) t^{n-k}$. Equivalently, if $\alpha_1 < \cdots < \alpha_n$ are the roots of $P \in \mathscr{D}_n$, we have $\sigma_0(P) = 1$ and

$$\sigma_k(P) = \sum_{1 \le i_1 < \cdots < i_k \le n} \alpha_{i_1} \cdots \alpha_{i_k}$$

for $k = 1, \ldots, n$. The map $(P \mapsto (\sigma_1(P), \ldots, \sigma_n(P))) : \mathscr{D}_n \to \mathbb{R}^n$ inverts the map $(x \mapsto P_x) : \{x \in \mathbb{R}^n \mid P_x \in \mathscr{D}_n\} \to \mathscr{D}_n$. Let $\widetilde{\mathscr{D}}_n \subset \mathbb{R}^n$ be the open set consisting of n-tuples $(\alpha_1, \ldots, \alpha_n)$ such that $\alpha_1 < \cdots < \alpha_n$. Finally, for $P \in \mathscr{D}_n$ with roots $\alpha = (\alpha_1 < \cdots < \alpha_n)$, we set $D(P) = \prod_{i<j} (\alpha_j - \alpha_i)^2 = \Delta(\alpha)^2$.

Lemma 2.5.11 *For $k, \ell = 1, \ldots, n$ and $\alpha = (\alpha_1, \ldots, \alpha_n) \in \widetilde{\mathscr{D}}_n$ put*

$$\tau_k = \tau_k(\alpha_1, \ldots, \alpha_n) = \sum_{1 \le i_1 < \cdots < i_k \le n} \alpha_{i_1} \cdots \alpha_{i_k}, \quad \tau_{k,\ell} = \frac{\partial \tau_k}{\partial \alpha_\ell}.$$

Then

$$\left| \det_{k,\ell=1}^n \tau_{k,\ell} \right| = \prod_{1 \le i < j \le n} |\alpha_i - \alpha_j| = |\Delta(\alpha)|. \tag{2.5.16}$$

Proof We have

$$\tau_{k,\ell} = \sigma_{k-1} \left(\prod_{i \in \{1,\ldots,n\} \setminus \{\ell\}} (t - \alpha_i) \right),$$

whence follows the identity

$$\sum_{m=1}^n (-1)^{m-1} \alpha_k^{n-m} \tau_{m,\ell} = \delta_{k\ell} \prod_{i \in \{1,\ldots,n\} \setminus \{\ell\}} (\alpha_\ell - \alpha_i).$$

This last is equivalent to a matrix identity $AB = C$ where $\det A$ up to a sign equals the Vandermonde determinant $\det_{i,j=1}^n \alpha_j^{n-i}$, $\det B$ is the determinant we want to calculate, and $\det C$ up to a sign equals $(\det A)^2$. Formula (2.5.16) follows. □

(See Exercise 2.5.16 for an alternative proof of Lemma 2.5.11.)

We can now rewrite (2.5.9).

Lemma 2.5.12 *The left side of (2.5.9) equals*

$$\int_{\mathscr{D}_n(0,1)} |P(0)|^{a-1}|P(1)|^{b-1}D(P)^{c-1/2}d\ell_n(P)\,. \tag{2.5.17}$$

Proof We prove a slightly more general statement: for any nonnegative ℓ_n-measurable function f on $\mathscr{D}_n$, we have

$$\int_{\mathscr{D}_n} f d\ell_n = \int_{\widetilde{\mathscr{D}}_n} f(\prod_{i=1}^{n}(t-\alpha_i))\Delta(\alpha)d\alpha_1\cdots d\alpha_n\,, \tag{2.5.18}$$

from which (2.5.17) follows by taking $f(P) = |P(0)|^{a-1}|P(1)|^{b-1}D(P)^{c-1/2}$. To see (2.5.18), put $g(x) = f(P_x)$ for $x \in \mathbb{R}^n$ such that $P_x \in \mathscr{D}_n$. Then, the left side of (2.5.18) equals

$$\int_{\{x\in\mathbb{R}^n|P_x\in\mathscr{D}_n\}} g(x_1,\dots,x_n)dx_1\cdots dx_n = \int_{\widetilde{\mathscr{D}}_n} g(\tau_1,\dots,\tau_n)\left|\det_{k,\ell=1}^{n}\tau_{k,\ell}\right|d\alpha_1\dots d\alpha_n\,, \tag{2.5.19}$$

by the usual formula for changing variables in a multivariable integral. The left sides of (2.5.18) and (2.5.19) are equal by definition; the right sides are equal by (2.5.16). □

We next transform some naturally occurring integrals on $\mathscr{D}_n$ to beta integrals, see Lemma 2.5.15 below. This involves some additional notation. Let $\mathscr{E}_n \subset \mathscr{D}_n \times \mathscr{D}_{n+1}$ be the subset consisting of pairs (P,Q) such that the roots $\alpha_1 < \cdots < \alpha_n$ of P and the roots $\beta_1 < \cdots < \beta_{n+1}$ of Q are *interlaced*, that is, $\alpha_i \in (\beta_i, \beta_{i+1})$ for $i = 1,\dots,n$. More generally, given an interval $I \subset \mathbb{R}$, let $\mathscr{E}_n I = \mathscr{E}_n \cap (\mathscr{D}_n I \times \mathscr{D}_{n+1} I)$.

Lemma 2.5.13 *Fix $Q \in \mathscr{D}_{n+1}$ with roots $\beta_1 < \cdots < \beta_{n+1}$. Fix real numbers $\gamma_1,\dots,\gamma_{n+1}$ and let $P(t)$ be the unique polynomial in t of degree $\leq n$ with real coefficients such that the partial fraction expansion*

$$\frac{P(t)}{Q(t)} = \sum_{i=1}^{n+1}\frac{\gamma_i}{t-\beta_i}$$

holds. Then the following statements are equivalent:

(I) $(P,Q) \in \mathscr{E}_n$.
(II) $\min_{i=1}^{n+1}\gamma_i > 0$ *and* $\sum_{i=1}^{n+1}\gamma_i = 1$.

Proof (I⇒II) The numbers $P(\beta_i)$ do not vanish and their signs alternate. Similarly, the numbers $Q'(\beta_i)$ do not vanish and their signs alternate. By L'Hôpital's rule, we have $\gamma_i = P(\beta_i)/Q'(\beta_i)$ for $i = 1,\dots,n+1$. Thus all the quantities γ_i are nonzero and have the same sign. The quantity $P(t)/Q'(t)$ depends continuously on t in the interval $[\beta_{n+1},\infty)$, does not vanish in that interval, and tends to $1/(n+1)$ as $t \to +\infty$. Thus γ_{n+1} is positive. Since the signs of $P(\beta_i)$ alternate, and so do the signs of $Q'(\beta_i)$, it follows that $\gamma_i = P(\beta_i)/Q'(\beta_i) > 0$ for all i. Because $P(t)$ is monic, the numbers γ_i sum to 1. Thus condition (II) holds.

(II⇒I) Because the signs of the numbers $Q'(\beta_i)$ alternate, we have sufficient information to force $P(t)$ to change sign $n+1$ times, and thus to have n distinct real roots interlaced with the roots of $Q(t)$. And because the numbers γ_i sum to 1, the polynomial $P(t)$ must be monic in t. Thus condition (I) holds. □

Lemma 2.5.14 *Fix $Q \in \mathscr{D}_{n+1}$ with roots $\beta_1 < \cdots < \beta_{n+1}$. Then we have*

$$\ell_n(\{P \in \mathscr{D}_n \mid (P,Q) \in \mathscr{E}_n\}) = \frac{1}{n!}\prod_{j=1}^{n+1}|Q'(\beta_j)|^{1/2} = \frac{D(Q)^{1/2}}{n!}. \tag{2.5.20}$$

Proof Consider the set

$$A = \{x \in \mathbb{R}^n \mid (P_x, Q) \in \mathscr{E}_n\}.$$

By definition the left side of (2.5.20) equals the Lebesgue measure of A. Consider the polynomials $Q_j(t) = Q(t)/(t-\beta_j)$ for $j = 1,\ldots,n+1$. By Lemma 2.5.13, for all $x \in \mathbb{R}^n$, we have $x \in A$ if and only if $P_x(t) = \sum_{i=1}^{n+1}\gamma_i Q_i(t)$ for some real numbers γ_i such that $\min \gamma_i > 0$ and $\sum \gamma_i = 1$, or equivalently, A is the interior of the convex hull of the points

$$\big(\tau_{2,j}(\beta_1,\ldots,\beta_{n+1}),\ldots,\tau_{n+1,j}(\beta_1,\ldots,\beta_{n+1})\big) \in \mathbb{R}^n \quad \text{for } j = 1,\ldots,n+1,$$

where the τs are defined as in Lemma 2.5.11 (but with n replaced by $n+1$). Noting that $\tau_{1,\ell} \equiv 1$ for $\ell = 1,\ldots,n+1$, the Lebesgue measure of A equals the absolute value of $\frac{1}{n!}\det_{k,\ell=1}^{n+1}\tau_{k,\ell}(\beta_1,\ldots,\beta_{n+1})$ by the determinantal formula for computing the volume of a simplex in $\mathbb{R}^n$. Finally, we get the claimed result by (2.5.16). □

Lemma 2.5.15 *Fix $Q \in \mathscr{D}_{n+1}$ with roots $\beta_1 < \cdots < \beta_{n+1}$. Fix positive numbers $s_1,\ldots,s_{n+1}$. Then we have*

$$\int_{\{P\in\mathscr{D}_n|(P,Q)\in\mathscr{E}_n\}}\prod_{i=1}^{n+1}|P(\beta_i)|^{s_i-1}d\ell_n(P) = \frac{\prod_{i=1}^{n+1}|Q'(\beta_i)|^{s_i-1/2}\Gamma(s_i)}{\Gamma(\sum_{i=1}^{n+1}s_i)}. \tag{2.5.21}$$

Proof For P in the domain of integration in the left side of (2.5.21), define $\gamma_i = \gamma_i(P) = P(\beta_i)/Q'(\beta_i)$, $i = 1,\ldots,n+1$. By Lemma 2.5.13, $\gamma_i > 0$, $\sum_{i=1}^{n+1}\gamma_i = 1$, and further $P \mapsto (\gamma_i)_{i=1}^n$ is a bijection from $\{P \in \mathscr{D}_n \mid (P,Q) \in \mathscr{E}_n\}$ to the domain of integration in the right side of (2.5.15). Further, the map $x \mapsto \gamma(P_x)$ is linear.

Hence

$$\int_{\{P\in\mathscr{D}_n|(P,Q)\in\mathscr{E}_n\}} \prod_{i=1}^{n+1}\left|\frac{P(\beta_i)}{Q'(\beta_i)}\right|^{s_i-1} d\ell_n(P)$$

equals, up to a constant multiple C independent of $\{s_i\}$, the right side of (2.5.15). Finally, by evaluating the left side of (2.5.21) for $s_1 = \cdots = s_{n+1} = 1$ by means of Lemma 2.5.14 (and recalling that $\Gamma(n+1) = n!$) we find that $C = 1$. □

We may now complete the proof of Theorem 2.5.8. Recall that the integral on the left side of (2.5.9), denoted as above by $S_n(a,b,c)$, can be represented as the integral (2.5.17). Consider the double integral

$$K_n(a,b,c) = \int_{\mathscr{E}_n(0,1)} |Q(0)|^{a-1}|Q(1)|^{b-1}|R(P,Q)|^{c-1} d\ell_n(P) d\ell_{n+1}(Q)\,,$$

where $R(P,Q)$ denotes the resultant of P and Q, see Appendix A.4. We will apply Fubini's Theorem in both possible ways. On the one hand, we have

$$\begin{aligned} K_n(a,b,c) &= \int_{\mathscr{D}_{n+1}(0,1)} |Q(0)|^{a-1}|Q(1)|^{b-1} \\ &\quad \times \left(\int_{\{P\in\mathscr{D}_n(0,1)|(P,Q)\in\mathscr{E}_n\}} |R(P,Q)|^{c-1} d\ell_n(P)\right) d\ell_{n+1}(Q) \\ &= S_{n+1}(a,b,c)\frac{\Gamma(c)^{n+1}}{\Gamma((n+1)c)}\,, \end{aligned}$$

via Lemma 2.5.15. On the other hand, writing $\tilde{P} = t(t-1)P$, we have

$$\begin{aligned} K_n(a,b,c) &= \int_{\mathscr{D}_n(0,1)} \left(\int_{\{Q\in\mathscr{D}_{n+1}|(Q,\tilde{P})\in\mathscr{E}_{n+2}\}} \right. \\ &\qquad \left. |Q(0)|^{a-1}|Q(1)|^{b-1}|R(P,Q)|^{c-1} d\ell_{n+1}(Q)\right) d\ell_n(P) \\ &= \int_{\mathscr{D}_n(0,1)} |\tilde{P}'(0)|^{a-1/2}|\tilde{P}'(1)|^{b-1/2}|R(P,\tilde{P}')|^{c-1/2} d\ell_n(P)\frac{\Gamma(a)\Gamma(b)\Gamma(c)^n}{\Gamma(a+b+nc)} \\ &= S_n(a+c,b+c,c)\frac{\Gamma(a)\Gamma(b)\Gamma(c)^n}{\Gamma(a+b+nc)}\,, \end{aligned}$$

by another application of Lemma 2.5.15. This proves (2.5.9) by induction on n; the induction base $n=1$ is an instance of (2.5.15). □

Exercise 2.5.16 Provide an alternative proof of Lemma 2.5.11 by noting that the determinant in the left side of (2.5.16) is a polynomial of degree $n(n-1)/2$ that vanishes whenever $x_i = x_j$ for some $i \neq j$, and thus, must equal a constant multiple of $\Delta(x)$.

2.5.4 Joint distribution of eigenvalues: alternative formulation

It is sometimes useful to represent the formulas for the joint distribution of eigenvalues as integration formulas for functions that depend only on the eigenvalues. We develop this correspondence now.

Let $f : \mathscr{H}_N^{(\beta)} \to [0,\infty]$ be a Borel function such that $f(H)$ depends only on the sequence of eigenvalues $\lambda_1(H) \leq \cdots \leq \lambda_N(H)$. In this situation, for short, we say that $f(H)$ depends only on the eigenvalues of H. (Note that the definition implies that f is a *symmetric* function of the eigenvalues of H.) Let $X \in \mathscr{H}_N^{(\beta)}$ be random with law $P_N^{(\beta)}$. Assuming the validity of Theorem 2.5.2, we have

$$Ef(X) = \frac{\int_{-\infty}^{\infty} \cdots \int_{-\infty}^{\infty} f(x_1,\ldots,x_N)|\Delta(x)|^\beta \prod_{i=1}^N e^{-\beta x_i^2/4} dx_i}{\int_{-\infty}^{\infty} \cdots \int_{-\infty}^{\infty} |\Delta(x)|^\beta \prod_{i=1}^N e^{-\beta x_i^2/4} dx_i}, \tag{2.5.22}$$

where $f(x_1,\ldots,x_N)$ denotes the value of f at the diagonal matrix with diagonal entries $x_1,\ldots,x_N$. Conversely, assuming (2.5.22), we immediately verify that (2.5.3) is proportional to the joint density of the eigenvalues $\lambda_1(X),\ldots,\lambda_N(X)$ by taking $f(H) = 1_{(\lambda_1(H),\ldots,\lambda_N(H))\in A}$ where $A \subset \mathbb{R}^N$ is any Borel set. In turn, to prove (2.5.22), it suffices to prove the general integration formula

$$\int f(H)\ell_N^{(\beta)}(dH) = C_N^{(\beta)} \int_{-\infty}^{\infty} \cdots \int_{-\infty}^{\infty} f(x_1,\ldots,x_N)|\Delta(x)|^\beta \prod_{j=1}^N dx_i, \tag{2.5.23}$$

where

$$C_N^{(\beta)} = \begin{cases} \dfrac{1}{N!} \displaystyle\prod_{k=1}^N \frac{\Gamma(1/2)^k}{\Gamma(k/2)} & \text{if } \beta = 1, \\[2ex] \dfrac{1}{N!} \displaystyle\prod_{k=1}^N \frac{\pi^{k-1}}{(k-1)!} & \text{if } \beta = 2, \end{cases}$$

and as in (2.5.22), the integrand $f(H)$ is nonnegative, Borel measurable, and depends only on the eigenvalues of H. Moreover, assuming the validity of (2.5.23), it follows by taking $f(H) = \exp(-a\mathrm{tr}(H^2)/2)$ with $a > 0$ and using Gaussian integration that

$$\begin{aligned} &\frac{1}{N!} \int_{-\infty}^{\infty} \cdots \int_{-\infty}^{\infty} |\Delta(x)|^\beta \prod_{i=1}^N e^{-ax_i^2/2} dx_i \\ &= (2\pi)^{N/2} a^{-\beta N(N-1)/4 - N/2} \prod_{j=1}^N \frac{\Gamma(j\beta/2)}{\Gamma(\beta/2)} =: \frac{1}{N!\bar{C}_N^{(\beta)}}. \end{aligned} \tag{2.5.24}$$

Thus, Theorem 2.5.2 is equivalent to the integration formula (2.5.23).

2.5.5 Superposition and decimation relations

The goal of this short subsection is to show how the eigenvalues of the GUE can be coupled (that is, constructed on the same probability space) with the eigenvalues of the GOE. As a by-product, we also discuss the eigenvalues of the GSE. Besides the obvious probabilistic interest in such a construction, the coupling will actually save us some work in the analysis of limit distributions for the maximal eigenvalue of the GOE and the GSE.

To state our results, we introduce some notation. For a finite subset $A \subset \mathbb{R}$ with $|A| = n$, we define $\mathrm{Ord}(A)$ to be the vector in $\mathbb{R}^n$ whose entries are the elements of A, ordered, that is

$$\mathrm{Ord}(A) = (x_1, \ldots, x_n) \quad \text{with } x_i \in A \text{ and } x_1 \le x_2 \le \ldots \le x_n\,.$$

For a vector $\mathbf{x} = (x_1, \ldots, x_n) \in \mathbb{R}^n$, we define $\mathrm{Dec}(\mathbf{x})$ as the even-location decimated version of $\mathbf{x}$, that is

$$\mathrm{Dec}(\mathbf{x}) = (x_2, x_4, \ldots, x_{2\lfloor n/2 \rfloor})\,.$$

Note that if $\mathbf{x}$ is ordered, then $\mathrm{Dec}(\mathbf{x})$ erases from $\mathbf{x}$ the smallest entry, the third smallest entry, etc.

The main result of this section is the following.

Theorem 2.5.17 *For $N > 0$ integer, let A_N and B_{N+1} denote the (collection of) eigenvalues of two independent random matrices distributed according to GOE(N) and GOE(N+1), respectively. Set*

$$(\eta_1^N, \ldots, \eta_N^N) = \eta^N = \mathrm{Dec}(\mathrm{Ord}(A_N \cup B_{N+1}))\,, \tag{2.5.25}$$

and

$$(\theta_1^N, \ldots, \theta_N^N) = \theta^N = \mathrm{Dec}(\mathrm{Ord}(A_{2N+1}))\,. \tag{2.5.26}$$

Then, $\{\eta^N\}$ (resp., $\{\theta^N\}$) is distributed as the eigenvalues of GUE(N) (resp., GSE(N)).

The proof of Theorem 2.5.17 goes through an integration relation that is slightly more general than our immediate needs. To state it, let $L = (a, b) \subset \mathbb{R}$ be a nonempty open interval, perhaps unbounded, and let f and g be positive real-valued infinitely differentiable functions defined on L. We will use the following assumption on the triple (L, f, g).

Assumption 2.5.18 *For (L, f, g) as above, for each integer $k \ge 0$, write $f_k(x) = x^k f(x)$ and $g_k(x) = x^k g(x)$ for $x \in L$. Then the following hold.*

(I) *There exists a matrix* $M^{(n)} \in \mathrm{Mat}_{n+1}(\mathbb{R})$, *independent of* x, *such that* $\det M^{(n)} > 0$ *and*

$$M^{(n)}(f_0, f_1, \dots, f_n)^{\mathrm{T}} = (g_0', g_1', \dots, g_{n-1}', f_0)^{\mathrm{T}}.$$

(II) $\int_a^b |f_n(x)| dx < \infty$.

(III) $\lim_{x \downarrow a} g_n(x) = 0$ *and* $\lim_{x \uparrow b} g_n(x) = 0$.

For a vector $\mathbf{x}_n = (x_1, \dots, x_n)$, recall that $\Delta(\mathbf{x}_n) = \prod_{1 \le i < j \le n}(x_j - x_i)$ is the Vandermonde determinant associated with $\mathbf{x}_n$, noting that if $\mathbf{x}_n$ is ordered then $\Delta(\mathbf{x}_n) \ge 0$. For an ordered vector $\mathbf{x}_n$ and an ordered collection of indices $I = \{i_1 < i_2 < \dots < i_{|I|}\} \subset \{1, \dots, n\}$, we write $\mathbf{x}_I = (x_{i_1}, x_{i_2}, \dots, x_{i_{|I|}})$. The key to the proof of Theorem 2.5.17 is the following proposition.

Proposition 2.5.19 *Let Assumption 2.5.18 hold for a triple* (L, f, g) *with* $L = (a, b)$. *For* $\mathbf{x}_{2n+1} = (x_1, \dots, x_{2n+1})$, *set*

$$\mathbf{x}_n^{(\mathrm{e})} = \mathrm{Dec}(\mathbf{x}_{2n+1}) = (x_2, x_4, \dots, x_{2n}), \text{ and } \mathbf{x}_{n+1}^{(\mathrm{o})} = (x_1, x_3, \dots, x_{2n+1}).$$

Let

$$\mathscr{I}_{2n+1} = \{(I,J) : I, J \subset \{1, \dots, 2n+1\}, |I| = n, |J| = n+1, I \cap J = \emptyset\}.$$

Then for each positive integer n *and* $\mathbf{x}_n^{(\mathrm{e})} \in L^n$, *we have the integration identities*

$$\int_a^{x_2} \int_{x_2}^{x_4} \cdots \int_{x_{2n}}^b \left(\sum_{(I,J) \in \mathscr{I}_{2n+1}} \Delta(\mathbf{x}_I)\Delta(\mathbf{x}_J) \right) \left(\prod_{i=1}^{2n+1} f(x_i) \right) dx_{2n+1} \cdots dx_3 dx_1$$
$$= \frac{2^n \left(\Delta(\mathbf{x}_n^{(\mathrm{e})})\right)^2 \left(\int_a^b f(x)dx\right) \left(\prod_{i=1}^n f(x_{2i})\right) \left(\prod_{i=1}^n g(x_{2i})\right)}{\det M^{(n)}}, \tag{2.5.27}$$

and

$$\int_a^{x_2} \int_{x_2}^{x_4} \cdots \int_{x_{2n}}^b \Delta(\mathbf{x}_{2n+1}) \left(\prod_{i=1}^{2n+1} f(x_i) \right) dx_{2n+1} \cdots dx_3 dx_1$$
$$= \frac{\left(\int_a^b f(x)dx\right) \left(\Delta(\mathbf{x}_n^{(\mathrm{e})})\right)^4 \left(\prod_{i=1}^n g(x_{2i})\right)^2}{\det M^{(2n)}}. \tag{2.5.28}$$

Assumption 2.5.18(II) guarantees the finiteness of the integrals in the proposition. The value of the positive constant $\det M^{(n)}$ will be of no interest in applications.

The proof of Proposition 2.5.19 will take up most of this section, after we complete the

Proof of Theorem 2.5.17 We first check that Assumption 2.5.18 with $L = (-\infty, \infty)$,

$f(x) = g(x) = e^{-x^2/4}$ holds, that is we verify that a matrix $M^{(n)}$ as defined there exists. Define $\tilde{M}^{(n)}$ as the solution to

$$\tilde{M}^{(n)}(f_0, f_1, \ldots, f_n)^{\mathrm{T}} = (f_0, f_0', f_1', \ldots, f_{n-1}')^{\mathrm{T}}.$$

Because f_i' is a polynomial of degree $i+1$ multiplied by $e^{-x^2/4}$, with leading coefficient equal $-1/2$, we have that $\tilde{M}^{(n)}$ is a lower triangular matrix, with $\tilde{M}^{(n)}_{1,1} = -1/2$ for $i > 1$ and $\tilde{M}^{(n)}_{1,1} = 1$, and thus $\det \tilde{M}^{(n)} = (-1/2)^n$. Since $M^{(n)}$ is obtained from $\tilde{M}^{(n)}$ by a cyclic permutation (of length $n+1$, and hence sign equal to $(-1)^n$), we conclude that $\det M^{(n)} = (1/2)^n > 0$, as needed.

To see the statement of Theorem 2.5.17 concerning the GUE, one applies equation (2.5.27) of Proposition 2.5.19 with the above choices of (L, f, g) and $M^{(n)}$, together with Theorem 2.5.2. The statement concerning the GSE follows with the same choice of (L, f, g), this time using (2.5.28). □

In preparation for the proof of Proposition 2.5.19, we need three lemmas. Only the first uses Assumption 2.5.18 in its proof. To compress notation, write

$$[A_{ij}]_{n,N} = \begin{bmatrix} A_{11} & \ldots & A_{1N} \\ \vdots & & \vdots \\ A_{n1} & \ldots & A_{nN} \end{bmatrix}.$$

Lemma 2.5.20 *For positive integers n and N, we have*

$$M^{(n)} \left[\int_{x_{j-1}}^{x_j} f_{i-1}(x)dx \right]_{n+1,N+1} [\mathbf{1}_{i \le j}]_{N+1,N+1}$$
$$= \left[\begin{cases} g_{i-1}(x_j) & \text{if } i < n+1 \text{ and } j < N+1, \\ 0 & \text{if } i < n+1 \text{ and } j = N+1, \\ \int_a^{x_j} f_0(x)dx & \text{if } i = n+1 \end{cases} \right]_{n+1,N+1} \qquad (2.5.29)$$

for all $a = x_0 < x_1 < \cdots < x_N < x_{N+1} = b$.

The left side of (2.5.29) is well-defined by Assumptions 2.5.18(I,II).

Proof Let $h_i = g_i'$ for $i = 0, \ldots, n-1$ and put $h_n = f_0$. The left side of (2.5.29) equals $[\int_a^{x_j} h_{i-1}(x)dx]_{n+1,N+1}$ and this in turn equals the right side of (2.5.29) by Assumption 2.5.18(III). □

Lemma 2.5.21 *For every positive integer n and $\mathbf{x} \in L^n$, we have*

$$(\Delta(\mathbf{x}))^4 \left(\prod_{i=1}^n g(x_i) \right)^2 = \det \left[\begin{cases} g_{i-1}(x_{(j+1)/2}) & \text{if } j \text{ is odd} \\ g_{i-1}'(x_{j/2}) & \text{if } j \text{ is even} \end{cases} \right]_{2n,2n}. \qquad (2.5.30)$$

The case $g = 1$ is the classical *confluent alternant identity*.

Proof Write $\mathbf{y}_{2n} = (y_1,\dots,y_{2n})$. Set

$$G(\mathbf{y}_{2n}) = \det([g_{i-1}(y_j)]_{2n,2n}) = \Delta(\mathbf{y})\prod_{i=1}^{2n} g(y_i)\,. \tag{2.5.31}$$

Dividing $G(\mathbf{y}_{2n})$ by $\prod_{i=1}^{n}(y_{2i}-y_{2i-1})$ and substituting $y_{2i-1} = y_{2i} = x_i$ for $i = 1,\dots,n$ give the left side of (2.5.30). On the other hand, let u_j denote the jth column of $[g_{i-1}(y_j)]_{2n,2n}$. (Thus, $G(\mathbf{y}_{2n}) = \det[u_1,\dots,u_{2n}]$.) Since it is a determinant, $G(\mathbf{y}_{2n}) = \det[u_1, u_2-u_1, u_3, u_4-u_3,\dots,u_{2n-1}, u_{2n}-u_{2n-1}]$ and thus

$$\frac{G(\mathbf{y}_{2n})}{\prod_{i=1}^{n}(y_{2i}-y_{2i-1})} = \det\left[u_1, \frac{u_2-u_1}{y_2-y_1},\dots,u_{2n-1}, \frac{u_{2n}-u_{2n-1}}{y_{2n}-y_{2n-1}}\right].$$

Applying L'Hôpital's rule thus shows that the last expression evaluated at $y_{2i-1} = y_{2i} = x_i$ for $i = 1,\dots,n$ equals the right side of (2.5.30). □

Lemma 2.5.22 *For every positive integer n and $\mathbf{x}_{2n+1} = (x_1,\dots,x_{2n+1})$ we have an identity*

$$2^n\Delta(\mathbf{x}_{n+1}^{(o)})\Delta(\mathbf{x}_n^{(e)}) = \sum_{(I,J)\in\mathscr{I}_{2n+1}} \Delta(\mathbf{x}_I)\Delta(\mathbf{x}_J)\,. \tag{2.5.32}$$

Proof Given $I = \{i_1 < \cdots < i_r\} \subset \{1,\dots,2n+1\}$, we write $\Delta_I = \Delta(\mathbf{x}_I)$. Given a polynomial $P = P(x_1,\dots,x_{2n+1})$ and a permutation $\tau \in S_{2n+1}$, let τP be defined by the rule

$$(\tau P)(x_1,\dots,x_{2n+1}) = P(x_{\tau(1)},\dots,x_{\tau(2n+1)})\,.$$

Given a permutation $\tau \in S_{2n+1}$, let $\tau I = \{\tau(i) \mid i \in I\}$. Now let $\Delta_I\Delta_J$ be a term appearing on the right side of (2.5.32) and let $\tau = (ij) \in S_{2n+1}$ be a transposition. We claim that

$$\frac{\tau(\Delta_I\Delta_J)}{\Delta_{\tau I}\Delta_{\tau J}} = \begin{cases} -1 & \text{if } \{i,j\}\subset I \text{ or } \{i,j\}\subset J,\\ (-1)^{|i-j|+1} & \text{otherwise.}\end{cases} \tag{2.5.33}$$

To prove (2.5.33), since the cases $\{i,j\}\subset I$ and $\{i,j\}\subset J$ are trivial, and we may allow i and j to exchange roles, we may assume without loss of generality that $i \in I$ and $j \in J$. Let k (resp., ℓ) be the number of indices in the set I (resp., J) strictly between i and j. Then

$$k+\ell = |i-j|-1,\quad \tau\Delta_I/\Delta_{\tau I} = (-1)^k,\quad \tau\Delta_J/\Delta_{\tau J} = (-1)^\ell,$$

which proves (2.5.33). It follows that if i and j have the same parity, the effect of applying τ to the right side of (2.5.32) is to multiply by -1, and therefore $(x_i - x_j)$ divides the right side. On the other hand, the left side of (2.5.32) equals 2^n times the product of $(x_i - x_j)$ with $i < j$ of the same parity. Therefore, because the polynomial functions on both sides of (2.5.32) are homogeneous of the same total degree in the variables $x_1,\dots,x_{2n+1}$, the left side equals the right side times some constant factor. Finally, the constant factor has to be 1 because the monomial $\prod_{i=1}^{n+1} x_{2i-1}^{i-1} \prod_{i=1}^{n} x_{2i}^{i-1}$ appears with coefficient 2^n on both sides. □

We can now provide the

Proof of Proposition 2.5.19 Let $x_0 = a$ and $x_{2n+2} = b$. To prove (2.5.27), use (2.5.32) to rewrite the left side multiplied by $\det M^{(n)}$ as

$$2^n \Delta(\mathbf{x}_n^{(\mathrm{e})}) \det \left(M^{(n)} \left[\int_{x_{2j-2}}^{x_{2j}} f_{i-1}(x)dx \right]_{n+1.n+1} \right) \prod_{i=1}^{n} f(x_{2i}),$$

and then evaluate using (2.5.29) and the second equality in (2.5.31). To prove (2.5.28), rewrite the left side multiplied by $\det M^{(2n)}$ as

$$\det \left(M^{(2n)} \left[\begin{cases} \int_{x_{j-1}}^{x_{j+1}} f_{i-1}(x)dx & \text{if } j \text{ is odd} \\ f_{i-1}(x_j) & \text{if } j \text{ is even} \end{cases} \right]_{2n+1,2n+1} \right),$$

and then evaluate using (2.5.29) and (2.5.30). □

Exercise 2.5.23 Let $\alpha, \gamma > -1$ be real constants. Show that each of the following triples (L, f, g) satisfies Assumption 2.5.18:
(a) $L = (0,\infty)$, $f(x) = x^\alpha e^{-x}$, $g(x) = x^{\alpha+1} e^{-x}$ (the Laguerre ensembles);
(b) $L = (0,1)$, $f(x) = x^\alpha (1-x)^\gamma$, $g(x) = x^{\alpha+1}(1-x)^{\gamma+1}$ (the Jacobi ensembles).

2.6 Large deviations for random matrices

In this section, we consider N random variables $(\lambda_1,\cdots,\lambda_N)$ with law

$$P_{V,\beta}^N(d\lambda_1,\cdots,d\lambda_N) = (Z_{V,\beta}^N)^{-1} |\Delta(\lambda)|^\beta e^{-N\sum_{i=1}^N V(\lambda_i)} \prod_{i=1}^N d\lambda_i, \tag{2.6.1}$$

for a $\beta > 0$ and a continuous function $V : \mathbb{R} \to \mathbb{R}$ such that, for some $\beta' > 1$ satisfying $\beta' \geq \beta$,

$$\liminf_{|x|\to\infty} \frac{V(x)}{\beta' \log|x|} > 1. \tag{2.6.2}$$

Here, $\Delta(\lambda) = \prod_{1\leq i<j\leq N} (\lambda_i - \lambda_j)$ and

$$Z_{V,\beta}^N = \int_{\mathbb{R}} \cdots \int_{\mathbb{R}} |\Delta(\lambda)|^\beta e^{-N\sum_{i=1}^N V(\lambda_i)} \prod_{i=1}^N d\lambda_i. \tag{2.6.3}$$

When $V(x) = \beta x^2/4$, and $\beta = 1, 2$, we saw in Section 2.5 that $P_{\beta x^2/4,\beta}^N$ is the law of the (rescaled) eigenvalues of a GOE(N) matrix when $\beta = 1$, and of a GUE(N) matrix when $\beta = 2$. It also follows from the general results in Section 4.1 that the case $\beta = 4$ corresponds to another matrix ensemble, namely the GSE(N). In view of these and applications to certain problems in physics, we consider in this section the slightly more general model. We emphasize, however, that the distribution (2.6.1) precludes us from considering random matrices with independent non-Gaussian entries.

We have proved earlier in this chapter (for the GOE, see Section 2.1, and for the GUE, see Section 2.2) that the empirical measure $L_N = N^{-1}\sum_{i=1}^N \delta_{\lambda_i}$ converges in probability (and almost surely, under appropriate moment assumptions), and we studied its fluctuations around its mean. We have also considered the convergence of the top eigenvalue λ_N^N. Such results did not depend much on the Gaussian nature of the entries.

We address here a different type of question. Namely, we study the probability that L_N, or λ_N^N, take a very unlikely value. This was already considered in our discussion of concentration inequalities, see Section 2.3, where the emphasis was put on obtaining upper bounds on the probability of deviation. In contrast, the purpose of the analysis here is to exhibit a precise estimate of these probabilities, or at least of their logarithmic asymptotics. The appropriate tool for handling such questions is large deviation theory, and we give in Appendix D a concise introduction to that theory and related definitions, together with related references.

2.6.1 Large deviations for the empirical measure

We endow $M_1(\mathbb{R})$ with the usual weak topology, compatible with the Lipschitz bounded metric, see (C.1). Our goal is to estimate the probability $P_{V,\beta}^N(L_N \in A)$, for measurable sets $A \subset M_1(\mathbb{R})$. Of particular interest is the case where A does not contain the limiting distribution of L_N.

Define the *noncommutative entropy* $\Sigma : M_1(\mathbb{R}) \to [-\infty, \infty)$ as

$$\Sigma(\mu) = \begin{cases} \iint \log|x-y| d\mu(x) d\mu(y) & \text{if } \int \log(|x|+1) d\mu(x) < \infty, \\ -\infty & \text{otherwise}, \end{cases} \tag{2.6.4}$$

and the function $I_\beta^V : M_1(\mathbb{R}) \to [0, \infty]$ as

$$I_\beta^V(\mu) = \begin{cases} \int V(x) d\mu(x) - \frac{\beta}{2}\Sigma(\mu) - c_\beta^V & \text{if } \int V(x) d\mu(x) < \infty, \\ \infty & \text{otherwise}, \end{cases} \tag{2.6.5}$$

where

$$c_\beta^V = \inf_{\nu \in M_1(\mathbb{R})} \{ \int V(x) d\nu(x) - \frac{\beta}{2} \Sigma(\nu) \} \in (-\infty, \infty). \tag{2.6.6}$$

(Lemma 2.6.2 below and its proof show that both Σ and I_β^V are well defined, and that c_β^V is finite.)

Theorem 2.6.1 *Let $L_N = N^{-1} \sum_{i=1}^N \delta_{\lambda_i^N}$ where the random variables $\{\lambda_i^N\}_{i=1}^N$ are distributed according to the law $P_{V,\beta}^N$ of (2.6.1), with potential V satisfying (2.6.2). Then, the family of random measures L_N satisfies, in $M_1(\mathbb{R})$ equipped with the weak topology, a large deviation principle with speed N^2 and good rate function I_β^V. That is,*

(a) $I_\beta^V : M_1(\mathbb{R}) \to [0, \infty]$ possesses compact level sets $\{\nu : I_\beta^V(\nu) \le M\}$ for all $M \in \mathbb{R}_+$,

(b) for any open set $O \subset M_1(\mathbb{R})$,

$$\liminf_{N\to\infty} \frac{1}{N^2} \log P_{\beta,V}^N (L_N \in O) \ge -\inf_O I_\beta^V, \tag{2.6.7}$$

(c) for any closed set $F \subset M_1(\mathbb{R})$,

$$\limsup_{N\to\infty} \frac{1}{N^2} \log P_{\beta,V}^N (L_N \in F) \le -\inf_F I_\beta^V. \tag{2.6.8}$$

The proof of Theorem 2.6.1 relies on the properties of the function I_β^V collected in Lemma 2.6.2 below. Define the *logarithmic capacity* of a measurable set $A \subset \mathbb{R}$ as

$$\gamma(A) := \exp\left\{ - \inf_{\nu \in M_1(A)} \int\int \log \frac{1}{|x-y|} d\nu(x) d\nu(y) \right\}.$$

Lemma 2.6.2

(a) $c_\beta^V \in (-\infty, \infty)$ *and I_β^V is well defined on $M_1(\mathbb{R})$, taking its values in $[0, +\infty]$.*

(b) $I_\beta^V(\mu)$ *is infinite as soon as μ satisfies one of the following conditions*

(b.1) $\int V(x) d\mu(x) = +\infty$.

(b.2) *There exists a set $A \subset \mathbb{R}$ of positive μ mass but null logarithmic capacity, i.e. a set A such that $\mu(A) > 0$ but $\gamma(A) = 0$.*

(c) I_β^V *is a good rate function.*

(d) I_β^V *is a strictly convex function on $M_1(\mathbb{R})$.*

(e) I_β^V *achieves its minimum value at unique $\sigma_\beta^V \in M_1(\mathbb{R})$. The measure σ_β^V is compactly supported, and is characterized by the equality*

$$V(x) - \beta \langle \sigma_\beta^V, \log|\cdot - x| \rangle = C_\beta^V, \quad \text{for } \sigma_\beta^V\text{-almost every } x, \tag{2.6.9}$$

and inequality

$$V(x) - \beta\langle \sigma_\beta^V, \log|\cdot - x|\rangle > C_\beta^V, \quad \text{for all } x \notin \operatorname{supp}(\sigma_\beta^V), \tag{2.6.10}$$

for some constant C_β^V. *Necessarily,* $C_\beta^V = 2c_\beta^V - \langle \sigma_\beta^V, V\rangle$.

As an immediate corollary of Theorem 2.6.1 and of part (e) of Lemma 2.6.2 we have the following.

Corollary 2.6.3 *Under* $P_{V,\beta}^N$, L_N *converges almost surely towards* σ_β^V.

Proof of Lemma 2.6.2 For all $\mu \in M_1(\mathbb{R})$, $\Sigma(\mu)$ is well defined and $< \infty$ due to the bound

$$\log|x-y| \le \log(|x|+1) + \log(|y|+1). \tag{2.6.11}$$

Further, $c_\beta^V < \infty$ as can be checked by taking ν as the uniform law on $[0,1]$.

Set

$$f(x,y) = \frac{1}{2}V(x) + \frac{1}{2}V(y) - \frac{\beta}{2}\log|x-y|. \tag{2.6.12}$$

Note that (2.6.2) implies that $f(x,y)$ goes to $+\infty$ when x,y do since (2.6.11) yields

$$f(x,y) \ge \frac{1}{2}(V(x) - \beta\log(|x|+1)) + \frac{1}{2}(V(y) - \beta\log(|y|+1)). \tag{2.6.13}$$

Further, $f(x,y)$ goes to $+\infty$ when x,y approach the diagonal $\{x=y\}$. Therefore, for all $L>0$, there exists a constant $K(L)$ (going to infinity with L) such that, with $B_L := \{(x,y) : |x-y| < L^{-1}\} \cup \{(x,y) : |x| > L\} \cup \{(x,y) : |y| > L\}$,

$$B_L \subset \{(x,y) : f(x,y) \ge K(L)\}. \tag{2.6.14}$$

Since f is continuous on the compact set B_L^c, we conclude that f is bounded below on $\mathbb{R}^2$, and denote by $b_f > -\infty$ a lower bound. It follows that $c_\beta^V \ge b_f > -\infty$. Thus, because V is bounded below by (2.6.2), we conclude that I_β^V is well defined and takes its values in $[0,\infty]$, completing the proof of part (a). Further, since for any measurable subset $A \subset \mathbb{R}$,

$$\begin{aligned}
I_\beta^V(\mu) &= \iint (f(x,y) - b_f)d\mu(x)d\mu(y) + b_f - c_\beta^V \\
&\ge \int_A\int_A (f(x,y) - b_f)d\mu(x)d\mu(y) + b_f - c_\beta^V \\
&\ge \frac{\beta}{2}\int_A\int_A \log|x-y|^{-1}d\mu(x)d\mu(y) + \inf_{x\in\mathbb{R}} V(x)\mu(A)^2 - |b_f| - c_\beta^V \\
&\ge -\frac{\beta}{2}\mu(A)^2\log(\gamma(A)) - |b_f| - c_\beta^V + \inf_{x\in\mathbb{R}} V(x)\mu(A)^2,
\end{aligned}$$

one concludes that if $I_\beta^V(\mu) < \infty$, and A is a measurable set with $\mu(A) > 0$, then $\gamma(A) > 0$. This completes the proof of part (b).

We now show that I_V^β is a good rate function, and first that its level sets $\{I_V^\beta \leq M\}$ are closed, that is that I_V^β is lower semicontinuous. Indeed, by the monotone convergence theorem,

$$\begin{aligned} I_\beta^V(\mu) &= \iint f(x,y)d\mu(x)d\mu(y) - c_\beta^V \\ &= \sup_{M\geq 0} \iint (f(x,y)\wedge M)d\mu(x)d\mu(y) - c_\beta^V . \end{aligned}$$

But $f^M = f \wedge M$ is bounded continuous and so, for $M < \infty$,

$$I_\beta^{V,M}(\mu) = \iint (f(x,y)\wedge M)d\mu(x)d\mu(y)$$

is bounded continuous on $M_1(\mathbb{R})$. As a supremum of the continuous functions $I_\beta^{V,M}$, I_β^V is lower semicontinuous.

To complete the proof that I_β^V is a good rate function, we need to show that the set $\{I_\beta^V \leq L\}$ is compact. By Theorem C.9, to see the latter it is enough to show that $\{I_\beta^V \leq L\}$ is included in a compact subset of $M_1(\mathbb{R})$ of the form

$$K_\varepsilon = \bigcap_{B\in\mathbb{N}} \{\mu \in M_1(\mathbb{R}) : \mu([-B,B]^c) \leq \varepsilon(B)\},$$

with a sequence $\varepsilon(B)$ going to zero as B goes to infinity. Arguing as in (2.6.14), there exist constants $K'(L)$ going to infinity as L goes to infinity, such that

$$\{(x,y) : |x| > L, |y| > L\} \subset \{(x,y) : f(x,y) \geq K'(L)\}. \tag{2.6.15}$$

Therefore, for any large positive L,

$$\begin{aligned} \mu(|x| > L)^2 &= \mu \otimes \mu(|x| > L, |y| > L) \\ &\leq \mu \otimes \mu(f(x,y) \geq K'(L)) \\ &\leq \frac{1}{K'(L) - b_f} \iint (f(x,y) - b_f)d\mu(x)d\mu(y) \\ &= \frac{1}{K'(L) - b_f}(I_\beta^V(\mu) + c_\beta^V - b_f). \end{aligned}$$

Hence, taking $\varepsilon(B) = [\sqrt{(M + c_\beta^V - b_f)_+}/\sqrt{(K'(B) - b_f)_+}] \wedge 1$, which goes to zero when B goes to infinity, one has that $\{I_\beta^V \leq M\} \subset K_\varepsilon$. This completes the proof of part (c).

Since I_β^V is a good rate function, it achieves its minimal value. Let σ_β^V be a minimizer. Let us derive some consequences of minimality. For any signed measure $\bar{\nu}(dx) = \phi(x)\sigma_\beta^V(dx) + \psi(x)dx$ with two bounded measurable compactly supported functions (ϕ, ψ) such that $\psi \geq 0$ and $\bar{\nu}(\mathbb{R}) = 0$, for $\varepsilon > 0$ small enough, $\sigma_\beta^V + \varepsilon\bar{\nu}$ is a probability measure so that

$$I_\beta^V(\sigma_\beta^V + \varepsilon\bar{\nu}) \geq I_\beta^V(\sigma_\beta^V), \tag{2.6.16}$$

which implies

$$\int \left(V(x) - \beta \int \log|x-y| d\sigma_\beta^V(y) \right) d\bar{\nu}(x) \geq 0.$$

Taking $\psi = 0$, we deduce (using $\pm\phi$) that there is a constant C_β^V such that

$$V(x) - \beta \int \log|x-y| d\sigma_\beta^V(y) = C_\beta^V, \quad \sigma_\beta^V \text{ a.s.}, \tag{2.6.17}$$

which implies that σ_β^V is compactly supported (because $V(x) - \beta \int \log|x-y| d\sigma_\beta^V(y)$ goes to infinity when x does by (2.6.13)). Taking $\phi = -\int \psi(y)dy$ on the support of σ_β^V, we then find that

$$V(x) - \beta \int \log|x-y| d\sigma_\beta^V(y) \geq C_\beta^V, \tag{2.6.18}$$

Lebesgue almost surely, and then everywhere outside of the support of σ_β^V by continuity. Integrating (2.6.17) with respect to σ_β^V then shows that

$$C_\beta^V = 2c_\beta^V - \langle \sigma_\beta^V, V \rangle,$$

proving (2.6.9) and (2.6.10), with the strict inequality in (2.6.10) following from the uniqueness of σ_β^V, since the later implies that the inequality (2.6.16) is strict as soon as $\bar{\nu}$ is nontrivial. Finally, integrating (2.6.9) with respect to σ_β^V reveals that the latter must be a minimizer of I_β^V, so that (2.6.9) characterizes σ_β^V.

The claimed uniqueness of σ_β^V, and hence the completion of the proof of part (e), will follow from the strict convexity claim (part (d) of the lemma), which we turn to next. Note first that, extending the definition of Σ to signed measures in evident fashion when the integral in (2.6.4) is well defined, we can rewrite I_β^V as

$$I_\beta^V(\mu) = -\frac{\beta}{2}\Sigma(\mu - \sigma_\beta^V) + \int \left(V(x) - \beta \int \log|x-y| d\sigma_\beta^V(y) - C_\beta^V \right) d\mu(x).$$

The fact that I_β^V is strictly convex will follow as soon as we show that Σ is strictly concave. Toward this end, note the formula

$$\log|x-y| = \int_0^\infty \frac{1}{2t} \left(\exp\{-\frac{1}{2t}\} - \exp\{-\frac{|x-y|^2}{2t}\} \right) dt, \tag{2.6.19}$$

which follows from the equality

$$\frac{1}{z} = \frac{1}{2z}\int_0^\infty e^{-u/2}du$$

by the change of variables $u \mapsto z^2/t$ and integration of z from 1 to $|x-y|$. Now, (2.6.19) implies that for any $\mu \in M_1(\mathbb{R})$,

$$\Sigma(\mu - \sigma_\beta^V) = -\int_0^\infty \frac{1}{2t}\left(\iint \exp\{-\frac{|x-y|^2}{2t}\}d(\mu - \sigma_\beta^V)(x)d(\mu - \sigma_\beta^V)(y)\right)dt\,. \tag{2.6.20}$$

Indeed, one may apply Fubini's Theorem when μ, σ_β^V are supported in $[-\frac{1}{2}, \frac{1}{2}]$ since then $\mu \otimes \sigma_\beta^V(\exp\{-\frac{1}{2t}\} - \exp\{-\frac{|x-y|^2}{2t}\} \leq 0) = 1$. One then deduces (2.6.20) for any compactly supported probability measure μ by scaling and finally for all probability measures by approximations. The fact that, for all $t \geq 0$,

$$\begin{aligned}&\iint \exp\{-\frac{|x-y|^2}{2t}\}d(\mu - \sigma_\beta^V)(x)d(\mu - \sigma_\beta^V)(y)\\ &\quad = \sqrt{\frac{t}{2\pi}}\int_{-\infty}^{+\infty}\left|\int \exp\{i\lambda x\}d(\mu - \sigma_\beta^V)(x)\right|^2 \exp\{-\frac{t\lambda^2}{2}\}d\lambda\,,\end{aligned}$$

therefore entails that Σ is concave since $\mu \to \left|\int \exp\{i\lambda x\}d(\mu - \sigma_\beta^V)(x)\right|^2$ is convex for all $\lambda \in \mathbb{R}$. Strict convexity comes from the fact that

$$\Sigma(\alpha\mu + (1-\alpha)\nu) - (\alpha\Sigma(\mu) + (1-\alpha)\Sigma(\nu)) = (\alpha^2 - \alpha)\Sigma(\mu - \nu)\,,$$

which vanishes for $\alpha \in (0,1)$ if and only if $\Sigma(\nu - \mu) = 0$. The latter equality implies that all the Fourier transforms of $\nu - \mu$ vanish, and hence $\mu = \nu$. This completes the proof of part (d) and hence of the lemma. □

Proof of Theorem 2.6.1 With f as in (2.6.12),

$$P_{V,\beta}^N(d\lambda_1, \cdots, d\lambda_N) = (Z_N^{\beta,V})^{-1}e^{-N^2\iint_{x\neq y} f(x,y)dL_N(x)dL_N(y)}\prod_{i=1}^N e^{-V(\lambda_i)}d\lambda_i\,.$$

(No typo here: indeed, no N before $V(\lambda_i)$.) Hence, if

$$\mu \to \int_{x\neq y} f(x,y)d\mu(x)d\mu(y)$$

were a bounded continuous function, the proof would follow from a standard application of Varadhan's Lemma, Theorem D.8. The main point will therefore be to overcome the singularities of this function, with the most delicate part being to overcome the singularity of the logarithm.

Following Appendix D (see Corollary D.6 and Definition D.3), a full large deviation principle can be proved by proving that exponential tightness holds, as well as estimating the probability of small balls. We follow these steps below.

Exponential tightness

Observe that, by Jensen's inequality, for some constant C,

$$\begin{aligned}\log Z_N^{\beta,V} &\geq N\log\int e^{-V(x)}dx \\ &\quad -N^2\int\left(\int_{x\neq y} f(x,y)dL_N(x)dL_N(y)\right)\prod_{i=1}^{N}\frac{e^{-V(\lambda_i)}d\lambda_i}{\int e^{-V(x)}dx} \geq -CN^2 .\end{aligned}$$

Moreover, by (2.6.13) and (2.6.2), there exist constants $a>0$ and $c>-\infty$ so that

$$f(x,y)\geq a|V(x)|+a|V(y)|+c,$$

from which one concludes that for all $M\geq 0$,

$$P_{V,\beta}^N\left(\int|V(x)|dL_N\geq M\right)\leq e^{-2aN^2M+(C-c)N^2}\left(\int e^{-V(x)}dx\right)^N . \tag{2.6.21}$$

Since V goes to infinity at infinity, $K_M=\{\mu\in M_1(\mathbb{R}):\int|V|d\mu\leq M\}$ is a compact set for all $M<\infty$, so that we have proved that the law of L_N under $P_{V,\beta}^N$ is exponentially tight.

A large deviation upper bound

Recall that d denotes the Lipschitz bounded metric, see (C.1). We prove here that for any $\mu\in M_1(\mathbb{R})$, if we set $\bar{P}_{V,\beta}^N=Z_N^{\beta,V}P_{V,\beta}^N$,

$$\lim_{\varepsilon\to 0}\limsup_{N\to\infty}\frac{1}{N^2}\log\bar{P}_{V,\beta}^N\left(d(L_N,\mu)\leq\varepsilon\right)\leq-\int f(x,y)d\mu(x)d\mu(y). \tag{2.6.22}$$

(We will prove the full LDP for $P_{V,\beta}^N$ as a consequence of both the upper and lower bounds on $\bar{P}_{V,\beta}^N$, see (2.6.28) below.) For any $M\geq 0$, set $f_M(x,y)=f(x,y)\wedge M$. Then the bound

$$\bar{P}_{V,\beta}^N\left(d(L_N,\mu)\leq\varepsilon\right)\leq\int_{d(L_N,\mu)\leq\varepsilon}e^{-N^2\int_{x\neq y}f_M(x,y)dL_N(x)dL_N(y)}\prod_{i=1}^N e^{-V(\lambda_i)}d\lambda_i$$

holds. Since under the product Lebesgue measure, the λ_is are almost surely distinct, it holds that $L_N\otimes L_N(x=y)=N^{-1}$, $\bar{P}_{V,\beta}^N$ almost surely. Thus we deduce that

$$\int f_M(x,y)dL_N(x)dL_N(y)=\int_{x\neq y}f_M(x,y)dL_N(x)dL_N(y)+MN^{-1},$$

and so

$$\bar{P}^N_{V,\beta}(d(L_N,\mu)\le\varepsilon)$$

$$\le e^{MN}\int_{d(L_N,\mu)\le\varepsilon} e^{-N^2\int f_M(x,y)dL_N(x)dL_N(y)}\prod_{i=1}^N e^{-V(\lambda_i)}d\lambda_i\,.$$

Since f_M is bounded and continuous, $I_\beta^{V,M}:\nu\mapsto\int f_M(x,y)d\nu(x)d\nu(y)$ is a continuous functional, and therefore we deduce that

$$\lim_{\varepsilon\to 0}\limsup_{N\to\infty}\frac{1}{N^2}\log\bar{P}^N_{V,\beta}(d(L_N,\mu)\le\varepsilon)\le -I_\beta^{V,M}(\mu)\,.$$

We finally let M go to infinity and conclude by the monotone convergence theorem. Note that the same argument shows that

$$\limsup_{N\to\infty}\frac{1}{N^2}\log Z_N^{\beta,V}\le -\inf_{\mu\in M_1(\mathbb{R})}\int f(x,y)d\mu(x)d\mu(y)\,. \qquad (2.6.23)$$

A large deviation lower bound

We prove here that for any $\mu\in M_1(\mathbb{R})$,

$$\lim_{\varepsilon\to 0}\liminf_{N\to\infty}\frac{1}{N^2}\log\bar{P}^N_{V,\beta}(d(L_N,\mu)\le\varepsilon)\ge -\int f(x,y)d\mu(x)d\mu(y)\,. \qquad (2.6.24)$$

Note that we can assume without loss of generality that $I_\beta^V(\mu)<\infty$, since otherwise the bound is trivial, and so, in particular, we may and will assume that μ has no atoms. We can also assume that μ is compactly supported since if we consider $\mu_M=\mu([-M,M])^{-1}1_{|x|\le M}d\mu(x)$, clearly μ_M converges towards μ and by the monotone convergence theorem, one checks that, since f is bounded below,

$$\lim_{M\uparrow\infty}\int f(x,y)d\mu_M(x)d\mu_M(y)=\int f(x,y)d\mu(x)d\mu(y)\,,$$

which ensures that it is enough to prove the lower bound for $(\mu_M, M\in\mathbb{R}_+, I_\beta^V(\mu)<\infty)$, and so for compactly supported probability measures with finite entropy.

The idea is to localize the eigenvalues $(\lambda_i)_{1\le i\le N}$ in small sets and to take advantage of the fast speed N^2 of the large deviations to neglect the small volume of these sets. To do so, we first remark that, for any $\nu\in M_1(\mathbb{R})$ with no atoms, if we set

$$\begin{aligned} x^{1,N} &= \inf\left\{x:\ \nu((-\infty,x])\ge\frac{1}{N+1}\right\},\\ x^{i+1,N} &= \inf\left\{x\ge x^{i,N}:\nu\left((x^{i,N},x]\right)\ge\frac{1}{N+1}\right\},\qquad 1\le i\le N-1, \end{aligned}$$

for any real number η, there exists a positive integer $N(\eta)$ such that, for any N larger than $N(\eta)$,

$$d\left(\nu, \frac{1}{N}\sum_{i=1}^{N}\delta_{x^{i,N}}\right) < \eta .$$

In particular, for $N \geq N(\frac{\delta}{2})$,

$$\left\{(\lambda_i)_{1\leq i\leq N} \mid |\lambda_i - x^{i,N}| < \frac{\delta}{2}\ \forall i \in [1,N]\right\} \subset \{(\lambda_i)_{1\leq i\leq N} \mid d(L_N,\nu) < \delta\},$$

so that we have the lower bound

$$\begin{aligned}
&\bar{P}^N_{V,\beta}\left(d(L_N,\mu)\leq\varepsilon\right)\\
&\geq \int_{\cap_i\{|\lambda_i - x^{i,N}|<\frac{\delta}{2}\}} e^{-N^2\int_{x\neq y} f(x,y)dL_N(x)dL_N(y)}\prod_{i=1}^{N} e^{-V(\lambda_i)}d\lambda_i\\
&= \int_{\cap_i\{|\lambda_i|<\frac{\delta}{2}\}} \prod_{i<j}|x^{i,N}-x^{j,N}+\lambda_i-\lambda_j|^{\beta} e^{-N\sum_{i=1}^{N}V(x^{i,N}+\lambda_i)}\prod_{i=1}^{N}d\lambda_i\\
&\geq \left(\prod_{i+1<j}|x^{i,N}-x^{j,N}|^{\beta}\prod_i |x^{i,N}-x^{i+1,N}|^{\frac{\beta}{2}}e^{-N\sum_{i=1}^{N}V(x^{i,N})}\right)\\
&\quad\times\left(\int_{\substack{\cap_i\{|\lambda_i|<\frac{\delta}{2}\}\\ \lambda_i<\lambda_{i+1}}}\prod_i|\lambda_i-\lambda_{i+1}|^{\frac{\beta}{2}}e^{-N\sum_{i=1}^{N}[V(x^{i,N}+\lambda_i)-V(x^{i,N})]}\prod_{i=1}^{N}d\lambda_i\right)\\
&=: P_{N,1}\times P_{N,2}, \qquad (2.6.25)
\end{aligned}$$

where we used the fact that $|x^{i,N}-x^{j,N}+\lambda_i-\lambda_j| \geq |x^{i,N}-x^{j,N}| \vee |\lambda_i-\lambda_j|$ when $\lambda_i \geq \lambda_j$ and $x^{i,N} \geq x^{j,N}$. To estimate $P_{N,2}$, note that since we assumed that μ is compactly supported, the $(x^{i,N}, 1\leq i\leq N)_{N\in\mathbb{N}}$ are uniformly bounded and so, by continuity of V,

$$\lim_{N\to\infty}\sup_{N\in\mathbb{N}}\sup_{1\leq i\leq N}\sup_{|x|\leq\delta}|V(x^{i,N}+x)-V(x^{i,N})| = 0.$$

Moreover, writing $u_1 = \lambda_1$, $u_{i+1} = \lambda_{i+1}-\lambda_i$,

$$\int_{\substack{|\lambda_i|<\frac{\delta}{2}\ \forall i\\ \lambda_i<\lambda_{i-1}}}\prod_i|\lambda_i-\lambda_{i+1}|^{\frac{\beta}{2}}\prod_{i=1}^{N}d\lambda_i \geq \int_{0<u_i<\frac{\delta}{2N}}\prod_{i=2}^{N}u_i^{\frac{\beta}{2}}\prod_{i=1}^{N}du_i \geq \left(\frac{\delta}{(\beta+2)N}\right)^{N(\frac{\beta}{2}+1)}.$$

Therefore,

$$\lim_{\delta\to 0}\liminf_{N\to\infty}\frac{1}{N^2}\log P_{N,2}\geq 0. \qquad (2.6.26)$$

To handle the term $P_{N,1}$, the uniform boundedness of the $x^{i,N}$s and the convergence of their empirical measure towards μ imply that

$$\lim_{N\to\infty}\frac{1}{N}\sum_{i=1}^{N}V(x^{i,N})=\int V(x)d\mu(x). \tag{2.6.27}$$

Finally since $x\to\log(x)$ increases on $\mathbb{R}^+$, we notice that

$$\begin{aligned}
&\int_{x^{1,N}\le x<y\le x^{N,N}}\log(y-x)d\mu(x)d\mu(y)\\
&\quad\le \sum_{1\le i\le j\le N-1}\log(x^{j+1,N}-x^{i,N})\int_{\substack{x\in[x^{i,N},x^{i+1,N}]\\ y\in[x^{j,N},x^{j+1,N}]}}1_{x<y}d\mu(x)d\mu(y)\\
&\quad= \frac{1}{(N+1)^2}\sum_{i<j}\log|x^{i,N}-x^{j+1,N}|+\frac{1}{2(N+1)^2}\sum_{i=1}^{N-1}\log|x^{i+1,N}-x^{i,N}|.
\end{aligned}$$

Since $\log|x-y|$ is upper-bounded when x,y are in the support of the compactly supported measure μ, the monotone convergence theorem implies that the left side in the last display converges towards $\frac{1}{2}\Sigma(\mu)$. Thus, with (2.6.27), we have proved

$$\liminf_{N\to\infty}\frac{1}{N^2}\log P_{N,1}\ge\beta\int_{x<y}\log(y-x)d\mu(x)d\mu(y)-\int V(x)d\mu(x),$$

which concludes, with (2.6.25) and (2.6.26), the proof of (2.6.24).

Conclusion of the proof of Theorem 2.6.1

By (2.6.24), for all $\mu\in M_1(\mathbb{R})$,

$$\begin{aligned}
\liminf_{N\to\infty}\frac{1}{N^2}\log Z^N_{\beta,V} &\ge \lim_{\varepsilon\to0}\liminf_{N\to\infty}\frac{1}{N^2}\log\bar{P}^N_{V,\beta}(d(L_N,\mu)\le\varepsilon)\\
&\ge -\iint f(x,y)d\mu(x)d\mu(y),
\end{aligned}$$

and so, optimizing with respect to $\mu\in M_1(\mathbb{R})$ and with (2.6.23),

$$\lim_{N\to\infty}\frac{1}{N^2}\log Z^N_{\beta,V}=-\inf_{\mu\in M_1(\mathbb{R})}\{\int f(x,y)d\mu(x)d\mu(y)\}=-c^V_\beta.$$

Thus, (2.6.24) and (2.6.22) imply the weak large deviation principle, i.e. that for all $\mu\in M_1(\mathbb{R})$,

$$\begin{aligned}
&\lim_{\varepsilon\to0}\liminf_{N\to\infty}\frac{1}{N^2}\log P^N_{V,\beta}(d(L_N,\mu)\le\varepsilon)\\
&\quad=\lim_{\varepsilon\to0}\limsup_{N\to\infty}\frac{1}{N^2}\log P^N_{V,\beta}(d(L_N,\mu)\le\varepsilon)=-I^V_\beta(\mu).
\end{aligned} \tag{2.6.28}$$

This, together with the exponential tightness property proved above, completes the proof of Theorem 2.6.1. □

Exercise 2.6.4 [Proof #5 of Wigner's Theorem] Take $V(x) = \beta x^2/4$ and apply Corollary 2.6.3 together with Lemma 2.6.2 to provide a proof of Wigner's Theorem 2.1.1 in the case of GOE or GUE matrices.
Hint: It is enough to check (2.6.9) and (2.6.10), that is to check that

$$\int \log|x-y|\sigma(dy) \le \frac{x^2}{4} - \frac{1}{2},$$

with equality for $x \in [-2,2]$, where σ is the semicircle law. Toward this end, use the representation of the Stieltjes transform of σ, see (2.4.6).

2.6.2 Large deviations for the top eigenvalue

We consider next the large deviations for the maximum $\lambda_N^* = \max_{i=1}^N \lambda_i$, of random variables that possess the joint law (2.6.1). These will be obtained under the following assumption.

Assumption 2.6.5 *The normalization constants $Z_{V,\beta}^N$ satisfy*

$$\lim_{N\to\infty} \frac{1}{N} \log \frac{Z_{NV/(N-1),\beta}^{N-1}}{Z_{V,\beta}^N} = \alpha_{V,\beta}\,. \tag{2.6.29}$$

It is immediate from (2.5.11) that if $V(x) = \beta x^2/4$ then Assumption 2.6.5 holds, with $\alpha_{V,\beta} = -\beta/2$.

Assumption 2.6.5 is crucial in deriving the following LDP.

Theorem 2.6.6 *Let $(\lambda_1^N, \dots, \lambda_N^N)$ be distributed according to the joint law $P_{V,\beta}^N$ of (2.6.1), with continuous potential V that satisfies (2.6.2) and Assumption 2.6.5. Let σ_β^V be the minimizing measure of Lemma 2.6.2, and set $x^* = \max\{x : x \in \operatorname{supp}\sigma_\beta^V\}$. Then, $\lambda_N^* = \max_{i=1}^N \lambda_i^N$ satisfies the LDP in $\mathbb{R}$ with speed N and good rate function*

$$J_\beta^V(x) = \begin{cases} \beta \int \log|x-y|\sigma_\beta^V(dy) - V(x) - \alpha_{V,\beta} & \text{if } x \ge x^*, \\ \infty & \text{otherwise}. \end{cases}$$

Proof of Theorem 2.6.6 Since $J_\beta^V(\cdot)$ is continuous on (x^*, ∞) and $J_\beta^V(x)$ increases to infinity as $x \to \infty$, it is a good rate function. Therefore, the stated LDP follows as soon as we show that

$$\text{for any } x < x^*, \quad \limsup_{N\to\infty} \frac{1}{N} \log P_{V,\beta}^N(\lambda_N^* \le x) = -\infty, \tag{2.6.30}$$

$$\text{for any } x > x^*, \quad \limsup_{N\to\infty} \frac{1}{N} \log P^N_{V,\beta}(\lambda^*_N \geq x) \leq -J^V_\beta(x) \tag{2.6.31}$$

and

$$\text{for any } x > x^*, \quad \lim_{\delta\to 0} \liminf_{N\to\infty} \frac{1}{N} \log P^N_{V,\beta}(\lambda^*_N \in (x-\delta, x+\delta)) \geq -J^V_\beta(x). \tag{2.6.32}$$

The limit (2.6.30) follows immediately from the LDP (at speed N^2) for the empirical measure, Theorem 2.6.1; indeed, the event $\lambda^*_N \leq x$ implies that $L_N((x,x^*]) = 0$. Hence, one can find a bounded continuous function f with support in $(x,x^*]$, independent of N, such that $\langle L_N, f\rangle = 0$ but $\langle \sigma^V_\beta, f\rangle > 0$. Theorem 2.6.1 implies that this event has probability that decays exponentially (at speed N^2), whence (2.6.30) follows.

The following lemma, whose proof is deferred, will allow for a proper truncation of the top and bottom eigenvalues. (The reader interested only in the GOE or GUE setups can note that Lemma 2.6.7 is then a consequence of Exercise 2.1.30.)

Lemma 2.6.7 *Under the assumptions of Theorem 2.6.6, we have*

$$\limsup_{N\to\infty} \frac{1}{N} \log \frac{Z^{N-1}_{V,\beta}}{Z^N_{V,\beta}} < \infty. \tag{2.6.33}$$

Further,

$$\lim_{M\to\infty} \limsup_{N\to\infty} \frac{1}{N} \log P^N_{V,\beta}(\lambda^*_N > M) = -\infty \tag{2.6.34}$$

and, with $\lambda^*_1 = \min_{i=1}^N \lambda^N_i$,

$$\lim_{M\to\infty} \limsup_{N\to\infty} \frac{1}{N} \log P^N_{V,\beta}(\lambda^*_1 < -M) = -\infty. \tag{2.6.35}$$

Equipped with Lemma 2.6.7, we may complete the proof of Theorem 2.6.6. We begin with the upper bound (2.6.31). Note that for any $M > x$,

$$P^N_{V,\beta}(\lambda^*_N \geq x) \leq P^N_{V,\beta}(\lambda^*_N > M) + P^N_{V,\beta}(\lambda^*_N \in [x,M]). \tag{2.6.36}$$

By choosing M large enough and using (2.6.34), the first term in the right side of (2.6.36) can be made smaller than $e^{-NJ^V_\beta(x)}$, for all N large. In the sequel, we fix an M such that the above is satisfied, the analogous bound with $-\lambda^*_1$ also holds, and further

$$\left[\beta \int \log|x-y|\sigma^V_\beta(dy) - V(x)\right] > \sup_{z\in[M,\infty)} \left[\beta \int \log|z-y|\sigma^V_\beta(dy) - V(z)\right]. \tag{2.6.37}$$

Set, for $z \in [-M, M]$ and μ supported on $[-M, M]$,

$$\Phi(z,\mu) = \beta \int \log|z-y|\mu(dy) - V(z) \le \beta \log(2M) + V_- =: \Phi_M ,$$

where $V_- = -\inf_{x\in\mathbb{R}} V(x) < \infty$. Setting $B(\delta)$ as the ball of radius δ around σ_β^V, $B_M(\delta)$ as those probability measures in $B(\delta)$ with support in $[-M, M]$, and writing

$$\zeta_N = \frac{Z^{N-1}_{NV/(N-1),\beta}}{Z^N_{V,\beta}}, \quad I_M = [-M,M]^{N-1},$$

we get

$$\begin{aligned}
&P^N_{V,\beta}(\lambda_N^* \in [x,M]) \\
&\quad \le \; P^N_{V,\beta}(\lambda_1^* < -M) \\
&\qquad + N\zeta_N \int_x^M d\lambda_N \int_{I_M} e^{(N-1)\Phi(\lambda_N, L_{N-1})} P^{N-1}_{NV/(N-1),\beta}(d\lambda_1,\ldots,d\lambda_{N-1}) \\
&\quad \le \; P^N_{V,\beta}(\lambda_1^* < -M) + N\zeta_N \Big[\int_x^M e^{(N-1)\sup_{\mu\in B_M(\delta)}\Phi(z,\mu)} dz \\
&\qquad\qquad + (M-x) e^{(N-1)\Phi_M} P^{N-1}_{NV/(N-1),\beta}(L_{N-1} \notin B(\delta)) \Big] . \qquad (2.6.38)
\end{aligned}$$

(The choice of metric in the definition of $B(\delta)$ plays no role in our argument, as long as it is compatible with weak convergence.) Noting that the perturbation involving the multiplication of V by $N/(N-1)$ introduces only an exponential in N factor, see (2.6.33), we get from the LDP for the empirical measure, Theorem 2.6.1, that

$$\limsup_{N\to\infty} \frac{1}{N^2} \log P^{N-1}_{NV/(N-1),\beta}(L_{N-1} \notin B(\delta)) < 0 ,$$

and hence, for any fixed $\delta > 0$,

$$\limsup_{N\to\infty} \frac{1}{N} \log P^{N-1}_{NV/(N-1),\beta}(L_{N-1} \notin B(\delta)) = -\infty . \qquad (2.6.39)$$

We conclude from (2.6.38) and (2.6.39) that

$$\begin{aligned}
\limsup_{N\to\infty} \frac{1}{N} P^N_{V,\beta}(\lambda_N^* \in [x,M]) &\le \limsup_{N\to\infty} \frac{1}{N} \log \zeta_N + \lim_{\delta\to 0} \sup_{z\in[x,M],\mu\in B_M(\delta)} \Phi(z,\mu) \\
&= \alpha_{V,\beta} + \lim_{\delta\to 0} \sup_{z\in[x,M],\mu\in B_M(\delta)} \Phi(z,\mu) . \qquad (2.6.40)
\end{aligned}$$

Since $\Phi(z,\mu) = \inf_{\eta>0}[\beta \int \log(|z-y| \vee \eta)\mu(dy) - V(z)$, it holds that $(z,\mu) \mapsto \Phi(z,\mu)$ is upper semicontinuous on $[-M,M] \times M_1([-M,M])$. Therefore, using (2.6.37) in the last equality,

$$\lim_{\delta\to 0} \sup_{z\in[x,M],\mu\in B_M(\delta)} \Phi(z,\mu) = \sup_{z\in[x,M]} \Phi(z,\sigma_\beta^V) = \sup_{z\in[x,\infty)} \Phi(z,\sigma_\beta^V) .$$

Combining the last equality with (2.6.40) and (2.6.36), we obtain (2.6.31).

We finally prove the lower bound (2.6.32). Let $2\delta < x - x^*$ and fix $r \in (x^*, x - 2\delta)$. Then, with $I_r = (-M, r)^{N-1}$,

$$\begin{aligned} & P^N_{V,\beta}(\lambda^*_N \in (x-\delta, x+\delta)) \\ & \quad \geq P^N_{V,\beta}(\lambda_N \in (x-\delta, x+\delta), \lambda_i \in (-M, r), i = 1, \ldots, N-1) \qquad (2.6.41) \\ & \quad = \zeta_N \int_{x-\delta}^{x+\delta} d\lambda_N \int_{I_r} e^{(N-1)\Phi(\lambda_N, L^N_{N-1})} P^{N-1}_{NV/(N-1),\beta}(d\lambda_1, \ldots, d\lambda_{N-1}) \\ & \quad \geq 2\delta\zeta_N \exp\Big((N-1) \inf_{\substack{z \in (x-\delta, x+\delta) \\ \mu \in B_{r,M}(\delta)}} \Phi(z, \mu)\Big) P^{N-1}_{NV/(N-1),\beta}(L_{N-1} \in B_{r,M}(\delta)), \end{aligned}$$

where $B_{r,M}(\delta)$ denotes those measures in $B(\delta)$ with support in $[-M, r]$. Recall from the upper bound (2.6.31) together with (2.6.35) that

$$\limsup_{N\to\infty} P^{N-1}_{NV/(N-1),\beta}(\lambda_i \notin (-M, r) \text{ for some } i \in \{1, \ldots, N-1\}) = 0.$$

Combined with (2.6.39) and the strict inequality in (2.6.10) of Lemma 2.6.2, we get by substituting in (2.6.41) that

$$\begin{aligned} \lim_{\delta\to 0} \liminf_{N\to\infty} \frac{1}{N} \log P^N_{V,\beta}(\lambda^*_N \in (x-\delta, x+\delta)) & \geq \alpha_{V,\beta} + \lim_{\delta\to 0} \inf_{\substack{z \in (x-\delta, x+\delta) \\ \mu \in B_{r,M}(\delta)}} \Phi(z, \mu) \\ & = \alpha_{V,\beta} + \Phi(x, \sigma^\beta_V), \end{aligned}$$

where in the last step we used the continuity of $(z, \mu) \mapsto \Phi(z, \mu)$ on $[x-\delta, x+\delta] \times M_1([-M, r])$. The bound (2.6.32) follows. □

Proof of Lemma 2.6.7 We first prove (2.6.33). Note that, for any $\delta > 0$ and all N large,

$$\frac{Z^{N-1}_{V,\beta}}{Z^N_{V,\beta}} = \frac{Z^{N-1}_{V,\beta}}{Z^{N-1}_{NV/(N-1),\beta}} \cdot \frac{Z^{N-1}_{NV/(N-1),\beta}}{Z^N_{V,\beta}} \leq \frac{Z^{N-1}_{V,\beta}}{Z^{N-1}_{NV/(N-1),\beta}} \cdot e^{N(\alpha_{V,\beta}+\delta)}, \qquad (2.6.42)$$

by (2.6.29). On the other hand,

$$\frac{Z^{N-1}_{V,\beta}}{Z^{N-1}_{NV/(N-1),\beta}} = \int e^{N\langle L_{N-1}, V\rangle} dP_{N-1, NV/(N-1)} \,. \qquad (2.6.43)$$

By the LDP for L_{N-1} (at speed N^2, see Theorem 2.6.1), Lemma 2.6.2 and (2.6.21), the last integral is bounded above by $e^{N(\langle \sigma^V_\beta, V\rangle + \delta)}$. Substituting this in (2.6.43) and (2.6.42) yields (2.6.33).

For $|x| > M$, M large and $\lambda_i \in \mathbb{R}$, for some constants a_β, b_β,

$$|x - \lambda_i|^\beta e^{-V(\lambda_i)} \leq a_\beta(|x|^\beta + |\lambda_i|^\beta) e^{-V(\lambda_i)} \leq b_\beta |x|^\beta \leq b_\beta e^{V(x)} \,.$$

Therefore,

$$\begin{aligned} P^N_{V,\beta}(\lambda_N^* > M) &\leq N\frac{Z^{N-1}_{V,\beta}}{Z^N_{V,\beta}}\int_M^\infty e^{-NV(\lambda_N)}d\lambda_N \int_{\mathbb{R}^{N-1}} \prod_{i=1}^{N-1}\left(|x-\lambda_i|^\beta e^{-V(\lambda_i)}\right) dP^{N-1}_{V,\beta} \\ &\leq Nb_\beta^{N-1}e^{-NV(M)/2}\frac{Z^{N-1}_{V,\beta}}{Z^N_{V,\beta}}\int_M^\infty e^{-V(\lambda_N)}d\lambda_N\,, \end{aligned}$$

implying, together with (2.6.33), that

$$\lim_{M\to\infty}\limsup_{N\to\infty}\frac{1}{N}\log P^N_{V,\beta}(\lambda_N^* > M) = -\infty\,.$$

This proves (2.6.34). The proof of (2.6.35) is similar. □

2.7 Bibliographical notes

Wigner's Theorem was presented in [Wig55], and proved there using the method of moments developed in Section 2.1. Since then, this result has been extended in many directions. In particular, under appropriate moment conditions, an almost sure version holds, see [Arn67] for an early result in that direction. Relaxation of moment conditions, requiring only the existence of third moments of the variables, is described by Bai and co-workers, using a mixture of combinatorial, probabilistic and complex-analytic techniques. For a review, we refer to [Bai99]. It is important to note that one cannot hope to forgo the assumption of finiteness of second moments, because without this assumption the empirical measure, properly rescaled, converges toward a noncompactly supported measure, see [BeG08].

Regarding the proof of Wigner's Theorem that we presented, there is a slight ambiguity in the literature concerning the numbering of Catalan numbers. Thus, [Aig79, p. 85] denotes by c_k what we denote by C_{k-1}. Our notation follows [Sta97]. Also, there does not seem to be a clear convention as to whether the Dyck paths we introduced should be called Dyck paths of length $2k$ or of length k. Our choice is consistent with our notion of length of Bernoulli walks. Finally, we note that the first part of the proof of Lemma 2.1.3 is an application of the reflection principle, see [Fel57, Ch. III.2].

The study of Wigner matrices is closely related to the study of Wishart matrices, discussed in Exercises 2.1.18 and 2.4.8. The limit of the empirical measure of eigenvalues of Wishart matrices (and generalizations) can be found in [MaP67], [Wac78] and [GrS77]. Another similar model is given by band matrices, see [BoMP91]. In fact, both Wigner and Wishart matrices fall under the class of the general band matrices discussed in [Shl96], [Gui02] (for the Gaussian case) and [AnZ05], [HaLN06].

Another promising combinatorial approach to the study of the spectrum of random Wigner matrices, making a direct link with orthogonal polynomials, is presented in [Sod07].

The rate of convergence toward the semicircle distribution has received some attention in the literature, see, e.g., [Bai93a], [Bai93b], [GoT03].

Lemma 2.1.19 first appears in [HoW53]. In the proof we mention that permutation matrices form the extreme points of the set of doubly stochastic matrices, a fact that is is usually attributed to G. Birkhoff. See [Chv83] for a proof and a historical discussion which attributes this result to D. Konig. The argument we present (that bypasses this characterization) was kindly communicated to us by Hongjie Dong. The study of the distribution of the maximal eigenvalue of Wigner matrices by combinatorial techniques was initiated by [Juh81], and extended by [FuK81] (whose treatment we essentially follow; see also [Vu07] for recent improvements). See also [Gem80] for the analogous results for Wishart matrices. The method was widely extended in the papers [SiS98a], [SiS98b], [Sos99] (with symmetric distribution of the entries) and [PeS07] (in the general case), allowing one to derive much finer behavior on the law of the largest eigenvalue, see the discussion in Section 3.7. Some extensions of the Füredi–Komlós and Sinai–Soshnikov techniques can also be found in [Kho01]. Finally, conditions for the almost sure convergence of the maximal eigenvalue of Wigner matrices appear in [BaY88].

The study of maximal and minimal eigenvalues for Wishart matrices is of fundamental importance in statistics, where they are referred to as sample covariance matrices, and has received a great deal of attention recently. See [SpT02], [BeP05], [LiPRTJ05], [TaV09a], [Rud08], [RuV08] for a sample of recent developments.

The study of central limit theorems for traces of powers of random matrices can be traced back to [Jon82], in the context of Wishart matrices (an even earlier announcement appears in [Arh71], without proofs). Our presentation follows to a large extent Jonsson's method, which allows one to derive a CLT for polynomial functions. A by-product of [SiS98a] is a CLT for $\mathrm{tr} f(X_N)$ for analytic f, under a symmetry assumption on the moments. The paper [AnZ05] generalizes these results, allowing for differentiable functions f and for nonconstant variance of the independent entries. See also [AnZ08a] for a different version of Lemma 2.1.34. For functions of the form $f(x) = \sum a_i/(z_i - x)$ where $z_i \in \mathbb{C} \setminus \mathbb{R}$, and matrices of Wigner type, CLT statements can be found in [KhKP96], with somewhat sketchy proofs. A complete treatment for f analytic in a domain including the support of the limit of the empirical distribution of eigenvalues is given in [BaY05] for matrices of Wigner type, and in [BaS04] for matrices of Wishart type under a certain restriction on fourth moments. Finally, an approach based on Fourier transforms and interpolation was recently proposed in [PaL08].

Much more is known concerning the CLT for restricted classes of matrices: [Joh98] uses an approach based on the explicit joint density of the eigenvalues, see Section 2.5. (These results apply also to a class of matrices with dependent entries.) For Gaussian matrices, an approach based on the stochastic calculus introduced in Section 4.3 can be found in [Cab01] and [Gui02]. Recent extensions and reinterpretation of this work, using the notion of second order freeness, can be found in [MiS06] (see Chapter 5 for the notion of freeness and its relation to random matrices).

The study of spectra of random matrices via the Stieltjes transform (resolvent functions) was pioneered by Pastur co-workers, and greatly extended by Bai and co-workers. See [MaP67] for an early reference, and [Pas73] for a survey of the literature up to 1973. Our derivation is based on [KhKP96], [Bai99] and [SiB95].

We presented in Section 2.3 a very brief introduction to concentration inequalities. This topic is picked up again in Section 4.4, to which we refer the reader for a complete introduction to different concentration inequalities and their application in RMT, and for full bibliographical notes. Good references for the logarithmic Sobolev inequalities used in Section 2.3 are [Led01] and [AnBC$^+$00]. Our treatment is based on [Led01] and [GuZ00]. Lemma 2.3.2 is taken from [BoL00, Proposition 3.1]. We note in passing that, on $\mathbb{R}$, a criterion for a measure to satisfy the logarithmic Sobolev inequality was developed by Bobkov and Götze [BoG99]. In particular, any probability measure on $\mathbb{R}$ possessing a bounded above and below density with respect to the measures $\nu(dx) = Z^{-1}e^{-|x|^\alpha}dx$ for $\alpha \geq 2$, where $Z = \int e^{-|x|^\alpha}dx$, satisfies the LSI, see [Led01], [GuZ03, Property 4.6]. Finally, in the Gaussian case, estimates on the expectation of the maximal eigenvalue (or minimal and maximal singular values, in the case of Wishart matrices) can be obtained from Slepian's and Gordon's inequalities, see [LiPRTJ05] and [DaS01]. In particular, these estimates are useful when using, in the Gaussian case, (2.3.10) with $k = N$.

The basic results on joint distribution of eigenvalues in the GOE and GUE presented in Section 2.5, as well as an extensive list of integral formulas similar to (2.5.4) are given in [Meh91], [For05]. We took, however, a quite different approach to all these topics based on the elementary proof of the Selberg integral formula [Sel44], see [AnAR99], given in [And91]. The proof of [And91] is based on a similar proof [And90] of some trigonometric sum identities, and is also similar in spirit to the proofs in [Gus90] of much more elaborate identities. For a recent review of the importance of the Selberg integral, see [FoO08], where in particular it is pointed out that Lemma 2.5.15 seems to have first appeared in [Dix05].

We follow [FoR01] in our treatment of "superposition and decimation" (Theorem 2.5.17). We remark that triples (L, f, g) satisfying Assumption 2.5.18, and hence the conclusions of Proposition 2.5.19, can be classified, see [FoR01], to which we refer for other classical examples where superposition and decimation relations hold. An early precursor of such relations can be traced to [MeD63].

Theorem 2.6.1 is stated in [BeG97, Theorem 5.2] under the additional assumption that V does not grow faster than exponentially and proved there in detail when $V(x) = x^2$. In [HiP00b], the same result is obtained when the topology over $M_1(\mathbb{R})$ is taken to be the weak topology with respect to polynomial test functions instead of bounded continuous functions. Large deviations for the empirical measure of random matrices with complex eigenvalues were considered in [BeZ98] (where non self-adjoint matrices with independent Gaussian entries were studied) and in [HiP00a] (where Haar unitary distributed matrices are considered). This strategy can also be used when one is interested in discretized versions of the law $P^N_{\beta,V}$ as they appear in the context of Young diagrams, see [GuM05]. The LDP for the maximal eigenvalue described in Theorem 2.6.6 is based on [BeDG01]. We mention in passing that other results discussed in this chapter have analogs for the law $P^N_{\beta,V}$. In particular, the CLT for linear statistics is discussed in [Joh98], and concentration inequalities for V convex are a consequence of the results in Section 4.4.

Models of random matrices with various degrees of dependence between entries have also be treated extensively in the literature. For a sample of existing results, see [BodMKV96], [ScS05] and [AnZ08b]. Random Toeplitz, Hankel and Markov matrices have been studied in [BrDJ06] and [HaM05].

Many of the results described in this chapter (except for Sections 2.3, 2.5 and 2.6) can also be found in the book [Gir90], a translation of a 1975 Russian edition, albeit with somewhat sketchy and incomplete proofs.

We have restricted attention in this chapter to Hermitian matrices. A natural question concerns the *complex* eigenvalues of a matrix X_N where all are i.i.d. In the Gaussian case, the joint distribution of the eigenvalues was derived by [Gin65]. The analog of the semicircle law is now the circular law: the empirical measure of the (rescaled) eigenvalues converges to the circular law, i.e. the measure uniform on the unit disc in the complex plane. This is stated in [Gir84], with a sketchy proof. A full proof for the Gaussian case is provided in [Ede97], who also evaluated the probability that exactly k eigenvalues are real. Large deviations for the empirical measure in the Gaussian case are derived in [BeZ98]. For non-Gaussian entries whose law possesses a density and finite moments of order at least 6, a full proof, based on Girko idea's, appears in [Bai97]. The problem was recently settled in full generality, see [TaV08a], [TaV08b], [GoT07]; the extra ingredients in the proof are closely related to the study of the minimal singular value of XX^* discussed above.

3

Hermite polynomials, spacings and limit distributions for the Gaussian ensembles

In this chapter, we present the analysis of asymptotics for the joint eigenvalue distribution for the Gaussian ensembles: the GOE, GUE and GSE. As it turns out, the analysis takes a particularly simple form for the GUE, because then the process of eigenvalues is a *determinantal process*. (We postpone to Section 4.2 a discussion of general determinantal processes, opting to present here all computations "with bare hands".) In keeping with our goal of making this chapter accessible with minimal background, in most of this chapter we consider the GUE, and discuss the other Gaussian ensembles in Section 3.9. Generalizations to other ensembles, refinements and other extensions are discussed in Chapter 4 and in the bibliographical notes.

3.1 Summary of main results: spacing distributions in the bulk and edge of the spectrum for the Gaussian ensembles

We recall that the N eigenvalues of the GUE/GOE/GSE are spread out on an interval of width roughly equal to $4\sqrt{N}$, and hence the spacing between adjacent eigenvalues is expected to be of order $1/\sqrt{N}$.

3.1.1 Limit results for the GUE

Using the determinantal structure of the eigenvalues $\{\lambda_1^N, \dots, \lambda_N^N\}$ of the GUE, developed in Sections 3.2–3.4, we prove the following.

Theorem 3.1.1 (Gaudin–Mehta) *For any compact set $A \subset \mathbb{R}$,*

$$\lim_{N\to\infty} P[\sqrt{N}\lambda_1^N,\dots,\sqrt{N}\lambda_N^N \notin A] = 1+\sum_{k=1}^{\infty}\frac{(-1)^k}{k!}\int_A\cdots\int_A \det_{i,j=1}^k K_{\text{sine}}(x_i,x_j)\prod_{j=1}^k dx_j\,, \quad (3.1.1)$$

where

$$K_{\text{sine}}(x,y)=\begin{cases}\frac{1}{\pi}\frac{\sin(x-y)}{x-y}, & x\neq y,\\ \frac{1}{\pi}, & x=y.\end{cases}$$

(Similar results apply to the sets $A+c\sqrt{n}$ with $|c|<2$, see Exercise 3.7.5.)

As a consequence of Theorem 3.1.1, we will show that the theory of integrable systems applies and yields the following fundamental result concerning the behavior of spacings between eigenvalues *in the bulk*.

Theorem 3.1.2 (Jimbo–Miwa–Môri–Sato) *One has*

$$\lim_{N\to\infty} P[\sqrt{N}\lambda_1^N,\dots,\sqrt{N}\lambda_N^N \notin (-t/2,t/2)] = 1-F(t),$$

with

$$1-F(t)=\exp\left(\int_0^t \frac{\sigma(x)}{x}dx\right) \textit{ for } t\geq 0,$$

with σ the solution of

$$(t\sigma'')^2+4(t\sigma'-\sigma)(t\sigma'-\sigma+(\sigma')^2)=0,$$

so that

$$\sigma=-\frac{t}{\pi}-\frac{t^2}{\pi^2}-\frac{t^3}{\pi^3}+O(t^4) \text{ as } t\downarrow 0. \quad (3.1.2)$$

The differential equation satisfied by σ is the *σ-form* of Painlevé V. Note that Theorem 3.1.2 implies that $F(t)\to_{t\to 0} 0$. Additional analysis (see Remark 3.6.5 in Subsection 3.6.3) yields that also $F(t)\to_{t\to\infty} 1$, showing that F is the distribution function of a probability distribution on $\mathbb{R}_+$.

We now turn our attention to the edge of the spectrum.

Definition 3.1.3 The *Airy function* is defined by the formula

$$\mathrm{Ai}(x)=\frac{1}{2\pi i}\int_C e^{\zeta^3/3-x\zeta}d\zeta\,, \quad (3.1.3)$$

where C is the contour in the ζ-plane consisting of the ray joining $e^{-\pi i/3}\infty$ to the origin plus the ray joining the origin to $e^{\pi i/3}\infty$ (see Figure 3.1.1).

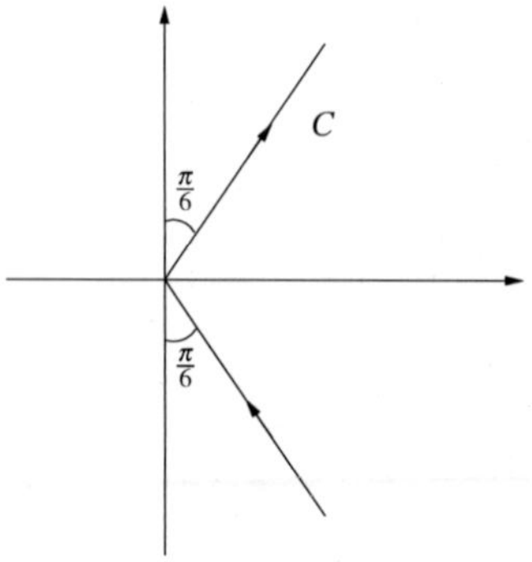

Fig. 3.1.1. Contour of integration for the Airy function

The *Airy kernel* is defined by

$$K_{\mathrm{Airy}}(x,y) = A(x,y) := \frac{\mathrm{Ai}(x)\,\mathrm{Ai}'(y) - \mathrm{Ai}'(x)\,\mathrm{Ai}(y)}{x-y},$$

where the value for $x = y$ is determined by continuity.

By differentiating under the integral and then integrating by parts, it follows that $\mathrm{Ai}(x)$, for $x \in \mathbb{R}$, satisfies the *Airy equation*:

$$\frac{d^2y}{dx^2} - xy = 0. \tag{3.1.4}$$

Various additional properties of the Airy function and kernel are summarized in Subsection 3.7.3.

The fundamental result concerning the eigenvalues of the GUE at the edge of the spectrum is the following.

Theorem 3.1.4 *For all* $-\infty < t \le t' \le \infty$,

$$\lim_{N\to\infty} P\left[N^{2/3}\left(\frac{\lambda_i^N}{\sqrt{N}} - 2\right) \notin [t,t'], i = 1,\dots,N\right]$$
$$= 1 + \sum_{k=1}^{\infty} \frac{(-1)^k}{k!} \int_t^{t'} \cdots \int_t^{t'} \det_{i,j=1}^{k} A(x_i,x_j) \prod_{j=1}^{k} dx_j, \tag{3.1.5}$$

with A the Airy kernel. In particular,

$$\lim_{N\to\infty} P\left[N^{2/3}\left(\frac{\lambda_N^N}{\sqrt{N}} - 2\right) \le t\right]$$
$$= 1 + \sum_{k=1}^{\infty} \frac{(-1)^k}{k!} \int_t^{\infty} \cdots \int_t^{\infty} \det_{i,j=1}^{k} A(x_i,x_j) \prod_{j=1}^{k} dx_j =: F_2(t). \tag{3.1.6}$$

Note that the statement of Theorem 3.1.4 does not ensure that F_2 is a distribution function (and in particular, does not ensure that $F_2(-\infty) = 0$), since it only implies the vague convergence, not the weak convergence, of the random variables $\lambda_N^N/\sqrt{N} - 2$. The latter convergence, as well as a representation of F_2, are contained in the following.

Theorem 3.1.5 (Tracy–Widom) *The function $F_2(\cdot)$ is a distribution function that admits the representation*

$$F_2(t) = \exp\left(-\int_t^\infty (x-t)q(x)^2 dx\right), \tag{3.1.7}$$

where q satisfies

$$q'' = tq + 2q^3, \quad q(t) \sim \mathrm{Ai}(t), \text{as } t \to +\infty. \tag{3.1.8}$$

The function $F_2(\cdot)$ is the *Tracy–Widom* distribution. Equation (3.1.8) is the *Painlevé II* equation. Some information on its solutions is collected in Remark 3.8.1 below.

3.1.2 Generalizations: limit formulas for the GOE and GSE

We next state the results for the GOE and GSE in a concise way that allows easy comparison with the GUE. Most of the analysis will be devoted to controlling the influence of the departure from a determinantal structure in these ensembles.

For $\beta = 1,2,4$, let $\lambda^{(\beta,n)} = (\lambda_1^{(\beta,n)}, \ldots, \lambda_n^{(\beta,n)})$ be a random vector in $\mathbb{R}^n$ with the law $\mathscr{P}_n^{(\beta)}$, see (2.5.6), possessing a density with respect to Lebesgue measure proportional to $|\Delta(x)|^\beta e^{-\beta|x|^2/4}$. (Thus, $\beta = 1$ corresponds to the GOE, $\beta = 2$ to the GUE and $\beta = 4$ to the GSE.) Consider the limits

$$\begin{aligned} 1 - F_{\beta,\text{bulk}}(t) &= \lim_{n\to\infty} P(\{\sqrt{n}\lambda^{(\beta,n)}\} \cap (-t/2, t/2)\} = \emptyset), \\ &\quad \text{for } t > 0, \end{aligned} \tag{3.1.9}$$

$$\begin{aligned} F_{\beta,\text{edge}}(t) &= \lim_{n\to\infty} P\left(\left\{n^{1/6}(\lambda^{(\beta,n)} - 2\sqrt{n})\right\} \cap (t,\infty) = \emptyset\right), \\ &\quad \text{for all real } t. \end{aligned} \tag{3.1.10}$$

The existence of these limits for $\beta = 2$ follows from Theorems 3.1.2 and 3.1.4, together with Corollary 3.1.5. Further, from Lemma 3.6.6 below, we also have

$$1 - F_{2,\text{bulk}}(t) = \exp\left(-\frac{t}{\pi} - \int_0^t (t-x)r(x)^2 dx\right),$$

where

$$t^2((tr)'' + (tr))^2 = 4(tr)^2((tr)^2 + ((tr)')^2), \quad r(t) = \frac{1}{\pi} + \frac{t}{\pi^2} + O_{t\downarrow 0}(t^2).$$

The following is the main result of the analysis of spacings for the GOE and GSE.

Theorem 3.1.6 *The limits* $1-F_{\beta,\text{bulk}}$ $(\beta=1,4)$ *exist and are as follows:*

$$\frac{1-F_{1,\text{bulk}}(t)}{\sqrt{1-F_{2,\text{bulk}}(t)}} = \exp\left(-\frac{1}{2}\int_0^t r(x)dx\right), \tag{3.1.11}$$

$$\frac{1-F_{4,\text{bulk}}(t/2)}{\sqrt{1-F_{2,\text{bulk}}(t)}} = \cosh\left(-\frac{1}{2}\int_0^t r(x)dx\right). \tag{3.1.12}$$

Theorem 3.1.7 *The limits* $F_{\beta,\text{edge}}$ $(\beta=1,4)$ *exist and are as follows:*

$$\frac{F_{1,\text{edge}}(t)}{\sqrt{F_{2,\text{edge}}(t)}} = \exp\left(-\frac{1}{2}\int_t^\infty q(x)dx\right), \tag{3.1.13}$$

$$\frac{F_{4,\text{edge}}(t/2^{2/3})}{\sqrt{F_{2,\text{edge}}(t)}} = \cosh\left(-\frac{1}{2}\int_t^\infty q(x)dx\right). \tag{3.1.14}$$

The proofs of Theorems 3.1.6 and 3.1.7 appear in Section 3.9.

3.2 Hermite polynomials and the GUE

In this section we show why orthogonal polynomials arise naturally in the study of the law of the GUE. The relevant orthogonal polynomials in this study are the Hermite polynomials and the associated oscillator wave-functions, which we introduce and use to derive a Fredholm determinant representation for certain probabilities connected with the GUE.

3.2.1 The GUE and determinantal laws

We now show that the joint distribution of the eigenvalues following the GUE has a nice *determinantal* form, see Lemma 3.2.2 below. We then use this formula in order to deduce a *Fredholm determinant* expression for the probability that no eigenvalues belong to a given interval, see Lemma 3.2.4.

Throughout this section, we shall consider the eigenvalues of GUE matrices with complex Gaussian entries of unit variance as in Theorem 2.5.2, and later normalize the eigenvalues to study convergence issues. We shall be interested in symmetric statistics of the eigenvalues. For $p \le N$, recalling the joint distributions $\mathscr{P}_N^{(2)}$ of the unordered eigenvalues of the GUE, see Remark 2.5.3, we call its marginal $\mathscr{P}_{p,N}$ on p coordinates the *distribution of* p *unordered eigenvalues of the GUE*. More explicitly, $\mathscr{P}^{(2)}_{p,N}$ is the probability measure on $\mathbb{R}^p$ so that, for any $f \in C_b(\mathbb{R}^p)$,

$$\int f(\theta_1,\cdots,\theta_p)d\mathscr{P}^{(2)}_{p,N}(\theta_1,\cdots,\theta_p) = \int f(\theta_1,\cdots,\theta_p)d\mathscr{P}^{(2)}_{N}(\theta_1,\cdots,\theta_N)$$

(recall that $\mathscr{P}^{(2)}_N$ is the law of the *unordered* eigenvalues). Clearly, one also has

$$\int f(\theta_1,\cdots,\theta_p)d\mathscr{P}^{(2)}_{p,N}(\theta_1,\cdots,\theta_p)$$
$$= \frac{(N-p)!}{N!}\sum_{\sigma\in S_{p,N}}\int f(\theta_{\sigma(1)},\cdots,\theta_{\sigma(p)})d\mathscr{P}^{(2)}_{N}(\theta_1,\cdots,\theta_N),$$

where $S_{p,N}$ is the set of injective maps from $\{1,\cdots,p\}$ into $\{1,\cdots,N\}$. Note that we automatically have $\mathscr{P}^{(2)}_{N,N} = \mathscr{P}^{(2)}_{N}$.

We now introduce the Hermite polynomials and associated normalized (harmonic) oscillator wave-function.

Definition 3.2.1 (a) The nth *Hermite polynomial* $\mathfrak{H}_n(x)$ is defined as

$$\mathfrak{H}_n(x) := (-1)^n e^{x^2/2}\frac{d^n}{dx^n}e^{-x^2/2}. \tag{3.2.1}$$

(b) The nth *normalized oscillator wave-function* is the function

$$\psi_n(x) = \frac{e^{-x^2/4}\mathfrak{H}_n(x)}{\sqrt{\sqrt{2\pi}\,n!}}.$$

(Often, in the literature, $(-1)^n e^{x^2}\frac{d^n}{dx^n}e^{-x^2}$ is taken as the definition of the nth Hermite polynomial. We find (3.2.1) more convenient.)

For our needs, the most important property of the oscillator wave-functions is their orthogonality relations

$$\int \psi_k(x)\psi_\ell(x)dx = \delta_{k\ell}. \tag{3.2.2}$$

We will also use the monic property of the Hermite polynomials, that is

$$\mathfrak{H}_n(x) \text{ is a polynomial of degree } n \text{ with leading term } x^n. \tag{3.2.3}$$

The proofs of (3.2.2) and (3.2.3) appear in Subsection 3.2.2, see Lemmas 3.2.7 and 3.2.5.

We are finally ready to describe the determinantal structure of $\mathscr{P}^{(2)}_{p,N}$. (See Section 4.2 for more information on implications of this determinantal structure.)

Lemma 3.2.2 *For any $p \le N$, the law $\mathscr{P}^{(2)}_{p,N}$ is absolutely continuous with respect to Lebesgue measure, with density*

$$\rho^{(2)}_{p,N}(\theta_1,\cdots,\theta_p) = \frac{(N-p)!}{N!} \det_{k,l=1}^{p} K^{(N)}(\theta_k,\theta_l)\,,$$

where

$$K^{(N)}(x,y) = \sum_{k=0}^{N-1} \psi_k(x)\psi_k(y)\,. \tag{3.2.4}$$

Proof Theorem 2.5.2 shows that $\rho^{(2)}_{p,N}$ exists and equals

$$\rho^{(2)}_{p,N}(\theta_1,\cdots,\theta_p) = C_{p,N}\int |\Delta(x)|^2 \prod_{i=1}^{N} e^{-x_i^2/2} \prod_{i=p+1}^{N} d\zeta_i\,, \tag{3.2.5}$$

where $x_i = \theta_i$ for $i \le p$ and ζ_i for $i > p$, and $C_{p,N}$ is a normalization constant. The fundamental remark is that this density depends on the Vandermonde determinant

$$\Delta(x) = \prod_{1\le i<j\le N}(x_j - x_i) = \det_{i,j=1}^{N} x_i^{j-1} = \det_{i,j=1}^{N} \mathfrak{H}_{j-1}(x_i)\,, \tag{3.2.6}$$

where we used (3.2.3) in the last equality.

We begin by considering $p = N$, writing $\rho^{(2)}_N$ for $\rho^{(2)}_{N,N}$. Then

$$\begin{aligned} \rho^{(2)}_N(\theta_1,\cdots,\theta_N) &= C_{N,N}\left(\det_{i,j=1}^{N} \mathfrak{H}_{j-1}(\theta_i)\right)^2 \prod_{i=1}^{N} e^{-\theta_i^2/2} \\ &= \tilde{C}_{N,N}\left(\det_{i,j=1}^{N} \psi_{j-1}(\theta_i)\right)^2 = \tilde{C}_{N,N} \det_{i,j=1}^{N} K^{(N)}(\theta_i,\theta_j)\,, \end{aligned} \tag{3.2.7}$$

where in the last line we used the fact that $\det(AB) = \det(A)\det(B)$ with $A = B^* = (\psi_{j-1}(\theta_i))_{i,j=1}^N$. Here, $\tilde{C}_{N,N} = \prod_{k=0}^{N-1}(\sqrt{2\pi}k!)C_{N,N}$.

Of course, from (2.5.4) we know that $C_{N,N} = \bar{C}^{(2)}_N$. We provide now yet another direct evaluation of the normalization constant, following [Meh91]. We introduce a trick that will be very often applied in the sequel.

Lemma 3.2.3 *For any square-integrable functions $f_1,\ldots,f_n$ and $g_1,\ldots,g_n$ on the real line, we have*

$$\begin{aligned} &\frac{1}{n!}\int\cdots\int \det_{i,j=1}^{n}\left(\sum_{k=1}^{n} f_k(x_i)g_k(x_j)\right)\prod_{i=1}^{n} dx_i \\ &\quad = \frac{1}{n!}\int\cdots\int \det_{i,j=1}^{n} f_i(x_j)\cdot \det_{i,j=1}^{n} g_i(x_j)\prod_{i=1}^{n} dx_i = \det_{i,j=1}^{n}\int f_i(x)g_j(x)dx\,. \end{aligned} \tag{3.2.8}$$

Proof Using the identity $\det(AB) = \det(A)\det(B)$ applied to $A = \{f_k(x_i)\}_{ik}$ and $B = \{g_k(x_j)\}_{kj}$, we get

$$\int\cdots\int \det_{i,j=1}^{n}\left(\sum_{k=1}^{n} f_k(x_i)g_k(x_j)\right)\prod_{i=1}^{n} dx_i = \int\cdots\int \det_{i,j=1}^{n} f_i(x_j)\cdot \det_{i,j=1}^{n} g_i(x_j)\prod_{i=1}^{n} dx_i\,,$$

which equals, by expanding the determinants involving the families $\{g_i\}$ and $\{f_i\}$,

$$\begin{aligned} &\sum_{\sigma,\tau\in S_n} \varepsilon(\sigma)\varepsilon(\tau)\int\cdots\int \prod_{i=1}^{n} f_{\sigma(i)}(x_i)\cdot g_{\tau(i)}(x_i))\prod_{i=1}^{n} dx_i \\ = &\sum_{\sigma,\tau\in S_n} \varepsilon(\sigma)\varepsilon(\tau)\prod_{i=1}^{n}\int f_{\sigma(i)}(x)g_{\tau(i)}(x)dx \\ = &\; n!\sum_{\sigma\in S_n}\varepsilon(\sigma)\prod_{i=1}^{n}\int f_i(x)g_{\sigma(i)}(x)dx = n!\det_{i,j=1}^{n}\int f_i(x)g_j(x)dx. \end{aligned}$$

□

Substituting $f_i = g_i = \psi_{i-1}$ and $n = N$ in Lemma 3.2.3, and using the orthogonality relations (3.2.2), we deduce that

$$\int \det_{i,j=1}^{N} K^{(N)}(\theta_i,\theta_j)\prod_{i=1}^{N} d\theta_i = N!\,, \tag{3.2.9}$$

which completes the evaluation of $C_{N,N}$ and the proof of Lemma 3.2.2 for $p = N$.

For $p < N$, using (3.2.5) and (3.2.6) in a manner similar to (3.2.7), we find that for some constant $\tilde{C}_{p,N}$, with $x_i = \theta_i$ if $i \le p$ and ζ_i otherwise,

$$\begin{aligned} \rho^{(2)}_{p,N}(\theta_1,\cdots,\theta_p) &= \tilde{C}_{p,N}\int (\det_{i,j=1}^{N}\psi_{j-1}(x_i))^2\prod_{i=p+1}^{N} d\zeta_i \\ &= \tilde{C}_{p,N}\sum_{\sigma,\tau\in S_N}\varepsilon(\sigma)\varepsilon(\tau)\int\prod_{j=1}^{N}\psi_{\sigma(j)-1}(x_j)\psi_{\tau(j)-1}(x_j)\prod_{i=p+1}^{N} d\zeta_i\,. \end{aligned}$$

Therefore, letting $\mathscr{S}(p,\nu)$ denote the bijections from $\{1,\cdots,p\}$ to $\{\nu_1,\cdots,\nu_p\}$ $:=\nu$, we get

$$\begin{aligned} &\rho^{(2)}_{p,N}(\theta_1,\cdots,\theta_p) \\ &= \tilde{C}_{p,N}\sum_{1\le\nu_1<\cdots<\nu_p\le N}\;\sum_{\sigma,\tau\in\mathscr{S}(p,\nu)}\varepsilon(\sigma)\varepsilon(\tau)\prod_{i=1}^{p}\psi_{\sigma(i)-1}(\theta_i)\psi_{\tau(i)-1}(\theta_i) \\ &= \tilde{C}_{p,N}\sum_{1\le\nu_1<\cdots<\nu_p\le N}\left(\det_{i,j=1}^{p}\psi_{\nu_j-1}(\theta_i)\right)^2, \end{aligned} \tag{3.2.10}$$

where in the first equality we used the orthogonality of the family $\{\psi_j\}$ to conclude that the contribution comes only from permutations of $\mathscr{S}_N$ for which $\tau(i) =$

$\sigma(i)$ for $i > p$, and we put $\{\nu_1, \cdots, \nu_p\} = \{\tau(1), \cdots, \tau(p)\} = \{\sigma(1), \cdots, \sigma(p)\}$. Using the Cauchy–Binet Theorem A.2 with $A = B^*$ (of dimension $p \times N$) and $A_{i,j} = \psi_{j-1}(\theta_i)$, we get that

$$\rho^{(2)}_{p,N}(\theta_1, \cdots, \theta_p) = \tilde{C}_{p,N} \det_{i,j=1}^{p} (K^{(N)}(\theta_i, \theta_j)).$$

To compute $\tilde{C}_{p,N}$, note that, by integrating both sides of (3.2.10), we obtain

$$1 = \tilde{C}_{p,N} \sum_{1 \le \nu_1 < \cdots < \nu_p \le N} \int \left(\det_{i,j=1}^{p} \psi_{\nu_j - 1}(\theta_i) \right)^2 d\theta_1 \cdots d\theta_p, \tag{3.2.11}$$

whereas Lemma 3.2.3 implies that for all $\{\nu_1, \ldots, \nu_p\}$,

$$\int \left(\det_{i,j=1}^{p} \psi_{\nu_j - 1}(\theta_i) \right)^2 d\theta_1 \cdots d\theta_p = p!.$$

Thus, since there are $(N!)/((N-p)!p!)$ terms in the sum at the right side of (3.2.11), we conclude that $\tilde{C}_{p,N} = (N-p)!/N!$. □

Now we arrive at the main point, on which the study of the local properties of the GUE will be based.

Lemma 3.2.4 *For any measurable subset A of $\mathbb{R}$,*

$$P^{(2)}_N(\bigcap_{i=1}^{N} \{\lambda_i \in A\}) = 1 + \sum_{k=1}^{\infty} \frac{(-1)^k}{k!} \int_{A^c} \cdots \int_{A^c} \det_{i,j=1}^{k} K^{(N)}(x_i, x_j) \prod_{i=1}^{k} dx_i. \tag{3.2.12}$$

(The proof will show that the sum in (3.2.12) is actually finite.) The last expression appearing in (3.2.12) is a *Fredholm determinant*. The latter are discussed in greater detail in Section 3.4.

Proof By using Lemmas 3.2.2 and 3.2.3 in the first equality, and the orthogonality relations (3.2.2) in the second equality, we have

$$\begin{aligned} &P^{(2)}_N[\lambda_i \in A, i = 1, \ldots, N] \\ &\quad = \det_{i,j=0}^{N-1} \int_A \psi_i(x)\psi_j(x)dx = \det_{i,j=0}^{N-1} \left(\delta_{ij} - \int_{A^c} \psi_i(x)\psi_j(x)dx \right) \\ &\quad = 1 + \sum_{k=1}^{N} (-1)^k \sum_{0 \le \nu_1 < \cdots < \nu_k \le N-1} \det_{i,j=1}^{k} \left(\int_{A^c} \psi_{\nu_i}(x)\psi_{\nu_j}(x)dx \right), \end{aligned}$$

Therefore,

$$
\begin{aligned}
P_N^{(2)}[\lambda_i \in A, i=1,\ldots,N] \\
&= 1+\sum_{k=1}^{N}\frac{(-1)^k}{k!}\int_{A^c}\cdots\int_{A^c}\sum_{0\le\nu_1<\cdots<\nu_k\le N-1}\left(\det_{i,j=1}^{k}\psi_{\nu_i}(x_j)\right)^2\prod_{i=1}^{k}dx_i \\
&= 1+\sum_{k=1}^{N}\frac{(-1)^k}{k!}\int_{A^c}\cdots\int_{A^c}\det_{i,j=1}^{k}K^{(N)}(x_i,x_j)\prod_{i=1}^{k}dx_i \\
&= 1+\sum_{k=1}^{\infty}\frac{(-1)^k}{k!}\int_{A^c}\cdots\int_{A^c}\det_{i,j=1}^{k}K^{(N)}(x_i,x_j)\prod_{i=1}^{k}dx_i\,,
\end{aligned}
\tag{3.2.13}
$$

where the first equality uses (3.2.8) with $g_i(x)=f_i(x)=\psi_{\nu_i}(x)\mathbf{1}_{A^c}(x)$, the second equality uses the Cauchy–Binet Theorem A.2, and the last step is trivial since the determinant $\det_{i,j=1}^{k}K^{(N)}(x_i,x_j)$ has to vanish identically for $k>N$ because the rank of $\{K^{(N)}(x_i,x_j)\}_{i,j=1}^{k}$ is at most N. □

3.2.2 Properties of the Hermite polynomials and oscillator wave-functions

Recall the definition of the Hermite polynomials, Definition 3.2.1. Some properties of the Hermite polynomials are collected in the following lemma. Throughout, we use the notation $\langle f,g\rangle_{\mathscr{G}}=\int_{\mathbb{R}}f(x)g(x)e^{-x^2/2}dx$. In anticipation of further development, we collect much more information than was needed so far. Thus, the proof of Lemma 3.2.5 may be skipped at first reading. Note that (3.2.3) is the second point of Lemma 3.2.5.

Lemma 3.2.5 *The sequence of polynomials $\{\mathfrak{H}_n(x)\}_{n=0}^{\infty}$ has the following properties.*

1. $\mathfrak{H}_0(x)=1$, $\mathfrak{H}_1(x)=x$ and $\mathfrak{H}_{n+1}(x)=x\mathfrak{H}_n(x)-\mathfrak{H}_n'(x)$.
2. $\mathfrak{H}_n(x)$ is a polynomial of degree n with leading term x^n.
3. $\mathfrak{H}_n(x)$ is even or odd according as n is even or odd.
4. $\langle x,\mathfrak{H}_n^2\rangle_{\mathscr{G}}=0$.
5. $\langle \mathfrak{H}_k,\mathfrak{H}_\ell\rangle_{\mathscr{G}}=\sqrt{2\pi}\,k!\,\delta_{k\ell}$.
6. $\langle f,\mathfrak{H}_n\rangle_{\mathscr{G}}=0$ for all polynomials $f(x)$ of degree $<n$.
7. $x\mathfrak{H}_n(x)=\mathfrak{H}_{n+1}(x)+n\mathfrak{H}_{n-1}(x)$ for $n\ge 1$.
8. $\mathfrak{H}_n'(x)=n\mathfrak{H}_{n-1}(x)$.
9. $\mathfrak{H}_n''(x)-x\mathfrak{H}_n'(x)+n\mathfrak{H}_n(x)=0$.
10. For $x\neq y$,

$$\sum_{k=0}^{n-1}\frac{\mathfrak{H}_k(x)\mathfrak{H}_k(y)}{k!}=\frac{(\mathfrak{H}_n(x)\mathfrak{H}_{n-1}(y)-\mathfrak{H}_{n-1}(x)\mathfrak{H}_n(y))}{(n-1)!(x-y)}\,.$$

Property 2 shows that $\{\mathfrak{H}_n\}_{n\geq 0}$ is a basis of polynomial functions, whereas property 5 implies that it is an orthogonal basis for the scalar product $\langle f,g\rangle_{\mathscr{G}}$ defined on $L^2(e^{-x^2/2}dx)$ (since the polynomial functions are dense in the latter space).

Remark 3.2.6 Properties 7 and 10 are the *three-term recurrence* and the *Christoffel–Darboux identity* satisfied by the Hermite polynomials, respectively.

Proof of Lemma 3.2.5 Properties 1, 2 and 3 are clear. To prove property 5, use integration by parts to get that

$$\begin{aligned}\int \mathfrak{H}_k(x)\mathfrak{H}_l(x)e^{-x^2/2}dx &= (-1)^l \int \mathfrak{H}_k(x)\frac{d^l}{dx^l}(e^{-x^2/2})dx \\ &= \int \left[\frac{d^l}{dx^l}\mathfrak{H}_k(x)\right] e^{-x^2/2}dx\end{aligned}$$

vanishes if $l > k$ (since the degree of $\mathfrak{H}_k$ is strictly less than l), and is equal to $\sqrt{2\pi}k!$ if $k = l$, by property 2. Then, we deduce property 4 since, by property 3, $\mathfrak{H}_n^2$ is an even function and so is the function $e^{-x^2/2}$. Properties 2 and 5 suffice to prove property 6. To prove property 7, we proceed by induction on n. By properties 2 and 5 we have, for $n \geq 1$,

$$x\mathfrak{H}_n(x) = \sum_{k=0}^{n+1} \frac{\langle x\mathfrak{H}_n,\mathfrak{H}_k\rangle_{\mathscr{G}}}{\langle \mathfrak{H}_k,\mathfrak{H}_k\rangle_{\mathscr{G}}}\mathfrak{H}_k(x).$$

By property 6 the kth term on the right vanishes unless $|k-n| \leq 1$, by property 4 the nth term vanishes, and by property 2 the $(n+1)$st term equals 1. To get the $(n-1)$st term we observe that

$$\frac{\langle x\mathfrak{H}_n,\mathfrak{H}_{n-1}\rangle_{\mathscr{G}}}{\langle \mathfrak{H}_{n-1},\mathfrak{H}_{n-1}\rangle_{\mathscr{G}}} = \frac{\langle x\mathfrak{H}_n,\mathfrak{H}_{n-1}\rangle_{\mathscr{G}}}{\langle \mathfrak{H}_n,\mathfrak{H}_n\rangle_{\mathscr{G}}}\frac{\langle \mathfrak{H}_n,\mathfrak{H}_n\rangle_{\mathscr{G}}}{\langle \mathfrak{H}_{n-1},\mathfrak{H}_{n-1}\rangle_{\mathscr{G}}} = 1\cdot n = n$$

by induction on n and property 5. Thus property 7 is proved. Property 8 is a direct consequence of properties 1 and 7, and property 9 is obtained by differentiating the last identity in property 1 and using property 8. To prove property 10, call the left side of the claimed identity $F(x,y)$ and the right side $G(x,y)$. Using properties 2 and 5, followed by integration by parts and property 8, one sees that the integral

$$\int\int e^{-x^2/2-y^2/2}\mathfrak{H}_k(x)\mathfrak{H}_\ell(y)F(x,y)(x-y)dxdy$$

equals the analogous integral with $G(x,y)$ replacing $F(x,y)$; we leave the details to the reader. Equality of these integrals granted, property 10 follows since $\{\mathfrak{H}_k\}_{k\geq 0}$ being a basis of the set of polynomials, it implies almost sure equality and hence equality by continuity of F, G. Thus the claimed properties of Hermite polynomials are proved. □

Recall next the oscillator wave-functions, see Definition 3.2.1. Their basic properties are contained in the following lemma, which is an easy corollary of Lemma 3.2.5. Note that (3.2.2) is just the first point of the lemma.

Lemma 3.2.7 *The oscillator wave-functions satisfy the following.*

1. $\int \psi_k(x)\psi_\ell(x)dx = \delta_{k\ell}\,.$
2. $x\psi_n(x) = \sqrt{n+1}\,\psi_{n+1}(x) + \sqrt{n}\,\psi_{n-1}(x)\,.$
3. $\sum_{k=0}^{n-1} \psi_k(x)\psi_k(y) = \sqrt{n}(\psi_n(x)\psi_{n-1}(y) - \psi_{n-1}(x)\psi_n(y))/(x-y)\,.$
4. $\psi_n'(x) = -\frac{x}{2}\psi_n(x) + \sqrt{n}\,\psi_{n-1}(x)\,.$
5. $\psi_n''(x) + (n + \frac{1}{2} - \frac{x^2}{4})\psi_n(x) = 0\,.$

We remark that the last relation above is the one-dimensional *Schrödinger equation* for the eigenstates of the one-dimensional quantum-mechanical harmonic oscillator. This explains the terminology.

3.3 The semicircle law revisited

Let $X_N \in \mathcal{H}_N^{(2)}$ be a random Hermitian matrix from the GUE with eigenvalues $\lambda_1^N \leq \cdots \leq \lambda_N^N$, and let

$$L_N = (\delta_{\lambda_1^N/\sqrt{N}} + \cdots + \delta_{\lambda_N^N/\sqrt{N}})/N \tag{3.3.1}$$

denote the empirical distribution of the eigenvalues of the rescaled matrix $X_N/\sqrt{N}$. L_N thus corresponds to the eigenvalues of a Gaussian Wigner matrix.

We are going to make the average empirical distribution $\bar{L}_N$ explicit in terms of Hermite polynomials, calculate the moments of $\bar{L}_N$ explicitly, check that the moments of $\bar{L}_N$ converge to those of the semicircle law, and thus provide an alternative proof of Lemma 2.1.7. We also derive a recursion for the moments of $\bar{L}_N$ and estimate the order of fluctuation of the renormalized maximum eigenvalue $\lambda_N^N/\sqrt{N}$ above the spectrum edge, an observation that will be useful in Section 3.7.

3.3.1 Calculation of moments of $\bar{L}_N$

In this section, we derive the following explicit formula for $\langle \bar{L}_N, e^{s\cdot} \rangle$.

Lemma 3.3.1 *For any $s \in \mathbb{R}$, any $N \in \mathbb{N}$,*

$$\langle \bar{L}_N, e^{s\cdot} \rangle = e^{s^2/(2N)} \sum_{k=0}^{N-1} \frac{1}{k+1} \binom{2k}{k} \frac{(N-1)\cdots(N-k)}{N^k} \frac{s^{2k}}{(2k)!}. \tag{3.3.2}$$

Proof By Lemma 3.2.2,

$$\langle \bar{L}_N, \phi \rangle = \frac{1}{N} \int_{-\infty}^{\infty} \phi\left(\frac{x}{\sqrt{N}}\right) K^{(N)}(x,x) dx = \int_{-\infty}^{\infty} \phi(x) \frac{K^{(N)}(\sqrt{N}x, \sqrt{N}x)}{\sqrt{N}} dx. \tag{3.3.3}$$

This last identity shows that $\bar{L}_N$ is absolutely continuous with respect to Lebesgue measure, with density $K^{(N)}(\sqrt{N}x, \sqrt{N}x)/\sqrt{N}$.

Using points 3 and 5 of Lemma 3.2.7, we obtain that, for any n,

$$K^{(n)}(x,y)/\sqrt{n} = \frac{\psi_n(x)\psi_{n-1}(y) - \psi_{n-1}(x)\psi_n(y)}{x-y}$$

and hence by L'Hôpital's rule

$$K^{(n)}(x,x)/\sqrt{n} = \psi_n'(x)\psi_{n-1}(x) - \psi_{n-1}'(x)\psi_n(x).$$

Therefore

$$\frac{d}{dx} K^{(n)}(x,x)/\sqrt{n} = \psi_n''(x)\psi_{n-1}(x) - \psi_{n-1}''(x)\psi_n(x) = -\psi_n(x)\psi_{n-1}(x). \tag{3.3.4}$$

By (3.3.3) the function $K^{(N)}(\sqrt{N}x, \sqrt{N}x)/\sqrt{N}$ is the Radon–Nikodym derivative of $\bar{L}_N$ with respect to Lebesgue measure and hence we have the following representation of the moment-generating function of $\bar{L}_N$:

$$\langle \bar{L}_N, e^{s\cdot} \rangle = \frac{1}{N} \int_{-\infty}^{\infty} e^{sx/\sqrt{N}} K^{(N)}(x,x) dx. \tag{3.3.5}$$

Integrating by parts once and then applying (3.3.4), we find that

$$\langle \bar{L}_N, e^{s\cdot} \rangle = \frac{1}{s} \int_{-\infty}^{\infty} e^{sx/\sqrt{N}} \psi_N(x)\psi_{N-1}(x) dx. \tag{3.3.6}$$

Thus the calculation of the moment generating function of $\bar{L}_N$ boils down to the problem of evaluating the integral on the right.

By Taylor's theorem it follows from point 8 of Lemma 3.2.5 that, for any n,

$$\mathfrak{H}_n(x+t) = \sum_{k=0}^{n} \binom{n}{k} \mathfrak{H}_{n-k}(x) t^k = \sum_{k=0}^{n} \binom{n}{k} \mathfrak{H}_k(x) t^{n-k}.$$

Let $S_t^n =: \int_{-\infty}^{\infty} e^{tx}\psi_n(x)\psi_{n-1}(x)dx$. By the preceding identity and orthogonality we have

$$\begin{aligned} S_t^n &= \frac{\sqrt{n}}{n!\sqrt{2\pi}}\int_{-\infty}^{\infty} \mathfrak{H}_n(x)\mathfrak{H}_{n-1}(x)e^{-x^2/2+tx}dx \\ &= \frac{\sqrt{n}e^{t^2/2}}{n!\sqrt{2\pi}}\int_{-\infty}^{\infty} \mathfrak{H}_n(x+t)\mathfrak{H}_{n-1}(x+t)e^{-x^2/2}dx \\ &= e^{t^2/2}\sqrt{n}\sum_{k=0}^{n-1}\frac{k!}{n!}\binom{n}{k}\binom{n-1}{k}t^{2n-1-2k}. \end{aligned}$$

Changing the index of summation in the last sum from k to $n-1-k$, we then get

$$\begin{aligned} S_t^n &= e^{t^2/2}\sqrt{n}\sum_{k=0}^{n-1}\frac{(n-1-k)!}{n!}\binom{n}{n-1-k}\binom{n-1}{n-1-k}t^{2k+1} \\ &= e^{t^2/2}\sqrt{n}\sum_{k=0}^{n-1}\frac{(n-1-k)!}{n!}\binom{n}{k+1}\binom{n-1}{k}t^{2k+1}. \end{aligned}$$

From the last calculation combined with (3.3.6) and after a further bit of rearrangement we obtain (3.3.2). □

We can now present another

Proof of Lemma 2.1.7 (for Gaussian Wigner matrices) We have written the moment generating function in the form (3.3.2), making it obvious that as $N \to \infty$ the moments of $\bar{L}_N$ tend to the moments of the semicircle distribution. □

3.3.2 The Harer–Zagier recursion and Ledoux's argument

Recall that, throughout this chapter, λ_N^N denotes the maximal eigenvalue of a GUE matrix. Our goal in this section is to provide the proof of the following lemma.

Lemma 3.3.2 (Ledoux's bound) *There exist positive constants c' and C' such that*

$$P\left(\frac{\lambda_N^N}{2\sqrt{N}} \geq e^{N^{-2/3}\varepsilon}\right) \leq C'e^{-c'\varepsilon}, \tag{3.3.7}$$

for all $N \geq 1$ and $\varepsilon > 0$.

Roughly speaking, the last inequality says that fluctuations of the rescaled top eigenvalue $\tilde{\lambda}_N^N := \lambda_N^N/2\sqrt{N} - 1$ above 0 are of order of magnitude $N^{-2/3}$. This is an *a priori* indication that the random variables $N^{2/3}\tilde{\lambda}_N^N$ converge in distribution, as stated in Theorems 3.1.4 and 3.1.5. In fact, (3.3.7) is going to play a role in the proof of Theorem 3.1.4, see Subsection 3.7.1.

The proof of Lemma 3.3.2 is based on a recursion satisfied by the moments of $\bar{L}_N$. We thus first introduce this recursion in Lemma 3.3.3 below, prove it, and then show how to deduce from it Lemma 3.3.2. Write

$$\langle \bar{L}_N, e^{s\cdot} \rangle = \sum_{k=0}^{\infty} \frac{b_k^{(N)}}{k+1} \binom{2k}{k} \frac{s^{2k}}{(2k)!}.$$

Lemma 3.3.3 (Harer–Zagier recursions) *For any integer numbers k and N,*

$$b_{k+1}^{(N)} = b_k^{(N)} + \frac{k(k+1)}{4N^2} b_{k-1}^{(N)}, \tag{3.3.8}$$

where if $k = 0$ we ignore the last term.

Proof of Lemma 3.3.3 Define the (hypergeometric) function

$$F_n(t) = F\left(\begin{array}{c} 1-n \\ 2 \end{array} \middle| t\right) := \sum_{k=0}^{\infty} \frac{(-1)^k}{(k+1)!} \binom{n-1}{k} t^k, \tag{3.3.9}$$

and note that

$$\left(t\frac{d^2}{dt^2} + (2-t)\frac{d}{dt} + (n-1)\right) F_n(t) = 0. \tag{3.3.10}$$

By rearranging (3.3.2) it follows from (3.3.9) that

$$\langle \bar{L}_N, e^{s\cdot} \rangle = \Phi_N\left(-\frac{s^2}{N}\right), \tag{3.3.11}$$

where

$$\Phi_n(t) = e^{-t/2} F_n(t).$$

From (3.3.10) we find that

$$\left(4t\frac{d^2}{dt^2} + 8\frac{d}{dt} + 4n - t\right) \Phi_n(t) = 0. \tag{3.3.12}$$

Write next $\Phi_n(t) = \sum_{k=0}^{\infty} a_k^{(n)} t^k$. By (3.3.12) we have

$$0 = 4(k+2)(k+1)a_{k+1}^{(n)} + 4na_k^{(n)} - a_{k-1}^{(n)},$$

where if $k = 0$ we ignore the last term. Clearly we have, taking $n = N$,

$$\frac{(-1)^k a_k^{(N)} (2k)!}{N^k} = \frac{b_k^{(N)}}{k+1} \binom{2k}{k} = \langle \bar{L}_N, x^{2k} \rangle.$$

The lemma follows. □

Proof of Lemma 3.3.2 From (3.3.8) and the definitions we obtain the inequalities

$$0 \le b_k^{(N)} \le b_{k+1}^{(N)} \le \left(1 + \frac{k(k+1)}{4N^2}\right) b_k^{(N)}$$

for $N \ge 1$, $k \ge 0$. As a consequence, we deduce that

$$b_k^{(N)} \le e^{c\frac{k^3}{N^2}}, \tag{3.3.13}$$

for some finite constant $c > 0$. By Stirling's approximation (2.5.12) we have

$$\sup_{k=0}^{\infty} \frac{k^{3/2}}{2^{2k}(k+1)} \binom{2k}{k} < \infty.$$

It follows from (3.3.13) and the last display that, for appropriate positive constants c and C,

$$\begin{aligned} P\left(\frac{\lambda_N^N}{2\sqrt{N}} \ge e^{\varepsilon}\right) &\le E\left(\frac{\lambda_N^N}{2\sqrt{N}e^{\varepsilon}}\right)^{2k} \qquad (3.3.14)\\ &\le \frac{e^{-2\varepsilon k} N b_k^{(N)}}{2^{2k}(k+1)} \binom{2k}{k} \le CNt^{-3/2} e^{-2\varepsilon t + ct^3/N^2}, \end{aligned}$$

for all $N \ge 1$, $k \ge 0$ and real numbers $\varepsilon, t > 0$ such that $k = \lfloor t \rfloor$, where $\lfloor t \rfloor$ denotes the largest integer smaller than or equal to t. Taking $t = N^{2/3}$ and substituting $N^{-2/3}\varepsilon$ for ε yields the lemma. □

Exercise 3.3.4 Prove that, in the setup of this section, for every integer k it holds that

$$\lim_{N\to\infty} E\langle L_N, x^k\rangle^2 = \lim_{N\to\infty} \langle \bar{L}_N, x^k\rangle^2. \tag{3.3.15}$$

Using the fact that the moments of $\bar{L}_N$ converge to the moments of the semicircle distribution, complete yet another proof of Wigner's Theorem 2.1.1 in the GUE setup.

Hint: Deduce from (3.3.3) that

$$\langle \bar{L}_N, x^k\rangle = \frac{1}{N^{k/2+1}} \int x^k K^{(N)}(x,x)dx.$$

Also, rewrite $E\langle L_N, x^k\rangle^2$ as

$$\begin{aligned} &= \frac{1}{N^{2+k}} \frac{1}{N!} \int \cdots \int \left(\sum_{i=1}^N x_i^k\right)^2 \det_{i,j=1}^N K^{(N)}(x_i,x_j) \prod_{j=1}^N dx_j \\ &\overset{!}{=} \frac{1}{N^{k+2}} \int\int x^{2k} K^{(N)}(x,y)^2 dxdy + \frac{1}{N^{k+2}} \left(\int x^k K^{(N)}(x,x)dx\right)^2 \\ &= \langle \bar{L}_N, x^k\rangle^2 + I_k^{(N)}, \end{aligned}$$

where $I_k^{(N)}$ is equal to

$$\frac{1}{N^{k+3/2}}\int\int\frac{x^{2k}-x^k y^k}{x-y}(\psi_N(x)\psi_{N-1}(y)-\psi_{N-1}(x)\psi_N(y))K(x,y)dxdy.$$

To prove the equality marked with the exclamation point, show that

$$\int_{-\infty}^{\infty}K^{(n)}(x,t)K^{(n)}(t,y)dt=K^{(n)}(x,y),$$

while the expression for $I_k^{(N)}$ uses the Christoffel–Darboux formula (see Section 3.2.1). To complete the proof of (3.3.15), show that $\lim_{N\to\infty} I_k^{(N)}=0$, expanding the expression

$$\frac{x^{2k}-x^k y^k}{x-y}(\psi_N(x)\psi_{N-1}(y)-\psi_{N-1}(x)\psi_N(y))$$

as a linear combination of the functions $\psi_\ell(x)\psi_m(y)$ by exploiting the three-term recurrence (see Section 3.2.1) satisfied by the oscillator wave-functions.

Exercise 3.3.5 With the notation of Lemma 3.3.2, show that there exist $c', C' > 0$ so that, for all $N\geq 1$, if $\varepsilon > 1$ then

$$P\left(\frac{\lambda_N^N}{2\sqrt{N}}\geq e^{N^{-2/3}\varepsilon}\right)\leq C'\frac{1}{\varepsilon^{\frac{3}{4}}}e^{-c'\varepsilon^{\frac{3}{2}}}.$$

This bound improves upon (3.3.7) for large ε.
Hint: optimize differently over the parameter t at the end of the proof of Lemma 3.3.2, replacing ε there by $\varepsilon N^{-2/3}$.

Exercise 3.3.6 The function $F_n(t)$ defined in (3.3.9) is a particular case of the general *hypergeometric function*, see [GrKP94]. Let

$$x^{\bar{k}}=x(x+1)\cdots(x+k-1)$$

be the ascending factorial power. The general hypergeometric function is given by the rule

$$F\left(\begin{array}{ccc}a_1 & \cdots & a_p\\ b_1 & \cdots & b_q\end{array}\middle|\, t\right)=\sum_{k=0}^{\infty}\frac{a_1^{\bar{k}}\cdots a_p^{\bar{k}}}{b_1^{\bar{k}}\cdots b_q^{\bar{k}}}\frac{t^k}{k!}.$$

(i) Verify the following generalization of (3.3.10):

$$\frac{d}{dt}\left(t\frac{d}{dt}+b_1-1\right)\cdots\left(t\frac{d}{dt}+b_q-1\right)F\left(\begin{array}{ccc}a_1 & \cdots & a_p\\ b_1 & \cdots & b_q\end{array}\middle|\, t\right)$$

$$=\left(t\frac{d}{dt}+a_1\right)\cdots\left(t\frac{d}{dt}+a_p\right)F\left(\begin{array}{ccc}a_1 & \cdots & a_p\\ b_1 & \cdots & b_q\end{array}\middle|\, t\right).$$

(ii) (Proposed by D. Stanton) Check that $F_n(t)$ in (3.3.9) is a Laguerre polynomial.

3.4 Quick introduction to Fredholm determinants

We have seen in Lemma 3.2.4 that a certain gap probability, i.e. the probability that a set does not contain any eigenvalue, is given by a Fredholm determinant. The asymptotic study of gap probabilities thus involves the analysis of such determinants. Toward this end, in this section we review key definitions and facts concerning Fredholm determinants. We make no attempt to achieve great generality. In particular we do not touch here on any functional analytic aspects of the theory of Fredholm determinants. The reader interested only in the proof of Theorem 3.1.1 may skip Subsection 3.4.2 in a first reading.

3.4.1 The setting, fundamental estimates and definition of the Fredholm determinant

Let X be a locally compact Polish space, with $\mathscr{B}_X$ denoting its Borel σ-algebra. Let ν be a complex-valued measure on $(X, \mathscr{B}_X)$, such that

$$\|\nu\|_1 = \int_X |\nu(dx)| < \infty. \tag{3.4.1}$$

(In many applications, $X = \mathbb{R}$, and ν will be a scalar multiple of the Lebesgue measure on a bounded interval).

Definition 3.4.1 A *kernel* is a Borel measurable, complex-valued function $K(x,y)$ defined on $X \times X$ such that

$$\|K\| := \sup_{(x,y)\in X\times X} |K(x,y)| < \infty. \tag{3.4.2}$$

The *trace* of a kernel $K(x,y)$ (with respect to ν) is

$$\mathrm{tr}(K) = \int K(x,x)d\nu(x). \tag{3.4.3}$$

Given two kernels $K(x,y)$ and $L(x,y)$, define their *composition* (with respect to ν) as

$$(K \star L)(x,y) = \int K(x,z)L(z,y)d\nu(z). \tag{3.4.4}$$

The trace in (3.4.3) and the composition in (3.4.4) are well defined because $\|\nu\|_1 < \infty$ and $\|K\| < \infty$, and further, $K \star L$ is itself a kernel. By Fubini's Theorem, for any three kernels K, L and M, we have

$$\mathrm{tr}(K \star L) = \mathrm{tr}(L \star K) \text{ and } (K \star L) \star M = K \star (L \star M).$$

Warning We do not restrict K in Definition 3.4.1 to be continuous. Thus, we may have situations where two kernels K, K' satisfy $K = K'$, $\nu \times \nu$- a.e., but $\mathrm{tr}(K) \neq \mathrm{tr}(K')$.

We turn next to a basic estimate.

Lemma 3.4.2 *Fix $n > 0$. For any two kernels $F(x,y)$ and $G(x,y)$ we have*

$$\left| \det_{i,j=1}^{n} F(x_i, y_j) - \det_{i,j=1}^{n} G(x_i, y_j) \right| \leq n^{1+n/2} \|F - G\| \cdot \max(\|F\|, \|G\|)^{n-1} \tag{3.4.5}$$

and

$$\left| \det_{i,j=1}^{n} F(x_i, y_j) \right| \leq n^{n/2} \|F\|^{n}. \tag{3.4.6}$$

The factor $n^{n/2}$ in (3.4.5) and (3.4.6) comes from Hadamard's inequality (Theorem A.3). In view of Stirling's approximation (2.5.12), it is clear that the Hadamard bound is much better than the bound $n!$ we would get just by counting terms.

Proof Define

$$H_i^{(k)}(x,y) = \begin{cases} G(x,y) & \text{if } i < k, \\ F(x,y) - G(x,y) & \text{if } i = k, \\ F(x,y) & \text{if } i > k, \end{cases}$$

noting that, by the linearity of the determinant with respect to rows,

$$\det_{i,j=1}^{n} F(x_i, y_j) - \det_{i,j=1}^{n} G(x_i, y_j) = \sum_{k=1}^{n} \det_{i,j=1}^{n} H_i^{(k)}(x_i, y_j). \tag{3.4.7}$$

Considering the vectors $v_i = v_i^{(k)}$ with $v_i(j) = H_i^{(k)}(x_i, y_j)$, and applying Hadamard's inequality (Theorem A.3), one gets

$$\left| \det_{i,j=1}^{n} H_i^{(k)}(x_i, y_j) \right| \leq n^{n/2} \|F - G\| \cdot \max(\|F\|, \|G\|)^{n-1}.$$

Substituting in (3.4.7) yields (3.4.5). Noting that the summation in (3.4.7) involves only one nonzero term when $G = 0$, one obtains (3.4.6). □

We are now finally ready to define the Fredholm determinant associated with a kernel $K(x,y)$. For $n > 0$, put

$$\Delta_n = \Delta_n(K, \nu) = \int \cdots \int \det_{i,j=1}^{n} K(\xi_i, \xi_j) d\nu(\xi_1) \cdots d\nu(\xi_n), \tag{3.4.8}$$

setting $\Delta_0 = \Delta_0(K,\nu) = 1$. We have, by (3.4.6),

$$\left|\int \cdots \int \det_{i,j=1}^{n} K(\xi_i,\xi_j) d\nu(\xi_1)\cdots d\nu(\xi_n)\right| \leq \|\nu\|_1^n \|K\|^n n^{n/2}. \tag{3.4.9}$$

So, Δ_n is well defined.

Definition 3.4.3 The *Fredholm determinant* associated with the kernel K is defined as

$$\Delta(K) = \Delta(K,\nu) = \sum_{n=0}^{\infty} \frac{(-1)^n}{n!} \Delta_n(K,\nu).$$

(As in (3.4.8) and Definition 3.4.3, we often suppress the dependence on ν from the notation for Fredholm determinants.) In view of Stirling's approximation (2.5.12) and estimate (3.4.9), the series in Definition 3.4.3 converges absolutely, and so $\Delta(K)$ is well defined. The reader should not confuse the Fredholm determinant $\Delta(K)$ with the Vandermonde determinant $\Delta(x)$: in the former, the argument is a kernel while, in the latter, it is a vector.

Remark 3.4.4 Here is some motivation for calling $\Delta(K)$ a determinant. Let $f_1(x),\ldots,f_N(x)$, $g_1(x),\ldots,g_N(x)$ be given. Put

$$K(x,y) = \sum_{i=1}^{N} f_i(x) g_i(y).$$

Assume further that $\max_i \sup_x f_i(x) < \infty$ and $\max_j \sup_y g_j(y) < \infty$. Then $K(x,y)$ is a kernel and so fits into the theory developed thus far. Paraphrasing the proof of Lemma 3.2.4, we have that

$$\Delta(K) = \det_{i,j=1}^{N} \left(\delta_{ij} - \int f_i(x) g_j(x) d\nu(x)\right). \tag{3.4.10}$$

For this reason, one often encounters the notation $\det(I-K)$ for the Fredholm determinant of K.

The determinants $\Delta(K)$ inherit good continuity properties with respect to the $\|\cdot\|$ norm.

Lemma 3.4.5 *For any two kernels $K(x,y)$ and $L(x,y)$ we have*

$$|\Delta(K) - \Delta(L)| \leq \left(\sum_{n=1}^{\infty} \frac{n^{1+n/2}\|\nu\|_1^n \cdot \max(\|K\|,\|L\|)^{n-1}}{n!}\right) \cdot \|K-L\|. \tag{3.4.11}$$

Proof Sum the estimate (3.4.5). □

In particular, with K held fixed, and with L varying in such a way that $\|K-L\| \to 0$, it follows that $\Delta(L) \to \Delta(K)$. This is the only thing we shall need to obtain the convergence in law of the spacing distribution of the eigenvalues of the GUE, Theorem 3.1.1. On the other hand, the next subsections will be useful in the proof of Theorem 3.1.2.

3.4.2 Definition of the Fredholm adjugant, Fredholm resolvent and a fundamental identity

Throughout, we fix a measure ν and a kernel $K(x,y)$. We put $\Delta = \Delta(K)$. All the constructions under this heading depend on K and ν, but we suppress reference to this dependence in the notation in order to control clutter. Define, for any integer $n \geq 1$,

$$K\left(\begin{array}{ccc} x_1 & \dots & x_n \\ y_1 & \dots & y_n \end{array}\right) = \det_{i,j=1}^{n} K(x_i, y_j), \tag{3.4.12}$$

set

$$H_n(x,y) = \int \cdots \int K\left(\begin{array}{cccc} x & \xi_1 & \dots & \xi_n \\ y & \xi_1 & \dots & \xi_n \end{array}\right) d\nu(\xi_1) \cdots d\nu(\xi_n) \tag{3.4.13}$$

and

$$H_0(x,y) = K(x,y).$$

We then have from Lemma 3.4.2 that

$$|H_n(x,y)| \leq \|K\|^{n+1} \|\nu\|_1^n (n+1)^{(n+1)/2}. \tag{3.4.14}$$

Definition 3.4.6 The *Fredholm adjugant* of the kernel $K(x,y)$ is the function

$$H(x,y) = \sum_{n=0}^{\infty} \frac{(-1)^n}{n!} H_n(x,y). \tag{3.4.15}$$

If $\Delta(K) \neq 0$ we define the *resolvent* of the kernel $K(x,y)$ as the function

$$R(x,y) = \frac{H(x,y)}{\Delta(K)}. \tag{3.4.16}$$

By (3.4.14), the series in (3.4.15) converges absolutely and uniformly on $X \times X$. Therefore $H(\cdot)$ is well defined (and continuous on X^{2p} if K is continuous on $X \times X$). The main fact to bear in mind as we proceed is that

$$\sup |F(x,y)| < \infty \tag{3.4.17}$$

for $F = K, H, R$. These bounds are sufficient to guarantee the absolute convergence of all the integrals we will encounter in the remainder of Section 3.4. Also it bears emphasizing that the two-variable functions $H(x,y)$ (resp., $R(x,y)$ if defined) are kernels.

We next prove a fundamental identity relating the Fredholm adjugant and determinant associated with a kernel K.

Lemma 3.4.7 (The fundamental identity) *Let $H(x,y)$ be the Fredholm adjugant of the kernel $K(x,y)$. Then,*

$$\begin{aligned}\int K(x,z)H(z,y)d\nu(z) &= H(x,y) - \Delta(K)\cdot K(x,y) \\ &= \int H(x,z)K(z,y)d\nu(z), \qquad (3.4.18)\end{aligned}$$

and hence (equivalently)

$$K \star H = H - \Delta(K)\cdot K = H \star K. \qquad (3.4.19)$$

Remark 3.4.8 Before proving the fundamental identity (3.4.19), we make some amplifying remarks. If $\Delta(K) \neq 0$ and hence the resolvent $R(x,y) = H(x,y)/\Delta(K)$ of $K(x,y)$ is well defined, then the fundamental identity takes the form

$$\int K(x,z)R(z,y)d\nu(z) = R(x,y) - K(x,y) = \int R(x,z)K(z,y)d\nu(z) \qquad (3.4.20)$$

and hence (equivalently)

$$K \star R = R - K = R \star K.$$

It is helpful if not perfectly rigorous to rewrite the last formula as the operator identity

$$1 + R = (1-K)^{-1}.$$

Rigor is lacking here because we have not taken the trouble to associate linear operators with our kernels. Lack of rigor notwithstanding, the last formula makes it clear that $R(x,y)$ deserves to be called the resolvent of $K(x,y)$. Moreover, this formula is useful for discovering composition identities which one can then verify directly and rigorously.

Proof of Lemma 3.4.7 Here are two reductions to the proof of the fundamental identity. Firstly, it is enough just to prove the first of the equalities claimed in (3.4.18) because the second is proved similarly. Secondly, proceeding term by term, since $H_0 = K$ and $\Delta_0 = 1$, it is enough to prove that, for $n > 0$,

$$\frac{(-1)^{n-1}}{(n-1)!}\int K(x,z)H_{n-1}(z,y)d\nu(z) = \frac{(-1)^n}{n!}(H_n(x,y) - \Delta_n \cdot K(x,y))$$

or, equivalently,

$$H_n(x,y) = \Delta_n \cdot K(x,y) - n\int K(x,z)H_{n-1}(z,y)d\nu(z)\,, \tag{3.4.21}$$

where $\Delta_n = \Delta_n(K)$.

Now we can quickly give the proof of the fundamental identity (3.4.19). Expanding by minors of the first row, we find that

$$\begin{aligned} &K\begin{pmatrix} x & \xi_1 & \dots & \xi_n \\ y & \xi_1 & \dots & \xi_n \end{pmatrix} \\ &= K(x,y)K\begin{pmatrix} \xi_1 & \dots & \xi_n \\ \xi_1 & \dots & \xi_n \end{pmatrix} \\ &\quad + \sum_{j=1}^n (-1)^j K(x,\xi_j)K\begin{pmatrix} \xi_1 & \dots & \xi_{j-1} & \xi_j & \xi_{j+1} & \dots & \xi_n \\ y & \xi_1 & \dots & \xi_{j-1} & \xi_{j+1} & \dots & \xi_n \end{pmatrix} \\ &= K(x,y)K\begin{pmatrix} \xi_1 & \dots & \xi_n \\ \xi_1 & \dots & \xi_n \end{pmatrix} \\ &\quad - \sum_{j=1}^n K(x,\xi_j)K\begin{pmatrix} \xi_j & \xi_1 & \dots & \xi_{j-1} & \xi_{j+1} & \dots & \xi_n \\ y & \xi_1 & \dots & \xi_{j-1} & \xi_{j+1} & \dots & \xi_n \end{pmatrix}. \end{aligned}$$

Integrating out the variables $\xi_1, \dots, \xi_n$ in evident fashion, we obtain (3.4.21). Thus the fundamental identity is proved. □

We extract two further benefits from the proof of the fundamental identity. Recall from (3.4.8) and Definition 3.4.3 the abbreviated notation $\Delta_n = \Delta_n(K)$ and $\Delta(K)$.

Corollary 3.4.9 *(i) For all $n \geq 0$,*

$$\frac{(-1)^n}{n!}H_n(x,y) = \sum_{k=0}^n \frac{(-1)^k}{k!}\Delta_k \cdot (\underbrace{K\star\cdots\star K}_{n+1-k})(x,y)\,. \tag{3.4.22}$$

(ii) Further,

$$\frac{(-1)^n}{n!}\Delta_{n+1} = \sum_{k=0}^n \frac{(-1)^k}{k!}\Delta_k \cdot \mathrm{tr}(\underbrace{K\star\cdots\star K}_{n+1-k})\,. \tag{3.4.23}$$

In particular, the sequence of numbers

$$\mathrm{tr}(K),\ \ \mathrm{tr}(K\star K),\ \ \mathrm{tr}(K\star K\star K),\ \dots$$

uniquely determines the Fredholm determinant $\Delta(K)$.

Proof Part (i) follows from (3.4.21) by employing an induction on n. We leave the details to the reader. Part (ii) follows by putting $x = \xi$ and $y = \xi$ in (3.4.22), and integrating out the variable ξ. □

Multiplicativity of Fredholm determinants

We now prove a result needed for our later analysis of GOE and GSE. A reader interested only in GUE can skip this material.

Theorem 3.4.10 *Fix kernels* $K(x,y)$ *and* $L(x,y)$ *arbitrarily. We have*

$$\Delta(K + L - L \star K) = \Delta(K)\Delta(L)\,. \tag{3.4.24}$$

In the sequel we refer to this relation as the *multiplicativity* of the Fredholm determinant construction.

Proof Let t be a complex variable. We are going to prove multiplicativity by studying the entire function

$$\varphi_{K,L}(t) = \Delta(K + t(L - L \star K))$$

of t. We assume below that $\varphi_{K,L}(t)$ does not vanish identically, for otherwise there is nothing to prove. We claim that

$$\begin{aligned} \varphi'_{K,L}(0) &= -\Delta(K)\operatorname{tr}(L - L\star K) + \operatorname{tr}((L - L\star K)\star H) \\ &= -\Delta(K)\operatorname{tr}(L)\,, \end{aligned} \tag{3.4.25}$$

where H is the Fredholm adjugant of K, see equation (3.4.15). The first step is justified by differentiation under the integral; to justify the exchange of limits one notes that for any entire analytic function $f(z)$ and $\varepsilon > 0$ one has $f'(0) = \frac{1}{2\pi i}\int_{|z|=\varepsilon} \frac{f(z)}{z^2} dz$, and then uses Fubini's Theorem. The second step follows by the fundamental identity, see Lemma 3.4.7. This completes the proof of (3.4.25).

Since $\varphi_{0,L}(t) = \Delta(tL)$ equals 1 for $t = 0$, the product $\varphi_{0,L}(t)\varphi_{K,L}(t)$ does not vanish identically. Arbitrarily fix a complex number t_0 such that $\varphi_{0,L}(t_0)\varphi_{K,L}(t_0) \neq 0$. Note that the resolvant S of $t_0 L$ is defined. One can verify by straightforward calculation that the kernels

$$\tilde{K} = K + t_0(L - L\star K), \quad \tilde{L} = L + L\star S\,, \tag{3.4.26}$$

satisfy the composition identity

$$K + (t_0 + t)(L - L\star K) = \tilde{K} + t(\tilde{L} - \tilde{L}\star\tilde{K})\,. \tag{3.4.27}$$

With $\tilde{K}$ and $\tilde{L}$ as in (3.4.26), we have $\varphi_{\tilde{K},\tilde{L}}(t) = \varphi_{K,L}(t+t_0)$ by (3.4.27) and hence

$$\frac{d}{dt}\log\varphi_{K,L}(t)\bigg|_{t=t_0} = -\operatorname{tr}(\tilde{L})$$

by (3.4.25). Now the last identity holds also for $K=0$ and the right side is independent of K. It follows that the logarithmic derivatives of the functions $\varphi_{0,L}(t)$ and $\varphi_{K,L}(t)$ agree wherever neither has a pole, and so these logarithmic derivatives must be identically equal. Integrating and exponentiating once we obtain an identity $\varphi_{K,L}(t) = \varphi_{K,L}(0)\varphi_{0,L}(t)$ of entire functions of t. Finally, by evaluating the last relation at $t=1$, we recover the multiplicativity relation (3.4.24). □

3.5 Gap probabilities at 0 and proof of Theorem 3.1.1

In the remainder of this chapter, we let $X_N \in \mathscr{H}_N^{(2)}$ be a random Hermitian matrix from the GUE with eigenvalues $\lambda_1^N \leq \cdots \leq \lambda_N^N$. We initiate in this section the study of the *spacings* between eigenvalues of X_N. We focus on those eigenvalues that lie near 0, and seek, for a fixed $t>0$, to evaluate the limit

$$\lim_{N\to\infty} P[\sqrt{N}\lambda_1^N,\ldots,\sqrt{N}\lambda_N^N \notin (-t/2,t/2)]\,, \tag{3.5.1}$$

see the statement of Theorem 3.1.1. We note that *a priori*, because of Theorems 2.1.1 and 2.1.22, the limit in (3.5.1) has some chance of being nondegenerate because the N random variables $\sqrt{N}\lambda_1^N,\ldots,\sqrt{N}\lambda_N^N$ are spread out over an interval very nearly of length $4N$. As we will show in Section 4.2, the computation of the limit in (3.5.1) allows one to evaluate other limits, such as the limit of the empirical measure of the spacings in the bulk of the spectrum.

As in (3.2.4), set

$$K^{(n)}(x,y) = \sum_{k=0}^{n-1}\psi_k(x)\psi_k(y) = \sqrt{n}\,\frac{\psi_n(x)\psi_{n-1}(y)-\psi_{n-1}(x)\psi_n(y)}{x-y}\,,$$

where the $\psi_k(x)$ are the normalized oscillator wave-functions introduced in Definition 3.2.1. Set

$$S^{(n)}(x,y) = \frac{1}{\sqrt{n}}K^{(n)}\left(\frac{x}{\sqrt{n}},\frac{y}{\sqrt{n}}\right).$$

A crucial step in the proof of Theorem 3.1.1 is the following lemma, whose proof, which takes most of the analysis in this section, is deferred.

Lemma 3.5.1 *With the above notation, it holds that*

$$\lim_{n\to\infty} S^{(n)}(x,y) = \frac{1}{\pi}\frac{\sin(x-y)}{x-y}\,, \tag{3.5.2}$$

uniformly on each bounded subset of the (x,y)-plane.

Proof of Theorem 3.1.1 Recall that by Lemma 3.2.4,

$$
\begin{aligned}
P[\sqrt{n}\lambda_1^{(n)},\dots,\sqrt{n}\lambda_n^{(n)} \notin A] \\
&= 1+\sum_{k=1}^{\infty}\frac{(-1)^k}{k!}\int_{\sqrt{n}^{-1}A}\cdots\int_{\sqrt{n}^{-1}A}\det_{i,j=1}^{k} K^{(n)}(x_i,x_j)\prod_{j=1}^{k}dx_j \\
&= 1+\sum_{k=1}^{\infty}\frac{(-1)^k}{k!}\int_A\cdots\int_A\det_{i,j=1}^{k} S^{(n)}(x_i,x_j)\prod_{j=1}^{k}dx_j\,.
\end{aligned}
$$

(The scaling of Lebesgue's measure in the last equality explains the appearance of the scaling by $1/\sqrt{n}$ in the definition of $S^{(n)}(x,y)$.) Lemma 3.5.1 together with Lemma 3.4.5 complete the proof of the theorem. □

The proof of Lemma 3.5.1 takes up the rest of this section. We begin by bringing, in Subsection 3.5.1, a quick introduction to Laplace's method for the evaluation of asymptotics of integrals, which will be useful for other asymptotic computations, as well. We then apply it in Subsection 3.5.2 to conclude the proof.

Remark 3.5.2 We remark that one is naturally tempted to guess that the random variable W_N ="width of the largest open interval symmetric about the origin containing none of the eigenvalues $\sqrt{N}\lambda_1^N,\dots,\sqrt{N}\lambda_N^N$" should possess a limit in distribution. Note however that we do not *a priori* have tightness for that random variable. But, as we show in Section 3.6, we do have tightness (see (3.6.34) below) *a posteriori*. In particular, in Section 3.6 we prove Theorem 3.1.2, which provides an explicit expression for the limit distribution of W_N.

3.5.1 The method of Laplace

Laplace's method deals with the asymptotic (as $s\to\infty$) evaluation of integrals of the form

$$\int f(x)^s g(x)dx\,.$$

We will be concerned with the situation in which the function f possesses a global maximum at some point a, and behaves quadratically in a neighborhood of that maximum. More precisely, let $f:\mathbb{R}\mapsto\mathbb{R}_+$ be given, and for some constant a and positive constants s_0,K,L,M, let $\mathcal{G}=\mathcal{G}(a,\varepsilon_0,s_0,f(\cdot),K,L,M)$ be the class of measurable functions $g:\mathbb{R}\mapsto\mathbb{R}$ satisfying the following conditions:

(i) $|g(a)|\le K$;

(ii) $\sup_{0<|x-a|\leq\varepsilon_0}\left|\frac{g(x)-g(a)}{x-a}\right|\leq L$;
(iii) $\int f(x)^{s_0}|g(x)|dx \leq M$.

We then have the following.

Theorem 3.5.3 (Laplace) *Let $f:\mathbb{R}\to\mathbb{R}_+$ be a function such that, for some $a\in\mathbb{R}$ and some positive constants ε_0, c, the following hold.*

(a) $f(x)\leq f(x')$ if either $a-\varepsilon_0\leq x\leq x'\leq a$ or $a\leq x'\leq x\leq a+\varepsilon_0$.
(b) For all $\varepsilon<\varepsilon_0$, $\sup_{|x-a|>\varepsilon} f(x)\leq f(a)-c\varepsilon^2$.
(c) $f(x)$ has two continuous derivatives in the interval $(a-2\varepsilon_0, a+2\varepsilon_0)$.
(d) $f''(a)<0$.

Then, for any function $g\in\mathscr{G}(a,\varepsilon_0,s_0,f(\cdot),K,L,M)$, we have

$$\lim_{s\to\infty}\sqrt{s}f(a)^{-s}\int f(x)^s g(x)dx=\sqrt{-\frac{2\pi f(a)}{f''(a)}}\,g(a)\,, \tag{3.5.3}$$

and moreover, for fixed $f,a,\varepsilon_0,s_0,K,L,M$, the convergence is uniform over the class $\mathscr{G}(a,\varepsilon_0,s_0,f(\cdot),K,L,M)$.

Note that by point (b) of the assumptions, $f(a)>0$. The intuition here is that as s tends to infinity the function $(f(x)/f(a))^s$ near $x=a$ peaks more and more sharply and looks at the microscopic level more and more like a bell-curve, whereas $f(x)^s$ elsewhere becomes negligible. Formula (3.5.3) is arguably the simplest nontrivial application of Laplace's method. Later we are going to encounter more sophisticated applications.

Proof of Theorem 3.5.3 Let $\varepsilon(s)$ be a positive function defined for $s\geq s_0$ such that $\varepsilon(s)\to_{s\to\infty}0$ and $s\varepsilon(s)^2\to_{s\to\infty}\infty$, while $\varepsilon_0=\sup_{s\geq s_0}\varepsilon(s)$. For example, we could take $\varepsilon(s)=\varepsilon_0\cdot(s_0/s)^{1/4}$. For $s\geq s_0$, write

$$\int f(x)^s g(x)dx = g(a)I_1+I_2+I_3\,,$$

where

$$\begin{aligned}
I_1 &= \int_{|x-a|\leq\varepsilon(s)} f(x)^s dx\,,\\
I_2 &= \int_{|x-a|\leq\varepsilon(s)} f(x)^s(g(x)-g(a))dx\,,\\
I_3 &= \int_{|x-a|>\varepsilon(s)} f(x)^s g(x)dx\,.
\end{aligned}$$

For $|t|<2\varepsilon_0$, put

$$h(t)=\int_0^1(1-r)(\log f)''(a+rt)dr\,,$$

thus defining a continuous function of t such that $h(0) = f''(a)/2f(a)$ and which by Taylor's Theorem satisfies

$$f(x) = f(a)\exp(h(x-a)(x-a)^2)$$

for $|x-a| < 2\varepsilon_0$. We then have

$$I_1 = \frac{f(a)^s}{\sqrt{s}} \int_{|t| \le \varepsilon(s)\sqrt{s}} \exp\left(h\left(\frac{t}{\sqrt{s}}\right)t^2\right) dt\,,$$

and hence

$$\lim_{s\to\infty} \sqrt{s} f(a)^{-s} I_1 = \int_{-\infty}^{\infty} \exp\left(h(0)t^2\right) dt = \sqrt{-\frac{2\pi f(a)}{f''(a)}}\,.$$

We have $|I_2| \le L\varepsilon(s) I_1$ and hence

$$\lim_{s\to\infty} \sqrt{s} f(a)^{-s} I_2 = 0\,.$$

We have, since $\varepsilon(s) < \varepsilon_0$,

$$|I_3| \le M \sup_{x:|x-a|>\varepsilon(s)} |f(x)|^{s-s_0} \le M f(a)^{s-s_0} \left(1 - \frac{c\varepsilon(s)^2}{f(a)}\right)^{s-s_0},$$

and hence

$$\lim_{s\to\infty} \sqrt{s} f(a)^{-s} I_3 = 0\,.$$

This is enough to prove that the limit formula (3.5.3) holds and enough also to prove the uniformity of convergence over all functions $g(x)$ of the class $\mathscr{G}$. □

3.5.2 Evaluation of the scaling limit: proof of Lemma 3.5.1

The main step in the proof of Lemma 3.5.1 is the following uniform convergence result, whose proof is deferred. Let

$$\Psi_\nu(t) = n^{\frac{1}{4}} \psi_\nu\left(\frac{t}{\sqrt{n}}\right),$$

with ν a quantity whose difference from n is fixed (in the proof of Lemma 3.5.1, we will use $\nu = n, n-1, n-2$).

Lemma 3.5.4 *Uniformly for t in a fixed bounded interval,*

$$\lim_{n\to\infty} |\Psi_\nu(t) - \frac{1}{\sqrt{\pi}} \cos\left(t - \frac{\pi\nu}{2}\right)| = 0\,. \tag{3.5.4}$$

With Lemma 3.5.4 granted, we can complete the

Proof of Lemma 3.5.1 Recall that

$$S^{(n)}(x,y) = \sqrt{n}\frac{\psi_n(\frac{x}{\sqrt{n}})\psi_{n-1}(\frac{y}{\sqrt{n}}) - \psi_{n-1}(\frac{x}{\sqrt{n}})\psi_n(\frac{y}{\sqrt{n}})}{x-y}.$$

In order to prove the claimed uniform convergence, it is useful to get rid of the division by $(x-y)$ in $S^{(n)}(x,y)$. Toward this end, noting that for any differentiable functions f,g on $\mathbb{R}$,

$$\begin{aligned}
&\frac{f(x)g(y)-f(y)g(x)}{x-y} \\
&= \left(\frac{f(x)-f(y)}{x-y}\right)g(y) + f(y)\left(\frac{g(y)-g(x)}{x-y}\right) \\
&= g(y)\int_0^1 f'(tx+(1-t)y)dt - f(y)\int_0^1 g'(tx+(1-t)y)dt\,, \qquad (3.5.5)
\end{aligned}$$

we deduce

$$\begin{aligned}
S^{(n)}(x,y) &= \psi_{n-1}(\frac{y}{\sqrt{n}})\int_0^1 \psi_n'(t\frac{x}{\sqrt{n}}+(1-t)\frac{y}{\sqrt{n}})dt \\
&\quad -\psi_n(\frac{y}{\sqrt{n}})\int_0^1 \psi_{n-1}'(t\frac{x}{\sqrt{n}}+(1-t)\frac{y}{\sqrt{n}})dt \qquad (3.5.6) \\
&= \psi_{n-1}(\frac{y}{\sqrt{n}})\int_0^1 (\sqrt{n}\psi_{n-1}(z)-\frac{z}{2}\psi_n(z))|_{z=t\frac{x}{\sqrt{n}}+(1-t)\frac{y}{\sqrt{n}}}dt \\
&\quad -\psi_n(\frac{y}{\sqrt{n}})\int_0^1 (\sqrt{n-1}\psi_{n-2}(z)-\frac{z}{2}\psi_{n-1}(z))|_{z=t\frac{x}{\sqrt{n}}+(1-t)\frac{y}{\sqrt{n}}}dt\,,
\end{aligned}$$

where we used in the last equality point 4 of Lemma 3.2.7. Using (3.5.4) (in the case $\nu = n, n-1, n-2$) in (3.5.6) and elementary trigonometric formulas shows that

$$\begin{aligned}
S^{(n)}(x,y) &\sim \frac{1}{\pi}\Bigg(\cos(y-\frac{\pi(n-1)}{2})\int_0^1 \cos\left(tx+(1-t)y-\frac{\pi(n-1)}{2}\right)dt \\
&\quad -\cos(y-\frac{\pi n}{2})\int_0^1 \cos\left(tx+(1-t)y-\frac{\pi(n-2)}{2}\right)dt\Bigg) \\
&\sim \frac{1}{\pi}\frac{\sin(x-y)}{x-y}\,,
\end{aligned}$$

which, Lemma 3.5.4 granted, completes the proof of Lemma 3.5.1. □

Proof of Lemma 3.5.4 Recall the Fourier transform identity

$$e^{-x^2/2} = \frac{1}{\sqrt{2\pi}}\int e^{-\xi^2/2-\mathrm{i}\xi x}d\xi\,.$$

Differentiating under the integral, we find that

$$\mathfrak{H}_n(x)e^{-x^2/2} = (-1)^n \frac{d^n}{dx^n} e^{-x^2/2} = \frac{1}{\sqrt{2\pi}} \int (\mathrm{i}\xi)^n e^{-\xi^2/2-\mathrm{i}\xi x} d\xi \,,$$

or equivalently

$$\psi_\nu(x) = \frac{\mathrm{i}^\nu e^{x^2/4}}{(2\pi)^{3/4}\sqrt{\nu!}} \int \xi^\nu e^{-\xi^2/2-\mathrm{i}\xi x} d\xi \,. \tag{3.5.7}$$

We use the letter ν here instead of n to help avoid confusion at the next step. As a consequence, setting $C_{\nu,n} = \sqrt{n}/(2\pi)$, we have

$$\begin{aligned}
\Psi_\nu(t) &= \frac{\mathrm{i}^\nu e^{t^2/(4n)} n^{1/4}}{(2\pi)^{3/4}\sqrt{\nu!}} \int \xi^\nu e^{-\xi^2/2-\mathrm{i}\xi t/\sqrt{n}} d\xi \\
&= \frac{(2\pi)^{1/4} C_{\nu,n} e^{t^2/(4n)} n^{1/4+\nu/2}}{\sqrt{\nu!}} \int (\xi e^{-\xi^2/2})^n \mathrm{i}^\nu e^{-\mathrm{i}\xi t} \xi^{\nu-n} d\xi \\
&\sim \frac{(2\pi)^{1/4} C_{\nu,n} n^{1/4+n/2}}{\sqrt{n!}} \int (\xi e^{-\xi^2/2})^n \mathrm{i}^\nu e^{-\mathrm{i}\xi t} \xi^{\nu-n} d\xi \\
&\sim C_{\nu,n} e^{n/2} \int |\xi e^{-\xi^2/2}|^n \Re[(\mathrm{i}\,\mathrm{sign}\xi)^\nu e^{-\mathrm{i}\xi t}] |\xi^{\nu-n}| d\xi \,,
\end{aligned}$$

where Stirling's approximation (2.5.12) and the fact that $\Psi_\nu(t)$ is real were used in the last line. Using symmetry, we can rewrite the last expressions as

$$2C_{\nu,n} e^{n/2} \int_{-\infty}^{\infty} f(\xi)^n g_t(\xi) d\xi \,,$$

with $f(x) = xe^{-x^2/2}\mathbf{1}_{x\geq 0}$ and $g(x) = g_t(x) = \cos(xt - \frac{\pi\nu}{2})x^{\nu-n}$.

Consider t as fixed, and let $n \to \infty$ in one of the four possible ways such that $g(\cdot)$ does not depend on n (recall that $\nu - n$ does not depend on n). Note that $f(x)$ achieves its maximal value at $x = 1$ and

$$f(1) = e^{-1/2}, \quad f'(1) = 0, \quad f''(1) = -2e^{-1/2}.$$

Hence, we can apply Laplace's method (Theorem 3.5.3) to find that

$$\Psi_\nu(t) \to_{n\to\infty} \frac{1}{\sqrt{\pi}} \cos\left(t - \frac{\pi\nu}{2}\right).$$

Moreover, the convergence here is uniform for t in a fixed bounded interval, as follows from the uniformity asserted for convergence in limit formula (3.5.3). □

Exercise 3.5.5 Use Laplace's method (Theorem 3.5.3) with $a = 1$ to prove (2.5.12): as $s \to \infty$ along the positive real axis,

$$\Gamma(s) = \int_0^\infty x^s e^{-x} \frac{dx}{x} = s^s \int_0^\infty (xe^{-x})^s \frac{dx}{x} \sim \sqrt{2\pi}\, s^{s-1/2} e^{-s} \,.$$

This recovers in particular Stirling's approximation (2.5.12).

3.5.3 A complement: determinantal relations

Let integers $\ell_1,\dots,\ell_p \geq 0$ and bounded disjoint Borel sets $A_1,\dots,A_p$ be given. Put

$$P_N(\ell_1,\dots,\ell_p;A_1,\dots,A_p) = P\left[\ell_i = \left|\{\sqrt{N}\lambda_1^N,\dots,\sqrt{N}\lambda_N^N\} \cap A_i\right|, \text{ for } i=1,\dots,p\right].$$

We have the following.

Lemma 3.5.6 *Let $s_1,\dots,s_p$ be independent complex variables and let*

$$\varphi = (1-s_1)\mathbf{1}_{A_1} + \cdots + (1-s_p)\mathbf{1}_{A_p}.$$

Then, the limit

$$P(\ell_1,\dots,\ell_p;A_1,\dots,A_p) = \lim_{N\to\infty} P_N(\ell_1,\dots,\ell_p;A_1,\dots,A_p) \tag{3.5.8}$$

exists and satisfies

$$\sum_{\ell_1=0}^{\infty}\cdots\sum_{\ell_p=0}^{\infty} P(\ell_1,\dots,\ell_p;A_1,\dots,A_p)s_1^{\ell_1}\cdots s_p^{\ell_p}$$

$$= 1+\sum_{k=1}^{\infty}\frac{(-1)^k}{k!}\int\cdots\int \det_{i,j=1}^k \frac{1}{\pi}\frac{\sin(x_i-x_j)}{x_i-x_j}\prod_{j=1}^k \varphi(x_j)dx_j. \tag{3.5.9}$$

That is, the generating function in the left side of (3.5.8) can be represented in terms of a Fredholm determinant. We note that this holds in greater generality, see Section 4.2.

Proof The proof is a slight modification of the method presented in Subsection 3.5.2. Note that the right side of (3.5.9) defines, by the fundamental estimate (3.4.9), an entire function of the complex variables $s_1,\dots,s_p$, whereas the left side defines a function analytic in a domain containing the product of p copies of the unit disc centered at the origin. Clearly we have

$$E\prod_{i=1}^N\left(1-\varphi\left(\sqrt{N}\lambda_i^N\right)\right) = \sum_{\substack{\ell_1,\dots,\ell_p\geq 0\\ \ell_1+\cdots+\ell_p\leq N}} P_N(\ell_1,\dots,\ell_p;A_1,\dots,A_p)s_1^{\ell_1}\cdots s_p^{\ell_p}. \tag{3.5.10}$$

The function of $s_1,\dots,s_p$ on the right is simply a polynomial, whereas the expectation on the left can be represented as a Fredholm determinant. From this, the lemma follows after representing the probability $P_N(\ell_1,\dots,\ell_p;A_1,\dots,A_p)$ as a p-dimensional Cauchy integral. □

3.6 Analysis of the sine-kernel

Our goal in this section is to derive differential equations (in the parameter t) for the probability that no eigenvalue of the (properly rescaled) GUE lies in the interval $(-t/2, t/2)$. We will actually derive slightly more general systems of differential equations that can be used to evaluate expressions like (3.5.9).

3.6.1 General differentiation formulas

Recalling the setting of our general discussion of Fredholm determinants in Section 3.4, we fix a bounded open interval $(a,b) \subset \mathbb{R}$, real numbers

$$a < t_1 < \cdots < t_n < b$$

in the interval (a,b) and complex numbers

$$s_1, \ldots, s_{n-1}, \quad s_0 = 0 = s_n.$$

Set

$$\eta = s_1 \mathbf{1}_{(t_1,t_2)} + \cdots + s_{n-1} \mathbf{1}_{(t_{n-1},t_n)},$$

and define ν so that it has density η with respect to the Lebesgue measure on $X = \mathbb{R}$. We then have, for $f \in L^1[(a,b)]$,

$$\langle f, \nu \rangle = \int f(x) d\nu(x) = \sum_{i=1}^{n-1} s_i \int_{t_i}^{t_{i+1}} f(x) dx.$$

Motivated by Theorem 3.1.1, we fix the function

$$S(x,y) = \frac{\sin(x-y)}{\pi(x-y)} \tag{3.6.1}$$

on $(a,b)^2$ as our kernel. As usual $\Delta = \Delta(S)$ denotes the Fredholm determinant associated with S and the measure ν. We assume that $\Delta \neq 0$ so that the Fredholm resolvent $R(x,y)$ is also defined.

Before proceeding with the construction of a system of differential equations, we provide a description of the main ideas, disregarding in this sketch issues of rigor, and concentrating on the most important case of $n = 2$. View the kernels S and R as operators on $L^1[(a,b)]$, writing multiplication instead of the $\star$ operation. As noted in Remark 3.4.8, we have, with $\tilde{S}(x,y) = (x-y)S(x,y)$ and $\tilde{R}(x,y) = (x-y)R(x,y)$, that

$$(1-S)^{-1} = 1 + R, \quad \tilde{S} = [M,S], \quad \tilde{R} = [M,R],$$

where M is the operation of multiplication by x and the bracket $[A,B] = AB - BA$ is the commutator of the operators A, B. Note also that under our special assumptions

$$\tilde{S}(x,y) = (\sin x \cos y - \sin y \cos x)/\pi,$$

and hence the operator $\tilde{S}$ is of rank 2. But we have

$$\begin{aligned} \tilde{R} &= [M,R] = [M,(1-S)^{-1}] \\ &= -(1-S)^{-1}[M,1-S](1-S)^{-1} = (1+R)\tilde{S}(1+R), \end{aligned}$$

and hence $\tilde{R}$ is also of rank 2. Letting $P(x) = (1+R)\cos(x)/\sqrt{\pi}$ and $Q(x) = (1+R)\sin(x)/\sqrt{\pi}$, we then obtain $\tilde{R} = Q(x)P(y) - Q(y)P(x)$, and thus

$$R(x,y) = \frac{Q(x)P(y) - Q(y)P(x)}{x-y}. \tag{3.6.2}$$

(See Lemma 3.6.2 below for the precise statement and proof.) One checks that differentiating with respect to the endpoints t_1, t_2 the function $\log \Delta(S)$ yields the functions $R(t_i,t_i)$, $i = 1,2$, which in turn may be related to derivatives of P and Q by a careful differentiation, using (3.6.2). The system of differential equations thus obtained, see Theorem 3.6.2, can then be simplified, after specialization to the case $t_2 = -t_1 = t/2$, to yield the Painlevé V equation appearing in Theorem 3.1.2.

Turning to the actual derivation, we consider the parameters $t_1, \ldots, t_n$ as variable, whereas we consider the kernel $S(x,y)$ and the parameters $s_1, \ldots, s_{n-1}$ to be fixed. Motivated by the sketch above, set $f(x) = (\sin x)/\sqrt{\pi}$ and

$$Q(x) = f(x) + \int R(x,y) f(y)\, d\nu(y), \quad P(x) = f'(x) + \int R(x,y) f'(y)\, d\nu(y). \tag{3.6.3}$$

We emphasize that $P(x)$, $Q(x)$ and $R(x,y)$ depend on $t_1, \ldots, t_n$ (through ν), although the notation does not show it. The main result of this section, of which Theorem 3.1.2 is an easy corollary, is the following system of differential equations.

Theorem 3.6.1 *With the above notation, put, for $i, j = 1, \ldots, n$,*

$$p_i = P(t_i), \quad q_i = Q(t_i), \quad R_{ij} = R(t_i,t_j).$$

Then, for $i, j = 1, \ldots, n$ with $i \neq j$, we have the following equations:

$$\begin{aligned}
R_{ij} &= (q_i p_j - q_j p_i)/(t_i - t_j)\,,\\
\partial q_j/\partial t_i &= -(s_i - s_{i-1}) \cdot R_{ji} q_i\,,\\
\partial p_j/\partial t_i &= -(s_i - s_{i-1}) \cdot R_{ji} p_i\,,\\
\partial q_i/\partial t_i &= +p_i + \sum_{k \neq i} (s_k - s_{k-1}) \cdot R_{ik} q_k\,,\\
\partial p_i/\partial t_i &= -q_i + \sum_{k \neq i} (s_k - s_{k-1}) \cdot R_{ik} p_k\,,\\
R_{ii} &= p_i \partial q_i/\partial t_i - q_i \partial p_i/\partial t_i\,,\\
(\partial/\partial t_i) \log \Delta &= (s_i - s_{i-1}) \cdot R_{ii}\,.
\end{aligned} \tag{3.6.4}$$

The proof of Theorem 3.6.1 is completed in Subsection 3.6.2. In the rest of this subsection, we derive a fundamental differentiation formula, see (3.6.10), and derive several relations concerning the functions P, Q introduced in (3.6.3), and the resolvent R.

In the sequel, we write $\int_{I_i}$ for $\int_{t_i}^{t_{i+1}}$. Recall from (3.4.8) that

$$\Delta_\ell = \sum_{i_1=1}^{n-1} \cdots \sum_{i_\ell=1}^{n-1} s_{i_1} \cdots s_{i_\ell} \int_{I_{i_1}} \cdots \int_{I_{i_\ell}} S\begin{pmatrix} \xi_1 & \ldots & \xi_\ell \\ \xi_1 & \ldots & \xi_\ell \end{pmatrix} d\xi_1 \cdots d\xi_\ell\,.$$

Therefore, by the fundamental theorem of calculus,

$$\begin{aligned}
&\frac{\partial}{\partial t_i} \Delta_\ell(x,y)\\
&= -\sum_{j=1}^{\ell} \sum_{i_1=1}^{n-1} \cdots \sum_{i_{j-1}=1}^{n-1} \sum_{i_{j+1}=1}^{n-1} \cdots \sum_{i_\ell=1}^{n-1} s_{i_1} \cdots s_{i_{j-1}} s_{i_{j+1}} \cdots s_{i_k} (s_i - s_{i-1})\\
&\quad \times \int_{I_{i_1}} \cdots \int_{I_{i_{j-1}}} \int_{I_{i_{j+1}}} \cdots \int_{I_{i_\ell}} S\begin{pmatrix} \xi_1 & \ldots & \xi_{i-1} & t_i & \xi_{i+1} & \ldots & \xi_\ell \\ \xi_1 & \ldots & \xi_{i-1} & t_i & \xi_{i+1} & \ldots & \xi_\ell \end{pmatrix} \prod_{\substack{j=1\\ j \neq i}}^{\ell} d\xi_j\\
&= -\ell(s_i - s_{i-1}) H_{\ell-1}(t_i, t_i)\,,
\end{aligned} \tag{3.6.5}$$

with $H_{\ell-1}$ as in (3.4.13). Multiplying by $(-1)^\ell/\ell!$ and summing, using the estimate (3.4.9) and dominated convergence, we find that

$$\frac{\partial}{\partial t_i} \Delta = (s_i - s_{i-1}) H(t_i, t_i)\,. \tag{3.6.6}$$

From (3.6.6) we get

$$\frac{\partial}{\partial t_i} \log \Delta = (s_i - s_{i-1}) R(t_i, t_i)\,. \tag{3.6.7}$$

We also need to be able to differentiate $R(x,y)$. From the fundamental identity (3.4.20), we have

$$\frac{\partial}{\partial t_i}R(x,y) = -(s_i - s_{i-1})R(x,t_i)S(t_i,y) + \int S(x,z)\frac{\partial R(z,y)}{\partial t_i}\nu(dz). \quad (3.6.8)$$

Substituting $y = z'$ in (3.6.8) and integrating against $R(z',y)$ with respect to $\nu(dz')$ gives

$$\int \frac{\partial R(x,z')}{\partial t_i}R(z',y)\nu(dz') = -(s_i - s_{i-1})R(x,t_i)\int S(t_i,z')R(z',y)\nu(dz')$$
$$+\int\int S(x,z)\frac{\partial R(z,z')}{\partial t_i}R(z',y)\nu(dz)\nu(dz'). \quad (3.6.9)$$

Summing (3.6.8) and (3.6.9) and using again the fundamental identity (3.4.20) then yields

$$\frac{\partial}{\partial t_i}R(x,y) = (s_{i-1} - s_i)R(x,t_i)R(t_i,y). \quad (3.6.10)$$

The next lemma will play an important role in the proof of Theorem 3.6.1.

Lemma 3.6.2 *The functions P,Q,R satisfy the following relations:*

$$R(x,y) = \frac{Q(x)P(y) - Q(y)P(x)}{x-y} = R(y,x), \quad (3.6.11)$$

$$R(x,x) = Q'(x)P(x) - Q(x)P'(x), \quad (3.6.12)$$

$$\frac{\partial}{\partial t_i}Q(x) = (s_{i-1} - s_i)R(x,t_i)Q(t_i), \quad (3.6.13)$$

and similarly

$$\frac{\partial}{\partial t_i}P(x) = (s_{i-1} - s_i)R(x,t_i)P(t_i). \quad (3.6.14)$$

Proof We rewrite the fundamental identity (3.4.19) in the abbreviated form

$$R \star S = R - S = S \star R. \quad (3.6.15)$$

To abbreviate notation further, put

$$\tilde{R}(x,y) = (x-y)R(x,y), \quad \tilde{S}(x,y) = (x-y)S(x,y).$$

From (3.6.15) we deduce that

$$\tilde{R}\star S + R\star\tilde{S} = \tilde{R} - \tilde{S}.$$

Applying the operation $(\cdot)\star R$ on both sides, we get

$$\tilde{R}\star(R-S)+R\star\tilde{S}\star R=\tilde{R}\star R-\tilde{S}\star R\,.$$

Adding the last two relations and making the obvious cancellations and rearrangements, we get

$$\tilde{R}=(1+R)\star\tilde{S}\star(1+R).$$

Together with the trigonometric identity

$$\sin(x-y)=\sin x\cos y-\sin y\cos x$$

as well as the symmetry

$$S(x,y)=S(y,x),\quad R(x,y)=R(y,x)\,,$$

this yields (3.6.11). An application of L'Hôpital's rule then yields (3.6.12). Finally, by (3.6.10) and the definitions we obtain

$$\begin{aligned}\frac{\partial}{\partial t_i}Q(x) &= (s_{i-1}-s_i)R(x,t_i)\left(f(t_i)+\int R(t_i,y)f(y)d\nu(y)\right)\\ &= (s_{i-1}-s_i)R(x,t_i)Q(t_i)\,,\end{aligned}$$

yielding (3.6.13). Equation (3.6.14) is obtained similarly. □

Exercise 3.6.3 An alternative to the elementary calculus used in deriving (3.6.5) and (3.6.6), which is useful in obtaining higher order derivatives of the determinants, resolvents and adjugants, is sketched in this exercise.

(i) Let D be a domain (connected open subset) in $\mathbb{C}^n$. With X a measure space, let $f(x,\zeta)$ be a measurable function on $X\times D$, depending analytically on ζ for each fixed x and satisfying the condition

$$\sup_{\zeta\in K}\int|f(x,\zeta)|d\mu(x)<\infty$$

for all compact subsets $K\subset D$. Prove that the function

$$F(\zeta)=\int f(x,\zeta)d\mu(x)$$

is analytic in D and that for each index $i=1,\dots,n$ and all compact $K\subset D$,

$$\sup_{\zeta\in K}\int\left|\frac{\partial}{\partial\zeta_i}f(x,\zeta)\right|d\mu(x)<\infty.$$

Further, applying Cauchy's Theorem to turn the derivative into an integral, and then Fubini's Theorem, prove the identity of functions analytic in D:

$$\frac{\partial}{\partial\zeta_i}F(\zeta)=\int\left(\frac{\partial}{\partial\zeta_i}f(x,\zeta)\right)d\mu(x)\,.$$

(ii) Using the fact that the kernel S is an entire function, extend the definitions of H_ℓ, H and Δ in the setup of this section to analytic functions in the parameters $t_1,\ldots,t_n,s_1,\ldots,s_{n-1}$.

(iii) View the signed measure ν as defining a family of distributions η (in the sense of Schwartz) on the interval (a,b) depending on the parameters $t_1,\ldots,t_n$, by the formula

$$\langle \varphi, \eta \rangle = \sum_{i=1}^{n-1} s_i \int_{t_i}^{t_{i+1}} \varphi(x)dx\,,$$

valid for any smooth function $\varphi(x)$ on (a,b). Show that $\partial\eta/\partial t_i$ is a distribution satisfying

$$\frac{\partial}{\partial t_i}\eta = (s_{i-1} - s_i)\delta_{t_i} \tag{3.6.16}$$

for $i = 1,\ldots,n$, and that the distributional derivative $(d/dx)\eta$ of η satisfies

$$\frac{d}{dx}\eta = \sum_{i=1}^{n}(s_i - s_{i-1})\delta_{t_i} = -\sum_{i=1}^{n}\frac{\partial\eta}{\partial t_i}\,. \tag{3.6.17}$$

(iv) Use (3.6.16) to justify (3.6.5) and step (i) to justify (3.6.6).

3.6.2 Derivation of the differential equations: proof of Theorem 3.6.1

To proceed farther we need means for differentiating $Q(x)$ and $P(x)$ both with respect to x and with respect to the parameters $t_1,\ldots,t_n$. To this end we introduce the further abbreviated notation

$$S'(x,y) = \left(\frac{\partial}{\partial x} + \frac{\partial}{\partial y}\right)S(x,y) = 0,\quad R'(x,y) = \left(\frac{\partial}{\partial x} + \frac{\partial}{\partial y}\right)R(x,y)$$

and

$$(F\star' G)(x,y) = \int F(x,z)G(z,y)d\nu'(z) := \sum_{i=1}^{n}(s_i - s_{i-1})F(x,t_i)G(t_i,y)\,,$$

which can be taken as the definition of ν'. Below we persist for a while in writing S' instead of just automatically putting $S' = 0$ everywhere in order to keep the structure of the calculations clear. From the fundamental identity (3.4.19),

$$R\star S = R - S = S\star R\,,$$

we deduce, after integrating by parts, that

$$R'\star S + R\star' S + R\star S' = R' - S'\,.$$

Applying the operation $\star R$ on both sides of the last equation we find that

$$R' \star (R - S) + R \star' (R - S) + R \star S' \star R = R' \star R - S' \star R.$$

Adding the last two equations and then making the obvious cancellations (including now the cancellation $S' = 0$) we find that

$$R' = R \star' R.$$

Written out "in longhand" the last equation says that

$$\left(\frac{\partial}{\partial x} + \frac{\partial}{\partial y}\right) R(x,y) = \sum_{i=1}^{n} (s_i - s_{i-1}) R(x,t_i) R(t_i,y). \tag{3.6.18}$$

Now we can differentiate $Q(x)$ and $P(x)$. We have from the last identity

$$\begin{aligned} Q'(x) &= f'(x) + \int \frac{\partial}{\partial x} R(x,y) f(y) d\nu(y) \\ &= f'(x) - \int \frac{\partial}{\partial y} R(x,y) f(y) d\nu(y) \\ &\quad + \int \left(\int R(x,t) R(t,y) d\nu'(t) \right) f(y) d\nu(y). \end{aligned}$$

Integrating by parts and then rearranging the terms, we get

$$\begin{aligned} Q'(x) &= f'(x) + \int R(x,y) f'(y) d\nu(y) + \int R(x,y) f(y) d\nu'(y) \\ &\quad + \int \left(\int R(x,t) R(t,y) \eta(t) dt \right) f(y) d\nu(y) \\ &= f'(x) + \int R(x,y) f'(y) d\nu(y) \\ &\quad + \int R(x,t) \left(f(t) + \int R(t,y) f(y) d\nu(y) \right) d\nu'(t) \\ &= P(x) + \sum_{k=1}^{n} (s_k - s_{k-1}) R(x,t_k) Q(t_k), \end{aligned} \tag{3.6.19}$$

and similarly

$$P'(x) = -Q(x) + \sum_{k=1}^{n} (s_k - s_{k-1}) R(x,t_k) P(t_k). \tag{3.6.20}$$

Observing now that

$$\frac{\partial}{\partial t_i} Q(t_i) = Q'(t_i) + \left.\frac{\partial}{\partial t_i} Q(x)\right|_{x=t_i}, \quad \frac{\partial}{\partial t_i} P(t_i) = P'(t_i) + \left.\frac{\partial}{\partial t_i} P(x)\right|_{x=t_i},$$

and adding (3.6.19) and (3.6.13), we have

$$\frac{\partial}{\partial t_i}Q(t_i) = P(t_i) + \sum_{k=1,k\neq i}^{n} (s_k - s_{k-1})R(t_i,t_k)Q(t_k)\,. \tag{3.6.21}$$

Similarly, by adding (3.6.20) and (3.6.14) we have

$$\frac{\partial}{\partial t_i}P(t_i) = -Q(t_i) + \sum_{k=1,k\neq i}^{n} (s_k - s_{k-1})R(t_i,t_k)P(t_k). \tag{3.6.22}$$

It follows also via (3.6.12) and (3.6.13) that

$$R(t_i,t_i) = P(t_i)\frac{\partial}{\partial t_i}Q(t_i) - Q(t_i)\frac{\partial}{\partial t_i}P(t_i)\,. \tag{3.6.23}$$

(Note that the terms involving $\partial Q(x)/\partial t_i|_{x=t_i}$ cancel out to yield the above equality.) Unraveling the definitions, this completes the proof of (3.6.4) and hence of Theorem 3.6.1. □

3.6.3 Reduction to Painlevé V

In what follows, we complete the proof of Theorem 3.1.2. We take in Theorem 3.6.1 the values $n = 2, s_1 = s$. Our goal is to figure out the ordinary differential equation we get by reducing still farther to the case $t_1 = -t/2$ and $t_2 = t/2$. Recall the sine-kernel S in (3.6.1), set $\eta = \frac{d\nu}{dx} = s\mathbf{1}_{(-t/2,t/2)}$ and write $\Delta = \Delta(S)$ for the Fredholm determinant of S with respect to the measure ν. Finally, set $\sigma = \sigma(t) = t\frac{d}{dt}\log\Delta$. We now prove the following.

Lemma 3.6.4 *With notation as above,*

$$(t\sigma'')^2 + 4(t\sigma' - \sigma)(t\sigma' - \sigma + (\sigma')^2) = 0\,, \tag{3.6.24}$$

and, for each fixed s, Δ is analytic in $t \in \mathbb{C}$, with the following expansions as $t \to 0$:

$$\Delta = 1 - \left(\frac{s}{\pi}\right)t + O(t^4),\quad \sigma = -\left(\frac{s}{\pi}\right)t - \left(\frac{s}{\pi}\right)^2 t^2 - \left(\frac{s}{\pi}\right)^3 t^3 + O(t^4)\,. \tag{3.6.25}$$

Proof We first consider the notation of Theorem 3.6.1 specialized to $n = 2$, writing $\Delta(t_1,t_2)$ for the Fredholm determinant there. (Thus, $\Delta = \Delta(t_1,t_2)|_{t_1=-t_2=t/2}$.) Recall that

$$R_{21} = (q_2p_1 - q_1p_2)/(t_2 - t_1) = R_{12}\,.$$

From Theorem 3.6.1 specialized to $n = 2$ we have

$$\begin{aligned}
\frac{1}{2}(\partial/\partial t_2 - \partial/\partial t_1)\log\Delta(t_1,t_2) &= -\frac{1}{2}s(p_1^2+q_1^2+p_2^2+q_2^2)+s^2(t_2-t_1)R_{21}^2\,,\\
\frac{1}{2}(\partial q_1/\partial t_2 - \partial q_1/\partial t_1) &= -p_1/2+sR_{12}q_2\,,\\
\frac{1}{2}(\partial p_1/\partial t_2 - \partial p_1/\partial t_1) &= +q_1/2+sR_{12}p_2\,.
\end{aligned} \tag{3.6.26}$$

We now analyze symmetry. Temporarily, we write

$$p_1(t_1,t_2),\quad q_1(t_1,t_2),\quad p_2(t_1,t_2),\quad q_2(t_1,t_2),$$

in order to emphasize the roles of the parameters t_1 and t_2. To begin with, since

$$S(x+c,y+c) = S(x,y)\,,$$

for any constant c we have

$$\Delta(t_1,t_2) = \Delta(t_1+c,t_2+c) = \Delta(-t_2,-t_1)\,. \tag{3.6.27}$$

Further, we have (recall that $f(x) = (\sin x)/\sqrt{\pi}$)

$$\begin{aligned}
p_1(t_1,t_2) &= f'(t_1)+\frac{1}{\Delta(t_1,t_2)}\sum_{n=0}^{\infty}\frac{(-1)^n s^{n+1}}{n!}\\
&\quad\times\int_{t_1}^{t_2}\cdots\int_{t_1}^{t_2} S\begin{pmatrix} t_1 & x_1 & \dots & x_n\\ y & x_1 & \dots & x_n\end{pmatrix} f'(y)\,dx_1\cdots dx_n dy\\
&= f'(-t_1)+\frac{1}{\Delta(-t_2,-t_1)}\sum_{n=0}^{\infty}\frac{(-1)^n s^{n+1}}{n!}\\
&\quad\times\int_{t_1}^{t_2}\cdots\int_{t_1}^{t_2} S\begin{pmatrix} -t_1 & -x_1 & \dots & -x_n\\ -y & -x_1 & \dots & -x_n\end{pmatrix} f'(y)\,dx_1\cdots dx_n dy\\
&= f'(-t_1)+\frac{1}{\Delta(-t_2,-t_1)}\sum_{n=0}^{\infty}\frac{(-1)^n s^{n+1}}{n!}\\
&\quad\times\int_{-t_2}^{-t_1}\cdots\int_{-t_2}^{-t_1} S\begin{pmatrix} -t_1 & x_1 & \dots & x_n\\ y & x_1 & \dots & x_n\end{pmatrix} f'(y)\,dx_1\cdots dx_n dy\\
&= p_2(-t_2,-t_1)\,.
\end{aligned} \tag{3.6.28}$$

Similarly, we have

$$q_1(t_1,t_2) = -q_2(-t_2,-t_1)\,. \tag{3.6.29}$$

Now we are ready to reduce to the one-dimensional situation. We specialize as follows. Put

$$\begin{aligned} p &= p(t) = p_1(-t/2,t/2) = p_2(-t/2,t/2), \\ q &= q(t) = q_1(-t/2,t/2) = -q_2(-t/2,t/2), \\ r &= r(t) = R_{12}(-t/2,t/2) = -2pq/t, \\ \sigma &= \sigma(t) = t\frac{d}{dt}\log\Delta(-t/2,t/2). \end{aligned} \tag{3.6.30}$$

Note that, by the symmetry relations, writing $'$ for differentiation with respect to t, we have

$$\begin{aligned} p'(t) &= \frac{1}{2}(\partial p_1/\partial t_2 - \partial p_1/\partial t_1)|_{t_2=-t_1=t/2}, \\ q'(t) &= \frac{1}{2}(\partial q_1/\partial t_2 - \partial q_1/\partial t_1)|_{t_2=-t_1=t/2}, \end{aligned}$$

while

$$\sigma(t) = \frac{t}{2}(\partial/\partial t_2 - \partial/\partial t_1)\log\Delta(t_1,t_2)|_{t_2=-t_1=t/2}.$$

From (3.6.26) and the above we get

$$\begin{aligned} \sigma &= -st(p^2+q^2)+4s^2q^2p^2, \\ q' &= -p/2+2spq^2/t, \\ p' &= +q/2-2sp^2q/t, \end{aligned} \tag{3.6.31}$$

while differentiating σ (twice) and using these relations gives

$$\begin{aligned} \sigma' &= -s(p^2+q^2), \\ t\sigma'' &= 4s^2(p^3q-q^3p). \end{aligned} \tag{3.6.32}$$

Using (3.6.32) together with the equation for σ from (3.6.31) to eliminate the variables p,q, we obtain finally

$$4t(\sigma')^3+4t^2(\sigma')^2-4\sigma(\sigma')^2+4\sigma^2+(t\sigma'')^2-8t\sigma\sigma' = 0, \tag{3.6.33}$$

or equivalently, we get (3.6.24). Note that the differential equation is independent of s.

Turning to the proof of the claimed analyticity of Δ and of (3.6.25), we write

$$\begin{aligned} \Delta &= 1+\sum_{k=1}^{\infty}\frac{(-s)^k}{k!}\int_{-t/2}^{t/2}\cdots\int_{-t/2}^{t/2}\det_{i,j=1}^{k}\frac{\sin(x_i-x_j)}{\pi(x_i-x_j)}\prod_{j=1}^{k}dx_j \\ &= 1+\lim_{n\to\infty}\sum_{k=1}^{n}\frac{(-st)^k}{k!}\int_{-1/2}^{1/2}\cdots\int_{-1/2}^{1/2}\det_{i,j=1}^{k}\frac{\sin(tx_i-tx_j)}{\pi(tx_i-tx_j)}\prod_{j=1}^{k}dx_j. \end{aligned}$$

Each of the terms inside the limit in the last display is an entire function in t, and the convergence (in n) is uniform due to the boundedness of the kernel and the Hadamard inequality, see Lemma 3.4.2. The claimed analyticity of Δ in t follows.

We next explicitly compute a few terms of the expansion of Δ in powers of t. Indeed,

$$\int_{-t/2}^{t/2} dx = t, \quad \int_{-t/2}^{t/2} \cdots \int_{-t/2}^{t/2} \det_{i,j=1}^{k} \frac{\sin(x_i - x_j)}{\pi(x_i - x_j)} \prod_{j=1}^{k} dx_j = O(t^4) \text{ for } k \geq 2,$$

and hence the part of (3.6.25) dealing with Δ follows. With more computational effort, which we omit, one verifies the other part of (3.6.25). □

Proof of Theorem 3.1.2 We use Lemma 3.6.4. Take $s = 1$ and set

$$F(t) = 1 - \Delta = 1 - \exp\left(\int_0^t \frac{\sigma(u)}{u} du\right) \text{ for } t \geq 0.$$

Then by (3.1.1) we have

$$1 - F(t) = \lim_{N \to \infty} P[\sqrt{N}\lambda_1^N, \ldots, \sqrt{N}\lambda_N^N \notin (-t/2, t/2)],$$

completing the proof of the theorem. □

Remark 3.6.5 We emphasize that we have not yet proved that the function $F(\cdot)$ in Theorem 3.1.2 is a distribution function, that is, we have not shown tightness for the sequence of gaps around 0. From the expansion at 0 of $\sigma(t)$, see (3.1.2), it follows immediately that $\lim_{t \to 0} F(t) = 0$. To show that $F(t) \to 1$ as $t \to \infty$ requires more work. One approach, that uses careful and nontrivial analysis of the resolvent equation, see [Wid94] for the first rigorous proof, shows that in fact

$$\sigma(t) \sim -t^2/4 \text{ as } t \to +\infty, \tag{3.6.34}$$

implying that $\lim_{t \uparrow \infty} F(t) = 1$. An easier approach, which does not however yield such precise information, proceeds from the CLT for determinantal processes developed in Section 4.2; indeed, it is straightforward to verify, see Exercise 4.2.40, that for the determinantal process determined by the sine-kernel, the expected number of points in an interval of length L around 0 increases linearly in L, while the variance increases only logarithmically in N. This is enough to show that with $A = [-t/2, t/2]$, the right side of (3.1.1) decreases to 0 as $t \to \infty$, which implies that $\lim_{t \uparrow \infty} F(t) = 1$. In particular, it follows that the random variable giving the width of the largest open interval centered at the origin in which no eigenvalue of $\sqrt{N}X_N$ appears is weakly convergent as $N \to \infty$ to a random variable with distribution F.

We finally present an alternative formulation of Theorem 3.1.2 that is useful in comparing with the limit results for the GOE and GSE. Recall the function $r = r(t) = R_{12}(-t/2, t/2)$, see (3.6.30).

Lemma 3.6.6 *With $F(\cdot)$ as in Theorem 3.1.2, we have*

$$1 - F(t) = \exp\left(-\frac{t}{\pi} - \int_0^t (t-x)r(x)^2 dx\right), \tag{3.6.35}$$

and furthermore the differential equation

$$t^2((tr)'' + (tr))^2 = 4(tr)^2((tr)^2 + ((tr)')^2) \tag{3.6.36}$$

is satisfied with boundary conditions

$$r(t) = \frac{1}{\pi} + \frac{t}{\pi^2} + O_{t\downarrow 0}(t^2). \tag{3.6.37}$$

The function $r(t)$ has a convergent expansion in powers of t valid for small t.

Proof Recall p and q from (3.6.30). We have

$$-\frac{\sigma}{t} = p^2 + q^2 - \frac{4p^2q^2}{t}, \quad tr = -2pq, \quad p' = q/2 - 2p^2q/t, \quad q' = -p/2 + 2pq^2/t,$$

hence (3.6.36) holds and furthermore

$$\frac{d}{dt}\left(\frac{\sigma}{t}\right) = -r^2, \tag{3.6.38}$$

as one verifies by straightforward calculations. From the analyticity of Δ it follows that it is possible to extend both $r(t)$ and $\sigma(t)$ to analytic functions defined in a neighborhood of $[0,\infty)$ in the complex plane, and thus in particular both functions have convergent expansions in powers of t valid for small t. It is clear that

$$\lim_{t\downarrow 0} r(t) = \frac{1}{\pi}. \tag{3.6.39}$$

Thus (3.6.35) and (3.6.37) follow from (3.6.33), (3.6.38), (3.6.39) and (3.6.25). □

3.7 Edge-scaling: proof of Theorem 3.1.4

Our goal in this section is to study the spacing of eigenvalues at the edge of the spectrum. The main result is the proof of Theorem 3.1.4, which is completed in Subsection 3.7.1 (some technical estimates involving the steepest descent method are postponed to Subsection 3.7.2). For the proof of Theorem 3.1.4, we need the following *a priori* estimate on the Airy kernel. Its proof is postponed to Subsection 3.7.3, where additional properties of the Airy function are studied.

Lemma 3.7.1 *For any* $x_0 \in \mathbb{R}$,

$$\sup_{x,y \geq x_0} e^{x+y}|A(x,y)| < \infty. \tag{3.7.1}$$

3.7.1 *Vague convergence of the largest eigenvalue: proof of Theorem 3.1.4*

Again we let $X_N \in \mathscr{H}_N^{(2)}$ be a random Hermitian matrix from the GUE with eigenvalues $\lambda_1^N \leq \cdots \leq \lambda_N^N$. We now present the

Proof of Theorem 3.1.4 As before put

$$K^{(n)}(x,y) = \sqrt{n}\frac{\psi_n(x)\psi_{n-1}(y) - \psi_{n-1}(x)\psi_n(y)}{x-y},$$

where the $\psi_n(x)$ is the normalized oscillator wave-function. Define

$$A^{(n)}(x,y) \quad = \quad \frac{1}{n^{1/6}}K^{(n)}\left(2\sqrt{n}+\frac{x}{n^{1/6}}, 2\sqrt{n}+\frac{y}{n^{1/6}}\right). \tag{3.7.2}$$

In view of the basic estimate (3.4.9) in the theory of Fredholm determinants and the crude bound (3.7.1) for the Airy kernel we can by dominated convergence integrate to the limit on the right side of (3.1.5). By the bound (3.3.7) of Ledoux type, if the limit

$$\lim_{t' \to +\infty} \lim_{N \to \infty} P\left[N^{2/3}\left(\frac{\lambda_i^N}{\sqrt{N}} - 2\right) \notin (t,t') \text{ for } i = 1,\dots,N\right] \tag{3.7.3}$$

exists then the limit (3.1.6) also exists and both limits are equal. Therefore we can take the limit as $t' \to \infty$ on the left side of (3.1.5) inside the limit as $n \to \infty$ in order to conclude (3.1.6). We thus concentrate in the sequel on proving (3.1.5) for $t' < \infty$.

We begin by extending by analyticity the definition of $K^{(n)}$ and $A^{(n)}$ to the complex plane $\mathbb{C}$. Our goal will be to prove the convergence of $A^{(n)}$ to A on compact sets of $\mathbb{C}$, which will imply also the convergence of derivatives. Recall that by part 4 of Lemma 3.2.7,

$$K^{(n)}(x,y) = \frac{\psi_n(x)\psi_n'(y) - \psi_n(y)\psi_n'(x)}{x-y} - \frac{1}{2}\psi_n(x)\psi_n(y),$$

so that if we set

$$\Psi_n(x) := n^{1/12}\psi_n(2\sqrt{n}+\frac{x}{n^{1/6}}),$$

then

$$A^{(n)}(x,y) = \frac{\Psi_n(x)\Psi_n'(y) - \Psi_n(y)\Psi_n'(x)}{x-y} - \frac{1}{2n^{1/3}}\Psi_n(x)\Psi_n(y).$$

The following lemma plays the role of Lemma 3.5.1 in the study of the spacing in the bulk. Its proof is rather technical and takes up most of Subsection 3.7.2.

Lemma 3.7.2 *Fix a number $C > 1$. Then,*

$$\lim_{n\to\infty} \sup_{u\in\mathbb{C}:|u|<C} |\Psi_n(u) - \mathrm{Ai}(u)| = 0. \tag{3.7.4}$$

Since the functions Ψ_n are entire, the convergence in Lemma 3.7.2 entails the uniform convergence of Ψ_n' to Ai$'$ on compact subsets of $\mathbb{C}$. Together with Lemma 3.4.5, this completes the proof of the theorem. □

Remark 3.7.3 An analysis similar to, but more elaborate than, the proof of Theorem 3.1.4 shows that

$$\lim_{N\to\infty} P\left[N^{2/3}\left(\frac{\lambda^N_{N-\ell}}{\sqrt{N}} - 2\right) \le t\right]$$

exists for each positive integer ℓ and real number t. In other words, the suitably rescaled ℓth largest eigenvalue converges vaguely and in fact weakly. Similar statements can be made concerning the joint distribution of the rescaled top ℓ eigenvalues.

3.7.2 Steepest descent: proof of Lemma 3.7.2

In this subsection, we use the steepest descent method to prove Lemma 3.7.2. The steepest descent method is a general, more elaborate version of the method of Laplace discussed in Subsection 3.5.1, which is inadequate when oscillatory integrands are involved. Indeed, consider the evaluation of integrals of the form

$$\int f(x)^s g(x)dx,$$

see (3.5.3), in the situation where f and g are analytic functions and the integral is a contour integral. The oscillatory nature of f prevents the use of Laplace's method. Instead, the oscillatory integral is tamed by modifying the contour of integration in such a way that f can be written along the contour as $e^{\tilde f}$ with $\tilde f$ *real*, and the oscillations of g at a neighborhood of the critical points of $\tilde f$ are slow. In practice, one needs to consider slightly more general versions of this example, in which g itself may depend (weakly) on s.

Proof of Lemma 3.7.2 Throughout, we let

$$x = 2n^{1/2} + \frac{u}{n^{1/6}} = 2n^{1/2}\left(1 + \frac{u}{2n^{2/3}}\right), \quad \Psi_n(u) = n^{1/12}\psi_n(x).$$

We assume throughout the proof that n is large enough so that $|u| < C < n^{2/3}$.

Let ζ be a complex variable. By reinterpreting formula (3.5.7) above as a contour integral we get the formula

$$\psi_n(x) = \frac{e^{x^2/4}}{\mathrm{i}(2\pi)^{3/4}\sqrt{n!}} \int_{-\mathrm{i}\infty}^{\mathrm{i}\infty} \zeta^n e^{\zeta^2/2 - \zeta x} d\zeta. \tag{3.7.5}$$

The main effort in the proof is to modify the contour integral in the formula above in such a way that the leading asymptotic order of all terms in the integrand match, and then keep track of the behavior of the integrand near its critical point. To carry out this program, note that, by Cauchy's Theorem, we may replace the contour of integration in (3.7.5) by any straight line in the complex plane with slope of absolute value greater than 1 oriented so that height above the real axis is increasing (the condition on the slope is to ensure that no contribution appears from the contour near ∞). Since $\Re(x) > 0$ under our assumptions concerning u and n, we may take the contour of integration in (3.7.5) to be the perpendicular bisector of the line segment joining x to the origin, that is, replace ζ by $(x/2)(1+\zeta)$, to obtain

$$\psi_n(x) = \frac{e^{-x^2/8}(x/2)^{n+1}}{\mathrm{i}(2\pi)^{3/4}\sqrt{n!}} \int_{-\mathrm{i}\infty}^{\mathrm{i}\infty} (1+\zeta)^n e^{(x/2)^2(\zeta^2/2 - \zeta)} d\zeta. \tag{3.7.6}$$

Let $\log\zeta$ be the principal branch of the logarithm, that is, the branch real on the interval $(0,\infty)$ and analytic in the complement of the interval $(-\infty, 0]$, and set

$$F(\zeta) = \log(1+\zeta) + \zeta^2/2 - \zeta. \tag{3.7.7}$$

Note that the leading term in the integrand in (3.7.6) has the form $e^{nF(\zeta)}$, where $\Re(F)$ has a maximum along the contour of integration at $\zeta = 0$, and a Taylor expansion starting with $\zeta^3/3$ in a neighborhood of that point (this explains the particular scaling we took for u). Put

$$\omega = \left(\frac{x}{2}\right)^{2/3}, \quad u' = \omega^2 - n/\omega,$$

where to define fractional powers of complex numbers such as that figuring in the definition of ω we follow the rule that $\zeta^a = \exp(a\log\zeta)$ whenever ζ is in the domain of our chosen branch of the logarithm. We remark that as $n \to \infty$ we have $u' \to u$ and $\omega \sim n^{1/3}$, uniformly for $|u| < C$. Now rearrange (3.7.6) to the form

$$\Psi_n(u) = \frac{(2\pi)^{1/4} n^{1/12} (x/2)^{n+1/3} e^{-x^2/8}}{\sqrt{n!}} I_n(u), \tag{3.7.8}$$

where

$$I_n(u) = \frac{1}{2\pi \mathrm{i}} \int_{-\mathrm{i}\infty}^{\mathrm{i}\infty} \omega e^{\omega^3 F(\zeta) - u'\omega \log(1+\zeta)} d\zeta . \tag{3.7.9}$$

To prove (3.7.4) it is enough to prove that

$$\lim_{n\to\infty} \sup_{|u|<C} |I_n(u) - \mathrm{Ai}(u)| = 0, \tag{3.7.10}$$

because we have

$$\log \frac{n^{1/12}(x/2)^{n+1/3}e^{-x^2/8}}{e^{-n/2}n^{n/2+1/4}} = \left(n + \frac{1}{3}\right) \log\left(1 + \frac{u}{2n^{2/3}}\right) - \frac{n^{1/3}u}{2} - \frac{u^2}{8n^{1/3}}$$

and hence

$$\lim_{n\to\infty} \sup_{|u|<C} \left| \frac{(2\pi)^{1/4} n^{1/12} (x/2)^{n+1/3} e^{-x^2/8}}{\sqrt{n!}} - 1 \right| = 0,$$

by Stirling's approximation (2.5.12) and some calculus.

To prove (3.7.10), we proceed by a saddle point analysis near the critical point $\zeta = 0$ of $\Re(F)(\zeta)$. The goal is to replace complex integration with real integration. This is achieved by making a change of contour of integration so that F is real along that contour. Ideally, we seek a contour so that the maximum of F is achieved at a unique point along the contour. We proceed to find such a contour now, noting that since the maximum of $\Re(F)(\zeta)$ along the imaginary axis is 0 and is achieved at $\zeta = 0$, we may seek contours that pass through 0 and such that F is strictly negative at all other points of the contour.

Turning to the actual construction, consider the wedge-shaped closed set

$$S = \{re^{\mathrm{i}\theta} | r \in [0,\infty), \theta \in [\pi/3, \pi/2]\}$$

in the complex plane with "corner" at the origin. For each $\rho > 0$ let S_ρ be the intersection of S with the closed disc of radius ρ centered at the origin and let ∂S_ρ be the boundary of S_ρ. For each $t > 0$ and all sufficiently large ρ, the curve $F(\partial S_\rho)$ winds exactly once about the point $-t$. Since, by the argument principle of complex analysis, the winding number equals the difference between the number of zeros and the number of poles of the function $F(\cdot) + t$ in the domain S_ρ, and the function $F(\cdot) + t$ does not possess poles there, it follows that there exists a unique solution $\gamma(t) \in S$ of the equation $F(\zeta) = -t$ (see Figure 3.7.1). Clearly $\gamma(0) = 0$ is the unique solution of the equation $F(\zeta) = 0$ in S. We have the following.

Lemma 3.7.4 *The function $\gamma : [0,\infty) \to S$ has the following properties.*
(i) $\lim_{t\to\infty} |\gamma(t)| = \infty$.

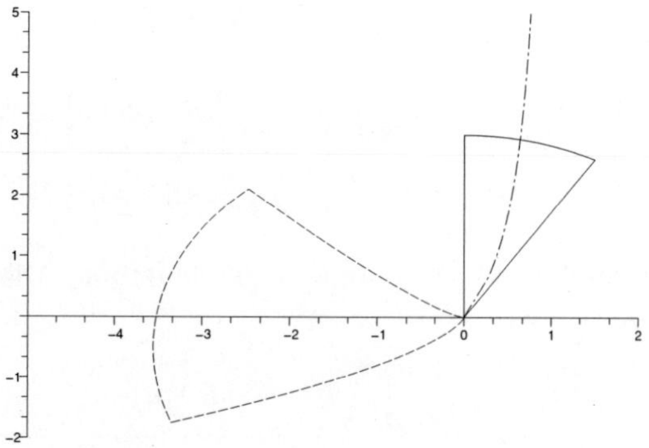

Fig. 3.7.1. The contour ∂S_3 (solid), its image $F(\partial S_3)$ (dashed), and the curve $\gamma(\cdot)$ (dash and dots).

(ii) $\gamma(t)$ *is continuous for* $t \geq 0$ *and real analytic for* $t > 0$.
(iii)

$$\begin{aligned}
\gamma(t) &= O(t^{1/2}) && \text{as } t \uparrow \infty, \\
\gamma'(t) &= O(t^{-1/2}) && \text{as } t \uparrow \infty, \\
\gamma(t) &= e^{\pi i/3} 3^{1/3} t^{1/3} + O(t^{4/3}) && \text{as } t \downarrow 0, \\
\gamma'(t) &= e^{\pi i/3} 3^{-2/3} t^{-2/3} + O(t^{1/3}) && \text{as } t \downarrow 0.
\end{aligned}$$

Proof (i) follows by noting that F restricted to S is proper, that is for any sequence $z_n \in S$ with $|z_n| \to \infty$ as $n \to \infty$, it holds that $|F(z_n)| \to \infty$. The real analyticity claim in (ii) follows from the implicit function theorem. (iii) follows from a direct computation, and together with $\gamma(0) = 0$ implies the continuity claim in (ii). □

From Lemma 3.7.4 we obtain the formula

$$I_n(u) = \frac{1}{2\pi i} \int_0^\infty \omega e^{-\omega^3 t} \left((1+\gamma(t))^{-\omega u'} \gamma'(t) - (1+\bar{\gamma}(t))^{-\omega u'} \bar{\gamma}'(t) \right) dt \,,$$

by deforming the contour $-i\infty \to i\infty$ in (3.7.9) to $\gamma - \bar{\gamma}$. After replacing t by $t^3/3n$ in the integral above we obtain the formula

$$I_n(u) = \frac{1}{2\pi i} \int_0^\infty (A_n(t,u) - B_n(t,u)) dt \,, \tag{3.7.11}$$

where

$$\begin{aligned}
A_n(t,u) &= \omega \exp\left(-\frac{\omega^3 t^3}{3n}\right) \left(1+\gamma\left(\frac{t^3}{3n}\right)\right)^{-\omega u'} \gamma'\left(\frac{t^3}{3n}\right) \frac{t^2}{n}, \\
B_n(t,u) &= \omega \exp\left(-\frac{\omega^3 t^3}{3n}\right) \left(1+\bar{\gamma}\left(\frac{t^3}{3n}\right)\right)^{-\omega u'} \bar{\gamma}'\left(\frac{t^3}{3n}\right) \frac{t^2}{n}.
\end{aligned}$$

Put

$$\begin{aligned} A(t,u) &= \exp\left(-\frac{t^3}{3} - e^{\pi \mathrm{i}/3} tu + \pi \mathrm{i}/3\right), \\ B(t,u) &= \exp\left(-\frac{t^3}{3} - e^{-\pi \mathrm{i}/3} tu - \pi \mathrm{i}/3\right). \end{aligned}$$

By modifying the contour of integration in the definition of the Airy function $\mathrm{Ai}(x)$, see (3.7.16), we have

$$\mathrm{Ai}(u) = \frac{1}{2\pi \mathrm{i}} \int_0^\infty (A(t,u) - B(t,u))dt. \tag{3.7.12}$$

A calculus exercise reveals that, for any positive constant c and each $t_0 \geq 0$,

$$\lim_{n\to\infty} \sup_{0\leq t\leq t_0} \sup_{|u|<c} \left| \frac{A_n(t,u)}{A(t,u)} - 1 \right| = 0 \tag{3.7.13}$$

and clearly the analogous limit formula linking $B_n(t,u)$ to $B(t,u)$ holds also. There exist positive constants c_1 and c_2 such that

$$|\log(1+\gamma(t))| \leq c_1 t^{1/3}, \quad |\gamma'(t)| \leq c_2 \max(t^{-2/3}, t^{-1/2})$$

for all $t > 0$. There exists a positive constant n_0 such that

$$\Re(\omega^3) \geq n/2, \quad |\omega| \leq 2n^{1/3}, \quad |u'| < 2c$$

for all $n \geq n_0$ and $|u| < c$. Also there exists a positive constant c_3 such that

$$e^{c_3 t^{1/3}} \geq t^{1/6}$$

for $t \geq 1$. Consequently there exist positive constants c_4 and c_5 such that

$$|\omega e^{\omega^3 t}(1+\gamma(t))^{-\omega u'} \gamma'(t)| \leq c_4 n^{1/3} e^{-nt/2 + c_5 n^{1/3} t} t^{-2/3},$$

hence

$$|A_n(t,u)| \leq c_4 e^{-t^3/6 + c_5 t} \tag{3.7.14}$$

for all $n \geq n_0$, $t > 0$ and $|u| < c$. Clearly we have the same majorization for $|B_n(t,u)|$. Integral formula (3.7.12), uniformity of convergence (3.7.13) and majorization (3.7.14) together are enough to finish the proof of limit formula (3.7.10) and hence of limit formula (3.7.4). □

Exercise 3.7.5 Set

$$S^{(n)}_{z_n}(x,y) = \frac{1}{\sqrt{n}} K^{(n)}(z_n + x/\sqrt{n}, z_n + y/\sqrt{n}).$$

Apply the steepest descent method to show that if $z_n/\sqrt{n} \to_{n\to\infty} c$ with $|c| < 2$, then $S^{(n)}_{z_n}(x,y)$ converges to the rescaled sine-kernel $\sin[g(c)(x-y)]/(\pi(x-y))$,

uniformly in x,y in compacts, where $g(c) = \pi\sigma(c) = \sqrt{4-c^2}/2$ and $\sigma(\cdot)$ is the semicircle density, see (2.1.3).
Hint: use (3.7.6) and note the different behavior of the function F at 0 when $c < 2$.

3.7.3 Properties of the Airy functions and proof of Lemma 3.7.1

Throughout this subsection, we will consider various contours in the complex plane. We introduce the following convenient notation: for complex numbers a,b, we let $[a,b]$ denote the contour joining a to b along the segment connecting them, i.e. the contour $(t \mapsto (1-t)a+tb) : [0,1] \to \mathbb{C}$. We also write $[a,c\infty)$ for the ray emanating from a in the direction c, that is the contour $(t \mapsto a+ct) : [0,\infty) \to \mathbb{C}$, and write $(c\infty,a] = -[a,c\infty)$. With this notation, and performing the change of variables $\zeta \mapsto -w$, we can rewrite (3.1.3) as

$$\mathrm{Ai}(x) = \frac{1}{2\pi \mathrm{i}} \int_{(e^{-2\pi \mathrm{i}/3}\infty,0]+[0,e^{2\pi \mathrm{i}/3}\infty)} e^{xw-w^3/3} dw. \qquad (3.7.15)$$

Note that the rapid decay of the integrand in (3.7.15) along the indicated contour ensures that $\mathrm{Ai}(x)$ is well defined and depends holomorphically on x. By parametrizing the contour appearing in (3.7.15) in evident fashion, we also obtain the formula

$$\mathrm{Ai}(x) = \frac{1}{2\pi \mathrm{i}} \int_0^\infty \exp\left(-\frac{t^3}{3}\right)\left(\exp\left(-xte^{\frac{\pi \mathrm{i}}{3}} + \frac{\pi \mathrm{i}}{3}\right) - \exp\left(-xte^{-\frac{\pi \mathrm{i}}{3}} - \frac{\pi \mathrm{i}}{3}\right)\right) dt. \qquad (3.7.16)$$

In the statement of the next lemma, we use the notation $x \uparrow \infty$ to mean that x goes to ∞ along the real axis. Recall also the definition of Euler's Gamma function, see (2.5.5): $\Gamma(s) = \int_0^\infty e^{-x}x^{s-1}dx$, for s with positive real part.

Lemma 3.7.6 (a) *For any integer $\nu \geq 0$, the derivative $\mathrm{Ai}^{(\nu)}(x)$ satisfies*

$$\mathrm{Ai}^{(\nu)}(x) \to 0, \quad \text{as } x \uparrow \infty. \qquad (3.7.17)$$

(b) *The function $\mathrm{Ai}(x)$ is a solution of* (3.1.4) *that satisfies*

$$\mathrm{Ai}(0) = \frac{1}{3^{2/3}\Gamma(2/3)}, \quad \mathrm{Ai}'(0) = -\frac{1}{3^{1/3}\Gamma(1/3)}. \qquad (3.7.18)$$

(c) $\mathrm{Ai}(x) > 0$ *and* $\mathrm{Ai}'(x) < 0$ *for all* $x > 0$.

Proof For $x \geq 0$ real, $c \in \mathbb{C}$ satisfying $c^3 = 1$ and $k \geq 0$ integer, define

$$I(x,c,k) = \int_{[0,c\infty)} w^k e^{wx-w^3/3} dw = c^{k+1}\int_0^\infty t^k e^{xct-t^3/3} dt. \qquad (3.7.19)$$

As $x \uparrow \infty$ we have $I(x, e^{\pm 2\pi \mathrm{i}/3}, k) \to 0$ by dominated convergence. This proves (3.7.17). Next, (3.7.18) follows from (3.7.19) and the definition of $\Gamma(\cdot)$. We next prove that $\mathrm{Ai}(x) > 0$ for $x > 0$. Assume otherwise that for some $x_0 > 0$ one has $\mathrm{Ai}(x_0) \le 0$. By (3.7.29), if $\mathrm{Ai}(x_0) = 0$ then $\mathrm{Ai}'(x_0) \neq 0$. Thus, for some $x_1 > 0$, $\mathrm{Ai}(x_1) < 0$. Since $\mathrm{Ai}(0) = 0$ and $\mathrm{Ai}(x) \to 0$ as $x \uparrow \infty$, $\mathrm{Ai}(\cdot)$ possesses a global minimum at some $x_2 \in (0, \infty)$, and $\mathrm{Ai}''(x_2) \ge 0$, contradicting the Airy differential equation. □

We next evaluate the asymptotics of the Airy functions at infinity. For two functions f, g, we write $f \sim g$ as $x \uparrow \infty$ if $\lim_{x \uparrow \infty} f(x)/g(x) = 1$.

Lemma 3.7.7 *For $x \uparrow \infty$ we have the following asymptotic formulas:*

$$\mathrm{Ai}(x) \sim \pi^{-1/2} x^{-1/4} e^{-\frac{2}{3}x^{3/2}}/2. \tag{3.7.20}$$

$$\mathrm{Ai}'(x) \sim -\pi^{-1/2} x^{1/4} e^{-\frac{2}{3}x^{3/2}}/2. \tag{3.7.21}$$

Proof Making the substitution $w \mapsto x^{1/2}(u-1)$ and deforming the contour of integration in (3.7.15), we obtain

$$2\pi \mathrm{i} x^{1/4} e^{2x^{2/3}/3} \mathrm{Ai}(x) = x^{3/4} \int_{C'} e^{x^{3/2}(u^2 - u^3/3)} du, \tag{3.7.22}$$

where

$$C' = (e^{-2\pi \mathrm{i}/3}\infty, -\mathrm{i}\sqrt{3}] + [-\mathrm{i}\sqrt{3}, \mathrm{i}\sqrt{3}] + [\mathrm{i}\sqrt{3}, e^{2\pi \mathrm{i}/3}\infty) =: C_1' + C_2' + C_3'.$$

Since the infimum of the real part of $u^2 - u^3/3$ on the rays C_1' and C_3' is strictly negative, the contribution of the integral over C_1' and C_3' to the right side of (3.7.22) vanishes as $x \uparrow \infty$. The remaining integral (over C_2') gives

$$\mathrm{i} \int_{-\sqrt{3}x^{3/4}}^{\sqrt{3}x^{3/4}} e^{-t^2 + \mathrm{i}t^3 x^{-3/4}/3} dt \to \mathrm{i} \int_{-\infty}^{\infty} e^{-t^2} dt = \mathrm{i}\sqrt{\pi} \quad \text{as } x \uparrow \infty,$$

by dominated convergence. This completes the proof of (3.7.20). A similar proof gives (3.7.21). Further details are omitted. □

Proof of Lemma 3.7.1 Fix $x_0 \in \mathbb{R}$. By (3.7.20), (3.7.21) and the Airy differential equation (3.1.4), there exists a positive constant C (possibly depending on x_0) such that

$$\max(|\mathrm{Ai}(x)|, |\mathrm{Ai}'(x)|, |\mathrm{Ai}''(x)|) \le Ce^{-x}$$

for all real $x \ge x_0$ and hence for $x, y \ge x_0$,

$$|x-y| \ge 1 \Rightarrow |A(x,y)| \le 2C^2 e^{-x-y}.$$

But by the variant (3.5.5) of Taylor's Theorem noted above we also have, for $x,y \geq x_0$,

$$|x-y|<1 \Rightarrow |A(x,y)| \leq 2C^2e^2e^{-x-y}.$$

Thus the lemma is proved. □

Exercise 3.7.8 Show that $\int_0^\infty \mathrm{Ai}(x)dx = 1/3$.
Hint: for $\rho > 0$, let γ_ρ denote the path $(t \mapsto \rho e^{2\pi i t}) : [5/6, 7/6] \to \mathbb{C}$, and define the contour $C_\rho = (e^{2\pi i/3}\infty, \rho e^{2\pi i/3}] + \gamma_\rho + [\rho e^{-2\pi i/3}, e^{-2\pi i/3}\infty)$. Show that

$$\int_0^\infty \mathrm{Ai}(x)dx = \frac{1}{2\pi i}\int_{C_\rho} w^{-1}e^{-w^3/3}dw,$$

and take $\rho \to 0$ to conclude.

Exercise 3.7.9 Write $x \downarrow -\infty$ if $x \to -\infty$ along the real axis. Prove the asymptotics

$$\mathrm{Ai}(x) \sim \frac{\sin(\frac{2}{3}|x|^{3/2} + \frac{\pi}{4})}{\sqrt{\pi}|x|^{1/4}} \quad \text{as } x \downarrow -\infty \tag{3.7.23}$$

and

$$\mathrm{Ai}'(x) \sim -\frac{\cos(\frac{2}{3}|x|^{3/2} + \frac{\pi}{4})|x|^{1/4}}{\sqrt{\pi}} \quad \text{as } x \downarrow -\infty. \tag{3.7.24}$$

Conclude that Lemma 3.7.1 can be strengthened to the statement

$$\sup_{x,y\in\mathbb{R}} e^{x+y}|A(x,y)| < \infty. \tag{3.7.25}$$

Exercise 3.7.10 The proof of Lemma 3.7.7 as well as the asymptotics in Exercise 3.7.17 are based on finding an appropriate explicit contour of integration. An alternative to this approach utilizes the steepest descent method. Provide the details of the proof of (3.7.20), using the following steps.
(a) Replacing ζ by $x^{1/2}\zeta$ in (3.1.3), deduce the integral representation, for $x > 0$,

$$\mathrm{Ai}(x) = \frac{x^{1/2}}{2\pi i}\int_C e^{x^{3/2}H(\zeta)}d\zeta, \quad H(\zeta) = \zeta^3/3 - \zeta. \tag{3.7.26}$$

(b) Modify the contour C to another (implicitly defined) contour C', so that $\Im(H(C'))$ is constant, and the deformed contour C' "snags" the critical point $\zeta = 1$ of H, so that the image $H(C')$ runs on the real axis from $-\infty$ to $-2/3$ and back.
Hint: Consider the closed sets

$$S' = \{1 + re^{i\theta} | r \geq 0,\ \theta \in [\pi/3, \pi/2]\}$$

and the intersection of S' with the closed disc of radius ρ about 1, and apply a reasoning similar to the proof of Lemma 3.7.2 to find a curve $\gamma(t)$ such that

$$\mathrm{Ai}(x) = \frac{e^{-2x^{3/2}/3}x^{1/2}}{2\pi\mathrm{i}} \int_0^\infty e^{-x^{3/2}t}(\gamma'(t) - \bar{\gamma}'(t))dt \text{ for } x > 0. \tag{3.7.27}$$

Identify the asymptotics of $\gamma(t)$ and its derivative as $t \to 0$ and $t \to \infty$.
(c) Apply Laplace's method, Lemma D.9, to obtain (3.7.20).

Exercise 3.7.11 Another solution of (3.1.4), denoted $\mathrm{Bi}(x)$, is obtained by replacing the contour in (3.7.15) with the contour $(e^{-2\pi\mathrm{i}/3}\infty, 0] + [0,\infty) + (e^{2\pi\mathrm{i}/3}\infty, 0] + [0,\infty)$, that is

$$\mathrm{Bi}(x) = \frac{1}{2\pi} \int_{(e^{-2\pi\mathrm{i}/3}\infty,0]+2[0,\infty)+(e^{2\pi\mathrm{i}/3}\infty,0]} e^{xw-w^3/3}dw. \tag{3.7.28}$$

Show that $\mathrm{Bi}(x)$ satisfies (3.1.4) with the boundary conditions $[\mathrm{Bi}(0)\ \mathrm{Bi}'(0)] = \left[\frac{1}{3^{1/6}\Gamma(2/3)}\ \ \frac{3^{1/6}}{\Gamma(1/3)}\right]$. Show that for any $x \in \mathbb{R}$,

$$\det \begin{bmatrix} \mathrm{Ai}(x) & \mathrm{Ai}'(x) \\ \mathrm{Bi}(x) & \mathrm{Bi}'(x) \end{bmatrix} = \frac{1}{\pi}, \tag{3.7.29}$$

concluding that Ai and Bi are linearly independent solutions. Show also that $\mathrm{Bi}(x) > 0$ and $\mathrm{Bi}'(x) > 0$ for all $x > 0$. Finally, repeat the analysis in Lemma 3.7.7, using the substitution $w \mapsto x^{1/2}(u+1)$ and the (undeformed!) contour

$$C = (-e^{-2\pi\mathrm{i}/3}\infty, -1] + [-1,1] + [1,\infty) + e^{-2\pi\mathrm{i}/3}\infty, -1] + [-1,1] + [1,\infty),$$

and conclude that

$$\mathrm{Bi}(x) \sim \pi^{-1/2}x^{-1/4}e^{\frac{2}{3}x^{3/2}}, \tag{3.7.30}$$

$$\mathrm{Bi}'(x) \sim -\pi^{-1/2}x^{1/4}e^{\frac{2}{3}x^{3/2}}. \tag{3.7.31}$$

3.8 Analysis of the Tracy–Widom distribution and proof of Theorem 3.1.5

We will study the Fredholm determinant

$$\Delta = \Delta(t) := 1 + \sum_{k=1}^{\infty} \frac{(-1)^k}{k!} \int_t^\infty \cdots \int_t^\infty A\begin{pmatrix} x_1 & \dots & x_k \\ x_1 & \dots & x_k \end{pmatrix} \prod_{j=1}^k dx_j$$

where $A(x,y)$ is the Airy kernel and as before we write

$$A\begin{pmatrix} x_1 & \dots & x_k \\ y_1 & \dots & y_k \end{pmatrix} = \det_{i,j=1}^{k} A(x_i, y_j).$$

We are going to explain why $\Delta(t)$ is a distribution function, which, together with Theorem 3.1.4, will complete our proof of weak convergence of $n^{2/3}\left(\frac{\lambda_n^{(n)}}{\sqrt{n}}-2\right)$. Further, we are going to link $\Delta(t)$ to the Painlevé II differential equation.

We begin by putting the study of the Tracy–Widom distribution $\Delta(t)$ into a framework compatible with the general theory of Fredholm determinants developed in Section 3.4. Let ν denote the measure on the real line with density $d\nu/dx = \mathbf{1}_{(t,\infty)}(x)$ with respect to the Lebesgue measure (although ν depends on t, we suppress this dependence from the notation). We have then

$$\Delta = 1 + \sum_{k=1}^{\infty} \frac{(-1)^k}{k!} \int \cdots \int A\begin{pmatrix} x_1 & \dots & x_k \\ x_1 & \dots & x_k \end{pmatrix} \prod_{j=1}^{k} d\nu(x_j).$$

Put

$$H(x,y) = A(x,y) + \sum_{k=1}^{\infty} \frac{(-1)^k}{k!} \int \cdots \int A\begin{pmatrix} x & x_1 & \dots & x_k \\ y & x_1 & \dots & x_k \end{pmatrix} \prod_{j=1}^{k} d\nu(x_j).$$

In view of the basic estimate (3.4.9) and the crude bound (3.7.1) for the Airy kernel, we must have $\Delta(t) \to 1$ as $t \uparrow \infty$. Similarly, we have

$$\sup_{t \geq t_0} \sup_{x,y \in \mathbb{R}} e^{x+y} |H(x,y)| < \infty \tag{3.8.1}$$

for each real t_0 and

$$\lim_{t \uparrow \infty} \sup_{x,y \in \mathbb{R}} e^{x+y} |H(x,y) - A(x,y)| = 0. \tag{3.8.2}$$

Note that because Δ can be extended to a not-identically-vanishing entire analytic function of t, it follows that Δ vanishes only for isolated real values of t. Put

$$R(x,y) = H(x,y)/\Delta,$$

provided of course that $\Delta \neq 0$; a similar proviso applies to each of the following definitions since each involves $R(x,y)$. Put

$$\begin{aligned}
Q(x) &= \operatorname{Ai}(x) + \int R(x,y) \operatorname{Ai}(y) d\nu(y), \\
P(x) &= \operatorname{Ai}'(x) + \int R(x,y) \operatorname{Ai}'(y) d\nu(y), \\
q &= Q(t), \quad p = P(t), \; u = \int Q(x) \operatorname{Ai}(x) d\nu(x), \\
v &= \int Q(x) \operatorname{Ai}'(x) d\nu(x) = \int P(x) \operatorname{Ai}(x) d\nu(x),
\end{aligned} \tag{3.8.3}$$

the last equality by symmetry $R(x,y) = R(y,x)$. Convergence of all these integrals is easy to check. Note that each of the quantities q, p, u and v tends to 0 as $t \uparrow \infty$.

More precise information is also available. For example, from (3.8.1) and (3.8.2) it follows that

$$q(x)/\operatorname{Ai}(x) \to_{x\to\infty} 1\,, \tag{3.8.4}$$

because for x large, (3.7.20) implies that for some constant C independent of x,

$$\int_x^\infty R(x,y)\operatorname{Ai}(y)dy \le C\int_x^\infty e^{-x-y}\operatorname{Ai}(y)dy \le C\operatorname{Ai}(x)e^{-2x}\,.$$

3.8.1 *The first standard moves of the game*

We follow the trail blazed in the discussion of the sine-kernel in Section 3.6. The first few steps we can get through quickly by analogy. We have

$$\frac{\partial}{\partial t}\log\Delta = R(t,t)\,, \tag{3.8.5}$$

$$\frac{\partial}{\partial t}R(x,y) = -R(x,t)R(t,y)\,. \tag{3.8.6}$$

As before we have a relation

$$R(x,y) = \frac{Q(x)P(y)-Q(y)P(x)}{x-y} = R(y,x) \tag{3.8.7}$$

and hence by L'Hôpital's rule we have

$$R(x,x) = Q'(x)P(x)-Q(x)P'(x)\,. \tag{3.8.8}$$

We have the differentiation formulas

$$\frac{\partial}{\partial t}Q(x) = -R(x,t)Q(t) = -Q(t)R(t,x)\,, \tag{3.8.9}$$

$$\frac{\partial}{\partial t}P(x) = -R(x,t)P(t) = -P(t)R(t,x)\,. \tag{3.8.10}$$

Here the Airy function and its derivative are playing the roles previously played by sine and cosine, but otherwise to this point our calculation is running just as before. Actually the calculation to this point is simpler since we are focusing on a single interval of integration rather than on several.

3.8.2 *The wrinkle in the carpet*

As before we introduce the abbreviated notation

$$A'(x,y) = \left(\frac{\partial}{\partial x}+\frac{\partial}{\partial y}\right)A(x,y),\quad R'(x,y) = \left(\frac{\partial}{\partial x}+\frac{\partial}{\partial y}\right)R(x,y)\,,$$

$$(F \star' G)(x,y) = \int F(x,z)G(z,y)dv'(z) = F(x,t)G(t,y).$$

Here's the wrinkle in the carpet that changes the game in a critical way: A' does not vanish identically. Instead we have

$$A'(x,y) = -\operatorname{Ai}(x)\operatorname{Ai}(y), \tag{3.8.11}$$

which is an immediate consequence of the Airy differential equation $y'' - xy = 0$. Calculating as before but this time *not* putting A' to zero we find that

$$R' = R \star' R + A' + R \star A' + A' \star R + R \star A' \star R.$$

Written out "in longhand" the last equation says that

$$\left(\frac{\partial}{\partial x} + \frac{\partial}{\partial y}\right) R(x,y) = R(x,t)R(t,y) - Q(x)Q(y). \tag{3.8.12}$$

The wrinkle "propagates" to produce the extra term on the right. We now have

$$\begin{aligned}
Q'(x) &= \operatorname{Ai}(x) + \int \left(\frac{\partial}{\partial x} R(x,y)\right) \operatorname{Ai}(y)dv(y) \\
&= \operatorname{Ai}'(x) - \int \left(\frac{\partial}{\partial y} R(x,y)\right) \operatorname{Ai}(y)dv(y) \\
&\quad + R(x,t)\int R(t,y)\operatorname{Ai}(y)dv(y) - Q(x)u \\
&= \operatorname{Ai}(x) + \int R(x,y)\operatorname{Ai}'(y)dv(y) + \int R(x,y)\operatorname{Ai}(y)dv'(y) \\
&\quad + R(x,t)\int R(t,y)\operatorname{Ai}(y)dv(y) - Q(x)u \\
&= \operatorname{Ai}(x) + \int R(x,y)\operatorname{Ai}'(y)dv(y) \\
&\quad + R(x,t)(\operatorname{Ai}(t) + \int R(t,y)\operatorname{Ai}(y)dv(y)) - Q(x)u \\
&= P(x) + R(x,t)Q(t) - Q(x)u.
\end{aligned} \tag{3.8.13}$$

Similar manipulations yield

$$P'(x) = xQ(x) + R(x,t)P(t) + P(x)u - 2Q(x)v. \tag{3.8.14}$$

This is more or less in analogy with the sine-kernel case. But the wrinkle continues to propagate, producing the extra terms involving the quantities u and v.

3.8.3 Linkage to Painlevé II

The derivatives of the quantities p, q, u and v with respect to t we denote simply by a prime. We calculate these derivatives as follows. Observe that

$$q' = \frac{\partial}{\partial t} Q(x)\bigg|_{x=t} + Q'(t), \quad p' = \frac{\partial}{\partial t} P(x)\bigg|_{x=t} + P'(t).$$

By adding (3.8.9) to (3.8.13) and (3.8.10) to (3.8.14) we have

$$q' = p - qu, \quad p' = tq + pu - 2qv. \tag{3.8.15}$$

It follows also via (3.8.8) that

$$\frac{\partial}{\partial t} \log \Delta(t) = R(t,t) = q'p - p'q = p^2 - tq^2 - 2pqu + 2q^2 v. \tag{3.8.16}$$

We have

$$\begin{aligned} u' &= \int \left(\frac{\partial}{\partial t} Q(x)\right) \mathrm{Ai}(x) d\nu(x) + \int Q(x) \,\mathrm{Ai}(x) d\left(\frac{\partial \nu}{\partial t}\right)(x) \\ &= -Q(t) \int R(t,x) \,\mathrm{Ai}(x) d\nu(x) - Q(t) \,\mathrm{Ai}(t) = -q^2. \end{aligned}$$

$$\begin{aligned} v' &= \int \left(\frac{\partial}{\partial t} Q(x)\right) \mathrm{Ai}'(x) d\nu(x) + \int Q(x) \,\mathrm{Ai}'(x) d\left(\frac{\partial \nu}{\partial t}\right)(x) \\ &= -Q(t) \int R(t,x) \,\mathrm{Ai}'(x) d\nu(x) - Q(t) \,\mathrm{Ai}'(t) = -pq. \end{aligned}$$

We have a first integral

$$u^2 - 2v = q^2;$$

at least it is clear that the t-derivative here vanishes, but then the constant of integration has to be 0 because all the functions here tend to 0 as $t \uparrow \infty$. Finally,

$$\begin{aligned} q'' &= (p - qu)' = p' - q'u - qu' = tq + pu - 2qv - (p - qu)u - q(-q^2) \\ &= tq + pu - 2qv - pu + qu^2 + q^3 = tq + 2q^3, \end{aligned} \tag{3.8.17}$$

which is Painlevé II; that $q(t) \sim \mathrm{Ai}(t)$ as $t \to \infty$ was already proved in (3.8.4).

It remains to prove that the function F_2 defined in (3.1.6) is a distribution function. By adding equations (3.8.12) and (3.8.6) we get

$$\left(\frac{\partial}{\partial x} + \frac{\partial}{\partial y} + \frac{\partial}{\partial t}\right) R(x,y) = -Q(x)Q(y). \tag{3.8.18}$$

By evaluating both sides at $x = t = y$ and also using (3.8.5) we get

$$\frac{\partial^2}{\partial t^2} \log \Delta = -q^2. \tag{3.8.19}$$

Let us now write $q(t)$ and $\Delta(t)$ to emphasize the t-dependence. In view of the rapid decay of $\Delta(t)-1$, $(\log\Delta(t))'$ and $q(t)$ as $t\uparrow\infty$ we must have

$$\Delta(t)=\exp\left(-\int_t^\infty (x-t)q(x)^2dx\right), \tag{3.8.20}$$

whence the conclusion that $F_2(t)=\Delta(t)$ satisfies $F_2(\infty)=1$ and, because of the factor $(x-t)$ in (3.8.20) and the fact that $q(\cdot)$ does not identically vanish, also $F_2(-\infty)=0$. In other words, F_2 is a distribution function. Together with (3.8.17) and Theorem 3.1.4, this completes the proof of Theorem 3.1.5. □

Remark 3.8.1 The Painlevé II equation $q''=tq+2q^3$ has been studied extensively. The following facts, taken from [HaM80], are particularly relevant: any solution of Painlevé II that satisfies $q(t)\to_{t\to\infty}0$ satisfies also that as $t\to\infty$, $q(t)\sim\alpha\,\mathrm{Ai}(t)$ for some $\alpha\in\mathbb{R}$, and for each fixed α, such a solution exists and is unique. For $\alpha=1$, which is the case of interest to us, see (3.1.8), one then gets

$$q(t)\sim\sqrt{-t/2},\quad t\to-\infty. \tag{3.8.21}$$

We defer additional comments to the bibliographical notes.

Remark 3.8.2 The analysis in this section would have proceeded verbatim if the Airy kernel $A(x,y)$ were replaced by $sA(x,y)$ for any $s\in(0,1)$, the only difference being that the boundary condition for (3.1.8) would be replaced by $q(t)\sim s\,\mathrm{Ai}(t)$ as $t\to\infty$. On the other hand, by Corollary 4.2.23 below, the kernel $sA^{(n)}(x,y)$ replaces $A^{(n)}(x,y)$ if one erases each eigenvalue of the GUE with probability s. In particular, one concludes that for any k fixed,

$$\lim_{t\to\infty}\limsup_{N\to\infty}P(N^{1/6}(\lambda^N_{N-k}-2\sqrt{N})\le t)=0. \tag{3.8.22}$$

This observation will be useful in the proof of Theorem 3.1.7.

Exercise 3.8.3 Using (3.7.20), (3.8.4) and (3.8.21), deduce from the representation (3.1.7) of F_2 that

$$\begin{aligned}\lim_{t\to\infty}\frac{1}{t^{3/2}}\log[1-F_2(t)] &= -\frac{4}{3},\\ \lim_{t\to-\infty}\frac{1}{t^3}\log F_2(t) &= -\frac{1}{12},\end{aligned}$$

Note the different decay rate of the upper and lower tails of the distribution of the (rescaled) largest eigenvalue.

3.9 Limiting behavior of the GOE and the GSE

We prove Theorems 3.1.6 and 3.1.7 in this section, using the tools developed in Sections 3.4, 3.6 and 3.7, along with some new tools, namely, Pfaffians and matrix kernels. The multiplicativity of Fredholm determinants, see Theorem 3.4.10, also plays a key role.

3.9.1 Pfaffians and gap probabilities

We begin our analysis of the limiting behavior of the GOE and GSE by proving a series of integration identities involving Pfaffians; the latter are needed to handle the novel algebraic situations created by the factors $|\Delta(x)|^\beta$ with $\beta \in \{1,4\}$ appearing in the joint distribution of eigenvalues in the GOE and GSE, respectively. Then, with Remark 3.4.4 in mind, we use the Pfaffian integration identities to obtain determinant formulas for squared gap probabilities in the GOE and GSE.

Pfaffian integration formulas

Recall that $\mathrm{Mat}_{k\times\ell}(\mathbb{C})$ denotes the space of k-by-ℓ matrices with complex entries, with $\mathrm{Mat}_n(\mathbb{C}) = \mathrm{Mat}_{n\times n}(\mathbb{C})$ and $I_n \in \mathrm{Mat}_n(\mathbb{C})$ denoting the identity matrix. Let

$$\mathbf{J}_n = \begin{bmatrix} 0 & 1 & & & \\ -1 & 0 & & & \\ & & \ddots & & \\ & & & 0 & 1 \\ & & & -1 & 0 \end{bmatrix} \in \mathrm{Mat}_{2n}(\mathbb{C})$$

be the block-diagonal matrix consisting of n copies of $\begin{bmatrix} 0 & 1 \\ -1 & 0 \end{bmatrix}$ strung along the diagonal. Given a family of matrices

$$\{X(i,j) \in \mathrm{Mat}_{k\times\ell}(\mathbb{C}) : i = 1,\dots,m \,\text{and}\, j = 1,\dots,n\},$$

let

$$X(i,j)|_{m,n} = \begin{bmatrix} X(1,1) & \dots & X(1,n) \\ \vdots & & \vdots \\ X(m,1) & \dots & X(m,n) \end{bmatrix} \in \mathrm{Mat}_{km\times\ell n}(\mathbb{C}).$$

For example, $\mathbf{J}_n = \delta_{i,j} \begin{bmatrix} 0 & 1 \\ -1 & 0 \end{bmatrix} |_{n,n} \in \mathrm{Mat}_{2n}(\mathbb{C})$.

Next, recall a basic definition.

Definition 3.9.1 (Pfaffians) Let $X \in \mathrm{Mat}_{2n}(\mathbb{C})$ be antisymmetric, that is, $X^{\mathrm{T}} = -X, X_{j,i} = -X_{i,j}$. The *Pfaffian* of X is defined by the formula

$$\mathrm{Pf}X = \frac{1}{2^n n!} \sum_{\sigma \in S_{2n}} (-1)^\sigma \prod_{i=1}^n X_{\sigma(2i-1),\sigma(2i)},$$

where $(-1)^\sigma$ denotes the sign of the permutation σ.

For example, $\mathrm{Pf}\,\mathbf{J}_n = 1$, which explains the normalization $\frac{1}{2^n n!}$.

We collect without proof some standard facts related to Pfaffians.

Theorem 3.9.2 *Let $X \in \mathrm{Mat}_{2n}(\mathbb{C})$ be antisymmetric. The following hold:*
(i) $\mathrm{Pf}(Y^{\mathrm{T}}XY) = (\mathrm{Pf}X)(\det Y)$ for every $Y \in \mathrm{Mat}_{2n}(\mathbb{C})$;
(ii) $(\mathrm{Pf}X)^2 = \det X$;
(iii) $\mathrm{Pf}X = \sum_{i=1}^{2n-1}(-1)^{i+1}X_{i,2n}\,\mathrm{Pf}X^{\{i,2n\}}$, where $X^{\{i,2n\}}$ is the submatrix obtained by striking out the ith row, ith column, $(2n)$th row and $(2n)$th column.

We next give a general integration identity involving Pfaffians, which is the analog for $\beta \in \{1,4\}$ of Lemma 3.2.3.

Proposition 3.9.3 *Let $f_1,\dots,f_{2n}$ and $g_1,\dots,g_{2n}$ be $\mathbb{C}$-valued measurable functions on the real line. Assume that all products $f_i g_j$ are integrable. For $x \in \mathbb{R}$, put*

$$F(x) = [f_i(x)\ g_i(x)]|_{2n,1} \in \mathrm{Mat}_{2n\times 2}(\mathbb{C}).$$

Then, for all measurable sets $A \subset \mathbb{R}$,

$$\mathrm{Pf}\int_A F(x)\mathbf{J}_1 F(x)^{\mathrm{T}} dx = \frac{1}{n!}\int_A \cdots \int_A \det[F(x_j)]|_{1,n} \prod_{i=1}^n dx_i\,. \tag{3.9.1}$$

Here and throughout the discussion of Pfaffian integration identities, measurable means Lebesgue measurable.

Proof Expand the right side of (3.9.1) as

$$\frac{1}{2^n n!}\sum_{\sigma\in S_{2n}}(-1)^\sigma \int_A \cdots \int_A \prod_{i=1}^n \det\left[\begin{array}{cc} f_{\sigma(2i-1)}(x_i) & g_{\sigma(2i-1)}(x_i) \\ f_{\sigma(2i)}(x_i) & g_{\sigma(2i)}(x_i) \end{array}\right] \prod_{i=1}^n dx_i\,. \tag{3.9.2}$$

The (i,j) entry of the matrix appearing on the left side of (3.9.1) can be expressed as $\int_A \det\left[\begin{array}{cc} f_i(x) & g_i(x) \\ f_j(x) & g_j(x) \end{array}\right] dx$. Therefore, by Fubini's Theorem, the expansion (3.9.2) matches term for term the analogous expansion of the left side of (3.9.1) according to the definition of the Pfaffian. □

To evaluate gap probabilities in the GOE and GSE, we will specialize Proposition 3.9.3 in several different ways, varying both F and n. To begin the evaluation, let φ denote a function on the real line of the form $\varphi(x) = e^{C_1x^2+C_2x+C_3}$, where $C_1 < 0$, C_2 and C_3 are real constants, and let $\mathscr{O}_n$ denote the span over $\mathbb{C}$ of the set of functions $\{x^{i-1}\varphi(x)\}_{i=0}^{n-1}$. Later we will make use of specially chosen bases for $\mathscr{O}_n$ consisting of suitably modified oscillator wave-functions, but initially these are not needed. Recall that $\Delta(x) = \prod_{1\le i<j\le n}(x_j - x_i)$ for $x = (x_1,\dots,x_n) \in \mathbb{R}^n$.

The application of (3.9.1) to the GSE is the following.

Proposition 3.9.4 *Let $\{f_i\}_{i=1}^{2n}$ be any family of elements of $\mathscr{O}_{2n}$. For $x \in \mathbb{R}$, put*

$$F(x) = [\ f_i'(x) \quad f_i(x)\ ]|_{2n,1} \in \mathrm{Mat}_{2n\times 2}(\mathbb{C})\,.$$

Then, for all measurable sets $A \subset \mathbb{R}$,

$$\mathrm{Pf}\int_A F(x)\mathbf{J}_1F(x)^{\mathrm{T}}dx = c\int_A\cdots\int_A \Delta(x)^4\prod_{i=1}^n \varphi(x_i)^2dx_i\,, \tag{3.9.3}$$

where $c = c(\{f_i\})$ is a complex number depending only on the family $\{f_i\}$, not on A. Further, $c \neq 0$ if and only if $\{f_i\}_{i=1}^{2n}$ is a basis for $\mathscr{O}_{2n}$ over $\mathbb{C}$.

Proof By Theorem 3.9.2(i), we may assume without loss of generality that $f_i(x) = x^{i-1}\varphi(x)$, and it suffices to show that (3.9.3) holds with $c \neq 0$. By identity (3.9.1) and the confluent alternant identity (2.5.30), identity (3.9.3) does indeed hold for suitable nonzero c independent of A. □

The corresponding result for the GOE uses indefinite integrals of functions. To streamline the handling of the latter, we introduce the following notation, which is used throughout Section 3.9. For each integrable real-valued function f on the real line we define a continuous function εf by the formula

$$\begin{aligned}(\varepsilon f)(x) &= \int \frac{1}{2}\mathrm{sign}\,(x-y)f(y)dy = -\int_x^\infty f(y)dy + \frac{1}{2}\int f(y)dy \\ &= \int_0^x f(y)dy - \frac{1}{2}\int \mathrm{sign}(y)f(y)dx,\end{aligned} \tag{3.9.4}$$

where $\mathrm{sign}(x) = \mathbf{1}_{x>0} - \mathbf{1}_{x<0}$, and we write $\int f(x)dx = \int_{-\infty}^{\infty} f(x)dx$ to abbreviate notation. Note that $(\varepsilon f)'(x) = f(x)$ almost everywhere, that is, ε inverts differentiation. Note also that the operation ε reverses parity and commutes with translation.

The application of (3.9.1) to the GOE is the following.

Proposition 3.9.5 *Let $\{f_i\}_{i=1}^{n}$ be any family of elements of $\mathscr{O}_n$. Let $a \neq 0$ be a* *complex constant. For each measurable set $A \subset \mathbb{R}$ and $x \in \mathbb{R}$, put*

$$F_A^{\mathrm{e}}(x) = \left[\ f_i(x) \quad \varepsilon(\mathbf{1}_A f_i)(x)\ \right]|_{n,1} \in \mathrm{Mat}_{n\times 2}(\mathbb{C})\,.$$

If n is even, let $F_A(x) = F_A^{\mathrm{e}}(x) \in \mathrm{Mat}_{n'\times 2}(\mathbb{C})$. Otherwise, if n is odd, let $F_A(x) \in \mathrm{Mat}_{n'\times 2}(\mathbb{C})$ be the result of adjoining the row $[0\ \ a]$ at the bottom of $F_A^{\mathrm{e}}(x)$. Then, for all measurable sets $A \subset \mathbb{R}$,

$$\mathrm{Pf}\int_A F_A(x)\mathbf{J}_1 F_A(x)^{\mathrm{T}}dx = c\int_A \cdots \int_A |\Delta(x)|\prod_{i=1}^n \varphi(x_i)dx_i\,, \tag{3.9.5}$$

where $c = c(\{f_i\},a)$ is a complex number depending only on the data $(\{f_i\},a)$, not on A. Further, $c \neq 0$ if and only if $\{f_i\}_{i=1}^n$ is a basis for $\mathscr{O}_n$ over $\mathbb{C}$.

Proof By Theorem 3.9.2(i), we may assume without loss of generality that $f_i(x) = x^{i-1}\varphi(x)$, and it suffices to show that (3.9.5) holds with $c \neq 0$ independent of A. For $x \in \mathbb{R}$, let $f(x) = [f_i(x)]|_{n,1} \in \mathrm{Mat}_{n\times 1}(\mathbb{C})$. Let A_+^n be the subset of $A^n \subset \mathbb{R}^n$ consisting of n-tuples in strictly increasing order. Then, using the symmetry of the integrand of (3.9.5) and the Vandermonde determinant identity, one can verify that the integral $\int_{A_+^n} \det[f(y_j)]|_{1,n}\prod_1^n dy_i$ equals the right side of (3.9.5) with $c = 1/n!$. Put $r = \lfloor n/2 \rfloor$. Consider, for $z \in \mathbb{R}^r$, the $n \times n$ matrix

$$\Psi_A(z) = \begin{cases} \left[\ [\varepsilon(\mathbf{1}_A f_i)|_{-\infty}^{z_1}]|_{n,1} \quad [\ f_i(z_j) \quad \varepsilon(\mathbf{1}_A f_i)|_{z_j}^{z_{j+1}}\ ]|_{n,r}\ \right] & \text{if } n \text{ is odd,}\\ \qquad\qquad [\ f_i(z_j) \quad \varepsilon(\mathbf{1}_A f_i)|_{z_j}^{z_{j+1}}\ ]|_{n,r} & \text{if } n \text{ is even,}\end{cases}$$

where $z_{r+1} = \infty$, and $h|_s^t = h(t) - h(s)$. By integrating every other variable, we obtain a relation

$$\int_{A_+^r} \det \Psi_A(z)\prod_1^r dz_i = \int_{A_+^n} \det[f(y_j)]|_{1,n}\prod_1^n dy_i\,.$$

Consider, for $z \in \mathbb{R}^r$, the $n \times n$ matrix

$$\Phi_A(z) = \begin{cases} [[F_A(z_j)]|_{1,r}\ \ a\int_A f(x)dx] & \text{if } n \text{ is odd,}\\ [F_A(z_j)]|_{1,r} & \text{if } n \text{ is even.}\end{cases}$$

Because $\Phi_A(z)$ arises from $\Psi_A(z)$ by evident column operations, we deduce that $\det \Phi_A(z) = c_1 \det \Psi_A(z)$ for some nonzero complex constant c_1 independent of A and z. Since the function $\det \Phi_A(z)$ of $z \in \mathbb{R}^r$ is symmetric, we have

$$\int_{A_+^r} \det \Phi_A(z)\prod_1^r dz_i = \frac{1}{r!}\int_{A^r} \det \Phi_A(z)\prod_1^r dz_i\,.$$

If n is even, we conclude the proof by using the Pfaffian integration identity (3.9.1) to verify that the right side above equals the left side of (3.9.5).

Assume for the rest of the proof that n is odd. For $i = 1,\ldots,n$, let $F_A^{\mathrm{e},i}(x)$ be the result of striking out the ith row from $F_A^{\mathrm{e}}(x)$ and similarly, let $\Phi_A^i(z)$ be the result of striking the ith row and last column from $\Phi_A(z)$. Then we have expansions

$$\begin{aligned}
\mathrm{Pf}\left[\begin{array}{cc} \int_A F_A^{\mathrm{e}}(x)\mathbf{J}_1 F_A^{\mathrm{e}}(x)^{\mathrm{T}}dx & a\int_A f(x)dx \\ -a\int_A f(x)^{\mathrm{T}}dx & 0 \end{array}\right] & \\
&= a\sum_{i=1}^n (-1)^{i+1}\left(\int_A f_i(x)dx\right)\left(\mathrm{Pf}\int_A F_A^{\mathrm{e},i}(x)\mathbf{J}_1 F_A^{\mathrm{e},i}(x)^{\mathrm{T}}dx\right), \\
\det\Phi_A(z) &= a\sum_{i=1}^n (-1)^{i+n}\left(\int_A f_i(x)dx\right)\det\Phi_A^i(z),
\end{aligned}$$

obtained in the first case by Theorem 3.9.2(iii), and in the second by expanding the determinant by minors of the last column. Finally, by applying (3.9.1) term by term to the latter expansion, and comparing the resulting terms with those of the former expansion, one verifies that $\frac{1}{r!}\int_{A^r}\det\Phi_A(z)\prod_1^r dz_i$ equals the left side of (3.9.5). This concludes the proof in the remaining case of odd n. □

The next lemma gives further information about the structure of the antisymmetric matrix $\int_A F_A(x)\mathbf{J}_1 F_A(x)^{\mathrm{T}}dx$ appearing in Proposition 3.9.5. Let $\eta_n = \sqrt{2}I_n$ for even n, and $\eta_n = \left[\begin{array}{cc} \sqrt{2}I_n & 0 \\ 0 & 1/\sqrt{2}\end{array}\right]$ for odd n.

Lemma 3.9.6 *In the setup of Proposition 3.9.5, for all measurable sets $A\subset\mathbb{R}$,*

$$\int_A F_A(x)\mathbf{J}_1 F_A(x)^{\mathrm{T}}dx = \int F_{\mathbb{R}}(x)\mathbf{J}_1 F_{\mathbb{R}}(x)^{\mathrm{T}}dx - \int_{A^c}\eta_n F_{\mathbb{R}}(x)\mathbf{J}_1 F_A(x)^{\mathrm{T}}\eta_n dx. \quad (3.9.6)$$

Proof Let $L_{i,j}$ (resp., $R_{i,j}$) denote the (i,j) entry of the matrix on the left (resp., right). To abbreviate notation we write $\langle f,g\rangle = \int f(x)g(x)dx$. For $i,j<n+1$, using antisymmetry of the kernel $\frac{1}{2}\mathrm{sign}(x-y)$, we have

$$\begin{aligned}
\frac{1}{2}L_{i,j} &= \frac{1}{2}(\langle \mathbf{1}_A f_i, \varepsilon(\mathbf{1}_A f_j)\rangle - \langle \mathbf{1}_A f_j, \varepsilon(\mathbf{1}_A f_i)\rangle) = \langle \mathbf{1}_A f_i, \varepsilon(\mathbf{1}_A f_j)\rangle \\
&= \langle f_i, \varepsilon f_j\rangle - \langle \mathbf{1}_{A^c} f_i, \varepsilon f_j\rangle - \langle \mathbf{1}_A f_i, \varepsilon(\mathbf{1}_{A^c} f_j)\rangle \\
&= \langle f_i, \varepsilon f_j\rangle - \langle \mathbf{1}_{A^c} f_i, \varepsilon f_j\rangle + \langle \varepsilon(\mathbf{1}_A f_i), \mathbf{1}_{A^c} f_j\rangle = \frac{1}{2}R_{i,j},
\end{aligned}$$

which concludes the proof in the case of even n. In the case of odd n it remains only to consider the cases $\max(i,j)=n+1$. If $i=j=n+1$, then $L_{i,j}=0=R_{i,j}$. If $i<j=n+1$, then $L_{i,j}=a\langle \mathbf{1}_A, f_i\rangle = R_{i,j}$. If $j<i=n+1$, then $L_{i,j}=-a\langle \mathbf{1}_A, f_j\rangle = R_{i,j}$. The proof is complete. □

Determinant formulas for squared gap probabilities

By making careful choices for the functions f_i in Propositions 3.9.4 and 3.9.5, and applying Theorems 3.9.2(ii) and 2.5.2, we are going to obtain determinant formulas for squared gap probabilities. Toward that end, for fixed $\sigma > 0$ and real ξ, let

$$\phi_n(x) = \phi_{n,\sigma,\xi}(x) = \sigma^{1/2}\psi_n(\sigma^{-1}x+\xi), \tag{3.9.7}$$

and $\phi_{-1} \equiv 0$ for convenience. The functions ϕ_n are shifted and scaled versions of the oscillator wave-functions, see Definition 3.2.1.

We are ready to state the main results for gap probabilities in the GSE and GOE. These should be compared with Lemma 3.2.4 and Remark 3.4.4. The result for the GSE is as follows.

Proposition 3.9.7 *For $x \in \mathbb{R}$, put*

$$H(x) = \frac{1}{\sigma\sqrt{2}}\left[\begin{array}{cc} \phi_{2i-1}(x) & \phi'_{2i-1}(x) \\ \varepsilon\phi_{2i-1}(x) & \phi_{2i-1}(x) \end{array}\right]\Big|_{1,n} \in \mathrm{Mat}_{2\times 2n}(\mathbb{C}) \tag{3.9.8}$$

and $\widetilde{H}(x) = \mathbf{J}_1 H(x)\mathbf{J}_n^{-1}$. Then, for all measurable sets $A \subset \mathbb{R}$,

$$\det\left(I_{2n} - \int_A \widetilde{H}(x)^{\mathrm{T}} H(x)dx\right) = \left(\frac{\int_{A^c}\cdots\int_{A^c}\Delta(x)^4\prod_{i=1}^n\phi_0(x_i)^2dx_i}{\int\cdots\int\Delta(x)^4\prod_{i=1}^n\phi_0(x_i)^2dx_i}\right)^2. \tag{3.9.9}$$

To prove the proposition we will interpret H as the transpose of a matrix of the form F appearing in Proposition 3.9.4, which is possible because ε inverts differentiation.

The result for the GOE is as follows.

Proposition 3.9.8 *Let $r = \lfloor n/2 \rfloor$. Let $n' = n$ if n is even, and otherwise, if n is odd, let $n' = n+1$. Let $\ell \in \{1,2\}$ have the same parity as n. For $x \in \mathbb{R}$, and measurable sets $A \subset \mathbb{R}$, put*

$$G_A^{\mathrm{e}}(x) = \frac{1}{\sigma}\left[\begin{array}{cc} \phi_{2i-\ell}(x) & \phi'_{2i-\ell}(x) \\ \varepsilon(\mathbf{1}_A\phi_{2i-\ell})(x) & \varepsilon(\mathbf{1}_A\phi'_{2i-\ell})(x) \end{array}\right]\Big|_{1,r} \in \mathrm{Mat}_{2\times 2r}(\mathbb{C}).$$

If n is even, put $G_A(x) = G_A^{\mathrm{e}}(x) \in \mathrm{Mat}_{2\times n'}(\mathbb{C})$. Otherwise, if n is odd, let $G_A(x) \in \mathrm{Mat}_{2\times n'}(\mathbb{C})$ be obtained from $G_A^{\mathrm{e}}(x)$ by adjoining the block

$$\left[\begin{array}{cc} \phi_{n-1}(x) & 0 \\ \varepsilon(\mathbf{1}_A\phi_{n-1})(x) & 1/\langle\phi_{n-1},1\rangle \end{array}\right]$$

on the far right. Also put $\widetilde{G}_A(x) = \mathbf{J}_1 G_A(x)\mathbf{J}_{n'/2}^{-1}$. Then, for all measurable sets $A \subset \mathbb{R}$,

$$\det\left(I_{n'} - \int_A \widetilde{G}_{\mathbb{R}}(x)^{\mathrm{T}} G_{A^c}(x)dx\right) = \left(\frac{\int_{A^c}\cdots\int_{A^c}|\Delta(x)|\prod_{i=1}^n\phi_0(x_i)dx_i}{\int\cdots\int|\Delta(x)|\prod_{i=1}^n\phi_0(x_i)dx_i}\right)^2. \tag{3.9.10}$$

To prove the proposition we will interpret G_A as a matrix of the form $F_A^{\mathrm{T}}\eta_n$ appearing on the right side of (3.9.6) in Lemma 3.9.6.

Before commencing the proofs we record a series of elementary properties of the functions ϕ_i following immediately from Lemmas 3.2.5 and 3.2.7. These properties will be useful throughout Section 3.9. As above, we write $\langle f,g\rangle = \int f(x)g(x)dx$. Let $k,\ell,n \geq 0$ be integers. Let $\mathscr{O}_n = \mathscr{O}_{n,\sigma,\xi}$ denote the span of the set $\{\phi_i\}_{i=0}^{n-1}$ over $\mathbb{C}$.

Lemma 3.9.9 *The following hold:*

$$\begin{aligned}
\phi_0(x) &= \sigma^{1/2}(2\pi)^{-1/4}e^{-\frac{(\sigma^{-1}x+\xi)^2}{4}}, & (3.9.11)\\
\sup_x e^{\gamma|x|}|\phi_n(x)| &< \infty \textit{ for every real constant } \gamma, & (3.9.12)\\
\phi_n &= \varepsilon(\phi_n') = (\varepsilon\phi_n)', & (3.9.13)\\
\langle\phi_k,\phi_\ell\rangle &= \sigma^2\delta_{k\ell} = -\langle\varepsilon\phi_k,\phi_\ell'\rangle, & (3.9.14)\\
\langle\phi_k,\varepsilon\phi_\ell\rangle &= 0 \textit{ and } \langle\phi_k,\phi_\ell'\rangle = 0 \textit{ for } k+\ell \textit{ even}, & (3.9.15)\\
\langle\phi_n,1\rangle &= 0 \textit{ for } n \textit{ odd}, & (3.9.16)\\
\sigma\phi_n' &= -\frac{\sqrt{n+1}}{2}\phi_{n+1} + \frac{\sqrt{n}}{2}\phi_{n-1}, & (3.9.17)\\
\langle\phi_n,1\rangle &> 0 \textit{ for } n \textit{ even}, & (3.9.18)\\
\varepsilon\phi_n &\in \mathscr{O}_{n-1} \textit{ for } n \textit{ odd}, & (3.9.19)\\
(\sigma^{-1}x+\xi)\phi_n(x) &= \sqrt{n+1}\phi_{n+1}(x) + \sqrt{n}\phi_{n-1}(x), & (3.9.20)\\
\sum_{i=0}^{n-1}\frac{\phi_i(x)\phi_i(y)}{\sigma^2} &= \frac{\phi_n(x)\phi_n'(y)-\phi_n'(x)\phi_n(y)}{x-y} - \frac{\phi_n(x)\phi_n(y)}{2\sigma^2}, & (3.9.21)\\
\sigma^2\phi_n''(x) &= \left(\frac{(\sigma^{-1}x+\xi)^2}{4} - n - \frac{1}{2}\right)\phi_n(x). & (3.9.22)
\end{aligned}$$

Proof of Proposition 3.9.7 Using property (3.9.19), and recalling that ε inverts differentiation, we observe that, with $\varphi = \phi_0$ and $F(x) = H(x)^{\mathrm{T}}$, the integration identity (3.9.3) holds with a constant c independent of A. Further, we have $\int \widetilde{H}(x)^{\mathrm{T}}H(x)dx = I_{2n}$ by (3.9.14) and (3.9.15), and hence

$$\det\left(I_n - \int_A \widetilde{H}(x)^{\mathrm{T}}H(x)dx\right) = \left(\mathrm{Pf}\int_{A^c} F(x)\mathbf{J}_1F(x)^{\mathrm{T}}dx\right)^2,$$

after some algebraic manipulations using part (ii) of Theorem 3.9.2 and the fact that $\det\mathbf{J}_n = 1$. Thus, by (3.9.3) with A replaced by A^c, the integration identity (3.9.9) holds up to a constant factor independent of A. Finally, since (3.9.9) obviously holds for $A = \emptyset$, it holds for all A. $\square$

Proof of Proposition 3.9.8 Taking η_n as in Lemma 3.9.6, $\varphi = \phi_0$ and $F_A(x) = \eta_n^{-1} G_A(x)^{\mathrm{T}}$, the integration identity (3.9.5) holds with a constant c independent of A. Further, we have $\mathbf{I}_n = \int \mathbf{J}_{n'/2}^{-1} F_{\mathbb{R}}(x) \mathbf{J}_1 F_{\mathbb{R}}(x)^{\mathrm{T}} dx$ by (3.9.14), (3.9.15) and (3.9.16), and hence

$$\det\left(I_{n'} - \int_A \tilde{G}(x)^{\mathrm{T}} G_{A^c}(x) dx\right) = \left(\mathrm{Pf}\int_{A^c} F_{A^c}(x) \mathbf{J}_1 F_{A^c}(x)^{\mathrm{T}} dx\right)^2$$

by Lemma 3.9.6 with A replaced by A^c, after some algebraic manipulations using part (ii) of Theorem 3.9.2 and the fact that $\det \mathbf{J}_n = 1$. Thus, by (3.9.5) with A replaced by A^c, the integration identity (3.9.10) holds up to a constant factor independent of A. Finally, since (3.9.10) obviously holds for $A = \emptyset$, it holds for all A. □

3.9.2 *Fredholm representation of gap probabilities*

In this section, by reinterpreting formulas (3.9.9) and (3.9.10), we represent the square of a gap probability for the GOE or GSE as a Fredholm determinant of a matrix kernel, see Theorem 3.9.19.

Matrix kernels and a revision of the Fredholm setup

We make some specialized definitions to adapt Fredholm determinants as defined in Section 3.4 to the study of limits in the GOE and GSE.

Definition 3.9.10 For $k \in \{1,2\}$, let Ker_k denote the space of Borel-measurable functions $K : \mathbb{R} \times \mathbb{R} \to \mathrm{Mat}_k(\mathbb{C})$. We call elements of Ker_1 *scalar kernels*, elements of Ker_2 *matrix kernels*, and elements of $\mathrm{Ker}_1 \cup \mathrm{Ker}_2$ simply *kernels*. We often view a matrix kernel $K \in \mathrm{Ker}_2$ as a 2×2 matrix with entries $K_{i,j} \in \mathrm{Ker}_1$.

We are now using the term "kernel" in a sense somewhat differing from that in Section 3.4. On the one hand, usage is more general because boundedness is not assumed any more. On the other hand, usage is more specialized in that kernels are always functions defined on $\mathbb{R} \times \mathbb{R}$.

Definition 3.9.11 Given $K, L \in \mathrm{Ker}_k$, we define $K \star L$ by the formula

$$(K \star L)(x,y) = \int K(x,t) L(t,y) dt\,,$$

whenever $\int |K_{i,\ell}(x,t) L_{\ell,j}(t,y)| dt < \infty$ for all $x, y \in \mathbb{R}$ and $i, j, \ell \in \{1, \dots, k\}$.

Since the definition of Fredholm determinant made in Section 3.4 applies only to bounded kernels on measure spaces of finite total mass, to use it efficiently we have to make the next several definitions.

Given a real constant $\gamma \geq 0$, let $\mathbf{w}_\gamma(x) = \exp(\gamma|x+\gamma| - \gamma^2)$ for $x \in \mathbb{R}$. Note that $\mathbf{w}_\gamma(x) = e^{\gamma x}$ for $x > -\gamma$ and $\mathbf{w}_0(x) \equiv 1$.

Definition 3.9.12 (γ-twisting) Given $k \in \{1,2\}$, a kernel $K \in \mathrm{Ker}_k$, and a constant $\gamma \geq 0$, we define the γ-*twisted* kernel $K^{(\gamma)} \in \mathrm{Ker}_k$ by

$$K^{(\gamma)}(x,y) = \begin{cases} K(x,y)\mathbf{w}_\gamma(y) & \text{if } k=1, \\ \\ \begin{bmatrix} \mathbf{w}_\gamma(x)K_{11}(x,y) & \mathbf{w}_\gamma(x)K_{12}(x,y)\mathbf{w}_\gamma(y) \\ K_{21}(x,y) & K_{22}(x,y)\mathbf{w}_\gamma(y) \end{bmatrix} & \text{if } k=2\,. \end{cases}$$

We remark that $K \in \mathrm{Ker}_2^\gamma \Rightarrow K_{11}^{\mathrm{T}}, K_{22} \in \mathrm{Ker}_1^\gamma$ where $K_{11}^{\mathrm{T}}(x,y) = K_{11}(y,x)$.

As before, let Leb denote Lebesgue measure on the real line. For $\gamma \geq 0$, let $\mathrm{Leb}_\gamma(dx) = \mathbf{w}_\gamma(x)^{-1}\mathrm{Leb}(dx)$, noting that $\mathrm{Leb}_0 = \mathrm{Leb}$, and that Leb_γ has finite total mass for $\gamma > 0$.

Definition 3.9.13 Given $k \in \{1,2\}$, a kernel $K \in \mathrm{Ker}_k$, and a constant $\gamma \geq 0$, we write $K \in \mathrm{Ker}_k^\gamma$ if there exists some open set $U \subset \mathbb{R}$ and constant $c > 0$ such that $\mathrm{Leb}_\gamma(U) < \infty$ and $\max_{i,j} |(K^{(\gamma)})_{i,j}| \leq c\mathbf{1}_{U\times U}$.

Note that Ker_k^γ is closed under the operation $\star$ because, for $K, L \in \mathrm{Ker}_k^\gamma$, we have

$$(K \star L)^{(\gamma)}(x,y) = \int K^{(\gamma)}(x,t)L^{(\gamma)}(t,y)\mathrm{Leb}_\gamma(dt) \tag{3.9.23}$$

and hence $K \star L \in \mathrm{Ker}_k^\gamma$.

We turn next to the formulation of a version of the definition of Fredholm determinant suited to kernels of the class Ker_k^γ.

Definition 3.9.14 Given $k \in \{1,2\}$, $\gamma \geq 0$, and $L \in \mathrm{Ker}_k^\gamma$, we define $\mathrm{Fred}_k^\gamma(L)$ by specializing the setup of Section 3.4 as follows.

(i) Choose $U \subset \mathbb{R}$ open and $c > 0$ such that $\max_{i,j} |(L^{(\gamma)})_{i,j}| \leq c\mathbf{1}_{U\times U}$.
(ii) Let $X = U \times \mathscr{I}$, where $\mathscr{I} = \{1\}, \{1,2\}$ according as $k = 1,2$.
(iii) Let $\nu =$ (restriction of Leb_γ to U) $\otimes$ (counting measure on $\mathscr{I}$).
(iv) Let $K((s,i),(t,j)) = L^{(\gamma)}(s,t)_{i,j}$ for $(s,i),(t,j) \in X$.

Finally, we let $\mathrm{Fred}_k^\gamma(L) = \Delta(K)$, where the latter is given as in Definition 3.4.3, with inputs X, ν and K as defined above.

The complex number $\mathrm{Fred}_k^\gamma(L)$ is independent of the choice of U and c made in point (i) of the definition, and hence well defined. The definition is contrived so that if $L \in \mathrm{Ker}_k^{\gamma_i}$ for $i = 1,2$, then $\mathrm{Fred}_k^{\gamma_i}(L)$ is independent of i, as one verifies by comparing the expansions of these Fredholm determinants term by term.

Two formal properties of $\mathrm{Fred}_k^\gamma(\cdot)$ deserve emphasis.

Remark 3.9.15 If $K, L \in \mathrm{Ker}_k^\gamma$, then multiplicativity holds in the form

$$\mathrm{Fred}_k^\gamma(K + L - K \star L) = \mathrm{Fred}_k^\gamma(K)\mathrm{Fred}_k^\gamma(L),$$

by (3.9.23) and Theorem 3.4.10. Further, by Corollary 3.4.9, if $K \in \mathrm{Ker}_2^\gamma$ satisfies $K_{21} \equiv 0$ or $K_{12} \equiv 0$, then

$$\mathrm{Fred}_2^\gamma(K) = \mathrm{Fred}_1^\gamma(K_{11}^{\mathrm{T}})\mathrm{Fred}_1^\gamma(K_{22}).$$

The analog of Remark 3.4.4 in the present situation is the following.

Remark 3.9.16 Let $\gamma \geq 0$ be a constant. Let $U \subset \mathbb{R}$ be an open set such that $\mathrm{Leb}_\gamma(U) < \infty$. Let $G, \tilde{G} : \mathbb{R} \to \mathrm{Mat}_{2\times 2n}(\mathbb{C})$ be Borel-measurable. Assume further that all entries of the matrices

$$\begin{bmatrix} \mathbf{w}_\gamma(x) & 0 \\ 0 & 1 \end{bmatrix} G(x), \quad \begin{bmatrix} 1 & 0 \\ 0 & \mathbf{w}_\gamma(x) \end{bmatrix} \widetilde{G}(x)$$

are bounded for $x \in U$. Let

$$K(x,y) = G(x)\widetilde{G}(y)^{\mathrm{T}} \in \mathrm{Mat}_2(\mathbb{C})$$

for $x, y \in \mathbb{R}$. Let $A \subset U$ be a Borel set. Then $\mathbf{1}_{A\times A}K \in \mathrm{Ker}_2^\gamma$ and

$$\mathrm{Fred}_2^\gamma(\mathbf{1}_{A\times A}K) = \det\left(I_{2n} - \int_A \widetilde{G}(x)^{\mathrm{T}} G(x)dx\right).$$

If $K \in \mathrm{Ker}_k^\gamma$ and $\mathrm{Fred}_k^\gamma(K) \neq 0$, then one can adapt the Fredholm adjugant construction, see equation (3.4.15), to the present situation, and one can verify that there exists unique $R \in \mathrm{Ker}_k^\gamma$ such that the resolvent equation $R - K = K \star R = R \star K$ holds.

Definition 3.9.17 The kernel $R \in \mathrm{Ker}_k^\gamma$ associated as above with $K \in \mathrm{Ker}_k^\gamma$ is called the *resolvent* of K with respect to γ, and we write $R = \mathrm{Res}_k^\gamma(K)$.

This definition is contrived so that if $K \in \mathrm{Ker}_k^{\gamma_i}$ for $i = 1,2$, then $\mathrm{Res}_k^{\gamma_i}(K)$ is independent of i. In fact, we will need to use this definition only for $k = 1$, and the only resolvents that we will need are those we have already used to analyze GUE in the bulk and at the edge of the spectrum.

Finally, we introduce terminology pertaining to useful additional structure a kernel may possess.

Definition 3.9.18 We say that $K \in \mathrm{Ker}_k$ for $k \in \{1,2\}$ is *smooth* if K is infinitely differentiable. We say that $L \in \mathrm{Ker}_1$ is *symmetric* (resp., *antisymmetric*) if $L(x,y) \equiv L(y,x)$ (resp., $L(x,y) \equiv -L(y,x)$). We say that $K \in \mathrm{Ker}_2$ is *self-dual* if K_{21} and K_{12} are antisymmetric and $K_{11}(x,y) \equiv K_{21}(x,y)$. Given smooth $L \in \mathrm{Ker}_1$ and $K \in \mathrm{Ker}_2$, we say that K is the *differential extension* of L if

$$K(x,y) \equiv \left[\begin{array}{cc} \frac{\partial L}{\partial x}(x,y) & -\frac{\partial^2 L}{\partial x \partial y}(x,y) \\ L(x,y) & -\frac{\partial L}{\partial y}(x,y) \end{array}\right].$$

Note that if $K \in \mathrm{Ker}_2$ is smooth, K_{21} is antisymmetric, and K is the differential extension of K_{21}, then K is self-dual and $K_{21}(x,y) = \int_y^x K_{11}(t,y)dt$.

Main results

Fix real constants $\sigma > 0$ and ξ. With $\phi_n = \phi_{n,\sigma,\xi}$ as defined by formula (3.9.7), we put

$$K_{n,\sigma,\xi,2}(x,y) = \frac{1}{\sigma^2} \sum_{i=0}^{n-1} \phi_i(x)\phi_i(y). \tag{3.9.24}$$

The kernel $K_{n,\sigma,\xi,2}(x,y)$ is nothing new: we have previously studied it to obtain limiting results for the GUE.

We come to the novel definitions. We write $K_n = K_{n,\sigma,\xi,2}$ to abbreviate. Let

$$\begin{aligned} K_{n,\sigma,\xi,1}(x,y) &= \left[\begin{array}{cc} K_n(x,y) & -\frac{\partial K_n}{\partial y}(x,y) \\ -\frac{1}{2}\mathrm{sign}(x-y) + \int_y^x K_n(t,y)dt & K_n(x,y) \end{array}\right] \\ &+ \frac{\sqrt{n}}{2\sigma^3}\left[\begin{array}{cc} \phi_{n-1}(x)\varepsilon\phi_n(y) & -\phi_{n-1}(x)\phi_n(y) \\ (\int_y^x \phi_{n-1}(t)dt)\varepsilon\phi_n(y) & \varepsilon\phi_n(x)\phi_{n-1}(y) \end{array}\right] \\ &+ \begin{cases} \left[\begin{array}{cc} \frac{\phi_{n-1}(x)}{\langle \phi_{n-1},1\rangle} & 0 \\ \frac{\int_y^x \phi_{n-1}(t)dt}{\langle \phi_{n-1},1\rangle} & \frac{\phi_{n-1}(y)}{\langle \phi_{n-1},1\rangle} \end{array}\right] & \text{if } n \text{ is odd,} \\ 0 & \text{if } n \text{ is even,} \end{cases} \end{aligned} \tag{3.9.25}$$

and

$$K_{n,\sigma,\xi,4}(x,y) = \frac{1}{2}\begin{bmatrix} K_{2n+1}(x,y) & -\frac{\partial K_{2n+1}}{\partial y}(x,y) \\ \int_y^x K_{2n+1}(t,y)dt & K_{2n+1}(x,y) \end{bmatrix} \tag{3.9.26}$$
$$+\frac{\sqrt{2n+1}}{4\sigma^3}\begin{bmatrix} \phi_{2n}(x)\varepsilon\phi_{2n+1}(y) & -\phi_{2n}(x)\phi_{2n+1}(y) \\ (\int_y^x \phi_{2n}(t)dt)\varepsilon\phi_{2n+1}(y) & \varepsilon\phi_{2n+1}(x)\phi_{2n}(y) \end{bmatrix}.$$

We then have the following representations of squares of gap probabilities as Fredholm determinants of matrix kernels.

Theorem 3.9.19 *Let $\gamma \geq 0$ and a Borel set $A \subset \mathbb{R}$ be given. Assume either that $\gamma > 0$ or that A is bounded. Let $\beta \in \{1,4\}$. Then we have*

$$\left(\frac{\int_{A^c}\cdots\int_{A^c}|\Delta(x)|^\beta \prod_{i=1}^n \phi_{0,\sigma,\xi}(x_i)\sqrt{\beta}dx_i}{\int\cdots\int|\Delta(x)|^\beta \prod_{i=1}^n \phi_{0,\sigma,\xi}(x_i)\sqrt{\beta}dx_i}\right)^2 = \mathrm{Fred}_2^\gamma(\mathbf{1}_{A\times A}K_{n,\sigma,\xi,\beta}). \tag{3.9.27}$$

It is easy to check using Lemma 3.9.9 that the right side is defined. For comparison, we note that under the same hypotheses on γ and A we have

$$\frac{\int_{A^c}\cdots\int_{A^c}|\Delta(x)|^2 \prod_{i=1}^n \phi_{0,\sigma,\xi}(x_i)^2dx_i}{\int\cdots\int|\Delta(x)|^2 \prod_{i=1}^n \phi_{0,\sigma,\xi}(x_i)^2dx_i} = \mathrm{Fred}_1^\gamma(\mathbf{1}_{A\times A}K_{n,\sigma,\xi,2}). \tag{3.9.28}$$

The latter is merely a restatement in the present setup of Lemma 3.2.4.

Before commencing the proof we need to prove a Pfaffian analog of (3.9.21). For integers $n > 0$, put

$$L_n(x,y) = L_{n,\sigma,\xi}(x,y) = \sigma^{-2}\sum_{\substack{0\leq\ell<n \\ (-1)^\ell=(-1)^n}}\begin{vmatrix} \varepsilon\phi_\ell(x) & \varepsilon\phi_\ell(y) \\ \phi_\ell(x) & \phi_\ell(y) \end{vmatrix}.$$

Lemma 3.9.20

$$L_n(x,y) = \frac{\sqrt{n}}{2\sigma^3}\varepsilon\phi_{n-1}(x)\varepsilon\phi_n(y) + \frac{1}{\sigma^2}\sum_{i=0}^{n-1}\varepsilon\phi_i(x)\phi_i(y).$$

Proof In view of (3.9.13), it is enough to prove

$$\sum_{\substack{0\leq\ell<n \\ (-1)^\ell=(-1)^n}}\begin{vmatrix} \phi_\ell(x) & \phi_\ell(y) \\ \phi_\ell'(x) & \phi_\ell'(y) \end{vmatrix} = \frac{\sqrt{n}}{2\sigma}\phi_{n-1}(x)\phi_n(y) + \sum_{i=0}^{n-1}\phi_i(x)\phi_i'(y).$$

Let $F_1(x,y)$ and $F_2(x,y)$ denote the left and right sides of the equation above, respectively. Fix $\alpha \in \{1,2\}$ and integers $j,k \geq 0$ arbitrarily. By means of (3.9.14) and (3.9.17), one can verify that $\iint F_\alpha(x,y)\phi_j(x)\phi_k(y)dxdy$ is independent of α, which is enough by (3.9.14) to complete the proof. □

Proof of Theorem 3.9.19 Given smooth $L \in \mathrm{Ker}_1$, to abbreviate notation, let $L^{\mathrm{ext}} \in \mathrm{Ker}_2$ denote the differential extension of L, see Definition 3.9.18.

First we prove the case $\beta = 4$ pertaining to the GSE. Let $H(x)$ be as defined in Proposition 3.9.7. By straightforward calculation based on Lemma 3.9.20, one can verify that

$$H(x)\mathbf{J}_n^{-1}H(y)^{\mathrm{T}}\mathbf{J}_1 = \frac{1}{2}L^{\mathrm{ext}}_{2n+1,\sigma,\xi}(x,y) = K_{n,\sigma,\xi,4}(x,y)\,.$$

Then formula (3.9.27) in the case $\beta = 4$ follows from (3.9.9) and Remark 3.9.16.

We next prove the case $\beta = 1$ pertaining to the GOE. We use all the notation introduced in Proposition 3.9.8. One verifies by straightforward calculation using Lemma 3.9.20 that

$$G_{\mathbb{R}}(x)\mathbf{J}_1 G_{\mathbb{R}}(y)^{\mathrm{T}}\mathbf{J}_n^{-1} = L^{\mathrm{ext}}_{n,\sigma,\xi}(x,y) + M^{\mathrm{ext}}_{n,\sigma,\xi}(x,y)\,,$$

where

$$M_{n,\sigma,\xi}(x,y) = \begin{cases} \frac{\varepsilon\phi_{n-1}(x)-\varepsilon\phi_{n-1}(y)}{\langle 1,\phi_{n-1}\rangle} & \text{if } n \text{ is odd,} \\ 0 & \text{if } n \text{ is even.} \end{cases}$$

Further, with

$$Q(x,y) = G_{A^c}(x)\mathbf{J}_1 G_{\mathbb{R}}(y)^{\mathrm{T}}\mathbf{J}_n^{-1}, \quad E(x,y) = \begin{bmatrix} 0 & 0 \\ \frac{1}{2}\mathrm{sign}(x-y) & 0 \end{bmatrix}, \tag{3.9.29}$$

$Q_A = \mathbf{1}_{A\times A}Q$ and $E_A = \mathbf{1}_{A\times A}E$, we have

$$-E_A + Q_A + E_A \star Q_A = \mathbf{1}_{A\times A}K_{n,\sigma,\xi,1}\,.$$

Finally, formula (3.9.27) in the case $\beta = 1$ follows from (3.9.10) combined with Remarks 3.9.15 and 3.9.16. □

Remark 3.9.21 Because the kernel $L_{n,\sigma,\xi}$ is smooth and antisymmetric, the proof above actually shows that $K_{n,\sigma,\xi,4}$ is both self-dual and the differential extension of its entry in the lower left. Further, the proof shows the same for $K_{n,\sigma,\xi,1} + E$.

3.9.3 Limit calculations

In this section we evaluate various limits of the form $\lim_{n\to\infty} K^{(\gamma)}_{n,\sigma_n,\xi_n,\beta}$, paying strict attention to uniformity of the convergence, see Theorems 3.9.22 and 3.9.24 below. Implications of these to spacing probabilities are summarized in Corollaries 3.9.23 and 3.9.25 below.

Statements of main results

Recall the symmetric scalar kernels, see Theorem 3.1.1, and Definition 3.1.3,

$$K_{\text{sine}}(x,y) = K_{\text{sine},2}(x,y) = \frac{1}{\pi}\frac{\sin(x-y)}{x-y}, \tag{3.9.30}$$

$$K_{\text{Airy}}(x,y) = K_{\text{Airy},2}(x,y) = \frac{\text{Ai}(x)\,\text{Ai}'(y) - \text{Ai}'(x)\,\text{Ai}(y)}{x-y}. \tag{3.9.31}$$

It is understood that these kernels are defined for $x = y$ in the unique way making them continuous (and in fact infinitely differentiable). The subscript 2 refers to the β parameter for the GUE.

We define matrix variants of the sine-kernel, and state the main result on convergence toward these variants. Let

$$K_{\text{sine},1}(x,y) = \begin{bmatrix} K_{\text{sine}}(x,y) & -\frac{\partial K_{\text{sine}}}{\partial y}(x,y) \\ -\frac{1}{2}\text{sign}(x-y) + \int_y^x K_{\text{sine}}(t,y)dt & K_{\text{sine}}(x,y) \end{bmatrix}, \tag{3.9.32}$$

$$K_{\text{sine},4}(x,y) = \frac{1}{2}\begin{bmatrix} K_{\text{sine}}(x,y) & -\frac{\partial K_{\text{sine}}}{\partial y}(x,y) \\ \int_y^x K_{\text{sine}}(t,y)dt & K_{\text{sine}}(x,y) \end{bmatrix}. \tag{3.9.33}$$

The subscripts 1 and 4 refer to the β parameters for the GOE and GSE, respectively. Note that each of the kernels $K_{\text{sine},4}$ and, with E as in (3.9.29), $E + K_{\text{sine},1}$ is self-dual and the differential extension of its entry in the lower left. In other words, the kernels $K_{\text{sine},\beta}$ have properties analogous to those of $K_{n,\sigma,\xi,\beta}$ mentioned in Remark 3.9.18.

We will prove the following limit formulas.

Theorem 3.9.22 *For all bounded intervals $I \subset \mathbb{R}$,*

$$\lim_{n\to\infty} K_{n,\sqrt{n},0,1}(x,y) = K_{\text{sine},1}(x,y), \tag{3.9.34}$$

$$\lim_{n\to\infty} K_{n,\sqrt{n},0,2}(x,y) = K_{\text{sine},2}(x,y), \tag{3.9.35}$$

$$\lim_{n\to\infty} K_{n,\sqrt{2n},0,4}(x,y) = K_{\text{sine},4}(x,y), \tag{3.9.36}$$

uniformly for $x,y \in I$.

Limit formula (3.9.35) is merely a restatement of Lemma 3.5.1, and to the proof of the latter there is not much to add in order to prove the other two limit formulas. Using these we will prove the following concerning the bulk limits $F_{\text{bulk},\beta}(t)$ considered in Theorem 3.1.6.

Corollary 3.9.23 *For $\beta \in \{1,2,4\}$ and constants $t > 0$, the limits $F_{\text{bulk},\beta}(t)$ exist. More precisely, with $I = (-t/2, t/2) \subset \mathbb{R}$,*

$$\begin{aligned}(1 - F_{\text{bulk},1}(t))^2 &= \text{Fred}_2^0(\mathbf{1}_{I\times I}K_{\text{sine},1}), & (3.9.37)\\ 1 - F_{\text{bulk},2}(t) &= \text{Fred}_1^0(\mathbf{1}_{I\times I}K_{\text{sine},2}), & (3.9.38)\\ (1 - F_{\text{bulk},4}(t/2))^2 &= \text{Fred}_2^0(\mathbf{1}_{I\times I}K_{\text{sine},4}). & (3.9.39)\end{aligned}$$

Further, for $\beta \in \{1,2,4\}$,

$$\lim_{t\to\infty} F_{\text{bulk},\beta}(t) = 1. \qquad (3.9.40)$$

Formula (3.9.38) merely restates the limit formula in Theorem 3.1.1. Note that the limit formulas $\lim_{t\downarrow 0} F_{\text{bulk},\beta}(t) = 0$ for $\beta \in \{1,2,4\}$ hold automatically as a consequence of the Fredholm determinant formulas (3.9.37), (3.9.38) and (3.9.39), respectively. The case $\beta = 2$ of (3.9.40) was discussed previously in Remark 3.6.5. We will see that the cases $\beta \in \{1,4\}$ are easily deduced from the case $\beta = 2$ by using decimation and superposition, see Theorem 2.5.17.

We turn to the study of the edge of the spectrum. We introduce matrix variants of the Airy kernel K_{Airy} and then state limit results. Let

$$\begin{aligned}&K_{\text{Airy},1}(x,y)\\ &= \left[\begin{array}{cc} K_{\text{Airy}}(x,y) & -\frac{\partial K_{\text{Airy}}}{\partial y}(x,y) \\ -\frac{1}{2}\text{sign}(x-y) + \int_y^x K_{\text{Airy}}(t,y)dt & K_{\text{Airy}}(x,y)\end{array}\right]\\ &\quad +\frac{1}{2}\left[\begin{array}{cc} \text{Ai}(x)(1-\int_y^\infty \text{Ai}(t)dt) & -\text{Ai}(x)\,\text{Ai}(y) \\ (\int_y^x \text{Ai}(t)dt)(1-\int_y^\infty \text{Ai}(t)dt) & (1-\int_x^\infty \text{Ai}(t)dt)\,\text{Ai}(y)\end{array}\right], \qquad (3.9.41)\\ &K_{\text{Airy},4}(x,y)\\ &= \frac{1}{2}\left[\begin{array}{cc} K_{\text{Airy}}(x,y) & -\frac{\partial K_{\text{Airy}}}{\partial y}(x,y) \\ \int_y^x K_{\text{Airy}}(t,y)dt & K_{\text{Airy}}(x,y)\end{array}\right]\\ &\quad +\frac{1}{4}\left[\begin{array}{cc} -\text{Ai}(x)\int_y^\infty \text{Ai}(t)dt & -\text{Ai}(x)\,\text{Ai}(y) \\ -(\int_y^x \text{Ai}(t)dt)(\int_y^\infty \text{Ai}(t)dt) & -(\int_x^\infty \text{Ai}(t)dt)\,\text{Ai}(y)\end{array}\right]. \qquad (3.9.42)\end{aligned}$$

Although it is not immediately apparent, the scalar kernels appearing in the lower left of $K_{\text{Airy},\beta}$ for $\beta \in \{1,4\}$ are antisymmetric, as can be verified by using formula (3.9.58) below and integration by parts. More precisely, each of the kernels $K_{\text{Airy},4}$ and $E + K_{\text{Airy},1}$ (with E as in (3.9.29)) is self-dual and the differential extension of its entry in the lower left. In other words, the kernels $K_{\text{Airy},\beta}$ have properties analogous to those of $K_{n,\sigma,\xi,\beta}$ mentioned in Remark 3.9.18.

We will prove the following limit formulas.

Theorem 3.9.24 *For constants $\gamma \geq 0$ and intervals $I \subset \mathbb{R}$ bounded below,*

$$\lim_{n\to\infty} K^{(\gamma)}_{n,n^{1/6},2\sqrt{n},1}(x,y) = K^{(\gamma)}_{\mathrm{Airy},1}(x,y), \quad (3.9.43)$$

$$\lim_{n\to\infty} K^{(\gamma)}_{n,n^{1/6},2\sqrt{n},2}(x,y) = K^{(\gamma)}_{\mathrm{Airy},2}(x,y), \quad (3.9.44)$$

$$\lim_{n\to\infty} K^{(\gamma)}_{n,(2n)^{1/6},2\sqrt{2n},4}(x,y) = K^{(\gamma)}_{\mathrm{Airy},4}(x,y), \quad (3.9.45)$$

uniformly for $x,y \in I$.

The proofs of the limit formulas are based on a strengthening of Lemma 3.7.2 capable of handling intervals unbounded above, see Proposition 3.9.30. The limit formulas imply, with some extra arguments, the following results concerning the edge limits $F_{\mathrm{edge},\beta}(t)$ considered in Theorem 3.1.7.

Corollary 3.9.25 *For $\beta \in \{1,2,4\}$ and real constants t, the edge limits $F_{\mathrm{edge},\beta}(t)$ exist. More precisely, with $I = (t,\infty)$, and $\gamma > 0$ any constant,*

$$F_{\mathrm{edge},1}(t)^2 = \mathrm{Fred}_2^{\gamma}(\mathbf{1}_{I\times I} K_{\mathrm{Airy},1}), \quad (3.9.46)$$

$$F_{\mathrm{edge},2}(t) = \mathrm{Fred}_1^{\gamma}(\mathbf{1}_{I\times I} K_{\mathrm{Airy},2}), \quad (3.9.47)$$

$$F_{\mathrm{edge},4}(t/2^{2/3})^2 = \mathrm{Fred}_2^{\gamma}(\mathbf{1}_{I\times I} K_{\mathrm{Airy},4}). \quad (3.9.48)$$

Further, for $\beta \in \{1,2,4\}$,

$$\lim_{t\to-\infty} F_{\mathrm{edge},\beta}(t) = 0. \quad (3.9.49)$$

We will show below, see Lemma 3.9.33, that for $\gamma \geq 0$ and $\beta \in \{1,2,4\}$, the γ-twisted kernel $K^{(\gamma)}_{\mathrm{Airy},\beta}$ is bounded on sets of the form $I \times I$ with I an interval bounded below, and hence all Fredholm determinants on the right are defined. Note that the limits $\lim_{t\to+\infty} F_{\mathrm{edge},\beta}(t) = 1$ for $\beta \in \{1,2,4\}$ follow automatically from formulas (3.9.46), (3.9.47) and (3.9.48), respectively. In particular, formula (3.9.47) provides another route to the proof of Theorem 3.1.4 concerning edge-scaling in the GUE which, bypassing the Ledoux bound (Lemma 3.3.2), handles the "right-tightness" issue directly.

Proofs of bulk results

The proof of Theorem 3.9.22 is based on the following refinement of (3.5.4).

Proposition 3.9.26 *For all integers $k \geq 0$, integers δ, and bounded intervals I of real numbers, we have*

$$\lim_{n\to\infty} \left| \left(\frac{d}{dx}\right)^k \left(\phi_{n+\delta,\sqrt{n},0}(x) - \frac{\cos(x - \pi(n+\delta)/2)}{\sqrt{\pi}}\right)\right| = 0,$$

uniformly for $x \in I$.

Proof The case $k = 0$ of the proposition is exactly (3.5.4). Assume hereafter that $k > 0$. By (3.9.17) and (3.9.20) we have

$$\phi'_{n+\delta,\sqrt{n},0}(x) = \sqrt{\frac{n+\delta}{n}}\phi_{n+\delta-1,\sqrt{n},0}(x) - \frac{x\phi_{n+\delta,\sqrt{n},0}(x)}{2n}.$$

Repeated differentiation of the latter yields a relation which finishes the proof by induction on k. □

Proposition 3.9.27 *For $\delta, \kappa \in \{0,1\}$ and bounded intervals $I \subset \mathbb{R}$ we have*

$$\lim_{n\to\infty}\left(\frac{\partial}{\partial y}\right)^{\kappa} K_{n+\delta,\sqrt{n},0,2}(x,y) = \left(\frac{\partial}{\partial y}\right)^{\kappa} K_{\mathrm{sine},2}(x,y),$$

uniformly for $x, y \in I$.

The proof is a straightforward modification of the proof of Lemma 3.5.1, using Proposition 3.9.26 to justify differentiation under the integral. We omit the details.

The following elementary properties of the oscillator wave-functions will also be needed.

Proposition 3.9.28 *We have*

$$\lim_{\substack{n\to\infty\\ n:\mathrm{even}}} n^{1/4}\int_{-\infty}^{\infty}\psi_n(x)dx = 2. \tag{3.9.50}$$

In the bulk case only the order of magnitude established here is needed, but in the edge case we will need the exact value of the limit.

Proof By (3.9.11) in the case $\sigma = 1$ and $\xi = 0$ we have

$$\psi_0(x) = 2^{-1/4}\pi^{-1/4}e^{-x^2/4}, \quad \int \psi_0(x)dx = 2^{3/4}\pi^{1/4}. \tag{3.9.51}$$

By (3.9.17) in the case $\xi = 0$ and $\sigma = 1$ we have

$$\frac{\int\psi_n(x)dx}{\int\psi_0(x)dx} = \sqrt{\prod_{i=1}^{n/2}\frac{2i-1}{2i}} = \sqrt{\frac{n!}{2^n((n/2)!)^2}} \sim \sqrt[4]{\frac{2}{\pi n}},$$

by the Stirling approximation, see (2.5.12). Then (3.9.50) follows from (3.9.51). □

Proposition 3.9.29 *We have*

$$\sup_{\substack{n=1\\ n:\text{odd}}}^{\infty}\left|\int_0^\infty \psi_n(x)dx\right| < \infty. \tag{3.9.52}$$

Proof For odd positive integers n we have a recursion

$$\int_0^\infty \psi_{n+2}(x)dx = \frac{2}{\sqrt{n+2}}\psi_{n+1}(0) + \sqrt{\frac{n+1}{n+2}}\int_0^\infty \psi_n(x)dx,$$

which follows directly from (3.9.17) in the case $\xi = 0$ and $\sigma = 1$. Iterating, and using also the special case

$$\sqrt{n+1}\,\psi_{n+1}(0) = -\sqrt{n}\,\psi_{n-1}(0) \tag{3.9.53}$$

of (3.9.20), we obtain the relation

$$\begin{aligned}(-1)^{(n+5)/2}\int_0^\infty \psi_{n+4}(x)dx - \sqrt{\frac{n+3}{n+4}}\sqrt{\frac{n+1}{n+2}}(-1)^{(n+1)/2}\int_0^\infty \psi_n(x)dx\\ = (-1)^{(n+1)/2}\psi_{n+1}(0)\frac{2}{\sqrt{n+4}}\left(-\sqrt{\frac{n+2}{n+3}}+\sqrt{\frac{n+3}{n+2}}\right),\end{aligned}$$

for odd positive integers n. By (3.9.51) and (3.9.53), the right side is positive and in any case is $O(n^{-5/4})$. The bound (3.9.52) follows. □

Proof of Theorem 3.9.22 The equality (3.9.35) is the case $\kappa = 0$ of Proposition 3.9.27. To prove (3.9.34) and (3.9.36), in view of Propositions 3.9.26 and 3.9.27, we just have to verify the (numerical) limit formulas

$$\begin{aligned}\lim_{\substack{n\to\infty\\ n:\text{odd}}}\frac{1}{\langle\phi_{n-1,\sqrt{n},0},1\rangle} &= \lim_{\substack{n\to\infty\\ n:\text{odd}}}\frac{1}{n^{3/4}\langle\psi_{n-1},1\rangle} = 0,\\ \lim_{\substack{n\to\infty\\ n:\text{odd}}}\frac{\varepsilon\phi_{n,\sqrt{n},0}(0)}{n} &= \lim_{\substack{n\to\infty\\ n:\text{odd}}}\frac{1}{2n^{1/4}}\int_0^\infty \psi_n(x)dx = 0.\end{aligned}$$

These hold by Propositions 3.9.28 and 3.9.29, respectively. The proof of Theorem 3.9.22 is complete. □

Proof of Corollary 3.9.23 For $\beta \in \{1,2,4\}$, let $\lambda^{(\beta,n)} = (\lambda_1^{(\beta,n)},\ldots,\lambda_n^{(\beta,n)})$ be a random vector in $\mathbb{R}^n$ with law possessing a density with respect to Lebesgue measure proportional to $|\Delta(x)|^\beta e^{\beta|x|^2/4}$. We have by Theorem 3.9.19, formula (3.9.11) and the definitions that

$$\begin{aligned}P(\{\sigma(\lambda^{(1,n)}-\xi)\}\cap I=\emptyset)^2 &= \mathrm{Fred}_2^0(\mathbf{1}_{I\times I}K_{n,\sigma,\xi,1}),\\ P(\{\sigma(\lambda^{(2,n)}-\xi)\}\cap I=\emptyset) &= \mathrm{Fred}_1^0(\mathbf{1}_{I\times I}K_{n,\sigma,\xi,2}),\\ P(\{\sigma(\sqrt{2}\lambda^{(4,n)}-\xi)\}\cap I=\emptyset)^2 &= \mathrm{Fred}_2^0(\mathbf{1}_{I\times I}K_{n,\sigma,\xi,4}).\end{aligned}$$

The proofs of (3.9.37), (3.9.38) and (3.9.39) are completed by using Lemma 3.4.5 and Theorem 3.9.22. It remains only to prove the statement (3.9.40). For $\beta = 2$, it is a fact which can be proved in a couple of ways described in Remark 3.6.5. The case $\beta = 2$ granted, the cases $\beta \in \{1,4\}$ can be proved by using decimation and superposition, see Theorem 2.5.17. Indeed, consider first the case $\beta = 1$. To derive a contradiction, assume $\lim_{t\to\infty} F_{\text{bulk},1}(t) = 1-\delta$ for some $\delta > 0$. Then, by the decimation relation (2.5.25), $\lim_{t\to\infty} F_{\text{bulk},2}(t) \le 1-\delta^2$, a contradiction. Thus, $\lim_{t\to\infty} F_{\text{bulk},1}(t) = 1$. This also implies by symmetry that the probability that no (rescaled) eigenvalue of the GOE appears in $[0,t]$, denoted $\tilde{F}_1(t)$, decays to 0 as $t \to \infty$. By the decimation relation (2.5.26), we then have

$$1 - F_{\text{bulk},4}(t) \le 2\tilde{F}_1(2t) \to_{t\to\infty} 0\,.$$

This completes the proof of (3.9.40). □

Proofs of edge results

The proof of Theorem 3.9.24 is similar in structure to that of Theorem 3.9.22. We begin by refining Lemma 3.7.2.

Proposition 3.9.30 *For all constants $\gamma \ge 0$, integers $k \ge 0$, integers δ and intervals I bounded below we have*

$$\lim_{n\to\infty} e^{\gamma x}\phi^{(k)}_{n+\delta,n^{1/6},2\sqrt{n}}(x) = e^{\gamma x}\,\mathrm{Ai}^{(k)}(x) \tag{3.9.54}$$

uniformly for $x \in I$.

We first need to prove two lemmas. The first is a classical trick giving growth information about solutions of one-dimensional Schrödinger equations. The second applies the first to the Schrödinger equation (3.9.22) satisfied by oscillator wave-functions.

Lemma 3.9.31 *Fix real numbers $a < b$. Let ϕ and V be infinitely differentiable real-valued functions defined on the interval (a,∞) satisfying the following:*
(i) $\phi'' = V\phi$; (ii) $\phi > 0$ on $[b,\infty)$; (iii) $\lim_{x\to\infty}(\log\phi)'(x) = -\infty$;
(iv) $V > 0$ on $[b,\infty)$; (v) $V' \ge 0$ on $[b,\infty)$.
Then $(\log\phi)' \le -\sqrt{V}$ on $[b,\infty)$.

The differentiability assumptions, while satisfied in our intended application, are much stronger than needed.

Proof Suppose rather that the conclusion does not hold. After replacing b by some point of the interval (b,∞) we may assume that $\frac{\phi'}{\phi}(b) > -\sqrt{V(b)}$. After making a linear change of both independent and dependent variables, we may assume that $b = 0$, $V(0) = 1$ and hence $\frac{\phi'}{\phi}(0) > -1$. Consider the function $\theta(x) = \cosh x + \frac{\phi'}{\phi}(0) \sinh x$. Clearly we have $\theta(0) = 1$, $\frac{\theta'}{\theta}(0) = \frac{\phi'}{\phi}(0)$ and $\theta'' = \theta$. Further, because $\frac{\phi'}{\phi}(0) > -1$, we have $\theta > 0$ and $\frac{\theta'}{\theta} > -1$ on $[0,\infty)$. Finally, we have

$$(\theta\phi' - \theta'\phi)(0) = 0, \quad \frac{d}{dx}(\theta\phi' - \theta'\phi) = \theta\phi\,(V-1) \geq 0 \text{ on } [0,\infty),$$

and hence $\frac{\phi'}{\phi} \geq \frac{\theta'}{\theta} > -1$ on $[0,\infty)$, which is a contradiction. □

Lemma 3.9.32 *Fix $n > 0$ and put $\phi_n(x) = \phi_{n,n^{1/6},2\sqrt{n}}(x)$. Then for $x \geq 1$ we have $\phi_n(x) > 0$ and $(\log\phi_n)'(x) \leq -(x-1/2)^{1/2}$.*

Proof Let ζ be the rightmost of the finitely many zeroes of the function ϕ_n. Then ϕ_n does not change sign on (ζ,∞) and in fact is positive by (3.9.20). The logarithmic derivative of ϕ_n tends to $-\infty$ as $x \to +\infty$ because ϕ_n is a polynomial in x times a Gaussian density function of x. In the present case the Schrödinger equation (3.9.22) takes the form

$$\phi_n''(x) = (x + n^{-2/3}x^2/4 - 1/(2n^{1/3}))\,\phi_n(x)\,. \tag{3.9.55}$$

We finally apply Lemma 3.9.31 with $a = \max(1,\zeta) < b$, thus obtaining the estimate

$$(\log\phi_n)'(b) \leq -\sqrt{b-1/2} \text{ for } b \in (\zeta,\infty)\cap(1,\infty)\,.$$

This inequality forces one to have $\zeta < 1$ because the function of b on the left side tends to $+\infty$ as $b \downarrow \zeta$. □

Proof of Proposition 3.9.30 We write $\phi_{n,\delta}(x)$ instead of $\phi_{n+\delta,n^{1/6},2\sqrt{n}}(x)$ to abbreviate. We have

$$\phi_{n,\delta\pm1}(x) - \phi_{n,\delta}(x) = \frac{x\phi_{n,\delta}(x)}{2n^{1/6}\sqrt{n+\delta}} + \left(\sqrt{\frac{n}{n+\delta}} - 1\right)\phi_{n,\delta}(x) \mp \frac{n^{1/6}}{\sqrt{n+\delta}}\phi_{n,\delta}'(x)\,,$$

by (3.9.20) and (3.9.17), and by means of this relation we can easily reduce to the case $\delta = 0$. Assume that $\delta = 0$ hereafter and write simply $\phi_n = \phi_{n,0}$.

By Lemma 3.7.2, the limit (3.9.54) holds on bounded intervals I. Further, from Lemma 3.7.7 and the Airy equation $\mathrm{Ai}''(x) = x\mathrm{Ai}(x)$, we deduce that

$$e^{\gamma x}\,\mathrm{Ai}^{(k)}(x) \text{ is bounded on intervals bounded below}\,. \tag{3.9.56}$$

Thus it is enough to establish the following bound, for arbitrary constants $\gamma \geq 0$ and integers $k \geq 0$:

$$\sup_{n=1}^{\infty} \sup_{x \geq 1} |e^{\gamma x} \phi_n^{(k)}(x)| < \infty. \tag{3.9.57}$$

Since in any case $\sup_{n=1}^{\infty} \phi_n(1) < \infty$, we get the bound (3.9.57) for $k = 0, 1$ and all $\gamma \geq 0$ by Lemma 3.9.32. We then get (3.9.57) for $k \geq 2$ and all $\gamma \geq 0$ by (3.9.55) and induction on k. □

Growth of $K_{\mathrm{Airy},\beta}$ is under control in the following sense.

Lemma 3.9.33 *For $\beta \in \{1,2,4\}$, $\gamma \geq 0$ and intervals I bounded below, $K^{(\gamma)}_{\mathrm{Airy},\beta}$ is bounded on $I \times I$.*

Proof We have

$$K_{\mathrm{Airy}}(x,y) = \int_0^\infty \mathrm{Ai}(x+t)\,\mathrm{Ai}(y+t)dt\,. \tag{3.9.58}$$

To verify this formula, first apply $\frac{\partial}{\partial x} + \frac{\partial}{\partial y}$ to both sides, using (3.9.56) to justify differentiation under the integral, then apply the Airy equation $\mathrm{Ai}''(x) = x\mathrm{Ai}(x)$ to verify equality of derivatives, and finally apply (3.9.56) again to fix the constant of integration. By further differentiation under the integral, it follows that for all integers $k, \ell \geq 0$, constants $\gamma \geq 0$ and intervals I bounded below,

$$\sup_{x,y \in I} \left| e^{\gamma(x+y)} \frac{\partial^{k+\ell}}{\partial x^k \partial y^\ell} K_{\mathrm{Airy}}(x,y) \right| < \infty. \tag{3.9.59}$$

The latter is more than enough to prove the lemma. □

The following is the analog of Proposition 3.9.27.

Proposition 3.9.34 *For $\delta, \kappa \in \{0,1\}$, constants $\gamma \geq 0$ and intervals $I \subset \mathbb{R}$ bounded below, we have*

$$\lim_{n\to\infty} e^{\gamma(x+y)} \left(\frac{\partial}{\partial y}\right)^\kappa K_{n+\delta, n^{1/6}, 2\sqrt{n}, 2}(x,y) = e^{\gamma(x+y)} \left(\frac{\partial}{\partial y}\right)^\kappa K_{\mathrm{Airy},2}(x,y)\,, \tag{3.9.60}$$

uniformly for $x, y \in I$.

Proof To abbreviate we write $\phi_{n,\delta} = \phi_{n+\delta,n^{1/6},2\sqrt{n}}$. We have

$$\begin{aligned} K_{n+\delta,n^{1/6},2\sqrt{n},2}(x,y) & \\ = & \int_0^\infty \phi_{n,\delta}(x+t)\phi_{n,\delta}(y+t)dt \\ & + \frac{1}{4n^{2/3}} \int_0^\infty (x+y+2t)\phi_{n,\delta}(x+t)\phi_{n,\delta}(y+t)dt \qquad (3.9.61) \\ & + \frac{1}{2n^{1/3}} \int_0^\infty \Big(\phi'_{n,\delta}(x+t)\phi_{n,\delta}(y+t) + \phi_{n,\delta}(x+t)\phi'_{n,\delta}(y+t)\Big)\, dt\,. \end{aligned}$$

This is proved using (3.9.12), (3.9.21) and (3.9.22), following the pattern set in proving (3.9.58) above. In the case $\kappa = 0$ we then get the desired uniform convergence (3.9.50) by Proposition 3.9.30 and dominated convergence. After differentiating under the integrals in (3.9.58) and (3.9.61), we get the desired uniform convergence for $\kappa = 1$ in similar fashion. □

Proof of Theorem 3.9.24 The limit (3.9.44) follows from Proposition 3.9.34. To see (3.9.43) and (3.9.45), note that by definitions (3.9.41) and (3.9.42), and Propositions 3.9.30 and 3.9.34, we just have to verify the (numerical) limit formulas

$$\begin{aligned} \lim_{\substack{n\to\infty \\ n:\text{even}}} \frac{1}{4}\langle \phi_{n,n^{1/6},2\sqrt{n}}, 1\rangle &= \lim_{\substack{n\to\infty \\ n:\text{even}}} \frac{n^{1/4}}{4}\langle \psi_n, 1\rangle = \frac{1}{2}\,, \\ \lim_{\substack{n\to\infty \\ n:\text{odd}}} \frac{1}{\langle \phi_{n-1,n^{1/6},2\sqrt{n}}, 1\rangle} &= \lim_{\substack{n\to\infty \\ n:\text{odd}}} \frac{1}{n^{1/4}\langle \psi_{n-1}, 1\rangle} = \frac{1}{2}\,. \end{aligned}$$

These hold by Proposition 3.9.28. The proof of Theorem 3.9.24 is complete. □

Proof of Corollary 3.9.25 With the notation $\lambda^{(\beta,n)}$ as defined at the beginning of the proof of Corollary 3.9.23, we have by Theorem 3.9.19, formula (3.9.11) and the definitions that

$$\begin{aligned} P(\{\sigma(\lambda^{(1,n)} - \xi)\} \cap I = \emptyset)^2 &= \mathrm{Fred}_2^\gamma(\mathbf{1}_{I\times I} K_{n,\sigma,\xi,1})\,, \\ P(\{\sigma(\lambda^{(2,n)} - \xi)\} \cap I = \emptyset) &= \mathrm{Fred}_1^\gamma(\mathbf{1}_{I\times I} K_{n,\sigma,\xi,2})\,, \\ P(\{\sigma(\sqrt{2}\lambda^{(4,n)} - \xi)\} \cap I = \emptyset)^2 &= \mathrm{Fred}_2^\gamma(\mathbf{1}_{I\times I} K_{n,\sigma,\xi,4}))\,. \end{aligned}$$

To finish the proofs of (3.9.46), (3.9.47) and (3.9.48), use Lemma 3.4.5 and Theorem 3.9.24. The statement (3.9.49) holds for $\beta = 2$ by virtue of Theorem 3.1.5, and for $\beta = 1$ as a consequence of the decimation relation (2.5.25).

The argument for $\beta = 4$ is slightly more complicated. We use some information on determinantal processes as developed in Section 4.2. By (3.8:22), the sequence of laws of the second eigenvalue of the GUE, rescaled at the "edge scaling", is tight. Exactly as in the argument above concerning $\beta = 1$, this property is inherited by the sequence of laws of the (rescaled) second eigenvalue of the GOE. Using

(2.5.26), we conclude that the same applies to the sequence of laws of the largest eigenvalue of the GSE. □

Remark 3.9.35 An alternative to using the decimation relations (2.5.25) and (2.5.26) in the proof of lower tail tightness is to use the asymptotics of solutions of the Painlevé II equations, see Remark 3.8.1. It has the advantage of leading to more precise tail estimates on $F_{\mathrm{edge},\beta}$. We sketch the argument in Exercise 3.9.36.

Exercise 3.9.36 Using Exercise 3.8.3, (3.7.20), (3.8.4), (3.8.21) and Theorem 3.1.7, show that for $\beta = 1,2,4$,

$$\begin{aligned}\lim_{t\to\infty}\frac{1}{t^{3/2}}\log[1-F_{\mathrm{edge},\beta}(t)] &= -\frac{2\beta}{3},\\ \lim_{t\to-\infty}\frac{1}{t^{3}}\log F_{\mathrm{edge},\beta}(t) &= -\frac{\beta}{24}.\end{aligned}$$

Again, note the different rates of decay for the upper and lower tails of the distribution of the largest eigenvalue.

3.9.4 Differential equations

We derive differential equations for the ratios

$$\rho_{\mathrm{bulk},\beta}(t) = \frac{(1-F_{\mathrm{bulk},\beta}(t/2))^2}{1-F_{\mathrm{bulk},2}(t)},\quad \rho_{\mathrm{edge},\beta}(t) = \frac{F_{\mathrm{edge},\beta}(t/2^{2/3})^2}{F_{\mathrm{edge},2}(t)}, \tag{3.9.62}$$

for $\beta \in \{1,4\}$, thus finishing the proofs of Theorems 3.1.6 and 3.1.7.

Block matrix calculations

We aim to represent each of the quantities $\rho_{\mathrm{bulk},\beta}(t)$ and $\rho_{\mathrm{edge},\beta}(t)$ as a Fredholm determinant of a finite rank kernel. Toward that end we prove the following two lemmas.

Fix a constant $\gamma \geq 0$. Fix kernels

$$\begin{bmatrix} a & b\\ c & d\end{bmatrix},\begin{bmatrix} 0 & 0\\ e & 0\end{bmatrix} \in \mathrm{Ker}_2^\gamma,\quad \sigma, w \in \mathrm{Ker}_1^\gamma. \tag{3.9.63}$$

Assume that

$$d = \sigma + w,\quad \mathrm{Fred}_1^\gamma(\sigma) \neq 0. \tag{3.9.64}$$

Below, for brevity, we suppress $\star$, writing AB for $A\star B$. Put

$$\mathbf{K}_1 = \begin{bmatrix} a-be & (a-be)b \\ c-de & w+(c-de)b \end{bmatrix}, \tag{3.9.65}$$

$$\mathbf{K}_4 = \begin{bmatrix} \frac{a-be}{2} & \frac{(a-be)b}{4} \\ \frac{c-de+e(a-be)}{2} & w+\frac{eb-d}{2}+\frac{(c-de+e(a-be))b}{4} \end{bmatrix}, \tag{3.9.66}$$

$$\mathbf{R} = \begin{bmatrix} 0 & 0 \\ 0 & \mathrm{Res}_1^\gamma(\sigma) \end{bmatrix} \in \mathrm{Ker}_2^\gamma.$$

That $\mathbf{R}$ is well defined and belongs to Ker_2^γ follows from assumption (3.9.64). That $\mathbf{K}_1$ and $\mathbf{K}_4$ are well defined will be proved below. Recall that for $k\in\{1,2\}$ and $L_1,L_2\in\mathrm{Ker}_k^\gamma$, again $L_1L_2\in\mathrm{Ker}_k^\gamma$, by (3.9.23).

Lemma 3.9.37 *With data* (3.9.63) *and under assumption* (3.9.64)*, the kernels* $\mathbf{K}_1$ *and* $\mathbf{K}_4$ *are well defined, and have the following properties:*

$$\mathbf{K}_1,\mathbf{K}_4 \in \mathrm{Ker}_2^\gamma, \tag{3.9.67}$$

$$\mathrm{Fred}_2^\gamma(\mathbf{K}_1+\mathbf{K}_1\mathbf{R}) = \frac{\mathrm{Fred}_2^\gamma\left(\begin{bmatrix} a & b \\ -e+c & d \end{bmatrix}\right)}{\mathrm{Fred}_1^\gamma(\sigma)}, \tag{3.9.68}$$

$$\mathrm{Fred}_2^\gamma(\mathbf{K}_4+\mathbf{K}_4\mathbf{R}) = \frac{\mathrm{Fred}_2^\gamma\left(\frac{1}{2}\begin{bmatrix} a & b \\ c & d \end{bmatrix}\right)}{\mathrm{Fred}_1^\gamma(\sigma)}. \tag{3.9.69}$$

Proof Put

$$\mathbf{B} = \begin{bmatrix} 0 & b \\ 0 & 0 \end{bmatrix},\quad \mathbf{E} = \begin{bmatrix} 0 & 0 \\ e & 0 \end{bmatrix},\quad \mathbf{S} = \begin{bmatrix} 0 & 0 \\ 0 & \sigma \end{bmatrix}.$$

Note that $\mathbf{B},\mathbf{E},\mathbf{S}\in\mathrm{Ker}_2^\gamma$. Given $L_1,\dots,L_n\in\mathrm{Ker}_2^\gamma$ with $n\ge 2$, let

$$\begin{aligned} m(L_1,L_2) &= L_1+L_2-L_1L_2\in\mathrm{Ker}_2^\gamma, \\ m(L_1,\dots,L_n) &= m(m(L_1,\dots,L_{n-1}),L_n)\in\mathrm{Ker}_2^\gamma \text{ for } n>2. \end{aligned}$$

Put

$$\mathbf{L}_1 = m\left(\begin{bmatrix} a & b \\ -e+c & d \end{bmatrix},\mathbf{E},-\mathbf{B},-\mathbf{R}\right),$$

$$\mathbf{L}_4 = m\left(-\mathbf{E},\frac{1}{2}\begin{bmatrix} a & b \\ c & d \end{bmatrix},\mathbf{E},-\frac{1}{2}\mathbf{B},-\mathbf{R}\right).$$

Ones verifies that

$$\mathbf{K}_\beta = \mathbf{L}_\beta - \mathbf{L}_\beta\mathbf{S},\ \ \mathbf{L}_\beta = \mathbf{K}_\beta+\mathbf{K}_\beta\mathbf{R} \tag{3.9.70}$$

for $\beta \in \{1,4\}$ by straightforward calculation with 2×2 matrices in which one uses the first part of assumption (3.9.64), namely $d = \sigma + w$, and the resolvent identity $\mathbf{R}-\mathbf{S} = \mathbf{RS} = \mathbf{SR}$. Relation (3.9.70) establishes that $\mathbf{K}_1$ and $\mathbf{K}_4$ are well defined and proves (3.9.67). By Remark 3.9.15, we have

$$\mathrm{Fred}_2^\gamma(c\mathbf{B}) = 1, \ \mathrm{Fred}_2^\gamma(\pm\mathbf{E}) = 1, \ \mathrm{Fred}_2^\gamma(\mathbf{R})\mathrm{Fred}_1^\gamma(\sigma) = 1,$$

where c is any real constant, and for $L_1,\dots,L_n \in \mathrm{Ker}_2^\gamma$ with $n\geq 2$,

$$\mathrm{Fred}_2^\gamma(m(L_1,\dots,L_n)) = \mathrm{Fred}_2^\gamma(L_1)\cdots\mathrm{Fred}_2^\gamma(L_n).$$

We can now evaluate $\mathrm{Fred}_2^\gamma(\mathbf{L}_\beta)$, thus proving (3.9.68) and (3.9.69). □

The next lemma shows that $\mathbf{K}_\beta$ can indeed be of finite rank in cases of interest.

Lemma 3.9.38 *Let $K \in \mathrm{Ker}_2$ be smooth, self-dual, and the differential extension of its entry $K_{21} \in \mathrm{Ker}_1$ in the lower left. Let $I = (t_1,t_2)$ be a bounded interval. Let*

$$\begin{bmatrix} a(x,y) & b(x,y) \\ c(x,y) & d(x,y) \end{bmatrix} = \mathbf{1}_{I\times I}(x,y)K(x,y), \ e(x,y) = \frac{1}{2}\mathbf{1}_{I\times I}(x,y)\mathrm{sign}(x-y),$$

thus defining $a,b,c,d,e \in \mathrm{Ker}_1^0$. Let

$$\phi(x) = \frac{1}{2}(K_{11}(x,t_1)+K_{11}(x,t_2)), \tag{3.9.71}$$

$$\psi(x) = K_{11}(x,t_2)-K_{11}(x,t_1), \tag{3.9.72}$$

$$\Phi(x) = \frac{1}{2}\left(\int_{t_1}^x \phi(y)dy - \int_x^{t_2}\phi(y)dy\right). \tag{3.9.73}$$

Let $\mathbf{K}_\beta$ for $\beta\in\{1,4\}$ be as defined in (3.9.65) and (3.9.66), respectively, with $w=0$. Then

$$\mathbf{K}_1(x,y) = \mathbf{1}_{I\times I}(x,y)\begin{bmatrix} \phi(x) \\ \Phi(x) \end{bmatrix}\begin{bmatrix} 1 & \psi(y) \end{bmatrix}, \tag{3.9.74}$$

$$\mathbf{K}_4(x,y) = \mathbf{1}_{I\times I}(x,y)\begin{bmatrix} \phi(x)/2 & 0 \\ \Phi(x) & -1 \end{bmatrix}\begin{bmatrix} 1 & \psi(y)/2 \\ 0 & \phi(y)/2 \end{bmatrix}. \tag{3.9.75}$$

We omit the straightforward proof.

Proof of Theorem 3.1.6

We begin by recalling basic objects from the analysis of the GUE in the bulk of the spectrum. Reverting to the briefer notation introduced in equation (3.6.1), we write $S(x,y) = K_{\text{sine},2}(x,y)$ for the sine-kernel. Explicitly, equation (3.9.38) says that

$$1 - F_{\text{bulk},2}(t) = 1 + \sum_{n=1}^{\infty} \frac{(-1)^n}{n!} \int_{[-\frac{t}{2},\frac{t}{2}]^n} S\begin{pmatrix} x_1 & \dots & x_n \\ x_1 & \dots & x_n \end{pmatrix} \prod_{i=1}^{n} dx_i \,.$$

Let $R(x,y;t)$ be the resolvent kernel introduced in Section 3.6.1 (obtained from the sine-kernel with the choice $n = 2$, $s_0 = 0 = s_2$, $s_1 = 1$ and $t_2 = -t_1 = t/2$). Explicitly, $R(x,y;t)$ is given by

$$(1 - F_{\text{bulk},2}(t))R(x,y;t) = S(x,y) + \sum_{n=1}^{\infty} \frac{(-1)^n}{n!} \int_{[-\frac{t}{2},\frac{t}{2}]^n} S\begin{pmatrix} x & x_1 & \cdots & x_n \\ y & x_1 & \cdots & x_n \end{pmatrix} \prod_{i=1}^{n} dx_i \,,$$

and satisfies

$$S(x,y) + \int_{-t/2}^{t/2} S(x,z)R(z,y;t)dz = R(x,y;t) \tag{3.9.76}$$

by the fundamental identity, see Lemma 3.4.7. Recall the functions

$$\begin{aligned} Q(x;t) &= \frac{\sin x}{\sqrt{\pi}} + \int_{-t/2}^{t/2} R(x,y;t)\frac{\sin y}{\sqrt{\pi}} dy \,, \\ P(x;t) &= \frac{\cos x}{\sqrt{\pi}} + \int_{-t/2}^{t/2} R(x,y;t)\frac{\cos y}{\sqrt{\pi}} dy \,, \end{aligned}$$

which are as in definition (3.6.3) as specialized to the case $n = 2$, $s_0 = 0$, $s_1 = 1$, $s_2 = 0$, $t_1 = -t/2$ and $t_2 = t/2$ studied in Section 3.6.3. Finally, as in (3.6.30), let

$$p = p(t) = P(-t/2;t), \;\; q = q(t) = Q(-t/2;t) \,,$$

noting that

$$r = r(t) = -2pq/t \,, \tag{3.9.77}$$

is the function appearing in Theorem 3.1.6.

We introduce a systematic method for extracting useful functions of t from $R(x,y;t)$. A smooth (infinitely differentiable) function $\phi(x;t)$ defined for real x and positive t will be called a *test-function*. Given two test-functions ϕ_1 and ϕ_2, we define

$$\begin{aligned} \langle \phi_1 | \phi_2 \rangle_t &= t \int_{-1/2}^{1/2} \phi_1(tx;t)\phi_2(tx;t)dx \\ &\quad + t^2 \int_{-1/2}^{1/2}\int_{-1/2}^{1/2} \phi_1(tx;t)R(tx,ty;t)\phi_2(ty;t)dxdy \,. \end{aligned}$$

We call the resulting function of t an *angle bracket*. Because

$$R(x,y;t) \equiv R(y,x;t) \equiv R(-x,-y;t) \,, \tag{3.9.78}$$

the pairing $\langle\cdot|\cdot\rangle_t$ is symmetric and, furthermore,

$$\phi_1(-x;t)\phi_2(-x;t) \equiv -\phi_1(x;t)\phi_2(x;t) \Rightarrow \langle\phi_1|\phi_2\rangle_t \equiv 0. \tag{3.9.79}$$

Given a test-function $\phi = \phi(x;t)$, we also define

$$\phi^\pm = \phi^\pm(t) = \phi(\pm t/2, t), \quad \phi' = \phi'(x;t) = \frac{\partial\phi}{\partial x}(x;t).$$

Now consider the test-functions

$$\begin{aligned}
f(x;t) &= \frac{\sin x}{\sqrt{\pi}}, \\
g(x;t) &= \frac{1}{2}(S(x,t/2) + S(x,-t/2)), \\
h(x;t) &= \frac{1}{2}(S(x,t/2) - S(x,-t/2)), \\
G(x;t) &= \int_0^x g(z;t)dz.
\end{aligned}$$

By the resolvent identity (3.9.76) and the symmetry (3.9.78) we have

$$p(t) = f'^+(t) + \langle g|f'\rangle_t, \quad -q(t) = f^+(t) + \langle h|f\rangle_t. \tag{3.9.80}$$

It follows by (3.9.77) that $r(t)$ is also expressible in terms of angle brackets. To link the function $r(t)$ to the ratios (3.9.62) in the bulk case, we begin by expressing the latter in terms of angle brackets, as follows.

Lemma 3.9.39 *For each constant $t > 0$ we have*

$$\rho_{\text{bulk},1}(t) = 1 - 2G^+(t) - 2\langle h|G\rangle_t, \tag{3.9.81}$$

$$\rho_{\text{bulk},4}(t) = (1 - G^+(t) - \langle h|G\rangle_t)(1 + \frac{1}{2}\langle g|1\rangle_t). \tag{3.9.82}$$

Proof Let $I = (-t/2, t/2)$ and define inputs to Lemma 3.9.37 as follows:

$$\begin{aligned}
\begin{bmatrix} a(x,y) & b(x,y) \\ c(x,y) & d(x,y) \end{bmatrix} &= 2\mathbf{1}_{I\times I}(x,y)K_{\text{sine},4}(x,y), \\
e(x,y) &= \mathbf{1}_{I\times I}(x,y)\frac{1}{2}\text{sign}(x-y), \\
\sigma(x,y) &= \mathbf{1}_{I\times I}(x,y)S(x,y), \quad w = 0.
\end{aligned}$$

Then we have

$$\begin{aligned}
\mathbf{K}_1(x,y) &= \mathbf{1}_{I\times I}(x,y)\begin{bmatrix} g(x;t) \\ G(x;t) \end{bmatrix}\begin{bmatrix} 1 & 2h(x;t) \end{bmatrix}, \\
\mathbf{K}_4(x,y) &= \mathbf{1}_{I\times I}(x,y)\begin{bmatrix} g(x;t)/2 & 0 \\ G(x;t) & -1 \end{bmatrix}\begin{bmatrix} 1 & h(y;t) \\ 0 & g(x;t)/2 \end{bmatrix}, \\
\mathbf{R}(x,y) &= \mathbf{1}_{I\times I}(x,y)\begin{bmatrix} 0 & 0 \\ 0 & R(x,y;t) \end{bmatrix},
\end{aligned}$$

where the first two formulas can be checked using Lemma 3.9.38, and the last formula holds by the resolvent identity (3.9.76).

The right sides of (3.9.68) and (3.9.69) equal $\rho_{\text{bulk},\beta}(t)$ for $\beta \in \{1,4\}$, respectively, by Corollary 3.9.23. Using Remark 3.9.15, one can check that the left side of (3.9.68) equals the right side of (3.9.81), which concludes the proof of the latter. A similar argument shows that the left side of (3.9.69) equals

$$\det\left(I_2 - \begin{bmatrix} G^+(t) + \langle h|G\rangle_t & -\langle h|1\rangle_t \\ \frac{1}{2}\langle g|G\rangle_t & -\frac{1}{2}\langle g|1\rangle_t \end{bmatrix}\right).$$

But $\langle h|1\rangle_t$ and $\langle g|G\rangle_t$ are forced to vanish identically by (3.9.79). This concludes the proof of (3.9.82). □

Toward the goal of evaluating the logarithmic derivatives of the right sides of (3.9.81) and (3.9.82), we prove a final lemma. Given a test-function $\phi = \phi(x;t)$, let $\mathbf{D}\phi = (\mathbf{D}\phi)(x;t) = (x\frac{\partial}{\partial x} + t\frac{\partial}{\partial t})\phi(x;t)$. In the statement of the lemma and the calculations following we drop subscripts of t for brevity.

Lemma 3.9.40 *For all test-functions ϕ_1, ϕ_2 we have*

$$\begin{aligned}
&\langle \phi_1'|\phi_2\rangle + \langle \phi_1|\phi_2'\rangle = \qquad (3.9.83) \\
&(\phi_1^+ + \langle g+h|\phi_1\rangle)(\phi_2^+ + \langle g+h|\phi_2\rangle) - (\phi_1^- + \langle g-h|\phi_1\rangle)(\phi_2^- + \langle g-h|\phi_2\rangle), \\
&t\frac{d}{dt}\langle \phi_1|\phi_2\rangle = \\
&\langle \phi_1|\phi_2\rangle + \langle \mathbf{D}\phi_1|\phi_2\rangle + \langle \phi_1|\mathbf{D}\phi_2\rangle + \langle \phi_1|f\rangle\langle f|\phi_2\rangle + \langle \phi_1|f'\rangle\langle f'|\phi_2\rangle. \qquad (3.9.84)
\end{aligned}$$

Proof The resolvent identity (3.9.76) and the symmetry $S(x,y) \equiv S(y,x)$ yield the relation

$$\langle g \pm h|\phi\rangle_t = \int_{-t/2}^{t/2} R(\pm t/2, x;t)\phi(x)dx.$$

Formula (3.6.18) with $n = 2$, $s_0 = 0 = s_2$, $s_1 = 1$, $t_2 = -t_1 = t/2$ states that

$$\left(\frac{\partial}{\partial x} + \frac{\partial}{\partial y}\right) R(x,y;t) = R(x,-t/2;t)R(-t/2,y;t) - R(x,t/2;t)R(t/2,y;t).$$

These facts, along with the symmetry (3.9.78) and integration by parts, yield (3.9.83) after a straightforward calculation. Similarly, using the previously proved formulas for $\frac{\partial}{\partial t}R(x,y;t)$, $(x-y)R(x,y;t)$, $P'(x;t)$ and $Q'(x;t)$, see Section 3.6, along with the trick

$$\left(1+x\frac{\partial}{\partial x}+y\frac{\partial}{\partial y}\right)R=\frac{\partial}{\partial x}(x-y)R+y\left(\frac{\partial}{\partial x}+\frac{\partial}{\partial y}\right)R,$$

one gets

$$\left(1+x\frac{\partial}{\partial x}+y\frac{\partial}{\partial y}+t\frac{\partial}{\partial t}\right)R(x,y;t)=P(x;t)P(y;t)+Q(x;t)Q(y;t),$$

whence formula (3.9.83) by differentiation under the integral. □

To apply the preceding lemma we need the following identities for which the verifications are straightforward.

$$h+\mathbf{D}h=f^+f,\ g+\mathbf{D}g=f'^+f',\ \mathbf{D}G=f'^+f,\ t\frac{d}{dt}G^+=f'^+f^+. \tag{3.9.85}$$

The notation here is severely abbreviated. For example, the third relation written out in full reads $(\mathbf{D}G)(x;t)=f'^+(t)f(x)=f'(t/2)f(x)$. The other relations are interpreted similarly.

We are ready to conclude. We claim that

$$\begin{aligned}
&t\frac{d}{dt}(1-2G^+-2\langle h|G\rangle)\\
&\quad= -2(f^++\langle h|f\rangle)(f'^++\langle f|G\rangle)=2q(f'^++\langle f|G\rangle)\\
&\quad= 2q(f'^++\langle g|f'\rangle-2(f'^++\langle g|f'\rangle)(G^++\langle h|G\rangle))\\
&\quad= 2pq(1-2G^+-2\langle h|G\rangle)=-tr(1-2G^+-2\langle h|G\rangle).
\end{aligned} \tag{3.9.86}$$

At the first step we apply (3.9.79), (3.9.84) and (3.9.85). At the second and fourth steps we apply (3.9.80). At the third step we apply (3.9.83) with $\phi_1=-f'$ and $\phi_2=G$, using (3.9.79) to simplify. At the last step we apply (3.9.77). Thus the claim (3.9.86) is proved. The claim is enough to prove (3.1.11) since both sides of the latter tend to 1 as $t\downarrow 0$. Similarly, we have

$$t\frac{d}{dt}(1+\langle g|1\rangle)=p\langle f'|1\rangle=-2pq(1+\langle g|1\rangle)=tr(1+\langle g|1\rangle),$$

which is enough in conjunction with (3.1.11) to verify (3.1.12). The proof of Theorem 3.1.6 is complete. □

Proof of Theorem 3.1.7

The pattern of the proof of Theorem 3.1.6 will be followed rather closely, albeit with some extra complications. We begin by recalling the main objects from the analysis of the GUE at the edge of the spectrum. We revert to the abbreviated notation $A(x,y) = K_{\mathrm{Airy},2}(x,y)$. Explicitly, equation (3.9.47) says that

$$F_{\mathrm{edge},2}(t) = 1 + \sum_{n=1}^{\infty} \frac{(-1)^n}{n!} \int_{[t,\infty)^n} A\left(\begin{array}{ccc} x_1 & \dots & x_n \\ x_1 & \dots & x_n \end{array} \right) \prod_{i=1}^{n} dx_i \,.$$

Let $R(x,y;t)$ be the resolvent kernel studied in Section 3.8. Explicitly, $R(x,y;t)$ is given by

$$F_{\mathrm{edge},2}(t)R(x,y;t) = A(x,y) + \sum_{n=1}^{\infty} \frac{(-1)^n}{n!} \int_{(t,\infty)^n} A\left(\begin{array}{cccc} x & x_1 & \cdots & x_n \\ y & x_1 & \cdots & x_n \end{array} \right) \prod_{i=1}^{n} dx_i \,,$$

and by Lemma 3.4.7 satisfies

$$A(x,y) + \int_t^{\infty} A(x,z)R(z,y;t)dz = R(x,y;t)\,. \tag{3.9.87}$$

Recall the functions

$$Q(x;t) = \mathrm{Ai}(x) + \int_t^{\infty} R(x,y;t)\,\mathrm{Ai}(y)dy\,, \quad q = q(t) = Q(t;t)\,,$$

which are as in definition (3.8.3), noting that q is the function appearing in Theorem 3.1.7.

Given any smooth functions $\phi_1 = \phi_1(x;t)$ and $\phi_2 = \phi_2(x;t)$ defined on $\mathbb{R}^2$, we define

$$\begin{aligned} \langle \phi_1 | \phi_2 \rangle_t &= \int_0^{\infty} \phi_1(t+x;t)\phi_2(t+x;t)dx \\ &\quad + \int_0^{\infty}\int_0^{\infty} \phi_1(t+x;t)R(t+x,t+y;t)\phi_2(t+y;t)dxdy\,, \end{aligned}$$

provided that the integrals converge absolutely for each fixed t. We call the resulting function of t an *angle bracket*. Since the kernel $R(x,y;t)$ is symmetric in x and y, we have $\langle \phi_1 | \phi_2 \rangle_t = \langle \phi_2 | \phi_1 \rangle_t$.

We will only need finitely many explicitly constructed pairs (ϕ_1, ϕ_2) to substitute into $\langle \cdot | \cdot \rangle_t$. For each of these pairs it will be clear using the estimates (3.9.56) and (3.9.59) that the integrals above converge absolutely, and that differentiation under the integral is permissible.

We now define the finite collection of smooth functions of $(x,t) \in \mathbb{R}^2$ from which we will draw pairs to substitute into $\langle\cdot|\cdot\rangle_t$. Let

$$\begin{aligned} f = f(x;t) &= \mathrm{Ai}(x)\,, \\ g = g(x;t) &= A(t,x)\,, \\ F = F(x;t) &= -\int_x^\infty f(z)dz\,, \\ G = G(x;t) &= -\int_x^\infty g(z;t)dz\,. \end{aligned}$$

Given any smooth function $\phi = \phi(x;t)$, it is convenient to define

$$\begin{aligned} \phi' = \phi'(x;t) &= \frac{\partial\phi}{\partial x}(x;t)\,, \\ \phi^- = \phi^-(t) &= \phi(t;t)\,, \\ \mathbf{D}\phi = (\mathbf{D}\phi)(x;t) &= \left(\frac{\partial}{\partial x} + \frac{\partial}{\partial y}\right)\phi(x;t)\,. \end{aligned}$$

We have

$$\mathbf{D}f = f',\ \ \mathbf{D}F = F' = f\,,\ G' = g\,,\ \frac{d}{dt}F^- = f^-\,, \tag{3.9.88}$$

$$\mathbf{D}g = -f^- f\,,\ \mathbf{D}G = -f^- F\,,\ G^- = -(F^-)^2/2\,,\ \frac{d}{dt}G^- = -f^- F^-\,, \tag{3.9.89}$$

the first four relations clearly, and the latter four following from the integral representation (3.9.58) of $A(x,y)$. We further have

$$q = f^- + \langle f|g\rangle\,, \tag{3.9.90}$$

by (3.9.87). The next lemma links q to the ratios (3.9.62) in the edge case by expressing these ratios in terms of angle brackets. For $\beta \in \{1,4\}$ let

$$\begin{aligned} \begin{bmatrix} h_\beta \\ g_\beta \\ f_\beta \end{bmatrix} &= \begin{bmatrix} -1 & -\frac{1}{2}F^- \\ \frac{1}{2} & \frac{\delta_{\beta,1}}{2} + \frac{1}{4}F^- \\ 0 & 1 \end{bmatrix} \begin{bmatrix} g \\ f \end{bmatrix}, \\ \begin{bmatrix} G_\beta \\ F_\beta \end{bmatrix} &= \begin{bmatrix} \frac{1}{2} & -\frac{\delta_{\beta,1}}{4}F^- & \frac{\delta_{\beta,1}}{2} + \frac{1}{4}F^- \\ 0 & \frac{\delta_{\beta,1}}{2} & \frac{1}{2} \end{bmatrix} \begin{bmatrix} G \\ 1 \\ F \end{bmatrix}. \end{aligned}$$

Lemma 3.9.41 *For each real t we have*

$$\rho_{\text{edge},1}(t) = \det\left(I_2 - \begin{bmatrix} -F^-(t)/2 + \langle h_1|G_1\rangle_t & \langle h_1|F_1\rangle_t \\ \langle f_1|G_1\rangle_t & \langle f_1|F_1\rangle_t \end{bmatrix}\right), \tag{3.9.91}$$

$$\rho_{\text{edge},4}(t) = \det\left(I_3 - \begin{bmatrix} \langle h_4|G_4\rangle_t/2 & -\langle h_4|1\rangle_t/2 & \langle h_4|F_4\rangle_t/2 \\ \langle g_4|G_4\rangle_t/2 & -\langle g_4|1\rangle_t/2 & \langle g_4|F_4\rangle_t/2 \\ \langle f_4|G_4\rangle_t & -\langle f_4|1\rangle_t & \langle f_4|F_4\rangle_t \end{bmatrix}\right). \tag{3.9.92}$$

It is easy to check that all the angle brackets are well defined.

Proof We arbitrarily fix real t, along with $\beta \in \{1,4\}$ and $\gamma > 0$. Let $K = E + K_{\text{Airy},1}$ if $\beta = 1$ and otherwise let $K = 2K_{\text{Airy},4}$ if $\beta = 4$. Let $I = (t,\infty)$ and define inputs to Lemma 3.9.37 as follows.

$$\begin{bmatrix} a(x,y) & b(x,y) \\ c(x,y) & d(x,y) \end{bmatrix} = \mathbf{1}_{I\times I}(x,y)K(x,y),$$

$$e(x,y) = \mathbf{1}_{I\times I}(x,y)\frac{1}{2}\text{sign}(x-y),$$

$$\sigma(x,y) = \mathbf{1}_{I\times I}(x,y)A(x,y),$$

$$w(x,y) = \frac{1}{2}\left(\delta_{\beta,1} - \int_x^\infty \text{Ai}(z)dz\right)\text{Ai}(y).$$

Using Lemma 3.9.38 with $t_1 = t$ and $t_2 \to \infty$, one can verify after a straightforward if long calculation that if $\beta = 1$, then

$$\mathbf{K}_1(x,y) = \mathbf{1}_{I\times I}(x,y)\begin{bmatrix} g_1(y;t) & 0 \\ G_1(y;t) & F_1(y;t) \end{bmatrix}\begin{bmatrix} 1 & h_1(x;t) \\ 0 & f_1(x;t) \end{bmatrix},$$

whereas, if $\beta = 4$, then

$$\mathbf{K}_4(x,y) = \mathbf{1}_{I\times I}(x,y)\begin{bmatrix} g_4(x;t)/2 & 0 & 0 \\ G_4(x;t) & -1 & F_4(x;t) \end{bmatrix}\begin{bmatrix} 1 & h_4(y;t)/2 \\ 0 & g_4(y;t)/2 \\ 0 & f_4(y;t) \end{bmatrix}.$$

We also have

$$\mathbf{R}(x,y) = \mathbf{1}_{I\times I}(x,y)\begin{bmatrix} 0 & 0 \\ 0 & R(x,y;t) \end{bmatrix}.$$

The right sides of (3.9.68) and (3.9.69) equal $\rho_{\text{edge},\beta}(t)$ for $\beta \in \{1,4\}$, respectively, by Corollary 3.9.25. Using Remark 3.9.15, and the identity

$$\int_t^\infty g_\beta(x;t)dx = -\frac{\delta_{\beta,1}}{2}F^-(t),$$

which follows from (3.9.88) and the definitions, one can check that for $\beta = 1$ the left side of (3.9.68) equals the right side of (3.9.91), and that for $\beta = 4$, the left side of (3.9.69) equals the right side of (3.9.92). This completes the proof. □

One last preparation is required. For the rest of the proof we drop the subscript t, writing $\langle\phi_1|\phi_2\rangle$ instead of $\langle\phi_1|\phi_2\rangle_t$. For $\phi_1 \in \{f,g\}$ and $\phi_2 \in \{1,F,G\}$, we have

$$\frac{d}{dt}\langle\phi_1|\phi_2\rangle = \langle\mathbf{D}\phi_1|\phi_2\rangle + \langle\phi_1|\mathbf{D}\phi_2\rangle - \langle f|\phi_1\rangle\langle f|\phi_2\rangle\,, \tag{3.9.93}$$

$$\langle\phi_1'|\phi_2\rangle + \langle\phi_1|\phi_2'\rangle = -(\phi_1^- + \langle g|\phi_1\rangle)(\phi_2^- + \langle g|\phi_2\rangle) + \langle f|\phi_1\rangle\langle f|\phi_2\rangle\,, \tag{3.9.94}$$

as one verifies by straightforwardly applying the previously obtained formulas for $\left(\frac{\partial}{\partial x} + \frac{\partial}{\partial y}\right) R(x,y;t)$ and $\frac{\partial}{\partial t} R(x,y;t)$, see Section 3.8.

We now calculate using (3.9.88), (3.9.89), (3.9.90), (3.9.93) and (3.9.94). We have

$$\begin{aligned}
\frac{d}{dt}(1+\langle g|1\rangle) &= q(-\langle f|1\rangle)\,,\\
\frac{d}{dt}(-\langle f|1\rangle) &= -\langle f'|1\rangle + \langle f|f\rangle\langle 1|f\rangle = q(1+\langle g|1\rangle)\,,\\
\frac{d}{dt}(1-\langle f|F\rangle) &= -\langle f'|F\rangle - \langle f|f\rangle + \langle f|f\rangle\langle f|F\rangle = q(F^- + \langle g|F\rangle)\,,\\
\frac{d}{dt}(F^- + \langle g|F\rangle) &= q(1-\langle f|F\rangle)\,,\\
\langle g|1\rangle &= -(G^- + \langle g|G\rangle)(1+\langle g|1\rangle) + \langle f|G\rangle\langle f|1\rangle\,,\\
\langle g|F\rangle + \langle f|G\rangle &= -(G^- + \langle g|G\rangle)(F^- + \langle g|F\rangle) + \langle f|F\rangle\langle f|G\rangle\,.
\end{aligned}$$

The first four differential equations are easy to integrate, and moreover the constants of integration can be fixed in each case by noting that the angle brackets tend to 0 as $t \to +\infty$, as does q. In turn, the last two algebraic equations are easily solved for $\langle g|G\rangle$ and $\langle f|G\rangle$. Letting

$$\mathbf{x} = \mathbf{x}(t) = \exp\left(-\int_t^\infty q(x)dx\right),$$

we thus obtain the relations

$$\begin{bmatrix} \langle g|G\rangle & \langle g|1\rangle & \langle g|F\rangle \\ \langle f|G\rangle & \langle f|1\rangle & \langle f|F\rangle \end{bmatrix} \tag{3.9.95}$$
$$= \begin{bmatrix} \frac{\mathbf{x}+\mathbf{x}^{-1}}{2} - \frac{\mathbf{x}-\mathbf{x}^{-1}}{2}F^- + (F^-)^2/2 - 1 & \frac{\mathbf{x}+\mathbf{x}^{-1}}{2} - 1 & \frac{\mathbf{x}-\mathbf{x}^{-1}}{2} - F^- \\ \frac{\mathbf{x}+\mathbf{x}^{-1}}{2}F^- - \frac{\mathbf{x}-\mathbf{x}^{-1}}{2} & -\frac{\mathbf{x}-\mathbf{x}^{-1}}{2} & 1 - \frac{\mathbf{x}+\mathbf{x}^{-1}}{2} \end{bmatrix}.$$

It remains only to use these formulas to evaluate the determinants on the right sides of (3.9.91) and (3.9.92) in terms of $\mathbf{x}$ and F^-. The former determinant evaluates to $\mathbf{x}$ and the latter to $\frac{\mathbf{x}+2+\mathbf{x}^{-1}}{4}$. The proof of Theorem 3.1.7 is complete. □

Remark 3.9.42 The evaluations of determinants which conclude the proof above are too long to suffer through by hand. Fortunately one can organize them into manipulations of matrices with entries that are (Laurent) polynomials in variables $\mathbf{x}$ and F^-, and carry out the details with a computer algebra system.

3.10 Bibliographical notes

The study of spacings between eigenvalues of random matrices in the bulk was motivated by "Wigner's surmise" [Wig58], that postulated a density of spacing distributions of the form $Cse^{-s^2/4}$. Soon afterwords, it was realized that this was not the case [Meh60]. This was followed by the path-breaking work [MeG60], that established the link with orthogonal polynomials and the sine-kernel. Other relevant papers from that early period include the series [Dys62b], [Dys62c], [Dys62d] and [DyM63]. An important early paper concerning the orthogonal and symplectic ensembles is [Dys70]. Both the theory and a description of the history of the study of spacings of eigenvalues of various ensembles can be found in the treatise [Meh91]. The results concerning the largest eigenvalue are due to [TrW94a] for the GUE (with a 1992 ArXiv online posting), and [TrW96] for the GOE and GSE; a good review is in [TrW93]. These results have been extended in many directions; at the end of this section we provide a brief description and pointers to the relevant (huge) literature.

The book [Wil78] contains an excellent short introduction to orthogonal polynomials as presented in Section 3.2. Other good references are the classical [Sze75] and the recent [Ism05]. The three term recurrence and the Christoffel–Darboux identities mentioned in Remark 3.2.6 hold for any system of polynomials orthogonal with respect to a given weight on the real line.

Section 3.3.1 follows [HaT03], who proved (3.3.11) and observed that differential equation (3.3.12) implies a recursion for the moments of $\bar{L}_N$ discovered by [HaZ86] in the course of the latter's investigation of the moduli space of curves. Their motivation came from the following: at least formally, we have the expansion

$$\langle \bar{L}_N, e^{s\cdot} \rangle = \sum_{p\geq 0} \frac{s^{2p}}{2p!} \langle \bar{L}_N, x^{2p} \rangle .$$

Using graphical rules for the evaluation of expectations of products of Gaussian variables (Feynman's diagrams), one checks that $\langle \bar{L}_N, x^{2p} \rangle$ expands formally into

$$\sum_{g\geq 0} \frac{1}{N^{2g}} \mathcal{N}\mathcal{C}_{\mathrm{tr}(\mathbf{X}^{2p}),\mathbf{g}}(1)$$

with $\mathcal{N}\mathcal{C}_{\mathrm{tr}(\mathbf{X}^{2\mathbf{p}}),\mathbf{g}}(1)$ the number of perfect matchings on one vertex of degree $2p$ whose associated graph has genus g. Hence, computing $\langle \bar{L}_N, e^{s\cdot} \rangle$ as in Lemma 3.3.1 gives exact expressions for the numbers $\mathcal{N}\mathcal{C}_{\mathrm{tr}(\mathbf{X}^{2\mathbf{p}}),\mathbf{g}}(1)$. The link between random matrices and the enumeration of maps was first described in the physics context in [t'H74] and [BrIPZ78], and has since been enormously developed, also to situations involving multi-matrices, see [GrPW91], [FrGZJ95] for a description of the connection to quantum gravity. In these cases, matrices do not have in general independent entries but their joint distribution is described by a Gibbs measure. When this joint distribution is a small perturbation of the Gaussian law, it was shown in [BrIPZ78] that, at least at a formal level, annealed moments $\langle \bar{L}_N, x^{2p} \rangle$ expands formally into a generating function of the numbers of maps. For an accessible introduction, see [Zvo97], and for a discussion of the associated asymptotic expansion (in contrast with formal expansion), see [GuM06], [GuM07], [Mau06] and the discussion of Riemann–Hilbert methods below.

The sharp concentration estimates for $\lambda_{\max}$ contained in Lemma 3.3.2 are derived in [Led03].

Our treatment of Fredholm determinants in Section 3.4 is for the most part adapted from [Tri85]. The latter gives an excellent short introduction to Fredholm determinants and integral equations from the classical viewpoint.

The beautiful set of nonlinear partial differential equations (3.6.4), contained in Theorem 3.6.1, is one of the great discoveries reported in [JiMMS80]. Their work follows the lead of the theory of holonomic quantum fields developed by Sato, Miwa and Jimbo in the series of papers [SaMJ80]. The link between Toeplitz and Fredholm determinants and the Painlevé theory of ordinary differential equations was earlier discussed in [WuMTB76], and influenced the series [SaMJ80]. See the recent monograph [Pal07] for a discussion of these developments in the original context of the evaluation of correlations for two dimensional fields. To derive the equations (3.6.4) we followed the simplified approach of [TrW93], however we altered the operator-theoretic viewpoint of [TrW93] to a "matrix algebra" viewpoint consistent with that taken in our general discussion in Section 3.4 of Fredholm determinants. The differential equations have a Hamiltonian structure discussed briefly in [TrW93]. The same system of partial differential equations is discussed in [Mos80] in a wider geometrical context. See also [HaTW93].

Limit formula (3.7.4) appears in the literature as [Sze75, Eq. 8.22.14, p. 201] but is stated there without much in the way of proof. The relatively short self-contained proof of (3.7.4) presented in Section 3.7.2 is based on the ideas of [PlR29]; the latter paper is, however, devoted to the asymptotic behavior of the Hermite polynomials $\mathfrak{H}_n(x)$ for real positive x only.

In Section 3.8, we follow [TrW02] fairly closely. It is possible to work out a system of partial differential equations for the Fredholm determinant of the Airy kernel in the multi-interval case analogous to the system (3.6.4) for the sine-kernel. See [AdvM01] for a general framework that includes also non-Gaussian models. As in the case of the sine-kernel, there is an interpretation of the system of partial differential equations connected to the Airy kernel in the multi-interval case as an integrable Hamiltonian system, see [HaTW93] for details.

The statement contained in Remark 3.8.1, taken from [HaM80], is a solution of a *connection problem*. For another early solution to connection problems, see [McTW77]. The book [FoIKN06] contains a modern perspective on Painlevé equations and related connection problems, via the Riemann–Hilbert approach. Precise asymptotics on the Tracy–Widom distribution are contained in [BaBD08] and [DeIK08].

Section 3.9 borrows heavily from [TrW96] and [TrW05], again reworked to our "matrix algebra" viewpoint.

Our treatment of Pfaffians in Section 3.9.1 is classical, see [Jac85] for more information. We avoided the use of quaternion determinants; for a treatment based on these, see e.g. [Dys70] and [Meh91].

An analog of Lemma 3.2.2 exists for $\beta = 1,4$, see Theorem 6.2.1 and its proof in [Meh91] (in the language of quaternion determinants) and the exposition in [Rai00] (in the Pfaffian language).

As mentioned above, the results of this chapter have been extended in many directions, seeking to obtain *universality* results, stating that the limit distributions for spacings at the bulk and the edge of the GOE/GUE/GSE appear also in other matrix models, and in other problems. Four main directions for such universality occur in the literature, and we describe these next.

First, other classical ensembles have been considered (see Section 4.1 for what ensembles mean in this context). These involve the study of other types of orthogonal polynomials than the Hermite polynomials (e.g., Laguerre or Jacobi). See [For93], [For94], [TrW94b], [TrW00], [Joh00], [John01], [For06], and the book [For05].

Second, one may replace the entries of the random matrix by non-Gaussian entries. In that case, the invariance of the law under conjugation is lost, and no explicit expression for the joint distribution of the eigenvalues exist. It is, however, remarkable that it is still possible to obtain results concerning the top eigenvalue and spacings at the edge that are of the same form as Theorems 3.1.4 and 3.1.7, in the case that the law of the entries possesses good tail properties. The seminal work is [Sos99], who extended the combinatorial techniques in [SiS98b] to show that the dominant term in the evaluation of traces of large powers of random matrices does not depend on the law of the entry, as long as the mean is zero, the variance as in the GOE/GUE, and the distribution of the entries is symmetric. This has been extended to other models, and specifically to certain Wishart matrices, see [Sos02b] and [Péc09]. Some partial results relaxing the symmetry assumption can be found in [PeS07], [PeS08b], although at this time the universality at the edge of Wigner matrices with entries possessing non-symmetric distribution remains open. When the entries possess heavy tail, limit laws for the largest eigenvalue change, see [Sos04], [AuBP07]. Concerning the spacing in the bulk, universality was proved when the i.i.d. entries are complex and have a distribution that can be written as convolution with a Gaussian law, see [Joh01b] (for the complex Wigner case) and [BeP05] (for the complex Wishart case). The proof is based on an application of the Itzykson–Zuber–Harish-Chandra formula, see the bibliographical notes for Chapter 4. Similar techniques apply to the study of the largest eigenvalue of so called *spiked* models, which are matrices of the form XTX^* with X possessing i.i.d. complex entries and T a diagonal real matrix, all of whose entries except for a finite number equal to 1, and to small rank perturbations of Wigner matrices, see [BaBP05], [Péc06], [FeP07], [Kar07b] and [Ona08]. Finally, a wide ranging extension of the universality results in [Joh01b] to Hermitian matrices with independent entries on and above the diagonal appears in [ERSY09], [TaV09b] and [ERS$^+$09].

Third, one can consider joint distribution of eigenvalues of the form (2.6.1), for general potentials V. This is largely motivated by applications in physics. When deriving the bulk and edge asymptotics, one is naturally led to study the asymptotics of orthogonal polynomials associated with the weight e^{-V}. At this point, the powerful Riemann–Hilbert approach to the asymptotics of orthogonal polynomials and spacing distributions can be applied. Often, that approach yields the sharpest estimates, especially in situations where the orthogonal polynomials are not known explicitly, thereby proving universality statements for random matrices. Describing this approach in detail goes beyond the scope of this book (and bibliography notes). For the origins and current state of the art of this approach we refer the reader to the papers [FoIK92], [DeZ93], [DeZ95], [DeIZ97], [DeVZ97] [DeKM$^+$98], [DeKM$^+$99], [BlI99], to the books [Dei99], [DeG09] and to the lecture [Dei07]. See also [PaS08a].

Finally, expressions similar to the joint distribution of the eigenvalues of random matrices have appeared in the study of various combinatorial problems. Arguably, the most famous is the problem of the longest increasing subsequence of a random permutation, also known as *Ulam's problem*, which we now describe.

Let L_n denote the length of the longest increasing subsequence of a random permutation on $\{1,\dots,n\}$. The problem is to understand the asymptotics of the law of L_n. Based on his subadditive ergodic theorem, Hammersley [Ham72] showed that $L_n/\sqrt{n}$ converges to a deterministic limit and, shortly thereafter, [VeK77] and [LoS77] independently proved that the limit equals 2. It was conjectured (in analogy with conjectures for first passage percolation, see [AlD99] for some of the history and references) that $\tilde{L}_n := (L_n - 2\sqrt{n})/n^{1/6}$ has variance of order 1. Using a combinatorial representation, due to Gessel, of the distribution of L_n in terms of an integral over an expression resembling a joint distribution of eigenvalues (but with non-Gaussian potential V), [BaDJ99] applied the Riemann–Hilbert approach to prove that not only is the conjecture true, but in fact $\tilde{L}_n$ asymptotically is distributed according to the Tracy–Widom distribution F_2. Subsequently, direct proofs that do not use the Riemann–Hilbert approach (but do use the random matrices connection) emerged, see [Joh01a], [BoOO00] and [Oko00]. Certain growth models also fall in the same pattern, see [Joh00] and [PrS02]. Since then, many other examples of combinatorial problems leading to a universal behavior of the Tracy–Widom type have emerged. We refer the reader to the forthcoming book [BaDS09] for a thorough discussion.

We have not discussed, neither in the main text nor in these bibliographical notes, the connections between random matrices and number theory, more specifically the connections with the Riemann zeta function. We refer the reader to [KaS99] for an introduction to these links, and to [Kea06] for a recent account.

4

Some generalities

In this chapter, we introduce several tools useful in the study of matrix ensembles beyond GUE, GOE and Wigner matrices. We begin by setting up in Section 4.1 a general framework for the derivation of joint distribution of eigenvalues in matrix ensembles and then we use it to derive joint distribution results for several classical ensembles, namely, the GOE/GUE/GSE, the Laguerre ensembles (corresponding to Gaussian Wishart matrices), the Jacobi ensembles (corresponding to random projectors) and the unitary ensembles (corresponding to random matrices uniformly distributed in classical compact Lie groups). In Section 4.2, we study a class of point processes that are *determinantal*; the eigenvalues of the GUE, as well as those for the unitary ensembles, fall within this class. We derive a representation for determinantal processes and deduce from it a CLT for the number of eigenvalues in an interval, as well as ergodic consequences. In Section 4.3, we analyze time-dependent random matrices, where the entries are replaced by Brownian motions. The introduction of Brownian motion allows us to use the powerful theory of Ito integration. Generalizations of the Wigner law, CLTs, and large deviations are discussed. We then present in Section 4.4 a discussion of concentration inequalities and their applications to random matrices, substantially extending Section 2.3. Concentration results for matrices with independent entries, as well as for matrices distributed according to Haar measure on compact groups, are discussed. Finally, in Section 4.5, we introduce a tridiagonal model of random matrices, whose joint distribution of eigenvalues generalizes the Gaussian ensembles by allowing for any value of $\beta \geq 1$ in Theorem 2.5.3. We refer to this matrix model as the *beta ensemble*.

4.1 Joint distribution of eigenvalues in the classical matrix ensembles

In Section 2.5, we derived an expression for the joint distribution of eigenvalues of a GUE or GOE matrix which could be stated as an integration formula, see (2.5.22). Although we did not emphasize it in our derivation, a key point was that the distribution of the random matrices was invariant under the action of a group (orthogonal for the GOE, unitary for the GUE). A collection of matrices equipped with a probability measure invariant under a large group of symmetries is generally called an *ensemble*. It is our goal in this section to derive integration formulas, and hence joint distribution of eigenvalues, for several ensembles of matrices, in a unified way, by following in the footsteps of Weyl. The point of view we adopt is that of differential geometry, according to which we consider ensembles of matrices as manifolds embedded in Euclidean spaces. The prerequisites and notation are summarized in Appendix F.

The plan for Section 4.1 is as follows. In Section 4.1.1, after briefly recalling notation, we present the main results of Section 4.1, namely integration formulas yielding joint distribution of eigenvalues in three classical matrix ensembles linked to Hermite, Laguerre and Jacobi polynomials, respectively, and also Weyl's integration formulas for the classical compact Lie groups. We then state in Section 4.1.2 a special case of Federer's coarea formula and illustrate it by calculating the volumes of unitary groups. (A proof of the coarea formula in the "easy version" used here is presented in Appendix F.) In Section 4.1.3 we present a generalized Weyl integration formula, Theorem 4.1.28, which we prove by means of the coarea formula and a modest dose of Lie group theory. In Section 4.1.4 we verify the hypotheses of Theorem 4.1.28 in each of the setups discussed in Section 4.1.1, thus completing the proofs of the integration formulas by an updated version of Weyl's original method.

4.1.1 Integration formulas for classical ensembles

Throughout this section, we let $\mathbb{F}$ denote any of the (skew) fields $\mathbb{R}$, $\mathbb{C}$ or $\mathbb{H}$. (See Appendix E for the definition of the skew field of quaternions $\mathbb{H}$. Recall that $\mathbb{H}$ is a skew field, but not a field, because the product in $\mathbb{H}$ is not commutative.) We set $\beta = 1,2,4$ according as $\mathbb{F} = \mathbb{R},\mathbb{C},\mathbb{H}$, respectively. (Thus β is the dimension of $\mathbb{F}$ over $\mathbb{R}$.) We next recall matrix notation which in greater detail is set out in Appendix E.1. Let $\mathrm{Mat}_{p\times q}(\mathbb{F})$ be the space of $p \times q$ matrices with entries in $\mathbb{F}$, and write $\mathrm{Mat}_n(\mathbb{F}) = \mathrm{Mat}_{n\times n}(\mathbb{F})$. For each matrix $X \in \mathrm{Mat}_{p\times q}(\mathbb{F})$, let $X^* \in \mathrm{Mat}_{q\times p}(\mathbb{F})$ be the matrix obtained by transposing X and then applying the conjugation operation * to every entry. We endow $\mathrm{Mat}_{p\times q}(\mathbb{F})$ with the structure of Euclidean space (that is, with the structure of finite-dimensional real Hilbert space) by setting $X \cdot Y = \Re \operatorname{tr} X^*Y$. Let $\mathrm{GL}_n(\mathbb{F})$ be the group of invertible elements of $\mathrm{Mat}_n(\mathbb{F})$, and let $\mathrm{U}_n(\mathbb{F})$ be the subgroup of $\mathrm{GL}_n(\mathbb{F})$ consisting of unitary matrices; by definition $U \in \mathrm{U}_n(\mathbb{F})$ iff $UU^* = I_n$ iff $U^*U = I_n$.

The Gaussian ensembles

The first integration formula that we present pertains to the Gaussian ensembles, that is, to the GOE, GUE and GSE. Let $\mathscr{H}_n(\mathbb{F}) = \{X \in \mathrm{Mat}_n(\mathbb{F}) : X^* = X\}$. Let $\rho_{\mathscr{H}_n(\mathbb{F})}$ denote the volume measure on $\mathscr{H}_n(\mathbb{F})$. (See Proposition F.8 for the general definition of the volume measure ρ_M on a manifold M embedded in a Euclidean space.) Let $\rho_{\mathrm{U}_n(\mathbb{F})}$ denote the volume measure on $\mathrm{U}_n(\mathbb{F})$. (We will check below, see Proposition 4.1.14, that $\mathrm{U}_n(\mathbb{F})$ is a manifold.) The measures $\rho_{\mathscr{H}_n(\mathbb{F})}$ and $\rho_{\mathrm{U}_n(\mathbb{F})}$ are just particular normalizations of Lebesgue and Haar measure, respectively. Let $\rho[\mathrm{U}_n(\mathbb{F})]$ denote the (finite and positive) total volume of $\mathrm{U}_n(\mathbb{F})$. (For any manifold M embedded in a Euclidean space, we write $\rho[M] = \rho_M(M)$.) We will calculate $\rho[\mathrm{U}_n(\mathbb{F})]$ explicitly in Section 4.1.2. Recall that if $x = (x_1, \ldots, x_n)$, then we write $\Delta(x) = \prod_{1 \le i < j \le n}(x_j - x_i)$. The notion of eigenvalue used in the next result is defined for general $\mathbb{F}$ in a uniform way by Corollary E.12 and is the standard one for $\mathbb{F} = \mathbb{R}, \mathbb{C}$.

Proposition 4.1.1 *For every nonnegative Borel-measurable function φ on $\mathscr{H}_n(\mathbb{F})$ such that $\varphi(X)$ depends only on the eigenvalues of X, we have*

$$\int \varphi d\rho_{\mathscr{H}_n(\mathbb{F})} = \frac{\rho[\mathrm{U}_n(\mathbb{F})]}{(\rho[\mathrm{U}_1(\mathbb{F})])^n n!} \int_{\mathbb{R}^n} \varphi(x) |\Delta(x)|^\beta \prod_{i=1}^n dx_i \,, \tag{4.1.1}$$

where for every $x = (x_1, \ldots, x_n) \in \mathbb{R}^n$ we write $\varphi(x) = \varphi(X)$ for any $X \in \mathscr{H}_n(\mathbb{F})$ with eigenvalues $x_1, \ldots, x_n$.

According to Corollary E.12, the hypothesis that $\varphi(X)$ depends only on the eigenvalues of X could be restated as the condition that $\varphi(UXU^*) = \varphi(X)$ for all $X \in \mathscr{H}_n(\mathbb{F})$ and $U \in \mathrm{U}_n(\mathbb{F})$.

Suppose now that $X \in \mathscr{H}_n(\mathbb{F})$ is random. Suppose more precisely that the entries on or above the diagonal are independent; that each diagonal entry is (real) Gaussian of mean 0 and variance $2/\beta$; and that each above-diagonal entry is standard normal over $\mathbb{F}$. (We say that a random variable G with values in $\mathbb{F}$ is *standard normal* if, with $\{G_i\}_{i=1}^4$ independent real-valued Gaussian random variables of zero mean and unit variance, we have that G is distributed like

$$\begin{array}{ll} G_1 & \text{if } \mathbb{F}=\mathbb{R}, \\ (G_1+\mathbf{i}G_2)/\sqrt{2} & \text{if } \mathbb{F}=\mathbb{C}, \\ (G_1+\mathbf{i}G_2+\mathbf{j}G_3+\mathbf{k}G_4)/2 & \text{if } \mathbb{F}=\mathbb{H}.) \end{array} \tag{4.1.2}$$

Then for $\mathbb{F}=\mathbb{R}$ (resp., $\mathbb{F}=\mathbb{C}$) the matrix X is a random element of the GOE (resp., GUE), and in the case $\mathbb{F}=\mathbb{H}$ is by definition a random element of the *Gaussian Symplectic Ensemble* (GSE). Consider now the substitution $\varphi(X) = e^{-\beta \operatorname{tr} X^2/4} f(X)$ in (4.1.1), in conjunction with Proposition 4.1.14 below which computes volumes of unitary groups. For $\beta=1,2$, we recover Theorem 2.5.2 in the formulation given in (2.5.22). In the remaining case $\beta=4$ the substitution yields the joint distribution of the (unordered) eigenvalues in the GSE.

Remark 4.1.2 As in formula (4.1.1), all the integration formulas in this section involve normalization constants given in terms of volumes of certain manifolds. Frequently, when working with probability distributions, one bypasses the need to evaluate these volumes by instead using the Selberg integral formula, Theorem 2.5.8, and its limiting forms, as in our previous discussion of the GOE and GUE in Section 2.5.

We saw in Chapter 3 that the Hermite polynomials play a crucial role in the analysis of GUE/GOE/GSE matrices. For that reason we will sometimes speak of Gaussian/Hermite ensembles. In similar fashion we will tag each of the next two ensembles by the name of the associated family of orthogonal polynomials.

Laguerre ensembles and Wishart matrices

We next turn our attention to random matrices generalizing the Wishart matrices discussed in Exercise 2.1.18, in the case of Gaussian entries. Fix integers $0 < p \le q$ and put $n = p+q$. Let $\rho_{\mathrm{Mat}_{p\times q}(\mathbb{F})}$ be the volume measure on the Euclidean space $\mathrm{Mat}_{p\times q}(\mathbb{F})$. The analog of integration formula (4.1.1) for singular values of rectangular matrices is the following. The notion of singular value used here is defined for general $\mathbb{F}$ in a uniform way by Corollary E.13 and is the standard one for $\mathbb{F}=\mathbb{R},\mathbb{C}$.

Proposition 4.1.3 *For every nonnegative Borel-measurable function φ on $\mathrm{Mat}_{p\times q}(\mathbb{F})$ such that $\varphi(X)$ depends only on the singular values of X, we have*

$$\int \varphi d\rho_{\mathrm{Mat}_{p\times q}(\mathbb{F})} = \frac{\rho[\mathrm{U}_p(\mathbb{F})]\rho[\mathrm{U}_q(\mathbb{F})]2^{\beta p/2}}{\rho[\mathrm{U}_1(\mathbb{F})]^p\rho[\mathrm{U}_{q-p}(\mathbb{F})]2^{\beta pq/2}p!} \times \int_{\mathbb{R}_+^p} \varphi(x)\,|\Delta(x^2)|^\beta \prod_{i=1}^p x_i^{\beta(q-p+1)-1}dx_i\,, \tag{4.1.3}$$

where for every $x=(x_1,\dots,x_p)\in\mathbb{R}_+^p$ we write $x^2=(x_1^2,\dots,x_p^2)$, and $\varphi(x)=\varphi(X)$ for any $X\in\mathrm{Mat}_{p\times q}(\mathbb{F})$ with singular values $x_1,\dots,x_p$.

Here and in later formulas, by convention, $\rho[\mathrm{U}_0(\mathbb{F})]=1$. According to Corollary E.13, the hypothesis that $\varphi(X)$ depends only on the singular values of X could be restated as the condition that $\varphi(UXV)=\varphi(X)$ for all $U\in\mathrm{U}_p(\mathbb{F})$, $X\in\mathrm{Mat}_{p\times q}(\mathbb{F})$ and $V\in\mathrm{U}_q(\mathbb{F})$.

Suppose now that the entries of $X\in\mathrm{Mat}_{p\times q}(\mathbb{F})$ are i.i.d. standard normal. In the case $\mathbb{F}=\mathbb{R}$ the random matrix XX^* is an example of a Wishart matrix, the latter as studied in Exercise 2.1.18. In the case of general $\mathbb{F}$ we call XX^* a *Gaussian Wishart matrix* over $\mathbb{F}$. Proposition 4.1.3 implies that the distribution of the (unordered) eigenvalues of XX^* (which are the squares of the singular values of X) possesses a density on $(0,\infty)^p$ with respect to Lebesgue measure proportional to

$$|\Delta(x)|^\beta\cdot\prod_{i=1}^p e^{-\beta x_i/4}\cdot\prod_{i=1}^p x_i^{\beta(q-p+1)/2-1}\,.$$

Now the orthogonal polynomials corresponding to weights of the form $x^\alpha e^{-\gamma x}$ on $(0,\infty)$ are the Laguerre polynomials. In the analysis of random matrices of the form XX^*, the Laguerre polynomials and their asymptotics play a role analogous to that played by the Hermite polynomials and their asymptotics in the analysis of GUE/GOE/GSE matrices. For this reason we also call XX^* a random element of a *Laguerre ensemble* over $\mathbb{F}$.

Jacobi ensembles and random projectors

We first make a general definition. Put

$$\mathrm{Flag}_n(\lambda,\mathbb{F})=\{U\lambda U^*:U\in\mathrm{U}_n(\mathbb{F})\}\subset\mathscr{H}_n(\mathbb{F}), \tag{4.1.4}$$

where $\lambda\in\mathrm{Mat}_n$ is any real diagonal matrix. The compact set $\mathrm{Flag}_n(\lambda,\mathbb{F})$ is always a manifold, see Lemma 4.1.18 and Exercise 4.1.19.

Now fix integers $0<p\le q$ and put $n=p+q$. Also fix $0\le r\le q-p$ and write $q=p+r+s$. Consider the diagonal matrix $D=\mathrm{diag}(I_{p+r},0_{p+s})$, and the

corresponding space $\mathrm{Flag}_n(D,\mathbb{F})$ as defined in (4.1.4) above. (As in Appendix E.1, we will use the notation diag to form block-diagonal matrices as well as matrices diagonal in the usual sense.) Let $\rho_{\mathrm{Flag}_n(D,\mathbb{F})}$ denote the volume measure on $\mathrm{Flag}_n(D,\mathbb{F})$. Given $W \in \mathrm{Flag}_n(D,\mathbb{F})$, let $W^{(p)} \in \mathscr{H}_p(\mathbb{F})$ denote the upper left $p \times p$ block. Note that all eigenvalues of $W^{(p)}$ are in the unit interval $[0,1]$.

Proposition 4.1.4 *With notation as above, for all Borel-measurable nonnegative functions φ on $\mathscr{H}_p(\mathbb{F})$ such that $\varphi(X)$ depends only on the eigenvalues of X, we have*

$$\begin{aligned}\int \varphi(W^{(p)})d\rho_{\mathrm{Flag}_n(D,\mathbb{F})}(W) &= \frac{\rho[\mathrm{U}_p(\mathbb{F})]\rho[\mathrm{U}_q(\mathbb{F})]2^{\beta p/2}}{\rho[\mathrm{U}_1(\mathbb{F})]^p\rho[\mathrm{U}_r(\mathbb{F})]\rho[\mathrm{U}_s(\mathbb{F})]2^p p!} \\ &\times \int_{[0,1]^p} \varphi(x)\,|\Delta(x)|^\beta \cdot \prod_{i=1}^p (x_i^{(r+1)\beta/2-1}(1-x_i)^{(s+1)\beta/2-1}dx_i), \qquad (4.1.5)\end{aligned}$$

where for every $x = (x_1,\ldots,x_p) \in \mathbb{R}^p$ we write $\varphi(x) = \varphi(X)$ for any matrix $X \in \mathscr{H}_p(\mathbb{F})$ with eigenvalues $x_1,\ldots,x_p$.

The symmetry here crucial for the proof is that $\varphi(W^{(p)}) = \varphi((UWU^*)^{(p)})$ for all $U \in \mathrm{U}_n(\mathbb{F})$ commuting with $\mathrm{diag}(I_p, 0_q)$ and all $W \in \mathrm{Flag}_n(D,\mathbb{F})$.

Now up to a normalization constant, $\rho_{\mathrm{Flag}_n(D,\mathbb{F})}$ is the law of a random matrix of the form $U_nDU_n^*$, where $U_n \in \mathrm{U}_n(\mathbb{F})$ is Haar-distributed. (See Exercise 4.1.19 for evaluation of the constant $\rho[\mathrm{Flag}_n(D,\mathbb{F})]$.) We call such a random matrix $U_nDU_n^*$ a *random projector*. The joint distribution of eigenvalues of the submatrix $(U_nDU_n^*)^{(p)}$ is then specified by formula (4.1.5). Now the orthogonal polynomials corresponding to weights of the form $x^\alpha(1-x)^\gamma$ on $[0,1]$ are the Jacobi polynomials. In the analysis of random matrices of the form $(U_nDU_n)^{(p)}$, the Jacobi polynomials play a role analogous to that played by the Hermite polynomials in the analysis of GUE/GOE/GSE matrices. For this reason we call $(U_nDU_n^*)^{(p)}$ a random element of a *Jacobi ensemble* over $\mathbb{F}$.

The classical compact Lie groups

The last several integration formulas we present pertain to the classical compact Lie groups $\mathrm{U}_n(\mathbb{F})$ for $\mathbb{F} = \mathbb{R},\mathbb{C},\mathbb{H}$, that is, to the ensembles of orthogonal, unitary and symplectic matrices, respectively, equipped with normalized Haar measure. We set $R(\theta) = \begin{bmatrix} \cos\theta & \sin\theta \\ -\sin\theta & \cos\theta \end{bmatrix} \in \mathrm{U}_2(\mathbb{R})$ for $\theta \in \mathbb{R}$. More generally, for $\theta = (\theta_1,\ldots,\theta_n) \in \mathbb{R}^n$, we set $R_n(\theta) = \mathrm{diag}(R(\theta_1),\ldots,R(\theta_n)) \in \mathrm{U}_{2n}(\mathbb{R})$. We also write $\mathrm{diag}(\theta) = \mathrm{diag}(\theta_1,\ldots,\theta_n) \in \mathrm{Mat}_n$.

We define nonnegative functions A_n, B_n, C_n, D_n on $\mathbb{R}^n$ as follows:

$$\begin{aligned} A_n(\theta) &= \prod_{1\le i<j\le n} |e^{\mathbf{i}\theta_i} - e^{\mathbf{i}\theta_j}|^2, \quad D_n(\theta) = A_n(\theta) \prod_{1\le i<j\le n} \left|e^{\mathbf{i}\theta_i} - e^{-\mathbf{i}\theta_j}\right|^2, \\ B_n(\theta) &= D_n(\theta)\prod_{i=1}^n |e^{\mathbf{i}\theta_i} - 1|^2, \quad C_n(\theta) = D_n(\theta)\prod_{i=1}^n |e^{\mathbf{i}\theta_i} - e^{-\mathbf{i}\theta_i}|^2. \end{aligned}$$

(Recall that $\mathbf{i}$ equals the imaginary unit viewed as an element of $\mathbb{C}$ or $\mathbb{H}$.)

Remark 4.1.5 The choice of letters A, B, C, and D made here is consistent with the standard labeling of the corresponding root systems.

We say that a function φ on a group G is *central* if $\varphi(g)$ depends only on the conjugacy class of g, that is, if $\varphi(g_1g_2g_1^{-1}) = \varphi(g_2)$ for all $g_1, g_2 \in G$.

Proposition 4.1.6 (Weyl) *(Unitary case) For every nonnegative Borel-measurable central function φ on $\mathrm{U}_n(\mathbb{C})$, we have*

$$\int \varphi \frac{d\rho_{\mathrm{U}_n(\mathbb{C})}}{\rho[\mathrm{U}_n(\mathbb{C})]} = \frac{1}{n!}\int_{[0,2\pi]^n} \varphi(e^{\mathbf{i}\mathrm{diag}(\theta)})A_n(\theta)\prod_{i=1}^n\left(\frac{d\theta_i}{2\pi}\right). \tag{4.1.6}$$

(Odd orthogonal case) For odd $n = 2\ell+1$ and every nonnegative Borel-measurable central function φ on $\mathrm{U}_n(\mathbb{R})$, we have

$$\int \varphi \frac{d\rho_{\mathrm{U}_n(\mathbb{R})}}{\rho[\mathrm{U}_n(\mathbb{R})]} = \frac{1}{2^{\ell+1}\ell!}\int_{[0,2\pi]^\ell} \sum_{k=0}^{1} \varphi(\mathrm{diag}(R_\ell(\theta),(-1)^k))B_\ell(\theta)\prod_{i=1}^\ell\left(\frac{d\theta_i}{2\pi}\right). \tag{4.1.7}$$

(Symplectic case) For every nonnegative Borel-measurable central function φ on $\mathrm{U}_n(\mathbb{H})$, we have

$$\int \varphi \frac{d\rho_{\mathrm{U}_n(\mathbb{H})}}{\rho[\mathrm{U}_n(\mathbb{H})]} = \frac{1}{2^n n!}\int_{[0,2\pi]^n} \varphi(e^{\mathbf{i}\mathrm{diag}(\theta)})C_n(\theta)\prod_{i=1}^n\left(\frac{d\theta_i}{2\pi}\right). \tag{4.1.8}$$

(Even orthogonal case) For even $n = 2\ell$ and every nonnegative Borel-measurable central function φ on $\mathrm{U}_n(\mathbb{R})$ we have

$$\begin{aligned} &\int \varphi \frac{d\rho_{\mathrm{U}_n(\mathbb{R})}}{\rho[\mathrm{U}_n(\mathbb{R})]} \\ &= \frac{1}{2^\ell \ell!}\int_{[0,2\pi]^\ell} \varphi(R_\ell(\theta))D_\ell(\theta)\prod_{i=1}^\ell\left(\frac{d\theta_i}{2\pi}\right) \qquad (4.1.9) \\ &\quad + \frac{1}{2^\ell(\ell-1)!}\int_{[0,2\pi]^{\ell-1}} \varphi(\mathrm{diag}(R_{\ell-1}(\theta),1,-1))C_{\ell-1}(\theta)\prod_{i=1}^{\ell-1}\left(\frac{d\theta_i}{2\pi}\right). \end{aligned}$$

We will recover these classical results of Weyl in our setup in order to make it clear that all the results on joint distribution discussed in Section 4.1 fall within Weyl's circle of ideas.

Remark 4.1.7 Because we have

$$D_n(\theta) = \prod_{1\le i<j\le n} (2\cos\theta_i - 2\cos\theta_j)^2,$$

the process of eigenvalues of $\mathrm{U}_n(\mathbb{F})$ is determinantal (see Section 4.2.9 and in particular Lemma 4.2.50) not only for $\mathbb{F} = \mathbb{C}$ but also for $\mathbb{F} = \mathbb{R}, \mathbb{H}$. This is in sharp contrast to the situation with Gaussian/Hermite, Laguerre and Jacobi ensembles where, in the cases $\mathbb{F} = \mathbb{R}, \mathbb{H}$, the eigenvalue (singular value) processes are *not* determinantal. One still has tools for studying the latter processes, but they are Pfaffian- rather than determinant-based, of the same type considered in Section 3.9 to obtain limiting results for GOE/GSE.

4.1.2 Manifolds, volume measures and the coarea formula

Section 4.1.2 introduces the *coarea formula*, Theorem 4.1.8. In the specialized form of Corollary 4.1.10, the coarea formula will be our main tool for proving the formulas of Section 4.1.1. To allow for quick reading by the expert, we merely state the coarea formula here, using standard terminology; precise definitions, preliminary material and a proof of Theorem 4.1.8 are all presented in Appendix F. After presenting the coarea formula, we illustrate it by working out an explicit formula for $\rho[\mathrm{U}_n(\mathbb{F})]$.

Fix a smooth map $f : M \to N$ from an n-manifold to a k-manifold, with derivative at a point $p \in M$ denoted $\mathbb{T}_p(f) : \mathbb{T}_p(M) \to \mathbb{T}_{f(p)}(N)$. Let M_{crit}, M_{reg}, N_{crit} and N_{reg} be the sets of critical (regular) points (values) of f, see Definition F.3 and Proposition F.10 for the terminology. For $q \in N$ such that $M_{\mathrm{reg}} \cap f^{-1}(q)$ is nonempty (and hence by Proposition F.16 a manifold) we equip the latter with the volume measure $\rho_{M_{\mathrm{reg}}\cap f^{-1}(q)}$ (see Proposition F.8). Put $\rho_\emptyset = 0$ for convenience. Finally, let $J(\mathbb{T}_p(f))$ denote the generalized determinant of $\mathbb{T}_p(f)$, see Definition F.17.

Theorem 4.1.8 (The coarea formula) *With notation and setting as above, let φ be any nonnegative Borel-measurable function on M. Then:*
(i) the function $p \mapsto J(\mathbb{T}_p(f))$ on M is Borel-measurable;
(ii) the function $q \mapsto \int \varphi(p) d\rho_{M_{\mathrm{reg}}\cap f^{-1}(q)}(p)$ on N is Borel-measurable;

(iii) the integral formula

$$\int \varphi(p)J(\mathbb{T}_p(f))d\rho_M(p) = \int \left(\int \varphi(p)d\rho_{M_{\mathrm{reg}}\cap f^{-1}(q)}(p) \right) d\rho_N(q) \qquad (4.1.10)$$

holds.

Theorem 4.1.8 is in essence a version of Fubini's Theorem. It is also a particular case of the general coarea formula due to Federer. The latter formula at "full strength" (that is, in the language of Hausdorff measures) requires far less differentiability of f and is much harder to prove.

Remark 4.1.9 Since f in Theorem 4.1.8 is smooth, we have by Sard's Theorem (Theorem F.11) that for ρ_N almost every q, $M_{\mathrm{reg}} \cap f^{-1}(q) = f^{-1}(q)$. Thus, with slight abuse of notation, one could write the right side of (4.1.10) with $f^{-1}(q)$ replacing $M_{\mathrm{reg}} \cap f^{-1}(q)$.

Corollary 4.1.10 *We continue in the setup of Theorem 4.1.8. For every Borel-measurable nonnegative function ψ on N one has the integral formula*

$$\int \psi(f(p))J(\mathbb{T}_p(f))d\rho_M(p) = \int_{N_{\mathrm{reg}}} \rho[f^{-1}(q)]\psi(q)d\rho_N(q). \qquad (4.1.11)$$

Proof of Corollary 4.1.10 By (4.1.10) with $\varphi = \psi \circ f$, we have

$$\int \psi(f(p))J(\mathbb{T}_p(f))d\rho_M(p) = \int \rho[M_{\mathrm{reg}} \cap f^{-1}(q)]\psi(q)d\rho_N(q),$$

whence the result by Sard's Theorem (Theorem F.11), Proposition F.16, and the definitions. □

Let S^{n-1} be the unit sphere centered at the origin in $\mathbb{R}^n$. We will calculate $\rho[\mathrm{U}_n(\mathbb{F})]$ by relating it to $\rho[S^{n-1}]$. We prepare by proving two well-known lemmas concerning S^{n-1} and its volume. Their proofs provide templates for the more complicated proofs of Lemma 4.1.15 and Proposition 4.1.14 below.

Lemma 4.1.11 *S^{n-1} is a manifold and for every $x \in S^{n-1}$ we have $\mathbb{T}_x(S^{n-1}) = \{X \in \mathbb{R}^n : x \cdot X = 0\}$.*

Proof Consider the smooth map $f = (x \mapsto x \cdot x) : \mathbb{R}^n \to \mathbb{R}$. Let γ be a curve with $\gamma(0) = x \in \mathbb{R}^n$ and $\gamma'(0) = X \in \mathbb{T}_x(\mathbb{R}^n) = \mathbb{R}^n$. We have $(\mathbb{T}_x(f))(X) = (\gamma \cdot \gamma)'(0) = 2x \cdot X$. Thus 1 is a regular value of f, whence the result by Proposition F.16. □

Recall that $\Gamma(s) = \int_0^\infty x^{s-1}e^{-x}dx$ is Euler's Gamma function.

Proposition 4.1.12 *With notation as above, we have*

$$\rho[S^{n-1}] = \frac{2\pi^{n/2}}{\Gamma(n/2)}. \tag{4.1.12}$$

Proof Consider the smooth map

$$f = (x \mapsto x/\|x\|) : \mathbb{R}^n \setminus \{0\} \to S^{n-1}.$$

Let γ be a curve with $\gamma(0) = x \in \mathbb{R}^n \setminus \{0\}$ and $\gamma'(0) = X \in \mathbb{T}_x(\mathbb{R}^n \setminus \{0\}) = \mathbb{R}^n$. We have

$$(\mathbb{T}_x(f))(X) = (\gamma/\|\gamma\|)'(0) = \frac{X}{\|x\|} - \frac{x}{\|x\|}\left(\frac{X}{\|x\|}\cdot\frac{x}{\|x\|}\right),$$

and hence $J(\mathbb{T}_x(f)) = \|x\|^{1-n}$. Letting $\varphi(x) = \|x\|^{n-1}\exp(-\|x\|^2)$, we have

$$\int \cdots \int e^{-x\cdot x} dx_1 \cdots dx_n = \rho[S^{n-1}] \int_0^\infty r^{n-1} e^{-r^2} dr,$$

by Theorem 4.1.8 applied to f and φ. Formula (4.1.12) now follows. □

As further preparation for the evaluation of $\rho[\mathrm{U}_n(\mathbb{F})]$, we state without proof the following elementary lemma which allows us to consider transformations of manifolds by left (or right) matrix multiplication.

Lemma 4.1.13 *Let $M \subset \mathrm{Mat}_{n\times k}(\mathbb{F})$ be a manifold. Fix $g \in \mathrm{GL}_n(\mathbb{F})$. Let $f = (p \mapsto gp) : M \to gM = \{gp \in \mathrm{Mat}_{n\times k}(\mathbb{F}) : p \in M\}$. Then:*
(i) gM is a manifold and f is a diffeomorphism;
(ii) for every $p \in M$ and $X \in \mathbb{T}_p(M)$ we have $\mathbb{T}_p(f)(X) = gX$;
(iii) if $g \in \mathrm{U}_n(\mathbb{F})$, then f is an isometry (and hence measure-preserving).

The analogous statement concerning right-multiplication by an invertible matrix also holds. The lemma, especially part (iii) of it, will be frequently exploited throughout the remainder of Section 4.1.

Now we can state our main result concerning $\mathrm{U}_n(\mathbb{F})$ and its volume. Recall in what follows that $\beta = 1, 2, 4$ according as $\mathbb{F} = \mathbb{R}, \mathbb{C}, \mathbb{H}$.

Proposition 4.1.14 $\mathrm{U}_n(\mathbb{F})$ *is a manifold whose volume is*

$$\rho[\mathrm{U}_n(\mathbb{F})] = 2^{\beta n(n-1)/4} \prod_{k=1}^n \rho[S^{\beta k-1}] = \prod_{k=1}^n \frac{2(2\pi)^{\beta k/2}}{2^{\beta/2}\Gamma(\beta k/2)}. \tag{4.1.13}$$

The proof of Proposition 4.1.14 will be obtained by applying the coarea formula to the smooth map

$$f = (g \mapsto (\text{last column of } g)) : \mathrm{U}_n(\mathbb{F}) \to S^{\beta n-1} \tag{4.1.14}$$

where, abusing notation slightly, we make the isometric identification

$$S^{\beta n-1} = \{x \in \mathrm{Mat}_{n\times 1}(\mathbb{F}) : x^*x = 1\}$$

on the extreme right in (4.1.14).

Turning to the actual proof, we begin with the identification of $\mathrm{U}_n(\mathbb{F})$ as a manifold and the calculation of its tangent space at I_n.

Lemma 4.1.15 *$\mathrm{U}_n(\mathbb{F})$ is a manifold and $\mathbb{T}_{I_n}(\mathrm{U}_n(\mathbb{F}))$ is the space of anti-self-adjoint matrices in* $\mathrm{Mat}_n(\mathbb{F})$.

Proof Consider the smooth map

$$h = (X \mapsto X^*X) : \mathrm{Mat}_n(\mathbb{F}) \to \mathscr{H}_n(\mathbb{F}).$$

Let γ be a curve in $\mathrm{Mat}_n(\mathbb{F})$ with $\gamma(0) = I_n$ and $\gamma'(0) = X \in \mathbb{T}_{I_n}(\mathrm{Mat}_n(\mathbb{F})) = \mathrm{Mat}_n(\mathbb{F})$. Then, for all $g \in \mathrm{U}_n(\mathbb{F})$ and $X \in \mathrm{Mat}_n(\mathbb{F})$,

$$(\mathbb{T}_g(h))(gX) = ((g\gamma)^*(g\gamma))'(0) = X + X^*. \tag{4.1.15}$$

Thus I_n is a regular value of h, and hence $\mathrm{U}_n(\mathbb{F})$ is a manifold by Proposition F.16.

To find the tangent space $\mathbb{T}_{I_n}(\mathrm{U}_n(\mathbb{F}))$, consider a curve $\gamma(t) \in \mathrm{U}_n(\mathbb{F})$ with $\gamma(0) = I_n$. Then, because $XX^* = I_n$ on $\mathrm{U}_n(\mathbb{F})$ and thus the derivative of $h(\gamma(t))$ vanishes for $t = 0$, we deduce from (4.1.15) that $X + X^* = 0$, and hence $\mathbb{T}_{I_n}(\mathrm{U}_n(\mathbb{F}))$ is contained in the space of anti-self-adjoint matrices in $\mathrm{Mat}_n(\mathbb{F})$. Because the latter two spaces have the same dimension, the inclusion must be an equality. □

Recall the function f introduced in (4.1.14).

Lemma 4.1.16 *f is onto, and furthermore (provided that $n > 1$), for any $s \in S^{\beta n-1}$, the fiber $f^{-1}(s)$ is isometric to* $\mathrm{U}_{n-1}(\mathbb{F})$.

Proof The first claim (which should be obvious in the cases $\mathbb{F} = \mathbb{R}, \mathbb{C}$) is proved by applying Corollary E.8 with $k = 1$. To see the second claim, note first that for any $W \in \mathrm{U}_{n-1}(\mathbb{F})$, we have

$$\begin{bmatrix} W & 0 \\ 0 & 1 \end{bmatrix} \in \mathrm{U}_n(\mathbb{F}), \tag{4.1.16}$$

and that every $g \in \mathrm{U}_n(\mathbb{F})$ whose last column is the unit vector $e_n = (0,\dots,0,1)^{\mathrm{T}}$ is necessarily of the form (4.1.16). Therefore the fiber $f^{-1}(e_n)$ is isometric to $\mathrm{U}_{n-1}(\mathbb{F})$. To see the claim for other fibers, note that if $g, h \in \mathrm{U}_n(\mathbb{F})$, then $f(gh) = gf(h)$, and then apply part (iii) of Lemma 4.1.13. □

Lemma 4.1.17 *Let f be as in (4.1.14). Then:*
(i) $J(\mathbb{T}_g(f))$ is constant as a function of $g \in \mathrm{U}_n(\mathbb{F})$;
(ii) $J(\mathbb{T}_{I_n}(f)) = \sqrt{2}^{\beta(1-n)}$;
(iii) every value of f is regular.

Proof (i) Fix $h \in \mathrm{U}_n(\mathbb{F})$ arbitrarily. Let $e_n = (0,\dots,0,1)^{\mathrm{T}} \in \mathrm{Mat}_{n\times 1}$. The diagram

$$\begin{array}{ccccc} & \mathbb{T}_{I_n}(\mathrm{U}_n(\mathbb{F})) & \xrightarrow{\mathbb{T}_{I_n}(f)} & \mathbb{T}_{e_n}(S^{\beta n-1}) & \\ \mathbb{T}_{I_n}(g\mapsto hg) & \downarrow & & \downarrow & \mathbb{T}_{e_n}(x\mapsto hx) \\ & \mathbb{T}_h(\mathrm{U}_n(\mathbb{F})) & \xrightarrow{\mathbb{T}_h(f)} & \mathbb{T}_{f(h)}(S^{\beta n-1}) & \end{array}$$

commutes. Furthermore, its vertical arrows are, by part (ii) of Lemma 4.1.13, induced by left-multiplication by h, and hence are isometries of Euclidean spaces. Therefore we have $J(\mathbb{T}_h(f)) = J(\mathbb{T}_{I_n}(f))$.

(ii) Recall the notation $\mathbf{i},\mathbf{j},\mathbf{k}$ in Definition E.1. Recall the elementary matrices $e_{ij} \in \mathrm{Mat}_n(\mathbb{F})$ with 1 in position (i,j) and 0s elsewhere, see Appendix E.1. By Lemma 4.1.15 the collection

$$\{(ue_{ij} - u^* e_{ji})/\sqrt{2} : 1 \le i < j \le n,\ u \in \{1,\mathbf{i},\mathbf{j},\mathbf{k}\} \cap \mathbb{F}\} \\ \cup \quad \{ue_{ii} : 1 \le i \le n,\ u \in \{\mathbf{i},\mathbf{j},\mathbf{k}\} \cap \mathbb{F}\}$$

is an orthonormal basis for $\mathbb{T}_{I_n}(\mathrm{U}_n(\mathbb{F}))$. Let γ be a curve in $\mathrm{U}_n(\mathbb{F})$ with $\gamma(0) = I_n$ and $\gamma'(0) = X \in \mathbb{T}_{I_n}(\mathrm{U}_n(\mathbb{F}))$. We have

$$(\mathbb{T}_{I_n}(f))(X) = (\gamma e_n)'(0) = Xe_n,$$

hence the collection

$$\{(ue_{in} - u^* e_{ni})/\sqrt{2} : 1 \le i < n,\ u \in \{1,\mathbf{i},\mathbf{j},\mathbf{k}\} \cap \mathbb{F}\} \\ \cup \quad \{ue_{nn} : u \in \{\mathbf{i},\mathbf{j},\mathbf{k}\} \cap \mathbb{F}\}$$

is an orthonormal basis for $\mathbb{T}_{I_n}(\mathrm{U}_n(\mathbb{F})) \cap (\ker(\mathbb{T}_{I_n}(f)))^{\perp}$. An application of Lemma F.19 yields the desired formula.

(iii) This follows from the preceding two statements, since f is onto. □

Proof of Proposition 4.1.14 Assume at first that $n > 1$. We apply Corollary 4.1.10 to f with $\psi \equiv 1$. After simplifying with the help of the preceding two lemmas, we find the relation

$$\sqrt{2}^{\beta(1-n)} \rho[\mathrm{U}_n(\mathbb{F})] = \rho[\mathrm{U}_{n-1}(\mathbb{F})]\, \rho[S^{\beta n-1}].$$

By induction on n we conclude that formula (4.1.13) holds for all positive integers n; the induction base $n = 1$ holds because $S^{\beta-1} = \mathrm{U}_1(\mathbb{F})$. □

With an eye toward the proof of Proposition 4.1.4 about Jacobi ensembles, we prove the following concerning the spaces $\mathrm{Flag}_n(\lambda,\mathbb{F})$ defined in (4.1.4).

Lemma 4.1.18 *With p,q,n positive integers so that $p+q=n$, and $D=\mathrm{diag}(I_p,0_q)$, the collection $\mathrm{Flag}_n(D,\mathbb{F})$ is a manifold of dimension βpq.*

Proof In view of Corollary E.12 (the spectral theorem for self-adjoint matrices over $\mathbb{F}$), $\mathrm{Flag}_n(D,\mathbb{F})$ is the set of projectors in $\mathrm{Mat}_n(\mathbb{F})$ of trace p. Now consider the open set $O \subset \mathscr{H}_n(\mathbb{F})$ consisting of matrices whose p-by-p block in upper left is invertible, noting that $D \in O$. Using Corollary E.9, one can construct a smooth map from $\mathrm{Mat}_{p\times q}(\mathbb{F})$ to $O \cap \mathrm{Flag}_n(D,\mathbb{F})$ with a smooth inverse. Now let $P \in \mathrm{Flag}_n(D,\mathbb{F})$ be any point. By definition $P = U^*DU$ for some $U \in \mathrm{U}_n(D,\mathbb{F})$. By Lemma 4.1.13 the set $\{UMU^* \mid M \in O \cap \mathrm{Flag}_n(D,\mathbb{F})\}$ is a neighborhood of P diffeomorphic to $O \cap \mathrm{Flag}_n(D,\mathbb{F})$ and hence to $\mathrm{Mat}_{p\times q}(\mathbb{F})$. Thus $\mathrm{Flag}_n(D,\mathbb{F})$ is indeed a manifold of dimension βpq. □

Motivated by Lemma 4.1.18, we refer to $\mathrm{Flag}_n(D,\mathbb{F})$ as the *flag manifold* determined by D. In fact the claim in Lemma 4.1.18 holds for all real diagonal matrices D, see Exercise 4.1.19 below.

Exercise 4.1.19 Fix $\lambda_1,\dots,\lambda_n \in \mathbb{R}$ and put $\lambda = \mathrm{diag}(\lambda_1,\dots,\lambda_n)$. In this exercise we study $\mathrm{Flag}_n(\lambda,\mathbb{F})$. Write $\{\mu_1 < \cdots < \mu_\ell\} = \{\lambda_1,\dots,\lambda_n\}$ and let n_i be the number of indices j such that $\mu_i = \lambda_j$. (Thus, $n = n_1 + \cdots + n_\ell$.)
(a) Prove that $\mathrm{Flag}_n(\lambda,\mathbb{F})$ is a manifold of dimension equal to

$$\dim \mathrm{U}_n(\mathbb{F}) - \sum_{i=1}^{\ell} \dim \mathrm{U}_{n_i}(\mathbb{F}).$$

(b) Applying the coarea formula to the smooth map $f = (g \mapsto g\lambda g^{-1}) : \mathrm{U}_n(\mathbb{F}) \to \mathrm{Flag}_n(D,\mathbb{F})$, show that

$$\rho[\mathrm{Flag}_n(\lambda,\mathbb{F})] = \frac{\rho[\mathrm{U}_n(\mathbb{F})]}{\prod_{i=1}^{\ell} \rho[\mathrm{U}_{n_i}(\mathbb{F})]} \prod_{\substack{1\le i<j\le n \\ \lambda_i \neq \lambda_j}} |\lambda_i - \lambda_j|^\beta . \tag{4.1.17}$$

Exercise 4.1.20 We look at joint distribution of eigenvalues in the Gaussian ensembles (GUE/GOE/GSE) in yet another way. We continue with the notation of the previous exercise.
(a) Consider the smooth map $f = (A \mapsto (\mathrm{tr}(A), \mathrm{tr}(A^2)/2, \dots, \mathrm{tr}(A^n)/n)) : \mathscr{H}_n(\mathbb{F}) \to \mathbb{R}^n$. Show that $J(\mathbb{T}_A(f))$ depends only on the eigenvalues of $A \in \mathscr{H}_n(\mathbb{F})$, that $J(\mathbb{T}_\lambda(f)) = |\Delta(\lambda)|$, and that a point of $\mathbb{R}^n$ is a regular value of f if and only if it is of the form $f(X)$ for some $X \in \mathscr{H}_n(\mathbb{F})$ with distinct eigenvalues.
(b) Applying the coarea formula to f, prove that for any nonnegative Borel-

measurable function φ on $\mathscr{H}_n(\mathbb{F})$,

$$\int \varphi d\rho_{\mathscr{H}_n(\mathbb{F})} = \underbrace{\int \cdots \int}_{\substack{-\infty<\lambda_1<\cdots<\lambda_n<\infty \\ \lambda=\mathrm{diag}(\lambda_1,\ldots,\lambda_n)}} \left(\int \varphi d\rho_{\mathrm{Flag}_n(\lambda,\mathbb{F})} \right) d\lambda_1 \cdots d\lambda_n . \tag{4.1.18}$$

(c) Derive the joint distribution of eigenvalues in the GUE, GOE and GSE from (4.1.17) and (4.1.18).

Exercise 4.1.21 Fix $\lambda_1,\ldots,\lambda_n \in \mathbb{C}$ and put $\lambda = \mathrm{diag}(\lambda_1,\ldots,\lambda_n)$. Let $\mathrm{Flag}_n(\lambda,\mathbb{C})$ be the set of normal matrices with the same eigenvalues as λ. (When λ has real entries, then $\mathrm{Flag}_n(\lambda,\mathbb{C})$ is just as we defined it before.) Show that in this extended setting $\mathrm{Flag}_n(\lambda,\mathbb{C})$ is again a manifold and that formula (4.1.17), with $\mathbb{F}=\mathbb{C}$ and $\beta=2$, still holds.

4.1.3 An integration formula of Weyl type

For the rest of Section 4.1 we will be working in the setup of Lie groups, see Appendix F for definitions and basic properties. We aim to derive an integration formula of Weyl type, Theorem 4.1.28, in some generality, which encompasses all the results enunciated in Section 4.1.1.

Our immediate goal is to introduce a framework within which a uniform approach to derivation of joint eigenvalue distributions is possible. For motivation, suppose that G and M are submanifolds of $\mathrm{Mat}_n(\mathbb{F})$ and that G is a closed subgroup of $\mathrm{U}_n(\mathbb{F})$ such that $\{gmg^{-1} : m \in M, g \in G\} = M$. We want to "integrate out" the action of G. More precisely, given a submanifold $\Lambda \subset M$ which satisfies $M = \{g\lambda g^{-1} : g \in G,\ \lambda \in \Lambda\}$, and a function φ on M such that $\varphi(gmg^{-1}) = \varphi(m)$ for all $m \in M$ and $g \in G$, we want to represent $\int \varphi d\rho_M$ in a natural way as an integral on Λ. This is possible if we can control the set of solutions $(g,\lambda) \in G \times \Lambda$ of the equation $g\lambda g^{-1} = m$ for all but a negligible set of $m \in M$. Such a procedure was followed in Section 2.5 when deriving the law of the eigenvalues of the GOE. However, as was already noted in the derivation of the law of the eigenvalues of the GUE, decompositions of the form $m = g\lambda g^{-1}$ are not unique, and worse, the set $\{(g,\lambda) \in G \times \Lambda : g\lambda g^{-1} = m\}$ is in general not discrete. Fortunately, however, it typically has the structure of compact manifold. These considerations (and hindsight based on familiarity with classical matrix ensembles) motivate the following definition.

Definition 4.1.22 A *Weyl quadruple* (G,H,M,Λ) consists of four manifolds G,

H, M and Λ with common ambient space $\mathrm{Mat}_n(\mathbb{F})$ satisfying the following conditions:

(I) (a) G is a closed subgroup of $\mathrm{U}_n(\mathbb{F})$,
(b) H is a closed subgroup of G, and
(c) $\dim G - \dim H = \dim M - \dim \Lambda$.

(II) (a) $M = \{g\lambda g^{-1} : g \in G,\ \lambda \in \Lambda\}$,
(b) $\Lambda = \{h\lambda h^{-1} : h \in H,\ \lambda \in \Lambda\}$,
(c) for every $\lambda \in \Lambda$ the set $\{h\lambda h^{-1} : h \in H\}$ is finite, and
(d) for all $\lambda, \mu \in \Lambda$ we have $\lambda^*\mu = \mu\lambda^*$.

(III) There exists $\Lambda' \subset \Lambda$ such that

(a) Λ' is open in Λ,
(b) $\rho_\Lambda(\Lambda \setminus \Lambda') = 0$, and
(c) for every $\lambda \in \Lambda'$ we have $H = \{g \in G : g\lambda g^{-1} \in \Lambda\}$.

We say that a subset $\Lambda' \subset \Lambda$ for which (IIIa,b,c) hold is *generic*.

We emphasize that by conditions (Ia,b), the groups G and H are compact, and that by Lemma 4.1.13(iii), the measures ρ_G and ρ_H are Haar measures. We also remark that we make no connectedness assumptions concerning G, H, M and Λ. (In general, we do not require manifolds to be connected, although we do assume that all tangent spaces of a manifold are of the same dimension.) In fact, in practice, H is usually not connected.

In the next proposition we present the simplest example of a Weyl quadruple. We recall, as in Definition E.4, that a matrix $h \in \mathrm{Mat}_n(\mathbb{F})$ is *monomial* if it factors as the product of a diagonal matrix and a permutation matrix.

Proposition 4.1.23 *Let $G = \mathrm{U}_n(\mathbb{F})$ and let $H \subset G$ be the subset consisting of monomial elements. Let $M = \mathscr{H}_n(\mathbb{F})$, let $\Lambda \subset M$ be the subset consisting of (real) diagonal elements, and let $\Lambda' \subset \Lambda$ be the subset consisting of matrices with distinct diagonal entries. Then (G, H, M, Λ) is a Weyl quadruple with ambient space* $\mathrm{Mat}_n(\mathbb{F})$ *for which the set Λ' is generic, and furthermore*

$$\frac{\rho[G]}{\rho[H]} = \frac{\rho[\mathrm{U}_n(\mathbb{F})]}{n!\rho[\mathrm{U}_1(\mathbb{F})]^n}. \tag{4.1.19}$$

This Weyl quadruple and the value of the associated constant $\rho[G]/\rho[H]$ will be used to prove Proposition 4.1.1.

Proof Of all the conditions imposed by Definition 4.1.22, only conditions (Ic),

(IIa) and (IIIc) require special attention, because the others are clear. To verify condition (Ic), we note that

$$\dim M = n + \beta n(n-1)/2,\ \dim \Lambda = n,$$
$$\dim G = (\beta - 1)n + \beta n(n-1)/2,\ \dim H = (\beta - 1)n.$$

The first two equalities are clear since M and Λ are real vector spaces. By Lemma 4.1.15 the tangent space $\mathbb{T}_{I_n}(G)$ consists of the collection of anti-self-adjoint matrices in $\mathrm{Mat}_n(\mathbb{F})$, and thus the third equality holds. So does the fourth because $\mathbb{T}_{I_n}(H)$ consists of the diagonal elements of $\mathbb{T}_{I_n}(G)$. Thus condition (Ic) holds. To verify condition (IIa), we have only to apply Corollary E.12(i) which asserts the possibility of diagonalizing a self-adjoint matrix. To verify condition (IIIc), arbitrarily fix $\lambda \in \Lambda'$, $\mu \in \Lambda$ and $g \in G$ such that $g\lambda g^{-1} = \mu$, with the goal to show that $g \in H$. In any case, by Corollary E.12(ii), the diagonal entries of μ are merely a rearrangement of those of λ. After left-multiplying g by a permutation matrix (the latter belongs by definition to H), we may assume that $\lambda = \mu$, in which case g commutes with λ. Then, because the diagonal entries of λ are distinct, it follows that g is diagonal and thus belongs to H. Thus (IIIc) is proved. Thus (G,H,M,Λ) is a Weyl quadruple for which Λ' is generic.

We turn to the verification of formula (4.1.19). It is clear that the numerator on the right side of (4.1.19) is correct. To handle the denominator, we observe that H is the disjoint union of $n!$ isometric copies of the manifold $\mathrm{U}_1(\mathbb{F})^n$, and then apply Proposition F.8(vi). Thus (4.1.19) is proved. □

Note that condition (IIa) of Definition 4.1.22 implies that $gmg^{-1} \in M$ for all $m \in M$ and $g \in G$. Thus the following definition makes sense.

Definition 4.1.24 Given a Weyl quadruple (G,H,M,Λ) and a function φ on M (resp., a subset $A \subset M$), we say that φ (resp., A) is *G-conjugation-invariant* if $\varphi(gmg^{-1}) = \varphi(m)$ (resp., $\mathbf{1}_A(gmg^{-1}) = \mathbf{1}_A(m)$) for all $g \in G$ and $m \in M$.

Given a Weyl quadruple (G,H,M,Λ) and a G-conjugation-invariant nonnegative Borel-measurable function φ on M, we aim now to represent $\int \varphi d\rho_M$ as an integral on Λ. Our strategy for achieving this is to apply the coarea formula to the smooth map

$$f = (g \mapsto g\lambda g^{-1}) : G \times \Lambda \to M. \tag{4.1.20}$$

For the calculation of the factor $J(\mathbb{T}_{(g,\lambda)}(f))$ figuring in the coarea formula for the map f we need to understand for each fixed $\lambda \in \Lambda$ the structure of the derivative at $I_n \in G$ of the map

$$f_\lambda = (g \mapsto g\lambda g^{-1}) : G \to M \tag{4.1.21}$$

obtained by "freezing" the second variable in f. For study of the derivative $\mathbb{T}_{I_n}(f_\lambda)$ the following *ad hoc* version of the Lie bracket will be useful.

Definition 4.1.25 Given $X,Y \in \mathrm{Mat}_n(\mathbb{F})$, let $[X,Y] = XY - YX$.

Concerning the derivative $T_{I_n}(f_\lambda)$ we then have the following key result.

Lemma 4.1.26 *Fix a Weyl quadruple* (G,H,M,Λ) *with ambient space* $\mathrm{Mat}_n(\mathbb{F})$ *and a point* $\lambda \in \Lambda$. *Let* f_λ *be as in (4.1.21). Then we have*

$$\mathbb{T}_{I_n}(f_\lambda)(\mathbb{T}_{I_n}(H)) = 0, \tag{4.1.22}$$

$$\mathbb{T}_{I_n}(f_\lambda)(X) = [X,\lambda], \tag{4.1.23}$$

$$\mathbb{T}_{I_n}(f_\lambda)(\mathbb{T}_{I_n}(G)) \subset \mathbb{T}_\lambda(M) \cap \mathbb{T}_\lambda(\Lambda)^\perp. \tag{4.1.24}$$

The proof will be given later.

Definition 4.1.27 Let (G,H,M,Λ) be a Weyl quadruple. Given $\lambda \in \Lambda$, let

$$D_\lambda : \mathbb{T}_{I_n}(G) \cap \mathbb{T}_{I_n}(H)^\perp \to \mathbb{T}_\lambda(M) \cap \mathbb{T}_\lambda(\Lambda)^\perp \tag{4.1.25}$$

be the linear map induced by $\mathbb{T}_{I_n}(f_\lambda)$. For each $\lambda \in \Lambda$ we define the *Weyl operator* Θ_λ to equal $D_\lambda^* \circ D_\lambda$.

The abbreviated notation D_λ and Θ_λ is appropriate because in applications below the corresponding Weyl quadruple (G,H,M,Λ) will be fixed, and thus need not be referenced in the notation. We emphasize that the source and target of the linear map D_λ have the same dimension by assumption (Ic). The determinant $\det\Theta_\lambda$, which is independent of the choice of basis used to compute it, is nonnegative because Θ_λ is positive semidefinite, and hence $\sqrt{\det\Theta_\lambda}$ is a well-defined nonnegative number. We show in formula (4.1.29) below how to reduce the calculation of Θ_λ to an essentially mechanical procedure. Remarkably, in all intended applications, we can calculate $\det\Theta_\lambda$ by exhibiting an orthogonal basis for $\mathbb{T}_{I_n}(G) \cap \mathbb{T}_{I_n}(H)^\perp$ simultaneously diagonalizing the whole family $\{\Theta_\lambda\}_{\lambda\in\Lambda}$.

We are now ready to state the generalized Weyl integration formula.

Theorem 4.1.28 (Weyl) *Let* (G,H,M,Λ) *be a Weyl quadruple. Then for every Borel-measurable nonnegative G-conjugation-invariant function* φ *on* M, *we have*

$$\int \varphi d\rho_M = \frac{\rho[G]}{\rho[H]} \int \varphi(\lambda)\sqrt{\det\Theta_\lambda}\, d\rho_\Lambda(\lambda).$$

The proof takes up the rest of Section 4.1.3. We emphasize that a Weyl quadruple (G,H,M,Λ) with ambient space $\mathrm{Mat}_n(\mathbb{F})$ is fixed now and remains so until the end of Section 4.1.3.

We begin with the analysis of the maps f and f_λ defined in (4.1.20) and (4.1.21), respectively.

Lemma 4.1.29 *The restricted function $f_\lambda|_H$ is constant on connected components of H, and* a fortiori *has identically vanishing derivative.*

Proof The function $f_\lambda|_H$ is continuous and by assumption (IIc) takes only finitely many values. Thus $f_\lambda|_H$ is locally constant, whence the result. □

Lemma 4.1.30 *Let $\Lambda' \subset \Lambda$ be generic. Then for every $g_0 \in G$ and $\lambda_0 \in \Lambda'$, the fiber $f^{-1}(g_0\lambda_0 g_0^{-1})$ is a manifold isometric to H.*

It follows from Lemma 4.1.30 and Proposition F.8(v) that $\rho[f^{-1}(g_0\lambda_0 g_0^{-1})] = \rho[H]$.

Proof We claim that

$$f^{-1}(g_0\lambda_0 g_0^{-1}) = \{(g_0 h, h^{-1}\lambda_0 h) \in G \times M : h \in H\}.$$

The inclusion $\supset$ follows from assumption (IIb). To prove the opposite inclusion $\subset$, suppose now that $g\lambda g^{-1} = g_0\lambda_0 g_0^{-1}$ for some $g \in G$ and $\lambda \in \Lambda$. Then we have $g^{-1}g_0 \in H$ by assumption (IIIc), hence $g_0^{-1}g = h$ for some $h \in H$, and hence $(g,\lambda) = (g_0 h, h^{-1}\lambda_0 h)$. The claim is proved. By assumptions (Ia,b) and Lemma 4.1.13(iii), the map

$$(h \mapsto g_0 h) : H \to g_0 H = \{g_0 h : h \in H\}$$

is an isometry of manifolds, and indeed is the restriction to H of an isometry of Euclidean spaces. In view of Lemma 4.1.29, the map

$$(h \mapsto (g_0 h, h^{-1}\lambda_0 h)) : H \to f^{-1}(g_0\lambda_0 g_0^{-1}) \qquad (4.1.26)$$

is also an isometry, which finishes the proof of Lemma 4.1.30. □

Note that we have *not* asserted that the map (4.1.26) preserves distances as measured in ambient Euclidean spaces, but rather merely that it preserves geodesic distances within the manifolds in question. For manifolds with several connected components (as is typically the case for H), distinct connected components are considered to be at infinite distance one from the other.

Proof of Lemma 4.1.26 The identity (4.1.22) follows immediately from Lemma 4.1.29.

We prove (4.1.23). Let γ be a curve in G with $\gamma(0) = I_n$ and $\gamma'(0) = X \in \mathbb{T}_{I_n}(G)$. Since $(\gamma^{-1})' = -\gamma^{-1}\gamma'\gamma^{-1}$, we have $\mathbb{T}_{I_n}(f_\lambda)(X) = (\gamma\lambda\gamma^{-1})'(0) = [X,\lambda]$. Thus (4.1.23) holds.

It remains to prove (4.1.24). As a first step, we note that

$$[\lambda^*, X] = 0 \text{ for } \lambda \in \Lambda \text{ and } X \in T_\lambda(\Lambda). \tag{4.1.27}$$

Indeed, let γ be a curve in Λ with $\gamma(0) = \lambda$ and $\gamma'(0) = X$. Then $[\lambda^*, \gamma]$ vanishes identically by Assumption (IId) and hence $[\lambda^*, X] = 0$.

We further note that

$$[X,\lambda]\cdot Y = X \cdot [Y, \lambda^*] \;\text{ for } X, Y \in \mathrm{Mat}_n(\mathbb{F}), \tag{4.1.28}$$

which follows from the definition $A \cdot B = \Re \mathrm{tr} X^* Y$ for any $A, B \in \mathrm{Mat}_n(\mathbb{F})$ and straightforward manipulations.

We now prove (4.1.24). Given $X \in \mathbb{T}_{I_n}(G)$ and $L \in T_\lambda(\Lambda)$, we have

$$\mathbb{T}_{I_n}(f_\lambda)(X) \cdot L = [X,\lambda]\cdot L = X \cdot [L, \lambda^*] = 0,$$

where the first equality follows from (4.1.23), the second from (4.1.28) and the last from (4.1.27). This completes the proof of (4.1.24) and of Lemma 4.1.26. □

Lemma 4.1.31 *Let $\Pi : \mathrm{Mat}_n(\mathbb{F}) \to \mathbb{T}_{I_n}(G) \cap \mathbb{T}_{I_n}(H)^\perp$ be the orthogonal projection. Fix $\lambda \in \Lambda$. Then the following hold:*

$$\Theta_\lambda(X) = \Pi([\lambda^*, [\lambda, X]]) \text{ for } X \in \mathbb{T}_{I_n}(G) \cap \mathbb{T}_{I_n}(H)^\perp, \tag{4.1.29}$$

$$J(\mathbb{T}_{(g,\lambda)}(f)) = \sqrt{\det \Theta_\lambda} \text{ for } g \in G. \tag{4.1.30}$$

Proof We prove (4.1.29). Fix $X, Y \in \mathbb{T}_{I_n}(G) \cap \mathbb{T}_{I_n}(H)^\perp$ arbitrarily. We have

$$\begin{aligned} \Theta_\lambda(X)\cdot Y &= D_\lambda^*(D_\lambda(X)) \cdot Y = D_\lambda(X)\cdot D_\lambda(Y) \\ &= \mathbb{T}_{I_n}(f_\lambda)(X) \cdot \mathbb{T}_{I_n}(f_\lambda)(Y) \\ &= [X,\lambda]\cdot[Y,\lambda] = [[X,\lambda],\lambda^*]\cdot Y = \Pi([[X,\lambda],\lambda^*])\cdot Y \end{aligned}$$

at the first step by definition, at the second step by definition of adjoint, at the third step by definition of D_λ, at the fourth step by (4.1.23), at the fifth step by (4.1.28) and at the last step trivially. Thus (4.1.29) holds.

Fix $h \in G$ arbitrarily. We claim that $J(\mathbb{T}_{(h,\lambda)}(f))$ is independent of $h \in G$. Toward that end consider the commuting diagram

$$\begin{array}{ccccc} & \mathbb{T}_{(I_n,\lambda)}(G\times\Lambda) & \xrightarrow{\mathbb{T}_{(I_n,\lambda)}(f)} & \mathbb{T}_\lambda(M) & \\ \mathbb{T}_{(I_n,\lambda)}((g,\mu)\mapsto(hg,\mu)) & \downarrow & & \downarrow & \mathbb{T}_\lambda(m\mapsto hmh^{-1}). \\ & \mathbb{T}_{(h,\lambda)}(G\times\Lambda) & \xrightarrow{\mathbb{T}_{(h,\lambda)}(f)} & \mathbb{T}_{hmh^{-1}}(M) & \end{array}$$

Since the vertical arrows are isometries of Euclidean spaces by assumption (Ia) and Lemma 4.1.13(ii), it follows that $J(\mathbb{T}_{(h,\lambda)}(f)) = J(\mathbb{T}_{(I_n,\lambda)}(f)$, and in particular is independent of h, as claimed.

We now complete the proof of (4.1.30), assuming without loss of generality that $g = I_n$. By definition

$$\mathbb{T}_{(I_n,\lambda)}(G\times\Lambda) = \mathbb{T}_{I_n}(G)\oplus\mathbb{T}_\lambda(\Lambda),$$

where we recall that the direct sum is equipped with Euclidean structure by declaring the summands to be orthogonal. Clearly we have

$$(\mathbb{T}_{(I_n,\lambda)}(f))(X\oplus L) = \mathbb{T}_{I_n}(f_\lambda)(X) + L \text{ for } X\in\mathbb{T}_{I_n}(G) \text{ and } L\in\mathbb{T}_\lambda(\Lambda). \quad (4.1.31)$$

By (4.1.24) and (4.1.31), the linear map $\mathbb{T}_{(I_n,\lambda)}(f)$ decomposes as the orthogonal direct sum of $\Sigma\circ\mathbb{T}_{I_n}(f_\lambda)$ and the identity map of $\mathbb{T}_\lambda(\Lambda)$ to itself. Consequently we have $J(\mathbb{T}_{I_n,\lambda}(f)) = J(\Sigma\circ\mathbb{T}_{I_n}(f_\lambda))$ by Lemma F.18. Finally, by assumption (Ic), formula (4.1.22) and Lemma F.19, we find that $J(\Sigma\circ\mathbb{T}_{I_n}(f_\lambda)) = \sqrt{\det\Theta_\lambda}$. □

Proof of Theorem 4.1.28 Let M_{reg} be the set of regular values of the map f. We have

$$\begin{aligned}\int_{M_{\text{reg}}} \rho[f^{-1}(m)]\varphi(m)d\rho_M(m) &= \int \varphi(\lambda)\sqrt{\det\Theta_\lambda}d\rho_{G\times\Lambda}(g,\lambda)\\ &= \rho[G]\cdot\int\varphi(\lambda)\sqrt{\det\Theta_\lambda}d\rho_\Lambda(\lambda). \qquad (4.1.32)\end{aligned}$$

The two equalities in (4.1.32) are justified as follows. The first holds by formula (4.1.30), the "pushed down" version (4.1.11) of the coarea formula, and the fact that $\varphi(f(g,\lambda)) = \varphi(\lambda)$ by the assumption that φ is G-conjugation-invariant. The second holds by Fubini's Theorem and the fact that $\rho_{G\times\Lambda} = \rho_G\times\rho_\Lambda$ by Proposition F.8(vi).

By assumption (IIa) the map f is onto, hence $M_{\text{reg}} = M\setminus M_{\text{crit}}$, implying by Sard's Theorem (Theorem F.11) that M_{reg} has full measure in M. For every $m\in M_{\text{reg}}$, the quantity $\rho[f^{-1}(m)]$ is positive (perhaps infinite). The quantity $\rho[G]$ is positive and also finite since G is compact. It follows by (4.1.32) that the claimed integration formula at least holds in the weak sense that a G-conjugation-invariant Borel set $A\subset M$ is negligible in M if the intersection $A\cap\Lambda$ is negligible in Λ.

Now put $M' = \{g\lambda g^{-1} : g\in G,\ \lambda\in\Lambda'\}$. Then M' is a Borel set. Indeed, by assumption (IIIa) the set Λ' is σ-compact, hence so is M'. By construction M' is G-conjugation-invariant. Now we have $\Lambda'\subset M'\cap\Lambda$, hence by assumption (IIIb) the intersection $M'\cap\Lambda$ is of full measure in Λ, and therefore by what we proved in the paragraph above, M' is of full measure in M. Thus, if we replace φ by $\varphi\mathbf{1}_{M'}$ in (4.1.32), neither the first nor the last integral in (4.1.32) changes and further, by Lemma 4.1.30, we can replace the factor $\rho_{f^{-1}(m)} f^{-1}(m)$ in the first integral by $\rho[H]$. Therefore we have

$$\rho[H]\int_{M'\cap M_{\mathrm{reg}}} \varphi(m)d\rho_M(m) = \rho[G]\int_{M'\cap\Lambda} \varphi(\lambda)\sqrt{\det\Theta_\lambda}\,d\rho_\Lambda(\lambda)\,.$$

Finally, since $M'\cap M_{\mathrm{reg}}$ is of full measure in M and $M'\cap\Lambda$ is of full measure in Λ, the desired formula holds. □

4.1.4 Applications of Weyl's formula

We now present the proofs of the integration formulas of Section 4.1.1. We prove each by applying Theorem 4.1.28 to a suitable Weyl quadruple.

We begin with the Gaussian/Hermite ensembles.

Proof of Proposition 4.1.1 Let (G,H,M,Λ) be the Weyl quadruple defined in Proposition 4.1.23. As in the proof of Lemma 4.1.17 above, and for a similar purpose, we use the notation $e_{ij}, \mathbf{i}, \mathbf{j}, \mathbf{k}$. By Lemma 4.1.15 we know that $\mathbb{T}_{I_n}(G)\subset \mathrm{Mat}_n(\mathbb{F})$ is the space of anti-self-adjoint matrices, and it is clear that $\mathbb{T}_{I_n}(H)\subset \mathbb{T}_{I_n}(G)$ is the subspace consisting of diagonal anti-self-adjoint matrices. Thus the set

$$\left\{ ue_{ij} - u^* e_{ji} \,\middle|\, u\in\{1,\mathbf{i},\mathbf{j},\mathbf{k}\}\cap\mathbb{F},\ 1\le i<j\le n\right\}$$

is an orthogonal basis for $\mathbb{T}_{I_n}(G)\cap\mathbb{T}_{I_n}(H)^\perp$. By formula (4.1.29), we have

$$\Theta_{\mathrm{diag}(x)}(ue_{ij}-u^*e_{ji}) = [\mathrm{diag}(x),[\mathrm{diag}(x),ue_{ij}-u^*e_{ji}]] = (x_i-x_j)^2(ue_{ij}-u^*e_{ji})$$

and hence

$$\sqrt{\det\Theta_{\mathrm{diag}(x)}} = |\Delta(x)|^\beta \ \text{ for } x\in\mathbb{R}^n.$$

To finish the bookkeeping, note that the map $x\mapsto \mathrm{diag}(x)$ sends $\mathbb{R}^n$ isometrically to Λ and hence pushes Lebesgue measure on $\mathbb{R}^n$ forward to ρ_Λ. Then the integration formula (4.1.1) follows from Theorem 4.1.28 combined with formula (4.1.19) for $\rho[G]/\rho[H]$. □

We remark that the orthogonal projection Π appearing in formula (4.1.29) is unnecessary in the Gaussian setup. In contrast, we will see that it does play a nontrivial role in the study of the Jacobi ensembles.

We turn next to the Laguerre ensembles. The following proposition provides the needed Weyl quadruples.

Proposition 4.1.32 *Fix integers $0 < p \leq q$ and put $n = p + q$. Let*

$$
\begin{aligned}
G &= \{\mathrm{diag}(U,V) : U \in \mathrm{U}_p(\mathbb{F}), V \in \mathrm{U}_q(\mathbb{F})\} \subset \mathrm{U}_n(\mathbb{F}), \\
H &= \{\mathrm{diag}(U,V',V'') : U,V' \in \mathrm{U}_p(\mathbb{F}), V'' \in \mathrm{U}_{q-p}(\mathbb{F}), \\
&\quad\ U,V' \textit{ are monomial, } U(V')^* \textit{ is diagonal, } (U(V')^*)^2 = I_p\} \subset G, \\
M &= \left\{ \begin{bmatrix} 0 & X \\ X^* & 0 \end{bmatrix} : X \in \mathrm{Mat}_{p\times q}(\mathbb{F}) \right\} \subset \mathscr{H}_n(\mathbb{F}), \\
\Lambda &= \left\{ \begin{bmatrix} 0 & x & 0 \\ x & 0 & 0 \\ 0 & 0 & 0_{q-p} \end{bmatrix} : x \in \mathrm{Mat}_p \textit{ is (real) diagonal} \right\} \subset M.
\end{aligned}
$$

Let $\Lambda' \subset \Lambda$ be the subset consisting of elements for which the corresponding real diagonal matrix x has nonzero diagonal entries with distinct absolute values. Then (G,H,M,Λ) is a Weyl quadruple with ambient space $\mathrm{Mat}_n(\mathbb{F})$ for which the set Λ' is generic and, furthermore,

$$
\frac{\rho[G]}{\rho[H]} = \frac{\rho[\mathrm{U}_p(\mathbb{F})]\rho[\mathrm{U}_q(\mathbb{F})]}{2^p p!(2^{(\beta-1)/2}\rho[\mathrm{U}_1(\mathbb{F})])^p \rho[\mathrm{U}_{q-p}(\mathbb{F})]}. \tag{4.1.33}
$$

We remark that in the case $p = q$ we are abusing notation slightly. For $p = q$ one should ignore V'' in the definition of H, and similarly modify the other definitions and formulas.

Proof Of the conditions imposed by Definition 4.1.22, only conditions (Ic), (IIa) and (IIIc) deserve comment. As in the proof of Proposition 4.1.23 one can verify (Ic) by means of Lemma 4.1.15. Conditions (IIa) and (IIIc) follow from Corollary E.13 concerning the singular value decomposition in $\mathrm{Mat}_{p\times q}(\mathbb{F})$, and specifically follow from points (i) and (iii) of that corollary, respectively. Thus (G,H,M,Λ) is a Weyl quadruple for which Λ' is generic.

Turning to the proof of (4.1.33), note that the group G is isometric to the product $\mathrm{U}_p(\mathbb{F}) \times \mathrm{U}_q(\mathbb{F})$. Thus the numerator on the right side of (4.1.33) is justified. The map $x \mapsto \mathrm{diag}(x,x)$ from $\mathrm{U}_1(\mathbb{F})$ to $\mathrm{U}_2(\mathbb{F})$ magnifies by a factor of $\sqrt{2}$. Abusing notation, we denote its image by $\sqrt{2}\mathrm{U}_1(\mathbb{F})$. The group H is the disjoint union of $2^p p!$ isometric copies of the manifold $(\sqrt{2}\mathrm{U}_1(\mathbb{F}))^p \times \mathrm{U}_{q-p}(\mathbb{F})$. This justifies the denominator on the right side of (4.1.33), and completes the proof. □

Proof of Proposition 4.1.3 Let (G,H,M,Λ) be the Weyl quadruple defined in Proposition 4.1.32. By Lemma 4.1.15, $\mathbb{T}_{I_n}(G)$ consists of matrices of the form $\mathrm{diag}(X,Y)$, where $X \in \mathrm{Mat}_p(\mathbb{F})$ and $Y \in \mathrm{Mat}_q(\mathbb{F})$ are anti-self-adjoint. By the same lemma, $\mathbb{T}_{I_n}(H)$ consists of matrices of the form $\mathrm{diag}(W,W,Z)$, where $W \in \mathrm{Mat}_p(\mathbb{F})$ is diagonal anti-self-adjoint and $Z \in \mathrm{Mat}_{q-p}(\mathbb{R})$ is anti-self-adjoint. Thus

$\mathbb{T}_{I_n}(G) \cap \mathbb{T}_{I_n}(H)^{\perp}$ may be described as the set of matrices of the form

$$\left[\left[\begin{array}{c} a \\ b \\ c \end{array}\right]\right] := \left[\begin{array}{ccc} a+b & 0 & 0 \\ 0 & a-b & c \\ 0 & -c^* & 0 \end{array}\right]$$

where $a, b \in \mathrm{Mat}_p(\mathbb{F})$ are anti-self-adjoint with a vanishing identically on the diagonal, and $c \in \mathrm{Mat}_{p\times q}(\mathbb{F})$. Given (real) diagonal $x \in \mathrm{Mat}_p$, we also put

$$\lambda(x) := \left[\begin{array}{ccc} 0 & x & 0 \\ x & 0 & 0 \\ 0 & 0 & 0_{q-p} \end{array}\right],$$

thus parametrizing Λ. By a straightforward calculation using formula (4.1.29), in which the orthogonal projection Π is again unnecessary, one verifies that

$$\Theta_{\lambda(x)} \left[\left[\begin{array}{c} a \\ b \\ c \end{array}\right]\right] = \left[\left[\begin{array}{c} x^2 a - 2xax + ax^2 \\ x^2 b + 2xbx + bx^2 \\ x^2 c \end{array}\right]\right],$$

and hence that

$$\sqrt{\det \Theta_{\lambda(\mathrm{diag}(x))}} = |\Delta(x^2)|^{\beta} \cdot \prod_{i=1}^{p} |2x_i|^{\beta-1} \cdot \prod_{i=1}^{p} |x_i|^{\beta(q-p)} \text{ for } x \in \mathbb{R}^p.$$

Now for $X \in \mathrm{Mat}_{p\times q}(\mathbb{F})$, put $X' = \left[\begin{array}{cc} 0 & X \\ X^* & 0 \end{array}\right] \in M$. With φ as in the statement of formula (4.1.3), let ψ be the unique function on M such that $\psi(X') = \varphi(X)$ for all $X \in \mathrm{Mat}_{p\times q}(\mathbb{F})$. By construction, ψ is G-conjugation-invariant, and in particular, $\psi(\lambda(\mathrm{diag}(x))$ depends only on the absolute values of the entries of x. Note also that the map $X \mapsto X'$ magnifies by a factor of $\sqrt{2}$. We thus have integration formulas

$$2^{\beta pq/2} \int \varphi d\rho_{\mathrm{Mat}_{p\times q}(\mathbb{F})} = \int \psi d\rho_M, \quad 2^{3p/2} \int_{\mathbb{R}_+^p} \varphi(x) \prod_{i=1}^{p} dx_i = \int \psi d\rho_{\Lambda}.$$

Integration formula (4.1.3) now follows from Theorem 4.1.28 combined with formula (4.1.33) for $\rho[G]/\rho[H]$. □

We turn next to the Jacobi ensembles. The next proposition provides the needed Weyl quadruples.

Proposition 4.1.33 *Fix integers* $0 < p \leq q$ *and put* $n = p + q$. *Fix* $0 \leq r \leq q - p$

and write $q = p+r+s$. Let

$$\begin{aligned}
G &= \{\mathrm{diag}(U,V) : U \in \mathrm{U}_p(\mathbb{F}), V \in \mathrm{U}_q(\mathbb{F})\} \subset \mathrm{U}_n(\mathbb{F}),\\
H &= \{\mathrm{diag}(U,V',V'',V''') : U,V' \in \mathrm{U}_p(\mathbb{F}), V'' \in \mathrm{U}_r(\mathbb{F}), V''' \in \mathrm{U}_s(\mathbb{F}),\\
&\qquad U,V' \textit{ are monomial}, U(V')^* \textit{ is diagonal}, (U(V')^*)^2 = I_p\} \subset G,\\
M &= \mathrm{Flag}_n(\mathrm{diag}(I_{p+r}, 0_{p+s}), \mathbb{F}),\\
\Lambda &= \{\mathrm{diag}(\begin{bmatrix} x & y \\ y & I_p - x \end{bmatrix}, I_r, 0_s) : x,y \in \mathrm{Mat}_p \textit{ are diagonal}\\
&\qquad \textit{and } x^2+y^2 = x\} \subset M.
\end{aligned}$$

Let $\Lambda' \subset \Lambda$ be the subset consisting of elements such that the absolute values of the diagonal entries of the corresponding diagonal matrix y belong to the interval $(0,1/2)$ and are distinct. Then (G,H,M,Λ) is a Weyl quadruple with ambient space $\mathrm{Mat}_n(\mathbb{F})$ for which Λ' is generic and, furthermore,

$$\frac{\rho[G]}{\rho[H]} = \frac{\rho[\mathrm{U}_p(\mathbb{F})]\rho[\mathrm{U}_q(\mathbb{F})]}{2^p p!(2^{(\beta-1)/2}\rho[\mathrm{U}_1(\mathbb{F})])^p \rho[\mathrm{U}_r(\mathbb{F})]\rho[\mathrm{U}_s(\mathbb{F})]}. \tag{4.1.34}$$

As in Proposition 4.1.32, we abuse notation slightly; one has to make appropriate adjustments to handle extreme values of the parameters p,q,r,s.

Proof As in the proof of Proposition 4.1.32, of the conditions imposed by Definition 4.1.22, only conditions (Ic), (IIa) and (IIIc) need be treated. One can verify (Ic) by means of Lemma 4.1.18 and Lemma 4.1.15.

We turn to the verification of condition (IIa). By Proposition E.14, for every $m \in M$, there exists $g \in G$ such that

$$gmg^{-1} = \mathrm{diag}(\begin{bmatrix} x & y \\ y & z \end{bmatrix}, w)$$

where $x,y,z \in \mathrm{Mat}_p$ and $w \in \mathrm{Mat}_{n-2p}$ are real diagonal and satisfy the relations dictated by the fact that gmg^{-1} squares to itself and has trace $p+r$. If we have $\mathrm{tr}\, w = r$, then after left-multiplying g by a permutation matrix in G we have $w = \mathrm{diag}(I_r, 0_s)$, and we are done. Otherwise $\mathrm{tr}\, w \neq r$. After left-multiplying g by a permutation matrix belonging to G, we can write $y = \mathrm{diag}(y',0)$ where $y' \in \mathrm{Mat}_{p'}$ has nonzero diagonal entries. Correspondingly, we write $x = \mathrm{diag}(x',x'')$ and $z = \mathrm{diag}(z',z'')$ with $x',z' \in \mathrm{Mat}_{p'}$ and $x'',z'' \in \mathrm{Mat}_{p-p'}$. We then have $z' = I_{p'} - x'$. Further, all diagonal entries of x'' and z'' belong to $\{0,1\}$, and finally, $\mathrm{tr}\, z'' + \mathrm{tr}\, w \geq r$. Thus, if we left-multiply g by a suitable permutation matrix in G we can arrange to have $\mathrm{tr}\, w = r$ and we are done.

We turn finally to the verification of condition (IIIc). Fix $\lambda \in \Lambda'$ and $g \in G$ such that $g\lambda g^{-1} \in \Lambda$. Let $x,y \in \mathrm{Mat}_p$ be the real diagonal matrices corresponding to λ as in the definition of Λ. By definition of Λ', no two of the four diagonal matrices x, $I_p - x$, I_r and 0_s have a diagonal entry in common, and hence $g = \mathrm{diag}(U,V,W,T)$ for some $U,V \in \mathrm{U}_p(\mathbb{F})$, $W \in \mathrm{U}_r(\mathbb{F})$ and $T \in \mathrm{U}_s(\mathbb{F})$. Also by definition of Λ', the diagonal entries of y have distinct nonzero absolute values, and hence we have $g \in H$ by Corollary E.13(iii) concerning the singular value decomposition. Thus (G,H,M,Λ) is a Weyl quadruple for which Λ' is generic.

A slight modification of the proof of formula (4.1.33) yields formula (4.1.34).

□

Proof of Proposition 4.1.4 Let (G,H,M,Λ) be the Weyl quadruple provided by Proposition 4.1.33. We follow the pattern established in the previous analysis of the Laguerre ensembles, but proceed more rapidly. We parametrize Λ and $\mathbb{T}_{I_n}(G) \cap \mathbb{T}_{I_n}(H)^{\perp}$, respectively, in the following way.

$$\lambda(x,y) \quad := \quad \mathrm{diag}\left(\begin{bmatrix} x & y \\ y & I_p - x \end{bmatrix}, I_r, 0_s\right),$$

$$\left[\left[\begin{matrix} a \\ b \\ c \\ d \\ e \end{matrix}\right]\right] \quad := \quad \begin{bmatrix} a+b & 0 & 0 & 0 \\ 0 & a-b & c & d \\ 0 & -c^* & 0 & e \\ 0 & -d^* & -e^* & 0 \end{bmatrix},$$

where:

- $x, y \in \mathrm{Mat}_p$ are real diagonal and satisfy $x^2 + y^2 = x$,
- $a, b \in \mathrm{Mat}_p(\mathbb{F})$ are anti-self-adjoint with a vanishing identically along the diagonal, and
- $c \in \mathrm{Mat}_{p\times r}(\mathbb{F})$, $d \in \mathrm{Mat}_{p\times s}(\mathbb{F})$ and $e \in \mathrm{Mat}_{r\times s}(\mathbb{F})$.

By a straightforward if rather involved calculation using formula (4.1.29), we have

$$\Theta_{\lambda(x,y)}\left[\left[\begin{matrix} a \\ b \\ c \\ d \\ e \end{matrix}\right]\right] = \left[\left[\begin{matrix} xa + ax - 2xax - 2yay \\ xb + bx - 2xbx + 2yby \\ xc \\ (I_p - x)d \\ e \end{matrix}\right]\right].$$

(Unlike in the proofs of Propositions 4.1.1 and 4.1.3, the orthogonal projection Π is used nontrivially.) We find that

$$\sqrt{\det \Theta_{\lambda(\mathrm{diag}(x),\mathrm{diag}(y))}} = |\Delta(x)|^{\beta} \cdot \prod_{i=1}^{p} (4x_i(1-x_i))^{(\beta-1)/2} \cdot \prod_{i=1}^{p} (x_i^r(1-x_i)^s)^{\beta/2}$$

for $x,y \in \mathbb{R}^p$ such that $x_i(1-x_i) = y_i^2$ (and hence $x_i \in [0,1]$) for $i = 1,\dots,p$. The calculation of the determinant is straightforward once it is noted that the identity

$$(x_1+x_2-2x_1x_2-2y_1y_2)(x_1+x_2-2x_1x_2+2y_1y_2) = (x_1-x_2)^2$$

holds if $x_i(1-x_i) = y_i^2$ for $i = 1,2$.

Now let φ be as it appears in formula (4.1.5). Note that Λ is an isometric copy of $\mathrm{Flag}_2(\mathrm{diag}(1,0),\mathbb{R})^p$ and that $\mathrm{Flag}_2(\mathrm{diag}(1,0),\mathbb{R})$ is a circle of circumference $\sqrt{2}\pi$. Note also that

$$\int_0^\pi f((1+\cos\theta)/2)d\theta = \int_0^1 \frac{f(x)dx}{\sqrt{x(1-x)}}.$$

We find that

$$\int \varphi(\lambda^{(p)})d\rho_\Lambda(\lambda) = 2^{p/2}\int_{[0,1]^p}\varphi(x)\prod_{i=1}^p \frac{dx_i}{\sqrt{x_i(1-x_i)}}.$$

Finally, note that the unique function ψ on M satisfying $\psi(W) = \varphi(W^{(p)})$ is G-conjugation invariant. We obtain (4.1.5) now by Theorem 4.1.28 combined with formula (4.1.34) for $\rho[G]/\rho[H]$. □

The next five propositions supply the Weyl quadruples needed to prove Proposition 4.1.6. All the propositions have similar proofs, with the last two proofs being the hardest. We therefore supply only the last two proofs.

Proposition 4.1.34 *Let $G = M = \mathrm{U}_n(\mathbb{C})$. Let $H \subset G$ be the set of monomial elements of G. Let $\Lambda \subset G$ be the set of diagonal elements of G, and let $\Lambda' \subset \Lambda$ be the subset consisting of elements with distinct diagonal entries. Then (G,H,M,Λ) is a Weyl quadruple with ambient space* $\mathrm{Mat}_n(\mathbb{C})$ *for which Λ' is generic and, furthermore,*

$$\rho[H]/\rho[\Lambda] = n!. \tag{4.1.35}$$

The proof of this proposition is an almost verbatim repetition of that of Proposition 4.1.23.

Put $\iota = \begin{bmatrix} 0 & 1 \\ 1 & 0 \end{bmatrix} \in \mathrm{Mat}_2$ and recall the notation $R_n(\theta)$ used in Proposition 4.1.6.

Proposition 4.1.35 *Let $n = 2\ell+1$ be odd. Let $G = M = \mathrm{U}_n(\mathbb{R})$. Let W_n be the group consisting of permutation matrices in* Mat_n *commuting with* $\mathrm{diag}(\iota,\dots,\iota,1)$. *Let*

$$\Lambda = \{\pm\mathrm{diag}(R_\ell(\theta),1) : \theta \in \mathbb{R}^\ell\},\ H = \{w\lambda : \lambda \in \Lambda,\ w \in W_n\}.$$

Let $\Lambda' \subset \Lambda$ be the subset consisting of elements with distinct (complex) eigenvalues. Then (G,H,M,Λ) is a Weyl quadruple with ambient space $\mathrm{Mat}_n(\mathbb{R})$ *for which Λ' is generic and, furthermore,*

$$\rho[H]/\rho[\Lambda] = 2^\ell \ell!. \tag{4.1.36}$$

Proposition 4.1.36 *Let $G = M = \mathrm{U}_n(\mathbb{H})$. Let $H \subset G$ be the set of monomial elements with entries in $\mathbb{C} \cup \mathbb{C}\mathbf{j}$. Let $\Lambda \subset G$ be the set of diagonal elements with entries in $\mathbb{C}$. Let $\Lambda' \subset \Lambda$ be the subset consisting of elements λ such that* $\mathrm{diag}(\lambda, \lambda^*)$ *has distinct diagonal entries. Then (G,H,M,Λ) is a Weyl quadruple with ambient space* $\mathrm{Mat}_n(\mathbb{H})$ *for which Λ' is generic and, furthermore,*

$$\rho[H]/\rho[\Lambda] = 2^n n!. \tag{4.1.37}$$

Proposition 4.1.37 *Let $n = 2\ell$ be even. Let $G = \mathrm{U}_n(\mathbb{R})$ and let $M \subset G$ be the subset on which* $\det = 1$. *Let $W_n^+ \subset G$ be the group consisting of permutation matrices commuting with* $\mathrm{diag}(\iota, \dots, \iota)$. *Put*

$$\Lambda = \{R_\ell(\theta) : \theta \in \mathbb{R}^\ell\} \subset M, \quad H = \{w\lambda : \lambda \in \Lambda, w \in W_n^+\} \subset G.$$

Let $\Lambda' \subset \Lambda$ be the subset consisting of elements with distinct (complex) eigenvalues. Then (G,H,M,Λ) is a Weyl quadruple with ambient space $\mathrm{Mat}_n(\mathbb{R})$ *such that Λ' is generic and, furthermore,*

$$\rho[H]/\rho[\Lambda] = 2^\ell \ell!. \tag{4.1.38}$$

Proposition 4.1.38 *Let $n = 2\ell$ be even. Let $G = \mathrm{U}_n(\mathbb{R})$ and let $M \subset G$ be the subset on which* $\det = -1$. *Put*

$$\begin{aligned} W_n^- &= \{\mathrm{diag}(w, \pm 1, \pm 1) : w \in W_{n-2}^+\} \subset G, \\ \Lambda &= \{\mathrm{diag}(R_{\ell-1}(\theta), 1, -1) : \theta \in \mathbb{R}^{\ell-1}\} \subset M, \\ H &= \{w\lambda : w \in W_n^-, \lambda \in \Lambda\} \subset G. \end{aligned}$$

Let $\Lambda' \subset \Lambda$ be the subset consisting of elements with distinct (complex) eigenvalues. Then (G,H,M,Λ) is a Weyl quadruple with ambient space $\mathrm{Mat}_n(\mathbb{R})$ *for which Λ' is generic and, furthermore,*

$$\rho[H]/\rho[\Lambda] = 2^{\ell+1}(\ell-1)!. \tag{4.1.39}$$

Proof of Proposition 4.1.37 Only conditions (IIa) and (IIIc) require proof. The other parts of the proposition, including formula (4.1.38), are easy to check.

To verify condition (IIa), fix $m \in M$ arbitrarily. After conjugating m by some element of G, we may assume by Theorem E.11 that m is block-diagonal with $\mathbb{R}$-standard blocks on the diagonal. Now the only orthogonal $\mathbb{R}$-standard blocks are

$\pm 1 \in \mathrm{Mat}_1$ and $R(\theta) \in \mathrm{Mat}_2$ for $0 < \theta < \pi$. Since we assume $\det m = 1$, there are even numbers of 1s and -1s along the diagonal of m, and hence after conjugating m by a suitable permutation matrix, we have $m \in \Lambda$ as required. Thus condition (IIa) is proved.

To verify condition (IIIc), we fix $\lambda \in \Lambda'$, $g \in G$ and $\mu \in \Lambda$ such that $g\lambda g^{-1} = \mu$, with the goal to show that $g \in H$. After conjugating λ by a suitably chosen element of W_n^+, we may assume that the angles $\theta_1, \dots, \theta_\ell$ describing λ, as in the definition of Λ, satisfy $0 < \theta_1 < \dots < \theta_\ell < \pi$. By another application of Theorem E.11, after replacing g by wg for suitably chosen $w \in W_n^+$, we may assume that $\lambda = \mu$. Then g commutes with λ, which is possible only if $g \in \Lambda$. Thus condition (IIIc) is proved, and the proposition is proved. □

Proof of Proposition 4.1.38 As in the proof of Proposition 4.1.37, only conditions (IIa) and (IIIc) require proof. To verify condition (IIa) we argue exactly as in the proof of Proposition 4.1.37, but this time, because $\det m = -1$, we have to pair off a 1 with a -1, and we arrive at the desired conclusion. To prove condition (IIIc), we again fix $\lambda \in \Lambda'$, $g \in G$ and $\mu \in \Lambda$ such that $g\lambda g^{-1} = \mu$, with the goal to show that $g \in H$; and arguing as before, we may assume that g commutes with λ. The hypothesis that λ has distinct complex eigenvalues then insures then $g = \mathrm{diag}(I_{n-2}, \pm 1, \pm 1)v$ for some $v \in \Lambda$, and hence $g \in H$. Thus condition (IIIc) is verified, and the proposition is proved. □

Proof of Proposition 4.1.6 It remains only to calculate $\sqrt{\det \Theta_\lambda}$ for each of the five types of Weyl quadruples defined above in order to complete the proofs of (4.1.6), (4.1.7), (4.1.8) and (4.1.9), for then we obtain each formula by invoking Theorem 4.1.28, combined with the formulas (4.1.35), (4.1.36), (4.1.37), (4.1.38) and (4.1.39), respectively, for the ratio $\rho[H]/\rho[\Lambda]$. Note that the last two Weyl quadruples are needed to handle the two terms on the right side of (4.1.9), respectively.

All the calculations are similar. Those connected with the proof of (4.1.9) are the hardest, and may serve to explain all the other calculations. In the following, we denote the Weyl quadruples defined in Propositions 4.1.37 and 4.1.38 by (G, H^+, M^+, Λ^+) and (G, H^-, M^-, Λ^-), respectively. We treat each quadruple in a separate paragraph below.

To prepare for the calculation it is convenient to introduce two special functions. Given real numbers α and β, let $D(\alpha, \beta)$ be the square-root of the absolute value of the determinant of the $\mathbb{R}$-linear operator

$$Z \mapsto R(-\alpha)(R(\alpha)Z - ZR(\beta)) - (R(\alpha)Z - ZR(\beta))R(-\beta)$$

on $\mathrm{Mat}_2(\mathbb{R})$, and let $C(\alpha)$ be the square-root of the absolute value of the determinant of the $\mathbb{R}$-linear operator

$$Z \mapsto R(-\alpha)(R(\alpha)Z - Z\kappa) - (R(\alpha)Z - Z\kappa)\kappa$$

on $\mathrm{Mat}_2(\mathbb{R})$, where $\kappa = \mathrm{diag}(1,-1)$. Actually both operators in question are nonnegative definite and hence have nonnegative determinants. One finds that

$$D(\alpha,\beta) = |(e^{\mathrm{i}\alpha} - e^{\mathrm{i}\beta})(e^{\mathrm{i}\alpha} - e^{-\mathrm{i}\beta})|, \;\; C(\alpha) = |e^{\mathrm{i}\alpha} - e^{-\mathrm{i}\alpha}|$$

by straightforward calculations.

Consider the Weyl quadruple (G,H^+,M^+,Λ^+) and for $\theta \in \mathbb{R}^\ell$ put $\lambda^+(\theta) = R_\ell(\theta)$. The space $\mathbb{T}_{I_n}(G) \cap \mathbb{T}_{I_n}(H^+)^\perp$ consists of real antisymmetric matrices $X \in \mathrm{Mat}_n$ such that $X_{2i,2i-1} = 0$ for $i = 1,\ldots,\ell$. Using formula (4.1.29), one finds that

$$\sqrt{\det \Theta_{\lambda^+(\theta)}} = \prod_{1\le i<j\le \ell} D(\theta_i,\theta_j) = D_\ell(\theta)$$

which proves (4.1.9) for all functions φ supported on M^+.

Consider next the Weyl quadruple (G,H^-,M^-,Λ^-) and for $\theta \in \mathbb{R}^{\ell-1}$ put $\lambda^-(\theta) = \mathrm{diag}(R_\ell(\theta),1,-1)$. The space $\mathbb{T}_{I_n}(G) \cap \mathbb{T}_{I_n}(H^-)^\perp$ consists of real antisymmetric matrices $X \in \mathrm{Mat}_n$ such that $X_{2i,2i-1} = 0$ for $i = 1,\ldots,\ell-1$. Using formula (4.1.29) one finds that

$$\sqrt{\det \Theta_{\lambda^-(\theta)}} = \prod_{1\le i<j\le \ell-1} D(\theta_i,\theta_j) \cdot \prod_{1\le i\le \ell-1} C(\theta_i) \cdot 2,$$

which proves (4.1.9) for all functions φ supported on M^-. (The last factor of 2 is accounted for by the fact that for $Z \in \mathrm{Mat}_2$ real antisymmetric, $[\kappa,[\kappa,Z]] = 4Z$.) This completes the proof of (4.1.9).

All the remaining details needed to complete the proof of Proposition 4.1.6, being similar, we omit. □

Exercise 4.1.39

Let $G = \mathrm{U}_n(\mathbb{C})$ and let $H \subset G$ be the subgroup consisting of monomial elements. Let $M \subset \mathrm{Mat}_n(\mathbb{C})$ be the set consisting of normal matrices with distinct eigenvalues, and let $\Lambda \subset M$ be the subset consisting of diagonal elements. Show that (G,H,M,Λ) is a Weyl quadruple. Show that $\sqrt{\det \Theta_\lambda} = \prod_{1\le i<j\le n} |\lambda_i - \lambda_j|^2$ for all $\lambda = \mathrm{diag}(\lambda_1,\ldots,\lambda_n) \in \Lambda$.

4.2 Determinantal point processes

The collection of eigenvalues of a random matrix naturally can be viewed as a configuration of points (on $\mathbb{R}$ or on $\mathbb{C}$), that is, as a *point process*. This section is devoted to the study of a class of point processes known as determinantal processes; such processes possess useful probabilistic properties, such as CLTs for occupation numbers, and, in the presence of approximate translation invariance, convergence to stationary limits. The point process determined by the eigenvalues of the GUE is, as we show below, a determinantal process. Further, determinantal processes occur as limits of the rescaled configuration of eigenvalues of the GUE, in the bulk and in the edge of the spectrum, see Section 4.2.5.

4.2.1 Point processes: basic definitions

Let Λ be a locally compact Polish space, equipped with a (necessarily σ-finite) positive Radon measure μ on its Borel σ-algebra (recall that a positive measure is *Radon* if $\mu(K) < \infty$ for each compact set K). We let $\mathscr{M}(\Lambda)$ denote the space of σ-finite Radon measures on Λ, and let $\mathscr{M}_+(\Lambda)$ denote the subset of $\mathscr{M}(\Lambda)$ consisting of positive measures.

Definition 4.2.1 (a) A *point process* is a random, integer-valued $\chi \in \mathscr{M}_+(\Lambda)$. (By random we mean that for any Borel $B \subset \Lambda$, $\chi(B)$ is an integer-valued random variable.)
(b) A point process χ is *simple* if

$$P(\exists x \in \Lambda : \chi(\{x\}) > 1) = 0\,. \tag{4.2.1}$$

Note that the event in (4.2.1) is measurable due to the fact that Λ is Polish. One may think about χ also in terms of configurations. Let $\mathscr{X}$ denote the space of locally finite configurations in Λ, and let $\mathscr{X}^{\neq}$ denote the space of locally finite configurations with no repetitions. More precisely, for $x_i \in \Lambda$, $i \in I$ an interval of positive integers (beginning at 1 if nonempty), with I finite or countable, let $[x_i]$ denote the equivalence class of all sequences $\{x_{\pi(i)}\}_{i\in I}$, where π runs over all permutations (finite or countable) of I. Then, set

$$\mathscr{X} = \mathscr{X}(\Lambda) = \{\mathbf{x} = [x_i]_{i=1}^{\kappa}, \quad \text{where } x_i \in \Lambda,\ \kappa \le \infty, \text{ and}$$
$$|\mathbf{x}_K| := \sharp\{i : x_i \in K\} < \infty \text{ for all compact } K \subset \Lambda\}$$

and

$$\mathscr{X}^{\neq} = \{\mathbf{x} \in \mathscr{X} : x_i \neq x_j \text{ for } i \neq j\}\,.$$

We endow $\mathscr{X}$ and $\mathscr{X}^{\neq}$ with the σ-algebra $\mathscr{C}_{\mathscr{X}}$ generated by the cylinder sets $C_n^B = \{\mathbf{x} \in \mathscr{X} : |\mathbf{x}_B| = n\}$, with B Borel with compact closure and n a nonnegative integer. Since $\chi = \sum_{i=1}^{\kappa} \delta_{y_i}$ for some (possibly random) $\kappa \le \infty$ and random y_i, each point process χ can be associated with a point in $\mathscr{X}$ (in $\mathscr{X}^{\neq}$ if χ is simple). The converse is also true, as is summarized in the following elementary lemma, where we let ν be a probability measure on the measure space $(\mathscr{X}, \mathscr{C}_{\mathscr{X}})$.

Lemma 4.2.2 *A ν-distributed random element $\mathbf{x}$ of $\mathscr{X}$ can be associated with a point process χ via the formula $\chi(B) = |\mathbf{x}_B|$ for all Borel $B \subset \Lambda$. If $\nu(\mathscr{X}^{\neq}) = 1$, then χ is a simple point process.*

With a slight abuse, we will therefore not distinguish between the point process χ and the induced configuration $\mathbf{x}$. In the sequel, we associate the law ν with the point process χ, and write E_ν for expectation with respect to this law.

We next note that if $\mathbf{x}$ is not simple, then one may construct a simple point process $\mathbf{x}^* = \{(x_j^*, N_j)\}_{j=1}^{\kappa^*} \in \mathscr{X}(\Lambda^*)$ on $\Lambda^* = \Lambda \times \mathbb{N}_+$ by letting κ^* denote the number of distinct entries in $\mathbf{x}$, introducing a many-to-one mapping $j(i) : \{1, \ldots, \kappa\} \mapsto \{1, \ldots, \kappa^*\}$ with $N_j = |\{i : j(i) = j\}|$ such that if $j(i) = j(i')$ then $x_i = x_{i'}$, and then setting $x_j^* = x_i$ if $j(i) = j$. In view of this observation, we only consider in the sequel simple point processes.

Definition 4.2.3 Let χ be a simple point process. Assume locally integrable functions $\rho_k : \Lambda^k \to [0, \infty)$, $k \geq 1$, exist such that for any mutually disjoint family of subsets $D_1, \cdots, D_k$ of Λ,

$$E_\nu[\prod_{i=1}^{k} \chi(D_i)] = \int_{\prod_{i=1}^{k} D_i} \rho_k(x_1, \cdots, x_k) d\mu(x_1) \cdots d\mu(x_k).$$

Then the functions ρ_k are called the *joint intensities* (or *correlation functions*) of the point process χ with respect to μ.

The term "correlation functions" is standard in the physics literature, while "joint intensities" is more commonly used in the mathematical literature.

Remark 4.2.4 By Lebesgue's Theorem, for μ^k almost every $(x_1, \ldots, x_k)$,

$$\lim_{\varepsilon \to 0} \frac{P(\chi(B(x_i, \varepsilon)) = 1, i = 1, \ldots, k)}{\prod_{i=1}^{k} \mu(B(x_i, \varepsilon))} = \rho_k(x_1, \ldots, x_k).$$

Further, note that $\rho_k(\cdot)$ is in general only defined μ^k-almost everywhere, and that $\rho_k(x_1, \ldots, x_k)$ is not determined by Definition 4.2.3 if there are $i \neq j$ with $x_i = x_j$. For consistency with Lemma 4.2.5 below and the fact that we consider simple processes only, we set $\rho_k(x_1, \ldots, x_k) = 0$ for such points.

The joint intensities, if they exist, allow one to consider overlapping sets, as well. In what follows, for a configuration $\mathbf{x} \in \mathscr{X}^{\neq}$, and k integer, we let $\mathbf{x}^{\wedge k}$ denote the set of ordered samples of k distinct elements from $\mathbf{x}$. (Thus, if $\Lambda = \mathbb{R}$ and $\mathbf{x} = \{1,2,3\}$, then $\mathbf{x}^{\wedge 2} = \{(1,2),(2,1),(1,3),(3,1),(2,3),(3,2)\}$.)

Lemma 4.2.5 *Let χ be a simple point process with intensities ρ_k.*
(a) *For any Borel set $B \subset \Lambda^k$ with compact closure,*

$$E_\nu\left(|\mathbf{x}^{\wedge k} \cap B|\right) = \int_B \rho_k(x_1,\cdots,x_k)d\mu(x_1)\cdots d\mu(x_k). \qquad (4.2.2)$$

(b) *If D_i, $i = 1,\ldots,r$, are mutually disjoint subsets of Λ contained in a compact set K, and if $\{k_i\}_{i=1}^r$ is a collection of positive integers such that $\sum_{i=1}^r k_i = k$, then*

$$E_\nu\left[\prod_{i=1}^r \binom{\chi(D_i)}{k_i} k_i!\right] = \int_{\prod D_i^{\times k_i}} \rho_k(x_1,\ldots,x_k)\mu(dx_1)\cdots\mu(dx_k). \qquad (4.2.3)$$

Proof of Lemma 4.2.5 Note first that, for any compact $Q \subset \Lambda$, there exists an increasing sequence of partitions $\{Q_i^n\}_{i=1}^n$ of Q such that, for any $x \in Q$,

$$\bigcap_n \bigcap_{i:x\in Q_i^n} Q_i^n = \{x\}.$$

We denote by $\mathscr{Q}_n^k$ the collection of (ordered) k-tuples of distinct elements of $\{Q_i^n\}$.
(a) It is enough to consider sets of the form $B = B_1 \times B_2 \times \cdots \times B_k$, with the sets B_i Borel of compact closure. Then

$$M_k^n := \sum_{(Q_1,\ldots,Q_k)\in\mathscr{Q}_n^k} |(Q_1 \times \cdots \times Q_k) \cap B \cap \mathbf{x}^{\wedge k}| = \sum_{(Q_1,\ldots,Q_k)\in\mathscr{Q}_n^k} \prod_{i=1}^k \chi(Q_i \cap B_i).$$

Thus

$$E_\nu(M_k^n) = \sum_{(Q_1,\ldots,Q_k)\in\mathscr{Q}_n^k} \int_{(Q_1\times\cdots\times Q_k)\cap B} \rho_k(x_1,\ldots,x_k)d\mu(x_1)\ldots d\mu(x_k). \qquad (4.2.4)$$

Note that M_k^n increases monotonically in n to $|\mathbf{x}^{\wedge k} \cap B|$. On the other hand, since $\mathbf{x}$ is simple, and by our convention concerning the intensities ρ_k, see Remark 4.2.4,

$$\limsup_{n\to\infty} \sum_{(Q_1,\ldots,Q_k)\in(\mathscr{Q}_n^1)^k\setminus\mathscr{Q}_n^k} \int_{(Q_1\times\cdots\times Q_k)\cap B} \rho_k(x_1,\ldots,x_k)d\mu(x_1)\ldots d\mu(x_k) = 0.$$

The conclusion follows from these facts, the fact that $\mathscr{X}$ is a Radon measure and (4.2.4).
(b) Equation (4.2.3) follows from (4.2.2) through the choice $B = \prod D_i^{\times k_i}$. □

Remark 4.2.6 Note that a system of nonnegative, measurable and symmetric functions $\{\rho_r : \Lambda^r \to [0,\infty]\}_{r=1}^\infty$ is a system of joint intensities for a simple point process that consists of exactly n points almost surely, if and only if $\rho_r = 0$ for $r > n$, ρ_1/n is a probability density function, and the family is consistent, that is, for $1 < r \le n$,

$$\int_\Lambda \rho_r(x_1,\ldots,x_r)d\mu(x_r) = (n-r+1)\rho_{r-1}(x_1,\ldots,x_{r-1}).$$

As we have seen, for a simple point process, the joint intensities give information concerning the number of points in disjoint sets. Let now D_i be given disjoint compact sets, with $D = \bigcup_{i=1}^L D_i$ be such that $E(z^{\chi(D)}) < \infty$ for z in a neighborhood of 1. Consider the Taylor expansion, valid for z_ℓ in a neighborhood of 1,

$$\prod_{\ell=1}^L z_\ell^{\chi(D_\ell)} = 1 + \sum_{n=1}^\infty \sum_{\substack{n_i \le \chi(D_i) \\ n_i \vdash_L n}} \prod_{i=1}^L \frac{\chi(D_i)!}{(\chi(D_i)-n_i)!n_i!} \prod_{i=1}^L (z_i - 1)^{n_i} \tag{4.2.5}$$

$$= 1 + \sum_{n=1}^\infty \sum_{n_i \vdash_L n} \prod_{i=1}^L \frac{(\chi(D_i)(\chi(D_i)-1)\cdots(\chi(D_i)-n_i+1))}{n_i!} \prod_{i=1}^L (z_i-1)^{n_i},$$

where

$$\{n_i \vdash_L n\} = \{(n_1,\ldots,n_L) \in \mathbb{N}_+^L : \sum_{i=1}^L n_i = n\}.$$

Then one sees that, under these conditions, the factorial moments in (4.2.3) determine the characteristic function of the collection $\{\chi(D_i)\}_{i=1}^L$. A more direct way to capture the distribution of the point process χ is via its Jánossy densities, that we define next.

Definition 4.2.7 Let $D \subset \Lambda$ be compact. Assume there exist symmetric functions $j_{D,k} : D^k \to \mathbb{R}_+$ such that, for any finite collection of mutually disjoint measurable sets $D_i \subset D$, $i = 1,\ldots,k$,

$$P(\chi(D) = k, \chi(D_i) = 1, i = 1,\ldots,k) = \int_{\prod_i D_i} j_{D,k}(x_1,\ldots,x_k) \prod_i \mu(dx_i). \tag{4.2.6}$$

Then we refer to the collection $\{j_{D,k}\}_{k=1}^\infty$ as the *Jánossy densities* of χ in D.

The following easy consequences of the definition are proved in the same way that Lemma 4.2.5 was proved.

Lemma 4.2.8 *For any compact $D \subset \Lambda$, if the Jánossy densities $j_{D,k}$, $k \ge 1$ exist then*

$$P(\chi(D) = k) = \frac{1}{k!} \int_{D^k} j_{D,k}(x_1,\ldots,x_k) \prod_i \mu(dx_i), \tag{4.2.7}$$

and, for any mutually disjoint measurable sets $D_i \subset D$, $i = 1, \ldots, k$ and any integer $r \geq 0$,

$$P(\chi(D) = k + r, \chi(D_i) = 1, i = 1, \ldots, k)$$
$$= \frac{1}{r!} \int_{\prod_{i=1}^k D_i \times D^r} j_{D,k+r}(x_1, \ldots, x_{k+r}) \prod_i \mu(dx_i) . \tag{4.2.8}$$

In view of (4.2.8) (with $r = 0$), one can naturally view the collection of Jánossy densities as a distribution on the space $\otimes_{k=0}^{\infty} D^k$.

Jánossy densities and joint intensities are (at least locally, i.e. restricted to a compact set D) equivalent descriptions of the point process χ, as the following proposition states.

Proposition 4.2.9 *Let χ be a simple point process on Λ and assume $D \subset \Lambda$ is compact.*
(a) Assume the Jánossy densities $j_{D,k}$, $k \geq 1$, exist, and that

$$\sum_k \int_{D^k} \frac{k^r j_{D,k}(x_1, \ldots, x_k)}{k!} \prod_i \mu(dx_i) < \infty, \quad \textit{for all } r \textit{ integer}. \tag{4.2.9}$$

Then χ restricted to D possesses the intensities

$$\rho_k(x_1, \ldots, x_k) = \sum_{r=0}^{\infty} \frac{j_{D,k+r}(x_1, \ldots, x_k, D, \ldots, D)}{r!}, \quad x_i \in D, \tag{4.2.10}$$

where

$$j_{D,k+r}(x_1, \ldots, x_k, D, \ldots, D) = \int_{D^r} j_{D,k+r}(x_1, \ldots, x_k, y_1, \ldots, y_r) \prod_{i=1}^r \mu(dy_i) .$$

(b) Assume the intensities $\rho_k(x_1, \ldots, x_k)$ exist and satisfy

$$\sum_k \int_{D^k} \frac{k^r \rho_k(x_1, \ldots, x_k)}{k!} \prod_i \mu(dx_i) < \infty, \quad \textit{for all } r \textit{ integer}. \tag{4.2.11}$$

Then the Jánossy densities $j_{D,k}$ exist for all k and satisfy

$$j_{D,k}(x_1, \ldots, x_k) = \sum_{r=0}^{\infty} \frac{(-1)^r \rho_{k+r}(x_1, \ldots, x_k, D, \ldots, D)}{r!}, \tag{4.2.12}$$

where

$$\rho_{k+r}(x_1, \ldots, x_k, D, \ldots, D) = \int_{D^r} \rho_{k+r}(x_1, \ldots, x_k, y_1, \ldots, y_r) \prod_{i=1}^r \mu(dy_i) .$$

The proof follows the same procedure as in Lemma 4.2.5: partition Λ and use dominated convergence together with the integrability conditions and the fact that χ is assumed simple. We omit further details. We note in passing that under a slightly stronger assumption of the existence of exponential moments, part (b) of the proposition follows from (4.2.5) and part (b) of Lemma 4.2.5.

Exercise 4.2.10 Show that, for the standard Poisson process of rate $\lambda > 0$ on $\Lambda = \mathbb{R}$ with μ taken as the Lebesgue measure, one has, for any compact $D \subset \mathbb{R}$ with Lebesgue measure $|D|$,

$$\rho_k(x_1,\ldots,x_k) = e^{\lambda|D|} j_{D,k}(x_1,\ldots,x_k) = \lambda^k .$$

4.2.2 Determinantal processes

We begin by introducing the general notion of a determinantal process.

Definition 4.2.11 A simple point process χ is said to be a *determinantal point process* with kernel K (in short: determinantal process) if its joint intensities ρ_k exist and are given by

$$\rho_k(x_1,\cdots,x_k) = \det_{i,j=1}^{k} (K(x_i,x_j)) . \tag{4.2.13}$$

In what follows, we will be mainly interested in certain locally trace-class operators on $L^2(\mu)$ (viewed as either a real or complex Hilbert space, with inner product denoted $\langle f,g\rangle_{L^2(\mu)}$), motivating the following definition.

Definition 4.2.12 An integral operator $\mathscr{K} : L^2(\mu) \to L^2(\mu)$ with kernel K given by

$$\mathscr{K}(f)(x) = \int K(x,y)f(y)d\mu(y), \quad f \in L^2(\mu)$$

is *admissible* (with admissible kernel K) if $\mathscr{K}$ is self-adjoint, nonnegative and locally trace-class, that is, with the operator $\mathscr{K}_D = \mathbf{1}_D \mathscr{K} \mathbf{1}_D$ having kernel $K_D(x,y) = \mathbf{1}_D(x)K(x,y)\mathbf{1}_D(y)$, the operators $\mathscr{K}$ and $\mathscr{K}_D$ satisfy:

$$\langle g,\mathscr{K}(f)\rangle_{L^2(\mu)} = \langle \mathscr{K}(g),f\rangle_{L^2(\mu)}, \quad f,g \in L^2(\mu), \tag{4.2.14}$$

$$\langle f,\mathscr{K}(f)\rangle_{L^2(\mu)} \geq 0, \quad f \in L^2(\mu), \tag{4.2.15}$$

For all compact sets $D \subset \Lambda$, the eigenvalues $(\lambda_i^D)_{i\geq 0} (\in \mathbb{R}^+)$ of $\mathscr{K}_D$ satisfy $\sum \lambda_i^D < \infty$. (4.2.16)

We say that $\mathscr{K}$ is *locally admissible* (with locally admissible kernel K) if (4.2.14) and (4.2.15) hold with $\mathscr{K}_D$ replacing $\mathscr{K}$.

The following standard result, which we quote from [Sim05b, Theorem 2.12] without proof, gives sufficient conditions for a (positive definite) kernel to be admissible.

Lemma 4.2.13 *Suppose $K : \Lambda \times \Lambda \to \mathbb{C}$ is a continuous, Hermitian and positive definite function, that is, $\sum_{i=1}^{n} z_i^* z_j K(x_i, x_j) \geq 0$ for any n, $x_1, \ldots, x_n \in \Lambda$ and $z_1, \ldots, z_n \in \mathbb{C}$. Then $\mathscr{K}$ is locally admissible.*

By standard results, see e.g. [Sim05b, Theorem 1.4], an integral compact operator $\mathscr{K}$ with admissible kernel K possesses the decomposition

$$\mathscr{K} f(x) = \sum_{k=1}^{n} \lambda_k \phi_k(x) \langle \phi_k, f \rangle_{L^2(\mu)}, \tag{4.2.17}$$

where the functions ϕ_k are orthonormal in $L^2(\mu)$, n is either finite or infinite, and $\lambda_k > 0$ for all k, leading to

$$K(x,y) = \sum_{k=1}^{n} \lambda_k \phi_k(x) \phi_k(y)^* . \tag{4.2.18}$$

(The last equality is to be understood in $L^2(\mu \times \mu)$.) If K is only locally admissible, K_D is admissible and compact for any compact D, and the relation (4.2.18) holds with K_D replacing K and the λ_k and ϕ_k depending on D.

Definition 4.2.14 An admissible (respectively, locally admissible) integral operator $\mathscr{K}$ with kernel K is *good* (with *good kernel* K) if the λ_k (respectively, λ_k^D) in (4.2.17) satisfy $\lambda_k \in (0,1]$.

We will later see (see Corollary 4.2.21) that if the kernel K in definition 4.2.11 of a determinantal process is (locally) admissible, then it must in fact be good.

The following example is our main motivation for discussing determinantal point processes.

Example 4.2.15 *Let $(\lambda_1^N, \cdots, \lambda_N^N)$ be the eigenvalues of the GUE of dimension N, and denote by χ_N the point process $\chi_N(D) = \sum_{i=1}^{N} 1_{\lambda_i^N \in D}$. By Lemma 3.2.2, χ_N is a determinantal process with (admissible, good) kernel*

$$K^{(N)}(x,y) = \sum_{k=0}^{N-1} \psi_k(x) \psi_k(y),$$

where the functions ψ_k are the oscillator wave-functions.

We state next the following extension of Lemma 4.2.5. (Recall, see Definition 3.4.3, that $\Delta(G)$ denotes the Fredholm determinant of a kernel G.)

Lemma 4.2.16 *Suppose χ is a ν-distributed determinantal point processes. Then, for mutually disjoint Borel sets D_ℓ, $\ell = 1,\dots,L$, whose closure is compact,*

$$E_\nu\Big(\prod_{\ell=1}^{L} z_\ell^{\chi(D_\ell)}\Big) = \Delta\left(\mathbf{1}_D \sum_{\ell=1}^{L}(1-z_\ell)K\mathbf{1}_{D_\ell}\right), \tag{4.2.19}$$

where $D = \bigcup_{\ell=1}^{L} D_\ell$ and the equality is valid for all $(z_\ell)_{\ell=1}^{L} \in \mathbb{C}^L$. In particular, the law of the restriction of simple determinantal processes to compact sets is completely determined by the intensity functions, and the restriction of a determinantal process to a compact set D is determinantal with admissible kernel $\mathbf{1}_D(x)K(x,y)\mathbf{1}_D(y)$.

Proof of Lemma 4.2.16 By our assumptions, the right side of (4.2.19) is well defined for any choice of $(z_\ell)_{\ell=1}^{L} \in \mathbb{C}^L$ as a Fredholm determinant (see Definition 3.4.3), and

$$\begin{aligned}
&\Delta\left(\mathbf{1}_D \sum_{\ell=1}^{L}(1-z_\ell)K\mathbf{1}_{D_\ell}\right) - 1 \\
&= \sum_{n=1}^{\infty}\frac{1}{n!}\int_D\cdots\int_D \det\left\{\sum_{\ell=1}^{L}(z_\ell-1)K(x_i,x_j)\mathbf{1}_{D_\ell}(x_j)\right\}_{i,j=1}^{n} \mu(dx_1)\cdots\mu(dx_L) \\
&= \sum_{n=1}^{\infty}\frac{1}{n!}\sum_{\ell_1,\dots,\ell_n=1}^{L}\prod_{k=1}^{n}(z_{\ell_k}-1) \qquad (4.2.20)\\
&\quad\times\int\cdots\int \det\left\{\mathbf{1}_D(x_i)K(x_i,x_j)\mathbf{1}_{D_{\ell_j}}(x_j)\right\}_{i,j=1}^{n}\mu(dx_1)\cdots\mu(dx_L).
\end{aligned}$$

On the other hand, recall the Taylor expansion (4.2.5). Using (4.2.3) we see that the ν-expectation of each term in the last power series equals the corresponding term in the power series in (4.2.20), which represents an entire function. Hence, by monotone convergence, (4.2.19) follows. □

Note that an immediate consequence of Definition 4.2.3 and Lemma 4.2.16 is that the restriction of a determinantal process with kernel $K(x,y)$ to a compact subset D is determinantal, with kernel $\mathbf{1}_{x\in D}K(x,y)\mathbf{1}_{y\in D}$.

4.2.3 Determinantal projections

A natural question is now whether, given a good kernel K, one may construct an associated determinantal point process. We will answer this question in the affirmative by providing an explicit construction of determinantal point processes. We begin, however, with a particular class of determinantal processes defined by projection kernels.

Definition 4.2.17 A good kernel K is called a *trace-class projection kernel* if all eigenvalues λ_k in (4.2.18) satisfy $\lambda_k = 1$, and $\sum_{k=1}^n \lambda_k < \infty$. For a trace-class projection kernel K, set $H_K = \mathrm{span}\{\phi_k\}$.

Lemma 4.2.18 *Suppose χ is a determinantal point process with trace-class projection kernel K. Then $\chi(\Lambda) = n$, almost surely.*

Proof By assumption, $n < \infty$ in (4.2.18). The matrix $\{K(x_i,x_j)\}_{i,j=1}^k$ has rank at most n for all k. Hence, by (4.2.3), $\chi(\Lambda) \le n$, almost surely. On the other hand,

$$E_\nu(\chi(\Lambda)) = \int \rho_1(x)d\mu(x) = \int K(x,x)d\mu(x) = \sum_{i=1}^n \int |\phi_i(x)|^2 d\mu(x) = n\,.$$

This completes the proof. □

Proposition 4.2.19 *Let K be a trace-class projection kernel. Then a simple determinantal point process with kernel K exists.*

A simple proof of Proposition 4.2.19 can be obtained by noting that the function $\det_{i,j=1}^n K(x_i,x_j)/n!$ is nonnegative, integrates to 1, and by a computation similar to Lemma 3.2.2, see in particular (3.2.10), its kth marginal is $(n-k)!\det_{i,j=1}^k K(x_i,x_j)/n!$. We present an alternative proof that has the advantage of providing an explicit construction of the resulting determinantal point process.

Proof For a finite-dimensional subspace H of $L^2(\mu)$ of dimension d, let $\mathcal{K}_H$ denote the projection operator into H and let K_H denote an associated kernel. That is, $K_H(x,y) = \sum_{k=1}^d \psi_k(x)\psi_k^*(y)$ for some orthonormal family $\{\psi_k\}_{k=1}^d$ in H. For $x \in \Lambda$, set $k_x^H(\cdot) = K_H(x,\cdot)$. (Formally, $k_x^H = \mathcal{K}_H \delta_x$, in the sense of distributions.) The function $k_x^H(\cdot) \in L^2(\mu)$ does not depend on the choice of basis $\{\psi_k\}$, for almost every x: indeed, if $\{\phi_k\}$ is another orthonormal basis in H, then there exist complex coefficients $\{a_{i,j}\}_{i,j=1}^k$ such that

$$\phi_k = \sum_{j=1}^d a_{k,j}\psi_j\,, \quad \sum_{j=1}^d a_{k,j}a_{k,j'}^* = \delta_{j,j'}\,.$$

Hence, for μ-almost every x,y,

$$\sum_{k=1}^d \phi_k(x)\phi_k^*(y) = \sum_{k,j,j'=1}^d a_{k,j}a_{k,j'}^*\psi_j(x)\psi_{j'}^*(y) = \sum_{j=1}^d \psi_j(x)\psi_j^*(y)\,.$$

We have that $K_H(x,x) = \|k_x^H\|^2$ belongs to $L^1(\mu)$ and that different choices of basis $\{\psi_k\}$ lead to the same equivalent class of functions in $L^1(\mu)$. Let μ_H be the measure on Λ defined by $d\mu_H/d\mu(x) = K_H(x,x)$.

By assumption, $n < \infty$ in (4.2.18). Thus the associated subspace H_K is finite-dimensional. We construct a sequence of random variables $Z_1,\dots,Z_n$ in Λ as follows. Set $H_n = H_K$ and $j = n$.

- If $j = 0$, stop.
- Pick a point Z_j distributed according to μ_{H_j}/j.
- Let H_{j-1} be the orthocomplement to the function $k_{Z_j}^{H_j}$ in H_j.
- Decrease j by one and iterate.

We now claim that the point process $\mathbf{x} = (Z_1,\dots,Z_n)$, of law ν, is determinantal with kernel K. To see that, note that

$$k_{Z_j}^{H_j} = \mathscr{K}_{H_j} k_{Z_j}^H, \quad \text{in } L^2(\mu),\ \nu\text{-a.s.}$$

Hence the density of the random vector $(Z_1,\dots,Z_n)$ with respect to $\mu^{\otimes n}$ equals

$$p(x_1,\dots,x_n) = \prod_{j=1}^n \frac{\|k_{x_j}^{H_j}\|^2}{j} = \prod_{j=1}^n \frac{\|\mathscr{K}_{H_j} k_{x_j}^H\|^2}{j}.$$

Since $H_j = H \cap (k_{x_{j+1}}^H,\dots,k_{x_n}^H)^\perp$, it holds that

$$V = \prod_{j=1}^n \|\mathscr{K}_{H_j} k_{x_j}^H\|$$

equals the volume of the parallelepiped determined by the vectors $k_{x_1}^H,\dots,k_{x_n}^H$ in the finite-dimensional subspace $H \subset L^2(\mu)$. Since $\int k_{x_i}^H(x) k_{x_j}^H(x) \mu(dx) = K(x_i,x_j)$, it follows that $V^2 = \det(K(x_i,x_j))_{i,j=1}^n$. Hence

$$p(x_1,\dots,x_n) = \frac{1}{n!} \det(K(x_i,x_j))_{i,j=1}^n.$$

Thus, the random variables $Z_1,\dots,Z_n$ are exchangeable, almost surely distinct, and the n-point intensity of the point process $\mathbf{x}$ equals $n!p(x_1,\dots,x_n)$. In particular, integrating and applying the same argument as in (3.2.10), all k-point intensities have the determinantal form for $k \le n$. Together with Lemma 4.2.18, this completes the proof. □

Projection kernels can serve as building blocks for trace-class determinantal processes.

Proposition 4.2.20 *Suppose χ is a determinantal process with good kernel K of the form (4.2.18), with $\sum_k \lambda_k < \infty$. Let $\{I_k\}_{k=1}^n$ be independent Bernoulli variables with $P(I_k = 1) = \lambda_k$. Set*

$$K_I(x,y) = \sum_{k=1}^n I_k \phi_k(x)\phi_k^*(y),$$

and let χ_I denote the determinantal process with (random) kernel K_I. Then χ and χ_I have the same distribution.

The statement in the proposition can be interpreted as stating that the mixture of determinental processes χ_I has the same distribution as χ.

Proof Assume first n is finite. We need to show that for all $m \le n$, the m-point joint intensities of χ and χ_I are the same, that is

$$\det_{i,j=1}^m (K(x_i,x_j)) = E[\det_{i,j=1}^m (K_I(x_i,x_j))].$$

But, with $A_{i,k} = I_k\phi_k(x_i)$ and $B_{k,i} = \phi_k^*(x_i)$ for $1 \le i \le m, 1 \le k \le n$, then

$$(K_I(x_i,x_j))_{i,j=1}^m = AB, \tag{4.2.21}$$

and by the Cauchy–Binet Theorem A.2,

$$\det_{i,j=1}^m (K_I(x_i,x_j)) = \sum_{1\le \nu_1<\cdots<\nu_m\le n} \det(A_{\{1,..,m\}\times\{\nu_1,\cdots,\nu_m\}}) \det(B_{\{\nu_1,\cdots,\nu_m\}\times\{1,..,m\}}).$$

Since $E(I_k) = \lambda_k$, we have

$$E[\det(A_{\{1,..,m\}\times\{\nu_1,\cdots,\nu_m\}})] = \det(C_{\{1,..,m\}\times\{\nu_1,\cdots,\nu_m\}})$$

with $C_{i,k} = \lambda_k\phi_k(x_i)$. Therefore,

$$\begin{aligned} E[\det_{i,j=1}^m (K_I(x_i,x_j))] &= \sum_{1\le \nu_1<\cdots<\nu_m\le n} \det(C_{\{1,..,m\}\times\{\nu_1,\cdots,\nu_m\}}) \det(B_{\{\nu_1,\cdots,\nu_m\}\times\{1,..,m\}}) \\ &= \det(CB) = \det_{i=1}^m (K(x_i,x_j)), \end{aligned} \tag{4.2.22}$$

where the Cauchy–Binet Theorem A.2 was used again in the last line.

Suppose next that $n = \infty$. Since $\sum \lambda_k < \infty$, we have that $\mathscr{I} := \sum I_k < \infty$ almost surely. Thus, χ_I is a well defined point process. Let χ_I^N denote the determinantal process with kernel $K_I^N = \sum_{k=1}^N I_k\phi_k(x)\phi_k^*(y)$. Then χ_I^N is a well defined point process, and arguing as in (4.2.21), we get, for every integer m,

$$\begin{aligned} \det_{i,j=1}^m (K_I^N(x_i,x_j)) &= \sum_{1\le \nu_1<\cdots<\nu_m\le N} \det(A_{\{1,..,m\}\times\{\nu_1,\cdots,\nu_m\}}) \det(B_{\{\nu_1,\cdots,\nu_m\}\times\{1,..,m\}}) \\ &= \sum_{1\le \nu_1<\cdots<\nu_m\le N} \mathbf{1}_{\{I_{\nu_j}=1, j=1,\ldots,m\}} |\det(B_{\{\nu_1,\cdots,\nu_m\}\times\{1,..,m\}})|^2. \end{aligned} \tag{4.2.23}$$

In particular, the left side of (4.2.23) increases in N. Taking expectations and using the Cauchy–Binet Theorem A.2 and monotone convergence, we get, with the same notation as in (4.2.22), that

$$\begin{aligned} E \det_{i,j=1}^{m} (K_I(x_i,x_j)) &= \lim_{N\to\infty} E \det_{i,j=1}^{m} (K_I^N(x_i,x_j)) \\ &= \lim_{N\to\infty} \sum_{1\le v_1<\cdots<v_m\le N} \det(C_{\{1,..,m\}\times\{v_1,\cdots,v_m\}}) \det(B_{\{v_1,\cdots,v_m\}\times\{1,..,m\}}) \\ &= \lim_{N\to\infty} \det_{i,j=1}^{m} (K_N(x_i,x_j)) = \det_{i,j=1}^{m} (K(x_i,x_j)), \end{aligned} \tag{4.2.24}$$

where we write $K_N(x,y) = \sum_{k=1}^N \lambda_k \phi_k(x)\phi_k^*(y)$. □

We have the following.

Corollary 4.2.21 *Let $\mathscr{K}$ be admissible on $L^2(\mu)$, with trace-class kernel K. Then there exists a determinantal process χ with kernel K if and only if the eigenvalues of $\mathscr{K}$ belong to $[0,1]$.*

Proof From the definition, determinantal processes are determined by restriction to compact subsets, and the resulting process is determinantal too, see Lemma 4.2.16. Since the restriction of an admissible $\mathscr{K}$ to a compact subset is trace-class, it thus suffices to consider only the case where K is trace-class. Thus, the sufficiency is immediate from the construction in Proposition 4.2.20.

To see the necessity, suppose χ is a determinantal process with nonnegative kernel $K(x,y) = \sum \lambda_k \phi_k(x)\phi_k(y)$, with $\max \lambda_i = \lambda_1 > 1$. Let χ_1 denote the point process with each point x_i deleted with probability $1 - 1/\lambda_1$, independently. χ_1 is clearly a simple point process and, moreover, for disjoint subsets $D_1,\dots,D_k$ of Λ,

$$E_\nu[\prod_{i=1}^k \chi_1(D_i)] = \int_{\prod_{i=1}^k D_i} (1/\lambda_1)^k \rho_k(x_1,\cdots,x_k) d\mu(x_1)\cdots d\mu(x_k).$$

Thus, χ_1 is determinantal with kernel $K_1 = (1/\lambda_1)K$. Since χ had finitely many points almost surely (recall that K was assumed trace-class), it follows that $P(\chi_1(\Lambda) = 0) > 0$. But, the process χ_1 can be constructed by the procedure of Proposition 4.2.20, and since the top eigenvalue of K_1 equals 1, we obtain $P(\chi_1(\Lambda) \ge 1) = 1$, a contradiction. □

We also have the following corollaries.

Corollary 4.2.22 *Let K be a locally admissible kernel on Λ, such that for any compact $D \subset \Lambda$, the nonzero eigenvalues of K_D belong to $(0,1]$. Then K uniquely determines a determinantal point process on Λ.*

Proof By Corollary 4.2.21, a determinantal process is uniquely determined by K_D for any compact D. By the definition of the intensity functions, this sequence of laws of the processes is consistent, and hence they determine uniquely a determinantal process on Λ. □

Corollary 4.2.23 *Let χ be a determinantal process corresponding to an admissible trace class kernel K. Define the process χ_p by erasing, independently, each point with probability $(1-p)$. Then χ_p is a determinantal process with kernel pK.*

Proof Repeat the argument in the proof of the necessity part of Corollary 4.2.21. □

4.2.4 The CLT for determinantal processes

We begin with the following immediate corollary of Proposition 4.2.20 and Lemma 4.2.18. Throughout, for a good kernel K and a set $D \subset \Lambda$, we write $K_D(x,y) = \mathbf{1}_D(x)K(x,y)\mathbf{1}_D(y)$ for the restriction of K to D.

Corollary 4.2.24 *Let K be a good kernel, and let D be such that K_D is trace-class, with eigenvalues $\lambda_k, k \geq 1$. Then $\chi(D)$ has the same distribution as $\sum_k \xi_k$ where ξ_k are independent Bernoulli random variables with $P(\xi_k = 1) = \lambda_k$ and $P(\xi_k = 0) = 1 - \lambda_k$.*

The above representation immediately leads to a central limit theorem for occupation measures.

Theorem 4.2.25 *Let χ_n be a sequence of determinantal processes on Λ with good kernels K_n. Let D_n be a sequence of measurable subsets of Λ such that $(K_n)_{D_n}$ is trace class and $\mathrm{Var}(\chi_n(D_n)) \to_{n\to\infty} \infty$. Then*

$$Z_n = \frac{\chi_n(D_n) - E_\nu[\chi_n(D_n)]}{\sqrt{\mathrm{Var}(\chi_n(D_n))}}$$

converges in distribution towards a standard normal variable.

Proof We write K_n for the kernel $(K_n)_{D_n}$ and set $S_n = \sqrt{\mathrm{Var}(\chi_n(D_n))}$. By Corollary 4.2.24, $\chi_n(D_n)$ has the same distribution as the sum of independent Bernoulli variables ξ_k^n, whose parameters λ_k^n are the eigenvalues of K_n. In particular, $S_n^2 = \sum_k \lambda_k^n(1-\lambda_k^n)$. Since K_n is trace-class, we can write, for any θ real,

$$\begin{aligned}
\log E[e^{\theta Z_n}] &= \sum_k \log E[e^{\theta(\xi_k^n - \lambda_k^n)/S_n}] \\
&= -\frac{\theta \sum_k \lambda_k}{S_n} + \sum_k \log(1+\lambda_k^n(e^{\theta/S_n}-1)) \\
&= \frac{\theta^2 \sum_k \lambda_k^n(1-\lambda_k^n)}{2S_n^2} + o(\frac{\sum_k \lambda_k^n(1-\lambda_k^n)}{S_n^3}),
\end{aligned}$$

uniformly for θ in compacts. Since $\sum_k \lambda_k^n / S_n^3 \to_{n\to\infty} 0$, the conclusion follows.

□

We note in passing that, under the assumptions of Theorem 4.2.25,

$$\mathrm{Var}(\chi_n(D_n)) = \sum_k \lambda_k^n(1-\lambda_k^n) \le \sum_k \lambda_k^n = \int K_n(x,x)d\mu_n(x).$$

Thus, for $\mathrm{Var}(\chi_n(D_n))$ to go to infinity, it is necessary that

$$\lim_{n\to\infty} \int_{D_n} K_n(x,x)d\mu_n(x) = +\infty. \tag{4.2.25}$$

We also note that from (4.2.3) (with $r=1$ and $k=2$, and $\rho_k^{(n)}$ denoting the intensity functions corresponding to the kernel K_n from Theorem 4.2.25), we get

$$\mathrm{Var}(\chi_n(D_n)) = \int_{D_n} K_n(x,x)d\mu_n(x) - \int_{D_n\times D_n} K_n^2(x,y)d\mu_n(x)d\mu_n(y). \tag{4.2.26}$$

Exercise 4.2.26 Using (4.2.26), provide an alternative proof that a necessary condition for $\mathrm{Var}(\chi_n(D_n)) \to \infty$ is that (4.2.25) holds.

4.2.5 Determinantal processes associated with eigenvalues

We provide in this section several examples of point processes related to configurations of eigenvalues of random matrices that possess a determinantal structure. We begin with the eigenvalues of the GUE, and move on to define the sine and Airy processes, associated with the sine and Airy kernels.

The GUE

[Continuation of Example 4.2.15] Let $(\lambda_1^N, \cdots, \lambda_N^N)$ be the eigenvalues of the GUE of dimension N, and denote by χ_N the point process $\chi_N(D) = \sum_{i=1}^N 1_{\lambda_i^N \in D}$. Recall that, with the GUE scaling, the empirical measure of the eigenvalues is, with high probability, roughly supported on the interval $[-2\sqrt{N}, 2\sqrt{N}]$.

Corollary 4.2.27 *Let $D = [-a,b]$ with $a,b > 0$, $\alpha \in (-1/2,1/2)$, and set $D_N = N^\alpha D$. Then*

$$Z_N = \frac{\chi_N(D_N) - \mathbb{E}[\chi_N(D_N)]}{\sqrt{Var(\chi_N(D_N))}}$$

converges in distribution towards a standard normal variable.

Proof In view of Example 4.2.15 and Theorem 4.2.25, the only thing we need to check is that $\mathrm{Var}(\chi_N(D_N)) \to \infty$ as $N \to \infty$. Recalling that

$$\int_{\mathbb{R}} \left(K^{(N)}(x,y)\right)^2 dy = K^{(N)}(x,x),$$

it follows from (4.2.26) that for any $R > 0$, and all N large,

$$\begin{aligned}
\mathrm{Var}(\chi_N(D_N)) &= \int_{D_N}\int_{(D_N)^c} \left(K^{(N)}(x,y)\right)^2 dxdy \\
&= \int_{\sqrt{N}D_N}\int_{\sqrt{N}(D_N)^c} \left(\frac{1}{\sqrt{N}}K^{(N)}(\frac{x}{\sqrt{N}},\frac{y}{\sqrt{N}})\right)^2 dxdy \\
&\geq \int_{-R}^{0}\int_{0}^{R} S_{bN^\alpha}^{(N)}(x,y)dxdy, \qquad (4.2.27)
\end{aligned}$$

where

$$S_z^{(N)}(x,y) = \frac{1}{\sqrt{N}}K^{(N)}\left(z + \frac{x}{\sqrt{N}}, z + \frac{y}{\sqrt{N}}\right)$$

is as in Exercise 3.7.5, and $S_{bN^\alpha}^{(N)}(x,y)$ converges uniformly on compacts, as $N \to \infty$, to the sine-kernel $\sin(x-y)/(\pi(x-y))$. Therefore, there exists a constant $c > 0$ such that the right side of (4.2.27) is bounded below, for large N, by $c\log R$. Since R is arbitrary, the conclusion follows. □

Exercise 4.2.28 Using Exercise 3.7.5 again, prove that if $D_N = [-a\sqrt{N}, b\sqrt{N}]$ with $a,b \in (0,2)$, then Corollary 4.2.27 still holds.

Exercise 4.2.29 Prove that the conclusions of Corollary 4.2.27 and Exercise 4.2.28 hold when the GUE is replaced by the GOE.
Hint: Write $\chi^{(N)}(D_N)$ for the variable corresponding to $\chi_N(D_N)$ in Corollary 4.2.27, with the GOE replacing the GUE. Let $\chi^{(N)}(D_N)$ and $\chi^{(N+1)}(D_N)$ be independent.
(a) Use Theorem 2.5.17 to show that $\chi_N(D_N)$ can be constructed on the same probability space as $\chi^{(N)}(D_N), \chi^{(N+1)}(D_N)$ in such a way that, for any $\varepsilon > 0$, there is a C_ε so that

$$\limsup_{N\to\infty} P(|\chi_N(D_N) - (\chi^{(N)}(D_N) + \chi^{(N+1)}(D_N))/2| > C_\varepsilon) < \varepsilon.$$

(b) By writing a GOE($N+1$) matrix as a rank 2 perturbation of a GOE(N) matrix, show that the laws of $\chi^{(N)}(D_N)$ and $\chi^{(N+1)}(D_N)$ are close in the sense that a copy of $\chi^{(N)}(D_N)$ could be constructed on the same probability space as $\chi^{(N+1)}(D_N)$ in such a way that their difference is bounded by 4.

The sine process

Recall the sine-kernel

$$K_{\text{sine}}(x,y) = \frac{1}{\pi}\frac{\sin(x-y)}{x-y}.$$

Take $\Lambda = \mathbb{R}$ and μ to be the Lebesgue measure, and for $f \in L^2(\mathbb{R})$, define

$$\mathscr{K}_{\text{sine}} f(x) = \int K_{\text{sine}}(x-y) f(y) dy.$$

Writing $k_{\text{sine}}(z) = K_{\text{sine}}(x,y)|_{z=x-y}$, we see that $k_{\text{sine}}(z)$ is the Fourier transform of the function $\mathbf{1}_{[-1/2\pi,1/2\pi]}(\xi)$. In particular, for any $f \in L^2(\mathbb{R})$,

$$\langle f, \mathscr{K}_{\text{sine}} f\rangle = \int\int f(x)f(y)k_{\text{sine}}(x-y)dxdy = \int_{-1/2\pi}^{1/2\pi} |\hat{f}(\xi)|^2 d\xi \le \|f\|_2^2. \tag{4.2.28}$$

Thus, $K_{\text{sine}}(x,y)$ is positive definite, and by Lemma 4.2.13, $\mathscr{K}_{\text{sine}}$ is locally admissible. Further, (4.2.28) implies that all eigenvalues of restrictions of $\mathscr{K}_{\text{sine}}$ to any compact interval belong to the interval $[0,1]$. Hence, by Corollary 4.2.22, $\mathscr{K}_{\text{sine}}$ determines a determinantal point process on $\mathbb{R}$ (which is translation invariant in the terminology of Section 4.2.6 below).

The Airy process

Recall from Definition 3.1.3 the Airy function $\text{Ai}(x) = \frac{1}{2\pi i}\int_C e^{\zeta^3/3 - x\zeta} d\zeta$, where C is the contour in the ζ-plane consisting of the ray joining $e^{-\pi i/3}\infty$ to the origin plus the ray joining the origin to $e^{\pi i/3}\infty$, and the Airy kernel $K_{\text{Airy}}(x,y) = A(x,y) := (\text{Ai}(x)\,\text{Ai}'(y) - \text{Ai}'(x)\,\text{Ai}(y))/(x-y)$. Take $\Lambda = \mathbb{R}$ and μ the Lebesgue measure. Fix $L > -\infty$ and let $\mathscr{K}^L_{\text{Airy}}$ denote the operator on $L^2([L,\infty))$ determined by

$$\mathscr{K}^L_{\text{Airy}} f(x) = \int_L^\infty K_{\text{Airy}}(x,y) f(y) dy.$$

We now have the following.

Proposition 4.2.30 *For any $L > -\infty$, the kernel $K^L_{\text{Airy}}(x,y)$ is locally admissible. Further, all the eigenvalues of its restriction to compact sets belong to the interval $(0,1]$. In particular, K^L_{Airy} determines a determinantal point process.*

Proof We first recall, see (3.9.58), that

$$K_{\mathrm{Airy}}(x,y)=\int_0^\infty \mathrm{Ai}(x+t)\,\mathrm{Ai}(y+t)dt\,. \qquad (4.2.29)$$

In particular, for any $L>-\infty$ and functions $f,g\in L_2([L,\infty))$,

$$\langle f,\mathscr{K}^L_{\mathrm{Airy}}g\rangle=\langle g,\mathscr{K}^L_{\mathrm{Airy}}f\rangle=\int_L^\infty\int_L^\infty\int_0^\infty \mathrm{Ai}(x+t)\,\mathrm{Ai}(y+t)f(x)g(y)dtdxdy\,.$$

It follows that $\mathscr{K}^L_{\mathrm{Airy}}$ is self-adjoint on $L_2([L,\infty))$. Further, from this representation, by an application of Fubini's Theorem,

$$\langle f,\mathscr{K}^L_{\mathrm{Airy}}f\rangle=\int_0^\infty\left|\int_L^\infty f(x)\,\mathrm{Ai}(x+t)dx\right|^2dt\geq 0\,.$$

Together with Lemma 4.2.13, this proves that $\mathscr{K}^L_{\mathrm{Airy}}$ is locally admissible.

To complete the proof, as in the case of the sine process, we need an upper bound on the eigenvalues of restrictions of $\mathscr{K}_{\mathrm{Airy}}$ to compact subsets of $\mathbb{R}$. Toward this end, deforming the contour of integration in the definition of $\mathrm{Ai}(x)$ to the imaginary line, using integration by parts to control the contribution of the integral outside a large disc in the complex plane, and applying Cauchy's Theorem, we obtain the representation, for $x\in\mathbb{R}$,

$$\mathrm{Ai}(x)=\lim_{R\to\infty}\frac{1}{2\pi}\int_{-R}^R e^{\mathrm{i}(s^3/3+xs)}ds\,,$$

with the convergence uniform for x in compacts (from this, one can conclude that $\mathrm{Ai}(x)$ is the Fourier transform, in the sense of distributions, of $e^{\mathrm{i}s^3/3}/\sqrt{2\pi}$, although we will not use that). We now obtain, for continuous functions f supported on $[-M,M]\subset[L,\infty)$,

$$\langle f,\mathscr{K}_{\mathrm{Airy}}f\rangle=\int_0^\infty\left|\int_L^\infty f(x)\,\mathrm{Ai}(x+t)dx\right|^2dt\leq\int_{-\infty}^\infty\left|\int_{-M}^M f(x)\,\mathrm{Ai}(x+t)dx\right|^2dt\,. \qquad (4.2.30)$$

But, for any fixed $K>0$,

$$\begin{aligned}&\int_{-K}^K\left|\int_{-M}^M f(x)\,\mathrm{Ai}(x+t)dx\right|^2dt\\ &=\int_{-K}^K\left|\int_{-M}^M\lim_{R\to\infty}\frac{1}{2\pi}\int_{-R}^R e^{\mathrm{i}(s^3/3+ts)}e^{\mathrm{i}xs}dsf(x)dx\right|^2dt\\ &=\lim_{R\to\infty}\frac{1}{2\pi}\int_{-K}^K\left|\int_{-R}^R e^{\mathrm{i}(s^3/3+ts)}\hat{f}(-s)ds\right|^2dt\,,\end{aligned}$$

where $\hat{f}$ denotes the Fourier transform of f and we have used dominated convergence (to pull the limit out) and Fubini's Theorem in the last equality. Therefore,

$$\int_{-K}^{K}\left|\int_{-M}^{M} f(x)\,\mathrm{Ai}(x+t)dx\right|^2 dt = \lim_{R\to\infty}\int_{-K}^{K}\left|\frac{1}{\sqrt{2\pi}}\int_{-R}^{R} e^{-its}e^{-is^3/3}\hat{f}(s)ds\right|^2 dt$$
$$\leq \limsup_{R\to\infty}\int_{-\infty}^{\infty}\left|\frac{1}{\sqrt{2\pi}}\int_{-\infty}^{\infty} e^{-its}e^{-is^3/3}\mathbf{1}_{[-R,R]}(s)\hat{f}(s)ds\right|^2 dt$$
$$= \limsup_{R\to\infty}\int_{-\infty}^{\infty}\left|e^{-it^3/3}\mathbf{1}_{[-R,R]}(t)\hat{f}(t)dt\right|^2 dt \leq \int_{-\infty}^{\infty}\left|\hat{f}(t)\right|^2 dt = \|f\|_2^2,$$

where we used Parseval's Theorem in the two last equalities. Using (4.2.30), we thus obtain

$$\langle f, \mathscr{K}_{\mathrm{Airy}} f\rangle \leq \|f\|_2^2,$$

first for all compactly supported continuous functions f and then for all $f \in L_2([-L,\infty))$ by approximation. An application of Corollary 4.2.22 completes the proof. □

4.2.6 Translation invariant determinantal processes

In this section we specialize the discussion to determinantal processes on Euclidean space equipped with Lebesgue's measure. Thus, let $\Lambda = \mathbb{R}^d$ and let μ be the Lebesgue measure.

Definition 4.2.31 A determinantal process with $(\Lambda,\mu) = (\mathbb{R}^d, dx)$ is *translation invariant* if the associated kernel K is admissible and can be written as $K(x,y) = K(x-y)$ for some continuous function $K : \mathbb{R}^d \to \mathbb{R}$.

As we will see below after introducing appropriate notation, a determinantal process χ is translation invariant if its law is invariant under (spatial) shifts.

For translation invariant determinantal processes, the conditions of Theorem 4.2.25 can sometimes be simplified.

Lemma 4.2.32 *Assume that K is associated with a translation invariant determinantal process on $\mathbb{R}^d$. Then*

$$\lim_{L\to\infty}\frac{1}{(2L)^d}\mathrm{Var}(\chi([-L,L]^d)) = K(0) - \int_{\mathbb{R}^d} K(x)^2 dx. \tag{4.2.31}$$

Proof. By (4.2.26) with $D = [-L,L]^d$ and $\mathrm{Vol}(D) = (2L)^d$,

$$\mathrm{Var}(\chi(D)) = \mathrm{Vol}(D)K(0) - \int_{D\times D} K^2(x-y)dxdy.$$

In particular,

$$\mathrm{Vol}(D)K(0) \geq \iint_{D\times D} K^2(x-y)dxdy\,.$$

By monotone convergence, it then follows by taking $L \to \infty$ that $\int K^2(x)dx \leq K(0) < \infty$. Further, again from (4.2.26),

$$\mathrm{Var}(\chi(D)) \;=\; \mathrm{Vol}(D)(K(0) - \int_{\mathbb{R}^d} K(x)^2 dx) + \int_D dx \int_{y:y\notin D} K^2(x-y)dy.$$

Since $\int_{\mathbb{R}^d} K(x)^2 dx < \infty$, (4.2.31) follows from the last equality. □

We emphasize that the RHS in (4.2.31) can vanish. In such a situation, a more careful analysis of the limiting variance is needed. We refer to Exercise 4.2.40 for an example of such a situation in the (important) case of the sine-kernel.

We turn next to the ergodic properties of determinantal processes. It is natural to discuss these in the framework of the configuration space $\mathscr{X}$. For $t \in \mathbb{R}^d$, let T^t denote the shift operator, that is for any Borel set $A \subset \mathbb{R}^d$, $T^t A = \{x+t : x \in A\}$. We also write $T^t f(x) = f(x+t)$ for Borel functions. We can extend the shift to act on $\mathscr{X}$ via the formula $T^t \mathbf{x} = (x_i + t)_{i=1}^{\kappa}$ for $\mathbf{x} = (x_i)_{i=1}^{\kappa}$. T^t then extends to a shift on $\mathscr{C}_{\mathscr{X}}$ in the obvious way. Note that one can alternatively also define $T^t \chi$ by the formula $T^t \chi(A) = \chi(T^t A)$.

Definition 4.2.33 Let $\mathbf{x}$ be a point process in $(\mathscr{X}, \mathscr{C}_{\mathscr{X}}, \nu)$. We say that $\mathbf{x}$ is *ergodic* if for any $A \in \mathscr{C}_{\mathscr{X}}$ satisfying $T^t A = A$ for all real t, it holds that $\nu(A) \in \{0,1\}$. It is *mixing* if for any $A, B \in \mathscr{C}_{\mathscr{X}}$, $\nu(A \cap T^t B) \to_{|t|\to\infty} \nu(A)\nu(B)$.

By standard ergodic theory, if $\mathbf{x}$ is mixing then it is ergodic.

Theorem 4.2.34 *Let $\mathbf{x}$ be a translation invariant determinantal point process in $\mathbb{R}^d$, with good kernel K satisfying $K(|x|) \to_{|x|\to\infty} 0$. Then $\mathbf{x}$ is mixing, and hence ergodic.*

Proof Recall from Theorem 4.2.25 that $\int K^2(x)dx < \infty$. It is enough to check that for arbitrary collections of compact Borel sets $\{F_i\}_{i=1}^{L_1}$ and $\{G_j\}_{j=1}^{L_2}$ such that $F_i \cap F_{i'} = \emptyset$ and $G_j \cap G_{j'} = \emptyset$ for $i \neq i'$, $j \neq j'$, and with the notation $G_j^t = T^t G_j$, it holds that for any $\mathbf{z} = \{z_i\}_{i=1}^{L_1} \in \mathbb{C}^{L_1}$, $\mathbf{w} = \{w_j\}_{j=1}^{L_2} \in \mathbb{C}^{L_2}$,

$$E_\nu\left(\prod_{i=1}^{L_1} z_i^{\chi(F_i)} \prod_{j=1}^{L_2} w_j^{\chi(G_j^t)}\right) \to_{|t|\to\infty} E_\nu\left(\prod_{i=1}^{L_1} z_i^{\chi(F_i)}\right) E_\nu\left(\prod_{j=1}^{L_2} w_j^{\chi(G_j)}\right). \quad (4.2.32)$$

Define $F = \bigcup_{i=1}^{L_1} F_i$, $G^t = \bigcup_{j=1}^{L_2} G^t_j$. Let

$$\begin{aligned} K_1 &= \mathbf{1}_F \sum_{i=1}^{L_1} (1-z_i) K \mathbf{1}_{F_i}\,, \quad K_2^t = \mathbf{1}_{G^t} \sum_{j=1}^{L_2} (1-w_j) K \mathbf{1}_{G^t_j}\,, \\ K_{12}^t &= \mathbf{1}_F \sum_{j=1}^{L_2} (1-w_j) K \mathbf{1}_{G^t_j}\,, \quad K_{21}^t = \mathbf{1}_{G^t} \sum_{i=1}^{L_1} (1-z_i) K \mathbf{1}_{F_i}\,. \end{aligned}$$

By Lemma 4.2.16, the left side of (4.2.32) equals, for $|t|$ large enough so that $F \cap G^t = \emptyset$,

$$\Delta(K_1 + K_2^t + K_{12}^t + K_{21}^t)\,. \tag{4.2.33}$$

Note that, by assumption, $\sup_{x,y} K_{12}^t \to_{|t|\to\infty} 0$, $\sup_{x,y} K_{21}^t \to_{|t|\to\infty} 0$. Therefore, by Lemma 3.4.5, it follows that

$$\lim_{|t|\to\infty} |\Delta(K_1 + K_2^t + K_{12}^t + K_{21}^t) - \Delta(K_1 + K_2^t)| = 0\,. \tag{4.2.34}$$

Next, note that for $|t|$ large enough such that $F \cap G^t = 0$, $K_1 \star K_2^t = 0$ and hence, by the definition of the Fredholm determinant,

$$\Delta(K_1 + K_2^t) = \Delta(K_1)\Delta(K_2^t) = \Delta(K_1)\Delta(K_2)\,,$$

where $K_2 := K_2^0$ and the last equality follows from the translation invariance of K. Therefore, substituting in (4.2.33) and using (4.2.34), we get that the left side of (4.2.32) equals $\Delta(K_1)\Delta(K_2)$. Using Lemma 4.2.16 again, we get (4.2.32). □

Let χ be a nonzero translation invariant determinantal point process with good kernel K satisfying $K(|x|) \to_{|x|\to\infty} 0$. As a consequence of Theorem 4.2.34 and the ergodic theorem, the limit

$$c := \lim_{n\to\infty} \chi([-n,n]^d)/(2n)^d \tag{4.2.35}$$

exists and is strictly positive, and is called the *intensity* of the point process.

For stationary point processes, an alternative description can be obtained by considering configurations "conditioned to have a point at the origin". When specialized to one-dimensional stationary point processes, this point of view will be used in Subsection 4.2.7 when relating statistical properties of the gap around zero for determinantal processes to ergodic averages of spacings.

Definition 4.2.35 Let χ be a translation invariant point process, and let B denote a Borel subset of $\mathbb{R}^d$ of positive and finite Lebesgue measure. The *Palm distribution* Q associated with χ is the measure on $\mathscr{M}_+(\mathbb{R}^d)$ determined by the equation, valid for any measurable A,

$$Q(A) = E\left(\int_B \mathbf{1}_A(T^s\chi)\chi(ds)\right)/E(\chi(B)).$$

We then have:

Lemma 4.2.36 *The Palm distribution Q does not depend on the choice of the Borel set B.*

Proof We first note that, due to the stationarity, $E(\chi(B)) = c\mu(B)$ with μ the Lebesgue measure, for some constant c. (It is referred to as the *intensity* of χ, and for determinantal translation invariant point processes, it coincides with the previously defined notion of intensity, see (4.2.35)). It is obvious from the definition that the random measure

$$\chi_A(B) := \int_B \mathbf{1}_A(T^s\chi)\chi(ds)$$

is stationary, namely $\chi_A(T^tB)$ has the same distribution as $\chi_A(B)$. It follows that $E\chi_A(T^tB) = E\chi_A(B)$ for all $t \in \mathbb{R}^d$, implying that $E\chi_A(B) = c_A\mu(B)$ for some constant c_A, since the Lebesgue measure is (up to multiplication by scalar) the unique translation invariant measure on $\mathbb{R}^d$. The conclusion follows. □

Due to Lemma 4.2.36, we can speak of the point process χ^0 attached to the Palm measure Q, which we refer to as the *Palm process*. Note that χ^0 is such that $Q(\chi^0(\{0\}) = 1) = 1$, i.e. χ^0 is such that the associated configurations have a point at zero. It turns out that this analogy goes deeper, and in fact the law Q corresponds to "conditioning on an atom at the origin". Let V_{χ^0} denote the *Voronoi cell* associated with χ^0, i.e., with $B(a,r)$ denoting the Euclidean ball of radius r around a,

$$V_{\chi^0} = \{t \in \mathbb{R}^d : \chi^0(B(t,|t|)) = 0\}.$$

Proposition 4.2.37 *Let χ be a nonzero translation invariant point process with good kernel K satisfying $K(|x|) \to_{|x|\to\infty} 0$, with intensity c. Let χ^0 denote the associated Palm process. Then the law P of χ can be determined from the law Q of χ^0 via the formula, valid for any bounded measurable function f,*

$$Ef(\chi) = cE\int_{V_{\chi^0}} f(T^t\chi^0)dt, \tag{4.2.36}$$

where c is the intensity of χ.

Proof From the definition of χ^0 it follows that for any bounded measurable function g,

$$E\int_B g(T^s\chi)\chi(ds) = c\mu(B)Eg(\chi^0). \tag{4.2.37}$$

This extends by monotone class to jointly measurable nonnegative functions $h : \mathscr{M}_+(\mathbb{R}^d)\times\mathbb{R}^d \to \mathbb{R}$ as

$$E\int_{\mathbb{R}^d} h(T^t\chi,t)\chi(dt) = cE\int_{\mathbb{R}^d} h(\chi^0,t)dt.$$

Applying the last equality to $h(\chi,t) = g(T^{-t}\chi,t)$, we get

$$E\int_{\mathbb{R}^d} g(\chi,t)\chi(dt) = cE\int_{\mathbb{R}^d} g(T^{-t}\chi^0,t)dt = cE\int_{\mathbb{R}^d} g(T^t\chi^0,-t)dt. \tag{4.2.38}$$

Before proceeding, we note a particularly useful consequence of (4.2.38). Namely, let

$$\mathscr{D} := \{\chi : \text{there exist } t \neq t' \in \mathbb{R}^d \text{ with } \|t\| = \|t'\| \text{ and } \chi(\{t\})\cdot\chi(\{t'\}) = 1\}.$$

The measurability of $\mathscr{D}$ is immediate from the measurability of the set

$$\mathscr{D}' = \{(t,t') \in (\mathbb{R}^d)^2 : \|t\| = \|t'\|, t \neq t'\}.$$

Now, with $\mathscr{E}_t = \{\chi : \chi(y) = 1 \text{ for some } y \neq t \text{ with } \|y\| = \|t\|\}$,

$$\mathbf{1}_{\mathscr{D}} \le \int \mathbf{1}_{\mathscr{E}_t}\chi(dt).$$

Therefore, using (4.2.38),

$$P(\mathscr{D}) \le cE\int_{\mathbb{R}^d} \mathbf{1}_{T^{-t}\chi^0\in\mathscr{E}_t} dt.$$

Since all configurations are countable, the set of ts in the indicator in the inner integral on the right side of the last expression is contained in a countable collection of $(d-1)$-dimensional surfaces. In particular, its Lebesgue measure vanishes. One thus concludes that

$$P(\mathscr{D}) = 0. \tag{4.2.39}$$

Returning to the proof of the proposition, apply (4.2.38) with $g(\chi,t) = f(\chi)\mathbf{1}_{\chi(\{t\})=1,\chi(B(0,|t|))=0}$, and use the fact that $T^t\chi^0(B(0,|t|)) = 0$ iff $t \in V_{\chi^0}$ to conclude that

$$E\left(f(\chi)\int \mathbf{1}_{\chi(B(0,|t|))=0}\chi(dt)\right) = cE\left(\int_{V_{\chi^0}} f(T^t\chi^0)dt\right).$$

Since $P(\mathscr{D}) = 0$, it follows that $\int \mathbf{1}_{\chi(B(0,|t|))=0}\chi(dt) = 1$, for almost every χ. This yields (4.2.36).

□

Exercise 4.2.38 Let χ be a nonzero translation invariant determinantal point process with good kernel K. Show that the intensity c defined in (4.2.35) satisfies $c = K(0)$.

Exercise 4.2.39 Assume that K satisfies the assumptions of Lemma 4.2.32, and define the Fourier transform

$$\hat{K}(\lambda) = \int_{x\in\mathbb{R}^d} K(x)\exp(2\pi i x\cdot\lambda)dx \in L^2(\mathbb{R}^d).$$

Give a direct proof that the right side of (4.2.31) is nonnegative.
Hint: use the fact that, since K is a good kernel, it follows that $\|\hat{K}\|_\infty \leq 1$.

Exercise 4.2.40 [CoL95] Take $d = 1$ and check that the sine-kernel $K_{\text{sine}}(x) = \sin(x)/\pi x$ is a good translation invariant kernel for which the right side of (4.2.31) vanishes. Check that then, if $a < b$ are fixed,

$$E[\chi(L[a,b])] = L(b-a)/\pi,$$

whereas

$$\text{Var}(\chi(L[a,b])) = \frac{1}{\pi^2}\log L + O(1).$$

Hint: (a) Apply Parseval's Theorem and the fact that the Fourier transform of the function $\sin(x)/\pi x$ is the indicator over the interval $[-1/2\pi, 1/2\pi]$ to conclude that $\int_{-\infty}^{\infty} K^2(x)dx = 1/\pi = K(0)$.
(b) Note that, with $D = L[a,b]$ and $D_x = [La-x, Lb-x]$,

$$\int_D dx \int_{D^c} K^2(x-y)dy = \int_D dx \int_{D_x^c} K^2(u)du = \frac{1}{\pi^2}\int_D dx \int_{D_x^c} \frac{1-\cos(2u)}{2u^2}du,$$

from which the conclusion follows.

Exercise 4.2.41 Let $|V_{\chi^0}|$ denote the Lebesgue measure of the Voronoi cell for a Palm process χ^0 corresponding to a stationary determinantal process on $\mathbb{R}^d$ with intensity c. Prove that $E(|V_{\chi^0}|) = 1/c$.

4.2.7 *One-dimensional translation invariant determinantal processes*

We restrict attention in the sequel to the case of most interest to us, namely to dimension $d = 1$, in which case the results are particularly explicit. Indeed, when

$d = 1$, each configuration $\mathbf{x}$ of a determinantal process can be ordered, and we write $\mathbf{x} = (\ldots,x_{-1},x_0,x_1,\ldots)$ with the convention that $x_i < x_{i+1}$ for all i and $x_0 < 0 < x_1$ (by stationarity and local finiteness, $P(\chi(\{0\}) = 1) = 0$, and thus the above is well defined). We also use $\mathbf{x}^0 = (\ldots,x^0_{-1},0 = x^0_0,x^0_1,\ldots)$ to denote the configuration corresponding to the Palm process χ^0. The translation invariance of the point process χ translates then to stationarity for the Palm process increments, as follows.

Lemma 4.2.42 *Let $\mathbf{x}^0$ denote the Palm process associated with a determinantal translation invariant point process $\mathbf{x}$ on $\mathbb{R}$ with good kernel K satisfying $K(|x|) \to_{|x|\to\infty} 0$, and with intensity $c > 0$. Then the sequence $\mathbf{y}^0 := \{x^0_{i+1} - x^0_i\}_{i\in\mathbb{Z}}$ is stationary and ergodic.*

Proof Let $T\mathbf{y}^0 = \{y^0_{i+1}\}_{i\in\mathbb{Z}}$ denote the shift of $\mathbf{y}^0$. Consider g a Borel function on $\mathbb{R}^{2r}$ for some $r \geq 1$, and set $\bar{g}(\mathbf{y}^0) = g(y^0_{-r},\ldots,y^0_{r-1})$. For any configuration $\mathbf{x}$ with $x_i < x_{i+1}$ and $x_{-1} < 0 \leq x_0$, set $\mathbf{y} := \{x_{i+1} - x_i\}_{i\in\mathbb{Z}}$. Set $f(\mathbf{x}) = g(x_{-r+1} - x_{-r},\ldots,x_r - x_{r-1})$, and let $A_u = \{\chi : f(\mathbf{x}) \leq u\}$. A_u is clearly measurable, and by Definition 4.2.35 and Lemma 4.2.36, for any Borel B with positive and finite Lebesgue measure,

$$\begin{aligned} P(\bar{g}(\mathbf{y}^0) \leq u) &= Q(A_u) = E\left(\int_B \mathbf{1}_{A_u}(T^s\chi)\chi(ds)\right)/c\mu(B) \\ &= E\left(\sum_{i:x_i\in B} \mathbf{1}_{\bar{g}(T^i\mathbf{y})\leq u}\right)/c\mu(B)\,. \end{aligned} \tag{4.2.40}$$

(Note the different roles of the shifts T^s, which is a spatial shift, and T^i, which is a shift on the index set, i.e. on $\mathbb{Z}$.) Hence,

$$|P(\bar{g}(\mathbf{y}^0) \leq u) - P(\bar{g}(T\mathbf{y}^0) \leq u)| \leq 2/c\mu(B)\,.$$

Taking $B = B_n = [-n,n]$ and then $n \to \infty$, we obtain that the left side of the last expression vanishes. This proves the stationarity. The ergodicity (and in fact, mixing property) of the sequence $\mathbf{y}^0$ is proved similarly, starting from Theorem 4.2.34. □

We also have the following analog of Proposition 4.2.37.

Proposition 4.2.43 *Assume $\mathbf{x}$ is a nonzero stationary determinantal process on $\mathbb{R}$ with intensity c. Then for any bounded measurable function f,*

$$E(f(\mathbf{x})) = cE\int_0^{x^0_1} f(T^t\mathbf{x}^0)dt\,. \tag{4.2.41}$$

Proof Apply (4.2.38) with $g(\chi,t) = f(\mathbf{x})\mathbf{1}_{x_0(\chi)=-t}$. □

Proposition 4.2.43 gives an natural way to construct the point process χ starting from χ^0 (whose increments form a stationary sequence): indeed, it implies that χ is nothing but the *size biased* version of χ^0, where the size biasing is obtained by the value of x_1^0. More explicitly, let $\mathbf{x}$ denote a translation invariant determinantal process with intensity c, and let $\mathbf{x}^0$ denote the associated Palm process on $\mathbb{R}$. Consider the sequence $\mathbf{y}^0$ introduced in Lemma 4.2.42, and denote its law by Q^y. Let $\bar{\mathbf{y}}$ denote a sequence with law $\bar{Q}^y$ satisfying $d\bar{Q}^y/dQ^y(\mathbf{y}) = cy_0$, let $\bar{\mathbf{x}}^0$ denote the associated configuration, that is $\bar{x}_i^0 = \sum_{j=1}^{i-1} \bar{y}_j$, noting that $\bar{x}_0 = 0$, and let U denote a random variable distributed uniformly on $[0,1]$, independent of $\bar{\mathbf{x}}^0$. Set $\bar{\mathbf{x}} = T^{U\bar{x}_1^0}\bar{\mathbf{x}}^0$. We then have

Corollary 4.2.44 *The point process $\bar{\mathbf{x}}$ has the same law as* $\mathbf{x}$.

Proof By construction, for any bounded measurable f,

$$\begin{aligned} Ef(\bar{\mathbf{x}}) &= E\int_0^1 f(T^{u\bar{x}_1^0}\bar{\mathbf{x}}^0)du = E\int_0^{\bar{x}_1^0} f(T^t\bar{\mathbf{x}}^0)\frac{dt}{\bar{x}_1^0} \\ &= cE\int_0^{x_1^0} f(T^t\mathbf{x}^0)dt = Ef(\mathbf{x}), \end{aligned}$$

where Proposition 4.2.43 was used in the last step. □

Corollary 4.2.44 has an important implication to averages. Let $B_n = [0,n]$. For a bounded measurable function f and a point process $\mathbf{x}$ on $\mathbb{R}$, let

$$f_n(\mathbf{x}) = \frac{\sum_{x_i \in B_n} f(T^{x_i}\mathbf{x})}{|\{i : x_i \in B_n\}|}.$$

Corollary 4.2.45 *Let $\mathbf{x}$ be a translation invariant determinantal process with intensity c, and good kernel K satisfying $K(x) \to_{|x|\to\infty} 0$, and Palm measure Q. Then*

$$\lim_{n\to\infty} f_n(\mathbf{x}) = E_Q f\,, \text{almost surely}.$$

Proof The statement is immediate from the ergodic theorem and Lemma 4.2.42 for the functions $f_n(\mathbf{x}^0)$. Since, by Corollary 4.2.44, the law of $T^{x_1}\mathbf{x}$ is absolutely continuous with respect to that of $\mathbf{x}^0$, the conclusion follows by an approximation argument. □

Corollary 4.2.44 allows us to relate several quantities of interest in the study of determinantal processes. For a translation invariant determinantal point process $\mathbf{x}$, let $G_{\mathbf{x}} = x_1 - x_0$ denote the *gap* around 0. With Q_1 denoting the marginal on x_1^0 of the Palm measure, and with $\bar{Q}_1$ defined by $d\bar{Q}_1/dQ_1(u) = cu$, note that

$$P(G_{\mathbf{x}} \geq t) = P(\bar{x}_1^0 \geq t) = \int_t^\infty \bar{Q}_1(du) = c\int_t^\infty uQ_1(du).$$

Let $\bar{G}(t) = P(\{\mathbf{x}\} \cap (-t,t) = \emptyset)$ be the probability that the interval $(-t,t)$ does not contain any point of the configuration $\mathbf{x}$. Letting $D_t = \mathbf{1}_{(-t,t)}$ $K_t = \mathbf{1}_{D_t} K \mathbf{1}_{D_t}$, and $\chi_t = \chi(D_t)$, we have, using Lemma 4.2.16, that

$$\bar{G}(t) = P(\chi_t = 0) = \lim_{|z|\to 0} E(z^{\chi_t}) = \Delta(K_t), \tag{4.2.42}$$

that is, $\bar{G}(t)$ can be read off easily from the kernel K. Other quantities can be read off $\bar{G}$, as well. In particular, the following holds.

Proposition 4.2.46 *Let* $\mathbf{x}$ *be a translation invariant determinantal point process of intensity c. Then the function* $\bar{G}$ *is differentiable and*

$$\frac{\partial \bar{G}(t)}{\partial t} = -2c\int_{2t}^\infty Q_1(dw). \tag{4.2.43}$$

Proof By Corollary 4.2.44,

$$\begin{aligned}\bar{G}(t) &= 2\int_0^{1/2} P(u\bar{x}_1^0 \geq t)du = 2\int_0^{1/2}\int_{t/u}^\infty \bar{Q}_1(ds)du \\ &= 2t\int_{2t}^\infty dww^{-2}\int_w^\infty \bar{Q}_1(ds),\end{aligned}$$

where the change of variables $w = t/u$ was used in the last equality. Integrating by parts, using $\mathscr{V}(w) = -1/w$ and $\mathscr{U}(w) = \bar{Q}_1([w,\infty))$, we get

$$\begin{aligned}\bar{G}(t) &= \mathscr{U}(2t) - 2t\int_{2t}^\infty w^{-1}\bar{Q}_1(dw) \\ &= \mathscr{U}(2t) - 2ct\int_{2t}^\infty Q_1(dw) = \mathscr{U}(2t) - 2ctQ_1([2t,\infty)) \\ &= c\int_{2t}^\infty [w-2t]Q_1(dw).\end{aligned}$$

Differentiating in t, we then get (4.2.43). □

Finally, we describe an immediate consequence of Proposition 4.2.46, which is useful when relating different statistics related to the spacing of eigenvalues of random matrices. Recall the "spacing process" $\mathbf{y}$ associated with a stationary point process $\mathbf{x}$, i.e. $y_i = x_{i+1} - x_i$.

Corollary 4.2.47 *Let g be a bounded measurable function on $\mathbb{R}_+$ and define $g_n = \frac{1}{n}\sum_{i=1}^n g(y_i)$. Then*

$$g_n \to_{n\to\infty} E_{Q_1} g = \int_0^\infty g(w) Q_1(dw)\,, \text{ almost surely}.$$

In particular, with $g^t(w) = \mathbf{1}_{w>2t}$, we get

$$-\frac{\partial \bar{G}(t)}{\partial t} = 2c \lim_{n\to\infty} (g^t)_n\,, \text{ almost surely}. \tag{4.2.44}$$

4.2.8 Convergence issues

We continue to assume K is a good translation invariant kernel on $\mathbb{R}$ satisfying $K(|x|) \to_{|x|\to\infty} 0$. In many situations, the kernel K arises as a suitable limit of kernels $K_N(x,y)$ that are not translation invariant, and it is natural to relate properties of determinantal processes $\mathbf{x}^N$ (or χ^N) associated with K_N to those of the determinantal process $\mathbf{x}$ (or χ) associated with K.

We begin with a simple lemma that is valid for (not necessarily translation invariant) determinantal processes. Let K_N denote a sequence of good kernels corresponding to a determinantal process $\bar{x}^N$, and let K be a good kernel corresponding to a determinantal process $\bar{x}$. Set $\bar{G}(t) = P(\{\mathbf{x}\} \cap (-\mathbf{t},\mathbf{t}) = \emptyset)$ and $\bar{G}_N(t) = P(\{\mathbf{x}^N\} \cap (-t,t) = \emptyset)$.

Lemma 4.2.48 *Let D_ℓ denote disjoint compact subsets of $\mathbb{R}$. Suppose a sequence of good kernels K_N satisfy $K_N(x,y) \to K(x,y)$ uniformly on compact subsets of $\mathbb{R}$, where K is a good kernel. Then for any L finite, the random vector $(\chi^N(D_1),\ldots,\chi^N(D_L))$ converges to the random vector $(\chi(D_1),\ldots,\chi(D_L))$ in distribution. In particular, $\bar{G}_N(t) \to_{N\to\infty} \bar{G}(t)$.*

Proof It is clearly enough to check that

$$E\left(\prod_{\ell=1}^L z_\ell^{\chi^N(D_\ell)}\right) \to_{N\to\infty} E\left(\prod_{\ell=1}^L z_\ell^{\chi(D_\ell)}\right).$$

By Lemma 4.2.16, with $D = \bigcup_{\ell=1}^L$, the last limit would follow from the convergence

$$\Delta\left(\mathbf{1}_D \sum_{\ell=1}^L (1-z_\ell) K_N \mathbf{1}_{D_\ell}\right) \to_{N\to\infty} \Delta\left(\mathbf{1}_D \sum_{\ell=1}^L (1-z_\ell) K \mathbf{1}_{D_\ell}\right),$$

which is an immediate consequence of Lemma 3.4.5. □

In what follows, we assume that K is a good translation invariant kernel on $\mathbb{R}$

satisfying $K(|x|) \to_{|x|\to\infty} 0$. In many situations, the kernel K arises as a suitable limit of kernels $K_N(x,y)$ that are not translation invariant, and it is natural to relate properties of determinantal processes $\mathbf{x}^N$ (or χ^N) associated with K_N to those of the determinantal process $\mathbf{x}$ (or χ) associated with K.

We next discuss a modification of Corollary 4.2.47 that is applicable to the process $\mathbf{x}^N$ and its associated spacing process $\mathbf{y}^N$.

Theorem 4.2.49 *Let $g_t(x) = \mathbf{1}_{x>t}$, and define $g_{n,t}^N = \frac{1}{n}\sum_{i=1}^n g_t(y_i^N)$. Suppose further that $n = o(N) \to_{N\to\infty} \infty$ is such that for any constant $a > 0$,*

$$\limsup_{N\to\infty} \sup_{|x|+|y|\le 2an} |K_N(x,y) - K(x-y)| = 0. \tag{4.2.45}$$

Then

$$g_{n,t}^N \to_{N\to\infty} E_{Q_1} g_t = \int_t^\infty Q_1(dw), \text{ in probability}. \tag{4.2.46}$$

Proof In view of Corollary 4.2.47, it is enough to prove that $|g_{n,t}^N - g_{n,t}| \to_{N\to\infty} 0$, in probability. Let c denote the intensity of the process $\mathbf{x}$. For $a > 0$, let $D_{n,a} = [0, an]$. By Corollary 4.2.45, $\chi(D_{n,a})/n$ converges almost surely to a/c. We now claim that

$$\frac{\chi^N(D_{n,a})}{n} \to_{N\to\infty} \frac{a}{c}, \quad \text{in probability}. \tag{4.2.47}$$

Indeed, recall that by Lemma 4.2.5 and the estimate (4.2.45),

$$\frac{1}{n} E\chi^N(D_{n,a}) = \frac{1}{n}\int_0^{an} [K_N(x,x) - K(0)]dx + \frac{anK(0)}{n} \to_{N\to\infty} \frac{a}{c},$$

while, c.f. (4.2.26),

$$\mathrm{Var}\left(\frac{1}{n}\chi^N(D_{n,a})\right) \le \frac{1}{n^2}\int_0^{an} K_N(x,x)dx \to_{N\to\infty} 0,$$

proving (4.2.47).

In the sequel, fix $a > 0$ and let

$$C_N(s,n) = \frac{1}{n}\sum_{i=1}^\infty \mathbf{1}_{an > x_i^N, x_{i+1}^N - x_i^N > s}, \quad C(s,n) = \frac{1}{n}\sum_{i=1}^\infty \mathbf{1}_{an > x_i, x_{i+1} - x_i > s}.$$

In view of (4.2.47), in order to prove (4.2.46) it is enough to show that, for any $a, s > 0$,

$$|EC_N(s,n) - EC(s,n)| \to_{N\to\infty} 0, \quad |E(C_N(s,n))^2 - E(C(s,n))^2| \to_{N\to\infty} 0. \tag{4.2.48}$$

Fix $\delta > 0$, and divide the interval $[0, an)$ into $\lceil n/\delta \rceil$ disjoint intervals $D_i =$

$[(i-1)\delta, i\delta) \cap [0,n)$, each of length $\leq \delta$. Let $\chi_i^N = \chi^N(D_i)$ and $\chi_i = \chi(D_i)$. Set

$$S^N(s,\delta,n) = \frac{1}{n} \sum_{i=1}^{\lceil an/\delta \rceil} \mathbf{1}_{\chi_i^N \geq 1, \chi_j^N = 0, j=i+1,\ldots,i+\lfloor s/\delta \rfloor},$$

and

$$S(s,\delta,n) = \frac{1}{n} \sum_{i=1}^{\lceil an/\delta \rceil} \mathbf{1}_{\chi_i \geq 1, \chi_j^N = 0, j=i+1,\ldots,i+\lfloor s/\delta \rfloor}.$$

We prove below that, for any fixed s, δ,

$$|ES^N(s,\delta,n) - ES(s,\delta,n)| \to_{N\to\infty} 0, \tag{4.2.49}$$

$$|E(S^N(s,\delta,n)^2) - E(S(s,\delta,n)^2)| \to_{N\to\infty} 0, \tag{4.2.50}$$

from which (4.2.48) follows by approximation.

To see (4.2.49), note first that

$$\begin{aligned} ES^N(s,\delta,n) &= \frac{1}{n} \sum_{i=1}^{\lceil an/\delta \rceil} E\left(\mathbf{1}_{\chi_i^N \geq 1} \prod_{j=i+1}^{i+\lfloor s/\delta \rfloor} \chi_j^N\right) \\ &= \frac{1}{n} \sum_{i=1}^{\lceil an/\delta \rceil} E\left((1-\mathbf{1}_{\chi_i^N=0}) \prod_{j=i+1}^{i+\lfloor s/\delta \rfloor} \chi_j^N\right) \\ &= \frac{1}{n} \sum_{i=1}^{\lceil an/\delta \rceil} \lim_{\max_j |z_j| \to 0} \left[E\left(\prod_{j=i+1}^{i+\lfloor s/\delta \rfloor} z_j^{\chi_j^N}\right) - E\left(\prod_{j=i}^{i+\lfloor s/\delta \rfloor} z_j^{\chi_j^N}\right)\right] \\ &= \frac{1}{n} \sum_{i=1}^{\lceil an/\delta \rceil} \left[\Delta(\mathbf{1}_{B_i} K_N \mathbf{1}_{B_i}) - \Delta(\mathbf{1}_{B_i^+} K_N \mathbf{1}_{B_i^+})\right], \end{aligned}$$

where $B_i = \bigcup_{j=i+1}^{i+\lfloor s/\delta \rfloor} D_j$ and $B_i^+ = \bigcup_{j=i}^{i+\lfloor s/\delta \rfloor} D_j$, and we used Lemma 4.2.16 in the last equality. Similarly,

$$ES(s,\delta,n) = \frac{1}{n} \sum_{i=1}^{\lceil an/\delta \rceil} \left[\Delta(\mathbf{1}_{B_i} K \mathbf{1}_{B_i}) - \Delta(\mathbf{1}_{B_i^+} K \mathbf{1}_{B_i^+})\right],$$

Applying Corollary 4.2.45, (4.2.49) follows.

The proof of (4.2.50) is similar and omitted. □

4.2.9 Examples

We consider in this subsection several examples of determinantal processes.

The biorthogonal ensembles

In the setup of Subsection 4.2.1, let $(\psi_i, \phi_i)_{i\geq 0}$ be functions in $L^2(\Lambda, \mu)$. Let

$$g_{ij} = \int \psi_i(x)\phi_j(x)d\mu(x), 1 \leq i, j \leq N.$$

Define the measure μ^N on Λ^N by

$$\mu^N(dx_1, \cdots, dx_N) = \det_{i,j=1}^N(\phi_i(x_j)) \det_{i,j=1}^N(\psi_i(x_j)) \prod_{i=1}^N d\mu(x_i). \tag{4.2.51}$$

Lemma 4.2.50 *Assume that all principal minors of $G = (g_{ij})$ are not zero. Then the measure μ^N of (4.2.51) defines a determinantal simple point process with N points.*

Proof The hypothesis implies that G admits a Gauss decomposition, that is, it can be decomposed into the product of a lower triangular and an upper triangular matrix, with nonzero diagonal entries. Thus there exist matrices $L = (l_{ij})_{i,j=1}^N$ and $U = (u_{ij})_{i,j=1}^N$ so that $LGU = I$. Setting

$$\tilde{\phi} = U\phi \qquad \tilde{\psi} = L\psi,$$

it follows that, with respect to the scalar product in $L^2(\mu)$,

$$\langle \tilde{\phi}_i, \tilde{\psi}_j \rangle = \delta_{i,j}, \tag{4.2.52}$$

and, further,

$$\mu^N(dx_1, \cdots, dx_N) = C_N \det_{i,j=1}^N(\tilde{\phi}_i(x_j)) \det_{i,j=1}^N(\tilde{\psi}_i(x_j)) \prod_{i=1}^N d\mu(x_i)$$

for some constant C_N. Proceeding as in the proof of Lemma 3.2.2, we conclude that

$$\mu^N(dx_1, \cdots, dx_N) = C_N \det_{i,j=1}^N \sum_{k=1}^N \tilde{\phi}_k(x_i)\tilde{\psi}_k(x_j) \prod_{i=1}^N d\mu(x_i).$$

The proof of Lemma 4.2.50 is concluded by using (4.2.52) and computations similar to Lemma 3.2.2 in order to verify the property in Remark 4.2.6. □

Exercise 4.2.51 By using Remark 4.1.7, show that all joint distributions appearing in Weyl's formula for the unitary groups (Proposition 4.1.6) correspond to determinantal processes.

Birth–death processes conditioned not to intersect

Take Λ to be $\mathbb{Z}$, μ the counting measure and K_n a homogeneous (discrete time) Markov semigroup, that is, $K_n : \Lambda\times\Lambda\to\mathbb{R}^+$ so that, for any integers n,m,

$$K_{n+m}(x,y) = K_n \star K_m(x,y) = \int K_n(x,z)K_m(z,y)d\mu(z)\,,$$

and, further, $\int K_n(x,y)d\mu(y) = 1$. We assume $K_1(x,y) = 0$ if $|x-y| \neq 1$. We let $\{X_n\}_{n\geq 0}$ denote the Markov process with kernel K_1, that is for all $n < m$ integers,

$$P(X_m \in A|X_j, j\leq n) = P(X_m \in A|X_n) = \int_{y\in A} K_{m-n}(X_s,y)d\mu(y)\,.$$

Fix $\mathbf{x} = (x^1 < \cdots < x^N)$ with $x^i \in 2\mathbb{Z}$. Let $\{\mathbf{X}_n^{\mathbf{x}}\}_{n\geq 0} = \{(X_n^1,\ldots,X_n^N)\}_{n\geq 0}$ denote N independent copies of $\{X_n\}_{n\geq 0}$, with initial positions $(X_0^1,\ldots,X_0^N) = \mathbf{x}$. For integer T, define the event $\mathscr{A}_T = \bigcap_{0\leq k\leq T}\{X_k^1 < X_k^2 < \cdots < X_k^N\}$.

Lemma 4.2.52 (Gessel–Viennot) *With the previous notation, set* $\mathbf{y} = (y^1 < \cdots < y^N)$ *with* $y^i \in 2\mathbb{Z}$. *Then*

$$\begin{aligned} K_{2T}^N(\mathbf{x},\mathbf{y}) &\stackrel{\triangle}{=} P(\mathbf{X}_{2T}^{\mathbf{x}} = \mathbf{y}|\mathscr{A}_{2T}) \\ &= \frac{\det_{i,j=1}^N(K_{2T}(x^i,y^j))}{\int_{z^1<\cdots<z^N}\det_{i,j=1}^N(K_{2T}(x^i,z^j))\prod d\mu(z^j)}\,. \end{aligned}$$

Proof The proof is an illustration of the *reflection principle*. Let $\mathscr{P}_{2T}(x,y)$, $x,y\in 2\mathbb{Z}$, denote the collection of $\mathbb{Z}$-valued, nearest neighbor paths $\{\pi(\ell)\}_{\ell=0}^{2T}$ with $\pi(0) = x$, $\pi(2T) = y$ and $|\pi(\ell+1)-\pi(\ell)| = 1$. Let

$$\Pi_{2T}(\mathbf{x},\mathbf{y}) = \{\{\pi^i\}_{i=1}^N : \pi^i \in \mathscr{P}_{2T}(x^i,y^i)\}$$

denote the collection of N nearest neighbor paths, with the ith path connecting x^i and y^i. For any permutation $\sigma\in\mathscr{S}_N$, set $\mathbf{y}_\sigma = \{y^{\sigma(i)}\}_{i=1}^N$. Then

$$\det_{i,j=1}^N(K_{2T}(x^i,y^j)) = \sum_{\sigma\in\mathscr{S}_N}\varepsilon(\sigma)\sum_{\{\pi^i\}_{i=1}^N\in\Pi_{2T}(\mathbf{x},\mathbf{y}_\sigma)}\prod_{i=1}^N K_{2T}(\pi^i)\,, \tag{4.2.53}$$

where

$$K_{2T}(\pi^i) = K_1(x^i,\pi^i(2))\left(\prod_{k=2}^{2T-2}K_1(\pi^i(k),\pi^i(k+1))\right)K_1(\pi^i(2T-1),y^{\sigma(i)})\,.$$

On the other hand, let

$$\mathscr{N}C_{2T}^{\mathbf{x},\mathbf{y}} = \{\{\pi^i\}_{i=1}^N \in \Pi_{2T}(\mathbf{x},\mathbf{y}) : \{\pi^i\}\cap\{\pi^j\} = \emptyset \text{ if } i\neq j\}$$

denote the collection of disjoint nearest neighbor paths connecting $\mathbf{x}$ and $\mathbf{y}$. Then

$$P(\mathbf{X}^{\mathbf{x}}_{2T} = \mathbf{y}, \mathscr{A}_{2T}) = \sum_{\{\pi^i\}_{i=1}^N \in \mathscr{N}C^{\mathbf{x},\mathbf{y}}_{2T}} \prod_{i=1}^N K_{2T}(\pi^i) . \qquad (4.2.54)$$

Thus, to prove the lemma, it suffices to check that the total contribution in (4.2.53) of the collection of paths not belonging to $\mathscr{N}C^{\mathbf{x},\mathbf{y}}_{2T}$ vanishes. Toward this end, the important observation is that because we assumed $\mathbf{x}, \mathbf{y} \in 2\mathbb{Z}$, for any $n \le 2t$ and $i, j \le N$, any path $\pi \in \Pi_{2T}(x^i, y^j)$ satisfies $\pi(n) \in 2\mathbb{Z} + n$. In particular, if $\{\pi^i\}_{i=1}^N \in \bigcup_{\sigma \in \mathscr{S}_N} \Pi_{2T}(\mathbf{x}, \mathbf{y}_\sigma)$ and there is a time $n \le 2T$ and integers $i < j$ such that $\pi^i(n) \ge \pi^j(n)$, then there actually is a time $m \le n$ with $\pi^i(m) = \pi^j(m)$.

Now, suppose that in a family $\{\pi^i\}_{i=1}^N \in \Pi_{2T}(\mathbf{x}, \mathbf{y}_\sigma)$, there are integers $i < j$ so that $\pi^i(n) = \pi^j(n)$. Consider the path $\tilde{\pi}$ so that

$$\tilde{\pi}^k(\ell) = \begin{cases} \pi^j(\ell), & k = i, \ell > n \\ \pi^i(\ell), & k = j, \ell > n \\ \pi^k(\ell), & \text{otherwise.} \end{cases}$$

Then, obviously, $\prod_{i=1}^N K_{2T}(\pi^i) = \prod_{i=1}^N K_{2T}(\tilde{\pi}^i)$. Further, for some $\sigma' \in \mathscr{S}_N$, $\{\tilde{\pi}^i\}_{i=1}^N \in \Pi_{2T}(\mathbf{x}, \mathbf{y}_{\sigma'})$, with σ and σ' differing only by the transposition of i and j. In particular, $\varepsilon(\sigma) + \varepsilon(\sigma') = 0$.

We can now conclude: by the previous argument, the contribution in (4.2.53) of the collection of paths where π^1 intersects with any other path vanishes. On the other hand, for the collection of paths where π^1 does not intersect any other path (and thus $\pi^1(2T) = y^1$), one freezes a path π^1 and repeats the same argument to conclude that the sum over all other paths, restricted not to intersect the frozen path π^1 but to have π^2 intersect another path, vanishes. Proceeding inductively, one concludes that the sum in (4.2.53) over all collections $\{\pi^i\}_{i=1}^N \notin \mathscr{N}C^{\mathbf{x},\mathbf{y}}_{2T}$ vanishes. This completes the proof. □

Combining Lemma 4.2.52 with Lemma 4.2.50, we get the following.

Corollary 4.2.53 *In the setup of Lemma 4.2.52, let*

$$\mathscr{B}_{2T,\mathbf{y}} = \mathscr{A}_{2T} \bigcap_{i=1}^N \{X^i(2T) = y^i\} .$$

Conditioned on the event $\mathscr{B}_{2T,\mathbf{y}}$, the process $(X^1(n), \ldots, X^N(n))_{n \in [0,2T]}$ is a (time inhomogeneous) Markov process satisfying, with $\mathbf{z} = (z^1 < z^2 < \cdots < z^N)$ and $n < 2T$,

$$P(\mathbf{X}^{\mathbf{x}}_n = \mathbf{z} | \mathscr{A}_{2T}) = C_N(n, T, \mathbf{x}, \mathbf{y}) \det_{i,j=1}^N (K_n(x^i, z^j)) \det_{i,j=1}^N (K_{2T-n}(z^i, y^j))$$

with

$$C_N(n,T,\mathbf{x},\mathbf{y}) = \int \det_{i,j=1}^{N}(K_n(x^i,z^j)) \det_{i,j=1}^{N}(K_{2T-n}(z^i,y^j)) \prod_{i=1}^{N} d\mu(z^i).$$

At any time $n < 2T$, the configuration $(X^1(n),\ldots,X^N(n))$, conditioned on the event $\mathscr{B}_{2T,\mathbf{y}}$, is a determinantal simple point process.

We note that, in the proof of Lemma 4.2.52, it was enough to consider only the *first* time in which paths cross; the proof can therefore be adapted to cover diffusion processes, as follows. Take $\Lambda = \mathbb{R}$, μ the Lebesgue measure, and consider a time homogeneous, real valued diffusion process $(X_t)_{t\geq 0}$ with transition kernel $K_t(x,y)$ which is jointly continuous in (x,y). Fix $\mathbf{x} = (x^1 < \cdots < x^N)$ with $x^i \in \mathbb{R}$. Let $\{\mathbf{X}_t^{\mathbf{x}}\}_{t\geq 0} = \{(X_t^1,\ldots,X_t^N)\}_{t\geq 0}$ denote N independent copies of $\{X_t\}_{t\geq 0}$, with initial positions $(X_0^1,\ldots,X_0^N) = \mathbf{x}$. For real T, define the event $\mathscr{A}_T = \bigcap_{0\leq t\leq T}\{X_t^1 < X_t^2 < \cdots < X_t^N\}$.

Lemma 4.2.54 (Karlin–McGregor) *With the previous notation, the probability measure $P(\mathbf{X}_T^{\mathbf{x}} \in \cdot|\mathscr{A}_T)$ is absolutely continuous with respect to Lebesgue measure restricted to the set $\{\mathbf{y} = (y^1 < y^2 < \cdots < y^N)\} \subset \mathbb{R}^N$, with density $p_T^{\mathbf{x}}(\mathbf{y}|\mathscr{A}_T)$ satisfying*

$$p_T^{\mathbf{x}}(\mathbf{y}|\mathscr{A}_T) = \frac{\det_{i,j=1}^{N}(K_T(x^i,y^j))}{\int_{z^1<\cdots<z^N} \det_{i,j=1}^{N}(K_T(x^i,z^j))\prod dz^j}.$$

Exercise 4.2.55 Prove the analog of Corollary 4.2.53 in the setup of Lemma 4.2.54. Use the following steps.
(a) For $t < T$, construct the density $q_t^{N,T,\mathbf{x},\mathbf{y}}$ of $\mathbf{X}_t^{\mathbf{x}}$ "conditioned on $\mathscr{A}_T \cap \{\mathbf{X}_T^{\mathbf{x}} = \mathbf{y}\}$" so as to satisfy, for any Borel sets $A, B \subset \mathbb{R}^N$ and $t < T$,

$$P(\mathbf{X}_t^{\mathbf{x}} \in A, \mathbf{X}_T^{\mathbf{x}} \in B|\mathscr{A}_T) = \int_A \prod_{i=1}^{N} dz^i \int_B \prod_{i=1}^{N} dy^i q_t^{N,T,\mathbf{x},\mathbf{y}}(\mathbf{z}) p^{\mathbf{x}}(\mathbf{y}|\mathscr{A}_T).$$

(b) Show that the collection of densities $q_t^{N,T,\mathbf{x},\mathbf{y}}$ determine a Markov semigroup corresponding to a diffusion process, and

$$q_t^{N,T,\mathbf{x},\mathbf{y}}(\mathbf{z}) = C_{N,T}(t,\mathbf{x},\mathbf{y}) \det_{i,j=1}^{N}(K_t(x^i,z^j)) \det_{i,j=1}^{N}(K_{T-t}(z^i,y^j))$$

with

$$C_{N,T}(t,\mathbf{x},\mathbf{y}) = \int \det_{i,j=1}^{N}(K_t(x^i,z^j)) \det_{i,j=1}^{N}(K_{T-t}(z^i,y^j)) \prod_{i=1}^{N} d\mu(z^i),$$

whose marginal at any time $t < T$ corresponds to a determinantal simple point process with N points.

Exercise 4.2.56 (a) Use Exercise 4.2.55 and the heat kernel

$$K_1(x,y) = (2\pi)^{-1/2} e^{-(x-y)^2/2}$$

to conclude that the law of the (ordered) eigenvalues of the GOE coincides with the law of N Brownian motions run for a unit of time and conditioned not to intersect at positive times smaller than 1.
Hint: start the Brownian motion at locations $0 = x_1 < x_2 < \cdots < x_N$ and then take $x_N \to 0$, keeping only the leading term in $\mathbf{x}$ and noting that it is a polynomial in $\mathbf{y}$ that vanishes when $\Delta(\mathbf{y}) = 0$.
(b) Using part (a) and Exercise 4.2.55, show that the law of the (ordered) eigenvalues of the GUE coincides with the law of N Brownian motions at time 1, run for two units of time, and conditioned not to intersect at positive times less than 2, while returning to 0 at time 2.

4.3 Stochastic analysis for random matrices

In this section we introduce yet another effective tool for the study of Gaussian random matrices. The approach is based on the fact that a standard Gaussian variable of mean 0 and variance 1 can be seen as the value, at time 1, of a standard Brownian motion. (Recall that a Brownian motion W_t is a zero mean Gaussian process of covariance $E(W_t W_s) = t \wedge s$.) Thus, replacing the entries by Brownian motions, one gets a matrix-valued random process, to which stochastic analysis and the theory of martingales can be applied, leading to alternative derivations and extensions of laws of large numbers, central limit theorems, and large deviations for classes of Gaussian random matrices that generalize the Wigner ensemble of Gaussian matrices. As discussed in the bibliographical notes, Section 4.6, some of the later results, when specialized to fixed matrices, are currently only accessible through stochastic calculus.

Our starting point is the introduction of the symmetric and Hermitian Brownian motions; we leave the introduction of the symplectic Brownian motions to the exercises.

Definition 4.3.1 Let $(B_{i,j}, \tilde{B}_{i,j}, 1 \le i \le j \le N)$ be a collection of i.i.d. real valued standard Brownian motions. The *symmetric* (resp. *Hermitian*) *Brownian motion*, denoted $H^{N,\beta} \in \mathscr{H}_N^{\beta}$, $\beta = 1,2$, is the random process with entries $\{H_{i,j}^{N,\beta}(t), t \ge 0, i \le j\}$ equal to

$$H_{k,l}^{N,\beta} = \begin{cases} \frac{1}{\sqrt{\beta N}}(B_{k,l} + i(\beta-1)\tilde{B}_{k,l}), & \text{if } k < l, \\ \frac{\sqrt{2}}{\sqrt{\beta N}} B_{l,l}, & \text{if } k = l. \end{cases} \tag{4.3.1}$$

We will be studying the stochastic process of the (ordered) eigenvalues of $H^{N,\beta}$. In Subsection 4.3.1, we derive equations for the system of eigenvalues, and show that at all positive times, eigenvalues do not "collide". These stochastic equations are then used in Subsections 4.3.2, 4.3.3 and 4.3.4 to derive laws of large numbers, central limit theorems, and large deviation upper bounds, respectively, for the process of empirical measure of the eigenvalues.

4.3.1 Dyson's Brownian motion

We begin in this subsection our study of the process of eigenvalues of time-dependent matrices. Throughout, we let $(W_1,\ldots,W_N)$ be a N-dimensional Brownian motion in a probability space (Ω,P) equipped with a filtration $\mathscr{F}=\{\mathscr{F}_t,t\geq 0\}$. Let Δ_N denote the open simplex

$$\Delta_N=\{(x_i)_{1\leq i\leq N}\in\mathbb{R}^N : x_1<x_2<\cdots<x_{N-1}<x_N\},$$

with closure $\overline{\Delta_N}$. With $\beta\in\{1,2\}$, let $X^{N,\beta}(0)\in\mathscr{H}_N^\beta$ be a matrix with (real) eigenvalues $(\lambda_1^N(0),\ldots,\lambda_N^N(0))\in\overline{\Delta_N}$. For $t\geq 0$, let $\lambda^N(t)=(\lambda_1^N(t),\ldots,\lambda_N^N(t))\in\overline{\Delta_N}$ denote the ordered collection of (real) eigenvalues of

$$X^{N,\beta}(t)-X^{N,\beta}(0)+H^{N,\beta}(t), \tag{4.3.2}$$

with $H^{N,\beta}$ as in Definition 4.3.1. A fundamental observation (due to Dyson in the case $X^{N,\beta}(0)=0$) is that the process $(\lambda^N(t))_{t\geq 0}$ is a vector of semi-martingales, whose evolution is described by a stochastic differential system.

Theorem 4.3.2 (Dyson) *Let $\left(X^{N,\beta}(t)\right)_{t\geq 0}$ be as in (4.3.2), with eigenvalues $(\lambda^N(t))_{t\geq 0}$ and $\lambda^N(t)\in\overline{\Delta_N}$ for all $t\geq 0$. Then, the processes $(\lambda^N(t))_{t\geq 0}$ are semi-martingales. Their joint law is the unique distribution on $C(\mathbb{R}^+,\mathbb{R}^N)$ so that*

$$P\left(\forall t>0,\ (\lambda_1^N(t),\cdots,\lambda_N^N(t))\in\Delta_N\right)=1,$$

which is a weak solution to the system

$$d\lambda_i^N(t)=\frac{\sqrt{2}}{\sqrt{\beta N}}dW_i(t)+\frac{1}{N}\sum_{j:j\neq i}\frac{1}{\lambda_i^N(t)-\lambda_j^N(t)}dt\,,\quad i=1,\ldots,N, \tag{4.3.3}$$

with initial condition $\lambda^N(0)$.

We refer the reader to Appendix H, Definitions H.4 and H.3, for the notions of strong and weak solutions.

Note that, in Theorem 4.3.2, we do not assume that $\lambda^N(0) \in \Delta_N$. The fact that $\lambda^N(t) \in \Delta_N$ for all $t > 0$ is due to the natural repulsion of the eigenvalues. This repulsion will be fundamental in the proof of the theorem.

It is not hard to guess the form of the stochastic differential equation for the eigenvalues of $X^{N,\beta}(t)$, simply by writing $X^{N,\beta}(t) = (O^N)^*(t)\Lambda(t)O^N(t)$, with $\Lambda(t)$ diagonal and $(O^N)^*(t)O^N(t) = I_N$. Differentiating formally (using Itô's formula) then allows one to write the equations (4.3.3) and appropriate stochastic differential equations for $O^N(t)$. However, the resulting equations are singular, and proceeding this way presents several technical difficulties. Instead, our derivation of the evolution of the eigenvalues $\lambda^N(t)$ will be somewhat roundabout. We first show, in Lemma 4.3.3, that the solution of (4.3.3), when started at Δ_N, exists, is unique, and stays in Δ_N. Once this is accomplished, the proof that $(\lambda^N(t))_{t\geq 0}$ solves this system will involve routine stochastic analysis.

Lemma 4.3.3 *Let $\lambda^N(0) = (\lambda_1^N(0), \ldots, \lambda_N^N(0)) \in \Delta_N$. For any $\beta \geq 1$, there exists a unique strong solution $(\lambda^N(t))_{t\geq 0} \in C(\mathbb{R}^+, \Delta_N)$ to the stochastic differential system (4.3.3) with initial condition $\lambda^N(0)$. Further, the weak solution to (4.3.3) is unique.*

This result is extended to initial conditions $\lambda^N(0) \in \overline{\Delta_N}$ in Proposition 4.3.5.

Proof The proof is routine stochastic analysis, and proceeds in three steps. To overcome the singularity in the drift, one first introduces a cut-off, parametrized by a parameter M, thus obtaining a stochastic differential equation with Lipschitz coefficients. In a second step, a Lyapunov function is introduce that allows one to control the time T_M until the diffusion sees the cut-off; before that time, the solution to the system with cut-off is also a solution to the original system. Finally, taking $M \to \infty$ one shows that $T_M \to \infty$ almost surely, and thus obtains a solution for all times.

Turning to the proof, set, for $R > 0$,

$$\phi_R(x) = \begin{cases} x^{-1} & \text{if } |x| \geq R^{-1}, \\ R^2 x & \text{otherwise.} \end{cases}$$

Introduce the auxiliary system

$$d\lambda_i^{N,R}(t) = \sqrt{\frac{2}{\beta N}} dW_i(t) + \frac{1}{N} \sum_{j:j\neq i} \phi_R(\lambda_i^{N,R}(t) - \lambda_j^{N,R}(t))dt, \quad i = 1, \ldots, N, \tag{4.3.4}$$

with $\lambda_i^{N,R}(0) = \lambda_i^N(0)$ for $i = 1, \ldots, N$. Since ϕ_R is uniformly Lipschitz, it follows from Theorem H.6 that (4.3.4) admits a unique strong solution, adapted to the filtration $\mathscr{F}$, as well as a unique weak solution $P^{N,R}_{T,\lambda^N(0)} \in M_1(C([0,T],\mathbb{R}^N))$. Let

$$\tau_R := \inf\{t : \min_{i\neq j} |\lambda_i^{N,R}(t) - \lambda_j^{N,R}(t)| < R^{-1}\},$$

noting that τ_R is monotone increasing in R and

$$\lambda^{N,R}(t) = \lambda^{N,R'}(t) \text{ for all } t \leq \tau_R \text{ and } R < R'. \tag{4.3.5}$$

We now construct a solution to (4.3.3) by taking $\lambda^N(t) = \lambda^{N,R}(t)$ on the event $\tau_R > t$, and then showing that $\tau_R \to_{R\to\infty} \infty$, almost surely. Toward this end, consider the Lyapunov function, defined for $x = (x_1,\ldots,x_N) \in \Delta_N$,

$$f(x) = f(x_1,\ldots,x_N) = \frac{1}{N}\sum_{i=1}^N x_i^2 - \frac{1}{N^2}\sum_{i\neq j}\log|x_i - x_j|.$$

Using the fact that

$$\log|x-y| \leq \log(|x|+1) + \log(|y|+1) \quad \text{and} \quad x^2 - 2\log(|x|+1) \geq -4,$$

we find that, for all $i \neq j$,

$$f(x_1,\ldots,x_N) \geq 4, \quad -\frac{1}{N^2}\log|x_i - x_j| \leq f(x_1,\ldots,x_N) + 4. \tag{4.3.6}$$

For any $M > 0$ and $x = (x_1,\ldots,x_N) \in \Delta_N$, set

$$R = R(N,M) = e^{N^2(4+M)} \text{ and } T_M = \inf\{t \geq 0 : f(\lambda^{N,R}(t)) \geq M\}. \tag{4.3.7}$$

Since f is $C^\infty(\Delta_N,\mathbb{R})$ on sets where it is uniformly bounded (note here that f is bounded below uniformly), we have that $\{T_M > T\} \in \mathscr{F}_T$ for all $T \geq 0$, and hence T_M is a stopping time. Moreover, due to (4.3.6), on the event $\{T_M > T\}$, we get that, for all $t \leq T$,

$$|\lambda_i^{N,R}(t) - \lambda_j^{N,R}(t)| \geq R^{-1},$$

and thus on the event $\{T \leq T_M\}$, $(\lambda^{N,R}(t), t \leq T)$ provides an adapted strong solution to (4.3.3). For $i = 1,\ldots,N$ and $j = 1,2$, define the functions $u_{i,j} : \Delta_N \to \mathbb{R}$ by

$$u_{i,1}(x) = \sum_{k:k\neq i}\frac{1}{x_i - x_k}, \quad u_{i,2}(x) = \sum_{k:k\neq i}\frac{1}{(x_i - x_k)^2}.$$

Itô's Lemma (see Theorem H.9) gives

$$\begin{aligned} df(\lambda^{N,R}(t)) &= \frac{2}{N^2}\sum_{i=1}^N \left(\lambda_i^{N,R}(t) - \frac{1}{N}u_{i,1}(\lambda^{N,R}(t))\right)u_{i,1}(\lambda^{N,R}(t))dt \\ &\quad + \frac{2}{\beta N}\sum_{i=1}^N\left(1 + \frac{1}{N^2}u_{i,2}(\lambda^{N,R}(t))\right)dt + dM^N(t), \end{aligned} \tag{4.3.8}$$

with $M^N(t)$ the local martingale

$$dM^N(t) = \frac{2^{\frac{3}{2}}}{\beta^{\frac{1}{2}} N^{\frac{3}{2}}} \sum_{i=1}^{N} \left(\lambda_i^{N,R}(t) - \frac{1}{N} \sum_{k:k\neq i} \frac{1}{\lambda_i^{N,R}(t) - \lambda_k^{N,R}(t)} \right) dW_i(t) .$$

Observing that, for all $x = (x_1, \ldots, x_N) \in \Delta_N$,

$$\begin{aligned} \sum_{i=1}^{N} \left(u_{i,1}(x)^2 - u_{i,2}(x) \right) &= \sum_{\substack{k\neq i, l\neq i \\ k\neq l}} \frac{1}{x_i - x_k} \frac{1}{x_i - x_l} \\ &= \sum_{\substack{k\neq i, l\neq i \\ k\neq l}} \frac{1}{x_l - x_k} \left(\frac{1}{x_i - x_l} - \frac{1}{x_i - x_k} \right) = -2 \sum_{\substack{k\neq i, l\neq i \\ k\neq l}} \frac{1}{x_i - x_k} \frac{1}{x_i - x_l} , \end{aligned}$$

we conclude that, for $x \in \Delta_N$,

$$\sum_{i=1}^{N} \left(u_{i,1}(x)^2 - u_{i,2}(x) \right) = 0 .$$

Similarly,

$$\sum_{i=1}^{N} u_{i,1}(x) x_i = \frac{N(N-1)}{2} .$$

Substituting the last two equalities into (4.3.8), we get

$$df(\lambda^{N,R}(t)) = (1 + \frac{2}{\beta} - \frac{1}{N}) dt + \frac{2(1-\beta)}{\beta N^2} \sum_i u_{i,2}(\lambda^{N,R}(t)) dt + dM^N(t) .$$

Thus, for all $\beta \geq 1$, for all $M < \infty$, since $(M^N(t \wedge T_M), t \geq 0)$ is a martingale with zero expectation,

$$\mathbb{E}[f(\lambda^{N,R}(t \wedge T_M))] \;\leq\; 3\mathbb{E}[t \wedge T_M] + f(\lambda^{N,R}(0)).$$

Therefore, recalling (4.3.6),

$$\begin{aligned} (M+4)\mathbb{P}(T_M \leq t) &= \mathbb{E}[\left(f(\lambda^{N,R}(t \wedge T_M)) + 4\right) 1_{t \geq T_M}] \\ &\leq \mathbb{E}[f(\lambda^{N,R}(t \wedge T_M)) + 4] \leq 3\mathbb{E}[t \wedge T_M] + 4 + f(\lambda^{N,R}(0)) \\ &\leq 3t + 4 + f(\lambda^{N,R}(0)) , \end{aligned}$$

which proves that

$$\mathbb{P}(T_M \leq t) \leq \frac{3t + 4 + f(\lambda^{N,R}(0))}{M+4} .$$

Hence, the Borel–Cantelli Lemma implies that, for all $t \in \mathbb{R}^+$,

$$\mathbb{P}(\exists M \in \mathbb{N} : T_{M^2} \geq t) = 1,$$

and in particular, T_{M^2} goes to infinity almost surely. As a consequence, recalling that $M = -4 + (\log R)/N^2$, see (4.3.7), and setting $\lambda^N(t) = \lambda^{N,R}(t)$ for $t \le T_{M^2}$, gives, due to (4.3.5), a strong solution to (4.3.3), which moreover satisfies $\lambda^N(t) \in \Delta_N$ for all t. The strong (and weak) uniqueness of the solutions to (4.3.4), together with $\lambda^{N,R}(t) = \lambda^N(t)$ on $\{T \le T_M\}$ and the fact that $T_M \to \infty$ almost surely, imply the strong (and weak) uniqueness of the solutions to (4.3.3). □

Proof of Theorem 4.3.2 As a preliminary observation, note that the law of $H^{N,\beta}$ is invariant under the action of the orthogonal (when $\beta = 1$) or unitary (when $\beta = 2$) groups, that is, $(OH^{N,\beta}(t)O^*)_{t\ge 0}$ has the same distribution as $(H^{N,\beta}(t))_{t\ge 0}$ if O belongs to the orthogonal (if $\beta = 1$) or unitary (if $\beta = 2$) groups. Therefore, the law of $(\lambda^N(t))_{t\ge 0}$ does not depend on the basis of eigenvectors of $X^{N,\beta}(0)$ and we shall assume in the sequel, without loss of generality, that $X^{N,\beta}(0)$ is diagonal and real.

The proof we present goes "backward" by proposing a way to construct the matrix $X^{N,\beta}(t)$ from the solution of (4.3.3) and a Brownian motion on the orthogonal (resp. unitary) group. Its advantage with respect to a "forward" proof is that we do not need to care about justifying that certain quantities defined from $X^{N,\beta}$ are semi-martingales to insure that Itô's calculus applies.

We first prove the theorem in the case $\lambda_N(0) \in \Delta_N$. We begin by enlarging the probability space by adding to the independent Brownian motions $(W_i, 1 \le i \le N)$ an independent collection of independent Brownian motions $(w_{ij}, 1 \le i < j \le N)$, which are complex if $\beta = 2$ (that is, $w_{ij} = 2^{-\frac{1}{2}}(w^1_{ij} + \sqrt{-1}w^2_{ij})$ with two independent real Brownian motions w^1_{ij}, w^2_{ij}) and real if $\beta = 1$. We continue to use $\mathscr{F}_t$ to denote the enlarged sigma-algebra $\sigma(w_{ij}(s), 1 \le i < j \le N, W_i(s), 1 \le i \le N, s \le t)$.

Fix $M > 0$ and R as in (4.3.7). We consider the strong solution of (4.3.3), constructed with the Brownian motions $(W_i, 1 \le i \le N)$, till the stopping time T_M defined in (4.3.7). We set, for $i < j$,

$$dR^N_{ij}(t) = \frac{1}{\sqrt{N}} \frac{1}{\lambda^N_i(t) - \lambda^N_j(t)} dw_{ij}(t), \quad R^N_{ij}(0) = 0. \tag{4.3.9}$$

We let $R^N(t)$ be the skew-Hermitian matrix (i.e. $R^N(t) = -R^N(t)^*$) with such entries above the diagonal and null entries on the diagonal. Note that since $\lambda^N(t) \in \Delta_N$ for all t, the matrix-valued process $R^N(t)$ is well defined, and its entries are semi-martingales.

Recalling the notation for the bracket of semi-martingales, see (H.1), for A,B two semi-martingales with values in $\mathcal{M}_N$, we denote by $\langle A,B\rangle_t$ the matrix

$$(\langle A,B\rangle_t)_{ij} = \langle (AB)_{ij}\rangle_t = \sum_{k=1}^{N}\langle A_{ik},B_{kj}\rangle_t,\ 1\leq i,j\leq N.$$

Observe that for all $t\geq 0$, $\langle A,B\rangle_t^* = \langle B^*,A^*\rangle_t$. We set O^N to be the (strong) solution of

$$dO^N(t) = O^N(t)dR^N(t) - \frac{1}{2}O^N(t)d\langle (R^N)^*,R^N\rangle_t,\quad O^N(0) = I_N. \tag{4.3.10}$$

This solution exists and is unique since it is a linear equation in O^N and R^N is a well defined semi-martingale. In fact, as the next lemma shows, $O^N(t)$ describes a process in the space of unitary matrices (orthogonal if $\beta = 1$).

Lemma 4.3.4 *The solution of (4.3.10) satisfies*

$$O^N(t)O^N(t)^* = O^N(t)^*O^N(t) = I \ \textit{for all}\ t\geq 0.$$

Further, let $D(\lambda^N(t))$ denote a diagonal matrix with $D(\lambda^N(t))_{ii} = \lambda^N(t)_i$ and set $Y^N(t) = O^N(t)D(\lambda^N(t))O^N(t)^*$. Then

$$P(\forall t\geq 0,\quad Y^N(t)\in\mathcal{H}_N^{\beta}) = 1,$$

and the entries of the process $(Y^N(t))_{t\geq 0}$ are continuous martingales with respect to the filtration $\mathcal{F}$, with bracket

$$\langle Y_{ij}^N, Y_{kl}^N\rangle_t = N^{-1}(1_{ij=kl}(2-\beta)+1_{ij=lk})t.$$

Proof We begin by showing that $J^N(t) := O^N(t)^*O^N(t)$ equals the identity I_N for all time t. Toward this end, we write a differential equation for $K^N(t) := J^N(t) - I_N$ based on the fact that the process $(O^N(t))_{t\geq 0}$ is the strong solution of (4.3.10). We have

$$\begin{aligned}
(d\langle (O^N)^*,(O^N)\rangle_t)_{ij} &= \left(d\langle \int_0^{\cdot} d(R^N)^*(s)(O^N)^*(s), \int_0^{\cdot} O^N(s)dR^N(s)\rangle_t\right)_{ij}\\
&= \sum_{k=1}^{N} d\langle (\int_0^{\cdot}(dR^N)^*(s)(O^N)^*(s))_{ik}, (\int_0^{\cdot} O^N(s)dR^N(s))_{kj}\rangle_t\\
&= -\sum_{m,n=1}^{N}\sum_{k=1}^{N}\bar{O}_{km}^N(t)O_{kn}^N(t)d\langle \bar{R}_{mi}^N, R_{nj}^N\rangle_t\\
&= -\sum_{m,n=1}^{N} J_{mn}^N(t)d\langle \bar{R}_{mi}^N, R_{nj}^N\rangle_t,
\end{aligned}\tag{4.3.11}$$

where here and in the sequel we use $\int_0^{\cdot}$ to denote an indefinite integral viewed as a process. Therefore, setting $A.B = AB + BA$, we obtain

$$\begin{aligned} dK^N(t) &= J^N(t)[dR^N(t) - \frac{1}{2}d\langle (R^N)^*, R^N\rangle_t] \\ &\quad +[d(R^N)^*(t) - \frac{1}{2}d\langle (R^N)^*, R^N\rangle_t]J^N(t) + d\langle (O^N)^*, O^N\rangle_t \\ &= K^N(t).(dR^N(t) - \frac{1}{2}d\langle (R^N)^*, R^N\rangle_t) + dr^N(t), \end{aligned}$$

with $dr^N(t)_{ij} = -\sum_{m,n=1}^N K_{mn}^N(t) d\langle \bar{R}_{mi}^N, R_{nj}^N\rangle_t$. For any deterministic $M > 0$ and $0 \le S \le T$, set, with T_M given by (4.3.7),

$$\kappa(M,S,T) = \max_{1\le i,j\le N} \sup_{t\le S} |K_{ij}^N(t\wedge T_M)|^2,$$

and note that $E\kappa(M,S,T) < \infty$ for all M,S,T, and that it is nondecreasing in S. From the Burkholder–Davis–Gundy inequality (Theorem H.8), the equality $K_N(0) = 0$, and the fact that $(R^N(t\wedge T_M))_{t\le T}$ has a uniformly (in T) bounded martingale bracket, we deduce that there exists a constant $C(M) < \infty$ (independent of S,T) such that for all $S \le T$,

$$E\kappa(M,S,T) \le C(M)E\int_0^S \kappa(M,t,T)dt.$$

It follows that $E\kappa(M,T,T)$ vanishes for all T,M. Letting M going to infinity we conclude that $K^N(t) = 0$ almost surely, that is, $O^N(t)^*O^N(t) = I_N$.

We now show that Y^N has martingales entries and compute their martingale bracket. By construction,

$$\begin{aligned} dY^N(t) = dO^N(t)D(\lambda^N(t))O^N(t)^* + O^N(t)D(\lambda^N(t))dO^N(t)^* \\ +O^N(t)dD(\lambda^N(t))O^N(t)^* + d\langle O^N D(\lambda^N)(O^N)^*\rangle_t \end{aligned} \tag{4.3.12}$$

where for all $i,j \in \{1,\cdots,N\}$, we have denoted

$$\begin{aligned} &(d\langle O^N D(\lambda^N)(O^N)^*\rangle_t)_{ij} \\ &= \sum_{k=1}^N \left(\frac{1}{2}O_{ik}^N(t)d\langle \lambda_k^N, \bar{O}_{jk}^N\rangle_t + \lambda_k^N(t)d\langle O_{ik}^N, \bar{O}_{jk}^N\rangle_t + \frac{1}{2}O_{jk}^N(t)d\langle \lambda_k^N, \bar{O}_{ik}^N\rangle_t\right) \\ &= \sum_{k=1}^N \lambda_k^N(t)d\langle O_{ik}^N, \bar{O}_{jk}^N\rangle_t, \end{aligned}$$

and we used in the last equality the independence of $(w_{ij}, 1 \le i < j \le N)$ and $(W_i, 1 \le i \le N)$ to assert that the martingale bracket of λ^N and O^N vanishes. Setting

$$dZ^N(t) := O^N(t)^* dY^N(t) O^N(t), \tag{4.3.13}$$

we obtain from the left multiplication by $O^N(t)^*$ and right multiplication by $O^N(t)$ of (4.3.12) that

$$\begin{aligned} dZ^N(t) &= (O^N)^*(t) dO^N(t) D(\lambda^N(t)) + D(\lambda^N(t)) dO^N(t)^* O^N(t) \\ &\quad + dD(\lambda^N(t)) + O^N(t)^* d\langle O^N D(\lambda^N)(O^N)^* \rangle_t O^N(t). \end{aligned} \tag{4.3.14}$$

We next compute the last term in the right side of (4.3.14). For all $i, j \in \{1, \ldots, N\}^2$, we have

$$\begin{aligned} \left(d\langle O^N D(\lambda^N)(O^N)^*\rangle_t\right)_{ij} &= \sum_{k=1}^N \lambda_k^N(t) d\langle O_{ik}^N, \bar{O}_{jk}^N \rangle_t \\ &= \sum_{k,l,m=1}^N \lambda_k^N(t) O_{il}^N(t) \bar{O}_{jm}^N(t) d\langle R_{lk}^N, \bar{R}_{mk}^N \rangle_t . \end{aligned}$$

But, by the definition (4.3.9) of R^N,

$$d\langle R_{lk}^N, \bar{R}_{mk}^N \rangle_t = 1_{m=l} 1_{m \neq k} \frac{1}{N(\lambda_k^N(t) - \lambda_m^N(t))^2} dt, \tag{4.3.15}$$

and so we obtain

$$\left(d\langle O^N D(\lambda^N)(O^N)^*\rangle_t\right)_{ij} = \sum_{1 \le k \neq l \le N} \frac{\lambda_k^N(t)}{N(\lambda_k^N(t) - \lambda_l^N(t))^2} O_{il}^N(t) \bar{O}_{jl}^N(t) dt .$$

Hence, for all $i, j \in \{1, \ldots, N\}^2$,

$$[O^N(t)^* d\langle O^N D(\lambda^N)(O^N)^*\rangle_t O^N(t)]_{ij} = 1_{i=j} \sum_{\substack{1 \le k \le N \\ k \neq i}} \frac{\lambda_k^N(t)}{N(\lambda_i^N(t) - \lambda_k^N(t))^2} dt .$$

Similarly, recall that

$$O^N(t)^* dO^N(t) = dR^N(t) - 2^{-1} d\langle (R^N)^*, R^N \rangle_t ,$$

so that from (4.3.15) we get, for all $i, j \in \{1, \cdots, N\}^2$,

$$[O^N(t)^* dO^N(t)]_{ij} = dR_{ij}^N(t) - 2^{-1} 1_{i=j} \sum_{\substack{1 \le k \le N \\ k \neq i}} \frac{1}{N(\lambda_i^N(t) - \lambda_k^N(t))^2} dt .$$

Therefore, identifying the terms on the diagonal in (4.3.14) and recalling that R^N vanishes on the diagonal, we find, substituting in (4.3.13), that

$$dZ_{ii}^N(t) = \sqrt{\frac{2}{\beta N}} dW_i(t).$$

Away from the diagonal, for $i \neq j$, we get

$$dZ^N_{ij}(t) = [dR^N(t)D(\lambda^N(t)) + D(\lambda^N(t))dR^N(t)^*]_{ij} = \frac{1}{\sqrt{N}}dw_{ij}(t).$$

Hence, $(Z^N(t))_{t\geq 0}$ has the law of a symmetric (resp. Hermitian) Brownian motion. Thus, since $(O^N(t))_{t\geq 0}$ is adapted,

$$Y^N(t) = \int_0^t O^N(s)dZ^N(s)O^N(s)^*$$

is a continuous matrix-valued martingale whose quadratic variation $d\langle Y^N_{ij}, Y^N_{i'j'}\rangle_t$ is given by

$$\begin{aligned}
&\sum_{k,l,k',l'=1}^N O^N_{ik}(t)\bar{O}^N_{jl}(t)O^N_{i'k'}(t)\bar{O}^N_{j'l'}(t)d\langle Z^N_{kl}, Z^N_{k'l'}\rangle_t \\
&= \frac{1}{N}\sum_{k,l,k',l'=1}^N O^N_{ik}(t)\bar{O}^N_{jl}(t)O^N_{i'k'}(t)\bar{O}^N_{j'l'}(t)(1_{kl=l'k'} + 1_{\beta=1}1_{kl=k'l'})dt \\
&= \frac{1}{N}(1_{ij=j'i'} + 1_{\beta=1}1_{ij=i'j'})dt.
\end{aligned}$$

□

We return to the proof of Theorem 4.3.2. Applying Lévy's Theorem (Theorem H.2) to the entries of Y^N, we conclude that $(Y^N(t) - Y^N(0))_{t\geq 0}$ is a symmetric (resp. Hermitian) Brownian motion, and so $(Y^N(t))_{t\geq 0}$ has the same law as $(X^{N,\beta}(t))_{t\geq 0}$ since $X^N(0) = Y^N(0)$, which completes the proof of the theorem in the case $Y^N(0) \in \Delta_N$.

Consider next the case where $X^{N,\beta}(0) \in \overline{\Delta_N} \setminus \Delta_N$. Note that the condition $\lambda^N(t) \notin \Delta_N$ means that the discriminant of the characteristic polynomial of $X^{N,\beta}(t)$ vanishes. The latter discriminant is a polynomial in the entries of $X^{N,\beta}(t)$, that does not vanish identically. By the same argument as in the proof of Lemma 2.5.5, it follows that $\lambda^N(t) \in \Delta_N$, almost surely. Hence, for any $\varepsilon > 0$, the law of $(X^{N,\beta}(t))_{t\geq\varepsilon}$ coincides with the strong solution of (4.3.3) initialized at $X^{N,\beta}(\varepsilon)$. By Lemma 2.1.19, it holds that for all $s,t \in \mathbb{R}$,

$$\sum_{i=1}^N (\lambda_i^N(t) - \lambda_i^N(s))^2 \leq \frac{1}{N}\sum_{i,j=1}^N (H^{N,\beta}_{ij}(t) - H^{N,\beta}_{ij}(s))^2,$$

and thus the a.s. continuity of the Brownian motions paths results in the a.s. continuity of $t \to \lambda^N(t)$ for any given N. Letting $\varepsilon \to 0$ completes the proof of the theorem. □

Our next goal is to extend the statement of Lemma 4.3.3 to initial conditions belonging to $\overline{\Delta_N}$. Namely, we have the following.

Proposition 4.3.5 *Let $\lambda^N(0) = (\lambda_1^N(0), \dots, \lambda_N^N(0)) \in \overline{\Delta_N}$. For any $\beta \geq 1$, there exists a unique strong solution $(\lambda^N(t))_{t \geq 0} \in C(\mathbb{R}^+, \overline{\Delta_N})$ to the stochastic differential system (4.3.3) with initial condition $\lambda^N(0)$. Further, for any $t > 0$, $\lambda^N(t) \in \Delta_N$ and $\lambda^N(t)$ is a continuous function of $\lambda^N(0)$.*

When $\beta = 1,2,4$, Proposition 4.3.5 can be proved by using Theorem 4.3.2. Instead, we provide a proof valid for all $\beta \geq 1$, that does not use the random matrices representation of the solutions. As a preliminary step, we present a comparison between strong solutions of (4.3.3) with initial condition in Δ_N.

Lemma 4.3.6 *Let $(\lambda^N(t))_{t \geq 0}$ and $(\eta^N(t))_{t \geq 0}$ be two strong solutions of (4.3.3) starting, respectively, from $\lambda^N(0) \in \Delta_N$ and $\eta^N(0) \in \Delta_N$. Assume that $\lambda_i^N(0) < \eta_i^N(0)$ for all i. Then,*

$$P(\text{for all } t \geq 0 \text{ and } i = 1, \dots, N, \quad \lambda_i^N(t) < \eta_i^N(t)) = 1\,. \tag{4.3.16}$$

Proof of Lemma 4.3.6 We note first that $d(\sum_i \lambda_i^N(t) - \sum_i \eta_i^N(t)) = 0$. In particular,

$$\sum_i (\lambda_i^N(t) - \eta_i^N(t)) = \sum_i (\lambda_i^N(0) - \eta_i^N(0)) < 0\,. \tag{4.3.17}$$

Next, for all $i \in \{1, \dots, N\}$, we have from (4.3.3) and the fact that $\eta^N(t) \in \Delta_N$, $\lambda^N(t) \in \Delta_N$ for all t that

$$d(\lambda_i^N - \eta_i^N)(t) = \frac{1}{N} \sum_{j: j \neq i} \frac{(\eta_i^N - \lambda_i^N - \eta_j^N + \lambda_j^N)(t)}{(\eta_i^N(t) - \eta_j^N(t))(\lambda_i^N(t) - \lambda_j^N(t))} dt\,.$$

Thus, $\lambda_i^N - \eta_i^N$ is differentiable for all i and, by continuity, negative for small enough times. Let T be the first time at which $(\lambda_i^N - \eta_i^N)(t)$ vanishes for some $i \in \{1, \dots, N\}$, and assume $T < \infty$. Since $(\eta_i^N(t) - \eta_j^N(t))(\lambda_i^N(t) - \lambda_j^N(t))$ is strictly positive for all time, we deduce that $\partial_t (\lambda_i^N - \eta_i^N)|_{t=T}$ is negative (note that it is impossible to have $(\lambda_j^N - \eta_j^N)(T) = 0$ for all j because of (4.3.17)). This provides a contradiction since $(\lambda_i^N - \eta_i^N)(t)$ was strictly negative for $t < T$. □

We can now prove Proposition 4.3.5.

Proof of Proposition 4.3.5 Set $\lambda^N(0) = (\lambda_1^N(0), \dots, \lambda_N^N(0)) \in \overline{\Delta_N}$ and put for $n \in \mathbb{Z}$, $\lambda_i^{N,n}(0) = \lambda_i^N(0) + \frac{i}{n}$. We have $\lambda^{N,n}(0) \in \Delta_N$ and, further, if $n > 0$, $\lambda_i^{N,-n}(0) < \lambda_i^{N,-n-1}(0) < \lambda_i^{N,n+1}(0) < \lambda_i^{N,n}(0)$. Hence, by Lemma 4.3.6, the corresponding solutions to (4.3.3) satisfy almost surely and for all $t > 0$

$$\lambda_i^{N,-n}(t) < \lambda_i^{N,-n-1}(t) < \lambda_i^{N,n+1}(t) < \lambda_i^{N,n}(t)\,.$$

Since

$$\sum_{i=1}^{N}(\lambda^{N,n}(t)-\lambda^{N,-n}(t))=\sum_{i=1}^{N}(\lambda^{N,n}(0)-\lambda^{N,-n}(0)) \tag{4.3.18}$$

goes to zero as n goes to infinity, we conclude that the sequences $\lambda^{N,-n}$ and $\lambda^{N,n}$ converge uniformly to a limit, which we denote by λ^N. By construction, $\lambda^N \in C(\mathbb{R}^+,\overline{\Delta_N})$. Moreover, if we take any other sequence $\lambda^{N,p}(0)\in\Delta_N$ converging to $\lambda^N(0)$, the solution $\lambda^{N,p}$ to (4.3.3) also converges to λ^N (as can be seen by comparing $\lambda^{N,p}(0)$ with some $\lambda^{N,n}(0),\lambda^{N,-n}(0)$ for p large enough).

We next show that λ^N is a solution of (4.3.3). Toward that end it is enough to show that for all $t>0$, $\lambda^N(t)\in\Delta_N$, since then if we start at any positive time s we see that the solution of (4.3.3) starting from $\lambda^N(s)$ can be bounded above and below by $\lambda^{N,n}$ and $\lambda^{N,-n}$ for all large enough n, so that this solution must coincide with the limit $(\lambda^N(t),t\geq s)$. So let us assume that there is $t>0$ so that $\lambda^N(s)\in\overline{\Delta_N}\backslash\Delta_N$ for all $s\leq t$ and obtain a contradiction. We let I be the largest $i\in\{2,\ldots,N\}$ so that $\lambda_k^N(s)<\lambda_{k+1}^N(s)$ for $k\geq I$ but $\lambda_{I-1}^N(s)=\lambda_I^N(s)$ for $s\leq t$. Then, we find a constant C independent of n and ε_n going to zero with n so that, for n large enough,

$$|\lambda_k^{N,n}(s)-\lambda_{k+1}^{N,n}(s)|\geq C\, k\geq I, \qquad |\lambda_I^{N,n}(s)-\lambda_{I-1}^{N,n}(s)|\leq\varepsilon_n.$$

Since $\lambda^{N,n}$ solves (4.3.3), we deduce that for $s\leq t$

$$\lambda_{I-1}^{N,n}(s)\geq\lambda_{I-1}^{N,n}(0)+\frac{2}{\beta N}W_s^{I-1}+\frac{1}{N}(\varepsilon_n^{-1}-C(N-I))s.$$

This implies that $\lambda_{I-1}^{N,n}(s)$ goes to infinity as n goes to infinity, a.s. To obtain a contradiction, we show that with $C_N(n,t):=\frac{1}{N}\sum_{i=1}^N(\lambda_i^{N,n}(t))^2$, we have

$$\sup_n\sup_{s\in[0,t]}\sqrt{C_N(n,t)}<\infty,\ \text{a.s.} \tag{4.3.19}$$

With (4.3.19), we conclude that for all $t>0$, $\lambda^N(t)\in\Delta_N$, and in particular it is the claimed strong solution.

To see (4.3.19), note that since $\lambda_i^{N,n}(s)\geq\lambda_i^{N,n'}(s)$ for any $n\geq n'$ and all s by Lemma 4.3.6, we have that

$$\begin{aligned}|C_N(n,s)-C_N(n',s)| &= \frac{1}{N}\sum_{i=1}^N(\lambda_i^{N,n}(s)-\lambda_i^{N,n'}(s))|(\lambda_i^{N,n}(s)+\lambda_i^{N,n'}(s))|\\ &\leq \sum_{i=1}^N(\lambda_i^{N,n}(s)-\lambda_i^{N,n'}(s))\cdot\frac{1}{N}\sum_{i=1}^N(|(\lambda_i^{N,n}(s)|+|\lambda_i^{N,n'}(s)|)\\ &\leq (\sqrt{C_N(n,s)}+\sqrt{C_N(n',s)})\sum_{i=1}^N(\lambda^{N,n}(0)-\lambda^{N,n'}(0)),\end{aligned}$$

where (4.3.18) and the Cauchy–Schwarz inequality were used in the last inequality. It follows that

$$\sqrt{C_N(n,s)} \le \sqrt{C_N(n',s)} + \sum_{i=1}^{N}(\lambda^{N,n}(0) - \lambda^{N,n'}(0)),$$

and thus

$$\sup_{n\ge n'}\sup_{s\in[0,t]} \sqrt{C_N(n,s)} \le \sup_{s\in[0,t]} \sqrt{C_N(n',s)} + \sum_{i=1}^{N}(\lambda^{N,n}(0) - \lambda^{N,n'}(0)).$$

Thus, to see (4.3.19), it is enough to bound almost surely $\sup_{s\in[0,t]}\sqrt{C_N(n,t)}$ for a fixed n. From Itô's Lemma (see Lemma 4.3.12 below for a generalization of this particular computation),

$$C_N(n,t) = D_N(n,t) + \frac{2\sqrt{2}}{N\sqrt{\beta N}}\sum_{i=1}^{N}\int_0^t \lambda_i^{N,n}(s)dW_i(s)$$

with $D_N(n,t) := C_N(n,0) + (\frac{2}{\beta} + \frac{N-1}{N})t$. Define the stopping time $S_R = \inf\{s : C_N(n,s) \ge R\}$. Then, by the Burkholder–Davis–Gundy inequality (Theorem H.8) we deduce that

$$\begin{aligned}
&E[\sup_{s\in[0,t]} C_N(n,s\wedge S_R)^2] \\
&\quad\le\ 2[D_N(n,t)]^2 + 2N^{-2}\Lambda\int_0^t E[\sup_{s\in[0,u]} C_N(n,s\wedge S_R)]du \\
&\quad\le\ 2[D_N(n,t)]^2 + N^{-2}\Lambda t + N^{-2}\Lambda\int_0^t E[\sup_{s\in[0,u]} C_N(n,s\wedge S_R)^2]du,
\end{aligned}$$

where the constant Λ does not depend on R. Gronwall's Lemma then implies, with $E_N(n,t) := 2[D_N(n,t)]^2 + N^{-2}\Lambda t$, that

$$E[\sup_{s\in[0,t]} C_N(n,s\wedge S_R)^2] \le E_N(n,t) + \int_0^t e^{2N^{-2}\Lambda_1(s-t)}E_N(n,s)ds.$$

We can finally let R go to infinity and conclude that $E[\sup_{s\in[0,t]} C_N(n,s)]$ is finite and so $\sup_{s\in[0,t]}\sqrt{C_N(n,s)}$, and therefore $\sup_n\sup_{s\in[0,t]}\sqrt{C_N(n,s)}$, are finite almost surely, completing the proof of (4.3.19). □

Exercise 4.3.7 Let $H^{N,4} = \left(X_{ij}^{N,\beta}\right)$ be $2N\times 2N$ complex Gaussian Wigner matrices defined as the self-adjoint random matrices with entries

$$H_{kl}^{N,\beta} = \frac{\sum_{i=1}^4 g_{kl}^i e_\beta^i}{\sqrt{4N}}, \quad 1\le k<l\le N, \quad X_{kk}^{N,4} = \sqrt{\frac{1}{2N}}g_{kk}e_\beta^1, \quad 1\le k\le N,$$

where $(e^i_\beta)_{1\le i\le\beta}$ are the Pauli matrices

$$e_4^1=\begin{pmatrix}1&0\\0&1\end{pmatrix}, e_4^2=\begin{pmatrix}0&-1\\1&0\end{pmatrix}, e_4^3=\begin{pmatrix}0&-\mathrm{i}\\-\mathrm{i}&0\end{pmatrix}, e_4^4=\begin{pmatrix}\mathrm{i}&0\\0&-\mathrm{i}\end{pmatrix}.$$

Show that with $H^{N,4}$ as above, and $X^{N,4}(0)$ a Hermitian matrix with eigenvalues $(\lambda_1^N(0),\ldots,\lambda_{2N}^N(0))\in\overline{\Delta_N}$, the eigenvalues $(\lambda_1^N(t),\ldots,\lambda_{2N}^N(t))$ of $X^{N,4}(0)+H^{N,4}(t)$ satisfy the stochastic differential system

$$d\lambda_i^N(t)=\frac{1}{\sqrt{2N}}dW_i(t)+\frac{1}{N}\sum_{j\neq i}\frac{1}{\lambda_i^N(t)-\lambda_j^N(t)}dt\,,i=1,\ldots,2N\,. \tag{4.3.20}$$

Exercise 4.3.8 [Bru91] Let $V(t)$ be an $N\times M$ matrix whose entries are independent complex Brownian motions and let $V(0)$ be an $N\times M$ matrix with complex entries. Let $\lambda^N(0)=(\lambda^N(0),\ldots,\lambda_N^N(0))\in\overline{\Delta_N}$ be the eigenvalues of $V(0)V(0)^*$. Show that the law of the eigenvalues of $X(t)=V(t)^*V(t)$ is the weak solution to

$$d\lambda_i^N(t)=2\sqrt{\frac{\lambda_i^N(t)}{N}}dW_i(t)+2(\frac{M}{N}+\sum_{k\neq i}\frac{\lambda_k^N+\lambda_i^N}{\lambda_i^N-\lambda_k^N})dt\,,$$

with initial condition $\lambda^N(0)$.

Exercise 4.3.9 Let X^N be the matrix-valued process solution of the stochastic differential system $dX_t^N=dH_t^{N,\beta}-X_t^N dt$, with $D(X^N(0))\in\overline{\Delta_N}$.
(a) Show that the law of the eigenvalues of X_t^N is a weak solution of

$$d\lambda_i^N(t)=\frac{\sqrt{2}}{\sqrt{\beta N}}dW_i(t)+\frac{1}{N}\sum_{j\neq i}\frac{1}{\lambda_i^N(t)-\lambda_j^N(t)}dt-\lambda_i^N(t)dt\,. \tag{4.3.21}$$

(b) Show that if $X_0^N=H^{N,\beta}(1)$, then the law of X_t^N is the same law for all $t\ge 0$. Conclude that the law $P_N^{(\beta)}$ of the eigenvalues of Gaussian Wigner matrices is stationary for the process (4.3.21).
(c) Deduce that $P_N^{(\beta)}$ is absolutely continuous with respect to the Lebesgue measure, with density

$$1_{x_1\le\cdots\le x_N}\prod_{1\le i<j\le N}|x_i-x_j|^\beta\prod_{i=1}^N e^{-\beta x_i^2/4}\,,$$

as proved in Theorem 2.5.2. *Hint*: obtain a partial differential equation for the invariant measure of (4.3.21) and solve it.

4.3.2 A dynamical version of Wigner's Theorem

In this subsection, we derive systems of (deterministic) differential equations satisfied by the limits of expectation of $\langle L_N(t), g\rangle$, for nice test functions g and

$$L_N(t) = N^{-1}\sum \delta_{\lambda_i^N(t)}, \tag{4.3.22}$$

where $(\lambda_i^N(t))_{t\geq 0}$ is a solution of (4.3.3) for $\beta \geq 1$ (see Proposition 4.3.10). Specializing to $\beta = 1$ or $\beta = 2$, we will then deduce in Corollary 4.3.11 a dynamical proof of Wigner's Theorem, Theorem 2.1.1, which, while restricted to Gaussian entries, generalizes the latter theorem in the sense that it allows one to consider the sum of a Wigner matrix with an arbitrary, N-dependent Hermitian matrix, provided the latter has a converging empirical distribution. The limit law is then described as the law at time one of the solution to a complex Burgers equation, a definition which introduces already the concept of *free convolution* (with respect to the semicircle law) that we shall develop in Section 5.3.3. In Exercise 4.3.18, Wigner's Theorem is recovered from its dynamical version.

We recall that, for $T > 0$, we denote by $C([0,T], M_1(\mathbb{R}))$ the space of continuous processes from $[0,T]$ into $M_1(\mathbb{R})$ (the space of probability measures on $\mathbb{R}$, equipped with its weak topology). We now prove the convergence of the empirical measure $L_N(\cdot)$, viewed as an element of $C([0,T], M_1(\mathbb{R}))$.

Proposition 4.3.10 *Let $\beta \geq 1$ and let $\lambda^N(0) = (\lambda_1^N(0), \ldots, \lambda_N^N(0)) \in \overline{\Delta_N}$, be a sequence of real vectors so that $\lambda^N(0) \in \overline{\Delta_N}$,*

$$C_0 := \sup_{N\geq 0} \frac{1}{N}\sum_{i=1}^{N} \log(\lambda_i^N(0)^2 + 1) < \infty, \tag{4.3.23}$$

and the empirical measure $L_N(0) = \frac{1}{N}\sum_{i=1}^N \delta_{\lambda_k^N(0)}$ converges weakly as N goes to infinity towards a $\mu \in M_1(\mathbb{R})$.

Let $\lambda^N(t) = (\lambda_1^N(t), \ldots, \lambda_N^N(t))_{t\geq 0}$ be the solution of (4.3.3) with initial condition $\lambda^N(0)$, and set $L_N(t)$ as in (4.3.22). Then, for any fixed time $T < \infty$, $(L_N(t))_{t\in[0,T]}$ converges almost surely in $C([0,T], M_1(\mathbb{R}))$. Its limit is the unique measure-valued process $(\mu_t)_{t\in[0,T]}$ so that $\mu_0 = \mu$ and the function

$$G_t(z) = \int (z-x)^{-1} d\mu_t(x) \tag{4.3.24}$$

satisfies the equation

$$G_t(z) = G_0(z) - \int_0^t G_s(z)\partial_z G_s(z) ds \tag{4.3.25}$$

for $z \in \mathbb{C}\backslash\mathbb{R}$.

An immediate consequence of Proposition 4.3.10 is the following.

Corollary 4.3.11 *For $\beta = 1,2$, let $(X^{N,\beta}(0))_{N\in\mathbb{N}}$ be a sequence of real diagonal matrices, with eigenvalues $(\lambda_1^N(0),\dots,\lambda_N^N(0))$ satisfying the assumptions of Proposition 4.3.10. For $t \geq 0$, let $\lambda_i^N(t) = (\lambda_1^N(t),\dots,\lambda_N^N(t) \in \overline{\Delta_N}$ denote the eigenvalues of $X^{N,\beta}(t) = X^{N,\beta}(0) + H^{N,\beta}(t)$, and let $L_N(t)$ be as in (4.3.22). Then the measure-valued process $(L_N(t))_{t\geq 0}$ converges almost surely towards $(\mu_t)_{t\geq 0}$ in $C([0,T],M_1(\mathbb{R}))$.*

Proof of Proposition 4.3.10 We begin by showing that the sequence $(L_N(t))_{t\in[0,T]}$ is almost surely pre-compact in $C([0,T],M_1(\mathbb{R}))$ and then show that it has a unique limit point characterized by (4.3.25). The key step of our approach is the following direct application of Itô's Lemma, Theorem H.9, to the stochastic differential system (4.3.3), whose elementary proof we omit.

Lemma 4.3.12 *Under the assumptions of Proposition 4.3.10, for all $T > 0$, all $f \in C^2([0,T]\times\mathbb{R},\mathbb{R})$ and all $t \in [0,T]$,*

$$\begin{aligned}\langle f(t,\cdot),L_N(t)\rangle &= \langle f(0,\cdot),L_N(0)\rangle + \int_0^t \langle \partial_s f(s,\cdot),L_N(s)\rangle ds \qquad (4.3.26)\\ &+ \frac{1}{2}\int_0^t \iint \frac{\partial_x f(s,x) - \partial_y f(s,y)}{x-y} dL_N(s)(x)dL_N(s)(y)ds\\ &+ \left(\frac{2}{\beta}-1\right)\frac{1}{2N}\int_0^t \langle \partial_x^2 f(s,\cdot),L_N(s)\rangle ds + M_f^N(t),\end{aligned}$$

where M_f^N is the martingale given for $t \leq T$ by

$$M_f^N(t) = \frac{\sqrt{2}}{\sqrt{\beta}N^{\frac{3}{2}}}\sum_{i=1}^N \int_0^t \partial_x f(s,\lambda_i^N(s))dW_s^i.$$

We note that the bracket of the martingale M_f^N appearing in Lemma 4.3.12 is

$$\langle M_f^N\rangle_t = \frac{2}{\beta N^2}\int_0^t \langle(\partial_x f(s,x))^2,L_N(s)\rangle ds \leq \frac{2t\sup_{s\in[0,t]}\|\partial_x f(.,s)\|_\infty^2}{\beta N^2}.$$

We also note that the term multiplying $(2/\beta - 1)$ in (4.3.26) is coming from both the quadratic variation term in Itô's Lemma and the finite variation term where the terms on the diagonal $x = y$ were added. That it vanishes when $\beta = 2$ is a curious coincidence, and emphasizes once more that the Hermitian case ($\beta = 2$) is in many ways the simplest case.

We return now to the proof of Proposition 4.3.10, and begin by showing that the sequence $(L_N(t))_{t\in[0,T]}$ is a pre-compact family in $C([0,T],M_1(\mathbb{R}))$ for all $T<\infty$. Toward this end, we first describe a family of compact sets of $C([0,T],M_1(\mathbb{R}))$.

Lemma 4.3.13 *Let K be a an arbitrary compact subset of $M_1(\mathbb{R})$, let $(f_i)_{i\geq 0}$ be a sequence of bounded continuous functions dense in $C_0(\mathbb{R})$, and let C_i be compact subsets of $C([0,T],\mathbb{R})$. Then the sets*

$$\mathscr{K} := \{\forall t\in[0,T], \mu_t\in K\}\bigcap_{i\geq 0}\{t\to\mu_t(f_i)\in C_i\} \tag{4.3.27}$$

are compact subsets of $C([0,T],M_1(\mathbb{R}))$.

Proof of Lemma 4.3.13 The space $C([0,T],M_1(\mathbb{R}))$ being Polish, it is enough to prove that the set $\mathscr{K}$ is sequentially compact and closed. Toward this end, let $(\mu^n)_{n\geq 0}$ be a sequence in $\mathscr{K}$. Then, for all $i\in\mathbb{N}$, the functions $t\to\mu_t^n(f_i)$ belong to the compact sets C_i and hence we can find a subsequence $\phi_i(n)\to_{n\to\infty}\infty$ such that the sequence of bounded continuous functions $t\to\mu_t^{\phi_i(n)}(f_i)$ converges in $C[0,T]$. By a diagonalization procedure, we can find an i independent subsequence $\phi(n)\to_{n\to\infty}\infty$ such that for all $i\in\mathbb{N}$, the functions $t\to\mu_t^{\phi(n)}(f_i)$ converge towards some function $t\to\mu_t(f_i)\in C[0,T]$. Because $(f_i)_{i\geq 0}$ is convergence determining in $K\cap M_1(\mathbb{R})$, it follows that one may extract a further subsequence, still denoted $\phi(n)$, such that for a fixed dense countable subset of $[0,T]$, the limit μ_t belongs to M_1. The continuity of $t\to\mu_t(f_i)$ then shows that $\mu_t\in M_1(\mathbb{R})$ for all t, which completes the proof that $(\mu^n)_{n\geq 0}$ is sequentially compact. Since $\mathscr{K}$ is an intersection of closed sets, it is closed. Thus, $\mathscr{K}$ is compact, as claimed. □

We next prove the pre-compactness of the sequence $(L_N(t), t\in[0,T])$.

Lemma 4.3.14 *Under the assumptions of Proposition 4.3.10, fix $T\in\mathbb{R}^+$. Then the sequence $(L_N(t), t\in[0,T])$ is almost surely pre-compact in $C([0,T],M_1(\mathbb{R}))$.*

Proof We begin with a couple of auxiliary estimates. Note that from Lemma 4.3.12, for any function f that is twice continuously differentiable,

$$\iint \frac{f'(x)-f'(y)}{x-y}dL_N(s)(x)dL_N(s)(y)$$
$$= \iint\int_0^1 f''(\alpha x+(1-\alpha)y)d\alpha dL_N(s)(x)dL_N(s)(y). \tag{4.3.28}$$

Apply Lemma 4.3.12 with the function $f(x)=\log(1+x^2)$, which is twice continuously differentiable with second derivative uniformly bounded by 2, to deduce that

$$\sup_{t\leq T}|\langle f, L_N(t)\rangle| \leq |\langle f, L_N(0)\rangle| + T(1+\frac{1}{N}) + \sup_{t\leq T}|M_f^N(t)| \tag{4.3.29}$$

with M_f^N a martingale with bracket bounded by $2(\beta N^2)^{-1}$ since $|f'| \le 1$. By the Burkholder–Davis–Gundy inequality (Theorem H.8) and Chebyshev's inequality, we get that, for a universal constant Λ_1,

$$P(\sup_{t \le T} |M_f^N(t)| \ge \varepsilon) \le \frac{2\Lambda_1}{\varepsilon^2 \beta N^2}, \tag{4.3.30}$$

which, together with (4.3.29), proves that there exists $a = a(T) < \infty$ so that, for $M > T + C_0 + 1$,

$$P\left(\sup_{t \in [0,T]} \langle \log(x^2+1), L_N(t) \rangle \ge M \right) \le \frac{a}{(M - T - C_0 - 1)^2 N^2}. \tag{4.3.31}$$

We next need an estimate on the Hölder norm of the function $t \to \langle f, L_N(t) \rangle$, for any twice boundedly differentiable function f on $\mathbb{R}$, with first and second derivatives bounded by 1. We claim that there exists a constant $a = a(T)$ so that, for any $\delta \in (0,1)$ and $M > 2$,

$$P\left(\sup_{\substack{t,s \in [0,T] \\ |t-s| \le \delta}} |\langle f, L_N(t) \rangle - \langle f, L_N(s) \rangle| \ge M \delta^{\frac{1}{8}} \right) \le \frac{a \delta^{1/2}}{M^4 N^4}. \tag{4.3.32}$$

Indeed, apply Lemma 4.3.12 with $f(x,t) = f(x)$. Using (4.3.28), one deduces that for all $t \ge s$,

$$|\langle f, L_N(t) \rangle - \langle f, L_N(s) \rangle| \le ||f''||_\infty |s-t| + |M_f^N(t) - M_f^N(s)|, \tag{4.3.33}$$

where $M_f^N(t)$ is a martingale with bracket $2\beta^{-1} N^{-2} \int_0^t \langle (f')^2, L_N(u) \rangle du$. Now, cutting $[0,T]$ to intervals of length δ we get, with $J := [T\delta^{-1}]$,

$$\begin{aligned}
&P\left(\sup_{\substack{|t-s| \le \delta \\ t,s \le T}} \left| M_f^N(t) - M_f^N(s) \right| \ge (M-1)\delta^{1/8} \right) \\
&\le \sum_{k=1}^{J+1} P\left(\sup_{k\delta \le t \le (k+1)\delta} \left| M_f^N(t) - M_f^N(k\delta) \right| \ge (M-1)\delta^{1/8}/3 \right) \\
&\le \sum_{k=1}^{J+1} \frac{3^4}{\delta^{1/2}(M-1)^4} E\left(\sup_{k\delta \le t \le (k+1)\delta} \left| M_f^N(t) - M_f^N(k\delta) \right|^4 \right) \\
&\le \frac{4 \cdot 3^4 \Lambda_2 \delta^2}{\beta^2 N^4 \delta^{1/2} (M-1)^4} (J+1) ||f'||_\infty^2 =: \frac{a \delta^{\frac{1}{2}}}{N^2 (M-1)^4} ||f'||_\infty^2,
\end{aligned}$$

where again we used in the second inequality Chebyshev's inequality, and in the last the Burkholder–Davis–Gundy inequality (Theorem H.8) with $m = 2$. Combining this inequality with (4.3.33) completes the proof of (4.3.32).

We can now conclude the proof of the lemma. Setting

$$K_M = \{\mu \in M_1(\mathbb{R}) : \int \log(1+x^2)d\mu(x) \le M\},$$

Borel–Cantelli's Lemma and (4.3.31) show that

$$P\left(\bigcup_{N_0\ge 0}\bigcap_{N\ge N_0} \{\forall t\in[0,T],\, L_N(t)\in K_M\}\right) = 1\,. \tag{4.3.34}$$

Next, recall that by the Arzela–Ascoli Theorem, sets of the form

$$C = \bigcap_n \{g\in C([0,T],\mathbb{R}) : \sup_{\substack{t,s\in[0,T]\\ |t-s|\le\eta_n}} |g(t)-g(s)|\le \varepsilon_n,\ \sup_{t\in[0,T]}|g(t)|\le M\},$$

where $\{\varepsilon_n, n\ge 0\}$ and $\{\eta_n, n\ge 0\}$ are sequences of positive real numbers going to zero as n goes to infinity, are compact. For $f\in C^2(\mathbb{R})$ with derivatives bounded by 1, and $\varepsilon>0$, consider the subset of $C([0,T],M_1(\mathbb{R}))$ defined by

$$C_T(f,\varepsilon) := \bigcap_{n=1}^{\infty}\{\mu\in C([0,T],M_1(\mathbb{R})) : \sup_{|t-s|\le n^{-4}} |\mu_t(f)-\mu_s(f)| \le \frac{1}{\varepsilon\sqrt{n}}\}.$$

Then, by (4.3.32),

$$P(L_N\in C_T(f,\varepsilon)^c) \le \frac{a\varepsilon^4}{N^4}\,. \tag{4.3.35}$$

Choose a countable family f_k of twice continuously differentiable functions dense in $C_0(\mathbb{R})$, and set $\varepsilon_k = 1/k(\|f_k\|_\infty + \|f_k'\|_\infty + \|f_k''\|_\infty)^{\frac{1}{2}} < 2^{-1}$, with

$$\mathscr{K} = K_M \cap \bigcap_{k\ge 1} C_T(f_k,\varepsilon_k) \subset C([0,T],M_1(\mathbb{R}))\,. \tag{4.3.36}$$

Combining (4.3.34) and (4.3.35), we get from the Borel–Cantelli Lemma that

$$P\left(\bigcup_{N_0\ge 0}\bigcap_{N\ge N_0}\{L_N\in\mathscr{K}\}\right) = 1.$$

Since $\mathscr{K}$ is compact by Lemma 4.3.13, the claim follows. □

We return to the proof of Proposition 4.3.10. To characterize the limit points of L_N, we again use Lemma 4.3.12 with a general twice continuously differentiable function f with bounded derivatives. Exactly as in the derivation leading to (4.3.30), the Boreli–Cantelli Lemma and the Burkholder–Davis–Gundy inequality (Theorem H.8) yield the almost sure convergence of M_f^N towards zero, uniformly on compact time intervals. Therefore, any limit point $(\mu_t, t\in[0,T])$ of L_N satisfies the equation

$$\int f(t,x)d\mu_t(x) = \int f(0,x)d\mu_0(x) + \int_0^t \int \partial_s f(s,x)d\mu_s(x)ds + \frac{1}{2}\int_0^t \iint \frac{\partial_x f(s,x) - \partial_x f(s,y)}{x-y} d\mu_s(x)d\mu_s(y)ds. \quad (4.3.37)$$

Taking $f(x) = (z-x)^{-1}$ for some $z \in \mathbb{C}\backslash\mathbb{R}$, we deduce that the function $G_t(z) = \int (z-x)^{-1}d\mu_t(x)$ satisfies (4.3.24), (4.3.25). Note also that since the limit μ_t is a probability measure on the real line, $G_t(z)$ is analytic in z for $z \in \mathbb{C}_+$.

To conclude the proof of Proposition 4.3.10, we show below in Lemma 4.3.15 that (4.3.24), (4.3.25) possess a unique solution analytic on $z \in \mathbb{C}_+ := \{z \in \mathbb{C} : \Im(z) > 0\}$. Since we know *a priori* that the support of any limit point μ_t lives in $\mathbb{R}$ for all t, this uniqueness implies the uniqueness of the Stieltjes transform of μ_t for all t and hence, by Theorem 2.4.3, the uniqueness of μ_t for all t, completing the proof of Proposition 4.3.10. □

Lemma 4.3.15 *Let $\Gamma_{\alpha,\beta} = \{z \in \mathbb{C}_+ : \Im z \geq \alpha|\Re z|, |z| \geq \beta\}$ and for $t \geq 0$, set $\Lambda_t := \{z \in \mathbb{C}_+ : z + tG_0(z) \in \mathbb{C}_+\}$. For all $t \geq 0$, there exist positive constants $\alpha_t, \beta_t, \alpha'_t, \beta'_t$ such that $\Gamma_{\alpha_t,\beta_t} \subset \Lambda_t$ and the function $z \in \Gamma_{\alpha_t,\beta_t} \to z + tG_0(z) \in \Gamma_{\alpha'_t,\beta'_t}$ is invertible with inverse $H_t : \Gamma_{\alpha'_t,\beta'_t} \to \Gamma_{\alpha_t,\beta_t}$. Any solution of (4.3.24), (4.3.25) is the unique analytic function on $\mathbb{C}_+$ such that for all t and all $z \in \Gamma_{\alpha'_t,\beta'_t}$,*

$$G_t(z) = G_0(H_t(z)).$$

Proof We first note that since $|G_0(z)| \leq 1/|\Im z|$, $\Im(z + tG_0(z)) \geq \Im z - t/\Im z$ is positive for $t < (\Im z)^2$ and $\Im z > 0$. Thus, $\Gamma_{\alpha_t,\beta_t} \subset \Lambda_t$ for $t < (\alpha_t\beta_t)^2/(1+\alpha_t^2)$. Moreover, $|\Re G_0(z)| \leq 1/2|\Im z|$ from which we see that, for all $t \geq 0$, the image of $\Gamma_{\alpha_t,\beta_t}$ by $z + tG_0(z)$ is contained in some $\Gamma_{\alpha'_t,\beta'_t}$ provided β_t is large enough. Note that we can choose the $\Gamma_{\alpha_t,\beta_t}$ and $\Gamma_{\alpha'_t,\beta'_t}$ decreasing in time.

We next use the method of characteristics. Fix $G_.$ a solution of (4.3.24), (4.3.25). We associate with $z \in \mathbb{C}_+$ the solution $\{z_t, t \geq 0\}$ of the equation

$$\partial_t z_t = G_t(z_t), \quad z_0 = z. \quad (4.3.38)$$

We can construct a solution $z_.$ to this equation up to time $(\Im z)^2/4$ with $\Im z_t \geq \Im z/2$ as follows. We put for $\varepsilon > 0$,

$$G_t^\varepsilon(z) := \int \frac{\bar{z}-x}{|z-x|^2+\varepsilon} d\mu_t(x), \partial_t z_t^\varepsilon = G_t^\varepsilon(z_t^\varepsilon), \quad z_0^\varepsilon = z.$$

$z_.^\varepsilon$ exists and is unique since G_t^ε is uniformly Lipschitz. Moreover,

$$\frac{\partial_t \Im(z_t^\varepsilon)}{\Im(z_t^\varepsilon)} = -\int \frac{1}{|z_t - x|^2 + \varepsilon} d\mu_t(x) \in [-\frac{1}{|\Im(z_t^\varepsilon)|^2}, 0],$$

implies that $|\Im(z_t^\varepsilon)|^2 \in [|\Im(z)|^2 - 2t, |\Im(z)|^2]$ and

$$\partial_t \Re(z_t^\varepsilon) = \int \frac{\Re(z_t^\varepsilon) - x}{|z_t - x|^2 + \varepsilon} d\mu_t(x) \in [-\frac{1}{\sqrt{|\Im(z_t^\varepsilon)|^2 + \varepsilon}}, \frac{1}{\sqrt{|\Im(z_t^\varepsilon)|^2 + \varepsilon}}]$$

shows that $\Re(z_t^\varepsilon)$ stays uniformly bounded, independently of ε, up to time $(\Im z)^2/4$ as well as its time derivative. Hence, $\{z_t^\varepsilon, t \le (\Im z)^2/4\}$ is tight by Arzela–Ascoli's Theorem. Any limit point is a solution of the original equation and such that $\Im z_t \ge \Im z/2 > 0$. It is unique since G_t is uniformly Lipschitz on this domain.

Now, $\partial_t G_t(z_t) = 0$ implies that for $t \le (\Im z)^2/4$,

$$z_t = tG_0(z) + z, \; G_t(z + tG_0(z)) = G_0(z).$$

By the implicit function theorem, $z + tG_0(z)$ is invertible from $\Gamma_{\alpha_t,\beta_t}$ into $\Gamma_{\alpha_t',\beta_t'}$ since $1 + tG_0'(z) \neq 0$ (note that $\Im G_0'(z) \neq 0$) on $\Gamma_{\alpha_t,\beta_t}$. Its inverse H_t is analytic from $\Gamma_{\alpha_t',\beta_t'}$ into $\Gamma_{\alpha_t,\beta_t}$ and satisfies

$$G_t(z) = G_0(H_t(z)).$$

□

With a view toward later applications in Subsection 4.3.3 to the proof of central limit theorems, we extend the previous results to polynomial test functions.

Lemma 4.3.16 *Let $\beta \ge 1$. Assume that*

$$\tilde{C} = \sup_{N \in \mathbb{N}} \max_{1 \le i \le N} |\lambda_i^N(0)| < \infty.$$

With the same notation and assumptions as in Proposition 4.3.10, for any $T < \infty$, for any polynomial function q, the process $(\langle q, L_N(t)\rangle)_{t\in[0,T]}$ converges almost surely and in all L^p, towards the process $(\mu_t(q))_{t\in[0,T]}$, that is,

$$\limsup_{N\to\infty} \sup_{t\in[0,T]} |\langle q, L_N(t)\rangle - \langle q, \mu_t\rangle| = 0 \quad a.s.$$

and for all $p \in \mathbb{N}$,

$$\limsup_{N\to\infty} \mathbb{E}[\sup_{t\in[0,T]} |\langle q, L_N(t)\rangle - \langle q, \mu_t\rangle|^p] = 0.$$

A key ingredient in the proof is the following control of the moments of $\lambda_N^*(t) := \max_{1\le i\le N} |\lambda_i^N(t)| = \max(\lambda_N^N(t), -\lambda_1^N(t))$.

Lemma 4.3.17 *Let $\beta \ge 1$ and $\lambda_N(0) \in \Delta_N$. Then there exist finite constants $\alpha = \alpha(\beta) > 0, C = C(\beta)$, and for all $t \ge 0$ a random variable $\eta_N^*(t)$ with law independent of t, such that*

$$P(\eta_N^*(t) \ge x + C) \le e^{-\alpha N x}$$

and, further, the unique strong solution of (4.3.3) *satisfies, for all* $t \geq 0$,

$$\lambda_N^*(t) \leq \lambda_N^*(0) + \sqrt{t}\eta_N^*(t). \quad (4.3.39)$$

We note that for $\beta = 1,2,4$, this result can be deduced from the study of the maximal eigenvalue of $X^{N,\beta}(0) + H^{N,\beta}(t)$, since the spectral radius of $H^{N,\beta}(t)$ has the same law as the spectral radius of $\sqrt{t}H^{N,\beta}(1)$, that can be controlled as in Section 2.1.6. The proof we give below is based on stochastic analysis, and works for all $\beta \geq 1$. It is based on the comparison between strong solutions of (4.3.3) presented in Lemma 4.3.6.

Proof of Lemma 4.3.17 Our approach is to construct a stationary process $\eta^N(t) = (\eta_1^N(t),\ldots,\eta_N^N(t)) \in \Delta_N$, $t \geq 0$, with marginal distribution $P^N_{(\beta)} := P^N_{\beta x^2/4,\beta}$ as in (2.6.1), such that, with $\eta_N^*(t) = \max(\eta_N^N(t), -\eta_1^N(t))$, the bound (4.3.39) holds. We first construct this process (roughly corresponding to the process of eigenvalues of $H^{N,\beta}(t)/\sqrt{t}$ if $\beta = 1,2,4$) and then prove (4.3.39) by comparing solutions to (4.3.3) started from different initial conditions.

Fix $\varepsilon > 0$. Consider, for $t \geq \varepsilon$, the stochastic differential system

$$du_i^N(t) = \sqrt{\frac{2}{\beta N t}} dW_i(t) + \frac{1}{Nt}\sum_{j\neq i}\frac{1}{u_i^N(t) - u_j^N(t)}dt - \frac{1}{2t}u_i^N(t)dt. \quad (4.3.40)$$

Let P_N^β denote the rescaled version of $P_N^{(\beta)}$ from (2.5.1), that is, the law on Δ_N with density proportional to

$$\prod_{i<j}|\lambda_i - \lambda_j|^\beta \cdot \prod_i e^{-N\beta\lambda_i^2/4}.$$

Because $P_N^\beta(\Delta_N) = 1$, we may take $u^N(\varepsilon)$ distributed according to P_N^β, and the proof of Lemma 4.3.3 carries over to yield the strong existence and uniqueness of solutions to (4.3.40) initialized from such (random) initial conditions belonging to Δ_N.

Our next goal is to prove that P_N^β is a stationary distribution for the system (4.3.40) with this initial distribution, independently of ε. Toward this end, note that by Itô's calculus (Lemma 4.3.12), one finds that for any twice continuously differentiable function $f : \mathbb{R}^N \to \mathbb{R}$,

$$\begin{aligned}\partial_t E[f(u^N(t))] &= E[\frac{1}{2Nt}\sum_{i\neq j}\frac{\partial_i f(u^N(t)) - \partial_j f(u^N(t))}{u_i^N(t) - u_j^N(t)}] \\ &\quad -E[\frac{1}{2t}\sum_i u_i^N(t)\partial_i f(u^N(t))] + E[\frac{1}{\beta N t}\sum_i \partial_i^2 f(u^N(t))],\end{aligned}$$

where we used the notation $\partial_i f(x) = \partial_{x_i} f(x_1,\ldots,x_N)$. Hence, if at any time t,

$u^N(t)$ has law P_N^β, we see by integration by parts that $\partial_t E[f(u^N(t))]|$ vanishes for any twice continuously differentiable f. Therefore, $(u^N(t))_{t\geq\varepsilon}$ is a stationary process with marginal law P_N^β. Because the marginal P_N^β does not depend on ε, one may extend this process to a stationary process $(u^N(t))_{t\geq 0}$.

Set $u_N^*(t) = \max(u_N^N(t), -u_1^N(t))$. Recall that by Theorem 2.6.6 together with (2.5.11),

$$\lim_{N\to\infty} \frac{1}{N} \log P_N^\beta(\lambda_N \geq u) = -\inf_{s\geq u} J_\beta^{\beta x^2/4}(s),$$

with $J_\beta^{\beta x^2/4}(s) > 0$ for $s > 2$. Thus, there exist $C < \infty$ and $\alpha > 0$ so that for $x \geq C$, for all $N \in \mathbb{N}$,

$$P(u_N^*(t) \geq x) \leq 2P_N^\beta(\lambda_N \geq x) \leq e^{-\alpha N x}. \tag{4.3.41}$$

Define next $\lambda^{N,0}(t) = \sqrt{t}u^N(t)$. Clearly, $\lambda^{N,0}(0) = 0 \in \overline{\Delta_N}$. An application of Itô's calculus, Lemma 4.3.12, shows that $\lambda^{N,0}(t)$ is a continuous solution of (4.3.3) with initial data 0, and $\lambda^{N,0}(t) \in \Delta_N$ for all $t > 0$. For an arbitrary constant A, define $\lambda^{N,A}(t) \in \overline{\Delta_N}$ by $\lambda_i^{N,A}(t) = \lambda_i^{N,0}(t) + A$, noting that $(\lambda^{N,A}(t))_{t\geq 0}$ is again a solution of (4.3.3), starting from the initial data $(A,\ldots,A) \in \overline{\Delta_N}$, that belongs to Δ_N for all $t > 0$.

Note next that for any $\delta > 0$, $\lambda_i^{N,\delta+\lambda_N^*(0)}(0) > \lambda_i^N(0)$ for all i. Further, for t small, $\lambda_i^{N,\delta+\lambda_N^*(0)}(t) > \lambda_i^N(t)$ for all i by continuity. Therefore, we get from Lemma 4.3.6 that, for all $t > 0$,

$$\lambda_N^N(t) \leq \lambda_N^{N,\delta+\lambda_N^*(0)}(t) \leq \lambda_N^*(0) + \delta + \sqrt{t}u_N^*(t).$$

A similar argument shows that

$$-\lambda_1^N(t) \leq \lambda_N^*(0) + \delta + \sqrt{t}u_N^*(t).$$

Since $u^N(t)$ is distributed according to the law P_N^β, taking $\delta \to 0$ and recalling (4.3.41) completes the proof of the lemma. □

Proof of Lemma 4.3.16 We use the estimates on $\lambda_N^*(t)$ from Lemma 4.3.17 in order to approximate $\langle q, L_N(t)\rangle$ for polynomial functions q by similar expressions involving bounded continuous functions.

We begin by noting that, due to Lemma 4.3.17 and the Borel–Cantelli Lemma, for any fixed t,

$$\limsup_{N\to\infty} \lambda_N^*(t) \leq \lambda_N^*(0) + \sqrt{t}C \quad \text{a.s.} \tag{4.3.42}$$

Again from Lemma 4.3.17, we also have that, for any $p \geq 0$,

$$\begin{aligned} E[(\lambda_N^*(t))^p] &\leq 2^p \left((\lambda_N^*(0) + C\sqrt{t})^p + p t^{\frac{p}{2}} \int_0^\infty x^{p-1} e^{-\alpha N x} dx \right) \\ &= 2^p \left((\lambda_N^*(0) + C\sqrt{t})^p + \frac{p!}{(\alpha N)^p} t^{\frac{p}{2}} \right). \end{aligned} \tag{4.3.43}$$

As a consequence, there exists an increasing function $C(t)$, such that for any $T < \infty$, $C(T) = \sup_{t \leq T} C(t) < \infty$, and so that for all N sufficiently large, all $p \in [0, \alpha N]$,

$$E[(\lambda_N^*(t))^p] \leq (2C(t))^p. \tag{4.3.44}$$

Note that (4.3.42) implies that, under the current assumptions, the support of the limit μ_t, see Proposition 4.3.10, is contained in the compact set $[-A(t), A(t)]$, where $A(t) := \tilde{C} + C\sqrt{t}$.

We next improve (4.3.42) to uniform (in $t \leq T$) bounds. Fix a constant $\varepsilon < \min(\alpha/6, 1/T\sqrt{\Lambda_1})$, where Λ_1 is as in the Burkholder–Davis–Gundy inequality (Theorem H.8). We will show that, for all $T < \infty$ and $p \leq \varepsilon N$,

$$E[\sup_{t \in [0,T]} \langle |x|^p, L_N(t) \rangle] \leq C(T)^p. \tag{4.3.45}$$

This will imply that

$$E[\sup_{t \in [0,T]} \lambda_N^*(t)^p] \leq N C(T)^p, \tag{4.3.46}$$

and therefore, by Chebyshev's inequality, for any $\delta > 0$,

$$P(\sup_{t \in [0,T]} \lambda_N^*(t) > C(T) + \delta) \leq \frac{N C(T)^p}{(C(T) + \delta)^p}.$$

Taking $p = p(N) = (\log N)^2$, we conclude by the Borel–Cantelli Lemma that

$$\limsup_{N \to \infty} \sup_{0 \leq t \leq T} \lambda_N^*(t) \leq C(T) \quad \text{a.s.}$$

To prove (4.3.45), we use (4.3.26) with $f(t,x) = x^n$ and an integer $n > 0$ to get

$$\begin{aligned} \langle x^{n+2}, L_N(t) \rangle &= \langle x^{n+2}, L_N(0) \rangle + M_{n+2}^N(t) \\ &\quad + \frac{(n+1)(n+2)}{2N} \left(\frac{2}{\beta} - 1 \right) \int_0^t \langle x^n, L_N(s) \rangle ds \\ &\quad + \frac{(n+2)}{2} \sum_{\ell=0}^{n} \int_0^t \langle x^\ell, L_N(s) \rangle \langle x^{n-\ell}, L_N(s) \rangle ds, \end{aligned} \tag{4.3.47}$$

where M_{n+2}^N is a local martingale with bracket

$$\langle M_{n+2}^N \rangle_t = \frac{2(n+2)^2}{\beta N^2} \int_0^t \langle x^{2n+2}, L_N(s) \rangle ds.$$

Setting $n = 2p$ and using the Burkholder–Davis–Gundy inequality (Theorem H.8), one obtains

$$\begin{aligned} E[\sup_{t\in[0,T]} M^N_{2(p+1)}(t)^2] &\leq \frac{8\Lambda_1(p+1)^2}{\beta N^2} E[\int_0^T \langle x^{4p+2}, L_N(s)\rangle ds] \\ &\leq c\frac{\Lambda_1 p^2 \int_0^T C(t)^{(4p+2)m} dt}{N^2} \leq c\frac{\Lambda_1 p^2 T C(T)^{(4p+2)}}{N^2}, \end{aligned}$$

for some constant $c = c(\beta)$ independent of p or T, where we used (4.3.44) (and thus used that $4p+2 \leq \alpha N$). We set

$$\Lambda_t(p) := E[\sup_{t\in[0,T]} \langle |x|^p, L_N(t)\rangle],$$

and deduce from (4.3.47) and the last estimate that for $p \in [0, \varepsilon N/2]$ integer,

$$\begin{aligned} \Lambda_t(2(p+1)) &\leq \Lambda_0(2(p+1)) + \frac{(c\Lambda_1)^{\frac{1}{2}} p\sqrt{t} C(t)^{(2p+1)}}{N} \\ &\quad + (p+1)^2 \int_0^t E[(\lambda_N^*(t))^{2p}] ds \\ &\leq \tilde{C}^{2(p+1)} + \frac{(c\Lambda_1)^{\frac{1}{2}} p\sqrt{t} C(t)^{(2p+1)}}{N} + (\alpha N)^2 C(t)^{2p}. \end{aligned} \tag{4.3.48}$$

Taking $p = \varepsilon N/2$, we deduce that the left side is bounded by $(2C(T))^{\alpha N}$, for all N large. Therefore, by Jensen's inequality, we conclude

$$\Lambda_t(\ell) \leq \Lambda_t(\varepsilon N)^{\frac{\ell}{\varepsilon N}} \leq (2C(T))^\ell \text{ for all } \ell \in [0, \varepsilon N]. \tag{4.3.49}$$

We may now complete the proof of the lemma. For $\delta > 0$ and continuous function q, set

$$q_\delta(x) = q\left(\frac{x}{1+\delta x^2}\right).$$

By Proposition 4.3.10, for any $\delta > 0$, we have

$$\lim_{N\to\infty} \sup_{t\in[0,T]} |\langle q_\delta, L_N(t)\rangle - \langle q_\delta, \mu_t\rangle| = 0. \tag{4.3.50}$$

Further, since the collection of measures μ_t, $t \in [0,T]$, is uniformly compactly supported by the remark following (4.3.42), it follows that

$$\lim_{\delta\to 0} \sup_{t\in[0,T]} |\langle q_\delta, \mu_t\rangle - \langle q, \mu_t\rangle| = 0. \tag{4.3.51}$$

Now, if q is a polynomial of degree p, we find a finite constant C so that

$$|q(x) - q_\delta(x)| \leq C\delta(|x|^{p-1}+1)\frac{|x|^3}{1+\delta x^2} \leq C\delta(|x|^{p+2}+|x|^3).$$

Hence, (4.3.45) shows that, for any $A > 0$,

$$P\left(\sup_{t\in[0,T]} |\langle (q-q_\delta), L_N(t)\rangle| \geq AC\delta\right)$$
$$\leq \frac{1}{A^\ell} E[\sup_{t\in[0,T]} \langle (|x|^{(p+2)} + |x|^3)^\ell, L_N(t)\rangle] \leq \frac{1}{A^\ell}((2C(T))^{(p+2)} + (2C(T))^3)^\ell,$$

for any $\ell \leq \varepsilon N$. By the Borel–Cantelli Lemma, taking $\ell = (\log N)^2$ and A larger than $2C(T)$, we conclude that

$$\limsup_{N\to\infty} \sup_{t\in[0,T]} |\langle (q-q_\delta), L_N(t)\rangle| \leq [(2C(T))^{p+2} + (2C(T))^3]C\delta, \quad \text{a.s.}$$

Together with (4.3.50) and (4.3.51), this yields the almost sure uniform convergence of $\langle q, L_N(t)\rangle$ to $\langle q, \mu_t\rangle$. The proof of the L^p convergence is similar once we have (4.3.45). □

Exercise 4.3.18 Take $\mu_0 = \delta_0$. Show that the empirical measure $L_N(1)$ of the Gaussian (real) Wigner matrices converges almost surely. Show that

$$G_1(z) = \frac{1}{z} - G_1(z)^2$$

and conclude that the limit is the semicircle law, hence giving a new proof of Theorem 2.1.1 for Gaussian entries.
Hint: by the scaling property, show that $G_t(z) = t^{-1/2}G_1(t^{-1/2}z)$ and use Lemma 4.3.25.

Exercise 4.3.19 Using Exercise 4.3.7, extend Corollary 4.3.11 to the symplectic setup ($\beta = 4$).

4.3.3 Dynamical central limit theorems

In this subsection, we study the fluctuations of $(L_N(t))_{t\geq 0}$ on path space. We shall only consider the fluctuations of moments, the generalization to other test functions such as continuously differentiable functions is possible by using concentration inequalities, see Exercise 2.3.7.

We continue in the notation of Subsection 4.3.2. For any n-tuple of polynomial functions $P_1, \ldots, P_n \in \mathbb{C}[X]$ and $(\mu_t)_{t\in[0,T]}$ as in Lemma 4.3.16 with $\mu_0 = \mu$, set

$$G_{N,\mu}(P_1, \ldots, P_n)(t) = N\Big(\langle P_1, L_N(t) - \mu_t\rangle, \ldots, \langle P_n, L_N(t) - \mu_t\rangle\Big).$$

The main result of this subsection is the following.

Theorem 4.3.20 *Let $\beta \geq 1$ and $T < \infty$. Assume that*

$$\tilde{C} = \sup_{N \in \mathbb{N}} \max_{1 \leq i \leq N} |\lambda_i^N(0)| < \infty$$

and that $L_N(0)$ converges towards a probability measure μ in such a way that, for all $p \geq 2$,

$$\sup_{N \in \mathbb{N}} \mathbb{E}[|N(\langle x^n, L_N(0)\rangle - \langle x^n, \mu\rangle)|^p] < \infty.$$

Assume that for any $n \in \mathbb{N}$ and any $P_1, \ldots, P_n \in \mathbb{C}[X]$, $G_{N,\mu}(P_1, \ldots, P_n)(0)$ converges in law towards a random vector $(G(P_1)(0), \ldots, G(P_n)(0))$. Then
(a) there exists a process $(G(P)(t))_{t \in [0,T], P \in \mathbb{C}[X]}$, such that for any polynomial functions $P_1, \ldots, P_n \in \mathbb{C}[X]$, the process $(G_{N,\mu}(P_1, \ldots, P_n)(t))_{t \in [0,T]}$ converges in law towards $(G(P_1)(t), \ldots, G(P_n)(t))_{t \in [0,T]}$;
(b) the limit process $(G(P)(t))_{t \in [0,T], P \in \mathbb{C}[X]}$ is uniquely characterized by the following two properties.
(1) For all $P, Q \in \mathbb{C}[X]$ and $(\lambda, \alpha) \in \mathbb{R}^2$,

$$G(\lambda P + \alpha Q)(t) = \lambda G(P)(t) + \alpha G(Q)(t) \quad \forall t \in [0,T].$$

(2) For any $n \in \mathbb{N}$, $(G(x^n)(t))_{t \in [0,T], n \in \mathbb{N}}$ is the unique solution of the system of equations

$$G(1)(t) = 0, \quad G(x)(t) = G(x)(0) + G_t^1,$$

and, for $n \geq 2$,

$$\begin{aligned} G(x^n)(t) &= G(x^n)(0) + n \int_0^t \sum_{k=0}^{n-2} \mu_s(x^{n-k-2}) G(x^k)(s) ds \\ &\quad + \frac{2-\beta}{2\beta} n(n-1) \int_0^t \mu_s(x^{n-2}) ds + G_t^n, \end{aligned} \qquad (4.3.52)$$

where $(G_t^n)_{t \in [0,T], n \in \mathbb{N}}$ is a centered Gaussian process, independent of $(G(x^n)(0))_{n \in \mathbb{N}}$, such that, if $n_1, n_2 \geq 1$, then for all $s, t \geq 0$,

$$E[G_t^{n_1} G_s^{n_2}] = n_1 n_2 \int_0^{t \wedge s} \mu_u(x^{n_1+n_2-2}) du.$$

Note that a consequence of Theorem 4.3.20 is that if $(G(x^n)(0))_{n \in N}$ is a centered Gaussian process, then so is $(G(x^n)(t))_{t \in [0,T], n \in \mathbb{N}}$.

Proof of Theorem 4.3.20 The idea of the proof is to use (4.3.47) to show that the process $(G^N(x, \ldots, x^n)(t))_{t \in [0,T]}$ is the solution of a stochastic differential system whose martingale terms converge by Rebolledo's Theorem H.14 towards a Gaussian process.

It is enough to prove the theorem with $P_i = x^i$ for $i \in \mathbb{N}$. Set $G_i^N(t) := G^N(x^i)(t) =$

$N\langle x^i, (L_N(t) - \mu_t)\rangle$ to get, using (4.3.47) (which is still valid with obvious modifications if $i = 1$),

$$\begin{aligned} G_i^N(t) &= G_i^N(0) + i\sum_{k=0}^{i-2}\int_0^t G_k^N(s)\mu_s(x^{i-2-k})ds + M_i^N(t) \\ &+ \frac{2-\beta}{2\beta}i(i-1)\int_0^t \langle x^{i-2}, L_N(s)\rangle ds + \frac{i}{2N}\sum_{k=0}^{i-2}\int_0^t G_k^N(s)G_{i-2-k}^N(s)ds, \end{aligned} \quad (4.3.53)$$

where $(M_i^N, i \in \mathbb{N})$ are martingales with bracket

$$\langle M_i^N, M_j^N\rangle_t = \frac{2}{\beta}ij\int_0^t \langle x^{i+j-2}, L_N(s)\rangle ds.$$

(Note that by Lemma 4.3.16, the L^p norm of $\langle M_i^N\rangle$ is finite for all p, and so in particular M_i^N are martingales and not just local martingales.)

By Lemma 4.3.16, for all $t \geq 0$, $\langle M_i^N, M_j^N\rangle_t$ converges in L^2 and almost surely towards $\frac{2}{\beta}ij\int_0^t\langle x^{i+j-2}, \mu_s\rangle ds$. Thus, by Theorem H.14, and with the Gaussian process $(G_t^i)_{t\in[0,T],i\in\mathbb{N}}$ as defined in the theorem, we see that, for all $k \in \mathbb{N}$,

$$\begin{aligned} &(M_k^N(t), \ldots, M_1^N(t))_{t\in[0,T]} \text{ converges in law towards} \\ &\text{the } k\text{-dimensional Gaussian process } (G_t^k, G_t^{k-1}, \ldots, G_t^1)_{t\in[0,T]}. \end{aligned} \quad (4.3.54)$$

Moreover, $(G_t^k, G_t^{k-1}, \ldots, G_t^1)_{t\in[0,T]}$ is independent of $(G(x^n)(0))_{n\in\mathbb{N}}$ since the convergence in (4.3.54) holds given any initial condition such that $L_N(0)$ converges to μ. We next show by induction over p that, for all $q \geq 2$,

$$A_q^p := \max_{i\leq p}\sup_{N\in\mathbb{N}}\mathbb{E}[\sup_{t\in[0,T]}|G_i^N(t)|^q] < \infty. \quad (4.3.55)$$

To begin the induction, note that (4.3.55) holds for $p = 0$ since $G_0^N(t) = 0$. Assume (4.3.55) is verified for polynomials of degree strictly less than p and all q. Recall that, by (4.3.45) of Lemma 4.3.16, for all $q \in \mathbb{N}$,

$$B_q = \sup_{N\in\mathbb{N}}\sup_{t\in[0,T]}\mathbb{E}[\langle |x|^q, L_N(t)\rangle] < \infty. \quad (4.3.56)$$

Set $A_q^p(N,T) := E[\sup_{t\in[0,T]}|G_p^N(t)|^q]$. Using (4.3.56), Jensen's inequality in the form $E(x_1+x_2+x_3)^q \leq 3^{q-1}\sum_{i=1}^3 E|x_i|^q$, and the Burkholder–Davis–Gundy inequality (Theorem H.8), we obtain that, for all $\varepsilon > 0$,

$$\begin{aligned} A_q^p(N,T) &\leq 3^q[A_q^p(N,0) \\ &+ (pT)^q\sum_{k=0}^{p-2}(A_{q(1+\varepsilon)}^k(N,T))^{(1+\varepsilon)^{-1}}B_{(1+\varepsilon)\varepsilon^{-1}(p-2-k)q}^{\frac{\varepsilon}{1+\varepsilon}} \\ &+ (pN^{-1})^q T^{q-1}\Lambda_{q/2}E[\int_0^T \langle x^{2q(p-1)}, L_N(s)\rangle ds]. \end{aligned}$$

By the induction hypothesis ($A^k_{q(1+\varepsilon)}$ is bounded since $k < p$), the fact that we control $A^p_q(N,0)$ by hypothesis and the finiteness of B_q for all q, we conclude also that $A^p_q(N,T)$ is bounded uniformly in N for all $q \in \mathbb{N}$. This completes the induction and proves (4.3.55).

Set next, for $i \in \mathbb{N}$,

$$\varepsilon_N(i)(s) := iN^{-1} \sum_{k=0}^{i-2} \int_0^t G_k^N(s) G_{i-2-k}^N(s) ds .$$

Since

$$\sup_{s\in[0,T]} E[\varepsilon_N(i)(s)^q] \le N^{-q} i^{2q} (A^p_{2q})^{\frac{1}{2}} T ,$$

we conclude from (4.3.55) and the Borel–Cantelli Lemma that

$$\varepsilon_N(i)(\cdot) \to_{N\to\infty} 0, \quad \text{in all } L^q,\ q \ge 2, \text{ and a.s.} \tag{4.3.57}$$

Setting

$$Y_i^N(t) = G_i^N(t) - G_i^N(0) - i \sum_{k=0}^{i-2} \int_0^t G_k^N(s) \langle x^{i-2-k}, \mu_s \rangle ds,$$

for all $t \in [0,T]$, we conclude from (4.3.53), (4.3.54) and (4.3.57) that the processes $(Y_i^N(t), Y_{i-1}^N(t), \ldots, Y_1^N(t))_{t\ge 0}$ converge in law towards the centered Gaussian process $(G^i(t), \ldots, G^1(t))_{t\ge 0}$.

To conclude, we need to deduce the convergence in law of the G^Ns from that of the Y^Ns. But this is clear again by induction; G_1^N is uniquely determined from Y_1^N and $G_1^N(0)$, and so the convergence in law of Y_1^N implies that of G_1^N since $G_1^N(0)$ converges in law. By induction, if we assume the convergence in law of $(G_k^N, k \le p-2)$, we deduce that of G_{p-1}^N and G_p^N from the convergence in law of Y_p^N and Y_{p-1}^N. □

Exercise 4.3.21 Recover the results of Section 2.1.7 in the case of Gaussian Wigner matrices. by taking $X^{N,\beta}(0) = 0$, with $\mu_0 = 0$ and $G(x^n)(0) = 0$. Note that $m^n(t) := EG(x^n)(t) = t^{n/2} m^n(1)$ may not vanish.

Exercise 4.3.22 In each part of this exercise, check that the given initial data $X^N(0)$ fulfills the hypotheses of Theorem 4.3.20. (a) Let $X^N(0)$ be a diagonal matrix with entries on the diagonal $(\phi(\frac{i}{N}), 1 \le i \le N)$, with ϕ a continuously differentiable function on $[0,1]$. Show that

$$\mu_0(f) = \int_0^1 f(\phi(x)) dx, \quad G(x^p)(0) = \frac{1}{2}[\phi(1)^p - \phi(0)^p] \text{ for all } p,$$

and that $(G(x^p)(0), p \geq 0)$ are deterministic.
(b) Let $X^{N,\beta}(0)$ be a finite rank diagonal matrix, i.e. for some k fixed independently of N, $X_0^N = \mathrm{diag}(\eta_1, \ldots, \eta_k, 0, \ldots, 0)$, with the η_i's uniformly bounded. Check that

$$\mu_0 = \delta_0, \, G(x^p)(0) = \sum_{l=1}^{k} \eta_l^p \text{ for all } p,$$

and that $G(x^p)(0)$ is random if the η_is are.
(c) Let $X^{N,\beta}(0)$ be a diagonal matrix with entries $X^N(0)(ii) = \eta_i/\sqrt{N}$ for $1 \leq i \leq N$, with some i.i.d. centered bounded random variables η_i. Check that

$$\mu_0(f) = \delta_0, \quad G(x^p)(0) = 0 \text{ if } p \neq 1$$

but $G(x)(0)$ is a standard Gaussian variable.

4.3.4 Large deviation bounds

Fix $T \in \mathbb{R}_+$. We discuss in this subsection the derivation of large deviation estimates for the measure-valued process $\{L_N(t)\}_{t\in[0,T]}$. We will only derive exponential upper bounds, and refer the reader to the bibliographical notes for information on complementary lower bounds, applications and relations to spherical integrals.

We begin by introducing a candidate for a rate function on $C([0,T], M_1(\mathbb{R}))$. For any $f, g \in C_b^{2,1}(\mathbb{R}\times[0,T])$, $s \leq t \in [0,T]$ and $\nu_. \in C([0,T], M_1(\mathbb{R}))$, set

$$\begin{aligned} S^{s,t}(\nu, f) &= \int f(x,t) d\nu_t(x) - \int f(x,s) d\nu_s(x) \\ &\quad - \int_s^t \int \partial_u f(x,u) d\nu_u(x) du \\ &\quad - \frac{1}{2} \int_s^t \iint \frac{\partial_x f(x,u) - \partial_x f(y,u)}{x-y} d\nu_u(x) d\nu_u(y) du, \end{aligned} \quad (4.3.58)$$

$$\langle f, g \rangle_\nu^{s,t} = \int_s^t \int \partial_x f(x,u) \partial_x g(x,u) d\nu_u(x) du \quad (4.3.59)$$

and

$$\bar{S}^{s,t}(\nu, f) = S^{s,t}(\nu, f) - \frac{1}{2} \langle f, f \rangle_{s,t}^\nu . \quad (4.3.60)$$

Set, for any probability measure $\mu \in M_1(\mathbb{R})$,

$$S_\mu(\nu) := \begin{cases} +\infty, & \text{if } \nu_0 \neq \mu, \\ S^{0,T}(\nu) := \sup_{f \in C_b^{2,1}(\mathbb{R}\times[0,T])} \sup_{0\leq s\leq t\leq T} \bar{S}^{s,t}(\nu, f), & \text{otherwise.} \end{cases}$$

We now show that $S_\mu(\cdot)$ is a candidate for rate function, and that a large deviation upper bound holds with it.

Proposition 4.3.23 *(a) For any $\mu \in M_1(\mathbb{R})$, $S_\mu(\cdot)$ is a good rate function on $C([0,T], M_1(\mathbb{R}))$, that is, $\{\nu \in C([0,T], M_1(\mathbb{R})); S_\mu(\nu) \le M\}$ is compact for any $M \in \mathbb{R}^+$.*
(b) With assumptions as in Proposition 4.3.10, the sequence $(L_N(t))_{t\in[0,T]}$ satisfies a large deviation upper bound of speed N^2 and good rate function S_μ, that is, for all closed subsets F of $C([0,T], M_1(\mathbb{R}))$,

$$\limsup_{N\to\infty} \frac{1}{N^2} \log P(L_N(\cdot) \in F) \le -\inf_F S_\mu .$$

We note in passing that, since $S_\mu(\cdot)$ is a good rate function, the process $(L_N(t))_{t\in[0,T]}$ concentrates on the set $\{\nu. : S_\mu(\nu) = 0\}$. Exercise 4.3.25 below establishes that the latter set consists of a singleton, the solution of (4.3.25).

The proof of Proposition 4.3.23 is based on Itô's calculus and the introduction of exponential martingales. We first need to improve Lemma 4.3.14 in order to obtain exponential tightness.

Lemma 4.3.24 *Assume* (4.3.23). *Let $T \in \mathbb{R}^+$. Then, there exists $a(T) > 0$ and $M(T), C(T) < \infty$ so that:*
(a) for $M \ge M(T)$,

$$P\left(\sup_{t\in[0,T]} \langle \log(x^2+1), L_N(t)\rangle \ge M\right) \le C(T) e^{-a(T)MN^2} ;$$

(b) for any $L \in \mathbb{N}$, there exists a compact set $\mathscr{K}(L) \subset C([0,T], M_1(\mathbb{R}))$ so that

$$P(L_N(\cdot) \in \mathscr{K}(L)^c) \le e^{-N^2 L} .$$

It follows in particular from the second part of Lemma 4.3.24 that the sequence $(L_N(t), t \in [0,T])$ is almost surely pre-compact in $C([0,T], M_1(\mathbb{R}))$; compare with Lemma 4.3.14.

Proof The proof proceeds as in Lemma 4.3.14. Set first $f(x) = \log(x^2+1)$. Recalling (4.3.29) and Corollary H.13, we then obtain that, for all $L \ge 0$,

$$P\left(\sup_{s\le T} |M_f^N(s)| \ge L\right) \le 2e^{-\frac{\beta N^2 L^2}{16T}} ,$$

which combined with (4.3.29) yields the first part of the lemma.

For the second part of the lemma, we proceed similarly, by first noticing that if $f \in C^2(\mathbb{R})$ is bounded, together with its first and second derivatives, by 1, then from Corollary H.13 and (4.3.33) we have that

$$\sup_{i\delta \le s \le (i+1)\delta} |\langle f, L_N(s) - L_N(t_i)\rangle| \le 2\delta + \varepsilon ,$$

with probability greater than $1 - 2e^{-\frac{\beta N^2(\varepsilon)^2}{16\delta}}$. Using the compact sets $\mathscr{K} = \mathscr{K}_M$ of $C([0,T], M_1(\mathbb{R}))$ as in (4.3.36) with $\varepsilon_k = 1/kM(\|f_k\|_\infty + \|f_k'\|_\infty + \|f_k''\|_\infty)$, we then conclude that

$$P(L_N \notin \mathscr{K}_M) \le 2e^{-c_M N^2} ,$$

with $c_M \rightarrow_{M\rightarrow\infty} \infty$. Adjusting $M = M(L)$ completes the proof. □

Proof of Proposition 4.3.23 We first prove that $S_\mu(\cdot)$ is a good rate function. Then we obtain a weak large deviation upper bound, which gives, by the exponential tightness proved in the Lemma 4.3.24, the full large deviation upper bound.

(a) Observe first that, from Riesz' Theorem (Theorem B.11), $S_\mu(\nu)$ is also given, when $\nu_0 = \mu$, by

$$S_{\mu_D}(\nu) = \frac{1}{2} \sup_{f \in C_b^{2,1}(\mathbb{R}\times[0,T])} \sup_{0\le s\le t\le T} \frac{S^{s,t}(\nu, f)^2}{\langle f, f\rangle_\nu^{s,t}} . \tag{4.3.61}$$

Consequently, S_μ is nonnegative. Moreover, S_μ is obviously lower semicontinuous as a supremum of continuous functions. Hence, we merely need to check that its level sets are contained in relatively compact sets. By Lemma 4.3.13, it is enough to show that, for any $M > 0$:

(1) for any integer m, there is a positive real number L_m^M so that for any $\nu \in \{S_{\mu_D} \le M\}$,

$$\sup_{0\le s\le T} \nu_s(|x| \ge L_m^M) \le \frac{1}{m}, \tag{4.3.62}$$

proving that $\nu_s \in K_{L^M}$ for all $s \in [0,T]$;

(2) for any integer m and $f \in C_b^2(\mathbb{R})$, there exists a positive real number δ_m^M so that for any $\nu \in \{S_\mu(\cdot) \le M\}$,

$$\sup_{|t-s|\le \delta_m^M} |\nu_t(f) - \nu_s(f)| \le \frac{1}{m}, \tag{4.3.63}$$

showing that $s \rightarrow \nu_s(f) \in C_{\delta^M, \|f\|_\infty}$.

To prove (4.3.62), we consider, for $\delta > 0$, $f_\delta(x) = \log\left(x^2(1+\delta x^2)^{-1} + 1\right) \in C_b^{2,1}(\mathbb{R}\times[0,T])$. We observe that

$$C := \sup_{0<\delta\le 1} \|\partial_x f_\delta\|_\infty + \sup_{0<\delta\le 1} \|\partial_x^2 f_\delta\|_\infty$$

is finite and, for $\delta \in (0,1]$,

$$\left|\frac{\partial_x f_\delta(x) - \partial_x f_\delta(y)}{x-y}\right| \le C.$$

Hence, (4.3.61) implies, by taking $f = f_\delta$ in the supremum, that, for any $\delta \in (0,1]$, any $t \in [0,T]$, any $\mu_. \in \{S_{\mu_D} \le M\}$,

$$\mu_t(f_\delta) \le \mu_0(f_\delta) + 2Ct + 2C\sqrt{Mt}.$$

Consequently, we deduce by the monotone convergence theorem and letting δ decrease to zero that for any $\mu_. \in \{S_\mu(\cdot) \le M\}$,

$$\sup_{t\in[0,T]} \mu_t(\log(x^2+1)) \le \langle \mu, \log(x^2+1)\rangle + 2C(1+\sqrt{M}).$$

Chebyshev's inequality and (4.3.23) thus imply that for any $\mu_. \in \{S_\mu(\cdot) \le M\}$ and any $K \in \mathbb{R}^+$,

$$\sup_{t\in[0,T]} \mu_t(|x| \ge K) \le \frac{C_D + 2C(1+\sqrt{M})}{\log(K^2+1)},$$

which finishes the proof of (4.3.62).

The proof of (4.3.63) again relies on (4.3.61) which implies that for any $f \in C_b^2(\mathbb{R})$, any $\mu_. \in \{S_\mu(\cdot) \le M\}$ and any $0 \le s \le t \le T$,

$$|\langle f, \mu_t - \mu_s\rangle| \le \|f''\|_\infty |t-s| + 2\|f'\|_\infty \sqrt{M}\sqrt{|t-s|}. \tag{4.3.64}$$

We turn next to establishing the weak large deviation upper bound. Pick $\nu \in C([0,T], M_1(\mathbb{R}))$ and $f \in C^{2,1}([0,T]\times\mathbb{R})$. By Lemma 4.3.12, for any $s \ge 0$, the process $\{S^{s,t}(L_N, f), t \ge s\}$ is a martingale for the filtration of the Brownian motion W, which is equal to $\sqrt{2/\beta}N^{-3/2}\sum_{i=1}^N \int_s^t f'(\lambda_i^N(u))dW_u^i$. Its bracket is $\langle f, f\rangle_{L_N}^{s,t}$. As f' is uniformly bounded, we can apply Theorem H.10 to deduce that the process $\{M_N(L_N, f)(t), t \ge s\}$ is a martingale if for $\mu \in C([0,T], M_1(\mathbb{R}))$ we denote

$$M_N(\mu, f)(t) := \exp\{N^2 S^{s,t}(\mu, f) - \frac{N^2}{2}\langle f, f\rangle_\mu^{s,t} + N\varepsilon(f)_\mu^{s,t}\}$$

with

$$\varepsilon(f)_\mu^{s,t} := (\frac{1}{\beta} - \frac{1}{2})\int_s^t \int \partial_x^2 f(s,x) d\mu(x) du.$$

Moreover, $\mu \in C([0,T], M_1(\mathbb{R})) \to \bar{S}^{s,t}(\mu, f) := S^{s,t}(\mu, f) - \frac{1}{2}\langle f, f\rangle_\mu^{s,t}$ is continuous as f and its two first derivatives are bounded continuous whereas the function $\mu \mapsto \int_s^t \int \partial_x^2 f(s,x) d\mu(x) du$ is uniformly bounded by $T\|\partial_x^2 f\|_\infty$. Therefore, if we pick δ small enough so that $\bar{S}^{s,t}(.,f)$ varies by at most $\varepsilon > 0$ on the ball (for some metric d compatible with the weak topology on $C([0,T],M_1(\mathbb{R})))$ of radius δ around ν, we obtain, for all $s \le t \le T$,

$$\begin{aligned} P(d(L_N,\nu)<\delta) &= E[\frac{M_N(L_N,f)(t)}{M_N(L_N,f)(t)}1_{d(L_N,\nu)<\delta}] \\ &\le e^{N^2\varepsilon+N\|f''\|_\infty-N^2\bar{S}^{s,t}(\nu,f)}E[M_N(L_N,f)(t)1_{d(L_N,\nu)<\delta}] \\ &\le e^{N^2\varepsilon+N\|f''\|_\infty-N^2\bar{S}^{s,t}(\nu,f)}E[M_N(L_N,f)(t)] \\ &= e^{N^2\varepsilon+N\|f''\|_\infty-N^2\bar{S}^{s,t}(\nu,f)}, \end{aligned}$$

where we finally used the fact that $E[M_N(L_N,f)(t)] = E[M_N(L_N,f)(s)] = 1$ since the process $\{M_N(L_N,f)(t), t \ge s\}$ is a martingale. Hence,

$$\lim_{\delta\to 0}\lim_{N\to\infty}\frac{1}{N^2}\log P(d(L_N,\nu)<\delta) \le -\bar{S}^{s,t}(\nu,f)$$

for any $f \in C^{2,1}([0,T]\times\mathbb{R})$. Optimizing over f gives

$$\lim_{\delta\to 0}\lim_{N\to\infty}\frac{1}{N^2}\log P(d(L_N,\nu)<\delta) \le -S^{0,1}(\nu,f).$$

Since $L_N(0)$ is deterministic and converges to μ_A, if $\nu_0 \ne \mu_A$,

$$\lim_{\delta\to 0}\lim_{N\to\infty}\frac{1}{N^2}\log P(d(L_N,\nu)<\delta) = -\infty,$$

which allows us to conclude that

$$\lim_{\delta\to 0}\lim_{N\to\infty}\frac{1}{N^2}\log P(d(L_N,\nu)<\delta) \le -S_{\mu_A}(\nu,f).$$

□

Exercise 4.3.25 In this exercise, you prove that the set $\{\nu_\cdot : S_\mu(\nu) = 0\}$ consists of the unique solution of (4.3.25).
(a) By applying Riesz' Theorem, show that

$$S^{0,T}(\nu) := \sup_{f\in C_b^{2,1}(\mathbb{R}\times[0,T])}\sup_{0\le s\le t\le T}\frac{S^{s,t}(\nu,f)^2}{2\langle f,f\rangle^{s,t}}.$$

(b) Show that $S_\mu(\nu_\cdot) = 0$ iff $\nu_0 = \mu$ and $S^{s,t}(\nu,f) = 0$ for all $0 \le s \le t \le T$ and all $f \in C_b^{2,1}(\mathbb{R}\times[0,T])$. Take $f(x) = (z-x)^{-1}$ to conclude.

4.4 Concentration of measure and random matrices

We have already seen in Section 2.3 that the phenomenon of concentration of measure can be useful in the study of random matrices. In this section, we further expand on this theme by developing both concentration techniques and their applications to random matrices. To do so we follow each of two well-established routes. Taking the first route, we consider functionals of the empirical measure of a matrix as functions of the underlying entries. When enough independence is present, and for functionals that are smooth enough (typically, Lipschitz), concentration inequalities for product measures can be applied. Taking the second route, which applies to situations in which random matrix entries are no longer independent, we view ensembles of matrices as manifolds equipped with probability measures. When the manifold satisfies appropriate curvature constraints, and the measure satisfies coercivity assumptions, semigroup techniques can be invoked to prove concentration of measure results.

4.4.1 Concentration inequalities for Hermitian matrices with independent entries

We begin by considering Hermitian matrices X_N whose entries on-and-above the diagonal are independent (but not necessarily identically distributed) random variables. We will mainly be concerned with concentration inequalities for the random variable $\mathrm{tr} f(X_N)$, which is a Lipschitz function of the entries of X_N, see Lemma 2.3.1.

Remark 4.4.1 Wishart matrices, as well as matrices of the form $Y_N T_N Y_N^*$ with T_N diagonal and deterministic, and $Y_N \in \mathrm{Mat}_{M\times N}$ possessing independent entries, can be easily treated by the techniques of this section. For example, to treat Wishart matrices, fix $N \le M$ positive integers, and define the matrix $X_N \in \mathrm{Mat}_{N+M}$,

$$X_N = \begin{pmatrix} 0 & Y_N \\ Y_N^* & 0 \end{pmatrix}.$$

Now $(X_N)^2$ equals

$$\begin{pmatrix} Y_N Y_N^* & 0 \\ 0 & Y_N^* Y_N \end{pmatrix}$$

and therefore, for any continuous function f,

$$\mathrm{tr}(f(X_N^2)) = 2\mathrm{tr}(f(Y_N Y_N^*)) + (M-N)f(0).$$

Hence, concentration results for linear functionals of the empirical measure of the singular values of Y_N can be deduced from such results for the eigenvalues of X_N. For an example, see Exercise 4.4.9.

Entries satisfying Poincaré's inequality

Our first goal is to extend the concentration inequalities, Lemma 2.3.3 and Theorem 2.3.5, to Hermitian matrices whose independent entries satisfy a weaker condition than the LSI, namely to matrices whose entries satisfy a Poincaré type inequality.

Definition 4.4.2 (Poincaré inequality) A probability measure P on $\mathbb{R}^M$ satisfies the *Poincaré inequality* (PI) with constant $m > 0$ if, for all continuously differentiable functions f,

$$\mathrm{Var}_P(f) := E_P\left(|f(x) - E_P(f(x))|^2\right) \le \frac{1}{m} E_P(|\nabla f|^2).$$

It is not hard to check that if P satisfies an LSI with constant c, then it satisfies a PI with constant $m \ge c^{-1}$, see [GuZ03, Theorem 4.9]. However, there are probability measures which satisfy the PI but not the LSI such as $Z^{-1}e^{-|x|^a}dx$ for $a \in (1,2)$. Further, like the LSI, the PI tensorizes: if P satisfies the PI with constant m, $P^{\otimes M}$ also satisfies the PI with constant m for any $M \in \mathbb{N}$, see [GuZ03, Theorem 2.5]. Finally, if for some uniformly bounded function V we set $P_V = Z^{-1}e^{V(x)}dP(x)$, then P_V also satisfies the PI with constant bounded below by $e^{-\sup V + \inf V} m$, see [GuZ03, Property 2.6].

As we now show, probability measures on $\mathbb{R}^M$ satisfying the PI have sub-exponential tails.

Lemma 4.4.3 *Assume that P satisfies the PI on $\mathbb{R}^M$ with constant m. Then, for any differentiable function G on $\mathbb{R}^M$, for $|t| \le \sqrt{m}/\sqrt{2}\| \|\nabla G\|_2\|_\infty$,*

$$E_P(e^{t(G - E_P(G))}) \le K, \tag{4.4.1}$$

with $K = -\sum_{i\ge 0} 2^i \log(1 - 2^{-1}4^{-i})$. Consequently, for all $\delta > 0$,

$$P(|G - E_P(G)| \ge \delta) \le 2K e^{-\frac{\sqrt{m}}{\sqrt{2}\| \|\nabla G\|_2\|_\infty}\delta}. \tag{4.4.2}$$

Proof With G as in the statement, for $t^2 < m/\| \|\nabla G\|_2^2\|_\infty$, set $f = e^{tG}$ and note that

$$E_P(e^{2tG}) - \left(E_P(e^{tG})\right)^2 \le \frac{t^2}{m} \| \|\nabla G\|_2^2\|_\infty E_P(e^{2tG})$$

so that

$$E_P(e^{2tG}) \le \left(1 - \frac{t^2}{m\| \|\nabla G\|_2^2\|_\infty}\right)^{-1} \left(E_P(e^{tG})\right)^2.$$

Iterating, we deduce that

$$\log E_P(e^{2tG}) \leq -\sum_{i=0}^{n} 2^i \log(1 - \frac{4^{-i}t^2}{m} \| \|\nabla G\|_2^2 \|_\infty) + 2^{n+1} \log E_P(e^{2^{-n}tG}).$$

Since

$$\lim_{n\to\infty} 2^{n+1} \log E_P(e^{2^{-n}tG}) = 2tE_P(G)$$

and

$$D_t := -\sum_{i=0}^{\infty} 2^i \log(1 - \frac{4^{-i}t^2}{m} \| \|\nabla G\|_2^2 \|_\infty) < \infty$$

increases with $|t|$, we conclude that with $t_0 = \sqrt{m}/\sqrt{2}\| \|\nabla G\|_2\|_\infty$,

$$E_P(e^{2t_0(G-E_P(G))}) \leq D_{t_0} = K.$$

The estimate (4.4.2) then follows by Chebyshev's inequality. □

We can immediately apply this result in the context of large random matrices. Consider Hermitian matrices such that the laws of the independent entries $\{X_N(i,j)\}_{1\leq i\leq j\leq N}$ all satisfy the PI (over $\mathbb{R}$ or $\mathbb{R}^2$) with constant bounded below by Nm. Note that, as for the LSI, if P satisfies the PI with constant m, the law of ax under P satisfies it also with a constant bounded by a^2m^{-1}, so that our hypothesis includes the case where $X_N(i,j) = a_N(i,j)Y_N(i,j)$ with $Y_N(i,j)$ i.i.d. of law P satisfying the PI and $a(i,j)$ deterministic and uniformly bounded.

Corollary 4.4.4 *Under the preceding assumptions, there exists a universal constant $C > 0$ such that, for any differentiable function f, and any $\delta > 0$,*

$$P(|\mathrm{tr}(f(X_N)) - E[\mathrm{tr}(f(X_N))]| \geq \delta N) \leq Ce^{-\frac{\sqrt{Nm}}{C\| \|\nabla f\|_2\|_\infty}\delta}.$$

Exercise 4.4.5 Using an approximation argument similar to that employed in the proof of Herbst's Lemma 2.3.3, show that the conclusions of Lemma 4.4.3 and Corollary 4.4.4 remain true if G is only assumed Lipschitz continuous, with $|G|_{\mathscr{L}}$ replacing $\| \|\nabla G\|_2\|_\infty$.

Exercise 4.4.6 Let $\gamma(dx) = (2\pi)^{-1/2}e^{-\frac{x^2}{2}}dx$ be the standard Gaussian measure. Show that γ satisfies the Poincaré inequality with constant one, by following the following approaches.

- Use Lemma 2.3.2.
- Use the interpolation

$$\gamma((f-\gamma(f))^2) = -\int_0^1 \partial_\alpha \int \left(\int f(\sqrt{\alpha}x + \sqrt{1-\alpha}y)d\gamma(y)\right)^2 d\gamma(x)d\alpha,$$

integration by parts, the Cauchy–Schwarz inequality and the fact that, for any $\alpha \in [0,1]$, the law of $\sqrt{\alpha}x + \sqrt{1-\alpha}y$ is γ under $\gamma \otimes \gamma$.

Exercise 4.4.7 [GuZ03, Theorem 2.5] Show that the PI tensorizes: if P satisfies the PI with constant m then $P^{\otimes M}$ also satisfies the PI with constant m for any $M \in \mathbb{N}$.

Exercise 4.4.8 [GuZ03, Theorem 4.9] Show that if P satisfies an LSI with constant c, then it satisfies a PI with constant $m \geq c^{-1}$. *Hint*: Use the LSI with $f = 1 + \varepsilon g$ and $\varepsilon \to 0$.

Exercise 4.4.9 Show that Corollary 4.4.4 extends to the setup of singular values of the Wishart matrices introduced in Exercise 2.1.18. That is, in the setup described there, assume the entries $Y_N(i,j)$ satisfy the PI with constant bounded below by Nm, and set $X_N = (Y_N Y_N^T)^{1/2}$. Prove that, for a universal constant C, and all $\delta > 0$,

$$P\left(|\mathrm{tr}(f(X_N)) - E[\mathrm{tr}(f(X_N))]| \geq \delta(M+N)\right) \leq Ce^{-\frac{\sqrt{Nm}}{C\|\|\nabla f\|_2\|_\infty}\delta}.$$

Matrices with bounded entries and Talagrand's method

Recall that the *median* M_Y of a random variable Y is defined as the largest real number such that $P(Y \leq x) \leq 2^{-1}$. The following is an easy consequence of a theorem due to Talagrand, see [Tal96, Theorem 6.6].

Theorem 4.4.10 (Talagrand) *Let K be a convex compact subset of $\mathbb{R}$ with diameter $|K| = \sup_{x,y\in K}|x-y|$. Consider a convex real-valued function f defined on K^M. Assume that f is Lipschitz on K^M, with constant $|f|_{\mathscr{L}}$. Let P be a probability measure on K and let $X_1,\dots,X_M$ be independent random variables with law P. Then, if M_f is the median of $f(X_1,\dots,X_M)$, for all $\delta > 0$,*

$$P\left(|f(X_1,\dots,X_M) - M_f| \geq \delta\right) \leq 4e^{-\frac{\delta^2}{16|K|^2|f|_{\mathscr{L}}^2}}.$$

Under the hypotheses of Theorem 4.4.10,

$$\begin{aligned} E[|f(X_1,\dots,X_M) - M_f|] &= \int_0^\infty P\left(|f(X_1,\dots,X_M) - M_f| \geq t\right) dt \\ &\leq 4\int_0^\infty e^{-\frac{t^2}{16|K|^2|f|_{\mathscr{L}}^2}} dt = 16|K||f|_{\mathscr{L}}. \end{aligned}$$

Hence we obtain as an immediate corollary of Theorem 4.4.10 the following.

Corollary 4.4.11 *Under the hypotheses of Theorem 4.4.10, for all $t \in \mathbb{R}^+$,*

$$P(|f(X_1,\ldots,X_M) - E[f(X_1,\ldots,X_M)]| \geq (t+16)|K||f|_{\mathscr{L}}) \leq 4e^{-\frac{t^2}{16}}.$$

In order to apply Corollary 4.4.11 in the context of (Hermitian) random matrices X_N, we need to identify convex functions of the entries. Since

$$\lambda_1(X_N) = \sup_{v\in\mathbb{C}^N, |v|_2=1} \langle v, X_N v\rangle,$$

it is obvious that the top eigenvalue of a Hermitian matrix is a convex function of the real and imaginary parts of the entries. Somewhat more surprisingly, so is the trace of a convex function of the matrix.

Lemma 4.4.12 (Klein's Lemma) *Suppose that f is a real-valued convex function on $\mathbb{R}$. Then the function $X \mapsto \operatorname{tr} f(X)$ on the vector space $\mathscr{H}_N^{(2)}$ of N-by-N Hermitian matrices is convex.*

For f twice-differentiable and f'' bounded away from 0 we actually prove a sharper result, see (4.4.3) below.

Proof We denote by X (resp. Y) an $N \times N$ Hermitian matrix with eigenvalues $(x_i)_{1\leq i\leq N}$ (resp. $(y_i)_{1\leq i\leq N}$) and eigenvectors $(\xi_i)_{1\leq i\leq N}$ (resp. $(\zeta_i)_{1\leq i\leq N}$). Assume at first that f is twice continuously differentiable, and consider the Taylor remainder $R_f(x,y) = f(x) - f(y) - (x-y)f'(y)$. Since

$$f'' \geq c \geq 0$$

for some constant c, we have $R_f(x,y) \geq \frac{c}{2}(x-y)^2 = R_{\frac{c}{2}x^2}(x,y)$. Consider also the matrix $R_f(X,Y) = f(X) - f(Y) - (X-Y)f'(Y)$, noting that $\operatorname{tr}(R_{\frac{c}{2}x^2}(X,Y)) = \operatorname{tr}(\frac{c}{2}(X-Y)^2)$. For $i \in \{1,\ldots,N\}$, with $c_{ij} = |\langle \xi_i, \eta_j\rangle|^2$, and with summations on $j \in \{1,\ldots,N\}$, we have

$$\begin{aligned} \langle \xi_i, R_f(X,Y)\xi_i\rangle &= f(x_i) + \sum_j (-c_{ij} f(y_j) - x_i c_{ij} f'(y_j) + c_{ij} y_j f'(y_j)) \\ &= \sum_j c_{ij} R_f(x_i, y_j) \geq \sum_j c_{ij} R_{\frac{c}{2}x^2}(x_i, y_j), \end{aligned}$$

where at the middle step we use the fact that $\sum_j c_{ij} = 1$. After summing on $i \in \{1,\ldots,N\}$ we have

$$\operatorname{tr}(f(X) - f(Y) - (X-Y)f'(Y)) \geq \frac{c}{2}\operatorname{tr}(X-Y)^2 \geq 0. \qquad (4.4.3)$$

Now take successively $(X,Y) = (A,(A+B)/2),(B,(A+B)/2)$. After summing the resulting inequalities, we have for arbitrary $A,B \in \mathscr{H}_n^{(2)}$ that

$$\mathrm{tr}\left(f(\frac{1}{2}A+\frac{1}{2}B)\right) \le \frac{1}{2}\mathrm{tr}\,(f(A)) + \frac{1}{2}\mathrm{tr}\,(f(B))\,.$$

The result follows for general convex functions f by approximations. □

We can now apply Corollary 4.4.11 and Lemma 4.4.12 to the function $f(\{X_N(i,j)\}_{1\le i\le j\le N}) = \mathrm{tr}(f(X_N))$ to obtain the following.

Theorem 4.4.13 *Let $(P_{i,j}, i \le j)$ and $(Q_{i,j}, i < j)$ be probability measures supported on a convex compact subset K of $\mathbb{R}$. Let X_N be a Hermitian matrix, such that $\Re X_N(i,j)$, $i \le j$, is distributed according to $P_{i,j}$, and $\Im X_N(i,j)$, $i < j$, is distributed according to $Q_{i,j}$, and such that all these random variables are independent. Fix $\delta_1(N) = 8|K|\sqrt{\pi}a/N$. Then, for any $\delta \ge 4\sqrt{|K|\delta_1(N)}$, and any convex Lipschitz function f on $\mathbb{R}$,*

$$\begin{aligned} &P^N\left(|\mathrm{tr}(f(X_N)) - E^N[\mathrm{tr}(f(X_N))]| \ge N\delta\right) \\ &\quad \le \frac{32|K|}{\delta}\exp\left(-N^2\frac{1}{16|K|^2a^2}[\frac{\delta^2}{16|K||f|_{\mathscr{L}}^2} - \delta_1(N)]\right). \end{aligned} \tag{4.4.4}$$

4.4.2 Concentration inequalities for matrices with dependent entries

We develop next an approach to concentration inequalities based on semigroup theory. When working on $\mathbb{R}^m$, this approach is related to concentration inequalities for product measures, and in particular to the LSI. However, its great advantage is that it also applies to manifolds, through the Bakry–Emery criterion.

Our general setup will be concerned with a manifold M equipped with a measure μ. We will consider either $M = \mathbb{R}^m$ or M compact.

The setup with $M = \mathbb{R}^m$ and μ =Lebesgue measure

Let Φ be a smooth function from $\mathbb{R}^m$ into $\mathbb{R}$, with fast enough growth at infinity such that the measure

$$\mu_\Phi(dx) := \frac{1}{Z}e^{-\Phi(x_1,\dots,x_m)}dx_1\cdots dx_m$$

is a well defined probability measure. (Further assumptions of Φ will be imposed below.) We consider the operator $\mathscr{L}_\Phi$ on twice continuously differentiable functions defined by

$$\mathscr{L}_\Phi = \Delta - (\nabla\Phi)\cdot\nabla = \sum_{i=1}^{m}[\partial_i^2 - (\partial_i\Phi)\partial_i]\,.$$

Then, integrating by parts, we see that $\mathscr{L}_\Phi$ is symmetric in $L^2(\mu_\Phi)$, that is, for any compactly supported smooth functions f,g,

$$\int (f\mathscr{L}_\Phi g)\,d\mu_\Phi = \int (g\mathscr{L}_\Phi f)\,d\mu_\Phi\,.$$

In the rest of this section, we will use the notation $\mu_\Phi f = \int f d\mu_\Phi$.

Let $\mathscr{B}$ denote a Banach space of real functions on M, equipped with a partial order $<$, that contains $C_b(M)$, the Banach space of continuous functions on M equipped with the uniform norm, with the latter being dense in $\mathscr{B}$. We will be concerned in the sequel with $\mathscr{B} = L^2(\mu_\Phi)$.

Definition 4.4.14 A collection of operators $(P_t)_{t\geq 0}$ with $P_t : \mathscr{B}\to\mathscr{B}$ is a *Markov semigroup* with *infinitesimal generator* $\mathscr{L}$ if the following hold.
(i) $P_0 f = f$ for all $f \in \mathscr{B}$.
(ii) The map $t\to P_t$ is continuous in the sense that for all $f \in \mathscr{B}$, $t\to P_t f$ is a continuous map from $\mathbb{R}_+$ into $\mathscr{B}$.
(iii) For any $f \in \mathscr{B}$ and $(t,s) \in \mathbb{R}_+^2$, $P_{t+s}f = P_t P_s f$.
(iv) $P_t 1 = 1$ for $t \geq 0$, and P_t preserves positivity: for each $f \geq 0$ and $t \geq 0$, $P_t f \geq 0$.
(v) For any function f for which the limit exists,

$$\mathscr{L}(f) = \lim_{t\downarrow 0} t^{-1}(P_t f - f)\,. \tag{4.4.5}$$

The collection of functions for which the right side of (4.4.5) exists is the *domain* of $\mathscr{L}$, and is denoted $\mathscr{D}(\mathscr{L})$.

Property (iv) implies in particular that $\|P_t f\|_\infty \leq \|f\|_\infty$. Furthermore, P_t is reversible in $L^2(\mu_\Phi)$, i.e., $\mu_\Phi(fP_t g) = \mu_\Phi(gP_t f)$ for any smooth functions f,g. In particular, μ_Φ is invariant under P_t: that is, $\mu_\Phi P_t = \mu_\Phi$. It also follows immediately from the definition that, for any $f \in \mathscr{D}(\mathscr{L})$ and $t \geq 0$,

$$f \in \mathscr{D}(\mathscr{L}) \Rightarrow P_t f \in \mathscr{D}(\mathscr{L}),\ \mathscr{L}P_t f = P_t\mathscr{L}f\,. \tag{4.4.6}$$

In what follows we will be interested in the case where $\mathscr{L} = \mathscr{L}_\Phi$, at least as operators on a large enough class of functions. We introduce a family of bilinear forms Γ_n on smooth functions by setting $\Gamma_0(f,g) = fg$ and, for $n \geq 1$,

$$\Gamma_n(f,g) = \frac{1}{2}\left(\mathscr{L}_\Phi\Gamma_{n-1}(f,g) - \Gamma_{n-1}(f,\mathscr{L}_\Phi g) - \Gamma_{n-1}(g,\mathscr{L}_\Phi f)\right)\,.$$

We will only be interested in the cases $n = 1,2$. Thus, the *carré du champ operator* Γ_1 satisfies

$$\Gamma_1(f,g) = \frac{1}{2}\left(\mathscr{L}_\Phi fg - f\mathscr{L}_\Phi g - g\mathscr{L}_\Phi f\right), \tag{4.4.7}$$

and the *carré du champ itéré operator* Γ_2 satisfies

$$\Gamma_2(f,f) = \frac{1}{2}\{\mathscr{L}_\Phi \Gamma_1(f,f) - 2\Gamma_1(f,\mathscr{L}_\Phi f)\}. \tag{4.4.8}$$

We often write $\Gamma_i(f)$ for $\Gamma_i(f,f)$, $i = 1,2$. Simple algebra shows that $\Gamma_1(f) = \sum_{i=1}^m (\partial_i f)^2$, and

$$\Gamma_2(f,f) = \sum_{i,j=1}^m (\partial_i\partial_j f)^2 + \sum_{i,j=1}^m \partial_i f \mathrm{Hess}(\Phi)_{ij}\partial_j f, \tag{4.4.9}$$

with $\mathrm{Hess}(\Phi)_{ij} = \mathrm{Hess}(\Phi)_{ji} = \partial_i\partial_j\Phi$ the Hessian of Φ.

Remark 4.4.15 We introduced the forms $\Gamma_n(f,f)$ in a purely formal way. To motivate, note that, assuming all differentiation and limits can be taken as written, one has

$$\begin{aligned}\Gamma_n(f,g) &= \frac{1}{2}\frac{d}{dt}\left(P_t(\Gamma_{n-1}(f,g)) - \Gamma_{n-1}(P_t f, P_t g)\right)|_{t=0} \\ &= \frac{1}{2}\left(\mathscr{L}_\Phi \Gamma_{n-1}(f,g) - \Gamma_{n-1}(f,\mathscr{L}_\Phi g) - \Gamma_{n-1}(g,\mathscr{L}_\Phi f)\right).\end{aligned} \tag{4.4.10}$$

We will see below in Lemma 4.4.22 that indeed these manipulations are justified when f,g are sufficiently smooth.

Definition 4.4.16 We say that the *Bakry–Emery condition* (denoted BE) is satisfied if there exists a positive constant $c > 0$ such that

$$\Gamma_2(f,f) \geq \frac{1}{c}\Gamma_1(f,f) \tag{4.4.11}$$

for any smooth function f.

Note (by taking $f = \sum a_i x_i$ with a_i arbitrary constants) that the BE condition is equivalent to

$$\mathrm{Hess}(\Phi)(x) \geq \frac{1}{c}I \text{ for all } x \in \mathbb{R}^m, \text{ in the sense of the partial order on positive definite matrices}. \tag{4.4.12}$$

Theorem 4.4.17 *Assume that* $\Phi \in C^2(\mathbb{R}^m)$ *and that the BE condition* (4.4.12) *holds. Then, μ_Φ satisfies the logarithmic Sobolev inequality with constant c, that is, for any $f \in L^2(\mu_\Phi)$,*

$$\int f^2 \log \frac{f^2}{\int f^2 d\mu_\Phi} d\mu_\Phi \leq 2c \int \Gamma_1(f,f) d\mu_\Phi . \tag{4.4.13}$$

In the sequel, we let $C^\infty_{\text{poly}}(\mathbb{R}^m)$ denote the subset of $C^\infty(\mathbb{R}^m)$ that consists of functions all of whose derivatives have polynomial growth at infinity. The proof of Theorem 4.4.17 is based on the following result which requires stronger assumptions.

Theorem 4.4.18 *Assume the BE condition* (4.4.12). *Further assume that $\Phi \in C^\infty_{\text{poly}}(\mathbb{R}^m)$. Then μ_Φ satisfies the logarithmic Sobolev inequality with constant c.*

From Theorem 4.4.17, (4.4.9) and Lemma 2.3.3 of Section 2.3, we immediately get the following.

Corollary 4.4.19 *Under the hypotheses of Theorem 4.4.17,*

$$\mu_\Phi\left(|G - \int G(x)\mu_\Phi(dx)| \geq \delta\right) \leq 2e^{-\delta^2/2c|G|^2_{\mathscr{L}}} . \tag{4.4.14}$$

Proof of Theorem 4.4.17 (with Theorem 4.4.18 granted). Fix $\varepsilon > 0, M > 1$, and set $B(0,M) = \{x \in \mathbb{R}^m : \|x\|_2 \leq M\}$. We will construct below approximations of Φ by functions $\Phi_{M,\varepsilon} \in C^\infty_{\text{poly}}(\mathbb{R}^m)$ with the following properties:

$$\begin{aligned} \sup_{x \in B(0,M)} |\Phi_{M,\varepsilon}(x) - \Phi(x)| &\leq \varepsilon , \\ \text{Hess}(\Phi_M) &\geq \frac{1}{c+\varepsilon} I \text{ uniformly} . \end{aligned} \tag{4.4.15}$$

With such a construction, $\mu_{\Phi_{M,\varepsilon}}$ converges weakly (as M tends to infinity and ε tends to 0) toward μ_Φ, by bounded convergence. Further, by Theorem 4.4.18, for any M, ε as above, $\mu_{\Phi_{M,\varepsilon}}$ satisfies (4.4.13) with the constant $c + \varepsilon > 0$. For f^2 smooth, bounded below by a strictly positive constant, and constant outside a compact set, we deduce that μ_Φ satisfies (4.4.13) by letting M go to infinity and ε go to zero in this family of inequalities. We then obtain the bound (4.4.13) for all functions $f \in L^2(\mu_\Phi)$ with $\int \Gamma_1(f,f) d\mu_\Phi < \infty$ by density.

So it remains to construct a family $\Phi_{M,\varepsilon}$ satisfying (4.4.15). For $\delta > 0$, we let P_δ be a polynomial approximation of Φ on $B(0,2M)$ such that

$$\sup_{x \in B(0,2M)} \|\text{Hess}(P_\delta)(x) - \text{Hess}(\Phi)(x)\|_\infty < \frac{\delta}{4}, \; P_\delta(0) = \Phi(0), \; \nabla P_\delta(0) = \nabla\Phi(0)$$

with $\|\cdot\|_\infty$ the operator norm on $\mathrm{Mat}_m(\mathbb{R})$. Such an approximation exists by Weierstrass' Theorem. Note that

$$\begin{aligned}&\sup_{x\in B(0,2M)}|P_\delta(x)-\Phi(x)|\\ &\le \sup_{x\in B(0,2M)}\left|\int_0^1 \alpha d\alpha\langle x,(\mathrm{Hess}(P_\delta)(\alpha x)-\mathrm{Hess}(\Phi)(\alpha x))x\rangle\right|\le\frac{\delta M^2}{2}. \quad (4.4.16)\end{aligned}$$

With $c_\delta^{-1}=c^{-1}-\frac{\delta}{4}>0$ for δ small, note that $\mathrm{Hess}(P_\delta)(x)\ge c_\delta^{-1}I$ on $B(0,2M)$ and define $\tilde{P}_\delta$ as the function on $\mathbb{R}^m$ given by

$$\tilde{P}_\delta(x)=\sup_{y\in B(0,2M)}\left\{P_\delta(y)+\nabla P_\delta(y)\cdot(x-y)+\frac{1}{2c_\delta}\|x-y\|_2^2\right\}.$$

Note that $\tilde{P}_\delta=P_\delta$ on $B(0,2M)$ whereas $\mathrm{Hess}(\tilde{P}_\delta)\ge c_\delta^{-1}I$ almost everywhere since the map

$$x\to\sup_{y\in B(0,2M)}\left\{P_\delta(y)+\nabla P_\delta(y)\cdot(x-y)+\frac{1}{2c_\delta}\|x-y\|^2\right\}-\frac{1}{2c_\delta}\|x\|^2$$

is convex as a supremum of convex functions (and thus its Hessian, which is almost everywhere well defined, is nonnegative). Finally, to define a $C^\infty_{\mathrm{poly}}(\mathbb{R}^m)$-valued function we put, for some small t,

$$\Phi_{\delta,t}(x)=\int\tilde{P}_\delta(x+tz)d\mu(z)$$

with μ the standard centered Gaussian law. By (4.4.16) and since $\tilde{P}_\delta=P_\delta$ on $B(0,M)$, we obtain for $x\in B(0,M)$,

$$\begin{aligned}\Delta_M(\delta,t) &:= \sup_{x\in B(0,M)}|\Phi_{\delta,t}(x)-\Phi(x)|\\ &\le \sup_{x\in B(0,M)}\int|\tilde{P}_\delta(x+tz)-\tilde{P}_\delta(x)|d\mu(z)+\frac{\delta M^2}{2}.\end{aligned}$$

Thus, $\Delta_M(\delta,t)$ vanishes when δ and t go to zero and we choose these two parameters so that it is bounded by ε. Moreover, $\Phi_{\delta,t}$ belongs to $C^\infty_{\mathrm{poly}}(\mathbb{R}^m)$ since the density of the Gaussian law is C^∞ and $\tilde{P}_\delta$ has at most a quadratic growth at infinity. Finally, since $\mathrm{Hess}(\tilde{P}_\delta)\ge c_\delta^{-1}I$ almost everywhere, $\mathrm{Hess}\Phi_{\delta,t}\ge c_\delta^{-1}I$ everywhere. To conclude, we choose δ small enough so that $c_\delta\le c+\varepsilon$. □

Our proof of Theorem 4.4.18 proceeds via the introduction of the semigroup P_t associated with $\mathscr{L}_\Phi$ through the solution of the stochastic differential equation

$$dX_t^x=-\nabla\Phi(X_t^x)dt+\sqrt{2}dw_t\,,X_0^x=x, \qquad (4.4.17)$$

where w_t is an m-dimensional Brownian motion. We first verify the properties of the solutions of (4.4.17), and then deduce in Lemma 4.4.20 some analytical properties of the semigroup. The proof of Theorem 4.4.18 follows these preliminary steps.

Lemma 4.4.20 *With assumptions as in Theorem 4.4.18, for any $x \in \mathbb{R}^m$, the solution of* (4.4.17) *exists for all $t \in \mathbb{R}_+$. Further, the formula*

$$P_t f(x) = E(f(X_t^x)) \tag{4.4.18}$$

determines a Markov semigroup on $\mathscr{B} = L^2(\mu_\Phi)$, with infinitesimal generator $\mathscr{L}$ so that $\mathscr{D}(\mathscr{L})$ contains $C^\infty_{\text{poly}}(\mathbb{R}^m)$, and $\mathscr{L}$ coincides with $\mathscr{L}_\Phi$ on $C^\infty_{\text{poly}}(\mathbb{R}^m)$.

Proof Since the second derivatives of Φ are locally bounded, the coefficients of (4.4.17) are locally Lipschitz, and the solution exists and is unique up to (possibly) an explosion time. We now show that no explosion occurs, in a way similar to our analysis in Lemma 4.3.3. Let $T_n = \inf\{t : |X_t^x| > n\}$. Itô's Lemma and the inequality $x \cdot \nabla\Phi(x) \geq |x|^2/c - c'$ for some constant $c' > 0$ (consequence of (4.4.12)) imply that

$$\begin{aligned} E(|X^x_{t\wedge T_n}|^2) &= x^2 - E\left(\int_0^{t\wedge T_n} X_s \cdot \nabla\Phi(X_s)ds\right) + 2E(t\wedge T_n) \\ &\leq x^2 + \frac{1}{c}E\left(\int_0^{t\wedge T_n} |X_s|^2 ds\right) + (2+c')E(t\wedge T_n). \end{aligned} \tag{4.4.19}$$

Gronwall's Lemma then yields that

$$E(|X^x_{t\wedge T_n}|^2) \leq (x^2 + (2+c')t)e^{t/c}.$$

Since the right side of the last estimate does not depend on n, it follows from Fatou's Theorem that the probability that explosion occurs in finite time vanishes. That (4.4.18) determines a Markov semigroup is then immediate (note that P_t is a contraction on $L^2(\mu_\Phi)$ by virtue of Jensen's inequality).

To analyze the infinitesimal generator of P_t, we again use Itô's Lemma. First note that (4.4.19) implies that $E^x|X_t|^2 \leq C(t)(x^2+1)$ for some locally bounded $C(t)$. Repeating the same computation (with the function $|X^x_{t\wedge T_n}|^{2p}$, p positive integer) yields that $E^x|X_t|^{2p} \leq C(t,p)(x^{2p}+1)$. For $f \in C^\infty_{\text{poly}}(\mathbb{R}^m)$, we then get that

$$f(X^x_{t\wedge T_n}) - f(x) = \int_0^{t\wedge T_n} \mathscr{L}_\Phi f(X_s^x)ds + \int_0^{t\wedge T_n} g(X_s^x)dw_s, \tag{4.4.20}$$

where the function g has polynomial growth at infinity and thus, in particular,

$$E(\sup_{t\leq 1}(\int_0^t g(X_s^x)dw_s)^2) < \infty.$$

Arguing similarly with the term containing $\mathscr{L}_\Phi f(X_s^x)$, we conclude that all terms in (4.4.20) are uniformly integrable. Taking $n \to \infty$ and using the fact that $T_n \to \infty$ together with the above uniform integrability yields that

$$E(f(X_t^x)) - f(x) = E\int_0^t \mathscr{L}_\Phi f(X_s^x)ds\,.$$

Taking the limit as $t \to 0$ (and using again the uniform integrability together with the continuity $X_s^x \to_{s\to 0} x$ a.s.) completes the proof that $C^\infty_{\text{poly}}(\mathbb{R}^m) \subset \mathscr{D}(\mathscr{L})$. □

Remark 4.4.21 In fact, $\mathscr{D}(\mathscr{L})$ can be explicitly characterized: it is the subset of $L^2(\mu_\Phi)$ consisting of functions f that are locally in the Sobolev space $W^{2,2}$ and such that $\mathscr{L}_\Phi f \in L^2(\mu_\Phi)$ in the sense of distributions (see [Roy07, Theorem 2.2.27]). In the interest of providing a self-contained proof, we do not use this fact.

An important analytical consequence of Lemma 4.4.20 is the following.

Lemma 4.4.22 *With assumptions as in Theorem 4.4.18, we have the following.*
(i) If f is a Lipschitz(1) function on $\mathbb{R}^m$, then $P_t f$ is a Lipschitz($e^{-2t/c}$) function for all $t \in \mathbb{R}_+$.
(ii) If $f \in C_b^\infty(\mathbb{R}^m)$, then $P_t f \in C^\infty_{\text{poly}}(\mathbb{R}^m)$.
(iii) If $f, g \in C^\infty_{\text{poly}}(\mathbb{R}^m)$, then the equality (4.4.10) *with $n = 2$ holds.*

Proof (i) By applying Itô's Lemma we obtain that

$$\frac{d}{dt}|X_t^x - X_t^y|^2 = -2(X_t^x - X_t^y)(\nabla\Phi(X_t^x) - \nabla\Phi(X_t^y)) \le -\frac{2}{c}|X_t^x - X_t^y|^2\,.$$

In particular, $|X_t^x - X_t^y| \le |x-y|e^{-2t/c}$, and thus for f Lipschitz with Lipschitz constant equal to 1, we have $|f(X_t^x) - f(X_t^y)| \le |x-y|e^{-2t/c}$. Taking expectations completes the proof.
(ii) Since $f \in C^\infty_{\text{poly}}(\mathbb{R}^m)$, we have that $f \in \mathscr{D}(\mathscr{L})$ and $\mathscr{L}f = \mathscr{L}_\Phi f$. Therefore, also $P_t f \in \mathscr{D}(\mathscr{L})$, and $\mathscr{L}_\Phi P_t f = P_t \mathscr{L}_\Phi f \in L^2(\mu_\Phi)$ (since $\mathscr{L}_\Phi f \in L^2(\mu_\Phi)$ and P_t is a contraction on $L^2(\mu_\Phi)$). By part (i) of the lemma, $|\nabla P_t f|$ is uniformly bounded and, by assumption, $|\nabla\Phi|$ has at most polynomial growth. It follows that $\Delta P_t f$, which exists everywhere, satisfies

$$\Delta P_t f = g_t\,,$$

where the function $g_t \in L^2(\mu_\Phi)$ has at most polynomial growth at infinity. Standard estimates for the solutions of uniformly elliptic equations (for this version, see [GiT98, Theorem 4.8]) then imply that $P_t f \in C^\infty_{\text{poly}}(\mathbb{R}^m)$.
(iii) By assumption, $f, g \in C^\infty_{\text{poly}}(\mathbb{R}^m)$. Thus $\Gamma_1(f,g) \in C^\infty_{\text{poly}}(\mathbb{R}^m)$ and, in particular,

by Lemma 4.4.20, belongs to $\mathscr{D}(\mathscr{L})$ and so does $P_t\Gamma_1(f,g)$. The rest follows from the definitions. □

Proof of Theorem 4.4.18 Let h be a positive bounded continuous function so that $\int h d\mu_\Phi = 1$. We begin by proving that P_t is ergodic in the sense that

$$\lim_{t\to\infty} \mu_\Phi(P_t h - \mu_\Phi h)^2 = 0. \tag{4.4.21}$$

A direct proof can be given based on part (i) of Lemma 4.4.22. Instead, we present a slightly longer proof that allows us to derive useful intermediate estimates.

We first note that we can localize (4.4.21): because $P_t 1 = 1$ and $P_t f \geq 0$ for f positive continuous, it is enough to prove (4.4.21) for $h \in C_b(\mathbb{R}^m)$ that is compactly supported. Because $C_b^\infty(K)$ is dense in $C(K)$ for any compact K, it is enough to prove (4.4.21) for $h \in C_b^\infty(\mathbb{R}^m)$. To prepare for what follows, we will prove (4.4.21) for a function h satisfying $h = \phi(P_\theta g)$ for some $g \in C_b^\infty$, $\theta \geq 0$, and ϕ that is infinitely differentiable with bounded derivatives on the range of g (the immediate interest is with $\theta = 0$, $\phi(x) = x$).

Set $h_t = P_t h$ and for $s \in [0,t]$, define $\psi(s) = P_s\Gamma_1(h_{t-s}, h_{t-s})$. By part (ii) of Lemma 4.4.22, $\Gamma_1(h_{t-s}, h_{t-s}) \in \mathscr{D}(\mathscr{L})$. Therefore,

$$\frac{d}{ds}\psi(s) = 2P_s\Gamma_2(P_{t-s}h, P_{t-s}h) \geq \frac{2}{c}P_s\Gamma_1(P_{t-s}h, P_{t-s}h) = \frac{2}{c}\psi(s),$$

where we use the BE condition in the inequality. In particular,

$$\|\nabla h_t\|_2^2 = \Gamma_1(h_t, h_t) = \psi(0) \leq e^{-2t/c}\psi(t) = e^{-2t/c}P_t\Gamma_1(h,h). \tag{4.4.22}$$

The expression $\|\Gamma_1(h_t,h_t)\|_\infty$ converges to 0 as $t \to \infty$ because $\Gamma_1(h,h) = \|\nabla h\|_2^2$ is uniformly bounded. Further, since for any $x, y \in \mathbb{R}^m$,

$$\begin{aligned} |h_t(x) - h_t(y)| &= \left| \int_0^1 \langle \nabla h_t(\alpha x + (1-\alpha)y), (x-y)\rangle d\alpha \right| \\ &\leq \|x-y\|_2 \cdot \| \|\nabla h_t\|_2 \|_\infty \leq \|x-y\|_2 e^{-t/c} \| \|\nabla h\|_2 \|_\infty , \end{aligned}$$

it follows that $h_t(\cdot) - \mu_\Phi(h_t)$ converges almost everywhere to zero. Since $\mu_\Phi(h_t) = \mu_\Phi(h)$, we conclude that h_t converges almost everywhere and in $L^2(\mu_\Phi)$ to $\mu_\Phi(h)$, yielding (4.4.21).

We now prove Theorem 4.4.18 for $f^2 = h \in C_b^\infty$ that is uniformly bounded below by a strictly positive constant. Set

$$S_f(t) = \int (h_t \log h_t) d\mu_\Phi .$$

Since $h_t \log h_t$ is uniformly bounded and $h_t \in P_t C_b^\infty(\mathbb{R}^m)$, we have by (4.4.21) that

$S_h(t)$ converges to 0 as $t \to \infty$. Hence

$$S_f(0) = -\int_0^\infty dt \frac{d}{dt} S_f(t) = \int_0^\infty dt \int \Gamma_1(h_t, \log h_t) d\mu_\Phi \,, \tag{4.4.23}$$

where, in the second equality, we used (4.4.7) and the fact that $\int \mathscr{L}_\Phi(g) d\mu_\Phi = 0$ for any $g \in C^\infty_{\text{poly}}(\mathbb{R}^m)$ and, in particular, for $g = h_t \log h_t$.

Next, using the fact that P_t is symmetric together with the Cauchy–Schwarz inequality, we get

$$\begin{aligned}\int \Gamma_1(h_t, \log h_t) d\mu_\Phi &= \int \Gamma_1 (h, P_t(\log h_t))\, d\mu_\Phi \\ &\le \left(\int \frac{\Gamma_1(h,h)}{h} d\mu_\Phi\right)^{\frac{1}{2}} \left(\int h\Gamma_1(P_t \log h_t, P_t \log h_t) d\mu_\Phi\right)^{\frac{1}{2}}. \end{aligned} \tag{4.4.24}$$

Now, applying (4.4.22) with the function $\log h_t$ (note that since h_t is bounded below uniformly away from 0, $\log(\cdot)$ is indeed smooth on the range of h_t), we obtain

$$\begin{aligned}\int h\Gamma_1(P_t \log h_t, P_t \log h_t) d\mu_\Phi &\le \int h e^{-\frac{2}{c}t} P_t \Gamma_1(\log h_t, \log h_t) d\mu_\Phi \\ &= e^{-\frac{2}{c}t} \int h_t \Gamma_1(\log h_t, \log h_t) d\mu_\Phi = e^{-\frac{2}{c}t} \int \Gamma_1(h_t, \log h_t) d\mu_\Phi \,, \end{aligned} \tag{4.4.25}$$

where in the last equality we have used symmetry of the semigroup and the Leibniz rule for Γ_1. The inequalities (4.4.24) and (4.4.25) imply the bound

$$\int \Gamma_1(h_t, \log h_t) d\mu_\Phi \le e^{-\frac{2}{c}t} \int \frac{\Gamma_1(h,h)}{h} d\mu_\Phi = 4e^{-\frac{2}{c}t} \int \Gamma_1(h^{\frac{1}{2}}, h^{\frac{1}{2}}) d\mu_\Phi \,. \tag{4.4.26}$$

Using this, one arrives at

$$S_f(0) \le \int_0^\infty 4e^{-\frac{2t}{c}} dt \int \Gamma_1(h^{\frac{1}{2}}, h^{\frac{1}{2}}) d\mu_\Phi = 2c \int \Gamma_1(f,f) d\mu_\Phi \,,$$

which completes the proof of (4.4.13) when $f \in C_b^\infty$ is strictly bounded below.

To consider $f \in C_b^\infty$, apply the inequality (4.4.13) to the function $f_\varepsilon^2 = f^2 + \varepsilon$, noting that $\Gamma_1(f_\varepsilon, f_\varepsilon) \le \Gamma_1(f,f)$, and use monotone convergence. Another use of localization and dominated convergence is used to complete the proof for arbitrary $f \in L^2(\mu_\Phi)$ with $\Gamma_1(f,f) < \infty$. □

The setup with M a compact Riemannian manifold

We now consider the version of Corollary 4.4.19 applying to the setting of a compact connected manifold M of dimension m equipped with a Riemannian metric g and volume measure μ, see Appendix F for the notions employed.

We let Φ be a smooth function on M and define

$$\mu_\Phi(dx) = \frac{1}{Z} e^{-\Phi(x)} d\mu(x)$$

as well as the operator $\mathscr{L}_\Phi$ such that for all smooth functions $h, f \in C^\infty(M)$,

$$\mu_\Phi(f \mathscr{L}_\Phi h) = \mu_\Phi(h \mathscr{L}_\Phi f) = -\int_M g(\operatorname{grad} f, \operatorname{grad} h) d\mu_\Phi .$$

We have, for all $f \in C^\infty(M)$,

$$\mathscr{L}_\Phi f = \Delta f - g(\operatorname{grad} \Phi, \operatorname{grad} f),$$

where Δ is the Laplace–Beltrami operator. In terms of a local orthonormal frame $\{L_i\}$, we can rewrite the above as

$$\mathscr{L}_\Phi = \sum_i (L_i^2 - \nabla_{L_i} L_i - (L_i \Phi) L_i),$$

where ∇ is the Levi–Civita connection.

Remark 4.4.23 For the reader familiar with such language, we note that, in local coordinates,

$$\mathscr{L}_\Phi = \sum_{i,j=1}^m g^{ij} \partial_i \partial_j + \sum_{i=1}^m b_i^\Phi \partial_i$$

with

$$b_i^\Phi(x) = e^{\Phi(x)} \sum_j \partial_j \left(e^{-\Phi(x)} \sqrt{\det(g_x)} g_x^{ij} \right) .$$

We will not need to use this formula.

Given $f, h \in C^\infty(M)$ we define $\langle \operatorname{Hess} f, \operatorname{Hess} h \rangle \in C^\infty(M)$ by requiring that

$$\langle \operatorname{Hess} f, \operatorname{Hess} h \rangle = \sum_{i,j} (\operatorname{Hess} f)(L_i, L_j)(\operatorname{Hess} h)(L_i, L_j)$$

for all local orthonormal frames $\{L_i\}$.

We define Γ_n, for $n \geq 0$, as in (4.4.10). In particular, Γ_1 and Γ_2 are given by (4.4.7) and (4.4.8). We have $\Gamma_1(f,h) = g(\operatorname{grad} f, \operatorname{grad} h)$ or equivalently

$$\Gamma_1(f,h) = \sum_i (L_i f)(L_i h)$$

in terms of a local orthonormal frame $\{L_i\}$. The latter expression for Γ_1 may be verified by a straightforward manipulation of differential operators. The expression for Γ_2 is more complicated and involves derivatives of the metric g, reflecting the fact that the Levi–Civita connection does not preserve the Lie bracket. In other words, the curvature intervenes, as follows.

Lemma 4.4.24 (Bochner–Bakry–Emery)

$$\Gamma_2(f,f) = \langle \mathrm{Hess}\, f, \mathrm{Hess}\, f\rangle + (\mathrm{Ric} + \mathrm{Hess}\,\Phi)(\mathrm{grad}\, f, \mathrm{grad}\, f).$$

(See Appendix F for the definition of the Ricci tensor $\mathrm{Ric}(\cdot,\cdot)$.)

Proof Fix $p \in M$ arbitrarily and let $|_p$ denote evaluation at p. Let $L_1,\ldots,L_m$ be an orthonormal frame defined near $p \in M$. Write $\nabla_{L_i} L_j = \sum_k C^k_{ij} L_k$, where $C^k_{ij} = g(\nabla_{L_i} L_j, L_k)$. We assume that the frame $\{L_i\}$ is geodesic at p, see Definition F.26. After exploiting the simplifications made possible by use of a geodesic frame, it will be enough to prove that

$$\begin{aligned}\Gamma_2(f,f)|_p &= \sum_{ij}\left((L_iL_jf)^2 + L_iL_j\Phi)(L_if)(L_jf)\right)|_p \\ &\quad + \sum_{i,j,k}((L_iC^j_{kk} - L_kC^i_{jk})(L_if)(L_jf))|_p. \end{aligned} \tag{4.4.27}$$

To abbreviate write $A_i = L_i\Phi + \sum_k C^i_{kk}$. By definition, and after some trivial manipulations of differential operators, we have

$$\begin{aligned}\Gamma_2(f,f) &= \sum_{i,j}\left(\frac{1}{2}((L_i^2 - A_iL_i)f)(L_jf)^2 - (L_j(L_i^2 - A_iL_i)f)(L_jf)\right) \\ &= \sum_{i,j}((L_iL_jf)^2 + ([L_i,L_j]L_i + L_i[L_i,L_j] + [L_j,A_iL_i])f)(L_jf)).\end{aligned}$$

We have $[L_i,L_j] = \sum_k (C^k_{ij} - C^k_{ji})L_k$ because ∇ is torsion-free. We also have $[L_i,L_j]|_p = 0$ because $\{L_i\}$ is geodesic at p. It follows that

$$\begin{aligned}[L_i,L_j]L_if|_p &= 0, \\ L_i[L_i,L_j]f|_p &= \sum_k (L_iC^k_{ij} - L_iC^k_{ji})(L_kf)|_p, \\ ([L_j,A_iL_i]f)(L_jf)|_p &= \sum_k (L_jC^i_{kk} + L_jL_i\Phi)(L_if)(L_jf)|_p.\end{aligned}$$

We have $g(\nabla_{L_i}L_j, L_k) + g(L_j, \nabla_{L_i}L_k) = C^k_{ij} + C^j_{ik} = 0$ by orthonormality of $\{L_i\}$ and thus

$$\sum_{i,j}(L_i[L_i,L_j]f)(L_jf)|_p = -\sum_{i,j,k}(L_iC^k_{ji})(L_kf)(L_jf)|_p.$$

Therefore, after some relabeling of dummy indices, we can see that equation (4.4.27) holds. □

Rerunning the proofs of Theorem 4.4.18 and Lemma 2.3.3 (this time, not worrying about explosions, since the process lives on a compact manifold, and replacing throughout the space $C^\infty_{\mathrm{poly}}(\mathbb{R}^m)$ by $C^\infty_b(M)$), we deduce from Lemma 4.4.24 the following.

Corollary 4.4.25 *If for all $x \in M$ and $v \in \mathbb{T}_x M$,*

$$(\mathrm{Ric} + \mathrm{Hess}\,\Phi)_x(v,v) \geq c^{-1} g_x(v,v),$$

then μ_Φ satisfies the LSI (4.4.13) with constant c and, further, for any differentiable function G on M,

$$\mu_\Phi\left(|G - \int G(x)\mu_\Phi(dx)| \geq \delta\right) \leq 2e^{-\delta^2/2cE_{\mu_\Phi}\Gamma_1(G,G)}. \qquad (4.4.28)$$

Applications to random matrices

We begin by applying, in the setup $M = \mathbb{R}^m$ and μ =Lebesgue measure, the general concentration inequality of Corollary 4.4.19. For $X_N \in \mathscr{H}_N^{(\beta)}$ we write

$$d^\beta X_N = \prod_{i<j} dX_N(i,j) \prod_i dX_N(i,i),$$

for the product Lebesgue measure on the entries on-and-above the diagonal of X_N, where the Lebesgue measure on $\mathbb{C}$ is taken as the product of the Lebesgue measure on the real and imaginary parts.

Proposition 4.4.26 *Let $V \in C^\infty_{\mathrm{poly}}(\mathbb{R})$ be a strictly convex function satisfying $V''(x) \geq cI$ for all $x \in \mathbb{R}$ and some $c > 0$. Let $\beta = 1$ or $\beta = 2$, and suppose X_N^V is a random matrix distributed according to the probability measure*

$$\frac{1}{Z_N^V} e^{-N\mathrm{tr}(V(X_N))} d^\beta X_N.$$

Let P_N^V denote the law of the eigenvalues $(\lambda_1, \ldots, \lambda_N)$ of X_N^V. Then, for any Lipschitz function $f : \mathbb{R}^N \to \mathbb{R}$,

$$P_N^V\left(|f(\lambda_1, \ldots, \lambda_N) - P_N^V f| > \delta\right) \leq e^{-\frac{Nc\delta^2}{2|f|^2_{\mathscr{L}}}}.$$

Note that if $f(\lambda_1, \ldots, \lambda_N) = \frac{1}{N}\sum_{i=1}^N g(\lambda_i)$, then $|f|_{\mathscr{L}} = \sqrt{2N}^{-1}|g|_{\mathscr{L}}$.

Proof Take $m = N(N-1)\beta/2 + N$. Let $h : \mathscr{H}_N^{(\beta)} \to \mathbb{R}^m$ denote the one-to-one and onto mapping as defined in the beginning of Section 2.5.1, and let $\tilde{V}$ be the function on $\mathbb{R}^m$ defined by $\mathrm{tr}\,V(X) = \tilde{V}(h(X))$. Note that $\mathrm{tr}X^2 \geq \|h(X)\|^2$. For $X, Y \in \mathscr{H}_N^{(\beta)}$ we have

$$\mathrm{tr}\,(V(X) - V(Y) - (X-Y)V'(Y)) \geq \frac{c}{2}\|h(X) - h(Y)\|^2$$

by (4.4.3), and hence $\mathrm{Hess}\,\tilde{V} \geq cI_m$. Now the function f gives rise to a function $\tilde{f}(X) = f(\lambda_1, \ldots, \lambda_n)$ on $\mathbb{R}^m$, where the λ_i are the eigenvalues of $h^{-1}(X)$. By

Lemma 2.3.1, the Lipschitz constants of $\tilde{f}$ and f coincide. Applying Corollary 4.4.19 yields the proposition. □

We next apply, in the setup of compact Riemannian manifolds, the general concentration inequality of Corollary 4.4.25. We study concentration on orthogonal and unitary groups. We let $O(N)$ denote the N-by-N orthogonal group and $U(N)$ denote the N-by-N unitary group. (In the notation of Appendix E, $O(N) = \mathrm{U}_N(\mathbb{R})$ and $U(N) = \mathrm{U}_n(\mathbb{C})$.) We let $SU(N) = \{X \in U(N) : \det X = 1\}$ and $SO(N) = O(N) \cap SU(N)$. All the groups $O(N)$, $SO(N)$, $U(N)$ and $SU(N)$ are manifolds embedded in $\mathrm{Mat}_N(\mathbb{C})$. We consider each of these manifolds to be equipped with the Riemannian metric it inherits from $\mathrm{Mat}_N(\mathbb{C})$, the latter equipped with the inner product $X \cdot Y = \mathrm{tr} XY^*$. It is our aim is to get concentration results for $O(N)$ and $U(N)$ by applying Corollary 4.4.25 to $SO(N)$ and $SU(N)$.

We introduce some general notation. Given a compact group G, let ν_G denote the unique Haar probability measure on G. Given a compact Riemannian manifold M with metric g, and $f \in C^\infty(M)$, let $|f|_{\mathscr{L},M}$ be the maximum achieved by $g(\mathrm{grad} f, \mathrm{grad} f)^{1/2}$ on M.

Although we are primarily interested in $SO(N)$ and $SU(N)$, in the following result, for completeness, we consider also the Lie group $USp(N) = \mathrm{U}_N(\mathbb{H}) \subset \mathrm{Mat}_N(\mathbb{H})$.

Theorem 4.4.27 (Gromov) *Let $\beta \in \{1,2,4\}$. Let*

$$G_N = SO(N), SU(N), USp(N)$$

according as $\beta = 1,2,4$. Then, for all $f \in C^\infty(G_N)$ and $\delta \geq 0$, we have

$$\nu_{G_N}\left(|f - \nu_{G_N} f| \geq \delta\right) \leq 2e^{-\frac{\left(\frac{\beta(N+2)}{4}-1\right)\delta^2}{2|f|^2_{\mathscr{L},G_N}}}. \qquad (4.4.29)$$

Proof Recall from Appendix F, see (F.6), that the Ricci curvature of G_N is given by

$$\mathrm{Ric}_x(G_N)(X,X) = \left(\frac{\beta(N+2)}{4} - 1\right) g_x(X,X) \qquad (4.4.30)$$

for $x \in G_N$ and $X \in \mathbb{T}_x(G_N)$. Consider now the specialization of Corollary 4.4.25 to the following case:

- $M = G_N$, which is a connected manifold;
- $g =$ the Riemannian metric inherited from $\mathrm{Mat}_N(\mathbb{F})$, with $\mathbb{F} = \mathbb{R}, \mathbb{C}, \mathbb{H}$ according as $\beta = 1,2,4$;
- $\mu =$ the volume measure on M corresponding to g;

- $\Phi \equiv 0$ and (hence) $\mu_\Phi = \nu_{G_N}$.

Then the corollary yields the theorem. □

We next deduce a corollary with an elementary character which does not make reference to differential geometry.

Corollary 4.4.28 *Let $\beta \in \{1,2\}$. Let $G_N = O(N), U(N)$, according as $\beta = 1,2$. Put $SG_N = \{X \in G_N : \det X = 1\}$. Let f be a continuous real-valued function on G_N which, for some constant C and all $X,Y \in G_N$, satisfies*

$$|f(X) - f(Y)| \leq C \operatorname{tr}((X-Y)(X-Y)^*)^{1/2}. \tag{4.4.31}$$

Then we have

$$\sup_{X \in G_N} |\nu_{G_N} f - \int f(YX) d\nu_{SG_N}(Y)| \leq 2C, \tag{4.4.32}$$

and furthermore

$$\nu_{G_N}\left(|f(\cdot) - \int f(Y\cdot) d\nu_{SG_N}(Y)| \geq \delta\right) \leq 2e^{-\frac{\left(\frac{\beta(N+2)}{4}-1\right)\delta^2}{2C^2}} \tag{4.4.33}$$

for all $\delta > 0$.

For the proof we need a lemma which records some group-theoretical tricks. We continue in the setting of Corollary 4.4.28.

Lemma 4.4.29 *Let $H_N \subset G_N$ be the subgroup consisting of diagonal matrices with all diagonal entries equal to 1 except possibly the entry in the upper left corner. Let $H'_N \subset G_N$ be the subgroup consisting of scalar multiples of the identity. For any continuous real-valued function f on G_N, put*

$$\begin{aligned} (Sf)(X) &= \int f(YX) d\nu_{SG_N}(Y), \\ (Tf)(X) &= \int f(XZ) d\nu_{H_N}(Z), \\ (T'f)(X) &= \int f(XZ) d\nu_{H'_N}(Z). \end{aligned}$$

Then we have $TSf = STf = \nu_{G_N} f$. Furthermore, if $\beta = 2$ or N is odd, then we have $T'Sf = ST'f = \nu_{G_N} f$.

Proof It is clear that $TS = ST$. Since $G_N = \{XY : X \in SG_N, Y \in H_N\}$, and Haar measure on a compact group is both left- and right-invariant, it follows that TSf is constant, and hence that $TSf = \nu_{G_N} f$. The remaining assertions of the lemma are proved similarly. □

Proof of Corollary 4.4.28 From (4.4.31) it follows that $|f - Tf| \leq 2C$. The bound (4.4.32) then follows by applying the previous lemma. We turn to the proof of (4.4.33). By mollifying f as in the course of the proof of Lemma 2.3.3, we may assume for the rest of this proof that $f \in C^\infty(G_N)$. Now fix $Z \in H_N$ and define $f_Z \in C^\infty(SG_N)$ by $f_Z(Y) = f(YZ)$, noting that $\nu_{SG_N} f_Z = (Sf)(Z)$ and that the constant C bounds $|f_Z|_{\mathscr{L},SG_N}$. We obtain (4.4.33) by applying (4.4.29) to f_Z and then averaging over $Z \in H_N$. The proof is complete. □

We next describe a couple of important applications of Corollary 4.4.28. We continue in the setup of Corollary 4.4.28.

Corollary 4.4.30 *Let D be a constant and let $D_N, D'_N \in \mathrm{Mat}_N$ be real diagonal matrices with all entries bounded in absolute value by D. Let F be a Lipschitz function on $\mathbb{R}$ with Lipschitz constant $|F|_{\mathscr{L}}$. Set $f(X) = \mathrm{tr}(F(D'_N + XD_NX^*))$ for $X \in G_N$. Then for every $\delta > 0$ we have*

$$\nu_{G_N}(|f - \nu_{G_N} f| \geq \delta N) \leq 2\exp\left(-\frac{\left(\frac{\beta(N+2)}{4} - 1\right)N\delta^2}{16D^2\|F\|^2_{\mathscr{L}}}\right).$$

Proof To abbreviate we write $\|X\| = (\mathrm{tr}\, XX^*)^{1/2}$ for $X \in \mathrm{Mat}_N(\mathbb{C})$. For $X, Y \in G_N$ we have

$$|f(X) - f(Y)| \leq \sqrt{2N}\|F\|_{\mathscr{L}}\|XD'_NX^* - YD'_NY^*\| \leq 2\sqrt{2N}D\|X - Y\|.$$

Further, by Lemma 4.4.29, since $Tf = f$, we have $\nu_{G_N} f = Sf$. Plugging into Corollary 4.4.28, we obtain the result. □

In Chapter 5, we will need the following concentration result for noncommutative polynomials.

Corollary 4.4.31 *Let $X_i \in \mathrm{Mat}_N(\mathbb{C})$ for $i = 1, \ldots, k$ be a collection of nonrandom matrices and let D be a constant bounding all singular values of these matrices. Let $p = p(t_1, \ldots, t_{k+2})$ be a polynomial in $k+2$ noncommuting variables with complex coefficients, and for $X \in U(N)$, define $f(X) = \mathrm{tr}\, p(X, X^*, X_1, \ldots, X_k)$. Then there exist positive constants $N_0 = N_0(p)$ and $c = c(p, D)$ such that, for any $\delta > 0$ and $N > N_0(p)$,*

$$\nu_{U(N)}\left(|f - \nu_{U(N)} f| > \delta N\right) \leq 2e^{-cN^2\delta^2}. \tag{4.4.34}$$

Proof We may assume without loss of generality that $p = t_{i_1} \cdots t_{i_\ell}$ for some indices $i_1, \ldots, i_\ell \in \{1, \ldots, k+2\}$, and also that $N > \ell$. We claim first that, for all $X \in U(N)$,

$$\nu_{U(N)} f = \int f(YX) d\nu_{SU(N)}(Y) =: (Sf)(X). \tag{4.4.35}$$

For some integer a such that $|a| \le \ell$ we have $f(e^{i\theta}X) = e^{ia\theta}f(X)$ for all $\theta \in \mathbb{R}$ and $X \in U(N)$. If $a = 0$, then $Sf = \nu_{U(N)}f$ by Lemma 4.4.29. Otherwise, if $a > 0$, then $\nu_{U(N)}f = 0$, but also $Sf = 0$, because $f(e^{2\pi i/N}X) = e^{2\pi ia/N}f(X)$ and $e^{2\pi ia/N}I_N \in SU(N)$. This completes the proof of (4.4.35).

It is clear that f is a Lipschitz function, with Lipschitz constant depending only on ℓ and D. Thus, from Corollary 4.4.28 in the case $\beta = 2$ and the equality $\nu_{U(N)}f = Sf$, we obtain (4.4.34) for $p = t_{i_1} \cdots t_{i_\ell}$ with $N_0 = \ell$ and $c = c(\ell, D)$, which finishes the proof of Corollary 4.4.31. □

Exercise 4.4.32 Prove Lemma 2.3.2.
Hint: follow the approximation ideas used in the proof of Theorem 4.4.17, replacing V by an approximation $V_\varepsilon(x) = \int V(x+\varepsilon z)\mu(dz)$ with μ the normal distribution.

Exercise 4.4.33 In this exercise, you provide another proof of Proposition 4.4.26 by proving directly that the law

$$P_V^N(d\lambda_1, \ldots, d\lambda_N) = \frac{1}{Z_N^V} e^{-N\sum_{i=1}^N V(\lambda_i)} \Delta(\lambda_i)^\beta \prod_{i=1}^N d\lambda_i$$

on $\mathbb{R}^N$ satisfies the LSI with constant $(Nc)^{-1}$. This proof extends to the β-ensembles discussed in Section 4.5.
(i) Use Exercise 4.4.32 to show that Theorem 4.4.18 extends to the case where

$$\Phi(\lambda) = N \sum_{i=1}^N V(\lambda_i) - \frac{\beta}{2} \sum_{i \ne j} \log |\lambda_i - \lambda_j| .$$

(Alternatively, you may prove this directly by first smoothing Φ.)
(ii) Note that

$$\mathrm{Hess}\Big(-\frac{\beta}{2}\sum_{i\ne j}\log|\lambda_i-\lambda_j|\Big)_{kl} = \begin{cases} -\beta(\lambda_k-\lambda_l)^{-2} & \text{if } k \ne l, \\ \beta \sum_{j \ne k}(\lambda_k-\lambda_j)^{-2} & \text{otherwise}, \end{cases}$$

is a nonnegative matrix, and apply Theorem 4.4.18.

4.5 Tridiagonal matrix models and the β ensembles

We consider in this section a class of random matrices that are tridiagonal and possess joint distribution of eigenvalues that generalize the classical GOE, GUE and GSE matrices. The tridiagonal representation has some advantages, among them a link with the well-developed theory of random Schroedinger operators.

4.5.1 Tridiagonal representation of β ensembles

We begin by recalling the definition of χ random variables (with t degrees of freedom).

Definition 4.5.34 The density on $\mathbb{R}_+$

$$f_t(x) = \frac{2^{1-t/2} x^{t-1} e^{-x^2/2}}{\Gamma(t/2)}$$

is called the χ *distribution* with t degrees of freedom, and is denoted χ_t.

Here, $\Gamma(\cdot)$ is Euler's Gamma function, see (2.5.5). The reason for the name is that if t is integer and X is distributed according to χ_t, then X has the same law as $\sqrt{\sum_{i=1}^t \xi_i^2}$ where ξ_i are standard Gaussian random variables.

Let ξ_i be independent i.i.d. standard Gaussian random variables of zero mean and variance 1, and let $Y_i \sim \chi_{i\beta}$ be independent and independent of the variables $\{\xi_i\}$. Define the tridiagonal symmetric matrix $H_N \in \mathrm{Mat}_N(\mathbb{R})$ with entries $H_N(i,j) = 0$ if $|i-j| > 1$, $H_N(i,i) = \sqrt{2/\beta}\,\xi_i$ and $H_N(i,i+1) = Y_{N-i}/\sqrt{\beta}$, $i = 1,\dots,N$. The main result of this section is the following.

Theorem 4.5.35 (Edelman–Dumitriu) *The joint distribution of the eigenvalues of H_N is given by*

$$C_N(\beta)\Delta(\lambda)^\beta e^{-\frac{\beta}{4}\sum_{i=1}^N \lambda_i^2}\,, \tag{4.5.1}$$

where the normalization constant $C_N(\beta)$ can be read off (2.5.11).

We begin by performing a preliminary computation that proves Theorem 4.5.35 in the case $\beta = 1$ and also turns out to be useful in the proof of the theorem in the general case.

Proof of Theorem 4.5.35 ($\beta = 1$) Let X_N be a matrix distributed according to the GOE law (and in particular, its joint distribution of eigenvalues has the density (2.5.3) with $\beta = 1$, coinciding with (4.5.1)). Set $\xi_N = X_N(1,1)/\sqrt{2}$, noting that, due to the construction in Section 2.5.1, ξ_N is a standard Gaussian variable. Let $X_N^{(1,1)}$ denote the matrix obtained from X_N by striking the first column and row, and let $Z_{N-1}^{\mathrm{T}} = (X_N(1,2),\dots,X_N(1,N))$. Then Z_{N-1} is independent of $X_N^{(1,1)}$ and ξ_N. Let $\tilde{H}_N$ be an orthogonal $N-1$-by-$N-1$ matrix, measurable on $\sigma(Z_{N-1})$, such that $\tilde{H}_N Z_{N-1} = (\|Z_{N-1}\|_2, 0,\dots,0)$, and set $Y_{N-1} = \|Z_{N-1}\|_2$, noting that Y_{N-1} is independent of ξ_N and is distributed according to χ_{N-1}. (A particular choice of $\tilde{H}_N$ is the *Householder reflector* $\tilde{H}_N = I - 2uu^{\mathrm{T}}/\|u\|_2^2$, where $u = Z_{N-1} -$

$\|Z_{N-1}\|_2(1,\ldots,0)$.) Let

$$H_N = \begin{pmatrix} 1 & 0 \\ 0 & \tilde{H}_N \end{pmatrix}.$$

Then the law of eigenvalues of $H_N X_N H_N^{\mathrm{T}}$ is still (4.5.1), while

$$H_N X_N H_N^{\mathrm{T}} = \begin{pmatrix} \sqrt{2}\xi_N & Y_{N-1} & \mathbf{0}_{N-2} \\ Y_{N-1} & & \\ & X_{N-1} & \\ \mathbf{0}_{N-2} & & \end{pmatrix},$$

where X_{N-1} is again distributed according to the GOE and is independent of ξ_N and Y_{N-1}. Iterating this construction $N-1$ times (in the next step, with the Householder matrix corresponding to X_{N-1}), one concludes the proof (with $\beta = 1$). □

We next prove some properties of the eigenvalues and eigenvectors of tridiagonal matrices. Recall some notation from Section 2.5: $\mathscr{D}_N$ denotes the collection of diagonal N-by-N matrices with real entries, $\mathscr{D}_N^{\mathrm{d}}$ denotes the subset of $\mathscr{D}_N$ consisting of matrices with distinct entries, and $\mathscr{D}_N^{\mathrm{do}}$ denotes the subset of matrices with decreasing entries, that is $\mathscr{D}_N^{\mathrm{do}} = \{D \in \mathscr{D}_N^{\mathrm{d}} : D_{i,i} > D_{i+1,i+1}\}$. Recall also that $\mathscr{U}_N^{(1)}$ denotes the collection of N-by-N orthogonal matrices, and let $\mathscr{U}_N^{(1),+}$ denote the subset of $\mathscr{U}_N^{(1)}$ consisting of matrices whose first row has all elements strictly positive.

We parametrize tridiagonal matrices by two vectors of length N and $N-1$, $\mathbf{a} = (a_1,\ldots,a_N)$ and $\mathbf{b} = (b_1,\ldots,b_{N-1})$, so that if $H \in \mathscr{H}_N^{(1)}$ is tridiagonal then $H(i,i) = a_{N-i+1}$ and $H(i,i+1) = b_{N-i}$. Let $\mathscr{T}_N \subset \mathscr{H}_N^{(1)}$ denote the collection of tridiagonal matrices with all entries of $\mathbf{b}$ strictly positive.

Lemma 4.5.36 *The eigenvalues of any $H \in \mathscr{T}_N$ are distinct, and all eigenvectors $v = (v_1,\ldots,v_N)$ of H satisfy $v_1 \neq 0$.*

Proof The null space of any matrix $H \in \mathscr{T}_N$ is at most one dimensional. Indeed, suppose $Hv = 0$ for some nonzero vector $v = (v_1,\ldots,v_N)$. Because all entries of $\mathbf{b}$ are nonzero, it is impossible that $v_1 = 0$ (for then, necessarily all $v_i = 0$). So suppose $v_1 \neq 0$, and then $v_2 = -a_N/b_{N-1}$. By solving recursively the equation

$$b_{N-i}v_{i-1} + a_{N-i}v_i = -b_{N-i-1}v_{i+1},\ i = 2,\ldots,N-1, \tag{4.5.2}$$

which is possible because all entries of $\mathbf{b}$ are nonzero, all entries of v are determined. Thus, the null space of any $H \in \mathscr{T}_N$ is one dimensional at most. Since $H - \lambda I \in \mathscr{T}_N$ for any λ, the first part of the lemma follows. The second part follows because we showed that if $v \neq 0$ is in the null space of $H - \lambda I$, it is impossible to have $v_1 = 0$. □

Let $H \in \mathscr{T}_N$, with diagonals $\mathbf{a}$ and $\mathbf{b}$ as above, and write $H = UDU^{\mathrm{T}}$ with $D \in \mathscr{D}_N^{\mathrm{do}}$ and $U = [v^1, \ldots, v^N]$ orthogonal, such that the first row of U, denoted $\mathbf{v} = (v_1^1, \ldots, v_1^N)$, has nonnegative entries. (Note that $\|\mathbf{v}\|_2 = 1$.) Write $\mathbf{d} = (D_{1,1}, \ldots, D_{N,N})$. Let $\Delta_N^{\mathrm{c}} = \{(x_1, \ldots, x_N) : x_1 > x_2 \cdots > x_N\}$ and let

$$S_+^{N-1} = \{\mathbf{v} = (v_1, \ldots, v_N) \in \mathbb{R}^N : \|\mathbf{v}\|_2 = 1,\, v_i > 0\}.$$

(Note that Δ_N^{c} is similar to Δ_N, except that the ordering of coordinates is reversed.)

Lemma 4.5.37 *The map*

$$(\mathbf{a}, \mathbf{b}) \mapsto (\mathbf{d}, \mathbf{v}) : \mathbb{R}^N \times \mathbb{R}_+^{(N-1)} \to \Delta_N^{\mathrm{c}} \times S_+^{N-1} \tag{4.5.3}$$

is a bijection, whose Jacobian J is proportional to

$$\frac{\Delta(\mathbf{d})}{\prod_{i=1}^{N-1} b_i^{i-1}}\,. \tag{4.5.4}$$

Proof That the map in (4.5.3) is a bijection follows from the proof of Lemma 4.5.36, and in particular from (4.5.2) (the map $(\mathbf{d}, \mathbf{v}) \mapsto (\mathbf{a}, \mathbf{b})$ is determined by the relation $H = UDU^{\mathrm{T}}$).

To evaluate the Jacobian, we recall the proof of the $\beta = 1$ case of Theorem 4.5.35. Let X be a matrix distributed according to the GOE, consider the tridiagonal matrix with diagonals $\mathbf{a}, \mathbf{b}$ obtained from X by the successive Householder transformations employed in that proof. Write $X = UDU^*$ where U is orthogonal, D is diagonal (with elements $\mathbf{d}$), and the first row $\mathbf{u}$ of U consists of nonnegative entries (and strictly positive except on a set of measure 0). Note that, by Corollary 2.5.4, $\mathbf{u}$ is independent of D and, by Theorem 2.5.2, the density of the distribution of the vector $(\mathbf{d}, \mathbf{u})$ with respect to the product of the Lebesgue measure on Δ_N^{c} and the the uniform measure on S_+^{N-1} is proportional to $\Delta(\mathbf{d}) e^{-\sum_{i=1}^N d_i^2/4}$. Using Theorem 4.5.35 and the first part of the lemma, we conclude that the latter (when evaluated in the variables $\mathbf{a}, \mathbf{b}$) is proportional to

$$J e^{-\sum_{i=1}^N \frac{a_i^2}{4} - \sum_{i=1}^{N-1} \frac{b_i^2}{2}} \prod_{i=1}^{N-1} b_i^{i-1} = J e^{-\sum_{i=1}^N d_i^2/4} \prod_{i=1}^{N-1} b_i^{i-1}\,.$$

The conclusion follows. □

We will also need the following useful identity.

Lemma 4.5.38 *With notation as above, we have the identity*

$$\Delta(\mathbf{d}) = \frac{\prod_{i=1}^{N-1} b_i^i}{\prod_{i=1}^{N} v_1^i}. \tag{4.5.5}$$

Proof Write $H = UDU^{\mathrm{T}}$. Let $e_1 = (1,0,\dots,0)^{\mathrm{T}}$. Let w^1 be the first column of U^{T}, which is the vector made out of the first entries of $v^1,\dots,v^n$. One then has

$$\begin{aligned}\prod_{i=1}^{N-1} b_i^i &= \det[e_1, He_1, \dots, H^{N-1}e_1] = \det[e_1, UDU^{\mathrm{T}}e_1, \dots, UD^{N-1}U^{\mathrm{T}}e_1] \\ &= \pm\det[w^1, Dw^1, \dots, D^{N-1}w^1] = \pm\Delta(\mathbf{d})\prod_{i=1}^{N} v_1^i.\end{aligned}$$

Because all terms involved are positive by construction, the $\pm$ is actually a $+$, and the lemma follows. □

We can now conclude.

Proof of Theorem 4.5.35 (general $\beta > 0$) The density of the independent vectors **a** and **b**, together with Lemma 4.5.37, imply that the joint density of **d** and **v** with respect to the product of the Lebesgue measure on Δ_N^{c} and the uniform measure on S_+^{N-1} is proportional to

$$J\prod_{i=1}^{N-1} b_i^{i\beta-1} e^{-\frac{\beta}{4}\sum_{i=1}^{N} d_i^2}. \tag{4.5.6}$$

Using the expression (4.5.4) for the Jacobian, one has

$$J\prod_{i=1}^{N-1} b_i^{i\beta-1} = \Delta(\mathbf{d})\left(\prod_{i=1}^{N-1} b_i^i\right)^{\beta-1} = \Delta(\mathbf{d})^{\beta}\left(\prod_{i=1}^{N} v_1^i\right)^{\beta-1},$$

where (4.5.5) was used in the second equality. Substituting in (4.5.6) and integrating over the variables **v** completes the proof. □

4.5.2 Scaling limits at the edge of the spectrum

By Theorem 4.5.35, Corollary 2.6.3 and Theorem 2.6.6, we know that $\lambda_N/\sqrt{N}$, the maximal eigenvalue of $H_N/\sqrt{N}$, converges to 2 as $N \to \infty$. It is thus natural to consider the matrix $\tilde{H}_N = H_N - 2\sqrt{N}\mathbf{I}_N$, and study its top eigenvalue. For $\beta = 1,2,4$, we have seen in Theorems 3.1.4 and 3.1.7 that the top eigenvalue of $N^{1/6}\tilde{H}_N$ converges in distribution (to the Tracy–Widom distributions F_β). In this section, we give an alternative derivation, valid for all β, of the convergence in distribution, although the identification of the limit does not involve the Tracy–Widom distributions.

One of the advantages of the tridiagonal representation of Theorem 4.5.35 is that one can hope that scaling limits of tridiagonal matrices naturally relate to (second order) differential operators. We begin by providing a heuristic argument that allows one to guess both the correct scale and the form of the limiting operator. From the definition of χ variables with t degrees of freedom, such variables are asymptotically (for large t) equivalent to $\sqrt{t}+G/\sqrt{2}$ where G is a standard Gaussian random variable. Consider $\tilde{H}_N$ as an operator acting on column vectors $\psi=(\psi_1,\ldots,\psi_N)^{\mathrm{T}}$. We look for parameters α,γ such that, if one writes $n=[xN^\alpha]$ and $\psi_n=\Psi(x)$ for some "nice" function Ψ, the action of the top left corner of $N^\gamma\tilde{H}_N$ on ψ approximates the action of a second order differential operator on Ψ. (We consider the upper left corner because this is where the off-diagonal terms have largest order, and one expects the top of the spectrum to be related to that corner.) Toward this end, expand Ψ in a Taylor series up to second order, and write $\psi_{n\pm 1}\sim\psi_n\pm N^{-\alpha}\Psi'(x)+N^{-2\alpha}\Psi''(x)/2$. Using the asymptotic form of χ variables mentioned above, one gets, after neglecting small error terms, that, for $\alpha<1$ and x in some compact subset of $\mathbb{R}_+$,

$$
\begin{aligned}
(N^\gamma\tilde{H}_N\psi)(n) &\sim N^{\gamma+1/2-2\alpha}\Psi''(x)\\
&+\sqrt{\frac{1}{2\beta}}N^\gamma\left(2G_n^{(1)}+G_n^{(2)}+G_{n-1}^{(2)}\right)\Psi(x)-xN^{\alpha+\gamma-1/2}\Psi(x),
\end{aligned}
\tag{4.5.7}
$$

where $\{G_n^{(i)}\}$, $i=1,2$, are independent sequences of i.i.d. standard Gaussian variables. It is then natural to try to represent $G_n^{(i)}$ as discrete derivatives of independent Brownian motions: thus, let W_x, $\overline{W}_x$ denote standard Brownian motions and (formally) write $G_n^{(1)}=N^{-\alpha/2}W_x'$, $G_n^{(2)}=N^{-\alpha/2}\overline{W}_x'$ with the understanding that a rigorous definition will involve integration by parts. Substituting in (4.5.7) and writing $B_x=(W_x+\overline{W}_x)/\sqrt{2}$, we obtain formally

$$
(N^\gamma\tilde{H}_N\psi)(n)\sim N^{\gamma+1/2-2\alpha}\Psi''(x)+\frac{2N^{\gamma-\alpha/2}B_x'}{\sqrt{\beta}}\Psi(x)-xN^{\alpha+\gamma-1/2}\Psi(x), \tag{4.5.8}
$$

where (4.5.8) has to be understood after an appropriate integration by parts against smooth test functions. To obtain a scaling limit, one then needs to take α,γ so that

$$
\gamma+\frac{1}{2}-2\alpha=\gamma-\frac{\alpha}{2}=\alpha+\gamma-\frac{1}{2}=0\Rightarrow\alpha=\frac{1}{3},\gamma=\frac{1}{6}.
$$

In particular, we recover the Tracy–Widom scaling, and expect the top of the spectrum of $N^{1/6}\tilde{H}_N$ to behave like the top of the spectrum of the "stochastic Airy operator"

$$
\mathbf{H}_\beta:=\frac{d^2}{dx^2}-x+\frac{2}{\sqrt{\beta}}B_x'. \tag{4.5.9}
$$

The rest of this section is devoted to providing a precise definition of $\mathbf{H}_\beta$, developing some of its properties, and proving the convergence of the top eigenvalues of $N^{1/6}\tilde{H}_N$ to the top eigenvalues of $\mathbf{H}_\beta$. In doing so, the convergence of the quadratic forms associated with $N^{1/6}\tilde{H}_N$ toward a quadratic form associated with $\mathbf{H}_\beta$ plays an important role. We thus begin by providing some analytical machinery that will be useful in controlling this convergence.

On smooth functions of compact support in $(0,\infty)$, introduce the bilinear non-degenerate form

$$\langle f,g\rangle_* = \int_0^\infty f'(x)g'(x)dx + \int_0^\infty (1+x)f(x)g(x)dx\,.$$

Define $\mathscr{L}_*$ as the Hilbert space obtained by completion with respect to the inner product $\langle\cdot,\cdot\rangle_*$ (and norm $\|f\|_* = \sqrt{\langle f,f\rangle_*}$). Because of the estimate

$$|f(x)-f(y)| \le \sqrt{|x-y|}\|f\|_*\,, \tag{4.5.10}$$

elements of $\mathscr{L}_*$ are continuous functions, and vanish at the origin. Further properties of $\mathscr{L}_*$ are collected in Lemma 4.5.43 below.

Definition 4.5.39 A pair $(f,\lambda) \in \mathscr{L}_* \times \mathbb{R}$ is called an *eigenvector–eigenvalue pair* of $\mathbf{H}_\beta$ if $\|f\|_2 = 1$ and, for any compactly supported infinitely differentiable function ϕ,

$$\begin{aligned}\lambda\int_0^\infty \phi(x)f(x)dx &= \int_0^\infty [\phi''(x)f(x) - x\phi(x)f(x)]dx \\ &\quad -\frac{2}{\sqrt{\beta}}\left[\int_0^\infty \phi'(x)f(x)B_x dx + \int_0^\infty \phi(x)B_x f'(x)dx\right].\end{aligned} \tag{4.5.11}$$

Remark 4.5.40 Equation (4.5.11) expresses the following: (f,λ) is an eigenvector–eigenvalue pair of $\mathbf{H}_\beta$ if $\mathbf{H}_\beta f = \lambda f$ in the sense of Schwarz distributions, where we understand $f(x)B'_x$ as the Schwarz distribution that is the derivative of the continuous function $f(x)B_x - \int_0^x B_y f'(y)dy$.

Remark 4.5.41 Using the fact that $f \in \mathscr{L}_*$, one can integrate by parts in (4.5.11) and express all integrals as integrals involving ϕ' only. In this way, one obtains that (f,λ) is an eigenvector–eigenvalue pair of $\mathbf{H}_\beta$ if and only if, for Lebesgue almost every x and some constant C, $f'(x)$ exists and

$$f'(x) = C + \int_0^x (\lambda+\theta)f(\theta)d\theta - B_x f(x) + \int_0^x B_\theta f'(\theta)d\theta\,. \tag{4.5.12}$$

Since the right side is a continuous function, we conclude that f' can be taken continuous. (4.5.12) will be an important tool in deriving properties of eigenvector–

eigenvalue pairs, and in particular the nonexistence of two eigenvector–eigenvalue pairs sharing the same eigenvalue.

The main result of this section in the following.

Theorem 4.5.42 (Ramirez–Rider–Virag) *Fix $\beta > 0$ and let $\lambda_N^N > \lambda_{N-1}^N > \cdots$ denote the eigenvalues of H_N. For almost every Brownian path B_x, for each $k \geq 0$, the collection of eigenvalues of $\mathbf{H}_\beta$ possesses a well defined $k+1$st largest element λ_k. Further, the random vector $N^{1/6}(\lambda_{N-j}^N - 2\sqrt{N})_{j=0}^k$ converges in distribution to the random vector $(\lambda_j)_{j=0}^k$.*

The proof of Theorem 4.5.42 will take the rest of this section. It is divided into two main steps. We first study the operator $\mathbf{H}_\beta$ by associating with it a variational problem. We prove, see Corollary 4.5.45 and Lemma 4.5.47 below, that the eigenvalues of $\mathbf{H}_\beta$ are discrete, that they can be obtained from this variational problem and that the associated eigenspaces are simple. In a second step, we introduce a discrete quadratic form associated with $\hat{H}_N = N^{1/6}\tilde{H}_N$ and prove its convergence to that associated with $\mathbf{H}_\beta$, see Lemma 4.5.50. Combining these facts will then lead to the proof of Theorem 4.5.42.

We begin with some preliminary material related to the space $\mathscr{L}_*$.

Lemma 4.5.43 *Any $f \in \mathscr{L}_*$ is Hölder(1/2)-continuous and satisfies*

$$x^{1/4}|f(x)| \leq 2\|f\|_*, \quad x > 1. \tag{4.5.13}$$

Further, if $\{f_n\}$ is a bounded sequence in $\mathscr{L}_*$ then it possesses a subsequence that converges to some f in $\mathscr{L}_*$ in the following senses: (i) $f_n \to_{L^2} f$, (ii) $f_n' \to f'$ weakly in L^2, (iii) $f_n \to f$ uniformly on compacts, (iv) $f_n \to f$ weakly in $\mathscr{L}_*$.

Proof The Hölder continuity statement is a consequence of (4.5.10). The latter also implies that for any function f with derivative in L^2,

$$|f(y)| \geq \left(|f(x)| - \sqrt{|y-x|}\|f'\|_2\right)_+$$

and in particular, for any x,

$$f^2(x) \leq 2\|f\|_2\|f'\|_2 \,. \tag{4.5.14}$$

(Indeed, fix x and consider the set $A_x = \{y : |y-x| \leq f^2(x)/4\|f'\|_2^2\}$. On A_x, $|f(y)| \geq |f(x)|/2$. Writing $\|f\|_2^2 \geq \int_{A_x} f^2(y)dy$ then gives (4.5.14).) Since $\|f\|_*^2 \geq \int_z^\infty (1+x)f^2(x)dx \geq z\int_z^\infty f^2(x)dx$, applying the estimate (4.5.14) on the function $f(z)\mathbf{1}_{z\geq x}$ yields (4.5.13).

Points (ii) and (iv) in the statement of the lemma follow from the Banach–Alaoglu Theorem (Theorem B.8). Point (iii) follows from the uniform equicontinuity on compacts of the sequence f_n that is a consequence of the uniform Hölder estimate. Together with the uniform integrability $\sup_n \int x f_n^2(x)dx < \infty$, this gives (i). □

The next step is the introduction of a bilinear form on $\mathscr{L}_*$ associated with $\mathbf{H}_\beta$. Toward this end, note that if one interprets $-\mathbf{H}_\beta \phi$ for ϕ smooth in the sense of Schwarz distributions, then it can be applied (as a linear functional) again on ϕ, yielding the quadratic form

$$\langle \phi, \phi \rangle_{\mathbf{H}_\beta} := \|\phi'\|_2^2 + \|\sqrt{x}\phi(x)\|_2^2 + \frac{4}{\sqrt{\beta}} \int_0^\infty B_x \phi(x)\phi'(x)dx. \tag{4.5.15}$$

We seek to extend the quadratic form in (4.5.15) to functions in $\mathscr{L}_*$. The main issue is the integral

$$2\int_0^\infty B_x \phi(x)\phi'(x)dx = \int_0^\infty B_x (\phi(x)^2)' dx.$$

Since it is not true that $|B_x| < C\sqrt{x}$ for all large x, in order to extend the quadratic form in (4.5.15) to functions in $\mathscr{L}_*$, we need to employ the fact that B_x is itself regular in x. More precisely, define

$$\bar{B}_x = \int_x^{x+1} B_y dy.$$

For ϕ smooth and compactly supported, we can write $B_x = \bar{B}_x + (B_x - \bar{B}_x)$ and integrate by parts to obtain

$$\int_0^\infty B_x(\phi(x)^2)'dx = -\int_0^\infty (\bar{B}_x)'\phi^2(x)dx + 2\int_0^\infty (B_x - \bar{B}_x)\phi(x)\phi'(x)dx.$$

This leads us to define

$$\langle \phi, \phi \rangle_{\mathbf{H}_\beta} := \|\phi'\|_2^2 + \|\sqrt{x}\phi(x)\|_2^2 - \frac{2}{\sqrt{\beta}} \left[\int_0^\infty Q_x \phi^2(x)dx - 2\int_0^\infty R_x \phi(x)\phi'(x)dx \right], \tag{4.5.16}$$

where

$$Q_x = (\bar{B}_x)' = B_{x+1} - B_x, \quad R_x = B_x - \bar{B}_x. \tag{4.5.17}$$

As we now show, this quadratic form extends to $\mathscr{L}_*$.

Lemma 4.5.44 *(a) For each $\varepsilon > 0$ there exists a random constant C (depending on β, ε and $B_\cdot$ only) such that*

$$\frac{4}{\sqrt{\beta}} \sup_x \frac{|Q_x| \vee |R_x|}{C + \sqrt{x}} \le \varepsilon. \tag{4.5.18}$$

(b) The quadratic form $\langle \cdot,\cdot \rangle_{\mathbf{H}_\beta}$ *of* (4.5.16) *extends to a continuous symmetric bilinear form on* $\mathscr{L}_* \times \mathscr{L}_*$*: there exists a (random) constant* C'*, depending on the Brownian path* $B_\cdot$ *only, such that, almost surely,*

$$\frac{1}{2}\|f\|_*^2 - C'\|f\|_2^2 \le \langle f, f\rangle_{\mathbf{H}_\beta} \le C'\|f\|_*^2 . \tag{4.5.19}$$

Proof For part (a), note that

$$|Q_x| \vee |R_x| \le Z_{[x]} + Z_{[x]+1} ,$$

where $Z_i = \sup_{y\in[0,1]} |B_{i+y} - B_i|$. The random variables Z_i are i.i.d. and satisfy $P(Z_i > t) \le 4P(G > t)$ where G is a standard Gaussian random variable. From this and the Borel–Cantelli Lemma, (4.5.18) follows.

We turn to the proof of (b). The sum of the first two terms in the definition of $\langle f, f\rangle_{\mathbf{H}_\beta}$ equals $\|f\|_*^2 - \|f\|_2^2$. By the estimate (4.5.18) on Q with $\varepsilon = 1/10$, the third term can be bounded in absolute value by $\|f\|_*^2/10 + C_1\|f\|_2^2$ for some (random) constant C_1 (this is achieved by upper bounding $C(1+\sqrt{x})$ by $C_1 + x/10$). Similarly, the last term can be controlled as

$$\int_0^\infty (C + \frac{1}{10}\sqrt{x})|f(x)||f'(x)|dx \le C\|f\|\|f\|_* + \frac{1}{10}\|f\|_*^2 \le \frac{1}{5}\|f\|_*^2 + C_2\|f\|_2^2 .$$

Combining these estimates (and the fact that $\|f\|_*$ dominates $\|f\|_2$) yields (4.5.19). □

We can now consider the variational problem associated with the quadratic form $\langle \cdot,\cdot \rangle_{\mathbf{H}_\beta}$ of (4.5.16).

Corollary 4.5.45 *The infimum in the minimization problem*

$$\Lambda_0 := \inf_{f\in\mathscr{L}_*, \|f\|_2 = 1} \langle f, f\rangle_{\mathbf{H}_\beta} \tag{4.5.20}$$

is achieved at some $f \in \mathscr{L}_*$*, and* $(f, -\Lambda_0)$ *is an eigenvector–eigenvalue pair for* $\mathbf{H}_\beta$*, with* $-\Lambda_0 = \lambda_0$.

We will shortly see in Lemma 4.5.47 that the minimizer in Corollary 4.5.45 is unique.

Proof By the estimate (4.5.19), the infimum in (4.5.20) is finite. Let $\{f_n\}_n$ be a minimizing sequence, that is $\|f_n\|_2 = 1$ and $\langle f_n, f_n\rangle_{\mathbf{H}_\beta} \to \Lambda_0$. Again by (4.5.19), there is some (random) constant K so that $\|f_n\|_* \le K$ for all n. Write

$$\langle f_n, f_n\rangle_{\mathbf{H}_\beta} = \|f_n\|_*^2 - \|f_n\|_2^2 - \frac{2}{\sqrt{\beta}}\left[\int_0^\infty Q_x f_n^2(x)dx - 2\int_0^\infty R_x f_n(x) f_n'(x)dx\right] .$$

Let $f \in \mathscr{L}_*$ be a limit point of f_n (in all the senses provided by Lemma 4.5.43).

Then $1 = \|f_n\|_2 \to \|f\|_2$ and hence $\|f\|_2 = 1$, while $\liminf \|f_n\|_* \geq \|f\|_*$. Fix $\varepsilon > 0$. Then, by (4.5.18), there is a random variable X such that

$$\left| \frac{2}{\sqrt{\beta}} \left[\int_X^\infty Q_x f_n^2(x) dx - 2 \int_X^\infty R_x f_n(x) f_n'(x) dx \right] \right| \leq \varepsilon \|f_n\|_* .$$

The convergence of f_n to f uniformly on $[0,X]$ together with the boundedness of $\|f_n\|_*$ then imply that

$$\langle f, f \rangle_{\mathbf{H}_\beta} \leq \liminf_{n\to\infty} \langle f_n, f_n \rangle_{\mathbf{H}_\beta} + \varepsilon K = \Lambda_0 + \varepsilon K .$$

Since ε is arbitrary, it follows from the definition of Λ_0 that $\langle f, f \rangle_{\mathbf{H}_\beta} = \Lambda_0$, as claimed.

To see that $(f, -\Lambda_0)$ is an eigenvector–eigenvalue pair, fix $\varepsilon > 0$ and ϕ smooth of compact support, and set $f^{\varepsilon,\phi} = (f + \varepsilon\phi)/\|f + \varepsilon\phi\|_2$ (reduce ε if needed so that $\phi \neq f/\varepsilon$). Then

$$\begin{aligned}
&\langle f^\varepsilon, f^\varepsilon \rangle_{\mathbf{H}_\beta} - \langle f, f \rangle_{\mathbf{H}_\beta} \\
&= -2\varepsilon \langle f, f \rangle_{\mathbf{H}_\beta} \int_0^\infty f(x)\phi(x)dx + 2\varepsilon \int_0^\infty (f'(x)\phi'(x) + xf(x)\phi(x))dx \\
&\quad - \frac{4\varepsilon}{\sqrt{\beta}} \left[\int_0^\infty Q_x \phi(x) f(x) dx - \int_0^\infty R_x [\phi(x) f(x)]' dx \right] + O(\varepsilon^2) .
\end{aligned}$$

Thus, a necessary condition for f to be a minimizer is that the linear in ε term in the last equality vanishes for all such smooth and compactly supported ϕ. Using the fact that ϕ is compactly supported, one can integrate by parts the term involving Q and rewrite it in terms of B_x. Using also the fact that $\langle f, f \rangle_{\mathbf{H}_\beta} = \Lambda_0$, one gets from this necessary condition that $(f, -\Lambda_0)$ satisfies (4.5.11).

Finally, we note that by (4.5.11) and an integration by parts, if (g, λ) is an eigenvector–eigenvalue pair then for any compactly supported smooth ϕ,

$$\begin{aligned}
\lambda \int_0^\infty \phi(x) g(x) dx &= \int_0^\infty [\phi''(x) g(x) - x\phi(x) g(x)] dx \\
&\quad - \frac{4}{\sqrt{\beta}} \left[\int_0^\infty \phi(x) g(x) Q_x dx - \int_0^\infty R_x [\phi(x) g(x)]' dx \right] . \qquad (4.5.21)
\end{aligned}$$

Take a sequence $\{\phi_n\}$ of smooth, compactly supported functions, so that $\phi_n \to g$ in $\mathscr{L}_*$. Applying the same argument as in the proof of Lemma 4.5.44, one concludes that all terms in (4.5.21) (with ϕ_n replacing ϕ) converge to their value with f replacing ϕ. This implies that $\langle g, g \rangle_{\mathbf{H}_\beta} = -\lambda \|g\|_2^2$, and in particular that $\lambda \leq -\Lambda_0$. Since the existence of a minimizer f to (4.5.20) was shown to imply that $(f, -\Lambda_0)$ is an eigenvector–eigenvalue pair, we conclude that in fact $-\Lambda_0 = \lambda_0$. □

Remark 4.5.46 The collection of scalar multiples of minimizers in Corollary 4.5.45 forms a linear subspace $\mathscr{H}_0$. We show that $\mathscr{H}_0$ is finite dimensional: indeed, let $\{f_n\}$ denote an orthogonal (in L^2) basis of $\mathscr{H}_0$, and suppose that it is infinite dimensional. By Lemma 4.5.44, there is a constant C such that $\|f_n\|_* \leq C$. Switching to a subsequence if necessary, it follows from Lemma 4.5.43 that f_n converges to some f in L^2, with $\|f\|_2 = 1$, and in fact $f \in \mathscr{H}_0$. But on the other hand, f is orthogonal to all f_n in $\mathscr{H}_0$ and thus $f \notin \mathscr{H}_0$, a contradiction.

We can now repeat the construction of Corollary 4.5.45 inductively. For $k \geq 1$, with $\mathscr{H}_{k-1}^{\perp}$ denoting the ortho-complement of $\mathscr{H}_{k-1}$ in L^2, set

$$\Lambda_k := \inf_{f \in \mathscr{L}_*, \|f\|_2 = 1, f \in \mathscr{H}_k^{\perp}} \langle f, f \rangle_{\mathbf{H}_\beta} \,. \tag{4.5.22}$$

Mimicking the proof of Corollary 4.5.45, one shows that the infimum in (4.5.22) is achieved at some $f \in \mathscr{L}_*$, and $(f, -\Lambda_k)$ is an eigenvector–eigenvalue pair for $\mathbf{H}_\beta$, with $-\Lambda_k = \lambda_k$. We then denote by $\mathscr{H}_k$ the (finite dimensional) linear space of scalar multiples of minimizers in (4.5.22). It follows that the collection of eigenvalues of $\mathbf{H}_\beta$ is discrete and can be ordered as $\lambda_0 > \lambda_1 > \cdots$.

Our next goal is to show that the spaces $\mathscr{H}_k$ are one dimensional, i.e. that each eigenvalue is simple. This will come from the analysis of (4.5.12). We have the following.

Lemma 4.5.47 *For each given C, λ and continuous function $B_\cdot$, the solution to* (4.5.12) *is unique. As a consequence, the spaces $\mathscr{H}_k$ are all one-dimensional.*

Proof Integrating by parts, we rewrite (4.5.12) as

$$f'(x) = C + (\lambda + x)\int_0^x f'(\theta)d\theta - \int_0^x f'(\theta)d\theta - B_x\int_0^x f'(\theta)d\theta + \int_0^x B_\theta f'(\theta)d\theta \,. \tag{4.5.23}$$

By linearity, it is enough to show that solutions of (4.5.23) vanish when $C = 0$. But, for $C = 0$, one gets that for some bounded $C'(x) = C'(\lambda, B_\cdot, x)$ with $C'(x)$ increasing in x, $|f'(x)| \leq C' \int_0^x |f'(\theta)|d\theta$. An application of Gronwall's Lemma shows that $f'(x) = 0$ for all positive x. To see that $\mathscr{H}_k$ is one dimensional, note that if f satisfies (4.5.12) with constant C, then cf satisfies the same with constant cC. □

Another ingredient of the proof of Theorem 4.5.42 is the representation of the matrix $\hat{H}_N := N^{1/6}\tilde{H}_N$ as an operator on $\mathscr{L}_*$. Toward this end, define (for $x \in \mathbb{R}_+$)

$$y_{N,1}(x) = N^{-1/6}\sqrt{\frac{2}{\beta}}\sum_{i=1}^{[xN^{1/3}]} H_N(i,i), \tag{4.5.24}$$

$$y_{N,2}(x) = 2N^{-1/6}\sum_{i=1}^{[xN^{1/3}]} (\sqrt{N} - H_N(i,i+1)). \tag{4.5.25}$$

Standard estimates lead to the following.

Lemma 4.5.48 *There exists a probability space supporting the processes $y_{N,j}(\cdot)$ and two independent Brownian motions $B_{\cdot,j}$, $j = 1,2$, such that, with respect to the Skorohod topology, the following convergence holds almost surely:*

$$y_{N,j}(\cdot) \Rightarrow \sqrt{\frac{2}{\beta}}B_{x,j} + x^2(j-1)/2, \quad j = 1,2.$$

In the sequel, we work in the probability space determined by Lemma 4.5.48, and write $B_x = B_{x,1} + B_{x,2}$ (thus defining naturally a version of the operator $\mathbf{H}_\beta$ whose relation to the matrices $\hat{H}_N$ needs clarification). Toward this end, we consider the matrices $\hat{H}_N$ as operators acting on $\mathbb{R}^N$ equipped with the norm

$$\|v\|_{N,*}^2 = N^{1/3}\sum_{i=1}^{N}(v(i+1) - v(i))^2 + N^{-2/3}\sum_{i=1}^{N} iv(i)^2 + N^{-1/3}\sum_{i=1}^{N} v(i)^2,$$

where we set $v(N+1) = 0$. Write $\langle v, w\rangle_{N,2} = N^{-1/3}\sum_{i=1}^{N} v(i)w(i)$ and let $\|v\|_{N,2}$ denote the associated norm on $\mathbb{R}^N$. Recall the random variables Y_i appearing in the definition of the tridiagonal matrix H_N, see Theorem 4.5.35, and motivated by the scaling in Theorem 4.5.42, introduce

$$\eta_i = 2N^{-1/6}(\sqrt{N} - \frac{1}{\sqrt{\beta}}EY_{N-i}),$$

$$\gamma_i = 2N^{-1/6}\frac{1}{\sqrt{\beta}}(EY_{N-i} - Y_{N-i}).$$

It is straightforward to verify that $\eta_i \geq 0$ and that, for some constant κ independent of N,

$$\frac{i}{\kappa\sqrt{N}} - \kappa \leq \eta_i \leq \frac{\kappa i}{\sqrt{N}} + \kappa. \tag{4.5.26}$$

Also, with $w_k^{(1)} = \sqrt{2/\beta}N^{-1/6}\sum_{i=1}^{k}\xi_i$ and $w_k^{(2)} = \sum_{i=1}^{k}\gamma_i$, we have that for any $\varepsilon > 0$ there is a tight sequence of random variables $\kappa_{N,\varepsilon}$ satisfying

$$\sup_{i \leq k \leq i+N^{1/3}} |w_k^{(j)} - w_i^{(j)}|^2 \leq \varepsilon i N^{-1/3} + \kappa_{N,\varepsilon}. \tag{4.5.27}$$

We now have the following analog of (4.5.19).

Lemma 4.5.49 *There exists a tight sequence of random variables $c_i = c_i(N)$, $i = 1,2,3$, so that, for all N and v,*

$$c_1\|v\|_{N,*}^2 - c_2\|v\|_{N,2}^2 \le -\langle v, \hat{H}_N v\rangle_{N,2} \le c_3\|v\|_{N,*}^2 .$$

Proof Using the definitions, one gets (setting $v(N+1)=0$)

$$\begin{aligned} -\langle v, \hat{H}_N v\rangle_{N,2} &= N^{1/3}\sum_{i=1}^{N}(v(i+1)-v(i))^2 + 2N^{-1/6}\sum_{i=1}^{N}\eta_i v(i)v(i+1) \\ &\quad -\sqrt{\frac{2}{\beta}}N^{-1/6}\sum_{i=1}^{N}v^2(i)\xi_i + 2N^{-1/6}\sum_{i=1}^{N}\gamma_i v(i)v(i+1) \\ &=: S_1 + S_2 - S_3 + S_4 . \end{aligned} \tag{4.5.28}$$

One identifies S_1 with the first term in $\|v\|_{N,*}^2$. Next, we have

$$\sum_{i=1}^{N}\eta_i v(i)v(i+1) \le \sqrt{\sum_{i=1}^{N}\eta_i v(i)^2 \cdot \sum_{i=1}^{N}\eta_i v(i+1)^2} ,$$

and thus, together with the bound (4.5.26), we have that S_2 is bounded above by a constant multiple of the sum of the second and third terms in $\|v\|_{N,*}^2$. Similarly, we have from the bound $ab \ge -(a-b)^2/3 + a^2/4$ that

$$\eta_i v(i)v(i+1) \ge -\frac{1}{3}\eta(v_{i+1}-v_i)^2 + \frac{1}{4}\eta_i v_i^2 \ge -\frac{1}{3}\eta(v_{i+1}-v_i)^2 + \frac{1}{4\kappa}iv_i^2 - \frac{\kappa}{4}v_i^2$$

and using (4.5.26) again, we conclude that, for an appropriate constant $c(\kappa)$,

$$S_2 + S_1 \ge \frac{2}{3}\|v\|_{N,*}^2 - c(\kappa)\|v\|_2^2 . \tag{4.5.29}$$

We turn next to S_3. Write $\delta w_k^{(j)} = N^{-1/3}[w_{k+N^{1/3}}^{(j)} - w_k^{(j)}]$, $j = 1,2$. Summing by parts we get

$$\begin{aligned} S_3 &= \sum_{i=1}^{N}(w_{i+1}^{(1)} - w_i^{(1)} - \delta w_i^{(1)})v^2(i) + \sum_{i=1}^{N}\delta w_i^{(1)}v^2(i) \\ &= N^{-1/3}\sum_{i=1}^{N}\left(\sum_{\ell=i+1}^{i+N^{1/3}}(w_\ell^{(1)} - w_i^{(1)})\right)(v^2(i+1)-v^2(i)) + \sum_{i=1}^{N}\delta w_i^{(1)}v^2(i) \\ &=: S_{3,1} + S_{3,2} . \end{aligned} \tag{4.5.30}$$

Using (4.5.27) we find that

$$\begin{aligned}
|S_{3,1}| &\le \sum_{i=1}^{N} |v^2(i+1)-v^2(i)|\sqrt{\varepsilon i N^{-1/3}+\kappa_{N,\varepsilon}} \\
&\le \sqrt{\varepsilon}N^{1/3}\sum_{i=1}^{N}(v(i+1)-v(i))^2+\frac{1}{\sqrt{\varepsilon}}\sum_{i=1}^{N}(\varepsilon i N^{-2/3}+\kappa_{N,\varepsilon}N^{-1/3})v^2(i) \\
&\le \sqrt{\varepsilon}\|v\|_{N,*}^2+\frac{\kappa_{N,\varepsilon}}{\sqrt{\varepsilon}}\|v\|_2^2 .
\end{aligned}$$

Applying (4.5.27) again to estimate $S_{3,2}$, we conclude that

$$|S_3| \le (\sqrt{\varepsilon}+\varepsilon)\|v\|_{N,*}^2+(\frac{1}{\sqrt{\varepsilon}}+1)\kappa_{N,\varepsilon}\|v\|_2^2 .$$

A similar argument applies to S_4. Choosing ε small and combining with the estimate (4.5.29) then concludes the proof of the lemma. □

Because the family of random variables in Lemma 4.5.49 is tight, any subsequence $\{N_k\}$ possesses a further subsequence $\{N_{k_i}\}$ so that the estimates there hold with fixed random variables c_i (now independent of N). To prove Theorem 4.5.42, it is enough to consider such a subsequence. With some abuse of notation, we continue to write N instead of N_k.

Each vector $v \in \mathbb{R}^N$ can be identified with a piecewise constant function f_v by the formula $f_v(x) = v(\lceil N^{1/3}x\rceil)$ for $x \in [0,\lceil N^{2/3}\rceil]$ and $f_v(x) = 0$ for all other x. The collection of such functions (for a fixed N) forms a closed linear subspace of $L^2 := L^2(\mathbb{R}_+)$, denoted $L^{2,N}$, and $\hat{H}_N$ acts naturally on $L^{2,N}$. Let $\mathscr{P}_N$ denote the projection from L^2 to $L^{2,N} \subset L^2$. Then $\hat{H}_N$ extends naturally to an operator on L^2 by the formula $\hat{H}_N f = \hat{H}_N \mathscr{P}_N f$. The relation between the operators $\hat{H}_N$ and $\mathbf{H}_\beta$ is clarified in the following lemma.

Lemma 4.5.50 *(a) Let $f_N \in L^{2,N}$ and suppose $f_N \to f$ weakly in L^2, so that*

$$N^{1/3}(f_N(x+N^{-1/3})-f_N(x)) \to f'(x) \quad \text{weakly in } L^2 .$$

Then, for any compactly supported ϕ,

$$\langle \phi, \hat{H}_N f_N\rangle_2 \to \langle \phi, \phi\rangle_{\mathbf{H}_\beta} . \tag{4.5.31}$$

(b) Let $f_N \in L^{2,N}$ with $\|f_N\|_{N,*} \le c$ and $\|f_N\|_2 = 1$. Then there exists an $f \in \mathscr{L}_*$ and a subsequence $N_\ell \to \infty$ so that $f_{N_\ell} \to f$ in L^2 and, for all smooth, compactly supported ϕ, one has

$$\langle \phi, \hat{H}_{N_\ell} f_{N_\ell}\rangle_2 \to_{\ell\to\infty} \langle \phi, f\rangle_{\mathbf{H}_\beta} .$$

Proof The first part is an exercise in summation by parts that we omit. To see the second part, pick a subsequence such that both f_N and $N^{1/3}(f_N(x+N^{-1/3}) - f_N(x))$ converge weakly in L^2 to a limit (f,g), with $f(x) = \int_0^t g(s)ds$ (this is possible because $\|f_N\|_{N,*} < \infty$). An application of the first part of the lemma then completes the proof. □

We have now put in place all the analytic machinery needed to conclude.

Proof of Theorem 4.5.42 Write $\eta_{N,k} = N^{1/6}(\lambda_{N-k}^N - 2\sqrt{N})$. Then $\eta_{N,k}$ is the kth top eigenvalue of $\hat{H}_N$. Let $v_{N,k}$ denote the associated eigenvector, so that $\|f_{v_{N,k}}\|_2 = 1$. We first claim that $\bar{\eta}_k := \limsup \eta_{k,N} \leq \lambda_k$. Indeed, if $\bar{\eta}_k > -\infty$, one can find a subsequence, that we continue to denote by N, so that $(\eta_{N,1}, \dots, \eta_{N,k}) \to (\xi_1, \dots, \xi_k = \bar{\eta}_k)$. By Lemma 4.5.49, for $j = 1, \dots, k$, $\|v_{N,j}\|_{N,*}$ are uniformly bounded, and hence, on a further subsequence, $f_{v_{N,j}}$ converge in L^2 to a limit f_j, $j = 1, \dots, N$, and the f_j are eigenvectors of $\mathbf{H}_\beta$ with eigenvalue at least $\bar{\eta}_k$. Since the f_j are orthogonal in L^2 and the spaces $\mathscr{H}_j$ are one dimensional, it follows that $\lambda_k \geq \bar{\eta}_k$.

To see the reverse implication, that will complete the proof, we use an inductive argument. Suppose that $\eta_{N,j} \to \lambda_j$ and $f_{v_{N,j}} \to f_j$ in L^2 for $j = 1, \dots, k-1$, where (f_j, λ_j) is the jth eigenvector–eigenvalue pair for $\mathbf{H}_\beta$. Let (f_k, λ_k) be the kth eigenvector–eigenvalue pair for $\mathbf{H}_\beta$. Let f_k^ε be smooth and of compact support, so that $\|f_k - f_k^\varepsilon\|_* \leq \varepsilon$, and set

$$f_{N,k} = \mathscr{P}_N f_k^\varepsilon - \sum_{j-1}^{k-1} \langle v_{N,j}, \mathscr{P}_N f_k^\varepsilon \rangle v_{N,j}.$$

Since $\|v_{N,j}\|_{N,*} < c$ for some fixed c by Lemma 4.5.49, and $\|\mathscr{P}_N f_k^\varepsilon - f_{v_{N,k}}\|_2$ is bounded by 2ε for N large, it follows that $\|f_{N,k} - \mathscr{P}_N f_k^\varepsilon\|_{N,*} < c\varepsilon$ for some (random) constant c. Using Lemma 4.5.49 again, we get that

$$\liminf_{N\to\infty} \eta_{N,k} \geq \liminf_{N\to\infty} \frac{\langle f_{N,k}, \hat{H}_N f_{N,k} \rangle}{\langle f_{N,k}, f_{N,k} \rangle} = \liminf_{N\to\infty} \frac{\langle \mathscr{P}_N f_k^\varepsilon, \hat{H}_N \mathscr{P}_N f_k^\varepsilon \rangle}{\langle \mathscr{P}_N f_k^\varepsilon, \mathscr{P}_N f_k^\varepsilon \rangle} + s(\varepsilon), \tag{4.5.32}$$

where $s(\varepsilon) \to_{\varepsilon\to 0} 0$. Applying (4.5.31), we have that

$$\lim_{N\to\infty} \langle \mathscr{P}_N f_k^\varepsilon, \hat{H}_N \mathscr{P}_N f_k^\varepsilon \rangle = \langle f_k^\varepsilon, f_k^\varepsilon \rangle_{\mathbf{H}_\beta}.$$

Substituting in (4.5.32), we get that

$$\liminf_{N\to\infty} \eta_{N,k} \geq \frac{\langle f_k^\varepsilon, f_k^\varepsilon \rangle_{\mathbf{H}_\beta}}{\|f_k\|_2} + s'(\varepsilon),$$

where again $s'(\varepsilon) \to_{\varepsilon\to 0} 0$. This implies, after taking $\varepsilon \to 0$, that

$$\liminf_{N\to\infty} \eta_{N,k} \geq \lambda_k.$$

The convergence of $f_{\nu_{N,k}} \to f_k$ follows from point (b) of Lemma 4.5.50. □

4.6 Bibliographical notes

The background material on manifolds that we used in Section 4.1 can be found in [Mil97] and [Ada69]. The Weyl formula (Theorem 4.1.28) can be found in [Wey39]. A general version of the coarea formula, Theorem 4.1.8, is due to Federer and can be found in [Fed69], see also [Sim83] and [EvG92] for less intimidating descriptions.

The physical motivation for studying different ensembles of random matrices is discussed in [Dys62e]. We note that the Laguerre and Jacobi ensembles occur also through statistical applications (the latter under the name MANOVA, or multivariate analysis of variance), see [Mui81].

Our treatment of the derivation of joint distributions of eigenvalues was influenced by [Due04] (the latter relies directly on Weyl's formula) and [Mat97]. The book [For05] is an excellent recent reference on the derivation of joint distributions of eigenvalues of random matrices belonging to various ensembles; see also [Meh91] and the more recent [Zir96]. Note, however, that the circular ensembles *COE* and *CSE* do *not* correspond to random matrices drawn uniformly from the unitary ensembles as in Proposition 4.1.6. A representation theoretic approach to the study of the latter that also gives central limit theorems for moments is presented in [DiS94] and further developed in [DiE01]. The observation contained in Remark 4.1.7 is motivated by the discussion in [KaS99]. For more on the root systems mentioned in Remark 4.1.5 and their link to the Weyl integration formula, see [Bou05, Chapter 9, Section 2].

The theory of point processes and the concept of Palm measures apply to much more general situations than we have addressed in Section 4.2. A good treatment of the theory is contained in [DaVJ88]. Our exposition builds on [Kal02, Chapter 11].

Point processes $\mathbf{x}^0$ on $\mathbb{R}$ whose associated difference sequences $\mathbf{y}^0$ (see Lemma 4.2.42) are stationary with marginals of finite mean are called *cyclo-stationary*. It is a general fact, see [Kal02, Theorem 11.4], that all cyclo-stationary processes are in one-to-one correspondence with nonzero stationary simple point processes of finite intensity via the Palm recipe.

Determinantal point processes were studied in [Mac75], see also the survey [Sos00]. The representation of Proposition 4.2.20, as well as the observation that it leads to a simple proof of Corollary 4.2.21 and of the CLT of Corollary 4.2.23 (originally proved in [Sos02a]), is due to [HoKPV06]. See also [HoKPV09]. The

Jánossy densities of Definition 4.2.7 for determinantal processes were studied in [BoS03], see [Sos03] for the Pfaffian analog.

The argument in the proof of Proposition 4.2.30 was suggested to us by T. Suidan. Lemma 4.2.50 appears in [Bor99]. Lemma 4.2.52 is taken from [GeV85]. A version valid for continuous time processes was proved earlier in [KaM59]. The relation between non-intersecting random walks, Brownian motions and queueing systems was developed in [OcY01], [OcY02], [KoOR02] and [Oco03]. There is a bijection between paths conditioned not to intersect and certain tiling problems, see [Joh02], [Kra90] and references therein; thus, certain tiling problems are related to determinantal processes. The relation with spanning trees in graphs is described in [BuP93]. Finally, two-dimensional determinantal processes appear naturally in the study of zeroes of random analytic functions, as was discovered in [PeV05], see [HoKPV09].

The description of eigenvalues of the GUE as a diffusion process, that is, Theorem 4.3.2, was first stated by Dyson [Dys62a]. McKean [McK05, p.123] considered the symmetric Brownian motion and related its eigenvalues to Dyson's Brownian motion. A more general framework is developed in [NoRW86] in the context of Brownian motions of ellipsoids. The relation between paths conditioned not to intersect and the Dyson process is studied in [BiBO05] and [DoO05]. The ideas behind Lemma 4.3.6 come from [Śni02]. A version of Lemma 4.3.10 can be found in [RoS93]. When $\beta = 1, 2$, μ_t in that lemma is the asymptotic limit of the spectral measure of $X^{N,\beta}(0) + H^{N,\beta}(t)$. It is a special case of free convolution (of the law μ and the semicircle law with variance t) that we shall describe in Chapter 5. A refined study of the analytic properties of free convolution with a semicircle law that greatly expands on the results in Lemma 4.3.15 appears in [Bia97b].

The properly rescaled process of eigenvalues converges weakly to the *sine process* (in the bulk) and the *Airy process* (at the edge), see [TrW03], [Adl05] and [AdvM05]. The Airy process also appears as the limit of various combinatorial problems. For details, see [PrS02] or [Joh05]. Other processes occur in the study of rescaled versions of the eigenvalue processes of other random matrices. In particular, the Laguerre process arises as the scaling limit of the low-lying eigenvalues of Wishart matrices, see [Bru91], [KoO01] and [Dem07], and has the interpretation of Bessel processes conditioned not to intersect.

The use of stochastic calculus as in Theorem 4.3.20 to prove central limit theorems in the context of Gaussian random matrices was introduced in [Cab01]. This approach extends to the study of the fluctuations of words of two (or more) independent Wigner matrices, see [Gui02] who considered central limit theorems for words of a Gaussian band matrix and deterministic diagonal matrices.

Proposition 4.3.23 is due to [CaG01]. It was completed into a full large deviation principle in [GuZ02] and [GZ04]. By the contraction principle (Theorem D.7), it implies also the large deviations principle for $L_N(1)$, and in particular for the empirical measure of eigenvalues for the sum of a Gaussian Wigner matrix X_N and a deterministic matrix A_N whose empirical measure of eigenvalues converges and satisfies (4.3.23). For $A_N = 0$, this recovers the results of Theorem 2.6.1 in the Gaussian case.

As pointed out in [GuZ02] (see also [Mat94]), the large deviations for the empirical measure of the eigenvalues of $A_N + X_N$ are closely related to the Itzykson–Zuber–Harish-Chandra integral, also called spherical integral, given by

$$I_N^{(2)}(A,D) = \int e^{\frac{\beta N}{2}\mathrm{tr}(UDU^*A)} dm_N^{(\beta)}(U),$$

where the integral is with respect to the Haar measure on the orthogonal group (when $\beta = 1$) and unitary group (when $\beta = 2$). This integral appeared first in the work of Harish-Chandra [Har56] who proved that when $\beta = 2$,

$$I_N^{(2)}(A,D) = \frac{\det((e^{Nd_i a_j})_{1\le i,j\le N})}{\prod_{i<j}(a_i - a_j)\prod_{i<j}(d_i - d_j)},$$

where $(d_i)_{1\le i\le N}$ (resp. $(a_i)_{1\le i\le N}$) denote the eigenvalues of D (resp. A). Itzykson and Zuber [ItZ80] rederived this result, proved it using the heat equation, and gave some properties of $I_N^{(2)}(A,D)$ as N goes to infinity. The integral $I_N^{(2)}(A,D)$ is also related to Schur functions, see [GuM05].

Concentration inequalities have a long history, we refer to [Led01] for a modern and concise introduction. Theorem 4.4.13 is taken from [GuZ00], where analogous bounds are derived, via Talagrand's method [Tal96], for the case in which the entries of the matrix X_N are bounded uniformly by $c/\sqrt{N}$ for some constant c. Under boundedness assumptions, concentration inequalities for the s-largest eigenvalue are derived in [AlKV02]. The proof of Klein's Lemma 4.4.12 follows [Rue69, Page 26].

In [GuZ00] it is explained how Theorems 2.3.5 and 4.4.4 allow one to obtain concentration results for the empirical measure, with respect to the Wasserstein distance

$$d(\mu,\nu) = \sup_{f:\|f\|_\infty\le 1,\|f\|_{\mathcal{L}}\le 1} \left|\int f d\mu - \int f d\nu\right|, \quad \mu,\nu \in M_1(\mathbb{R}).$$

($d(\mu,\nu)$ is also called the Monge–Kantorovich–Rubinstein distance, see the historical comments in [Dud89, p. 341–342]).

Concentration inequalities for the Lebesgue measure on compact connected Riemannian manifold were first obtained, in the case of the sphere, in [Lév22] and then generalized to arbitrary compact connected Riemannian manifold of dimension n with Ricci curvature bounded below by $(n-1)R^2$ for some $R > 0$ in [GrMS86, p. 128]. Our approach in Section 4.4.2 follows Bakry and Emery [BaE85], who introduced the criterion that carries their names. The ergodicity of P_t invoked in the course of proving Theorem 4.4.18, see (4.4.21), does not depend on the BE criterion and holds in greater generality, as a consequence of the fact that Γ vanishes only on the constants, see [Bak94]. In much of our treatment, we follow [AnBC$^+$00, Ch. 5], [GuZ03, Ch. 4] and [Roy07], which we recommend for more details and other applications.

Concentration inequalities for the empirical measure and largest eigenvalue of Hermitian matrices with stable entries are derived in [HoX08].

The first derivation of tridiagonal matrix models for the β-Hermite and Laguerre ensembles is due to [DuE02]. These authors used the models to derive CLT results for linear statistics [DuE06]. In our derivation, we borrowed some tools from [Par80, Ch. 7]. Soon after, other three- and five-diagonal models for the β-Jacobi and circular ensembles were devised in [KiN04], explicitly linking to the theory of orthogonal polynomials on the unit circle and the canonical matrix form of unitary matrices introduced in [CaMV03]. The book [Sim05a] and the survey [Sim07] contains much information on the relations between the coefficients in the three term recursions for orthogonal polynomials on the unit circle with respect to a given measure (the Verblunsky coefficients) and the CMV matrices of [CaMV03]. In this language, the key observation of [KiN04] is that the Verblunsky coefficients corresponding to Haar-distributed unitaries are independent. See also [FoR06], [KiN07] and [BoNR08] for further developments in this direction.

The derivation in Section 4.5.2 of the asymptotics of the eigenvalues of the β-ensembles at the edge is due to [RaRV06], who followed a conjecture of Edelman and Sutton [EdS07]. (In [RaRV06], tail estimates on the top eigenvalue are deduced from the diffusion representation.) The results in [RaRV06] are more general than we have exposed here in that they apply to a large class of tridiagonal matrices, as long as properly rescaled coefficients converge to Brownian motion. Analogous results for the "hard edge" (as in the case of the bottom eigenvalue of Wishart matrices) are described in [RaR08]. A major challenge is to identify the Tracy–Widom distributions (and their β-analogs) from the diffusion in Theorem 4.5.42. The description of the process of eigenvalues in the bulk involves a different machinery, see [VaV07] (where it is called "Brownian carousel") and [KiS09].

5
Free probability

Citing D. Voiculescu, *"Around 1982, I realized that the right way to look at certain operator algebra problems was by imitating some basic probability theory. More precisely, in noncommutative probability theory a new kind of independence can be defined by replacing tensor products with free products and this can help understand the von Neumann algebras of free groups. The subject has evolved into a kind of parallel to basic probability theory, which should be called free probability theory."*

Thus, Voiculescu's first motivation to introduce free probability was the analysis of the von Neumann algebras of free groups. One of his central observations was that such groups can be equipped with tracial states (also called traces), which resemble expectations in classical probability, whereas the property of freeness, once properly stated, can be seen as a notion similar to independence in classical probability. This led him to the statement

free probability theory=noncommutative probability theory+ free independence.

These two components are the basis for a probability theory for noncommutative variables where many concepts taken from probability theory such as the notions of laws, convergence in law, independence, central limit theorem, Brownian motion, entropy and more can be naturally defined. For instance, the law of one self-adjoint variable is simply given by the traces of its powers (which generalizes the definition through moments of compactly supported probability measures on the real line), and the joint law of several self-adjoint noncommutative variables is defined by the collection of traces of words in these variables. Similarly to the classical notion of independence, freeness is defined by certain relations between traces of words. Convergence in law just means that the trace of any word in the noncommutative variables converges towards the right limit.

This chapter is devoted to free probability theory and some of its consequences for the study of random matrices.

5.1 Introduction and main results

The key relation between free probability and random matrices was discovered by Voiculescu in 1991 when he proved that the trace of any word in independent Wigner matrices converges toward the trace of the corresponding word in free semicircular variables. Roughly speaking, he proved the following (see Theorem 5.4.2 for a complete statement).

Theorem 5.1.1 *Let $(\Omega,\mathscr{B},P)$ be a probability space and N,p be positive integers. Let $X_i^N:\Omega\to\mathscr{H}_N^{(\beta)}$, $1\le i\le p$, be a family of independent Gaussian Wigner matrices following the (rescaled) GOE or GUE. Then, for any integer $k\ge 1$ and $i_1,\ldots,i_k\in\{1,\ldots,p\}$, $N^{-1}\mathrm{tr}(X_{i_1}^N\cdots X_{i_\ell}^N)$ converges almost surely (and in expectation) as $N\to\infty$ to a limit denoted $\sigma^{(p)}(X_{i_1}\cdots X_{i_p})$. $\sigma^{(p)}$ is a linear form on noncommutative polynomial functions which is called the* law of p free semicircular variables.

Laws of free variables are defined in Definition 5.3.1. These are noncommutative laws which are defined uniquely in terms of the laws of their variables, that is, in terms of their one-variable marginal distributions. In Theorem 5.1.1 all the one-variable marginals are the same, namely, the semicircle law. The statement of Theorem 5.1.1 extends to Hermitian or real symmetric Wigner matrices whose entries have finite moments, see Theorem 5.4.2. Another extension deals with words that include also deterministic matrices whose law converges, as in the following.

Theorem 5.1.2 *Let $\beta=1$ or 2 and let $(\Omega,\mathscr{B},P)$ be a probability space. Let $\mathbf{D}^N=\{D_i^N\}_{1\le i\le p}$ be a sequence of Hermitian deterministic matrices with uniformly bounded spectral radius, and let $\mathbf{X}^N=\{X_i^N\}_{1\le i\le p}$, $X_i^N:\Omega\to\mathscr{H}_N^{(\beta)}$, $1\le i\le p$, be self-adjoint independent Wigner matrices whose entries have zero mean and finite moments of all order. Assume that for any positive integer k and $i_1,\ldots,i_k\in\{1,\ldots,p\}$, $N^{-1}\mathrm{tr}(D_{i_1}^N\cdots D_{i_k}^N)$ converges to some number $\mu(D_{i_1}\cdots D_{i_k})$.*

Then, for any positive integer ℓ and polynomial functions $(Q_i,P_i)_{1\le i\le\ell}$,

$$\frac{1}{N}\mathrm{tr}\left(Q_1(\mathbf{D}^N)P_1(\mathbf{X}^N)Q_2(\mathbf{D}^N)\cdots P_\ell(\mathbf{X}^N)\right)$$

converges almost surely and in expectation to a limit denoted

$$\tau\left(Q_1(\mathbf{D})P_1(\mathbf{X})Q_2(\mathbf{D})\cdots P_\ell(\mathbf{X})\right).$$

Here, τ is the law of p free semicircular variables $\mathbf{X}$, free from the collection of noncommutative variables $\mathbf{D}$ of law μ.

(See Theorem 5.4.5 for the full statement and the proof.)

Theorems 5.1.1 and 5.1.2 are extremely useful in the study of random matrices. Indeed, many classical models of random matrices can be written as some polynomials in Wigner matrices and deterministic matrices. This is the case for Wishart matrices or, more generally, for band matrices (see Exercises 5.4.14 and 5.4.16).

The law of free variables appears also when one considers random matrices following Haar measure on the unitary group. The following summarizes Theorem 5.4.10.

Theorem 5.1.3 *Take $\mathbf{D}^N = \{D_i^N\}_{1\le i\le p}$ as in Theorem 5.1.2. Let $\mathbf{U}^N = \{U_i^N\}_{1\le i\le p}$ be a collection of independent Haar-distributed unitary matrices independent from $\{D_i^N\}_{1\le i\le p}$, and set $(\mathbf{U}^N)^* = \{(U_i^N)^*\}_{1\le i\le p}$. Then, for any positive integer ℓ and any polynomial functions $(Q_i,P_i)_{1\le i\le \ell}$,*

$$\lim_{N\to\infty}\frac{1}{N}\mathrm{tr}\left(Q_1(\mathbf{D}^N)P_1(\mathbf{U}^N,(\mathbf{U}^N)^*)Q_2(\mathbf{D}^N)\cdots P_\ell(\mathbf{U}^N,(\mathbf{U}^N)^*)\right)$$
$$=\tau\left(Q_1(\mathbf{D})P_1(\mathbf{U},\mathbf{U}^*)Q_2(\mathbf{D})\cdots P_\ell(\mathbf{U},\mathbf{U}^*)\right)\quad a.s.,$$

where τ is the law of p free variables $\mathbf{U}=(U_1,\ldots,U_p)$, free from the noncommutative variables $\mathbf{D}$ of law μ. The law of U_i, $1\le i\le p$, is such that

$$\tau((U_iU_i^*-1)^2)=0,\quad \tau(U_i^n)=\tau((U_i^*)^n)=\mathbf{1}_{n=0}.$$

Thus, free probability appears as the natural setting to study the asymptotics of traces of words in several (possibly random) matrices.

Adopting the point of view that traces of words in several matrices are fundamental objects is fruitful because it leads to the study of some general structure such as freeness (see Section 5.3); freeness in turns simplifies the analysis of convergence of moments. The drawback is that one needs to consider more general objects than empirical measures of eigenvalues converging towards a probability measure, namely, traces of noncommutative polynomials in random matrices converging towards a linear functional on such polynomials, called a tracial state. Analysis of such objects is then achieved using free probability tools.

In the first part of this chapter, Section 5.2, we introduce the setup of free probability theory (the few required notions from the theory of operator algebras are contained in Appendix G). We then define in Section 5.3 the property of freeness and discuss free cumulants and free convolutions. In Section 5.4, which can be read independently of the previous ones except for the description of the limiting quantities in terms of free variables, we show that the asymptotics of many classical models of random matrices satisfy the freeness property, and use that observation to evaluate limiting laws. Finally, Section 5.5 uses free probability tools to describe the behavior of spectral norms of noncommutative polynomials in independent random matrices taken from the GUE.

5.2 Noncommutative laws and noncommutative probability spaces

In this section, we introduce the notions of noncommutative laws and noncommutative probability spaces. An example that the reader should keep in mind concerns $N \times N$ matrices $(M_1, \ldots, M_p)$; a natural noncommutative probability space is then the algebra of $N \times N$ matrices, equipped with the normalized trace $N^{-1}\mathrm{tr}$, whereas the law (or empirical distribution) of $(M_1, \ldots, M_p)$ is given by the collection of the normalized traces of all words in these matrices.

5.2.1 Algebraic noncommutative probability spaces and laws

Basic algebraic notions are recalled in Appendix G.1.

Definition 5.2.1 A *noncommutative probability space* is a pair $(\mathscr{A}, \phi)$ where $\mathscr{A}$ is a unital algebra over $\mathbb{C}$ and ϕ is a linear functional $\phi : \mathscr{A} \to \mathbb{C}$ so that $\phi(1) = 1$. Elements $a \in \mathscr{A}$ are called *noncommutative random variables*.

Let us give some relevant examples of noncommutative probability spaces.

Example 5.2.2

(i) *Classical probability theory* Let $(X, \mathscr{B}, \mu)$ be a probability space and set $\mathscr{A} = L^\infty(X, \mathscr{B}, \mu)$. Take ϕ to be the expectation $\phi(a) = \int_X a(x)\mu(dx)$. Note that, for any $p < \infty$, the spaces $L^p(X, \mathscr{B}, \mu)$ are not algebras for the usual product. (But the intersection $\bigcap_{1 \le p < \infty} L^p(X, \mathscr{B}, \mu)$ is again an algebra.) To consider unbounded variables, we will introduce later the notion of affiliated operators, see Subsection 5.2.3.

(ii) *Discrete groups* Let G be a discrete group with identity e and let $\mathscr{A} = \mathbb{C}(G)$ denote the group algebra (see Definition G.1). Take ϕ to be the linear functional on $\mathscr{A}$ so that, for all $g \in G$, $\phi(g) = 1_{g=e}$.

(iii) *Matrices* Let N be a positive integer and $\mathscr{A} = \mathrm{Mat}_N(\mathbb{C})$. Let $\langle \cdot, \cdot \rangle$ denote the scalar product on $\mathbb{C}^N$ and fix $v \in \mathbb{C}^N$ such that $\langle v, v \rangle = 1$. We can take ϕ on $\mathscr{A}$ to be given by $\phi_v(a) = \langle av, v \rangle$, or by $\phi_N(a) = N^{-1}\mathrm{tr}(a)$.

(iv) *Random matrices* Let $(X, \mathscr{B}, \mu)$ be a probability space. Define $\mathscr{A} = L^\infty(X, \mu, \mathrm{Mat}_N(\mathbb{C}))$, the space of $N \times N$-dimensional complex random matrices with μ-almost surely uniformly bounded entries. Set

$$\phi_N(a) = \frac{1}{N}\int_X \mathrm{tr}(a(x))\mu(dx) = \frac{1}{N}\sum_{i=1}^N \int \langle a(x)e_i, e_i \rangle \mu(dx), \qquad (5.2.1)$$

where here the e_i are the standard basis vectors in $\mathbb{C}^N$. Alternatively, one can consider, with $v \in \mathbb{C}^N$ so that $\langle v, v \rangle = 1$,

$$\phi_v(a) = \int_X \langle a(x)v, v \rangle \mu(dx). \qquad (5.2.2)$$

(v) *Bounded operators on a Hilbert space* Let H be a Hilbert space with inner product $\langle \cdot, \cdot \rangle$ and $B(H)$ be the set of bounded linear operators on H. We set for $v \in H$ so that $\langle v, v \rangle = 1$ and $a \in B(H)$,

$$\phi_v(a) = \langle av, v \rangle.$$

The GNS construction discussed below will show that this example is in a certain sense universal. It is therefore a particularly important example to keep in mind.

We now describe the notion of *laws* of noncommutative variables. Hereafter, J denotes a subset of $\mathbb{N}$, and $\mathbb{C}\langle X_i | i \in J \rangle$ denotes the set of polynomials in noncommutative indeterminates $\{X_i\}_{i \in J}$, that is, the set of all finite $\mathbb{C}$-linear combinations of words in the variables X_i with the empty word identified to $1 \in \mathbb{C}$; in symbols,

$$\mathbb{C}\langle X_i | i \in J \rangle = \{\gamma_0 + \sum_{k=1}^m \gamma_k X_{i_1^k} \cdots X_{i_{p_k}^k}, \gamma_k \in \mathbb{C}, m \in \mathbb{N}, i_j^k \in J\}.$$

$\mathbb{C}[X] = \mathbb{C}\langle X \rangle$ denotes the set of polynomial functions in one variable.

Definition 5.2.3 Let $\{a_i\}_{i \in J}$ be a family of elements in a noncommutative probability space $(\mathscr{A}, \phi)$. Then, the *distribution* (or *law*) of $\{a_i\}_{i \in J}$ is the map $\mu_{\{a_i\}_{i \in J}} : \mathbb{C}\langle X_i | i \in J \rangle \to \mathbb{C}$ such that

$$\mu_{\{a_i\}_{i \in J}}(P) = \phi(P(\{a_i\}_{i \in J})).$$

This definition is reminiscent of the description of compactly supported probability measures (on a collection of random variables) by means of their (mixed) moments. Since linear functionals on $\mathbb{C}\langle X_i|i\in J\rangle$ are uniquely determined by their values on words $X_{i_1}\cdots X_{i_k}$, $(i_1,\ldots,i_k)\in J$, we can and often do think of laws as word-indexed families of complex numbers.

Example 5.2.4 Example 5.2.2 continued.

(i) *Classical probability theory* If $a\in L^\infty(X,\mathscr{B},\mu)$, we get by definition that

$$\mu_a(P)=\int P(a(x))d\mu(x)$$

and so μ_a is (the sequence of moments of) the law of a under μ (or equivalently the push-forward $a_\#\mu$ of μ by a).

(ii) *Discrete groups* Let G be a group with identity e and take $\phi(g)=1_{g=e}$. Fix $\{g_i\}_{1\le i\le n}\in G^n$. The law $\mu=\mu_{\{g_i\}_{1\le i\le n}}$ has then the following description: for any monomial $P=X_{i_1}X_{i_2}\cdots X_{i_k}$, we have $\mu(P)=1$ if $g_{i_1}\cdots g_{i_k}=e$ and $\mu(P)=0$ otherwise.

(iii) *One matrix* Let a be an $N\times N$ Hermitian matrix with eigenvalues $(\lambda_1,\ldots,\lambda_N)$. Then we have, for all polynomials $P\in\mathbb{C}[X]$,

$$\mu_a(P)=\frac{1}{N}\mathrm{tr}(P(a))=\frac{1}{N}\sum_{i=1}^N P(\lambda_i).$$

Thus, μ_a is (the sequence of moments of) the spectral measure of a, and thus (in effect) a probability measure on $\mathbb{R}$.

(iv) *One random matrix* In the setting of part (iv) of Example 5.2.2, if $a:X\to\mathscr{H}_N^{(\beta)}$, for $\beta=1$ or 2, has eigenvalues $(\lambda_1(x),\ldots,\lambda_N(x))_{x\in X}$, we have

$$\begin{aligned}\phi_N(P(a)) &= \frac{1}{N}\int_X \mathrm{tr}(P(a)(x))\mu(dx)=\frac{1}{N}\sum_{i=1}^N\int P(\lambda_i(x))\mu(dx)\\ &= \langle \bar{L}_N,P\rangle\,. \end{aligned}\qquad(5.2.3)$$

Thus, μ_a is (the sequence of moments of) the mean spectral measure of a.

(v) *Several matrices* (Setting of Example 5.2.2, parts (iii) and (iv)) If we are given $\{a_i\}_{i\in J}\in\mathrm{Mat}_N(\mathbb{C})$ so that $a_i=a_i^*$ for all $i\in J$, then for $P\in\mathbb{C}\langle X_i|i\in J\rangle$,

$$\mu_{\{a_i\}_{i\in J}}(P):=N^{-1}\mathrm{tr}\,(P(\{a_i\}_{i\in J}))$$

defines a distribution of noncommutative variables. $\mu_{\{a_i\}_{i\in J}}$ is called the *empirical distribution* or *law of the matrices* $\{a_i\}_{i\in J}$. Note that if $J=\{1\}$ and a_1 is self-adjoint, μ_{a_1} can be identified, by the previous example, as the empirical distribution of the eigenvalues of a_1. Observe that if the $\{a_i\}_{i\in J}$ are random and with the notation of Example 5.2.2, part 4, we may define their "quenched empirical distribution" $\hat{\mu}_{\{a_i(x)\}_{i\in J}}$ for almost all x, or their "annealed empirical distribution" $\int \hat{\mu}_{\{a_i(x)\}_{i\in J}} d\mu(x)$.

(vi) *Bounded operators on a Hilbert space* Let H be a Hilbert space and T a bounded normal linear operator on H with spectrum $\sigma(T)$ (see Appendix G, and in particular Section G.1, for definitions). According to the spectral theorem, Theorem G.6, if χ is the spectral resolution of T, for any polynomial function $P \in \mathbb{C}[X]$,

$$P(T) = \int_{\sigma(T)} P(\lambda) d\chi(\lambda).$$

Therefore, with $v \in H$ so that $\langle v, v\rangle = 1$, we find that

$$\phi_v(P(T)) = \langle P(T)v, v\rangle = \int_{\sigma(T)} P(\lambda) d\langle \chi(\lambda)v, v\rangle.$$

Hence, the law of $T \in (B(H), \phi_v)$ is (the sequence of moments of) the compactly supported complex measure $d\langle\chi(\lambda)v, v\rangle$.

(vii) *Tautological example* Let $\mathscr{A} = \mathbb{C}\langle X_i | i \in J\rangle$ and let $\phi \in \mathscr{A}'$ be any linear functional such that $\phi(1) = 1$. Then $(\mathscr{A}, \phi)$ is a noncommutative probability space and ϕ is identically equal to the law $\mu_{\{X_i\}_{i\in J}}$.

It is convenient to have a notion of convergence of laws. It is easiest to work with the weak*-topology. This leads us to the following definition.

Definition 5.2.5 Let $(\mathscr{A}_N, \phi_N)$, $N \in \mathbb{N} \cup \{\infty\}$, be noncommutative probability spaces, and let $\{a_i^N\}_{i\in J}$ be a sequence of elements of $\mathscr{A}_N$. Then $\{a_i^N\}_{i\in J}$ *converges in law* to $\{a_i^\infty\}_{i\in J}$ if and only if for all $P \in \mathbb{C}\langle X_i | i \in J\rangle$,

$$\lim_{N\to\infty} \mu_{\{a_i^N\}_{i\in J}}(P) = \mu_{\{a_i^\infty\}_{i\in J}}(P).$$

We also say in such a situation that $\{a_i^N\}_{i\in J}$ *converges in moments* to $\{a_i^\infty\}_{i\subset J}$.

Since a law is uniquely determined by its values on monomials in the noncommutative variables X_i, the notion of convergence introduced here is the same as "word-wise" convergence.

The tautological example mentioned in Example 5.2.4 underscores the point that the notion of law is purely algebraic and for that reason too broad to capture any flavor of analysis. We have to enrich the structure of a noncommutative probability space in various ways in order to put the analysis back. To begin to see what sort of additional structure would be useful, consider the case in which J is reduced to a single element. Then a law α is simply a linear functional $\alpha \in \mathbb{C}[X]'$ such that $\alpha(1) = 1$, or equivalently a sequence of complex numbers $\alpha_n = \alpha(X^n)$ indexed by positive integers n. Consider the following question.

Does there exist a probability measure μ on the real line such that $\alpha(P) = \int P(x)\mu(dx)$ for all $P \in \mathbb{C}[X]$?

This is a reformulation in the present setup of the Hamburger moment problem. It is well known that the problem has an affirmative solution if and only if all the moments α_n are real, and furthermore the matrices $\{\alpha_{i+j}\}_{i,j=0}^{n-1}$ are positive definite for all n. We can rephrase the latter conditions in our setup as follows. Given $P = \sum_i a_i X^i \in \mathbb{C}[X]$, $a_i \in \mathbb{C}$, put $P^* = \sum_i a_i^* X^i$. Then the Hamburger moment problem has an affirmative solution if and only if $\alpha(P^*P) \geq 0$ for all $P \in \mathbb{C}[X]$. This example underscores the important role played by positivity. Our next immediate goal is, therefore, to introduce the notion of positivity into the setup of noncommutative probability spaces, through the concept of states and C^*-probability spaces. We will then give sufficient conditions, see Proposition 5.2.14, for a linear functional $\tau \in \mathbb{C}\langle X_i | i \in J\rangle'$ to be written $\phi(P(\{a_i\}_{i\in J})) = \tau(P)$ for all polynomials $P \in \mathbb{C}\langle X_i | i \in J\rangle$, where $\{a_i\}_{i\in J}$ is a fixed family of elements of a C^*-algebra $\mathscr{A}$ and ϕ is a state on $\mathscr{A}$.

5.2.2 C^*-probability spaces and the weak*-topology

We first recall C^*-algebras, see Appendix G.1 for detailed definitions. We will restrict our discussion throughout to unital C^*-algebras (and C^*-subalgebras) without further mentioning it. Thus, in the following, a C^*-algebra $\mathscr{A}$ is a unital algebra equipped with a norm $\|\cdot\|$ and an involution $*$ so that

$$\|xy\| \leq \|x\|\|y\|, \quad \|a^*a\| = \|a\|^2.$$

Recall that $\mathscr{A}$ is complete under its norm.

An element a of $\mathscr{A}$ is said to be *self-adjoint* (respectively, *normal*) if $a^* = a$ (respectively, $a^*a = aa^*$). Let $\mathscr{A}_{\mathrm{sa}}$ (respectively, $\mathscr{A}_{\mathrm{n}}$) denote the set of self-adjoint (respectively, normal) elements of $\mathscr{A}$.

Example 5.2.6 The following are examples of C^*-algebras.

(i) *Function spaces* If X is a Polish space, the spaces $B(X)$ and $C_b(X)$, of $\mathbb{C}$-valued functions which are, respectively, bounded and bounded continuous, are unital C^*-algebras when equipped with the supremum norm and the conjugation operation. Note however that the space $C_0(\mathbb{R})$ of continuous functions vanishing at infinity is in general not a (unital) C^*-algebra, for it has no unit.

(ii) *Classical probability theory* Take $(X, \mathscr{B}, \mu)$ a measure space and set $\mathscr{A} =$

$L^\infty(X,\mu)$, with the norm

$$||f|| = \text{ess sup}_x |f(x)|\,.$$

(iii) *Matrices* An important example is obtained if one takes $\mathscr{A} = \mathrm{Mat}_N(\mathbb{C})$. It is a C^*-algebra when equipped with the standard involution

$$(A^*)_{ij} = \bar{A}_{ji}\,, \quad 1 \le i,j \le N$$

and the operator norm given by the spectral radius.

(iv) *Bounded operators on a Hilbert space* The previous example generalizes as follows. Take H a complex Hilbert space, and consider as $\mathscr{A}$ the space $B(H)$ of linear operators $T : H \to H$ which are bounded for the norm

$$||T||_{B(H)} = \sup_{||e||_H = 1} ||Te||_H\,.$$

Here, the multiplication operation is taken as composition. The adjoint T^* of $T \in B(H)$ is defined as the unique element of $B(H)$ such that $\langle Ty, x\rangle = \langle y, T^*x\rangle$ for all $x,y \in H$, see (G.3).

Part (iv) of Example 5.2.6 is, in a sense, generic: any C^*-algebra $\mathscr{A}$ is isomorphic to a sub C^*-algebra of $B(H)$ for some Hilbert space H (see e.g. [Rud91, Theorem 12.41]). We provide below a concrete example.

Example 5.2.7 Let μ be a probability measure on a Polish space X. The C^*-algebra $\mathscr{A} = L^\infty(X,\mu)$ can be identified as a subset of $B(H)$ with $H = L^2(X,\mu)$ as follows. For all $f \in L^\infty(X,\mu)$, we define the multiplication operator $M_f \in B(H)$ by $M_f g = f \cdot g$ (which is in H if $g \in H$). Then M maps $L^\infty(X,\mu)$ into $B(H)$.

In C^*-algebras, spectral analysis can be developed. We recall (see Appendix G.2) that the spectrum of a normal operator a in a C^*-algebra $\mathscr{A}$ is the compact set

$$\mathrm{sp}(a) = \{\lambda \in \mathbb{C} : \lambda e - a \text{ is not invertible }\} \subset \{z \in \mathbb{C} : |z| \le ||a||\}.$$

The same functional calculus we encountered in the context of matrices can be used in C^*-algebras, for such normal operators a. Suppose that f is continuous on $\mathrm{sp}(a)$. By the Stone–Weierstrass Theorem, f can be uniformly approximated on $\mathrm{sp}(a)$ by a sequence of polynomials p_n^f in a and a^*. Then, by part (iii) of Theorem G.7, the limit

$$f(a) = \lim_{n\to\infty} p_n^f(a,a^*)$$

always exists, does not depend on the sequence of approximations, and yields an element of $\mathscr{A}$. It can thus serve as the definition of $f : a \in \mathscr{A} \mapsto f(a) \in \mathscr{A}$ (one may alternatively use the spectral theorem, see Section G.2).

Remark 5.2.8 The smallest C^*-subalgebra $\mathscr{A}_a \subset \mathscr{A}$ containing a given self-adjoint operator a is given by $\mathscr{A}_a = \{f(a) : f \in C(\mathrm{sp}(a))\}$. Indeed, $\mathscr{A}_a$ contains $\{p(a) : p \in \mathbb{C}[X]\}$ and so, by functional calculus, contains $\{f(a) : f \in C(\mathrm{sp}(a))\}$. The conclusion follows from the fact that the latter is a C^*-algebra. The norm on $\mathscr{A}_a$ is necessarily the spectral radius by Theorem G.3. Observe that this determines an isomorphism of $C(\mathrm{sp}(a))$ into $\mathscr{A}$ that preserves linearity and involution. It is a theorem of Gelfand and Naimark (see e.g. [Rud91, Theorem 11.18]) that if a C^*-algebra $\mathscr{A}$ is commutative then it is isomorphic to the algebra $C(X)$ for some compact X; we will not need this fact.

To begin discussing probability, we need two more concepts: the first is positivity and the second is that of a state.

Definition 5.2.9 Let $(\mathscr{A}, \|\cdot\|, *)$ be a C^*-algebra.

(i) An element $a \in \mathscr{A}$ is *nonnegative* (denoted $a \geq 0$) if $a^* = a$ and its spectrum $\mathrm{sp}(a)$ is nonnegative.
(ii) A *state* is a linear map $\phi : \mathscr{A} \to \mathbb{C}$ with $\phi(e) = 1$ and $\phi(a) \geq 0$ if $a \geq 0$.
(iii) A state is *tracial* if $\phi(ab) = \phi(ba)$ for all $a, b \in \mathscr{A}$.

It is standard to check (see e.g. [Mur90, Theorem 2.2.4]) that

$$\{a \in \mathscr{A} : a \geq 0\} = \{aa^* : a \in \mathscr{A}\}. \tag{5.2.4}$$

Example 5.2.10 An important example is $\mathscr{A} = C(X)$ with X some compact space. Then, by the Riesz representation theorem, Theorem B.11, a state is a probability measure on X.

C^*-probability spaces

Definition 5.2.11 A quadruple $(\mathscr{A}, \|\cdot\|, *, \phi)$ is called a *C^*-probability space* if $(\mathscr{A}, \|\cdot\|, *)$ is a C^*-algebra and ϕ is a state.

As a consequence of Theorem 5.2.24 below, the law of a family of random variables $\{a_i\}_{i\in J}$ in a C^*-probability space can always be realized as the law of random variables $\{b_i\}_{i\in J}$ in a C^*-probability space of the form $(B(H), \|\cdot\|, *, a \mapsto \langle av, v\rangle)$, where H is a Hilbert space with inner product $\langle\cdot,\cdot\rangle$, $\|\cdot\|$ is the operator norm, and $v \in H$ is a unit vector.

We show next how all cases in Example 5.2.2 can be made to fit the definition of C^*-probability space.

Example 5.2.12 Examples 5.2.2 and 5.2.4 continued.

(i) *Classical probability theory* Let $(X,\mathscr{B},\mu)$ be a probability space and set $\mathscr{A} = L^\infty(X,\mathscr{B},\mu)$. Let $\phi(a) = \int_X a(x)\mu(dx)$ be the expectation operator. In this setup, use $H = L^2(X,\mathscr{B},\mu)$, consider each $a \in \mathscr{A}$ as an element of $B(H)$ by associating with it the multiplication operator $M_a f = af$ (for $f \in H$), and then write $\phi(a) = \langle M_a 1, 1\rangle$. $\mathscr{A}$ is equipped with a structure of C^*-algebra as in part (i) of Example 5.2.6. Note that if a is self-adjoint, it is just a real-valued element of $L^\infty(X,\mathscr{B},\mu)$, and the spectrum of M_a is a subset of $[\text{ess-inf}_{x\in X} a(x), \text{ess-sup}_{x\in X} a(x)]$. The spectral projections are then given by $E(\Delta) = M_{1_{a^{-1}(\Delta)}}$ for any Δ in that interval.

(ii) *Discrete groups* Let G be a discrete group. Consider an orthonormal basis $\{v_g\}_{g\in G}$ of $\ell^2(G)$, the set of sums $\sum_{g\in G} c_g v_g$ with $c_g \in \mathbb{C}$ and $\sum |c_g|^2 < \infty$. $\ell^2(G)$ is equipped with a scalar product

$$\langle \sum_{g\in G} c_g v_g, \sum_{g\in G} c'_g v_g \rangle = \sum_{g\in G} c_g \bar{c}'_g,$$

which turns it into a Hilbert space. The action of each $g' \in G$ on $\ell^2(G)$ becomes $\lambda(g')(\sum_g c_g v_g) = \sum_g c_g v_{g'g}$, yielding the left regular representation determined by G, which defines a family of unitary operators on $\ell^2(G)$. These operators are determined by $\lambda(g)v_h = v_{gh}$. The C^*-algebra associated with this representation is generated by the unitary operators $\{\lambda(g)\}_{g\in G}$, and coincides with the operator-norm closure of the linear span of $\{\lambda(g)\}_{g\in G}$ (the latter contains any sum $\sum c_g \lambda(g)$ when $\sum |c_g| < \infty$). It is in particular included in $B(\ell^2(G))$. Take as trace the function $\phi(a) = \langle a v_e, v_e\rangle$ where $e \in G$ is the unit. In particular, $\phi(\sum_g b_g \lambda(g)) = b_e$.

(iii) *Random matrices* In the setting of part (iv) of Example 5.2.2, consider $\mathscr{A} = L^\infty(X,\mu,\mathrm{Mat}_N(\mathbb{C}))$. The function

$$\phi_N(a) = \frac{1}{N}\int_X \mathrm{tr}(a(x))\mu(dx) = \frac{1}{N}\sum_{i=1}^{N}\int \langle a(x)e_i, e_i\rangle \mu(dx), \qquad (5.2.5)$$

on $\mathscr{A}$ is a tracial state. There are many other states on $\mathscr{A}$; for any vector $v \in \mathbb{C}^N$ with $||v|| = 1$,

$$\phi_v(a) = \int \langle a(x)v, v\rangle d\mu(x)$$

is a state.

We now consider the set of laws of variables $\{a_i\}_{i\in J}$ defined on a C^*-probability space.

Definition 5.2.13 Let $(\mathscr{A}, \|\cdot\|, *)$ be a C^*-algebra. Define $\mathscr{M}_{\mathscr{A}} = \mathscr{M}_{\mathscr{A},\|\cdot\|,*}$ to be the set of *states* on $\mathscr{A}$, i.e. the set of linear forms α on $\mathscr{A}$ so that, for all positive elements $a \in \mathscr{A}$,

$$\alpha(a) \geq 0, \qquad \alpha(1) = 1. \tag{5.2.6}$$

(By Lemma G.11, a state α automatically satisfies $\|\alpha\| \leq 1$, that is, $|\alpha(x)| \leq \|x\|$ for any $x \in \mathscr{A}$.) Note that by either Lemma G.11 or (5.2.4), equation (5.2.6) is equivalent to

$$\alpha(bb^*) \geq 0 \quad \forall b \in \mathscr{A}, \qquad \alpha(1) = 1. \tag{5.2.7}$$

In studying laws of random variables $\{a_i\}_{i\in J}$ in a C^*-algebra $\mathscr{A}$, we may restrict attention to self-adjoint variables, by writing for any $a \in \mathscr{A}$, $a = b + \mathrm{i}c$ with $b = (a+a^*)/2$ and $c = \mathrm{i}(a^* - a)/2$ both self-adjoint. Thus, in the sequel, we restrict ourselves to studying the law of self-adjoint elements. In view of this restriction, it is convenient to equip $\mathbb{C}\langle X_i | i \in J\rangle$ with the unique involution so that $X_i = X_i^*$, and, as a consequence,

$$(\lambda X_{i_1} \cdots X_{i_m})^* = \bar{\lambda} X_{i_m} \cdots X_{i_1}\,, \tag{5.2.8}$$

We now present a criterion for verifying that a given linear functional on $\mathbb{C}\langle X_i | i \in J\rangle$ represents the law of a family of (self-adjoint) random variables on some C^*-algebra. Its proof follows ideas that are also employed in the proof of the Gelfand–Naimark–Segal construction, Theorem 5.2.24 below.

Proposition 5.2.14 *Let J be a set of positive integers. Fix a constant $0 < R < \infty$. Let the involution on $\mathbb{C}\langle X_i | i \in J\rangle$ be as in (5.2.8). Then there exists a C^*-algebra $\mathscr{A} = \mathscr{A}(R, J)$ and a family $\{a_i\}_{i\in J}$ of self-adjoint elements of it with the following properties.*

(a) $\sup_{i\in J} \|a_i\| \leq R$.
(b) *$\mathscr{A}$ is generated by $\{a_i\}_{i\in J}$ as a C^*-algebra.*
(c) *For any C^*-algebra $\mathscr{B}$ and family of self-adjoint elements $\{b_i\}_{i\in J}$ of it satisfying $\sup_{i\in J} \|b_i\| \leq R$, we have $\|P(\{a_i\}_{i\in J})\| \geq \|P(\{b_i\}_{i\in J})\|$ for all polynomials $P \in \mathbb{C}\langle X_i | i \in J\rangle$.*
(d) *A linear functional $\alpha \in \mathbb{C}\langle X_i | i \in J\rangle'$ is the law of $\{a_i\}_{i\in J}$ under some state $\tau \in \mathscr{M}_{\mathscr{A}}$ if and only if $\alpha(1) = 1$,*

$$|\alpha(X_{i_1} \cdots X_{i_k})| \leq R^k \tag{5.2.9}$$

for all words $X_{i_1}, \ldots, X_{i_k}$, and $\alpha(P^*P) \geq 0$ for all $P \in \mathbb{C}\langle X_i | i \in J\rangle$.

(e) *Under the equivalent conditions stated in point* (d), *the state* τ *is unique, and furthermore* τ *is tracial if* $\alpha(PQ) = \alpha(QP)$ *for all* $P, Q \in \mathbb{C}\langle X_i | i \in J\rangle$.

Points (a), (b) and (c) of Proposition 5.2.14 imply that, for any C^*-algebra $\mathscr{B}$ and $\{b_i\}_{i\in J}$ as in point (c), there exists a unique continuous algebra homomorphism $\mathscr{A} \to \mathscr{B}$ commuting with $*$ sending a_i to b_i for $i \in J$. In this sense, $\mathscr{A}$ is the universal example of a C^*-algebra equipped with an R-bounded J-indexed family of self-adjoint elements.

Proof To abbreviate notation, we write

$$A = \mathbb{C}\langle X_i | i \in J\rangle .$$

First we construct $\mathscr{A}$ and $\{a_i\}_{i\in J}$ to fulfill the first three points of the proposition by completing A in a certain way. For $P = P(\{X_i\}_{i\in J}) \in A$, put

$$\|P\|_{R,J,C^*} = \sup_{\mathscr{B},\{b_i\}_{i\in J}} \|P(\{b_i\}_{i\in J})\| , \tag{5.2.10}$$

where $\mathscr{B}$ ranges over all C^*-algebras and $\{b_i\}_{i\in J}$ ranges over all families of self-adjoint elements of $\mathscr{B}$ such that $\sup_{i\in J} \|b_i\| \le R$. Put

$$L = \{P \in A : \|P\|_{R,J,C^*} = 0\}.$$

Now the function $\|\cdot\|_{R,J,C^*}$ is a seminorm on the algebra $\mathscr{A}$. It follows that L is a two-sided ideal of $\mathscr{A}$ and that $\|\cdot\|_{R,J,C^*}$ induces on the quotient A/L a norm. Furthermore $\|PP^*\|_{R,J,C^*} = \|P\|^2_{R,J,C^*}$, and hence $\|P^*\|_{R,J,C^*} = \|P\|_{R,J,C^*}$ for all $P \in A$. In particular, the involution $*$ passes to the quotient A/L and preserves the norm induced by $\|\cdot\|_{R,J,C^*}$. Now complete A/L with respect to the norm induced by $\|\cdot\|_{R,J,C^*}$, and equip it with the involution induced by $P \mapsto P^*$, thus obtaining a C^*-algebra. Call this completion $\mathscr{A}$ and let a_i denote the image of X_i in $\mathscr{A}$ for $i \in J$. Thus we obtain $\mathscr{A}$ and self-adjoint $\{a_i\}_{i\in J}$ fulfilling points (a), (b), (c).

Since the implication (d)($\Rightarrow$) is trivial, and point (e) is easy to prove by approximation arguments, it remains only to prove (d)($\Leftarrow$). Given $P = \sum_\xi c_\xi \xi \in A$, where the summation extends over all words ξ in the X_i (including the empty word) and all but finitely many of the coefficients $c_\xi \in \mathbb{C}$ vanish, we define

$$\|P\|_{R,J} = \sum |c_\xi| R^{\deg \xi} < \infty ,$$

where $\deg \xi$ denotes the length of the word ξ. One checks that $\|P\|_{R,J}$ is a norm on A and further, from assumption (5.2.9),

$$|\alpha(P)| \le \|P\|_{R,J} , \quad P \in A . \tag{5.2.11}$$

For $P \in A$ and $Q \in A$ satisfying $\alpha(Q^*Q) > 0$ we define

$$\alpha_Q(P) = \frac{\alpha(Q^*PQ)}{\alpha(Q^*Q)},$$

and we set

$$\|P\|_\alpha = \left(\sup_{\substack{Q \in A \\ \alpha(Q^*Q)>0}} \alpha_Q(P^*P) \right)^{1/2}.$$

By the continuity of α with respect to $\|\cdot\|_{R,J}$, see (5.2.11), and Lemma G.22, we have that $\|P\|_\alpha \le \|P^*P\|_{R,J}^{1/2}$. In particular, $\|X_i\|_\alpha \le R$ for all $i \in J$.

We check that $\|\cdot\|_\alpha$ is a seminorm on A satisfying $\|P^*P\|_\alpha = \|P\|_\alpha^2$ for all $P \in A$. Indeed, for $\lambda \in \mathbb{C}$, $\|\lambda P\|_\alpha = |\lambda| \cdot \|P\|_\alpha$ by definition. We verify next the sub-additivity of $\|\cdot\|_\alpha$. Since α_Q is a nonnegative linear form on A, we have from (G.6) that for any $S, T \in A$,

$$[\alpha_Q((S+T)^*(S+T))]^{1/2} \le [\alpha_Q(S^*S)]^{1/2} + [\alpha_Q(T^*T)]^{1/2},$$

from which $\|S+T\|_\alpha \le \|S\|_\alpha + \|T\|_\alpha$ follows by optimization over Q.

To prove the sub-multiplicativity of $\|\cdot\|$, note first that by the Cauchy–Schwarz inequality (G.5), for $Q, S, T \in A$ with $\alpha(Q^*Q) > 0$,

$$\alpha_Q(T^*S^*ST) \quad \text{vanishes if} \quad \alpha_Q(T^*T) = 0.$$

Then, assuming $\|T\|_\alpha > 0$,

$$\begin{aligned} \|ST\|_\alpha^2 &= \sup_{\substack{Q \in A \\ \alpha(Q^*Q)>0}} \alpha_Q(T^*S^*ST) \\ &= \sup_{\substack{Q \in A \\ \alpha(Q^*T^*TQ)>0}} \alpha_{TQ}(S^*S)\alpha_Q(T^*T) \le \|S\|_\alpha^2 \|T\|_\alpha^2. \end{aligned} \tag{5.2.12}$$

We conclude that $\|\cdot\|_\alpha$ is a seminorm on A.

To verify that $\|TT^*\|_\alpha = \|T\|_\alpha^2$, note that by the Cauchy–Schwarz inequality (G.5) and $\alpha_Q(1) = 1$, we have $|\alpha_Q(T^*T)|^2 \le \alpha_Q((T^*T)^2)$, hence $\|T\|_\alpha^2 \le \|T^*T\|_\alpha$. By (5.2.12), $\|T^*T\|_\alpha \le \|T\|_\alpha \|T^*\|_\alpha$ and therefore we get that $\|T\|_\alpha \le \|T^*\|_\alpha$. By symmetry, this implies $\|T^*\|_\alpha = \|T\|_\alpha = \|T^*T\|_\alpha^{1/2}$, as claimed.

Using again the quotient and completion process which we used to construct $\mathscr{A}$, but this time using the seminorm $\|\cdot\|_\alpha$, we obtain a C^*-algebra $\mathscr{B}$ and self-adjoint elements $\{b_i\}_{i\in J}$ satisfying $\sup_{i\in J} \|b_i\| \le R$ and $\|P\|_\alpha = \|P(\{b_i\}_{i\in J})\|$ for $P \in A$. But then by point (c) we have $\|P\|_\alpha \le \|P\|_{R,J,C^*}$ for $P \in A$, and thus $|\alpha(P)| \le \|P\|_{R,J,C^*}$. Let τ be the unique continuous linear functional on $\mathscr{A}$ such that $\tau(P(\{a_i\}_{i\in J})) = \alpha(P)$ for all $P \in A$. Since $\alpha(P^*P) \geq 0$ for $P \in A$, it follows, see (5.2.7), that τ is positive and hence a state on $\mathscr{A}$. The proof of point (d)($\Leftarrow$) is complete. □

Example 5.2.15 Examples 5.2.2 continued.

(i) *Classical probability* The set $M_1([-R,R])$ of probability measures on $[-R,R]$ can be recovered as the set $\mathscr{M}_{\mathscr{A}(R,\{1\})}$.

(ii) *Matrices* The study of noncommutative laws of matrices $\{a_i\}_{i\in J}$ belonging to $\mathrm{Mat}_N(\mathbb{C})$ with spectral radii bounded by R reduces, by the remark following (5.2.7), to the study of laws of Hermitian matrices. For the latter, the noncommutative law of k matrices whose spectral radii are bounded by R can be represented as elements of $\mathscr{M}_{\mathscr{A}(R,\{1,\ldots,k\})}$.

The examples above do not accommodate laws of unbounded variables. We will see in Section 5.2.3 that such laws can be defined using the notion of affiliated operators.

Weak-topology*

Recall that we endowed the set of noncommutative laws with its weak*-topology, see Definition 5.2.5.

Corollary 5.2.16 *For $N \in \mathbb{N}$, let $\{a_i^N\}_{i\in J}$ be self-adjoint elements of a C^*-probability space $(\mathscr{A}_N, \|\cdot\|_N, *_N, \phi_N)$. Assume that for all $P \in \mathbb{C}\langle X_i | i \in J\rangle$, $\phi_N(P(a_i^N, i \in J))$ converges to some $\alpha(P)$. Let $R > 0$ be given, with $\mathscr{A}(R,J)$ the universal C^*-algebra and $\{a_i\}_{i\in J}$ the elements of it defined in Proposition 5.2.14.*

(i) *If $\sup_{i\in J,N} \|a_i^N\|_N \leq R$, then there exists a collection of states ψ_N, ψ on $\mathscr{A}(R,J)$ so that, for any $P \in \mathbb{C}\langle X_i | i \in J\rangle$,*

$$\psi_N(P(\{a_i\}_{i\in J})) = \phi_N(P(\{a_i^N\}_{i\in J})), \quad \psi(P(\{a_i\}_{i\in J})) = \alpha(P).$$

(ii) *If there exists a finite R so that for all $k \in \mathbb{N}$ and all $(i_j)_{1\leq j\leq k} \in J^k$,*

$$|\alpha(X_{i_1}\cdots X_{i_k})| \leq R^k, \tag{5.2.13}$$

then there exists a state ψ on $\mathscr{A}(R,J)$ so that, for any $P \in \mathbb{C}\langle X_i | i \in J\rangle$,

$$\psi(P(\{a_i\}_{i\in J})) = \alpha(P).$$

Proof By the remark following Proposition 5.2.14, there exist for $N \in \mathbb{N}$ C^*-homomorphisms $h_N : \mathscr{A}(R,J) \to \mathscr{A}_N$ so that $a_i^N = h_N(a_i)$ and the state $\psi_N = \phi_N \circ h_N$ satisfies $\phi_N(P(\{a_i^N\}_{i\in J})) = \psi_N(P(\{a_i\}_{i\in J}))$ for each $P \in \mathbb{C}\langle X_i | i \in J\rangle$. By assumption, $\psi_N(P(\{a_i\}))$ converges to $\alpha(P)$, and thus $|\alpha(P)| \leq \|P(\{a_i\}_{i\in J})\|$ (the norm here is the norm on $\mathscr{A}(R,J)$). As a consequence, α extends to a state on $\mathscr{A}(R,J)$, completing the proof of the first part of the corollary.

The second part of the corollary is a direct consequence of part (d) of Proposition 5.2.14. □

We remark that a different proof of part (i) of Corollary 5.2.16 can be given directly by using part (d) of Proposition 5.2.14. A different proof of part (ii) is sketched in Exercise 5.2.20.

Example 5.2.17 Examples 5.2.2, parts (iii) and (iv), continued.

(i) *Matrices* Let $\{M_j^N\}_{j\in J} \in \mathrm{Mat}_N(\mathbb{C})$ be a sequence of Hermitian matrices and assume that there exists R finite so that

$$\limsup_{N\to\infty} |\mu_{\{M_j^N\}_{j\in J}}(X_{i_1}\cdots X_{i_k})| \leq R^k .$$

Assume that $\mu_{\{M_j^N\}_{j\in J}}(P)$ converge as N goes to infinity to some limit $\alpha(P)$ for all $P \in \mathbb{C}\langle X_i | i \in J\rangle$. Then, there exist noncommutative random variables $\{a_j\}_{j\in J}$ in a C^*-probability space so that $a_i = a_i^*$ and $\{M_j^N\}_{j\in J}$ converge in law to $\{a_j\}_{j\in J}$.

(ii) *Random matrices* Let $(\Omega, \mathscr{B}, \mu)$ be a probability space. For $j \in J$, let $M_j^N(\omega) \in \mathscr{H}_N^{(2)}$ be a collection of Hermitian random matrices. If the requirements of the previous example are satisfied for almost all $\omega \in \Omega$, then we can conclude similarly that $\{M_j^N(\omega)\}_{j\in J} \in \mathrm{Mat}_N(\mathbb{C})$ converges in law to some $\{a^j(\omega)\}_{j\in J}$. Alternatively, assume one can show the convergence of the moments of products of elements from $\{M_j^N(\omega)\}_{j\in J}$, in $L^1(\mu)$. In this case, we endow the C^*-algebra $(\mathrm{Mat}_N(\mathbb{C}), \|\cdot\|_N, *)$ with the tracial state $\phi_N = N^{-1}\mu \circ \mathrm{tr}$. Observe that ϕ_N is continuous with respect to $\|M\|_\infty^\mu := \mathrm{ess\,sup}\|M(\omega)\|_\infty$, but the latter unfortunately may be infinite. However, if we assume that for all $i_j \in J$, $\phi_N(M_{i_1}^N \cdots M_{i_k}^N)$ converges as N goes to infinity to $\alpha(X_{i_1}\cdots X_{i_k})$, and that there exists $R < \infty$ so that, for all $i_j \in J$,

$$\alpha(X_{i_1}\cdots X_{i_k})| \leq R^k ,$$

then it follows from Corollary 5.2.16 that there exists a state ϕ_α on the universal C^*-algebra $\mathscr{A}(R,J)$ and elements $\{a_i\}_{i\in J} \in \mathscr{A}(R,J)$ so that $\{M_i^N(\omega)\}_{i\in J}$ converges in expectation to $\{a_i\}_{i\in J}$, i.e.

$$\lim_{N\to\infty} \phi_N(P(M_i^N(\omega), i \in J)) = \phi_\alpha(P(a_i, i \in J)) \quad \forall P \in \mathbb{C}\langle X_i | i \in J\rangle .$$

This example applies in particular to collections of independent Wigner matrices.

The space $\mathscr{M}_{\mathscr{A}}$ possesses a nice topological property that we state next. The main part of the proof (which we omit) uses the Banach–Alaoglu Theorem, Theorem B.8.

Lemma 5.2.18 *Let $(\mathscr{A}, \|\cdot\|, *)$ be a C^*-algebra, with $\mathscr{A}$ separable. Then $\mathscr{M}_{\mathscr{A}}$ is compact and separable, hence metrizable.*

Thus, on $\mathscr{M}_{\mathscr{A}}$, sequential convergence determines convergence.

As we next show, the construction of noncommutative laws is such that any one-dimensional marginal distribution is a probability measure. This can be seen as a variant of the Riesz representation theorem, Theorem B.11.

Lemma 5.2.19 *Let $(\mathscr{A}, \|\cdot\|, *)$ be a C^*-algebra and μ a state on $(\mathscr{A}, \|\cdot\|, *)$. Let $F \in \mathscr{A}$, $F = F^*$. Then there exists a unique probability measure $\mu_F \in M_1(\mathbb{R})$ with moments $\int x^k \mu_F(dx) = \mu(F^k)$. The support of μ_F is included in $[-\|F\|_{\mathscr{A}}, \|F\|_{\mathscr{A}}]$. Further, the map $\mu \mapsto \mu_F$ from $\mathscr{M}_{\mathscr{A}}$ furnished with the weak*-topology, into $M_1(\mathbb{R})$, equipped with the weak topology, is continuous.*

Proof The uniqueness of μ_F with the prescribed properties is a standard consequence of the bound $|\mu(F^k)| \le \|F\|^k_{\mathscr{A}}$. To prove existence of μ_F, recall the functional calculus described in Remark 5.2.8 which provides us with a map $f \mapsto f(F)$ identifying the C^*-algebra $C(sp_{\mathscr{A}}(F))$ isometrically with the C^*-subalgebra $\mathscr{A}_F \subset \mathscr{A}$ generated by F. The composite map $f \mapsto \mu(f(F))$ is then a state on $C(sp_{\mathscr{A}}(F))$ and hence by Example 5.2.10 a probability measure on $sp_{\mathscr{A}}(F) \subset [-\|F\|_{\mathscr{A}}, \|F\|_{\mathscr{A}}]$. It is clear that this probability measure has the moments prescribed for μ_F. Existence of $\mu_F \in \mathscr{M}_1(\mathbb{R})$ with the prescribed moments follows. Abusing notation, for $f \in C_b(\mathbb{R})$, let $f(F) = g(F) \in \mathscr{A}$ where $g = f|_{sp_{\mathscr{F}}(F)}$ and note that $\mu_F(f) = \int f d\mu_F = \mu(f(F))$ by construction. Finally, to see the claimed continuity, if we take a sequence $\mu^n \in \mathscr{M}_{\mathscr{A}}$ converging to μ for the weak*-topology, for any $f \in C_b(\mathbb{R})$, $\mu^n_F(f)$ converges to $\mu_F(f)$ as n goes to infinity since $f(F) \in \mathscr{A}$. Therefore $\mu \mapsto \mu_F$ is indeed continuous. □

Exercise 5.2.20 In the setting of Corollary 5.2.16, show, without using part (d) of Proposition 5.2.14, that under the assumptions of part (ii) of the corollary, there exists a sequence of states ψ_N on $\mathscr{A}(R+1,J)$ so that $\psi_N(P)$ converges to $\alpha(P)$ for all $P \in \mathbb{C}\langle X_i | i \in J\rangle$. Conclude that α is a state on $\mathscr{A}(R+1,J)$.
Hint: set $f_R(x) = x \wedge (R+1) \vee (-(R+1))$, and define $a_i^{N,R} = f_R(a_i^N)$. Using the Cauchy–Schwarz inequality, show that $\phi_N(P(\{a_i^{N,R}\}_{i\in J}))$ converges to $\alpha(P)$ for all $P \in \mathbb{C}\langle X_i | i \in J\rangle$. Conclude by applying part (i) of the corollary.

5.2.3 W^*-probability spaces

In the previous section, we considered noncommutative probability measures defined on C^*-algebras. This is equivalent, in the classical setting, to defining probability measures as linear forms on the set of continuous bounded functions. However, in the classical setting, it is well known that one can define probability measures as linear forms, satisfying certain regularity conditions, on the set of *measurable* bounded functions. One can define a generalization to the notion of measurable functions in the noncommutative setting.

If one deals with a single (not necessarily bounded) self-adjoint operator b, it is possible by the spectral theorem G.6 to define $g(b)$ for any function g in the set $B(\mathrm{sp}(b))$ of bounded, Borel-measurable functions on $\mathrm{sp}(b)$. This extension is such that for any $x,y \in H$, there exists a compactly supported measure $\mu^b_{x,y}$ (which equals $\langle \chi_b x, y\rangle$ if χ_b is the resolution of the identity of b, see Appendix G.2) such that

$$\langle g(b)x,y\rangle = \int g(z)d\mu^b_{x,y}(z)\,. \tag{5.2.14}$$

In general, $g(b)$ may not belong to the C^*-algebra generated by b; it will, however, belong to a larger algebra that we now define.

Definition 5.2.21 A C^*-algebra $\mathscr{A} \subset B(H)$ for some Hilbert space H is a *von Neumann algebra* (or W^*-algebra) if it is closed with respect to the weak operator topology.

(Weak operator topology closure means that $b_\alpha \to b$ on a net α if, for any fixed $x,y \in H$, $\langle b_\alpha x, y\rangle$ converges to $\langle bx, y\rangle$. Recall, see Theorem G.14, that in Definition 5.2.21, the requirement of closure with respect to the weak operator topology is equivalent to closure with respect to the strong operator topology, i.e., with the previous notation, to $b_\alpha x$ converging to bx in H.)

Definition 5.2.22 A *W^*-probability space* is a pair $(\mathscr{A},\phi)$ where $\mathscr{A}$ is a W^*-algebra, subset of $B(H)$ for some Hilbert space H, and ϕ is a state that can be written as $\phi(a) = \langle a\xi, \xi\rangle$ for some unit vector $\xi \in H$.

Example 5.2.23

(i) We have seen in Remark 5.2.8 that the C^*-algebra $\mathscr{A}_b$ generated by a self-adjoint bounded operator b on a separable Hilbert space H is exactly $\{f(b), f \in C(\mathrm{sp}(b))\}$. It turns out that the von Neumann algebra generated by b is $\bar{\mathscr{A}_b} = \{f(b), f \in B(\mathrm{sp}(b))\}$. Indeed, by Lusin's Theorem, Theorem B.13, for all $x, y \in H$, for any bounded measurable function g, there exists a sequence g_n of uniformly bounded continuous functions converging in $\mu^b_{x,y}$ probability to g. Since we assumed that H is separable, we can, by a diagonalization argument, assume that this convergence holds for all $x, y \in H$ simultaneously. Therefore, the above considerations show that $g_n(b)$ converges weakly to $g(b)$. Thus the weak closure of $\mathscr{A}_b$ contains $\bar{\mathscr{A}_b}$. One sees that $\bar{\mathscr{A}_b}$ is a von Neumann algebra by the double commutant theorem, Theorem G.13, and the spectral theorem, Theorem G.7.

(ii) As a particular case of the previous example (take b to be the right multiplication operator by a random variable with law μ), $L^\infty(X, \mu)$ can be identified as a W^*-algebra. In fact, every commutative von Neumann algebra on a separable Hilbert space H can be represented as $L^\infty(X, \mu)$ for some $(X, \mathscr{B}, \mu)$. (Since we do not use this fact, the proof, which can be found in [Mur90, Theorem 4.4.4], is omitted.)

(iii) An important example of a W^*-algebra is $B(H)$ itself which is a von Neumann algebra since it is trivially closed.

We saw in Proposition 5.2.14 sufficient conditions for a linear functional on $\mathbb{C}\langle X_i | i \in J\rangle$ to be represented by a state in a C^*-algebra $(\mathscr{A}, \|\cdot\|, *)$. The following *GNS construction* gives a canonical way to represent the latter as a state on $B(H)$ for some Hilbert space H.

Theorem 5.2.24 (Gelfand–Naimark–Segal construction) *Let α be a state on a unital C^*-algebra $(\mathscr{A}, \|\cdot\|, *)$ generated by a countable family $\{a_i\}_{i\in J}$ of self-adjoint elements. Then there exists a separable Hilbert space H, equipped with a scalar product $\langle \cdot, \cdot \rangle$, a norm-decreasing $*$-homomorphism $\pi : \mathscr{A} \to B(H)$ and a vector $\xi_1 \in H$ so that the following hold.*

(a) *$\{\pi(a)\xi_1 : a \in \mathscr{A}\}$ is dense in H.*

(b) *Set $\phi_\alpha(x) = \langle \xi_1, x\xi_1\rangle$ for $x \in B(H)$. Then, for all a in $\mathscr{A}$,*

$$\alpha(a) = \phi_\alpha(\pi(a)) .$$

(c) *The noncommutative law of $\{a_i\}_{i\in J}$ in the C^*-probability space $(\mathscr{A}, \|\cdot\|, *, \alpha)$ equals the law of $\{\pi(a_i)\}_{i\in J}$ in the W^*-probability space $(B(H), \phi_\alpha)$.*

(d) *Let $W^*(\{a_i\}_{i\in J})$ denote the von Neumann algebra generated by $\{\pi(a_i) : i \in J\}$ in $B(H)$. If α is tracial, so is the restriction of the state ϕ_α to $W^*(\{a_i\}_{i\in J})$.*

Proof of Theorem 5.2.24 Let $L_\alpha = \{f \in \mathscr{A} | \alpha(f^* f) = 0\}$. As in the proof of

Proposition 5.2.14, L_α is a left ideal. It is closed due to the continuity of the map $f \mapsto \alpha(f^*f)$. Consider the quotient space $\mathscr{A}^\alpha := \mathscr{A} \setminus L_\alpha$. Denote by $\xi : a \mapsto \xi_a$ the map from $\mathscr{A}$ into $\mathscr{A}^\alpha$. Note that, by (G.6), $\alpha(x^*y)$ depends only on ξ_x, ξ_y, and put

$$\langle \xi_x, \xi_y \rangle = \alpha(x^*y), \quad \|\xi_x\|_\alpha := \langle \xi_x, \xi_x \rangle^{\frac{1}{2}},$$

which defines a pre-Hilbert structure on $\mathscr{A}^\alpha$. Let H be the (separable) Hilbert space obtained by completing $\mathscr{A}^\alpha$ with respect to the Hilbert norm $\|\cdot\|_\alpha$.

To construct the morphism π, we consider $\mathscr{A}$ as acting on $\mathscr{A}^\alpha$ by left multiplication and define, for $a \in \mathscr{A}$ and $b \in \mathscr{A}^\alpha$,

$$\pi(a)\xi_b := \xi_{ab} \in \mathscr{A}^\alpha.$$

By (G.7),

$$||\pi(a)\xi_b||_\alpha^2 = ||\xi_{ab}||_\alpha^2 = \alpha(b^*a^*ab) \le ||a||^2\alpha(b^*b) = ||a||^2||\xi_b||_\alpha^2,$$

and therefore $\pi(a)$ extends uniquely to an element of $B(H)$, still denoted $\pi(a)$, with operator norm bounded by $\|a\|$. π is a $*$-homomorphism from $\mathscr{A}$ into $B(H)$, that is, $\pi(ab) = \pi(a)\pi(b)$ and $\pi(a)^* = \pi(a^*)$. To complete the construction, we take ξ_1 as the image under ξ of the unit in $\mathscr{A}$.

We now verify the conclusions (a)–(c) of the theorem. Part (a) holds since H was constructed as the closure of $\{\pi(a)\xi_1 : a \in \mathscr{A}\}$. To see (b), observe that for all $a \in \mathscr{A}$, $\langle \xi_1, \pi(a)\xi_1 \rangle = \langle \xi_1, \xi_a \rangle = \alpha(a)$. Finally, since π is a morphism, $\pi(P(\{a_i\}_{i\in J})) = P(\{\pi(a_i)\}_{i\in J})$, which together with part (b), shows part (c).

To verify part (d), note that part (b) implies that for $a, b \in \mathscr{A}$,

$$\alpha(ab) = \phi_\alpha(\pi(ab)) = \phi_\alpha(\pi(a)\pi(b))$$

and thus, if α is tracial, one gets $\phi_\alpha(\pi(a)\pi(b)) = \phi_\alpha(\pi(b)\pi(a))$. The conclusion follows by a density argument, using the Kaplansky density theorem, Theorem G.15, to first reduce attention to self-adjoint operators and their approximation by a net, belonging to $\pi(\mathscr{A})$, of self-adjoint operators. □

The norm-decreasing $*$-homomorphism constructed by the theorem is in general not one-to-one. This defect can be corrected as follows.

Corollary 5.2.25 *In the setup of Theorem 5.2.24, there exists a separable Hilbert space $\tilde{H}$, a norm-preserving $*$-homomorphism $\tilde{\pi} : \mathscr{A} \to B(\tilde{H})$ and a unit vector $\tilde{\xi} \in \tilde{H}$ such that for all $a \in \mathscr{A}$, $\alpha(a) = \langle \tilde{\pi}(a)\tilde{\xi}, \tilde{\xi} \rangle$.*

Proof By Theorem G.5 there exists a norm-preserving $*$-homomorphism $\pi_{\mathscr{A}} : \mathscr{A} \to B(H_{\mathscr{A}})$ but $H_{\mathscr{A}}$ might be nonseparable. Using the separability of $\mathscr{A}$, it is routine to construct a separable Hilbert space $H_0 \subset H_{\mathscr{A}}$ stable under the action of $\mathscr{A}$ via $\pi_{\mathscr{A}}$ so that the induced representation $\pi_0 : \mathscr{A} \to B(H_0)$ is a norm-preserving $*$-homomorphism. Then, with $\pi : \mathscr{A} \to B(H)$ and ξ_1 as in Theorem 5.2.24, the direct sum $\tilde{\pi} = \pi_0 \oplus \pi : \mathscr{A} \to B(H_0 \oplus H)$ of representations and the unit vector $\tilde{\xi} = 0 \oplus \xi_1 \in H_0 \oplus H$ have the desired properties. □

We will see that the state ϕ_α of Theorem 5.2.24 satisfies additional properties that we now define. These properties will play an important role in our treatment of unbounded operators in subsection 5.2.3.

Definition 5.2.26 Let $\mathscr{A}$ be a von Neumann algebra.

- A state τ on $\mathscr{A}$ is *faithful* iff $\tau(xx^*) = 0$ implies $x = 0$.
- A state on $\mathscr{A}$ is *normal* iff for any monotone decreasing to zero net a_β of nonnegative elements of $\mathscr{A}$,

$$\inf_\beta \tau(a_\beta) = 0\,.$$

The normality assumption is an analog in the noncommutative setup of the regularity assumptions on linear functionals on measurable functions needed to ensure they are represented by measures. For some consequences of normality, see Proposition G.21.

We next show that the Gelfand–Naimark–Segal construction allows us, if α is tracial, to represent any joint law of noncommutative variables as the law of elements of a von Neumann algebra equipped with a faithful normal state. In what follows, we will always restrict ourselves to W^*-probability spaces equipped with a tracial state ϕ. The properties we list below often depend on this assumption.

Corollary 5.2.27 *Let α be a tracial state on a unital C^*-algebra satisfying the assumptions of Theorem 5.2.24. Then, the tracial state ϕ_α on $W^*(\{a_i\}_{i\in J})$ of Theorem 5.2.24 is normal and faithful.*

Proof We keep the same notation as in the proof of Theorem 5.2.24. We begin by showing that ϕ_α is faithful on $W^*(\{a_i\}_{i\in J}) \subset B(H)$. Take $x \in W^*(\{a_i\}_{i\in J})$ so that $\phi_\alpha(x^*x) = 0$. Then we claim that

$$x\pi(a)\xi_1 = 0\,, \quad \text{for all } a \in \mathscr{A}\,. \tag{5.2.15}$$

Indeed, we have

$$\begin{aligned} \|x\pi(a)\xi_1\|_H^2 &= \langle x\pi(a)\xi_1, x\pi(a)\xi_1\rangle = \langle \xi_1, \pi(a)^* x^* x\pi(a)\xi_1\rangle \\ &= \phi_\alpha(\pi(a)^* x^* x\pi(a)) = \phi_\alpha(x\pi(a)\pi(a^*)x^*)\,, \end{aligned}$$

where we used in the last equality the fact that ϕ_α is tracial on $W^*(\{a_i\}_{i\in J})$. Because π is a morphism we have $\pi(a)\pi(a^*) = \pi(aa^*)$, and because the operator norm of $\pi(aa^*) \in B(H)$ is bounded by the norm $\|aa^*\|$ in $\mathscr{A}$, we obtain from the last display

$$\|x\pi(a)\xi_1\|_H^2 = \langle \xi_1, x\pi(aa^*)x^*\xi_1\rangle \le \|aa^*\|\phi_\alpha(x^*x) = 0,$$

completing the proof of (5.2.15). Since $\pi(a)\xi_1$ is dense in H by part (a) of Theorem 5.2.24, and $x \in B(H)$, we conclude that $x\xi = 0$ for all $\xi \in H$, and therefore $x = 0$, completing the proof that ϕ_α is faithful in $W^*(\{a_i\}_{i\in J})$. By using Proposition G.21 with x the projection onto the linear vector space generated by ξ_1, we see that ϕ_α is normal. □

Laws of self-adjoint operators

So far, we have considered bounded operators. However, with applications to random matrices in mind, it is useful also to consider unbounded operators. The theory incorporates such operators via the notion of affiliated operators. Let $\mathscr{A}$ be a W^* -algebra, subset of $B(H)$ for some Hilbert space H.

Definition 5.2.28 A densely defined self-adjoint operator X on a Hilbert space H is said to be *affiliated to* $\mathscr{A}$ if, for any bounded Borel function f on the spectrum of X, $f(X) \in \mathscr{A}$. A closed densely defined operator Y is affiliated with $\mathscr{A}$ if its polar decomposition $Y = uX$ (see Lemma G.9) is such that $u \in \mathscr{A}$ is a partial isometry and X is a self-adjoint operator affiliated with $\mathscr{A}$. We denote by $\widetilde{\mathscr{A}}$ the collection of operators affiliated with $\mathscr{A}$.

(Here, $f(X)$ is defined by the spectral theorem, Theorem G.8, see Section G.2 for details.)

It follows from the definition that a self-adjoint operator X is affiliated with $\mathscr{A}$ iff $(1+zX)^{-1}X \in \mathscr{A}$ for one (or equivalently all) $z \in \mathbb{C}\backslash\mathbb{R}$. (Equivalently, iff all the spectral projections of X belong to $\mathscr{A}$.) By the double commutant theorem, Theorem G.13, this is also equivalent to saying that, for any unitary operator u in the commutant of $\mathscr{A}$, $uXu^* = X$.

Example 5.2.29 *Let μ be a probability measure on $\mathbb{R}$, $H = L^2(\mu)$ and $\mathscr{A} = B(H)$. Let X be the left multiplication by x with law μ, that is, $Xf := xf$, $f \in H$. Then X is a densely defined operator, affiliated with $\mathscr{A}$.*

We define below the noncommutative laws of affiliated operators and of polynomials in affiliated operators.

Definition 5.2.30 Let $(\mathscr{A},\tau)$ be a W^*-probability space and let T be a self-adjoint operator affiliated with $\mathscr{A}$. Then, the *law* μ_T *of* T is the unique probability measure on $\mathbb{R}$ such that $\tau(u(T)) = \int u(\lambda) d\mu_T(\lambda)$ for any bounded measurable function u. The associated *distribution function* is $F_T(x) := F_{\mu_T}(x) := \mu_T((-\infty,x])$, $x \in \mathbb{R}$.

(The uniqueness of μ_T follows from the Riesz representation theorem, Theorem B.11.) The spectral theorem, Theorem G.8, implies that $F_T(x) = \tau(\chi_T((-\infty,x]))$ if χ_T is the resolution of the identity of the operator T (this is well defined since the spectral projection $\chi_T((-\infty,x])$ belongs to $\mathscr{A}$).

Polynomials of affiliated operators are defined by the following algebraic rules: $(A+B)v := Av + Bv$ for any $v \in H$ belonging to the domains of both A and B, and similarly, $(AB)v := A(Bv)$ for v in the domain of B such that Bv is in the domain of A. One difficulty arising with such polynomials is that, in general, they are not closed, and therefore not affiliated. This difficulty again can be overcome by an appropriate completion procedure, which we now describe. Given a W^*-algebra $\mathscr{A}$ equipped with a normal faithful tracial state τ, introduce a topology by declaring the sets

$$N(\varepsilon,\delta) = \{a \in \mathscr{A} : \text{for some projection } p \in \mathscr{A}, \|ap\| \le \varepsilon, \tau(1-p) \le \delta\}$$

and their translates to be neighborhoods. Similarly, introduce neighborhoods in H by declaring the sets

$$O(\varepsilon,\delta) = \{h \in H : \text{for some projection } p \in \mathscr{A}, \|ph\| \le \varepsilon, \tau(1-p) \le \delta\}$$

to be a fundamental system of neighborhoods, i.e. their translates are also neighborhoods. Let $\widehat{\mathscr{A}}$ be the completion of vector space $\mathscr{A}$ with respect to the uniformity defined by the system $N(\varepsilon,\delta)$ of neighborhoods of origin. Let $\widehat{\mathscr{H}}$ be the analogous completion with respect to the system of neighborhoods $O(\varepsilon,\delta)$. A fundamental property of this completion is the following theorem, whose proof, which we skip, can be found in [Nel74].

Theorem 5.2.31 (Nelson) *Suppose $\mathscr{A}$ is a von Neumann algebra equipped with a normal faithful tracial state.*

(i) *The mappings $a \mapsto a^*$, $(a,b) \mapsto a+b$, $(a,b) \mapsto ab$, $(h,g) \mapsto h+g$, $(a,h) \mapsto ah$ with $a,b \in \mathscr{A}$ and $h,g \in H$ possess unique uniformly continuous extensions to $\widehat{\mathscr{A}}$ and $\widehat{H}$.*

(ii) *With $b \in \widehat{\mathscr{A}}$ associate a multiplication operator M_b, with domain $\mathscr{D}(M_b) = \{h \in H : bh \in H\}$, by declaring $M_b h = bh$ for $h \in \mathscr{D}(M_b)$. Then M_b is a closed, densely defined operator affiliated with $\mathscr{A}$, with $M_b^* = M_{b^*}$. Further, if $a \in \widetilde{\mathscr{A}}$, then there exists a unique $b \in \widehat{\mathscr{A}}$ so that $a = M_b$.*

The advantage of the operators M_b is that they recover an algebraic structure. Namely, while if $a, a' \in \widetilde{\mathscr{A}}$ then it is not necessarily the case that $a + a'$ or aa' belong to $\widetilde{\mathscr{A}}$, however, if $a = M_b$ and $a' = M_{b'}$ then $M_{b+b'}$ and $M_{bb'}$ are affiliated operators that equal the closure of $M_b + M_{b'}$ and $M_b M_{b'}$ (see [Nel74, Theorem 4]). Thus, with some standard abuse of notation, if $T_i \in \widetilde{\mathscr{A}}$, $i = 1, \ldots, k$, we say that for $Q \in \mathbb{C}\langle X_i | 1 \le i \le k \rangle$, $Q(T_1, \ldots, T_k) \in \widetilde{\mathscr{A}}$, meaning that with $T_i = M_{a_i}$, we have $M_{Q(a_1,\ldots,a_k)} \in \widetilde{\mathscr{A}}$.

The assumption of the existence of a normal faithful tracial state ensures Property G.18, which is crucial in the proof of the following proposition.

Proposition 5.2.32 *Let $(\mathscr{A}, \tau)$ be a W^*-probability space, subset of $B(H)$ for some separable Hilbert space H. Assume that τ is a normal faithful tracial state. Let $Q \in \mathbb{C}\langle X_i | 1 \le i \le k \rangle$ be self-adjoint. Let $T_1, \ldots, T_k \in \widetilde{\mathscr{A}}$ be self-adjoint, and let $Q(T_1, \ldots, T_k)$ be the self-adjoint affiliated operator described following Theorem 5.2.31. Then, for any sequence u_n of bounded measurable functions converging, as n goes to infinity, to the identity uniformly on compact subsets of $\mathbb{R}$, the law of $Q(u_n(T_1), \ldots, u_n(T_k))$ converges to the law of $Q(T_1, \ldots, T_k)$.*

The proof of Proposition 5.2.32 is based on the two following auxiliary lemmas.

Lemma 5.2.33 *Let $(\mathscr{A}, \tau)$ be as in Proposition 5.2.32. Let $T_1, \ldots, T_k$ be self-adjoint operators in $\widetilde{\mathscr{A}}$, and let $Q \in \mathbb{C}\langle X_i | 1 \le i \le k \rangle$. Then there exists a constant $m(Q) < \infty$, such that, for any projections $p_1, \ldots, p_k \in \mathscr{A}$ so that $T_i' = T_i p_i \in \mathscr{A}$ for $i = 1, 2, \ldots, k$, there exists a projection p such that*

- $Q(T_1, \ldots, T_k)p = Q(T_1', \ldots, T_k')p$,
- $\tau(p) \ge 1 - m(Q) \max_{1 \le i \le k}(1 - \tau(p_i))$.

Note that part of the statement is that $Q(T_1, \ldots, T_k)p \in \mathscr{A}$. In the proof of Proposition 5.2.32, we use Lemma 5.2.33 with projections $p_i = p_i^n := \chi_{T_i}([-n, n])$ on the domain of the T_i that ensure that $(T_1', \ldots, T_k')$ belong to $\mathscr{A}$. Since such projections can be chosen with traces arbitrarily close to 1, Lemma 5.2.33 will allow us to define the law of polynomials in affiliated operators by density, as a consequence of the following lemma.

Lemma 5.2.34 *Let $(\mathscr{A}, \tau)$ be as in Proposition 5.2.32. Let X, Y be two self-adjoint operators in $\widetilde{\mathscr{A}}$. Fix $\varepsilon > 0$. Assume that there exists a projection $p \in \mathscr{A}$ such that $pXp = pYp$ and $\tau(p) \ge 1 - \varepsilon$ for some $\varepsilon > 0$. Then*

$$\sup_{x \in \mathbb{R}} |F_X(x) - F_Y(x)| \le \varepsilon .$$

Note that the Kolmogorov–Smirnov distance

$$d_{KS}(\mu,\nu) := \max_{x\in\mathbb{R}} |F_\mu(x) - F_\nu(x)|$$

dominates the Lévy distance on $M_1(\mathbb{R})$ defined in Theorem C.8. Lemma 5.2.34 shows that, with X,Y,p,ε as in the statement, $d_{KS}(\mu_X,\mu_Y) \le \varepsilon$.

Proof of Lemma 5.2.33 The key to the proof is to show that if $Z \in \widetilde{\mathscr{A}}$ and p is a projection, then there exists a projection q such that

$$\tau(q) \ge \tau(p) \text{ and } Zq = pZq. \tag{5.2.16}$$

With (5.2.16) granted, we proceed by induction, as follows. Let $S_i \in \widetilde{\mathscr{A}}$ and p_i be projections so that $S_i' = S_i p_i \in \mathscr{A}$, $i = 1,2$. (To prepare for the induction argument, at this stage we do not assume that the S_i are self-adjoint.) Write $p_{12} = p_1 \wedge p_2$. By (5.2.16) (applied with $p = p_{12}$), there exist two projections q and q' such that $p_{12}S_1q = S_1q$, $p_{12}S_2q' = S_2q'$. Set $p := p_1 \wedge p_2 \wedge q \wedge q'$. We have that $p_2p = p$ and $q'p = p$, and thus $S_2p = S_2q'p$. The range of S_2q' belongs to the range of p_1 and of p_2 (because $p_{12}S_2q' = S_2q'$). Thus

$$S_2p = S_2q'p = p_1S_2q'p = p_1S_2p = p_1S_2p_2p. \tag{5.2.17}$$

Therefore

$$S_1S_2p = S_1'S_2'p, \tag{5.2.18}$$

where (5.2.17) was used in the last equality. Note that part of the equality is that the image of S_2p is in the domain of S_1 and so $S_1S_2p \in \mathscr{A}$. Moreover, $\tau(p) \ge 1 - 4\max\tau(1-p_i)$ by Property G.18. We proceed by induction. We first detail the next step involving the product $S_1S_2S_3$. Set $S = S_2S_3$ and let p be the projection as in (5.2.18), so that $Sp = S_2'S_3'p \in \mathscr{A}$. Repeat the previous step now with S and S_1, yielding a projection q so that $S_1S_2S_3pq = S_1'S_2'S_3'pq$. Proceeding by induction, we can thus find a projection p' so that $S_1\cdots S_np' = S_1'\cdots S_n'p'$ with $S_i' = S_ip_i$ and $\tau(p) \ge 1 - 2^n \max\tau(1-p_i)$. Similarly, $(S_1+\cdots+S_n)q' = (S_1'+\cdots+S_n')q'$ if $q' = p_1 \wedge p_2 \cdots \wedge p_n$. Iterating these two results, for any given polynomial Q, we find a finite constant $m(Q)$ such that for any $T_i' = T_ip_i$ with $\tau(p_i) \ge 1-\varepsilon$, $1 \le i \le k$, there exists p so that $Q(T_1,\ldots,T_k)p = Q(T_1',\ldots,T_k')p$ and $\tau(p) \ge 1 - m(Q)\varepsilon$.

To complete the argument by proving (5.2.16), we write the polar decomposition $(1-p)Z = uT$ (see G.9), with a self-adjoint nonnegative operator $T = |(1-p)Z|$ and u a partial isometry such that u vanishes on the ortho-complement of the range of T. Set $q = 1 - u^*u$. Noting that $uu^* \le 1-p$, we have $\tau(q) \ge \tau(p)$. Also, $qT = (1-u^*u)T = 0$ implies that $Tq = 0$ since T and q are self-adjoint, and therefore $(1-p)Zq = 0$. □

Proof of Lemma 5.2.34 We first claim that, given an unbounded self-adjoint operator T affiliated to $\mathscr{A}$ and a real number x, we have

$$F_T(x) = \sup\{\tau(q) : q^* = q^2 = q \in \mathscr{A},\ qTq \in \mathscr{A},\ qTq \le xq\}. \qquad (5.2.19)$$

More precisely, we now prove that the supremum is achieved for $c \downarrow -\infty$ with the projections $q_{T,c}(x) = \chi_T((c,x])$ provided by the spectral theorem. At any rate, it is clear that $F_T(x) = \tau(\chi_T((-\infty,x]))$ is a lower bound for the right side of (5.2.19). To show that $F_T(x)$ is also an upper bound, consider any projection $r \in \mathscr{A}$ such that $\tau(r) > F_T(x)$ with rTr bounded. Put $q = \chi_T((-\infty,x])$. We have $\tau(r) > \tau(q)$. We have $\tau(r - r\wedge q) = \tau(r \vee q - q) \ge \tau(r) - \tau(q) > 0$ using Proposition G.17. Therefore we can find a unit vector $v \in H$ such that $\langle rTrv, v\rangle > x$, thus ruling out the possibility that $\tau(r)$ belongs to the set of numbers on the right side of (5.2.19). This completes the proof of the latter equality.

Consider next the quantity

$$F_{T,p}(x) = \sup\{\tau(q) : q^* = q^2 = q \in \mathscr{A},\ qTq \in \mathscr{A},\ qTq \le xq, q \le p\}.$$

We claim that

$$F_T(x) - \varepsilon \le F_{T,p}(x) \le F_T(x). \qquad (5.2.20)$$

The inequality on the right of (5.2.20) is obvious. We get the lower equality by taking $q = q_{T,c}(x) \wedge p$ on the right side of the definition of $F_{T,p}(x)$ with c large and using Proposition G.17 again. Thus, (5.2.20) is proved.

To complete the proof of Lemma 5.2.34, simply note that $F_{X,p}(x) = F_{Y,p}(x)$ by hypothesis, and apply (5.2.20). □

Proof of Proposition 5.2.32 Put $T_i^n := T_i p_i^n$ with $p_i^n = \chi_{T_i}([-n,n])$. Define the multiplication operator $M_Q := M_{Q(T_1,\ldots,T_k)}$ as in Theorem 5.2.31. By Lemma 5.2.33, we can find a projection p^n such that

$$X^n := p^n Q(T_1^n,\ldots,T_k^n)p^n = p^n Q(T_1,\ldots,T_k)p^n = p^n M_Q p^n$$

and $\tau(p^n) \ge 1 - m(Q)\max_i \tau(1 - \chi_{T_i}([-n,n]))$. By Lemma 5.2.34,

$$d_{KS}(\mu_{M_Q}, \mu_{Q(T_1^n,\ldots,T_k^n)}) \le m(Q)\max_i \tau(1 - \chi_{T_i}([-n,n])),$$

implying the convergence of the law of $Q(T_1^n,\ldots,T_k^n)$ to the law of M_Q. Since also by construction $p_i^n T_i p_i^n = w^n(T_i)$ with $w^n(x) = x1_{|x|\le n}$, we see that we can replace now w^n by any other local approximation u^n of the identity since the difference

$$X^n - p^n Q(u^n(T_1),\ldots,u^n(T_k))p^n$$

is uniformly bounded by $c\sup_{|x|\le n}|w^n - u^n|(x)$ for some finite constant $c = c(n, \sup_{|x|\le n}|w^n(x)|, Q)$ and therefore goes to zero when $u^n(x)$ approaches the identity map on $[-n,n]$. □

5.3 Free independence

What makes free probability special is the notion of freeness that we define in Section 5.3.1. It is the noncommutative analog of independence in probability. In some sense, probability theory distinguishes itself from integration theory by the notions of independence and of random variables which are the basis to treat problems from a different perspective. Similarly, free probability differentiates from noncommutative probability by this very notion of freeness which makes it a noncommutative analog of classical probability.

5.3.1 Independence and free independence

Classical independence of random variables can be defined in the noncommutative context. We assume throughout that $(\mathscr{A},\phi)$ is a noncommutative probability space. Suppose $\{\mathscr{A}_i\}_{i\in I}$ is a family of subalgebras of $\mathscr{A}$, each containing the unit of $\mathscr{A}$. The family is called *independent* if the algebras $\mathscr{A}_i$ commute and $\phi(a_1\cdots a_n)=\phi(a_1)\cdots\phi(a_n)$ for $a_i\in\mathscr{A}_{k(i)}$ with $i\neq j\Rightarrow k(i)\neq k(j)$. This is the natural notion of independence when considering tensor products, as is the case in the classical probability example $L^\infty(X,\mathscr{B},\mu)$.

Free independence is a completely different matter.

Definition 5.3.1 Let $\{\mathscr{A}_j\}_{j\in I}$ be a family of subalgebras of $\mathscr{A}$, each containing the unit of $\mathscr{A}$. The family $\{\mathscr{A}_j\}_{j\in I}$ is called *freely independent* if for any positive integer n, indices $k(1)\neq k(2)$, $k(2)\neq k(3)$, ... , $k(n-1)\neq k(n)$ in I and any $a_j\in\mathscr{A}_{k(j)}$, $j=1,\ldots,n$, with $\phi(a_j)=0$, it holds that

$$\phi(a_1\cdots a_n)=0\,.$$

Let r, $(m_k)_{1\le k\le r}$ be positive integers. The sets $(X_{1,p},\ldots,X_{m_p,p})_{1\le p\le r}$ of noncommutative random variables are called *free* if the algebras they generate are free.

Note that, in contrast to the classical notion of independence, repetition of indices is allowed provided they are not consecutive; thus, free independence is a truly noncommutative notion. Note also that it is impossible to have $a_i=1$ in Definition 5.3.1 because of the condition $\phi(a_i)=0$.

Observe that we could have assumed that $\mathscr{A}$ as well as all members of the family $\{\mathscr{A}_i\}_{i\in I}$ are W^*-algebras. In that situation, if σ_i is a family of generators of the W^*-algebra $\mathscr{A}_i$, then the W^*-subalgebras $\{\mathscr{A}_i\}_{i\in I}$ are free iff the families of variables $\{\sigma_i\}_{i\in I}$ are free.

Remark 5.3.2

(i) Independence and free independence are quite different. Indeed, let X,Y be two self-adjoint elements of a noncommutative probability space $(\mathscr{A},\phi)$ such that $\phi(X)=\phi(Y)=0$ but $\phi(X^2)\neq 0$ and $\phi(Y^2)\neq 0$. If X,Y commute and are independent,

$$\phi(XY)=0, \quad \phi(XYXY)=\phi(X^2)\phi(Y^2)\neq 0,$$

whereas if X,Y are free, then $\phi(XY)=0$ but $\phi(XYXY)=0$.

(ii) The interest in free independence is that if the subalgebras $\mathscr{A}_i$ are freely independent, the restrictions of ϕ to the $\mathscr{A}_i$ are sufficient in order to compute ϕ on the subalgebra generated by all $\mathscr{A}_i$. To see that, note that it is enough to compute $\phi(a_1a_2\cdots a_n)$ for $a_i\in\mathscr{A}_{k(i)}$ and $k(i)\neq k(i+1)$. But, from the freeness condition,

$$\phi((a_1-\phi(a_i)1)(a_2-\phi(a_2)1)\cdots(a_n-\phi(a_n)1))=0. \tag{5.3.1}$$

Expanding the product (using linearity), one can inductively compute $\phi(a_1\cdots a_n)$ as a function of lower order terms. We will see a systematic way to perform such computations in Section 5.3.2.

(iii) The law of free sets of noncommutative variables is a continuous function of the laws of the sets. For example, let $\mathbf{X}_p=(X_{1,p},\ldots,X_{m,p})$ and $\mathbf{Y}_p=(Y_{1,p},\ldots,Y_{n,p})$ be sets of noncommutative variables for each p which are free. Assume that the law of $\mathbf{X}_p$ (respectively, $\mathbf{Y}_p$) converges as p goes to infinity towards the law of $\mathbf{X}=(X_1,\ldots,X_m)$ (respectively, $\mathbf{Y}=(Y_1,\ldots,Y_n)$).

(a) If the sets $\mathbf{X}$ and $\mathbf{Y}$ are free, then the joint law of $(\mathbf{X}_p,\mathbf{Y}_p)$ converges to the joint law of $(\mathbf{X},\mathbf{Y})$.

(b) If instead the joint law of $(\mathbf{X}_p,\mathbf{Y}_p)$ converge to the joint law of $(\mathbf{X},\mathbf{Y})$, then $\mathbf{X}$ and $\mathbf{Y}$ are free.

(iv) If the restriction of ϕ to each of the subalgebras $\{\mathscr{A}_i\}_{i\in I}$ is tracial, then the restriction of ϕ to the algebra generated by $\{\mathscr{A}_i\}_{i\in I}$ is also tracial.

The proof of some basic properties of free independence that are inherited by subalgebras is left to Exercise 5.3.8.

The following are standard examples of free variables.

Example 5.3.3

(i) *Free products of groups* (Continuation of Example 5.2.2, part (ii)) Suppose G is a group which is the free product of its subgroups G_i, that is, every element in G can be written as the product of elements in the G_i and

$g_1 g_2 \cdots g_n \neq e$ whenever $g_j \in G_{i(j)} \setminus \{e\}$ and $i(j) \neq i(j+1)$ for all j. In this setup, we may take as $\mathscr{A}$ the W^*-algebra generated by the left regular representation $\lambda(G)$, see part (ii) of Example 5.2.12, and we may take τ as the trace ϕ defined in that example. Take also as $\mathscr{A}_i$ the W^*-algebra generated by the left regular representations $\lambda(G_i)$. This coincides with those operators $\sum_g c_g \lambda(g)$ with $c(g) = 0$ for $g \notin G_i$ that form bounded operators. Now, if $a \in \mathscr{A}_i$ and $\phi(a) = 0$ then $c_e = \phi(a) = 0$. Thus, if $a_i \in \mathscr{A}_{k(i)}$ with $\phi(a_i) = 0$ and $k(i) \neq k(i+1)$, the resulting operator corresponding to $a_1 \cdots a_n$, denoted $\sum_g c_g \lambda(g)$, satisfies $c_g \neq 0$ only if $g = g_1 \cdots g_n$ for $g_i \in G_{k(i)} \setminus e$. In particular, since $g_1 \cdots g_n \neq e$, we have that $c_e = 0$, i.e. $\phi(a_1 \cdots a_n) = 0$, which proves the freeness of the $\mathscr{A}_i$. The converse is also true, that is, if the subalgebras $\mathscr{A}_i$ associated with the subgroups G_i are free, then the subgroups are algebraically free.

(ii) *Fock spaces.* Let H be a Hilbert space and define the *Boltzmann–Fock* space as

$$\mathscr{T} = \bigoplus_{n \geq 0} H^{\otimes n} . \tag{5.3.2}$$

(Here, $H^{\otimes 0} = \mathbb{C}1$ where 1 is an arbitrary unit vector in H). $\mathscr{T}$ is itself a Hilbert space (with the inner product determined from the inner product in H by (G.1) and (G.2)). If $\{e_i\}$ is an orthonormal basis in H, then $\{1\}$ is an orthonormal basis for $H^{\otimes 0}$, and $\{e_{i_1} \otimes \cdots \otimes e_{i_n}\}$ is an orthonormal basis for $H^{\otimes n}$. An orthonormal basis for $\mathscr{T}$ is constructed naturally from these bases.

For $h \in H$, define $\ell(h)$ to be the left creation operator, $\ell(h)g = h \otimes g$. On the algebra of bounded operators on $\mathscr{T}$, denoted $\mathscr{B}(\mathscr{T})$, consider the state given by the *vacuum*, $\phi(a) = \langle a1, 1 \rangle$. We next show that the family $\{\ell(e_i), \ell^*(e_i)\}$ is freely independent in $(\mathscr{B}(\mathscr{T}), \phi)$. Here, $\ell_i^* := \ell^*(e_i)$, the *left annihilation* operator, is the operator adjoint to $\ell_i := \ell(e_i)$. We have $\ell_i^* 1 = 0$. More generally,

$$\ell_i^* e_{i_1} \otimes e_{i_2} \otimes \cdots \otimes e_{i_n} = \delta_{i i_1} e_{i_2} \otimes \cdots \otimes e_{i_n}$$

because, for $g \in \mathscr{T}$ with $(n-1)$th term equal to g_{n-1},

$$\begin{aligned} \langle e_{i_1} \otimes e_{i_2} \otimes \cdots \otimes e_{i_n}, \ell_i g \rangle &= \langle e_{i_1} \otimes e_{i_2} \otimes \cdots \otimes e_{i_n}, e_i \otimes g_{n-1} \rangle \\ &= \delta_{i i_1} \langle e_{i_2} \otimes \cdots \otimes e_{i_n}, g_{n-1} \rangle . \end{aligned}$$

Note that even though $\ell_i \ell_i^*$ is typically not the identity, it does hold true that $\ell_i^* \ell_j = \delta_{ij} I$ with I the identity in $\mathscr{B}(\mathscr{T})$. Due to that, the algebra generated by (ℓ_i, ℓ_i^*, I) is generated by the terms $\ell_i^q (\ell_i^*)^p$, $p + q > 0$, and I. Note also that

$$\phi(\ell_i^q(\ell_i^*)^p) = \langle(\ell_i^*)^p 1, (\ell_i^*)^q 1\rangle = 0,$$

since at least one of p,q is nonzero. Thus, we need only to prove that if $p_k + q_k > 0$, $i_k \neq i_{k+1}$,

$$Z := \phi\left(\ell_{i_1}^{q_1}(\ell_{i_1}^*)^{p_1}\ell_{i_2}^{q_2}(\ell_{i_2}^*)^{p_2}\cdots\ell_{i_n}^{q_n}(\ell_{i_n}^*)^{p_n}\right) = 0.$$

But necessarily if $Z \neq 0$ then $q_1 = 0$ (for otherwise a term e_{i_1} pops out on the left of the expression which will then be annihilated in the scalar product with 1). Thus, $p_1 > 0$, and then one must have $q_2 = 0$, implying in turn $p_2 > 0$, etc., up to $p_n > 0$. But since $(\ell_{i_n}^*)^{p_n} 1 = 0$, we conclude that $Z = 0$.

In classical probability one can create independent random variables by forming products of probability spaces. Analogously, in free probability, one can create free random variables by forming free products of noncommutative probability spaces. More precisely, if $\{(\mathscr{A}_j, \phi_j)\}$ is a family of noncommutative probability spaces, one may construct a noncommutative probability space $(\mathscr{A}, \phi)$ equipped with injections $i_j : \mathscr{A}_j \to \mathscr{A}$ such that $\phi_j = \phi \circ i_j$ and the images $i_j(\mathscr{A}_j)$ are free in $\mathscr{A}$.

We now explain the construction of free products in a simplified setting sufficient for the applications we have in mind. We assume each noncommutative probability space $(\mathscr{A}_j, \phi_j)$ is a C^*-probability space, $\mathscr{A}_j$ is separable, and the family $\{(\mathscr{A}_j, \phi_j)\}$ is countable. By Corollary 5.2.25, we may assume that $\mathscr{A}_j$ is a C^*-subalgebra of $B(H_j)$ for some separable Hilbert space H_j, and that for some unit vector $\zeta_j \in H_j$ we have $\phi_j(a) = \langle a\zeta_j, \zeta_j\rangle$ for all $a \in \mathscr{A}_j$. Then the free product $(\mathscr{A}, \phi)$ we aim to construct will be a C^*-subalgebra of $B(\mathscr{H})$ for a certain separable Hilbert space $\mathscr{H}$, and we will have for some unit vector $\zeta \in \mathscr{H}$ that $\phi(a) = \langle a\zeta, \zeta\rangle$ for all $a \in \mathscr{A}$.

We construct $(\mathscr{H}, \zeta)$ as the *free product* of the pairs (H_j, ζ_j). Toward that end, given $f \in H_j$, let $\mathring{f} = f - \langle f, \zeta_j\rangle\zeta_j \in H_j$ and put $\mathring{H}_j = \{\mathring{f} : f \in H_j\}$. Then, for a unit vector ζ in some Hilbert space which is independent of j, put

$$\mathscr{H}(j) := \mathbb{C}\zeta \oplus \bigoplus_{n\geq 1}\left(\bigoplus_{\substack{j_1\neq j_2\cdots\neq j_n\\ j_1\neq j}} \mathring{H}_{j_1} \otimes \mathring{H}_{j_2} \otimes \cdots \otimes \mathring{H}_{j_n}\right). \tag{5.3.3}$$

Let $\mathscr{H}$ be defined similarly but without the restriction $j_1 \neq j$. Note that all the Hilbert spaces $\mathscr{H}(j)$ are closed subspaces of $\mathscr{H}$. We equip $B(\mathscr{H})$ with the state $\tau = (a \mapsto \langle a\zeta, \zeta\rangle)$, and hereafter regard it as a noncommutative probability space.

We need next for each fixed j to define an embedding of $B(H_j)$ in $B(\mathscr{H})$. Toward that end we define a Hilbert space isomorphism $V_j : H_j \otimes \mathscr{H}(j) \to \mathscr{H}$ as follows, where h_j denotes a general element of H_j.

$$\begin{aligned} \zeta_j \otimes \zeta &\mapsto \zeta, \\ \mathring{h}_j \otimes \zeta &\mapsto \mathring{h}_j, \\ \zeta_j \otimes (\mathring{h}_{j_1} \otimes \mathring{h}_{j_2} \otimes \cdots \otimes \mathring{h}_{j_n}) &\mapsto \mathring{h}_{j_1} \otimes \mathring{h}_{j_2} \otimes \cdots \otimes \mathring{h}_{j_n}, \\ \mathring{h}_j \otimes (\mathring{h}_{j_1} \otimes \mathring{h}_{j_2} \otimes \cdots \otimes \mathring{h}_{j_n}) &\mapsto \mathring{h}_j \otimes \mathring{h}_{j_1} \otimes \mathring{h}_{j_2} \otimes \cdots \otimes \mathring{h}_{j_n}. \end{aligned}$$

Then, given $T \in B(H_j)$, we define $\pi_j(T) \in B(\mathscr{H})$ by the formula

$$\pi_j(T) = V_j \circ (T \otimes I_{\mathscr{H}(j)}) \circ V_j^*$$

where $I_{\mathscr{H}(j)}$ denotes the identity mapping of $\mathscr{H}(j)$ to itself. Note that π_j is a norm-preserving $*$-homomorphism of $B(H_j)$ into $B(\mathscr{H})$. The crucial feature of the definition is that for $j \neq j_1 \neq j_2 \neq \cdots \neq j_m$,

$$\pi_j(T)(\mathring{h}_{j_1} \otimes \cdots \otimes \mathring{h}_{j_m}) = \phi_j(T)\mathring{h}_{j_1} \otimes \cdots \otimes \mathring{h}_{j_m} + (T\zeta_j)^\circ \otimes \mathring{h}_{j_1} \otimes \cdots \otimes \mathring{h}_{j_m}. \quad (5.3.4)$$

We have nearly reached our goal. The key point is the following.

Lemma 5.3.4 *In the noncommutative probability space $(B(\mathscr{H}), \tau)$, the subalgebras $\pi_j(B(H_j))$ are free.*

The lemma granted, we can quickly conclude the construction of the free product $(\mathscr{A}, \phi)$, as follows. We take $\mathscr{A}$ to be the C^*-subalgebra of $B(\mathscr{H})$ generated by the images $\pi_j(\mathscr{A}_j)$, ϕ to be the restriction of τ to $\mathscr{A}$, and i_j to be the restriction of π_j to $\mathscr{A}_j$. It is immediate that the images $i_j(\mathscr{A}_j)$ are free in $(\mathscr{A}, \phi)$.

Proof of Lemma 5.3.4 Fix $j_1 \neq j_2 \neq \cdots \neq j_m$ and operators $T_k \in B(H_{j_k})$ for $k = 1, \ldots, m$. Note that by definition $\tau(\pi_{j_k}(T_k)) = \langle T_k \zeta_{j_k}, \zeta_{j_k} \rangle$. Put $\mathring{T}_k = T_k - \langle T_k \zeta_{j_k}, \zeta_{j_k} \rangle I_{j_k}$, where I_{j_k} denotes the identity mapping of H_{j_k} to itself, noting that $\tau(\pi_{j_k}(\mathring{T}_k)) = 0$. By iterated application of (5.3.4) we have

$$\pi_{j_1}(\mathring{T}_1) \cdots \pi_{j_m}(\mathring{T}_m)\zeta = (\mathring{T}_1 \zeta_{j_1}) \otimes \cdots \otimes (\mathring{T}_m \zeta_{j_m}) \in \mathring{H}_{j_1} \otimes \mathring{H}_{j_2} \otimes \cdots \otimes \mathring{H}_{j_m}.$$

Since the space on the right is orthogonal to ζ, we have

$$\tau(\pi_{j_1}(\mathring{T}_1) \cdots \pi_{j_m}(\mathring{T}_m)) = 0.$$

Thus the C^*-subalgebras $\pi_j(B(H_j))$ are indeed free in $B(\mathscr{H})$ with respect to the state τ. □

Remark 5.3.5 In point (i) of Example 5.3.3 the underlying Hilbert space equipped with unit vector is the free product of the pairs $(\ell^2(G_i), v_{e_{G_i}})$, while in point (ii) it is the free product of the pairs $(\bigoplus_{n=0}^{\infty} \mathbb{C}e_i^{\otimes n}, 1)$.

Remark 5.3.6 The free product $(\mathscr{A}, \phi)$ of a family $\{(\mathscr{A}_j, \phi_j)\}$ can be constructed purely algebraically, using just the spaces $(\mathscr{A}_j, \phi_j)$ themselves, but it is less simple to describe precisely. Given $a \in \mathscr{A}_j$, put $\mathring{a} = a - \phi_j(a)1_{\mathscr{A}_j}$ and $\mathring{\mathscr{A}}_j = \{\mathring{a} : a \in \mathscr{A}_j\}$. At the level of vector spaces,

$$\mathscr{A} = \mathbb{C}1_{\mathscr{A}} \oplus \left(\bigoplus_{j_1 \neq j_2 \neq \cdots \neq j_m} \mathring{\mathscr{A}}_{j_1} \otimes \cdots \otimes \mathring{\mathscr{A}}_{j_m} \right).$$

The injection $i_j : \mathscr{A}_j \to \mathscr{A}$ is given by the formula

$$i_j(a) = \phi_j(a)1_{\mathscr{A}} \oplus \mathring{a} \in \mathbb{C}1_{\mathscr{A}} \oplus \mathring{\mathscr{A}}_j \subset \mathscr{A}$$

and the state ϕ is defined by

$$\phi(1_{\mathscr{A}}) = 1, \quad \phi\left(\mathring{\mathscr{A}}_{j_1} \otimes \cdots \otimes \mathring{\mathscr{A}}_{j_m}\right) = 0.$$

Multiplication in $\mathscr{A}$ is obtained, roughly, by simplifying as much as possible when elements of the same algebra $\mathscr{A}_j$ are juxtaposed. Since a rigorous definition takes some effort and is not needed, we do not describe it in detail.

Exercise 5.3.7 In the setting of part (ii) of Example 5.3.3, show that, for all $n \in \mathbb{N}$,

$$\phi[(\ell_1 + \ell_1^*)^n] = \frac{1}{\pi} \int_{-2}^{2} x^n \sqrt{4 - x^2} dx.$$

Hint: Expand the left side and show that $\phi(\ell^{p_1} \ell^{p_2} \cdots \ell^{p_n})$, with $p_i = 1$ or $*$, vanishes unless $\sum_{i=1}^n 1_{p_i=1} = \sum_{i=1}^n 1_{p_i=*}$. Deduce that the left side vanishes when n is odd. Show that when n is even, the only indices $(p_1, \ldots, p_n)$ contributing to the expansion are those for which the path $(X_i = X_{i-1} + 1_{p_i=1} - 1_{p_i=*})_{1 \le i \le n}$, with $X_0 = 0$, is a Dyck path. Conclude by using Section 2.1.3.

Exercise 5.3.8 (i) Show that freely independent algebras can be "piled up", as follows. Let $\{\mathscr{A}_i\}_{i \in I}$ be a family of freely independent subalgebras of $\mathscr{A}$. Partition I into subsets $\{I_j\}_{j \in J}$ and denote by $\mathscr{B}_j$ the subalgebra generated by the family $\{\mathscr{A}_i\}_{i \in I_j}$. Show that the family $\{\mathscr{B}_j\}_{j \in J}$ is freely independent. (ii) Show that freeness is preserved under (strong or weak) closures, as follows. Suppose that $(\mathscr{A}, \phi)$ is a C^*- or W^*-probability space. Let $\{\mathscr{A}_i\}_{i \in I}$ be a family consisting of unital subalgebras closed under the involution, and for each index $i \in I$ let $\widehat{\mathscr{A}_i}$ be the strong or weak closure of $\mathscr{A}_i$. Show that the family $\{\widehat{\mathscr{A}_i}\}_{i \in I}$ is still freely independent.

5.3.2 *Free independence and combinatorics*

The definition 5.3.1 of free independence is given in terms of the vanishing of certain moments of the variables. It is not particularly easy to handle for computation. We explore in this section the notion of cumulant, which is often much easier to handle.

Basic properties of non-crossing partitions

Whereas classical cumulants are related to moments via a sum on the whole set of partitions, free cumulants are defined with the help of non-crossing partitions (recall Definition 2.1.4). A pictorial description of non-crossing versus crossing partitions was given in Figure 2.1.1.

Before turning to the definition of free cumulants, we need to review key properties of non-crossing partitions. It is convenient to define, for any finite nonempty set J of positive integers, the set $NC(J)$ to be the family of non-crossing partitions of J. This makes sense because the non-crossing property of a partition is well defined in the presence of a total ordering. Also, we define an *interval* in J to be any nonempty subset consisting of consecutive elements of J. Given $\sigma, \pi \in NC(J)$ we say that σ *refines* π if every block of σ is contained in some block of π, and in this case we write $\sigma \leq \pi$. Equipped with this partial order, $NC(J)$ is a poset, that is, a partially ordered set. For $J = \{1, \ldots, n\}$, we simply write $NC(n) = NC(J)$. The unique maximal element of $NC(n)$, namely $\{\{1, \ldots, n\}\}$, we denote by $\mathbf{1}_n$.

Property 5.3.9 *For any finite nonempty family $\{\pi_i\}_{i \in J}$ of elements of $NC(n)$ there exists a greatest lower bound $\wedge_{i \in J} \pi_i \in NC(n)$ and a least upper bound $\vee_{i \in J} \pi_i \in NC(n)$ with respect to the refinement partial ordering.*

We remark that greatest lower bounds and least upper bounds in a poset are automatically unique. Below, we write $\wedge_{i \in \{1,2\}} \pi_i = \pi_1 \wedge \pi_2$ and $\vee_{i \in \{1,2\}} \pi_i = \pi_1 \vee \pi_2$.

Proof It is enough to prove existence of the greatest lower bound $\wedge_{i \in J} \pi_i$, for then $\vee_{i \in J} \pi_i$ can be obtained as $\wedge_{k \in K} \sigma_k$, where $\{\sigma_k\}_{k \in K}$ is the family of elements of $NC(n)$ coarser than π_i for all $i \in J$. (The family $\{\sigma_k\}$ is nonempty since $\mathbf{1}_n$ belongs to it.) It is clear that in the refinement-ordered family of all partitions of $\{1, \ldots, n\}$ there exists a greatest lower bound π for the family $\{\pi_i\}_{i \in J}$. Finally, it is routine to check that π is in fact non-crossing, and hence $\pi = \wedge_{i \in J} \pi_i$. □

Remark 5.3.10 As noted in the proof above, for $\pi, \sigma \in NC(n)$, the greatest lower bound of π and σ in the poset $NC(n)$ coincides with the greatest lower bound in the poset of all partitions of $\{1,\ldots,n\}$. But the analogous statement about least upper bounds is false in general.

Property 5.3.11 *Let π be a non-crossing partition of a finite nonempty set S of positive integers. Let $S_1,\ldots,S_m$ be an enumeration of the blocks of π. For $i = 1,\ldots,m$ let π_i be a partition of S_i. Then the partition $\bigcup_{i=1}^m \pi_i$ of S obtained by combining the π_i is non-crossing if and only if π_i is non-crossing for $i = 1,\ldots,m$.*

The proof is straightforward and so omitted. But this property bears emphasis because it is crucial for defining free cumulants.

Property 5.3.12 *If a partition π of a finite nonempty set S of positive integers is non-crossing, then there is at least one block of π which is an interval in S.*

Proof Let W be any block of π, let $W' \supset W$ be the interval in S bounded by the least and greatest elements of W, and put $S' = W' \setminus W$. If S' is empty, we are done. Otherwise S' is a union of blocks of π, by the non-crossing property. Let π' be the restriction of π to S'. By induction on the cardinality of S, some block V of π' is an interval of S', hence V is an interval in S and a block of π. □

Free cumulants and freeness

In classical probability, moments can be written as a sum over partitions of classical cumulants. A similar formula holds in free probability except that partitions have to be non-crossing. This relation between moments and free cumulants can be used to define free cumulants, as follows.

We pause to introduce some notation. Suppose we are given a collection $\{\ell_n : \mathscr{A}^n \to \mathbb{C}\}_{n=1}^\infty$ of multilinear functionals on a fixed complex algebra $\mathscr{A}$. We define $\ell_\pi(\{a_i\}_{i\in J}) \in \mathbb{C}$ for finite nonempty sets J of positive integers, families $\{a_i\}_{i\in J}$ of elements of $\mathscr{A}$ and $\pi \in NC(J)$ in two stages: first we write $J = \{i_1 < \cdots < i_m\}$ and define $\ell(\{a_i\}_{i\in J}) = \ell_m(a_{i_1},\ldots,a_{i_m})$; then we define $\ell_\pi(\{a_i\}_{i\in J}) = \prod_{V\in\pi} \ell(\{a_i\}_{i\in V})$.

Definition 5.3.13 Let $(\mathscr{A},\phi)$ be a noncommutative probability space. The *free cumulants* are defined as a collection of multilinear functionals

$$k_n : \mathscr{A}^n \to \mathbb{C} \quad (n \in \mathbb{N})$$

by the following system of equations:

$$\phi(a_1\cdots a_n) = \sum_{\pi\in NC(n)} k_\pi(a_1,\ldots,a_n). \tag{5.3.5}$$

Lemma 5.3.14 *The free cumulants are well defined.*

Proof We define $\phi_\pi(\{a_i\}_{i\in J}) \in \mathbb{C}$ for finite nonempty sets J of positive integers, families $\{a_i\}_{i\in J}$ of elements of $\mathscr{A}$ and $\pi \in NC(J)$ in two stages: first we write $J = \{i_1 < \cdots < i_m\}$ and define $\prod_{i\in J} a_i = a_{i_1}\cdots a_{i_m}$; then we define $\phi_\pi(\{a_i\}_{i\in J}) = \prod_{V\in\pi} \phi(\prod_{i\in V} a_i)$. If the defining relations (5.3.5) hold, then, more generally, we must have

$$\phi_\pi(a_1,\ldots,a_n) = \sum_{\substack{\sigma\in NC(n)\\ \sigma\leq\pi}} k_\sigma(a_1,\ldots,a_n) \tag{5.3.6}$$

for all n, $(a_1,\ldots,a_n) \in \mathscr{A}^n$ and $\pi \in NC(n)$, by Property 5.3.11. Since every partial ordering of a finite set can be extended to a linear ordering, the system of linear equations (5.3.6), for fixed n and $(a_1,\ldots,a_n) \in \mathscr{A}^n$, has (in effect) a square triangular coefficient matrix with 1s on the diagonal, and hence a unique solution. Thus, the free cumulants are indeed well defined. □

We now turn to the description of freeness in terms of cumulants, which is analogous to the characterization of independence by cumulants in classical probability.

Theorem 5.3.15 *Let $(\mathscr{A},\phi)$ be a noncommutative probability space and consider unital subalgebras $\mathscr{A}_1,\ldots,\mathscr{A}_m \subset \mathscr{A}$. Then, $\mathscr{A}_1,\ldots,\mathscr{A}_m$ are free if and only if, for all $n\geq 2$ and for all $a_i \in \mathscr{A}_{j(i)}$ with $1\leq j(1),\ldots,j(n)\leq m$,*

$$k_n(a_1,\ldots,a_n) = 0 \quad \textit{if there exist } 1\leq l,k\leq n \textit{ with } j(l)\neq j(k). \tag{5.3.7}$$

Before beginning the proof of the theorem, we prove a result which explains why the description of freeness by cumulants does not require any centering of the variables.

Proposition 5.3.16 *Let $(\mathscr{A},\phi)$ be a noncommutative probability space and assume $a_1,\ldots,a_n \in \mathscr{A}$ with $n\geq 2$. If there is $i\in\{1,\ldots,n\}$ so that $a_i = 1$, then*

$$k_n(a_1,\ldots,a_n) = 0.$$

As a consequence, for $n\geq 2$ and any $a_1,\ldots,a_n \in \mathscr{A}$,

$$k_n(a_1,\ldots,a_n) = k_n(a_1-\phi(a_1),a_2-\phi(a_2),\ldots,a_n-\phi(a_n)).$$

Proof We use induction on $n\geq 2$. To establish the induction base, for $n=2$ we have, since $k_1(a) = \phi(a)$,

$$\phi(a_1a_2) = k_2(a_1,a_2) + \phi(a_1)\phi(a_2)$$

and so, if $a_1 = 1$ or $a_2 = 1$, we deduce, since $\phi(1) = 1$, that $k_2(a_1, a_2) = 0$. For the rest of the proof we assume that $n > 2$. By induction we may assume that for $p \le n-1$, $k_p(b_1, \ldots, b_p) = 0$ if one of the b_i is the identity. Suppose now that $a_i = 1$. Then

$$\phi(a_1 \cdots a_n) = k_n(a_1, \ldots, a_n) + \sum_{\substack{\pi \in NC(n) \\ \pi \neq \mathbf{1}_n}} k_\pi(a_1, \ldots, a_n), \tag{5.3.8}$$

where by our induction hypothesis all the partitions π contributing to the above sum must be such that $\{i\}$ is a block. But then, by the induction hypothesis,

$$\begin{aligned} \sum_{\substack{\pi \in NC(n) \\ \pi \neq \mathbf{1}_n}} k_\pi(a_1, \ldots, a_n) &= \sum_{\pi \in NC(n-1)} k_\pi(a_1, \ldots, a_{i-1}, a_{i+1}, \ldots, a_n) \\ &= \phi(a_1 \cdots a_{i-1} a_{i+1} \cdots a_n) \\ &= \phi(a_1 \cdots a_n) - k_n(a_1, \ldots, a_n) \end{aligned}$$

where the second equality is due to the definition of cumulants and the third to (5.3.8). As a consequence, because $\phi(a_1 \cdots a_{i-1} a_{i+1} \cdots a_n) = \phi(a_1 \cdots a_n)$, we have proved that $k_n(a_1, \ldots, a_n) = 0$. □

Proof of the implication $\Leftarrow$ in Theorem 5.3.15 We assume that the cumulants vanish when evaluated at elements of different algebras $\mathcal{A}_1, \ldots, \mathcal{A}_m$ and consider, for $a_i \in \mathcal{A}_{j(i)}$ with $j(i) \neq j(i+1)$ for all $i \in \{1, \ldots, n-1\}$, the equation

$$\phi((a_1 - \phi(a_1)) \cdots (a_n - \phi(a_n))) = \sum_{\pi \in NC(n)} k_\pi(a_1, \ldots, a_n).$$

By our hypothesis, k_π vanishes as soon as a block of π contains $1 \le p, q \le n$ so that $j(p) \neq j(q)$. Therefore, since we assumed $j(p) \neq j(p+1)$ for all $p \in \{1, \ldots, n-1\}$, we see that the contribution in the above sum comes from partitions π whose blocks cannot contain two nearest neighbors $\{p, p+1\}$ for any $p \in \{1, \ldots, n-1\}$. On the other hand, by Property 5.3.12, π must contain an interval in $\{1, \ldots, n\}$, and the previous remark implies that this interval must be of the form $V = \{p\}$ for some $p \in \{1, \ldots, n-1\}$. But then k_π vanishes since $k_1 = 0$ by centering of the variables. Therefore, if for $1 \le p \le n-1$, $j(p) \neq j(p+1)$, we get

$$\phi((a_1 - \phi(a_1)) \cdots (a_n - \phi(a_n))) = 0,$$

and hence ϕ satisfies (5.3.1). □

The next lemma handles an important special case of the implication $\Rightarrow$ in Theorem 5.3.15.

Lemma 5.3.17 *If $\mathscr{A}_1,\ldots,\mathscr{A}_m$ are free, then for $n\geq 2$,*

$$k_n(a_1,\ldots,a_n)=0 \;\; \text{if } a_j\in\mathscr{A}_{j(i)} \;\; \text{with } j(1)\neq j(2)\neq\cdots\neq j(n). \tag{5.3.9}$$

Proof We proceed by induction on $n\geq 2$. We have

$$\begin{aligned} 0 &= \phi((a_1-\phi(a_1))\cdots(a_n-\phi(a_n)))=\sum_{\pi\in NC(n)} k_\pi(a_1-\phi(a_1),\ldots,a_n-\phi(a_n)) \\ &= \sum_{\substack{\pi\in NC(n)\\ \pi \text{ has no singleton blocks}}} k_\pi(a_1,\ldots,a_n), \end{aligned} \tag{5.3.10}$$

where the second equality is due to Proposition 5.3.16 and the vanishing $k_1(a_i-\phi(a_i))=0$. To finish the proof of (5.3.9) it is enough to prove that the last sum reduces to $k_n(a_1,\ldots,a_n)$. If $n=2$ this is clear; otherwise, for $n>2$, this holds by induction on n, using Property 5.3.12. □

The next lemma provides the inductive step needed to finish the proof of Theorem 5.3.15.

Lemma 5.3.18 *Fix $n\geq 2$ and $a_1,\ldots,a_n\in\mathscr{A}$. Fix $1\leq i\leq n-1$ and let $\sigma\in NC(n)$ be the non-crossing partition all blocks of which are singletons except for $\{i,i+1\}$. Then for all $\eta\in NC(n-1)$ we have that*

$$k_\eta(a_1,\ldots,a_ia_{i+1},\ldots,a_n)=\sum_{\substack{\pi\in NC(n)\\ \pi\vee\sigma=\eta}} k_\pi(a_1,\ldots,a_n). \tag{5.3.11}$$

Proof Fix $\zeta\in NC(n-1)$ arbitrarily. It will be enough to prove equality after summing both sides of (5.3.11) over $\eta\leq\zeta$. Let

$$f:\{1,\ldots,n\}\to\{1,\ldots,n-1\}$$

be the unique onto monotone increasing function such that $f(i)=f(i+1)$. Let $\zeta'\in NC(n)$ be the partition whose blocks are of the form $f^{-1}(V)$ with V a block of ζ. Summing the left side of (5.3.11) on $\eta\leq\zeta$ we get $\phi_\zeta(a_1,\ldots,a_ia_{i+1},\ldots,a_n)$ by (5.3.6). Now summing the right side of (5.3.11) on $\eta\leq\zeta$ is the same thing as replacing the sum already there by a sum over $\pi\in NC(n)$ such that $\pi\leq\zeta'$. Thus, summing the right side of (5.3.11) over $\eta\leq\zeta$, we get $\phi_{\zeta'}(a_1,\ldots,a_n)$ by another application of (5.3.6). But clearly

$$\phi_\zeta(a_1,\ldots,a_ia_{i+1},\ldots,a_n)=\phi_{\zeta'}(a_1,\ldots,a_n),$$

Thus (5.3.11) holds. □

Proof of the implication $\Rightarrow$ in Theorem 5.3.15 For $n\geq 2$, indices $j(1),\ldots,j(n)\in\{1,\ldots,m\}$ such that $\{j(1),\ldots,j(n)\}$ is a set of more than one element, and $a_i\in$

$\mathscr{A}_{j(i)}$ for $i = 1,\dots,m$, assuming $\mathscr{A}_1,\dots,\mathscr{A}_m$ are free in $\mathscr{A}$ with respect to ϕ, we have to prove that $k_n(a_1,\dots,a_n) = 0$. We proceed by induction on $n \geq 2$. The induction base $n = 2$ holds by (5.3.9). Assume for the rest of the proof that $n > 2$. Because of (5.3.9), we may assume there exists $i \in \{1,\dots,n-1\}$ such that $j(i) = j(i+1)$. Let $\sigma \in NC(n)$ be the unique partition all blocks of which are singletons except for the block $\{i,i+1\}$. In the special case $\eta = \mathbf{1}_{n-1}$, equation (5.3.11) after slight rearrangement takes the form

$$k_n(a_1,\dots,a_n) = k_{n-1}(a_1,\dots,a_i a_{i+1},\dots,a_n) - \sum_{\substack{\mathbf{1}_n \neq \pi \in NC(n) \\ \pi \vee \sigma = \mathbf{1}_n}} k_\pi(a_1,\dots,a_n). \quad (5.3.12)$$

In the present case the first of the terms on the right vanishes by induction on n. Now each $\pi \in NC(n)$ contributing on the right is of the form $\pi = \{V_i, V_{i+1}\}$ where $i \in V_i$ and $i+1 \in V_{i+1}$. Since the function $i \mapsto j(i)$ cannot be constant both on V_i and on V_{i+1} lest it be constant, it follows that every term in the sum on the far right vanishes by induction on n. We conclude that $k_n(a_1,\dots,a_n) = 0$. The proof of Theorem 5.3.15 is complete. □

Exercise 5.3.19 Prove that

$$\begin{aligned} k_3(a_1,a_2,a_3) &= \phi(a_1a_2a_3) - \phi(a_1)\phi(a_2a_3) - \phi(a_1a_3)\phi(a_2) \\ &\quad -\phi(a_1a_2)\phi(a_3) + 2\phi(a_1)\phi(a_2)\phi(a_3). \end{aligned}$$

5.3.3 Consequence of free independence: free convolution

We postpone giving a direct link between free independence and random matrices in order to first exhibit some consequence of free independence, often described as *free harmonic analysis*. We will consider two self-adjoint noncommutative variables a and b. Our goal is to determine the law of $a+b$ or of ab when a,b are free. Since the law of (a,b) with a,b free is uniquely determined by the laws μ_a of a and μ_b of b (see part (ii) of Remark 5.3.2), the law of their sum (respectively, product) is a function of μ_a and μ_b denoted by $\mu_a \boxplus \mu_b$ (respectively, $\mu_a \boxtimes \mu_b$). There are several approaches to these questions; we will detail first a purely combinatorial approach based on free cumulants and then mention an algebraic approach based on the Fock space representations (see part (ii) of Example 5.3.3). These two approaches concern the case where the probability measures μ_a, μ_b have compact support (that is, a and b are bounded). We will generalize the results to unbounded variables in Section 5.3.5.

Free additive convolution

Definition 5.3.20 Let a, b be two noncommutative variables in a noncommutative probability space $(\mathscr{A}, \phi)$ with law μ_a, μ_b respectively. If a, b are free, then the law of $a+b$ is denoted $\mu_a \boxplus \mu_b$.

We use $k_n(a) = k_n(a, \ldots, a)$ to denote the nth cumulant of the variable a.

Lemma 5.3.21 *Let a, b be two bounded operators in a noncommutative probability space $(\mathscr{A}, \phi)$. If a and b are free, then for all $n \geq 1$,*

$$k_n(a+b) = k_n(a) + k_n(b).$$

Proof The result is obvious for $n = 1$ by linearity of k_1. Moreover, for all $n \geq 2$, by multilinearity of the cumulants,

$$\begin{aligned} k_n(a+b) &= \sum_{\varepsilon_i = 0,1} k_n(\varepsilon_1 a + (1-\varepsilon_1)b, \ldots, \varepsilon_n a + (1-\varepsilon_n)b) \\ &= k_n(a) + k_n(b), \end{aligned}$$

where the second equality is a consequence of Theorem 5.3.15. □

Definition 5.3.22 For a bounded operator a the formal power series

$$R_a(z) = \sum_{n \geq 0} k_{n+1}(a) z^n$$

is called the *R-transform* of the law μ_a. We also write $R_{\mu_a} := R_a$ since R_a only depends on the law μ_a.

By Lemma 5.3.21, the R-transform is to free probability what the log-Fourier transform is to classical probability in the sense that it is linear for free additive convolution, as stated by the next corollary.

Corollary 5.3.23 *Let a, b be two bounded operators in a noncommutative probability space $(\mathscr{A}, \phi)$. If a and b are free, we have*

$$R_{\mu_a \boxplus \mu_b} = R_{\mu_a} + R_{\mu_b},$$

where the equalities hold between formal series.

We next provide a more tractable definition of the R-transform in terms of the Stieltjes transform. Let $\mu : \mathbb{C}[X] \to \mathbb{C}$ be a distribution in the sense of Definition

5.2.3 and define the formal power series

$$G_\mu(z) := \sum_{n\geq 0} \mu(X^n) z^{-(n+1)}. \tag{5.3.13}$$

Let $K_\mu(z)$ be the formal inverse of G_μ, i.e. $G_\mu(K_\mu(z)) = z$. The *formal* power series expansion of K_μ is

$$K_\mu(z) = \frac{1}{z} + \sum_{n=1}^{\infty} C_n z^{n-1}.$$

Lemma 5.3.24 *Let μ be a compactly supported probability measure. For $n \geq 1$ integer, $C_n = k_n$ and so we have equality in the sense of formal series*

$$R_\mu(z) = K_\mu(z) - 1/z.$$

Proof Consider the generating function of the cumulants as the formal power series

$$C_a(z) = 1 + \sum_{n=1}^{\infty} k_n(a) z^n$$

and the generating function of the moments as the formal power series

$$M_a(z) = 1 + \sum_{n=1}^{\infty} m_n(a) z^n$$

with $m_n(a) := \mu(a^n)$. We will prove that

$$C_a(zM_a(z)) = M_a(z). \tag{5.3.14}$$

The rest of the proof is pure algebra since

$$G_a(z) := G_{\mu_a}(z) = z^{-1} M_a(z^{-1}), \quad R_a(z) := z^{-1}(C_a(z) - 1)$$

then gives $C_a(G_a(z)) = zG_a(z)$ and so, by composition with K_a,

$$zR_a(z) + 1 = C_a(z) = zK_a(z).$$

This equality proves that $k_n = C_n$ for $n \geq 1$. To derive (5.3.14), we will first show that

$$m_n(a) = \sum_{s=1}^{n} \sum_{\substack{i_1,\ldots,i_s \in \{0,1,\ldots,n-s\} \\ i_1+\cdots+i_s = n-s}} k_s(a) m_{i_1}(a) \cdots m_{i_s}(a). \tag{5.3.15}$$

With (5.3.15) granted, (5.3.14) follows readily since

$$\begin{aligned} M_a(z) &= 1+\sum_{n=1}^{\infty} m_n(a)z^n \\ &= 1+\sum_{n=1}^{\infty}\sum_{s=1}^{n}\sum_{\substack{i_1,\dots,i_s\in\{0,1,\dots,n-s\}\\ i_1+\dots+i_s=n-s}} k_s(a)z^s m_{i_1}(a)z^{i_1}\cdots m_{i_s}(a)z^{i_s} \\ &= 1+\sum_{s=1}^{\infty} k_s(z)z^s\left(\sum_{i=0}^{\infty} z^i m_i(a)\right)^s = C_a(zM_a(z)). \end{aligned}$$

To prove (5.3.15), recall that, by definition of the cumulants,

$$m_n(a) = \sum_{\pi\in NC(n)} k_\pi(a).$$

Given a non-crossing partition $\pi = \{V_1,\dots,V_r\} \in NC(n)$, write $V_1 = (1,v_2,\dots,v_s)$ with $s = |V_1| \in \{1,\dots,n\}$. Since π is non-crossing, we see that for any $l \in \{2,\dots,r\}$, there exists $k \in \{1,\dots,s\}$ so that the elements of V_l lie between v_k and v_{k+1}. Here $v_{s+1} = n+1$ by convention. This means that π decomposes into V_1 and at most s other (non-crossing) partitions $\tilde{\pi}_1,\dots,\tilde{\pi}_s$. Therefore

$$k_\pi = k_s k_{\tilde{\pi}_1}\cdots k_{\tilde{\pi}_s}.$$

If we let i_k denote the number of elements in $\tilde{\pi}_k$, we thus have proved that

$$\begin{aligned} m_n(a) &= \sum_{s=1}^{n} k_s(a) \sum_{\substack{\tilde{\pi}_k\in NC(i_k),\\ i_1+\dots+i_s=n-s}} k_{\tilde{\pi}_1}(a)\cdots k_{\tilde{\pi}_s}(a) \\ &= \sum_{s=1}^{n} k_s(a) \sum_{\substack{i_1+\dots+i_s=n-s\\ i_k\ge 0}} m_{i_1}(a)\cdots m_{i_s}(a), \end{aligned}$$

where we used again the relation (5.3.5) between cumulants and moments. The proof of (5.3.15), and hence of the lemma, is thus complete. □

We now digress by rapidly describing the original proof of Corollary 5.3.23 due to Voiculescu. The idea is that since laws only depends on moments, one can choose a specific representation of the free noncommutative variables a,b with given marginal distribution to actually compute the law of $a+b$. A standard choice is then to use left creation and annihilation operators as described in part (ii) of Example 5.3.3. Let $\mathcal{T}$ denote the Fock space described in (5.3.2) and $\ell_i = \ell(e_i)$, $i = 1,2$, be two creation operators on $\mathcal{T}$.

Lemma 5.3.25 *Let* $(\alpha_{j,i}, i = 1,2, j \in \mathbb{N})$ *be complex numbers and consider the operators on* $\mathscr{T}$

$$a_i = \ell_i^* + \alpha_{0,i} I + \sum_{j=1}^{\infty} \alpha_{j,i} \ell_i^j, \quad i = 1,2.$$

Then, denoting in short $\ell_i^0 = I$ *for* $i = 1,2$*, we have that*

$$a_1 + a_2 = (\ell_1^* + \ell_2^*) + \sum_{j=0}^{\infty} \alpha_{j,1} \ell_1^j + \sum_{j=0}^{\infty} \alpha_{j,2} \ell_2^j \tag{5.3.16}$$

and

$$a_3 = \ell_1^* + \sum_{j=0}^{\infty} \alpha_{j,1} \ell_1^j + \sum_{j=0}^{\infty} \alpha_{j,2} \ell_1^j \tag{5.3.17}$$

possess the same distribution in the noncommutative probability space $(\mathscr{T}, \langle \cdot 1, 1 \rangle)$*.*

In the above lemma, infinite sums are formal. The law of the associated operators is still well defined since the $(\ell_i^j)_{j \geq M}$ will not contribute to moments of order smaller than M; thus, any finite family of moments is well defined.

Proof We need to show that the traces $\langle a_3^k 1, 1 \rangle$ and $\langle (a_1 + a_2)^k 1, 1 \rangle$ are equal for all positive integers k. Comparing (5.3.16) and (5.3.17), there is a bijection between each term in the sum defining $(a_1 + a_2)$ and the sum defining a_3, which extends to the expansions of a_3^k and $(a_1 + a_2)^k$. We thus only need to compare the vacuum expectations of individual terms; for $\langle a_3^k 1, 1 \rangle$ they are of the form $Z := \langle \ell_1^{w_1} \ell_1^{w_2} \cdots \ell_1^{w_n} 1, 1 \rangle$ where $w_i \in \{*, 1\}$, whereas the expansion of $\langle (a_1 + a_2)^k 1, 1 \rangle$ yields similar terms except that ℓ_1^* has to be replaced by $\ell_1^* + \ell_2^*$ and some of the ℓ_1^1 by ℓ_2^1. Note, however, that $Z \neq 0$ if and only if the sequence $w_1, w_2, \ldots, w_n$ is a Dyck path, i.e. the walk defined by it forms a positive excursion that returns to 0 at time n (replacing the symbol $*$ by -1). But, since $(\ell_1^* + \ell_2^*)\ell_i = 1 = \ell_i^* \ell_i$ for $i = 1,2$, the value of Z is unchanged under the rules described above, which completes the proof. □

To deduce another proof of Lemma 5.3.21 from Lemma 5.3.25, we next show that the cumulants of the distribution of an operator of the form

$$a = \ell^* + \sum_{j \geq 0} \alpha_j \ell^j,$$

for some creation operator ℓ on $\mathcal{T}$, are given by $k_i = \alpha_{i+1}$. To prove this point, we compute the moments of a. By definition,

$$\begin{aligned}\langle a^n 1, 1\rangle &= \langle \left(\ell^* + \sum_{j\geq 0} \alpha_j \ell^j\right)^n 1, 1\rangle \\ &= \sum_{i(1),\ldots,i(n)\in\{-1,0,\ldots,n-1\}} \langle \ell^{i(1)} \cdots \ell^{i(n)} 1, 1\rangle \alpha_{i(1)} \cdots \alpha_{i(n)},\end{aligned}$$

where for $j = -1$ we wrote ℓ^* for ℓ^j and set $\alpha_{-1} = 1$, and further observed that mixed moments vanish if some $i(l) \geq n$. Recall now that $\langle \ell^{i(1)} \cdots \ell^{i(n)} 1, 1\rangle$ vanishes except if the path $(i(1),\ldots,i(n))$ forms a positive excursion that returns to the origin at time n, that is,

$$i(1) + \cdots + i(m) \geq 0 \text{ for all } m \leq n, \text{ and } i(1) + \cdots + i(n) = 0. \tag{5.3.18}$$

(Such a path is not in general a Dyck path since the $(i(p), 1 \leq p \leq n)$ may take any values in $\{-1, 0, \ldots, n-1\}$.) We thus have proved that

$$\langle a^n 1, 1\rangle = \sum_{\substack{i(1),\ldots,i(n)\in\{-1,\ldots,n-1\},\\ \sum_{p=1}^m i(p)\geq 0, \sum_{p=1}^n i(p)=0}} \alpha_{i(1)} \cdots \alpha_{i(n)}. \tag{5.3.19}$$

Define next a bijection between the set of integers $(i(1),\ldots,i(n))$ satisfying (5.3.18) and non-crossing partitions $\pi = \{V_1,\ldots,V_r\}$ by $i(m) = |V_i| - 1$ if m is the first element of the block V_i, and $i(m) = -1$ otherwise. To see it is a bijection, being given a partition, the numbers $(i(1),\ldots,i(n))$ satisfy (5.3.18). Reciprocally, being given the numbers $(i(1),\ldots,i(n))$, we have a unique non-crossing partition $\pi = (V_1,\ldots,V_k)$ satisfying $|V_i| = i(m) + 1$ with m the first point of V_i. It is drawn inductively by removing block intervals which are sequences of indices such that $\{i(m) = p, i(m+k) = -1, 1 \leq k \leq p\}$ (including $p = 0$ in which case an interval is $\{i(m) = 0\}$). Such a block must exist by the second assumption in (5.3.18). Fixing such intervals as blocks of the partition, we can remove the corresponding indices and search for intervals in the corresponding subset S of $\{i(k), 1 \leq k \leq n\}$. The indices in S also satisfy (5.3.18), so that we can continue the construction until no indices are left.

This bijection allows us to replace the summation over the $i(k)$ in (5.3.19) by summation over non-crossing partitions to obtain

$$\langle a^n 1, 1\rangle = \sum_{\pi=(V_1,\ldots,V_r)} \alpha_{|V_1|-1} \cdots \alpha_{|V_r|-1}.$$

Thus, by the definition (5.3.5) of the cumulants, we deduce that, for all $i \geq 0$, $\alpha_{i-1} = k_i$, with k_i the ith cumulant. Therefore, Lemma 5.3.25 is equivalent to the additivity of the free cumulants of Lemma 5.3.21 and the rest of the analysis is similar.

Example 5.3.26 Consider the standard semicircle law $\nu_a(dx) = \sigma(x)dx$. By Lemma 2.1.3 and Remark 2.4.2,

$$G_a(z) = \frac{z - \sqrt{z^2 - 4}}{2}.$$

Thus, $K_a(z) = z^{-1} + z$. In particular, the R-transform of the semicircle is the linear function z, and summing two (freely independent) semicircular variables yields again a semicircular variable with a different variance. Indeed, repeating the computation above, the R-transform of a semicircle with support $[-\alpha, \alpha]$ (or equivalently with variance $\alpha^2/4$) is $\alpha^2 z/4$. Note here that the linearity of the R-transform is equivalent to $k_n(a) = 0$ except if $n = 2$, and $k_2(a) = \alpha^2/4 = \phi(a^2)$.

Exercise 5.3.27 (i) Let $\mu = \frac{1}{2}(\delta_{+1} + \delta_{-1})$. Show that $G_\mu(z) = (z^2 - 1)^{-1} z$ and

$$R_\mu(z) = \frac{\sqrt{1 + 4z^2} - 1}{2z}$$

with the appropriate branch of the square root. Deduce that $G_{\mu \boxplus \mu}(z) = \sqrt{z^2 - 4}^{-1}$. Recall that if σ is the standard semicircle law $d\sigma(x) = \sigma(x)dx$, $G_\sigma(x) = \frac{1}{2}(z - \sqrt{z^2 - 4})$. Deduce by derivations and integration by parts that

$$\frac{1}{2}(1 - zG_{\mu \boxplus \mu}(z)) = \int \frac{1}{z - x} \partial_x \sigma(x) dx.$$

Conclude that $\mu \boxplus \mu$ is absolutely continuous with respect to Lebesgue measure and with density proportional to $1_{|x| \le 2}(4 - x^2)^{-\frac{1}{2}}$.
(ii) (Free Poisson) Let $\alpha > 0$. Show that if one takes $p_n(dx) = (1 - \frac{\lambda}{n})\delta_0 + \frac{\lambda}{n}\delta_\alpha$, $p_n^{\boxplus n}$ converges to a limit p whose R-transform is given by

$$R(z) = \frac{\lambda \alpha}{1 - \alpha z}.$$

Deduce that p is the Marčenko–Pastur law given, if $\lambda > 1$, by

$$p(dx) = \tilde{p}(dx) = \frac{1}{2\pi\alpha x} \sqrt{4\lambda\alpha^2 - (x - \alpha(\lambda + 1))^2} dx,$$

and for $\lambda < 1$, $p = (1 - \lambda)\delta_0 + \lambda \tilde{p}$.

Multiplicative free convolution

We consider again two bounded self-adjoint operators a, b in a noncommutative probability space $(\mathscr{A}, \phi)$ with laws μ_a and μ_b, but now study the law of ab, that is, the collection of moments $\{\phi((ab)^n), n \in \mathbb{N}\}$. Note that ab does not need to be a self-adjoint operator. In the case where ϕ is tracial and a self-adjoint positive, we can, however, rewrite $\phi((ab)^n) = \phi((a^{\frac{1}{2}} b a^{\frac{1}{2}})^n)$ so that the law of ab coincides with the spectral measure of $a^{\frac{1}{2}} b a^{\frac{1}{2}}$ when b is self-adjoint. However, the following analysis of the family $\{\phi((ab)^n), n \in \mathbb{N}\}$ holds in a more general context where these quantities might not be related to a spectral measure.

Definition 5.3.28 Let a, b be two noncommutative variables in a noncommutative probability space $(\mathscr{A}, \phi)$ with laws μ_a and μ_b respectively. If a and b are free, the law of ab is denoted $\mu_a \boxtimes \mu_b$.

Denote by m_a the generating function of the moments, that is, the formal power series

$$m_a(z) := \sum_{m \geq 1} \phi(a^n) z^n = M_a(z) - 1\,.$$

When $\phi(a) \neq 0$, m_a is invertible as a formal power series. Denote by m_a^{-1} its (formal) inverse. We then define

Definition 5.3.29 Assume $\phi(a) \neq 0$. The *S-transform* of a is given by

$$S_a(z) := \frac{1+z}{z} m_a^{-1}(z).$$

We next prove that the S-transform plays the same role in free probability that the Mellin transform does in classical probability.

Lemma 5.3.30 *Let a, b be two free bounded operators in a noncommutative probability space $(\mathscr{A}, \phi)$, so that $\phi(a) \neq 0$, $\phi(b) \neq 0$. Then*

$$S_{ab}(z) = S_a(z) S_b(z)\,.$$

See Exercise 5.3.31 for extensions of Lemma 5.3.30 to the case where either $\phi(a)$ or $\phi(b)$ vanish.

Proof The idea is to use the structure of non-crossing partitions to relate the generating functions

$$M_{ab}(z) = \sum_{n \geq 0} \phi((ab)^n) z^n, \quad M_{cd}^d(z) = \sum_{n \geq 0} \phi(d(cd)^n) z^n\,,$$

where $(c,d) = (a,b)$ or (b,a). Note first that, from Theorem 5.3.15,

$$\begin{aligned}\phi((ab)^n) &= \phi(abab\cdots ab) = \sum_{\pi\in NC(2n)} k_\pi(a,b,\ldots,a,b)\\ &= \sum_{\substack{\pi_1\in NC(1,3,\ldots,2n-1)\pi_2\in NC(2,4,\ldots,2n)\\ \pi_1\cup\pi_2\in NC(2n)}} k_{\pi_1}(a)k_{\pi_2}(b)\,.\end{aligned}$$

The last formula is symmetric in a,b so that, even if ϕ is not tracial, $\phi((ab)^n) = \phi((ba)^n)$ for all $n \geq 1$. We use below the notation $\mathscr{P}(odd)$ and $\mathscr{P}(even)$ for the partitions on the odd, respectively, even, positive integers. Fix the first block $V_1 = \{v_1,\ldots,v_s\}$ in the partition π_1. We denote by $W_1,\ldots,W_s$ the intervals between the elements of $V_1 \cup \{2n\}$. For $k = 1,\ldots,s$, the sum over the non-crossing partitions of W_k corresponds to a word $b(ab)^{i_k}$ if $|W_k| = 2i_k + 1 = v_{k+1} - v_k - 1$. Therefore we have

$$\begin{aligned}\phi((ab)^n) &= \sum_{s=1}^{n} k_s(a) \sum_{\substack{i_1+\cdots+i_s=n-s\\ i_k\geq 0}} \prod_{k=1}^{s}\left(\sum_{\substack{\pi_1\in\mathscr{P}(odd),\pi_2\in\mathscr{P}(even)\\ \pi_1\cup\pi_2\in NC(\{1,\ldots,2i_k+1\})}} k_{\pi_1}(b)k_{\pi_2}(a)\right)\\ &= \sum_{s=1}^{n} k_s(a) \sum_{\substack{i_1+\cdots+i_s=n-s\\ i_k\geq 0}} \prod_{k=1}^{s}\phi(b(ab)^{i_k})\,. \qquad (5.3.20)\end{aligned}$$

Now we can do the same for $\phi(b(ab)^n)$ by fixing the first block $V_1 = (v_1,\ldots,v_s)$ in the partition of the bs (on the odd numbers); the corresponding first intervals are $\{v_k+1, v_{k+1}-1\}$ for $k \leq s-1$ (representing the words of the form $(ab)^{i_k}a$, with $i_k = 2^{-1}(v_{k+1} - v_k) - 1)$, whereas the last interval $\{v_s+1, 2n+1\}$ corresponds to a word of the form $(ab)^{i_0}$ with $i_0 = 2^{-1}(2n+1-v_s)$. Thus we get, for $n \geq 0$,

$$\phi(b(ab)^n) = \sum_{s=0}^{n} k_{s+1}(b) \sum_{\substack{i_0+\cdots+i_s=n-s\\ i_k\geq 0}} \phi((ab)^{i_0}) \prod_{k=1}^{s}\phi(a(ba)^{i_k})\,. \qquad (5.3.21)$$

Set $c_a(z) := \sum_{n\geq 1} k_n(a)z^n$. Summing (5.3.20) and (5.3.21) yields the relations

$$\begin{aligned}M_{ab}(z) &= 1 + c_a(zM^b_{ab}(z))\,,\\ M^b_{ab}(z) &= \sum_{s\geq 0} z^s k_{s+1}(b) M_{ab}(z) M^a_{ba}(z)^s = \frac{M_{ab}(z)}{zM^a_{ba}(z)} c_b(zM^a_{ba}(z))\,.\end{aligned}$$

Since $M_{ab} = M_{ba}$, we deduce that

$$M_{ab}(z) - 1 = c_a(zM^b_{ab}(z)) = c_b(zM^a_{ba}(z)) = \frac{zM^b_{ab}(z)M^a_{ba}(z)}{M_{ab}(z)}\,,$$

which yields, noting that c_a, c_b are invertible as formal power series since $k_1(a) = \phi(a) \neq 0$ and $k_1(b) = \phi(b) \neq 0$ by assumption,

$$c_a^{-1}(M_{ab}(z) - 1)c_b^{-1}(M_{ab}(z) - 1) = zM_{ab}(z)(M_{ab}(z) - 1)\,. \tag{5.3.22}$$

Finally, from the equality (5.3.14) (note here that $c_a = C_a - 1$), if $m_a = M_a - 1$, then

$$m_a(z) = c_a(z(1 + m_a(z))) \Rightarrow c_a^{-1}(z) = (1+z)m_a^{-1}(z) = zS_a(z)\,.$$

Therefore, (5.3.22) implies

$$z^2 S_a(z) S_b(z) = (1+z)zm_{ab}^{-1}(z) = z^2 S_{ab}(z)\,,$$

which completes the proof of the lemma. □

Exercise 5.3.31 In the case where a is a self-adjoint operator such that $\phi(a) = 0$ but $a \neq 0$, define m_a^{-1}, the inverse of m_a, as a formal power series in $\sqrt{z}$. Define the S-transform $S_a(z) = (z^{-1} + 1)m_a^{-1}(z)$ and extend Lemma 5.3.30 to the case where $\phi(a)$ or $\phi(b)$ may vanish.
Hint: Note that $\phi(a^2) \neq 0$ so that $m_a(z) = \phi(a^2)z^2 + \sum_{m\geq 3}\phi(a^m)z^m$ has formal inverse $m_a^{-1}(z) = \phi(a^2)^{-\frac{1}{2}}\sqrt{z} + (\phi(a^3)/2\phi(a^2)^2)z + \cdots$, which is a formal power series in $\sqrt{z}$.

5.3.4 Free central limit theorem

In view of the free harmonic analysis that we developed in the previous sections, which is analogous to the classical one, it is no surprise that standard results from classical probability can be generalized to the noncommutative setting. One of the most important such generalizations is the free central limit theorem.

Lemma 5.3.32 *Let $\{a_i\}_{i\in\mathbb{N}}$ be a family of free self-adjoint random variables in a noncommutative probability space with a tracial state ϕ. Assume that, for all $k \in \mathbb{N}$,*

$$\sup_j |\phi(a_j^k)| < \infty\,. \tag{5.3.23}$$

Assume $\phi(a_i) = 0$, $\phi(a_i^2) = 1$. Then

$$X_N = \frac{1}{\sqrt{N}}\sum_{i=1}^N a_i$$

converges in law as N goes to infinity to a standard semicircle distribution.

Proof Note that by (5.3.23) the cumulants of words in the a_i are well defined and finite. Moreover, by Lemma 5.3.21, for all $p \geq 1$, we have

$$k_p(X_N) = \sum_{k=1}^{N} k_p(\frac{a_i}{\sqrt{N}}) = \frac{1}{N^{\frac{p}{2}}} \sum_{k=1}^{N} k_p(a_i) .$$

Since, for each p, $\{k_p(a_i)\}_{i=1}^{\infty}$ are bounded uniformly in i, we get, for $p \geq 3$,

$$\lim_{N\to\infty} k_p(X_N) = 0 .$$

Moreover, since $\phi(a_i) = 0, \phi(a_i^2) = 1$, for any integer N, $k_1(X_N) = 0$ whereas $k_2(X_N) = 1$. Therefore, we see by definition 5.3.13 that, for all $p \in \mathbb{N}$,

$$\lim_{N\to\infty} \phi(X_N^p) = \begin{cases} 0 \text{ if } p \text{ is odd,} \\ \sharp\{\pi \in NC(p), \pi \text{ pair partition}\} \end{cases} .$$

Here we recall that a pair partition is a partition whose blocks have exactly two elements. The right side corresponds to the definition of the moments of the semicircle law, see Proposition 2.1.11. □

5.3.5 Freeness for unbounded variables

The notion of freeness was defined for bounded variables possessing all moments. It naturally extends to general unbounded variables thanks to the notion of *affiliated operators* defined in Section 5.2.3, as follows.

Definition 5.3.33 Self-adjoint operators $\{X_i\}_{1\leq i\leq p}$, affiliated with a von Neumann algebra $\mathscr{A}$, are called *freely independent*, or simply free, iff the algebras generated by $\{f(X_i) : f \text{ bounded measurable}\}_{1\leq i\leq p}$ are free.

Free unbounded variables can be constructed in a noncommutative space, even though it is not possible anymore to represent these variables as bounded operators, so that standard tools such as the GNS representation, Theorem 5.2.24, do not hold directly. However, we can construct free affiliated variables as follows.

Proposition 5.3.34 *Let $(\mu_1, \ldots, \mu_p)$ be probability measures on $\mathbb{R}$. Then there exist a W^*-probability space $(\mathscr{A}, \tau)$ with τ a normal faithful tracial state, and self-adjoint operators $\{X_i\}_{1\leq i\leq p}$ which are affiliated with $\mathscr{A}$, with laws μ_i, $1 \leq i \leq p$, and which are free.*

Proof Set $\mathscr{A}_i = B(H_i)$ with $H_i = L^2(\mu_i)$ and construct the free product $\mathscr{H}$ as in the discussion following (5.3.3), yielding a C^*-probability space $(\mathscr{A}, \phi)$ with a tracial state ϕ and a morphism π such that the algebras $(\pi(\mathscr{A}_i))_{1\le i\le p}$ are free. By the GNS construction, see Proposition 5.2.24 and Corollary 5.2.27, we can construct a normal faithful tracial state τ on a von Neumann algebra $\mathscr{B}$ and unbounded operators $(a_1,\dots,a_p)$ affiliated with $\mathscr{B}$, with marginal distribution $(\mu_1,\dots,\mu_p)$. They are free since since the algebras they generate are free (note that ϕ and τ satisfy the relations of Definition 5.3.1 according to Remark 5.3.2). □

From now on we assume that we are given a Hilbert space H as well as a W^*-algebra $\mathscr{A}\subset B(H)$ and self-adjoint operators affiliated with $\mathscr{A}$. The law of affiliated operators is given by their spectral measure and, according to Theorem 5.2.31 and Proposition 5.2.32, if $\{T_i\}_{1\le i\le k}$ are self-adjoint affiliated operators, the law of $Q(\{T_i\}_{1\le i\le k})$ is well defined for any polynomial Q.

The following corollary is immediate.

Corollary 5.3.35 *Let $\{T_i\}_{1\le i\le k}\in\widetilde{\mathscr{A}}$ be free self-adjoint variables with marginal distribution $\{\mu_i\}_{1\le i\le k}$ and let Q be a self-adjoint polynomial in k noncommuting variables. Then the law of $Q(\{T_i\}_{1\le i\le k})$ depends only on $\{\mu_i\}_{1\le i\le k}$ and it is continuous in these measures.*

Proof of Corollary 5.3.35 Let $u_n:\mathbb{R}\to\mathbb{R}$ be bounded continuous functions so that $u_n(x)=x$ for $|x|<n$ and $u_n(x)=0$ for $|x|>2n$. By Proposition 5.2.32, the law of $Q(\{T_i\}_{1\le i\le k})$ can be approximated by the law of $Q(\{u_n(T_i)\}_{1\le i\le k})$. To see the claimed continuity, note that if $\mu_i^p\to\mu_i$ converges weakly as $p\to\infty$ for $i=1,\dots,k$, then the sequences $\{\mu_i^p\}$ are tight, and thus for each $\varepsilon>0$ there exists an M independent of p so that $\mu_i^p(\{x:|x|>M\})<\varepsilon$. In particular, with T_i^p denoting the operators corresponding to the measures μ_i^p, it follows that the convergence of the law of $Q(\{u_n(T_i^p)\}_{1\le i\le k})$ to the law of $Q(\{T_i^p\}_{1\le i\le k})$ is uniform in p. Since, for each n, the law of $Q(\{u_n(T_i^p)\}_{1\le i\le k})$ converges to that of $Q(\{u_n(T_i)\}_{1\le i\le k})$, the claimed continuity follows. □

Free harmonic analysis can be extended to affiliated operators, that is, to laws with unbounded support. We consider here the additive free convolution. We first show that the R-transform can be defined as an analytic function, at least for arguments with large enough imaginary part, without using the existence of moments.

Lemma 5.3.36 *Let μ be a probability measure on $\mathbb{R}$. For $\alpha,\beta>0$, let $\Gamma_{\alpha,\beta}\subset\mathbb{C}^+$ be given by*

$$\Gamma_{\alpha,\beta}=\{z=x+iy\in\mathbb{C}^+:|x|<\alpha y, y>\beta\}.$$

Put, for $z \in \mathbb{C}\backslash\mathbb{R}$,

$$G_\mu(z) := \int \frac{1}{z-x} d\mu(x), \quad F_\mu(z) = 1/G_\mu(z). \tag{5.3.24}$$

For any $\alpha > 0$ and $\varepsilon \in (0,\alpha)$, there exists $\beta > 0$ so that:

(i) *F_μ is univalent on $\Gamma_{\alpha,\beta}$;*
(ii) *$F_\mu(\Gamma_{\alpha,\beta})$ contains $\Gamma_{\alpha-\varepsilon,\beta(1+\varepsilon)}$ and in particular, the inverse of F_μ, denoted F_μ^{-1}, satisfies $F_\mu^{-1} : \Gamma_{\alpha-\varepsilon,\beta(1+\varepsilon)} \to \Gamma_{\alpha,\beta}$;*
(iii) *F_μ^{-1} is analytic on $\Gamma_{\alpha-\varepsilon,\beta(1+\varepsilon)}$.*

Proof Observe that F_μ is analytic on $\Gamma_{\alpha,\beta}$ and

$$\lim_{|z|\to\infty, z\in\Gamma_{\alpha,\beta}} F'_\mu(z) = -1 .$$

In particular, the latter shows that $|F'_\mu(z)| > 1/2$ on $\Gamma_{\alpha,\beta}$ for β large enough. We can thus apply the implicit function theorem (also known in this context as the Lagrange inversion theorem) to deduce that F_μ is invertible, with an analytic inverse. The other claims follow by noting that F_μ is approximately the identity for β sufficiently large. □

Definition 5.3.37 Let $\Gamma_{\alpha,\beta}$ be as in Lemma 5.3.36. We define the *Voiculescu transform* of μ on $\Gamma_{\alpha,\beta}$ as

$$\phi_\mu(z) = F_\mu^{-1}(z) - z .$$

For $1/z \in \Gamma_{\alpha,\beta}$, we define the *R-transform* of μ as $R_\mu(z) := \phi_\mu(\frac{1}{z})$.

By Lemma 5.3.36, for β large enough, ϕ_μ is analytic on $\Gamma_{\alpha,\beta}$. As the following lemma shows, the analyticity extends to a full neighborhood of infinity (and to an analyticity of R_μ in a neighborhood of 0) as soon as μ is compactly supported.

Lemma 5.3.38 *If μ is compactly supported and $|z|$ is small enough, then $R_\mu(z)$ equals the absolutely convergent series $\sum_{n\geq 0} k_{n+1}(a) z^n$.*

Note that the definition of G_μ given in (5.3.24) is analytic (in the upper half plane), whereas it was defined as a formal power series in (5.3.13). However, when μ is compactly supported and z is large enough, the formal series (5.3.13) is absolutely convergent and is equal to the analytic definition (5.3.24), which justifies the use of the same notation. Similarly, Lemma 5.3.38 shows that the formal Definition 5.3.22 of R_μ can be strengthened into an analytic definition when μ is compactly supported.

Proof Let μ be supported in $[-M,M]$ for some $M < \infty$. Then observe that G_μ defined in (5.3.13) can be as well defined as an absolutely converging series for $|z| > M$, and the resulting function is analytic in this neighborhood of infinity. R_μ is then defined using Lemma 5.3.36 by applying the same procedure as in Lemma 5.3.24, but on analytic functions rather than formal series. □

By Property 5.3.34, we can always construct a Hilbert space H, a tracial state ϕ, and two free variables X_1, X_2 with laws μ_1 and μ_2, respectively, affiliated with $B(H)$. By Corollary 5.3.35, we may define the law of $X_1 + X_2$ which we denote $\mu_1 \boxplus \mu_2$.

Corollary 5.3.39 *Let μ_1 and μ_2 be probability measures on $\mathbb{R}$, and let $\mu = \mu_1 \boxplus \mu_2$. For each $\alpha > 0$, we have $\phi_\mu = \phi_{\mu_1} + \phi_{\mu_2}$ in $\Gamma_{\alpha,\beta}$ for β sufficiently large.*

Proof The proof is obtained by continuity from the bounded variables case. Indeed, Lemmas 5.3.23 and 5.3.24, together with the last point of Lemma 5.3.36, show that Corollary 5.3.39 holds when μ_1 and μ_2 are compactly supported. We will next show that

$$\begin{array}{l}\text{if } \mu_n \text{ converge to } \mu \text{ in the weak topology, then there exist} \\ \alpha, \beta > 0 \text{ such that } \phi_{\mu_n} \text{ converges to } \phi_\mu \text{ uniformly on} \\ \text{compacts subsets of } \Gamma_{\alpha,\beta}\,.\end{array} \qquad (5.3.25)$$

With (5.3.25) granted, put $d\mu_i^n = \mu_i([-n,n])^{-1} 1_{|x|\le n} d\mu_i$, note that μ_i^n converges to μ_i for $i = 1,2$, and observe that the law $\mu_1^n \boxplus \mu_2^n$ of $u_n(X_1) + u_n(X_2)$, with X_1, X_2 being two free affiliated variables, converges to $\mu_1 \boxplus \mu_2$ by Proposition 5.2.32. The convergence of ϕ_{μ^n} to ϕ_μ on the compacts of some $\Gamma_{\alpha,\beta}$ for $\mu = \mu_1$, μ_2 and $\mu_1 \boxplus \mu_2$, together with the corollary applied to the compactly supported μ_i^n, implying

$$\phi_{\mu_1^n \boxplus \mu_2^n} = \phi_{\mu_1^n} + \phi_{\mu_2^n}\,,$$

yield the corollary for arbitrary measures μ_i.

It remains to prove (5.3.25). Fix a probability measure μ and a sequence μ^n converging to μ. Then, F_μ converges to F_μ uniformly on compact sets of $\mathbb{C}^+$ (as well as its derivatives, since the functions F_{μ_n} are analytic). Since $|F'_{\mu_n}(z)| > 1/2$ on $\Gamma_{\alpha,\beta}$ for β sufficiently large, $|F'_{\mu_n}(z)| > 1/4$ uniformly in n large enough for z in compact subsets of $\Gamma_{\alpha,\beta}$ for β sufficiently large. Therefore, the implicit function theorem asserts that there exist $\alpha, \beta > 0$ such that F_{μ_n} has a right inverse $F_{\mu_n}^{-1}$ on $\Gamma_{\alpha,\beta}$, and thus the functions $(\phi_{\mu_n}, n \in \mathbb{N}, \phi_\mu)$ are well defined analytic functions on $\Gamma_{\alpha,\beta}$ and are such that $\phi_{\mu_n}(z) = o(z)$ uniformly in n as $|z|$ goes to infinity. Therefore, by Montel's Theorem, the family $\{\phi_{\mu_n}, n \in \mathbb{N}\}$ has subsequences that converge uniformly on compacts of $\Gamma_{\alpha,\beta}$. We claim that all limit points must be equal to ϕ_μ and hence ϕ_{μ_n} converges to ϕ_μ on $\Gamma_{\alpha,\beta}$. Indeed, assume $\phi_{\mu_{n_j}}$ converges to ϕ on a compact $K \subset \Gamma_{\alpha,\beta}$. We have

$$\begin{aligned}|F_\mu(\phi(z)+z)-z| &= |F_\mu(\phi(z)+z)-F_{\mu_{n_j}}(\phi_{\mu_{n_j}}(z)+z)|\\ &= |F_\mu(\phi(z)+z)-F_\mu(\phi_{\mu_{n_j}}(z)+z)|\\ &\quad +|F_\mu(\phi_{\mu_{n_j}}(z)+z)-F_{\mu_{n_j}}(\phi_{\mu_{n_j}}(z)+z)|\,.\end{aligned}$$

The first term in the right side goes to zero as j goes to infinity by continuity of F_μ and the second term goes to zero by uniform convergence of $F_{\mu_{n_j}}$ on $\Gamma_{\alpha,\beta}$. (Note that $\phi_{\mu_{n_j}}(z)$ is uniformly small compared to $|z|$ so that $z+\phi_{\mu_{n_j}}(z)$, $j \in \mathbb{N}$, stays in $\Gamma_{\alpha,\beta}$.) Thus, $z+\phi$ is a right inverse of F_μ, that is, $\phi = \phi_\mu$. □

The study of free convolution via the analytic functions ϕ_μ (or R_μ) is useful in deducing properties of free convolution and of free infinitely divisible laws (whose definition is analogous to the classical one, with free convolution replacing classical convolution). The following lemma sheds light on the special role of the semicircle law with respect to free convolution. For a measure $\mu \in M_1(\mathbb{R})$, we define the rescaled measure $\mu_{\#\frac{1}{\sqrt{2}}} \in M_1(\mathbb{R})$ by the relation

$$\langle \mu_{\#\frac{1}{\sqrt{2}}}, f\rangle = \int f(\frac{x}{\sqrt{2}})d\mu(x) \quad \text{for all bounded measurable functions } f\,.$$

Lemma 5.3.40 *Let μ be a probability measure on $\mathbb{R}$, so that $\langle \mu, x^2\rangle < \infty$. If*

$$\mu_{\#\frac{1}{\sqrt{2}}} \boxplus \mu_{\#\frac{1}{\sqrt{2}}} = \mu\,, \tag{5.3.26}$$

then μ is a scalar rescale of the semicircle law.

(The assumption of finite variance in Lemma 5.3.40 is superfluous, see Section 5.6. The statement we present has the advantage of possessing a short proof.)

Proof Below, we consider the definition of Voiculescu's transform of μ, see Definition 5.3.37. We deduce from (5.3.26) that

$$\phi_\mu(z) = 2\phi_{\mu_{\#\frac{1}{\sqrt{2}}}}(z)\,.$$

But

$$G_{\mu_{\#\frac{1}{\sqrt{2}}}}(z) = \sqrt{2}G_\mu(\sqrt{2}z) \Rightarrow \phi_\mu(z) = \sqrt{2}\phi_{\mu_{\#\frac{1}{\sqrt{2}}}}(z/\sqrt{2})\,,$$

and so we obtain

$$\phi_\mu(z/\sqrt{2}) = \sqrt{2}\phi_\mu(z)\,. \tag{5.3.27}$$

When $\langle \mu, x^2 \rangle < \infty$ and z has large imaginary part, since

$$G_\mu(z) = \frac{1}{z}\left(1 + \frac{\langle \mu, x\rangle}{z} + \frac{\langle \mu, x^2\rangle}{z^2} + o(|\Im z|^{-2})\right),$$

we get

$$\phi_\mu(z) = \langle \mu, x\rangle + \frac{\langle \mu, x^2\rangle - \langle \mu, x\rangle^2}{2z} + o(|\Im z|^{-1}). \qquad (5.3.28)$$

From (5.3.27) and (5.3.28), we deduce first that $\langle \mu, x\rangle = 0$ and then that, as $\Im z \to \infty$, $z\phi_\mu(z)$ converges to $\langle \mu, x^2\rangle/2$. Since 5.3.27 implies that $z\phi_\mu(z) = 2^{n/2}\phi_\mu(2^{n/2}z)$, it follows by letting n go to infinity that $z\phi_\mu(z) = \langle \mu, x^2\rangle/2$, for all z with $\Im z \neq 0$. From Example 5.3.26, we conclude that μ is a scalar rescale of the semicircle law. □

Exercise 5.3.41 Let $\varepsilon > 0$ and $p_\varepsilon(dx)$ be the Cauchy law

$$p_\varepsilon(dx) = \frac{\varepsilon}{\pi}\frac{1}{x^2 + \varepsilon^2}dx.$$

Show that for $z \in \mathbb{C}^+$, $G_{p_\varepsilon}(z) = 1/(z + i\varepsilon)$ and so $R_{p_\varepsilon}(z) = -i\varepsilon$ and therefore that for any probability measure μ on $\mathbb{R}$, $G_{\mu \boxplus p_\varepsilon}(z) = G_\mu(z + i\varepsilon)$. Show by the residue theorem that $G_{\mu * p_\varepsilon}(z) = G_\mu(z + i\varepsilon)$ and conclude that $\mu \boxplus p_\varepsilon = \mu * p_\varepsilon$, that is, the free convolution by a Cauchy law is the same as the standard convolution.

5.4 Link with random matrices

Random matrices played a central role in free probability since Voiculescu's seminal observation that independent Gaussian Wigner matrices converge in distribution as their size goes to infinity to free semicircular variables (see Theorem 5.4.2). This result can be extended to approximate any law of free variables by taking diagonal matrices and conjugating them by independent unitary matrices (see Corollary 5.4.11). In this section we aim at presenting these results and the underlying combinatorics.

Definition 5.4.1 A sequence of collections of noncommutative random variables

$$(\{a_i^N\}_{i \in J})_{N \in \mathbb{N}}$$

in noncommutative probability spaces $(\mathcal{A}_N, *, \phi_N)$ is called *asymptotically free* if it converges in law as N goes to infinity to a collection of noncommutative random variables $\{a_i\}_{i\in J}$ in a noncommutative probability space $(\mathcal{A}, *, \phi)$, where $\{a_i\}_{i\in J}$

is free. In other words, for any positive integer p and any $i_1,\dots,i_p \in J$,

$$\lim_{N\to\infty} \phi_N(a_{i_1}^N a_{i_2}^N \cdots a_{i_p}^N) = \phi(a_{i_1}\cdots a_{i_p})$$

and the noncommutative variables a_i, $i \in J$, are free in $(A,*,\phi)$.

We first prove that independent (not necessarily Gaussian) Wigner matrices are asymptotically free.

Theorem 5.4.2 *Let $(\Omega,\mathscr{B},P)$ be a probability space and N,p be positive integers. Let $\beta = 1$ or 2, and let $X_i^N : \Omega \to \mathscr{H}_N^{(\beta)}$, $1 \le i \le p$, be a family of random matrices such that $X_i^N/\sqrt{N}$ are Wigner matrices. Assume that, for all $k \in \mathbb{N}$,*

$$\sup_{N\in\mathbb{N}} \sup_{1\le i\le p} \sup_{1\le m\le \ell\le N} E[|X_i^N(m,\ell)|^k] \le c_k < \infty, \tag{5.4.1}$$

that $(X_i^N(m,\ell), 1\le m\le \ell\le N, 1\le i\le p)$ are independent, and that $E[X_i^N(m,\ell)] = 0$ and $E[|X_i^N(m,\ell)|^2] = 1$.

Then the empirical distribution $\hat\mu_N := \mu_{\{\frac{1}{\sqrt N}X_i^N\}_{1\le i\le p}}$ of $\{\frac{1}{\sqrt N}X_i^N\}_{1\le i\le p}$ converges almost surely and in expectation to the law of p free semicircular variables. In other words, the matrices $\{\frac{1}{\sqrt N}X_i^N\}_{1\le i\le p}$, viewed as elements of the noncommutative probability space $(\mathrm{Mat}_N(\mathbb{C}),*,\frac{1}{N}\mathrm{tr})$ (respectively, $(\mathrm{Mat}_N(\mathbb{C}),*,E[\frac{1}{N}\mathrm{tr}])$), are almost surely asymptotically free (respectively, asymptotically free) and their spectral measures almost surely converge (respectively, converge) to the semicircle law.

In the course of the proof of this theorem, we shall prove the following useful intermediate remark, which in particular holds when only one matrix is involved.

Remark 5.4.3 Under the hypotheses of Theorem 5.4.2, except that we do not require that $E[|X_i^N(m,l)|^2] = 1$ but only that it is bounded by 1, for all monomials $q \in \mathbb{C}\langle X_i, 1\le i\le p\rangle$ of degree k normalized so that $q(1,1,\dots,1) = 1$,

$$\limsup_{N\to\infty} |E[\hat\mu_N(q)]| \le 2^k .$$

Proof of Theorem 5.4.2 We first prove the convergence of $E[\hat\mu_N]$. The proof follows closely that of Lemma 2.1.6 (see also Lemma 2.2.3 in the case of complex entries). We need to show, for any monomial $q(\{X_i\}_{1\le i\le p}) = X_{i_1}\cdots X_{i_k} \in \mathbb{C}\langle X_i | 1 \le i \le p\rangle$, the convergence of

$$E[\hat\mu_N(q)] = \frac{1}{N^{\frac{k}{2}+1}} \sum_{\mathbf{j}} \bar{\mathbf{T}}_{\mathbf{j}}, \tag{5.4.2}$$

where $\mathbf{j} = (j_1, \ldots, j_k)$ and

$$\bar{\mathbf{T}}_{\mathbf{j}} := E\left(X_{i_1}^N(j_1, j_2) X_{i_2}^N(j_2, j_3) \cdots X_{i_k}^N(j_k, j_1)\right).$$

(Compare with (2.1.10).) By (5.4.1), $\bar{\mathbf{T}}_{\mathbf{j}}$ is uniformly bounded by c_k.

We use the language of Section 2.1.3. Consider the closed word $w = w_{\mathbf{j}} = j_1 \cdots j_k j_1$ and recall that its weight $\mathrm{wt}(w)$ is the number of distinct letters in w. Let $G_w = (V_w, E_w)$ be the graph as defined in the proof of Lemma 2.1.6. As there, we need to find out which set of indices contributes to the leading order of the sum in the right side of (5.4.2). Loosely speaking, $\bar{\mathbf{T}}_{\mathbf{j}}$ vanishes more often when one has independent matrices than when one always has the same matrix. Hence, the indices corresponding to graphs G_w which are not trees will be negligible. We will then only consider indices corresponding to graphs which are trees, for which $\bar{\mathbf{T}}_{\mathbf{j}}$ will be easily computed. Recall the following from the proof of Lemma 2.1.6 (see also Lemma 2.2.3 for complex entries).

(i) $\bar{\mathbf{T}}_{\mathbf{j}}$ vanishes if each edge in $E_{w_{\mathbf{j}}}$ is not repeated at least twice (i.e. $N_e^{w_j} \geq 2$ for each $e \in E_{w_{\mathbf{j}}}$); hence, $\mathrm{wt}(w_{\mathbf{j}}) \leq \frac{k}{2} + 1$ for all contributing indices;
(ii) the number of N-words in the equivalence class of a given N-word of weight t is $N(N-1)\cdots(N-t+1) \leq N^t$;
(iii) the number of equivalence classes of closed N-words w of length $k+1$ and weight t such that $N_e^w \geq 2$ for each $e \in E_w$ is bounded by $t^k \leq k^k$.

Therefore,

$$\left|\sum_{\mathbf{j}: wt_{\mathbf{j}} \leq \frac{k}{2}} \bar{\mathbf{T}}_{\mathbf{j}}\right| \leq \sum_{t \leq \frac{k}{2}} N^t c_k t^k \leq C(k) N^{\frac{k}{2}}$$

and, considering (5.4.2), we deduce

$$\left| E[\hat{\mu}_N(q)] - \frac{1}{N^{\frac{k}{2}+1}} \sum_{\mathbf{j}: wt_{\mathbf{j}} = \frac{k}{2}+1} \bar{\mathbf{T}}_{\mathbf{j}} \right| \leq C(k) N^{-1}, \tag{5.4.3}$$

where the set $\{\mathbf{j} : wt_{\mathbf{j}} = \frac{k}{2} + 1\}$ is empty if k is odd. This already shows that, if k is odd,

$$\lim_{N \to \infty} E[\hat{\mu}_N(q)] = 0. \tag{5.4.4}$$

If k is even, recall also that if $\mathrm{wt}(w_{\mathbf{j}}) = \frac{k}{2} + 1$, then $G_{w_{\mathbf{j}}}$ is a tree (see an explanation below Definition 2.1.10) and (by the cited definition) $w_{\mathbf{j}}$ is a Wigner word. This means that each (unoriented) edge of $G_{w_{\mathbf{j}}}$ is traversed exactly once in each direction by the walk $j_1 \cdots j_k j_1$. Hence, $\bar{\mathbf{T}}_{\mathbf{j}}$ will be a product of covariances of the entries, and therefore vanishes if these covariances involve two independent matrices. Also, when $c_2 \le 1$, $\bar{\mathbf{T}}_{\mathbf{j}}$ will be bounded above by one and therefore $\limsup_{N\to\infty}|E[\hat{\mu}_N(q)]|$ is bounded above by $|\mathscr{W}_{k,k/2+1}| \le 2^k$, where, as in Definition 2.1.10, $\mathscr{W}_{k,k/2+1}$ denotes a set of representatives for equivalence classes of Wigner words of length $k+1$, and (hence) $|\mathscr{W}_{k,k/2+1}|$ is equal to the Catalan number $\frac{1}{k/2+1}\binom{k}{k/2}$. This will prove Remark 5.4.3.

We next introduce a refinement of Definition 2.1.8 needed to handle the more complicated combinatorics of monomials in several independent Wigner matrices. (Throughout, we consider the set $\mathscr{S} = \{1,\ldots,N\}$ and omit it from the notation.)

Definition 5.4.4 Let $q = q(\{X_i\}_{1\le i\le p}) = X_{i_1}\cdots X_{i_k} \in \mathbb{C}\langle X_i|1\le i\le p\rangle$ be given, where k is even. Let $w = s_1\cdots s_k s_{k+1}$, $s_{k+1} = s_1$ be any Wigner word of length $k+1$ and let G_w be the tree associated with w. We say that w is *q-colorable* if, for $j,\ell = 1,\ldots,k$, equality of edges $\{s_j,s_{j+1}\} = \{s_\ell,s_{\ell+1}\}$ of the tree G_w implies equality of indices ("colors") $i_j = i_\ell$. With, as above, $\mathscr{W}_{k,k/2+1}$ denoting a set of representatives for the equivalence classes of Wigner words of length $k+1$, let $\mathscr{W}^q_{k,k/2+1}$ denote the subset of q-colorable such.

By the previous considerations, each index $\mathbf{j}$ contributing to the leading order in the evaluation of $E[\hat{\mu}_N(q)]$ corresponds to a tree $G_{w_{\mathbf{j}}}$, each edge of which is traversed exactly once in each direction by the walk $j_1\cdots j_k j_1$. Further, since $E[X^N_{i_\ell}(1,2)X^N_{i_{\ell'}}(2,1)] = \mathbf{1}_{\ell=\ell'}$, an index $\mathbf{j}$ contributes to the leading order of $E[\hat{\mu}_N(q)]$ if and only if it the associated Wigner word $w_{\mathbf{j}}$ is q-colorable, and hence equivalent to an element of $\mathscr{W}^q_{k,k/2+1}$. Therefore, for even k,

$$\lim_{N\to\infty} E[\hat{\mu}_N(q)] = |\mathscr{W}^q_{k,k/2+1}|\,. \tag{5.4.5}$$

Moreover, trivially,

$$|\mathscr{W}^q_{k,k/2+1}| \le |\mathscr{W}^{X_1^k}_{k,k/2+1}| = |\mathscr{W}_{k,k/2+1}|\,. \tag{5.4.6}$$

Recall that $\mathscr{W}_{k,k/2+1}$ is canonically in bijection with the set $NC_2(k)$ of non-crossing pair partitions of $\mathscr{K}_k = \{1,\ldots,k\}$ (see Proposition 2.1.11 and its proof). Similarly, for $q = X_{i_1}\cdots X_{i_k}$, the set $\mathscr{W}^q_{k,k/2+1}$ is canonically in bijection with the subset of $NC_2(k)$ consisting of non-crossing pair partitions π of $\mathscr{K}_k$ such that for every block $\{b,b'\}\in\pi$ one has $i_b = i_{b'}$. Thus, we can also write

$$\lim_{N\to\infty} E[\hat{\mu}_N(q)] = \sum_{\pi\in NC_2(k)} \prod_{(b,b')\in\pi} \mathbf{1}_{i_b=i_{b'}}\,,$$

where the product runs over all blocks $\{b,b'\}$ of the pair partition π. Recalling that $k_n(a_i) = \mathbf{1}_{n=2}$ for semicircular variables by Example 5.3.26 and (5.3.7), we can rephrase the above as

$$\lim_{N\to\infty} E[\hat{\mu}_N(q)] = \sum_{\pi\in NC(k)} k_\pi(a_{i_1},\dots,a_{i_k}),$$

with $k_\pi = 0$ if π is not a pair partition and $k_2(a_i,a_j) = \mathbf{1}_{i=j}$. The right side corresponds to the definition of the moments of free semicircular variables according to Theorem 5.3.15 and Example 5.3.26. This proves the convergence of $E[\hat{\mu}_N]$ to the law of m free semicircular variables.

We now prove the almost sure convergence. Continuing to adapt the ideas of the (first) proof of Theorem 2.1.1, we follow the proof of Lemma 2.1.7 closely. (Recall that we proved in Lemma 2.1.7 that the variance of $\langle L_N, x^k\rangle$ is of order N^{-2}. As in Exercise 2.1.16, this was enough, using Chebyshev's inequality and the Borel–Cantelli Lemma, to conclude the almost sure convergence in Wigner's Theorem, Theorem 2.1.1.) Here, we study the variance of $\hat{\mu}_N(q)$ for $q(X_1,\dots,X_p) = X_{i_1}\cdots X_{i_k}$ which is given by

$$\mathrm{Var}(\hat{\mu}_N(q)) = E[|\hat{\mu}_N(q) - E[\hat{\mu}_N(q)]|^2] = \frac{1}{N^{k+2}}\sum_{\mathbf{j},\mathbf{j}'} T_{\mathbf{j},\mathbf{j}'} \qquad (5.4.7)$$

with

$$\begin{aligned} T_{\mathbf{j},\mathbf{j}'} &= E[X_{i_1}(j_1,j_2)\cdots X_{i_k}(j_k,j_1)X_{i_k}(j'_1,j'_2)\cdots X_{i_1}(j'_k,j'_1)] \\ &\quad -E[X_{i_1}(j_1,j_2)\cdots X_{i_k}(j_k,j_1)]E[X_{i_k}(j'_1,j'_2)\cdots X_{i_1}(j'_k,j'_1)], \end{aligned}$$

where we observed that $\overline{\hat{\mu}_N(q)} = \hat{\mu}_N(q^*)$. We consider the sentence $w_{\mathbf{j},\mathbf{j}'} = (j_1\cdots j_k j_1, j'_1 j'_2\cdots j'_1)$ and its associated graph $G_{w_{\mathbf{j},\mathbf{j}'}} = (V_{w_{\mathbf{j},\mathbf{j}'}}, E_{w_{\mathbf{j},\mathbf{j}'}})$. As in the proof of Lemma 2.1.7, $T_{\mathbf{j},\mathbf{j}'}$ vanishes unless each edge in $E_{w_{\mathbf{j},\mathbf{j}'}}$ appears at least twice and the graph $G_{w_{\mathbf{j},\mathbf{j}'}}$ is connected. This implies that the number of distinct elements in $V_{w_{\mathbf{j},\mathbf{j}'}}$ is not more than $k+1$, and it was further shown in the proof of Lemma 2.1.7 that the case where it is equal to $k+1$ never happens. Hence, there are at most k different vertices and so at most N^k possible choices for them. Thus, since $T_{\mathbf{j},\mathbf{j}'}$ is uniformly bounded by $2c_{2k}$, we conclude that there exists a finite constant $c(k)$ such that

$$\mathrm{Var}(\hat{\mu}_N(q)) \le \frac{c(k)}{N^2}.$$

By Chebyshev's inequality we therefore find that

$$P(|\hat{\mu}_N(X_{i_1}\cdots X_{i_k}) - E[\hat{\mu}_N(X_{i_1}\cdots X_{i_k})]| \ge \delta) \le \frac{c(k)}{\delta^2 N^2}.$$

The Borel–Cantelli Lemma then yields that

$$\lim_{N\to\infty} |\hat{\mu}_N(X_{i_1}\cdots X_{i_k}) - \mathbb{E}[\hat{\mu}_N(X_{i_1}\cdots X_{i_k})]| = 0, \text{ a.s.} \qquad \square$$

We next show that Theorem 5.4.2 generalizes to the case of polynomials that may include some deterministic matrices.

Theorem 5.4.5 *Let $\beta = 1$ or 2 and let $(\Omega, \mathscr{B}, P)$ be a probability space. Let $\mathbf{D}^N = \{D_i^N\}_{1\le i\le p}$ be a sequence of Hermitian deterministic matrices and let $\mathbf{X}^N = \{X_i^N\}_{1\le i\le p}$, $X_i^N : \Omega \to \mathscr{H}_N^{(\beta)}$, $1 \le i \le p$, be matrices satisfying the hypotheses of Theorem 5.4.2. Assume that*

$$D := \sup_{k\in\mathbb{N}} \max_{1\le i\le p} \sup_N \frac{1}{N}\mathrm{tr}(|D_i^N|^k)^{\frac{1}{k}} < \infty, \tag{5.4.8}$$

and that the law of $\mathbf{D}^N$ in the noncommutative probability space $(\mathrm{Mat}_N(\mathbb{C}), *, \frac{1}{N}\mathrm{tr})$ converges to a noncommutative law μ. Then we have the following.

(i) *The noncommutative variables $\frac{1}{\sqrt{N}}\mathbf{X}^N$ and $\mathbf{D}^N$ in the noncommutative probability space $(\mathrm{Mat}_N(\mathbb{C}), *, E[\frac{1}{N}\mathrm{tr}])$ are asymptotically free.*
(ii) *The noncommutative variables $\frac{1}{\sqrt{N}}\mathbf{X}^N$ and $\mathbf{D}^N$ in the noncommutative probability space $(\mathrm{Mat}_N(\mathbb{C}), *, \frac{1}{N}\mathrm{tr})$ are almost surely asymptotically free.*

In particular, the empirical distribution of $\{\frac{1}{\sqrt{N}}\mathbf{X}^N, \mathbf{D}^N\}$ converges almost surely and in expectation to the law of $\{\mathbf{X}, \mathbf{D}\}$, $\mathbf{X}$ and $\mathbf{D}$ being free, $\mathbf{D}$ with law μ and $\mathbf{X}$ being p free semicircular variables.

To avoid repetition, we follow a different route than that used in the proof of Theorem 5.4.2 (even though similar arguments could be developed). We denote by $\mathbb{C}\langle D_i, X_i | 1 \le i \le p\rangle$ the set of polynomials in $\{D_i, X_i\}_{1\le i\le p}$, by $\hat{\mu}_N$ (respectively, $\bar{\mu}_N$) the quenched (respectively, annealed) empirical distribution of $\{\mathbf{D}^N, N^{-\frac{1}{2}}\mathbf{X}^N\} = \{D_i^N, N^{-\frac{1}{2}}X_i^N\}_{1\le i\le p}$ given, for $q \in \mathbb{C}\langle D_i, X_i | 1 \le i \le p\rangle$, by

$$\hat{\mu}_N(q) := \frac{1}{N}\mathrm{tr}\left(q(\frac{\mathbf{X}^N}{\sqrt{N}}, \mathbf{D}^N)\right), \quad \bar{\mu}_N(q) := E[\hat{\mu}_N(q)].$$

To prove the convergence of $\{\bar{\mu}_N\}_{N\in\mathbb{N}}$ we first show that this sequence is tight (see Lemma 5.4.6), and then show that any limit point satisfies the so-called Schwinger–Dyson, or master loop, equation which has a unique solution (see Lemma 5.4.7).

Lemma 5.4.6 *For $R, d \in \mathbb{N}$, we denote by $\mathbb{C}\langle X_i, D_i | 1 \le i \le p\rangle_{R,d}$ the set of monomials in $\mathbf{X} := \{X_i\}_{1\le i\le p}$ and $\mathbf{D} := \{D_i\}_{1\le i\le p}$ with total degree in the variables $\mathbf{X}$ (respectively, $\mathbf{D}$) less than R (respectively, d). Under the hypotheses of Theorem 5.4.5, except that instead of $E[|X_i^N(m,l)|^2] = 1$ we only require that it is bounded by 1, assuming without loss of generality that $D \ge 1$, we have that, for* *any $R,d\in\mathbb{N}$,*

$$\sup_{q\in\mathbb{C}\langle X_i,D_i|1\le i\le p\rangle_{R,d}}\limsup_{N\to\infty}|\bar{\mu}_N(q)|\le D^d2^R. \tag{5.4.9}$$

As a consequence, $\{\hat{\mu}_N(q),q\in\mathbb{C}\langle X_i,D_i|1\le i\le p\rangle_{R,d}\}_{N\in\mathbb{N}}$ is tight as a $\mathbb{C}^{C(R,d)}$-valued sequence, with $C(R,d)$ denoting the number of monomials in $\mathbb{C}\langle X_i,D_i|1\le i\le p\rangle_{R,d}$.

We next characterize the limit points of $\{\hat{\mu}_N(q),q\in\mathbb{C}\langle X_i,D_i|1\le i\le p\rangle_{R,d}\}_{N\in\mathbb{N}}$. To this end, let ∂_i be the noncommutative derivative with respect to the variable X_i which is defined as the linear map from $\mathbb{C}\langle X_i,D_i|1\le i\le p\rangle$ to $\mathbb{C}\langle X_i,D_i|1\le i\le p\rangle^{\otimes 2}$ which satisfies the Leibniz rule

$$\partial_i PQ=\partial_i P\times(1\otimes Q)+(P\otimes 1)\times\partial_i Q \tag{5.4.10}$$

and $\partial_i X_j=\mathbf{1}_{i=j}1\otimes 1,\partial_i D_j=0\otimes 0$. (Here, $A\otimes B\times C\otimes D=AC\otimes BD$). If q is a monomial, we have

$$\partial_i q=\sum_{q=q_1X_iq_2}q_1\otimes q_2\,,$$

where the sum runs over all possible decompositions of q as $q_1X_iq_2$.

Lemma 5.4.7 *For any $R,d\in\mathbb{N}$, the following hold under the hypotheses of Theorem 5.4.5.*

(i) *Any limit point τ of $\{\hat{\mu}_N(q),q\in\mathbb{C}\langle X_i,D_i|1\le i\le p\rangle_{R,d}\}_{N\in\mathbb{N}}$ satisfies the boundary and tracial conditions*

$$\tau|_{\mathbb{C}\langle D_i|1\le i\le p\rangle_{0,d}}=\mu|_{\mathbb{C}\langle D_i|1\le i\le p\rangle_{0,d}}\,,\ \tau(PQ)=\tau(QP)\,, \tag{5.4.11}$$

where the second equality in (5.4.11) holds for all monomials P,Q such that $PQ\in\mathbb{C}\langle X_i,D_i|1\le i\le p\rangle_{R,d}$. Moreover, for all $i\in\{1,\ldots,m\}$ and all $q\in\mathbb{C}\langle X_i,D_i|1\le i\le m\rangle_{R-1,d}$, we have

$$\tau(X_iq)=\tau\otimes\tau(\partial_i q)\,. \tag{5.4.12}$$

(ii) *There exists a unique solution $\{\tau_{R,d}(q),q\in\mathbb{C}\langle X_i,D_i|1\le i\le p\rangle_{R,d}\}$ to (5.4.11) and (5.4.12).*

(iii) *Set τ to be the linear functional on $\mathbb{C}\langle X_i,D_i|1\le i\le p\rangle$ so that $\tau(q)=\tau_{R,d}(q)$ for $q\in\mathbb{C}\langle X_i,D_i|1\le i\le p\rangle_{R,d}$, any $R,d\in\mathbb{N}$. Then τ is characterized as the unique solution of the system of equations (5.4.11) and (5.4.12) holding for $q,Q,P\in\mathbb{C}\langle X_i,D_i|1\le i\le p\rangle$. Further, τ is the law of p free semicircular variables, free with variables $\{D_i\}_{1\le i\le p}$ possessing law μ.*

Note here that $q \in \mathbb{C}\langle X_i, D_i | 1 \le i \le p\rangle_{R,d}$ implies that $q_1, q_2 \in \mathbb{C}\langle X_i, D_i | 1 \le i \le p\rangle_{R,d}$ for any decomposition of q into $q_1 X_i q_2$. Therefore, equation (5.4.12), which is given by

$$\tau(X_i q) = \sum_{q = q_1 X_i q_2} \tau(q_1)\tau(q_2),$$

makes sense for any $q \in \mathbb{C}\langle X_i, D_i | 1 \le i \le p\rangle_{R-1,d}$ if $\{\tau(q), q \in \mathbb{C}\langle X_i, D_i | 1 \le i \le p\rangle_{R,d}\}$ is well defined.

Remark 5.4.8 The system of equations (5.4.11) and (5.4.12) is often referred to in the physics literature as the *Schwinger–Dyson*, or *master loop*, equation.

We next show heuristically how, when $\{X_i^N\}_{1 \le i \le p}$ are taken from the GUE, the Schwinger–Dyson equation can be derived using Gaussian integration by parts, see Lemma 2.4.5. Toward this end, we introduce the derivative $\partial_z = (\partial_{\Re z} - i\partial_{\Im z})/2$ with respect to the complex variable $z = \Re z + i\Im z$, so that $\partial_z z = 1$ but $\partial_z \bar{z} = 0$. Using this definition for the complex variable $X_i^N(\ell, r)$ when $\ell \neq r$ (and otherwise the usual definition for the real variable $X_i^N(\ell, \ell)$), note that we have

$$\partial_{X_i^N(\ell,r)} X_{i'}^N(\ell', r') = \delta_{i,i'}\delta_{\ell,\ell'}\delta_{r,r'}. \tag{5.4.13}$$

Lemma 2.4.5 can be extended to standard complex Gaussian variables, as introduced in (4.1.2), by

$$\int \partial_z f(z,\bar{z}) e^{-|z|^2} dz = \int \bar{z} f(z,\bar{z}) e^{-|z|^2} dz. \tag{5.4.14}$$

Here, dz is the Lebesgue measure on $\mathbb{C}$, $dz = d\Re z d\Im z$. Applying (5.4.14) with $z = X_i^N(m,\ell)$ for $m \neq \ell$ and $f(\mathbf{X}^N)$ a smooth function of $\{X_i^N\}_{1 \le i \le p}$ of polynomial growth along with its derivatives, we have

$$E\left[X_i^N(\ell,m) f(\mathbf{X}^N)\right] = E\left[\partial_{X_i^N(m,\ell)} f(\mathbf{X}^N)\right]. \tag{5.4.15}$$

Using Lemma 2.4.5 directly, one verifies that (5.4.15) still holds for $m = \ell$. (One could just as well take (5.4.15) as the definition of $\partial_{X_i^N(m,\ell)}$.) Now let us consider (5.4.15) with the special choice of $f = P(\frac{\mathbf{X}^N}{\sqrt{N}}, \mathbf{D}^N)(j,k)$, where $P \in \mathbb{C}\langle X_i, D_i | 1 \le i \le p\rangle$ and $j,k \in \{1,\dots,N\}$. Some algebra reveals that, using the notation $(A \otimes B)(j,m,\ell,k) = A(j,m)B(\ell,k)$,

$$\partial_{X_i^N(m,\ell)} \left(P(\mathbf{X}^N, \mathbf{D}^N)\right)(j,k) = \left(\partial_i P(\mathbf{X}^N, \mathbf{D}^N)\right)(j,m,\ell,k). \tag{5.4.16}$$

Together with (5.4.15), and after summation over $j = m$ and $\ell = k$, this shows that

$$E\left[\hat{\mu}_N(X_i P) - \hat{\mu}_N \otimes \hat{\mu}_N(\partial_i P)\right] = 0.$$

We have thus seen that, as a consequence of Gaussian integration by parts, $\hat{\mu}_N$ satisfies the master loop equation in expectation. In order to prove that $\bar{\mu}_N$ satisfies asymptotically the master loop equation, that is, part (i) of Lemma 5.4.7, it is therefore enough to show that $\hat{\mu}_N$ self-averages (that is, it is close to its expectation). The latter point is the content of the following technical lemma, which is stated in the generality of Theorem 5.4.5. The proof of the lemma is postponed until after we derive Theorem 5.4.5 from the lemma.

Lemma 5.4.9 *Let q be a monomial in $\mathbb{C}\langle X_i, D_i | 1 \le i \le p\rangle$. Under the hypotheses of Theorem 5.4.5, except that instead of $E[|X_i^N(m,l)|^2] =\le 1$, we only require that it is bounded by 1, we have the following for any $\varepsilon > 0$.*
(i) For any positive integer k,

$$\limsup_{N\to\infty} N^{-\varepsilon} \max_{1\le i\le j\le N} E[|q(\frac{\mathbf{X}^N}{\sqrt{N}}, \mathbf{D}^N)(i,j)|^k] = 0\,. \qquad (5.4.17)$$

(ii) There exists a finite constant $C(q)$ such that, for all positive integers N,

$$E[|\hat{\mu}_N(q) - \bar{\mu}_N(q)|^2] \le \frac{C(q)}{N^{2-\varepsilon}}\,. \qquad (5.4.18)$$

We next give the proof of Theorem 5.4.5, with Lemmas 5.4.6, 5.4.7 and 5.4.9 granted.

Proof of Theorem 5.4.5 By Lemmas 5.4.6 and 5.4.7, $\{\bar{\mu}_N(q), q \in \mathbb{C}\langle X_i, D_i | 1 \le i \le p\rangle_{R,d}\}$ is tight and converges to the unique solution $\{\tau_{R,d}(q), q \in \mathbb{C}\langle X_i, D_i | 1 \le i \le p\rangle_{R,d}\}$ of the system of equations (5.4.11) and (5.4.12). As a consequence, $\tau_{R,d}(q) = \tau_{R',d'}(q)$ for $q \in \mathbb{C}\langle X_i, D_i | 1 \le i \le p\rangle_{R',d'}$, $R \ge R'$ and $d \ge d'$, and we can define $\tau(q) = \tau_{R,d}(q)$ for $q \in \mathbb{C}\langle X_i, D_i | 1 \le i \le p\rangle_{R,d}$. This completes the proof of the first point of Theorem 5.4.5 since τ is the law of p free semicircular variables, free with $\{D_i\}_{1\le i\le p}$ with law μ by part (iii) of Lemma 5.4.7.

The almost sure convergence asserted in the second part of the theorem is a direct consequence of (5.4.18), the Borel–Cantelli Lemma and the previous convergence in expectation. □

We now prove Lemmas 5.4.6, 5.4.7 and 5.4.9.

Proof of Lemma 5.4.6 We prove by induction over R a slightly stronger result, namely that for all $R, d \in \mathbb{N}$, with $|q| = \sqrt{qq^*}$,

$$\sup_{r\ge 0} \sup_{q\in\mathbb{C}\langle X_i, D_i | 1\le i\le p\rangle_{R,d}} \limsup_{N\to\infty} |\bar{\mu}_N(|q|^r)|^{\frac{1}{r}} \le D^d 2^R\,. \qquad (5.4.19)$$

If $R = 0$, this is obvious by (5.4.8). When $R = 1$, by using (G.10) twice, for any $q \in \mathbb{C}\langle X_i, D_i | 1 \le i \le p\rangle_{1,d}$,

$$|\bar{\mu}_N(|q|^r)|^{\frac{1}{r}} \le D^d \max_{1 \le i \le p} |\bar{\mu}_N(|X_i|^r)|^{\frac{1}{r}},$$

which yields (5.4.19) since by Remark 5.4.3, if $r \le 2p$ for some $p \in \mathbb{N}$,

$$\limsup_{N\to\infty} |\bar{\mu}_N(|X_i|^r)|^{\frac{1}{r}} \le \limsup_{N\to\infty} |\bar{\mu}_N((X_i)^{2p})|^{\frac{1}{2p}} \le 2.$$

We next proceed by induction and assume that (5.4.19) is true up to $R = K - 1$. We write $q = q' X_j p(\mathbf{D})$ with p a monomial of degree ℓ and $q' \in \mathbb{C}\langle X_i, D_i | 1 \le i \le p\rangle_{K-1,d-\ell}$. By (G.10) and the induction hypothesis, we have, for all $r \ge 0$,

$$\limsup_{N\to\infty} |\bar{\mu}_N(|q|^r)|^{\frac{1}{r}} \le D^{\ell} |\bar{\mu}_N(|X_j|^{2r})|^{\frac{1}{2r}} |\bar{\mu}_N(|q'|^{2r})|^{\frac{1}{2r}} \le 2D^{\ell} 2^{K-1} D^{d-\ell},$$

which proves (5.4.19) for $K = R$, and thus completes the proof of the induction step. Equation (5.4.9) follows. □

Proof of Lemma 5.4.9 Without loss of generality, we assume in what follows that $D \ge 1$. If q is a monomial in $\mathbb{C}\langle X_i, D_i | 1 \le i \le p\rangle_{R,d}$, and if $\lambda_{\max}(X)$ denotes the spectral radius of a matrix X and e_i the canonical orthonormal basis of $\mathbb{C}^N$,

$$|q(\frac{\mathbf{X}^N}{\sqrt{N}}, \mathbf{D}^N)(i,j)| = |\langle e_i, q(\frac{\mathbf{X}^N}{\sqrt{N}}, \mathbf{D}^N) e_j\rangle| \le D^{\sum_{i=1}^p d_i} \prod_{1 \le i \le p} \lambda_{\max}(\frac{X_i^N}{\sqrt{N}})^{\gamma_i},$$

where γ_i (respectively, d_i) is the degree of q_i in the variable X_i (respectively, D_i) (in particular $\sum \gamma_i \le R$ and $\sum d_i \le d$). As a consequence, we obtain the following bound, for any even positive integer k and any $s \ge 1$,

$$\begin{aligned} E[|q(\frac{\mathbf{X}^N}{\sqrt{N}}, \mathbf{D}^N)(i,j)|^k] &\le D^{kd} \prod_{1 \le i \le p} E[\lambda_{\max}(\frac{X_i^N}{\sqrt{N}})^{k\gamma_i}] \\ &\le D^{kd} \prod_{i=1}^p E\left\{ \operatorname{tr}((\frac{X_1^N}{\sqrt{N}})^{ks\gamma_i}) \right\}^{\frac{1}{s}} \le D^{kd} N^{\frac{p}{s}} E\left\{ \hat{\mu}_N((X_1^N)^{ksR}) \right\}^{\frac{1}{s}}, \end{aligned}$$

where the last term is bounded uniformly in N by Lemma 2.1.6 (see Exercise 2.1.17 in the case where the variances of the entries are bounded by one rather than equal to one, and recall that $D \ge 1$) or Remark 5.4.3. Choosing s large enough so that $\frac{p}{s} < \varepsilon$ completes the proof of (5.4.17). Note that this control holds uniformly on all Wigner matrices with normalized entries possessing ksR moments bounded above by some value.

To prove (5.4.18) we consider a lexicographical order $(X^r, 1 \le r \le pN(N+1)/2)$ of the (independent) entries $(X_k^N(i,j), 1 \le i \le j \le N, 1 \le k \le p)$ and denote by $\Sigma_k = \sigma\{X^r, r \le k\}$ the associated sigma-algebra. By convention we denote by Σ_0 the trivial algebra. Then we have the decomposition

$$\delta_N := E[|\hat{\mu}_N(q) - \bar{\mu}_N(q)|^2] = \sum_{r=1}^{pN(N+1)/2} \Theta_r, \tag{5.4.20}$$

with

$$\Theta_r := E[|E[\hat{\mu}_N(q)|\Sigma_r] - E[\hat{\mu}_N(q)|\Sigma_{r-1}]|^2].$$

By the properties of conditional expectation and the independence of the X^r, we can write $\Theta_r = E[|\vartheta_r|^2]$ with

$$\vartheta_r := E[\hat{\mu}_N(q)|\Sigma_r](\tilde{X}^r, X^{r-1}, \ldots, X^1) - E[\hat{\mu}_N(q)|\Sigma_r](X^r, X^{r-1}, \ldots, X^1)$$

and $(\tilde{X}^r, X^r)$ identically distributed and independent of each other and of $X^{r'}$, $r' \neq r$. If $X^r = X_s^N(i,j)$ for some $s \in \{1, \ldots, p\}$ and $i, j \in \{1, \ldots, N\}^2$, we denote by X_γ^r the interpolation

$$X_\gamma^r := (1-\gamma)X^r + \gamma\tilde{X}^r.$$

Taylor's formula then gives

$$\begin{aligned}
\vartheta_r &= \int_0^1 \partial_\gamma E[\hat{\mu}_N(q)|\Sigma_r](X_\gamma^r, X^{r-1}, \ldots, X^1) d\gamma \\
&= \frac{1}{N^{3/2}} \int_0^1 \partial_\gamma X_\gamma^r \sum_{q = q_1 X_s q_2} E[(q_2 q_1)(j,i)|\Sigma_r](X_\gamma^r, X^{r-1}, \ldots, X^1) d\gamma \\
&\quad + \frac{1}{N^{3/2}} \int_0^1 \partial_\gamma \bar{X}_\gamma^r \sum_{q = q_1 X_s q_2} E[(q_2 q_1)(i,j)|\Sigma_r](X_\gamma^r, X^{r-1}, \ldots, X^1) d\gamma,
\end{aligned}$$

where the sum runs over all decompositions of q into $q_1 X_s q_2$. Hence we obtain that there exists a finite constant $\bar{C}(q)$ such that

$$\Theta_r \le \frac{\bar{C}(q)}{N^3} \sum_{\substack{q = q_1 X_s q_2 \\ (k,\ell)=(i,j) \text{ or } (j,i)}} \int_0^1 E[|Y_s^N(k,\ell)|^2 |(q_2 q_1)(\frac{\mathbf{X}_{\gamma,r}^N}{\sqrt{N}}, \mathbf{D}^N)(\ell,k)|^2] d\gamma,$$

with $\mathbf{X}_{\gamma,r}^N$ the p-tuple of matrices where the (i,j) and (j,i) entries of the matrix s were replaced by the interpolation X_γ^r and its conjugate and $Y_s^N(i,j) = X_s^N(i,j) - \tilde{X}_s^N(i,j)$. We interpolate again with the p-tuple $\mathbf{X}_r^N$ where the entries (i,j) and (j,i) of the matrix s vanishes to obtain by the Cauchy–Schwarz inequality and independence of $\mathbf{X}_r^N$ with $Y_s^N(i,j)$ that, for some finite constants $\bar{C}(q)_1, \bar{C}(q)_2$,

$$\begin{aligned}\Theta_r \leq & \frac{\bar{C}(q)_1}{N^3} \sum_{\substack{q=q_1X_sq_2 \\ (k,\ell)=(i,j) \text{ or } (j,i)}} \Bigg(E[|(q_2q_1)(\frac{\mathbf{X}_r^N}{\sqrt{N}}, \mathbf{D}^N)(k,\ell)|^2] \\ & + \int_0^1 E[|(q_2q_1)(\frac{\mathbf{X}_r^N}{\sqrt{N}}, \mathbf{D}^N)(k,\ell) - (q_2q_1)(\frac{\mathbf{X}_{\gamma,r}^N}{\sqrt{N}}, \mathbf{D}^N)(k,\ell)|^4]^{\frac{1}{2}} d\gamma\Bigg) \\ \leq & \frac{\bar{C}(q)_2}{N^3} \sum_{\substack{q=q_1X_sq_2 \\ (k,\ell)=(i,j) \text{ or } (j,i)}} \Bigg(E[|(q_2q_1)(\frac{\mathbf{X}^N}{\sqrt{N}}, \mathbf{D}^N)(k,\ell)|^2] \\ & + E[|(q_2q_1)(\frac{\mathbf{X}^N}{\sqrt{N}}, \mathbf{D}^N)(k,\ell) - (q_2q_1)(\frac{\mathbf{X}_r^N}{\sqrt{N}}, \mathbf{D}^N)(k,\ell)|^2] \\ & + \int_0^1 E[|(q_2q_1)(\frac{\mathbf{X}_r^N}{\sqrt{N}}, \mathbf{D}^N)(i,j) - (q_2q_1)(\frac{\mathbf{X}_{\gamma,r}^N}{\sqrt{N}}, \mathbf{D}^N)(k,\ell)|^4]^{\frac{1}{2}} d\gamma\Bigg). \end{aligned} \quad (5.4.21)$$

To control the last two terms, consider two p-tuples of matrices $\tilde{\mathbf{X}}^N$ and $\mathbf{X}^N$ that differ only at the entries (i,j) and (j,i) of the matrix s and put $Y_s^N(i,j) = \tilde{X}_s^N(i,j) - X_s^N(i,j)$. Let q be a monomial and $1 \leq k, \ell \leq N$. Then, if we set $\mathbf{X}_\gamma^N = (1-\gamma)\mathbf{X}^N + \gamma\tilde{\mathbf{X}}^N$, we have

$$\begin{aligned}\Delta q(k,\ell) &:= q(\frac{\tilde{\mathbf{X}}^N}{\sqrt{N}}, \mathbf{D}^N)(k,\ell) - q(\frac{\mathbf{X}^N}{\sqrt{N}}, \mathbf{D}^N)(k,\ell) \\ &= -\sum_{\substack{(m,n)=(i,j) \\ \text{or } (j,i)}} \frac{Y_s^N(m,n)}{\sqrt{N}} \int_0^1 \sum_{q=p_1X_sp_2} p_1(\frac{\mathbf{X}_\gamma^N}{\sqrt{N}}, \mathbf{D}^N)(k,m) p_2(\frac{\mathbf{X}_\gamma^N}{\sqrt{N}}, \mathbf{D}^N)(n,\ell) d\gamma.\end{aligned}$$

Using (5.4.17), we deduce, that for all $\varepsilon, r > 0$,

$$\lim_{N\to\infty} N^{\frac{r}{2}} N^{-\varepsilon} \max_{1\leq i,j\leq N} \max_{1\leq k,\ell\leq N} E[|\Delta q(k,\ell)|^r] = 0. \quad (5.4.22)$$

As a consequence, the two last terms in (5.4.21) are at most of order $N^{-1+\varepsilon}$ and summing (5.4.21) over r, we deduce that there exist finite constants $\bar{C}(q)_3, \bar{C}(q)_4$ so that

$$\begin{aligned}\delta_N &\leq \frac{\bar{C}(q)_3}{N^3} \sum_{s=1}^p \sum_{q=q_1X_sq_2} \Bigg(E[\sum_{1\leq i,j\leq N} |(q_2q_1)(\frac{\mathbf{X}^N}{\sqrt{N}}, \mathbf{D}^N)(i,j)|^2] + N^{1+\varepsilon}\Bigg) \\ &= \frac{\bar{C}(q)_3}{N^2} \sum_{s=1}^p \sum_{q=q_1X_sq_2} \bar{\mu}_N(q_2q_1q_1^*q_2^*) + \frac{\bar{C}(q)_4}{N^{2-\varepsilon}}.\end{aligned}$$

Using (5.4.17) again, we conclude that $\delta_N \leq C(q)/N^{2-\varepsilon}$. □

Proof of Lemma 5.4.7 To derive the equations satisfied by a limiting point $\tau_{R,d}$ of $\bar{\mu}_N$, note that the first equality of (5.4.11) holds since we assumed that the law of

$\{D_i^N\}_{1\le i\le p}$ converges to μ, whereas the second equality is verified by $\bar{\mu}_N$ for each N, and therefore by all its limit points. To check that $\tau_{R,d}$ also satisfies (5.4.12), we write

$$\bar{\mu}_N(X_i q) = \frac{1}{N^{3/2}} \sum_{j_1,j_2=1}^{N} E[X_i^N(j_1,j_2) q(\frac{\mathbf{X}^N}{\sqrt{N}}, \mathbf{D}^N)(j_2,j_1)] = \sum_{\ell_1,\ell_2} I_{\ell_1,\ell_2}, \quad (5.4.23)$$

where ℓ_1 (respectively, ℓ_2) denotes the number of occurrences of the entry $X_i^N(j_1,j_2)$ (respectively, $X_i^N(j_2,j_1)$) in the expansion of q in terms of the entries of $\mathbf{X}^N$. $I_{0,0}$ in the right side of (5.4.23) vanishes by independence and centering. To show that the equation (5.4.15) leading to the master loop equation is approximately true, we will prove that $\sum_{(\ell_1,\ell_2)\neq(0,1)} I_{\ell_1,\ell_2}$ is negligible.

We evaluate separately the different terms in the right side of (5.4.23). Concerning $I_{0,1}$, we have

$$I_{0,1} = \frac{1}{N^2} \sum_{j_1,j_2} \sum_{q=q_1 X_i q_2} E[q_1(\frac{\tilde{\mathbf{X}}^N}{\sqrt{N}}, \mathbf{D})(j_1,j_1) q_2(\frac{\tilde{\mathbf{X}}^N}{\sqrt{N}}, \mathbf{D}^N)(j_2,j_2)],$$

where $\tilde{\mathbf{X}}^N$ is the p-tuple of matrices whose entries are the same as $\mathbf{X}^N$, except that $\tilde{X}_i^N(j_1,j_2) = \tilde{X}_i^N(j_2,j_1) = 0$. By (5.4.22), we can replace the matrices $\tilde{\mathbf{X}}^N$ by $\mathbf{X}^N$ up to an error of order $N^{\frac{1}{2}-\varepsilon}$ for any $\varepsilon > 0$, and therefore

$$\begin{aligned} I_{0,1} &= \sum_{q=q_1 X_i q_2} E[\hat{\mu}_N(q_1)\hat{\mu}_N(q_2)] + o(1) \\ &= \sum_{q=q_1 X_i q_2} E[\hat{\mu}_N(q_1)] E[\hat{\mu}_N(q_2)] + o(1), \end{aligned} \quad (5.4.24)$$

where we used (5.4.18) in the second equality.

We similarly find that

$$I_{1,0} = \frac{1}{N^2} \sum_{j_1,j_2=1}^{N} \sum_{q=q_1 X_i q_2} E[q_1(\frac{\tilde{\mathbf{X}}^N}{\sqrt{N}}, \mathbf{D})(j_2,j_1) q_2(\frac{\tilde{\mathbf{X}}^N}{\sqrt{N}}, \mathbf{D})(j_2,j_1)]$$

so that replacing $\tilde{\mathbf{X}}^N$ by $\mathbf{X}^N$ as above shows that

$$I_{1,0} = \frac{1}{N} \bar{\mu}_N(q_1 q_2^*) + o(1) \to_{N\to\infty} 0, \quad (5.4.25)$$

where (5.4.9) was used in the limit, and we used that $(zX_{i_1}\cdots X_{i_p})^* = \bar{z} X_{i_p}\cdots X_{i_1}$.

Finally, with $(\ell_1,\ell_2) \neq (1,0)$ or $(0,1)$, we find that

$$I_{\ell_1,\ell_2} = \frac{1}{N^{2+\frac{\ell_1+\ell_2-1}{2}}} \sum_{q=q_1 X_i q_2 \cdots X_i q_{k+1}} \sum_{j_1,j_2} \sum_{\sigma} I(j_1,j_2,\sigma)$$

with

$$I(j_1,j_2,\sigma) := E[q_1(\frac{\tilde{\mathbf{X}}^N}{\sqrt{N}})(\sigma(1),\sigma(2))\cdots q_{k+1}(\frac{\tilde{\mathbf{X}}^N}{\sqrt{N}})(\sigma(k+1),\sigma(1))],$$

where we sum over all possible maps $\sigma : \{1,\ldots,k+1\} \to \{j_1,j_2\}$ corresponding to ℓ_1 (respectively, ℓ_2) occurrences of the oriented edge (j_1,j_2) (respectively, (j_2,j_1)). Using Hölder's inequality and (5.4.17) we find that the above is at most of order $N^{-\frac{\ell_1+\ell_2-1}{2}+\varepsilon}$ for any $\varepsilon > 0$. Combined with (5.4.24) and (5.4.25), we have proved that

$$\lim_{N\to\infty}\left(\bar{\mu}_N(X_i q) - \sum_{q=q_1X_iq_2}\bar{\mu}_N(q_1)\bar{\mu}_N(q_2)\right) = 0. \tag{5.4.26}$$

Since if $q \in \mathbb{C}\langle X_i, D_i | 1 \le i \le p\rangle_{R-1,d}$, any q_1,q_2 such that $q = q_1X_iq_2$ also belong to this set, we conclude that any limit point $\tau_{R,d}$ of $\bar{\mu}_{\{\frac{1}{\sqrt{N}}X_i^N, D_i^N\}_{1\le i\le p}}$ restricted to $\mathbb{C}\langle X_i, D_i | 1 \le i \le p\rangle_{R,d}$ satisfies (5.4.12).

Since (5.4.12) together with (5.4.11) defines $\tau(P)$ uniquely for any $P \in \mathbb{C}\langle X_i, D_i | 1 \le i \le p\rangle_{R,d}$ by induction over the degree of P in the X_i, it follows that $\bar{\mu}_N$ converges as N goes to infinity towards a law τ which coincides with $\tau_{R,d}$ on $\mathbb{C}\langle X_i, D_i | 1 \le i \le p\rangle_{R,d}$ for all $R,d \ge 0$. Thus, to complete the proof of part (i) of Theorem 5.4.5, it only remains to check that τ is the law of free variables. This task is achieved by induction: we verify that the trace of

$$Q(\mathbf{X},\mathbf{D}) = q_1(\mathbf{X})p_1(\mathbf{D})q_2(\mathbf{X})p_2(\mathbf{D})\cdots p_k(\mathbf{D}) \tag{5.4.27}$$

vanishes for all polynomials q_i, p_i such that $\tau(p_i(\mathbf{D})) = \tau(q_j(\mathbf{X})) = 0$, $i \ge 1, j \ge 2$. By linearity, we can restrict attention to the case where q_i, p_i are monomials.

Let $\deg_{\mathbf{X}}(Q)$ denote the degree of Q in $\mathbf{X}$. We need only consider $\deg_{\mathbf{X}}(Q) \ge 1$. If $\deg_{\mathbf{X}}(Q) = 1$ (and thus $Q = p_1(\mathbf{D})X_ip_2(\mathbf{D})$) we have $\tau(Q) = \tau(X_ip_2p_1(\mathbf{D})) = 0$ by (5.4.12). We continue by induction: assume that $\tau(Q) = 0$ whenever $\deg_{\mathbf{X}}(Q) < K$ and $\tau(p_i(\mathbf{D})) = \tau(q_j(\mathbf{X})) = 0$, $i \ge 1, j \ge 2$. Consider now Q of the form (5.4.27) with $\deg_{\mathbf{X}}(Q) = K$ and $\tau(q_j(\mathbf{X})) = 0, j \ge 2$, $\tau(p_i) = 0, i \ge 1$. Using traciality, we can write $\tau(Q) = \tau(X_iq)$ with $\deg_{\mathbf{X}}(q) = K-1$ and q satisfies all assumptions in the induction hypothesis. Applying (5.4.12), we find that $\tau(Q) = \sum_{q=q_1X_iq_2}\tau(q_1)\tau(q_2)$, where q_1 (respectively, q_2) is a product of centered polynomials except possibly for the first or last polynomials in the X_i. The induction hypothesis now yields that $\tau(X_iq) = \sum_{q=q_1X_iq_2}\tau(q_1)\tau(q_2) = 0$, completing the proof of the claimed asymptotic freeness. The marginal distribution of the $\{X_i\}_{1\le i\le p}$ is given by Theorem 5.4.2. □

We now consider conjugation by unitary matrices following the Haar measure $\rho_{U(N)}$ on the set $U(N)$ of $N \times N$ unitary matrices (see Theorem F.13 for a definition).

Theorem 5.4.10 *Let $\mathbf{D}^N = \{D_i^N\}_{1\le i\le p}$ be a sequence of Hermitian (possibly random) $N \times N$ matrices. Assume that their empirical distribution converges to a noncommutative law μ. Assume also that there exists a deterministic $D < \infty$ such that, for all $k \in \mathbb{N}$ and all $N \in \mathbb{N}$,*

$$\frac{1}{N}\mathrm{tr}((D_i^N)^{2k}) \le D^{2k}, \ a.s.$$

Let $\mathbf{U}^N = \{U_i^N\}_{1\le i\le p}$ be independent unitary matrices with Haar law $\rho_{U(N)}$, independent from $\{D_i^N\}_{1\le i\le p}$. Then the subalgebras $\mathscr{U}_i^N$ generated by the matrices $\{U_i^N, (U_i^N)^*\}_{1\le i\le p}$, and the subalgebra $\mathscr{D}^N$ generated by the matrices $\{D_i^N\}_{1\le i\le p}$, in the noncommutative probability space $(\mathrm{Mat}_N(\mathbb{C}), *, E[\frac{1}{N}\mathrm{tr}])$ (respectively, $(\mathrm{Mat}_N(\mathbb{C}), *, \frac{1}{N}\mathrm{tr})$) are asymptotically free (respectively, almost surely asymptotically free). For all $i \in \{1, \ldots, p\}$, the limit law of $\{U_i^N, (U_i^N)^*\}$ is given as the element of $\mathscr{M}_{\mathbb{C}\langle U,U^*\rangle, \|\cdot\|_{1,*}}$ such that

$$\tau((UU^* - 1)^2) = 0, \quad \tau(U^n) = \tau((U^*)^n) = \mathbf{1}_{n=0}\,.$$

We have the following corollary.

Corollary 5.4.11 *Let $\{D_i^N\}_{1\le i\le p}$ be a sequence of uniformly bounded real diagonal matrices with empirical measure of diagonal elements converging to μ_i, $i = 1, \ldots, p$ respectively. Let $\{U_i^N\}_{1\le i\le p}$ be independent unitary matrices following the Haar measure, independent from $\{D_i^N\}_{1\le i\le p}$.*

(i) *The noncommutative variables $\{U_i^N D_i^N (U_i^N)^*\}_{1\le i\le p}$ in the noncommutative probability space $(\mathrm{Mat}_N(\mathbb{C}), *, E[\frac{1}{N}\mathrm{tr}])$ (respectively, $(\mathrm{Mat}_N(\mathbb{C}), *, \frac{1}{N}\mathrm{tr})$) are asymptotically free (respectively, almost surely asymptotically free), the law of the marginals being given by the μ_i.*

(ii) *The empirical measure of eigenvalues of of $D_1^N + U_N D_2^N U_N^*$ converges weakly almost surely to $\mu_1 \boxplus \mu_2$ as N goes to infinity.*

(iii) *Assume that D_1^N is nonnegative. Then, the empirical measure of eigenvalues of*

$$(D_1^N)^{\frac{1}{2}} U_N D_2^N U_N^* (D_1^N)^{\frac{1}{2}}$$

converges weakly almost surely to $\mu_1 \boxtimes \mu_2$ as N goes to infinity.

Corollary 5.4.11 provides a comparison between independence (respectively, standard convolution) and freeness (respectively, free convolution) in terms of random matrices. If D_1^N and D_2^N are two diagonal matrices whose eigenvalues are independent and equidistributed, the spectral measure of $D_1^N + D_2^N$ converges to a standard convolution. At the other extreme, if the eigenvectors of a matrix A_1^N are "very independent" from those of a matrix A_2^N in the sense that the joint distribution of the matrices can be written as the distribution of $(A_1^N, U^N A_2^N (U^N)^*)$, then free convolution will describe the limit law.

Proof of Theorem 5.4.10 We denote by $\hat{\mu}_N := \mu_{\{D_i^N, U_i^N, (U_i^N)^*\}_{1\le i\le p}}$ the joint empirical distribution of $\{D_i^N, U_i^N, (U_i^N)^*\}_{1\le i\le p}$, considered as an element of the algebraic dual of $\mathbb{C}\langle X_i, 1\le i\le n\rangle$ with $n=3p$, equipped with the involution such that $(\lambda X_{i_1}\cdots X_{i_n})^* = \bar{\lambda} X_{i_n}^* \cdots X_{i_1}^*$ if

$$X_{3i-2}^* = X_{3i-2},\ 1\le i\le p, \quad X_{3i-1}^* = X_{3i},\ 1\le i\le p.$$

The norm is the operator norm on matrices. We may and will assume that $D\ge 1$, and then our variables are bounded uniformly by D. Hence, $\hat{\mu}_N$ is a state on the universal C^*-algebra $\mathscr{A}(D,\{1,\cdots,3n\})$ as defined in Proposition 5.2.14 by an appropriate separation/completion construction of $\mathbb{C}\langle X_i, 1\le i\le n\rangle$. The sequence $\{E[\hat{\mu}_N]\}_{N\in\mathbb{N}}$ is tight for the weak*-topology according to Lemma 5.2.18. Hence, we can take converging subsequences and consider their limit points. The strategy of the proof will be to show, as in the proof of Theorem 5.4.5, that these limit points satisfy a Schwinger–Dyson equation. Of course, this Schwinger–Dyson equation will be slightly different from the equation obtained in Lemma 5.4.7 in the context of Gaussian random matrices. However, it will again be a system of equations defined by an appropriate noncommutative derivative, and will be derived from the invariance by multiplication of the Haar measure, replacing the integration by parts (5.4.15) (the latter could be derived from the invariance by translation of the Lebesgue measure). We will also show that the Schwinger–Dyson equation has a unique solution, implying the convergence of $(E[\hat{\mu}_N], N\in\mathbb{N})$. We will then show that this limit is exactly the law of free variables. Finally, concentration inequalities will allow us to extend the result to the almost sure convergence of $\{\hat{\mu}_N\}_{N\in\mathbb{N}}$.

• *Schwinger–Dyson equation* We consider a limit point τ of $\{E[\hat{\mu}_N]\}_{N\in\mathbb{N}}$. Because we have $\hat{\mu}_N((U_i(U_i)^*-1)^2)=0$ and $\hat{\mu}_N(PQ)=\hat{\mu}_N(QP)$ for any $P,Q\in\mathbb{C}\langle D_i, U_i, U_i^* | 1\le i\le p\rangle$, almost surely, we know by taking the large N limit that

$$\tau(PQ)=\tau(QP), \quad \tau((U_iU_i^*-1)^2)=0,\ 1\le i\le p. \tag{5.4.28}$$

Since τ is a tracial state by Proposition 5.2.16, the second equality in (5.4.28) implies that, in the C^*-algebra $(\mathbb{C}\langle D_i, U_i, U_i^* | 1\le i\le p\rangle, *, \|\cdot\|_\tau)$, $U_iU_i^*=1$ (note that this algebra was obtained by taking the quotient with $\{P : \tau(PP^*)=0\}$).

By definition, the Haar measure $\rho_{U(N)}$ is invariant under multiplication by a unitary matrix. In particular, if $P \in \mathbb{C}\langle D_i, U_i, U_i^* | 1 \le i \le p\rangle$, we have for all $k, l \in \{1, \ldots, N\}$,

$$\partial_t \int \left(P(D_i, U_i e^{tB_i}, e^{-tB_i} U_i^*)\right)(k,l) d\rho_{U(N)}(U_1) \cdots d\rho_{U(N)}(U_p) = 0$$

for any anti-Hermitian matrices B_i ($B_i^* = -B_i$), $1 \le i \le p$, since $e^{tB_i} \in U(N)$. Taking $B_i = 0$ except for $i = i_0$ and $B_{i_0} = 0$ except at the entries (q,r) and (r,q), we find that

$$\int (\partial_{i_0} P)(\{D_i, U_i, U_i^*\}_{1 \le i \le p})(k,r,q,l) d\rho_{U(N)}(U_1) \cdots d\rho_{U(N)}(U_p) = 0$$

with ∂_i the derivative which obeys the Leibniz rules

$$\begin{aligned} \partial_i(PQ) &= \partial_i P \times 1 \otimes Q + P \otimes 1 \times \partial_i Q, \\ \partial_i U_j &= 1_{j=i} U_j \otimes 1, \partial_i U_j^* = -1_{j=i} 1 \otimes U_j^*, \end{aligned}$$

where we used the notation $(A \otimes B)(k,r,q,l) := A(k,r)B(q,l)$. Taking $k = r$ and $q = l$ and summing over r, q gives

$$E\left[\hat{\mu}_N \otimes \hat{\mu}_N(\partial_i P)\right] = 0. \tag{5.4.29}$$

Using Corollary 4.4.31 inductively (on the number p of independent unitary matrices), we find that, for any polynomial $P \in \mathbb{C}\langle D_i, U_i, U_i^* | 1 \le i \le p\rangle$, there exists a positive constant $c(P)$ such that

$$\rho_{U(N)}^{\otimes p}\left(|\mathrm{tr}P(\{D_i^N, U_i^N, (U_i^N)^*\}_{1 \le i \le p}) - E\mathrm{tr}P| > \delta\right) \le 2e^{-c(P)\delta^2},$$

and therefore

$$E[|\mathrm{tr}P - E\mathrm{tr}P|^2] \le \frac{2}{c(P)}.$$

Writing $\partial_i P = \sum_{j=1}^M P_j \otimes Q_j$ for appropriate integer M and polynomials $P_j, Q_j \in \mathbb{C}\langle D_i, U_i, U_i^* | 1 \le i \le p\rangle$, we deduce by the Cauchy–Schwarz inequality that

$$\begin{aligned} &|E\left[(\hat{\mu}_N - E[\hat{\mu}_N]) \otimes (\hat{\mu}_N - E[\hat{\mu}_N])(\partial_i P)\right]| \\ &\quad \le \left|\sum_{j=1}^M E\left[(\hat{\mu}_N - E[\hat{\mu}_N])(P_j)(\hat{\mu}_N - E[\hat{\mu}_N])(Q_j)\right]\right| \\ &\quad \le \frac{2M}{N^2} \max_{1 \le j \le p} \max\{\frac{1}{c(P_j)}, \frac{1}{c(Q_j)}\} \to_{N \to \infty} 0. \end{aligned}$$

We thus deduce from (5.4.29) that

$$\lim_{N \to \infty} E\left[\hat{\mu}_N\right] \otimes E\left[\hat{\mu}_N\right](\partial_i P) = 0.$$

Therefore, the limit point τ satisfies the Schwinger–Dyson equation

$$\tau\otimes\tau(\partial_i P)=0, \tag{5.4.30}$$

for all $i\in\{1,\dots,p\}$ and $P\in\mathbb{C}\langle D_i,U_i,U_i^*|1\le i\le p\rangle$.

• *Uniqueness of the solution to* (5.4.30) Let τ be a solution to (5.4.28) and (5.4.30), and let P be a monomial in $\mathbb{C}\langle D_i,U_i,U_i^*|1\le i\le p\rangle$. We show by induction over the total degree n of P in the variables U_i and U_i^* that $\tau(P)$ is uniquely determined by (5.4.28) and (5.4.30). Note that if $P\in\mathbb{C}\langle D_i|1\le i\le p\rangle$, $\tau(P)=\mu(P)$ is uniquely determined. If $P\in\mathbb{C}\langle D_i,U_i,U_i^*|1\le i\le p\rangle\backslash\mathbb{C}\langle D_i|1\le i\le p\rangle$ is a monomial, we can always write $\tau(P)=\tau(QU_i)$ or $\tau(P)=\tau(U_i^*Q)$ for some monomial Q by the tracial property (5.4.28). We study the first case, the second being similar. If $\tau(P)=\tau(QU_i)$,

$$\partial_i(QU_i)=\partial_iQ\times 1\otimes U_i+(QU_i)\otimes 1,$$

and so (5.4.30) gives

$$\begin{aligned}\tau(QU_i) &= -\tau\otimes\tau(\partial_iQ\times 1\otimes U_i)\\ &= -\sum_{Q=Q_1U_iQ_2}\tau(Q_1U_i)\tau(Q_2U_i)+\sum_{Q=Q_1U_i^*Q_2}\tau(Q_1)\tau(Q_2),\end{aligned}$$

where we used the fact that $\tau(U_i^*Q_2U_i)=\tau(Q_2)$ by (5.4.28). Each term in the right side is the trace under τ of a polynomial of degree strictly smaller in U_i and U_i^* than QU_i. Hence, this relation defines τ uniquely by induction. In particular, taking $P=U_i^n$ we get, for all $n\ge 1$,

$$\sum_{k=1}^{n}\tau(U_i^k)\tau(U_i^{n-k})=0,$$

from which we deduce by induction that $\tau(U_i^n)=0$ for all $n\ge 1$ since $\tau(U_i^0)=\tau(1)=1$. Moreover, as τ is a state, $\tau((U_i^*)^n)=\tau(((U_i)^n)^*)=\overline{\tau(U_i^n)}=0$ for $n\ge 1$.

• *The solution is the law of free variables* It is enough to show by the previous point that the joint law μ of the two free p-tuples $\{U_i,U_i^*\}_{1\le i\le p}$ and $\{D_i\}_{1\le i\le p}$ satisfies (5.4.30). So take $P=U_{i_1}^{n_1}B_1\cdots U_{i_p}^{n_p}B_p$ with some B_ks in the algebra generated by $\{D_i\}_{1\le i\le p}$ and $n_i\in\mathbb{Z}\backslash\{0\}$ (where we observed that $U_i^*=U_i^{-1}$). We wish to show that, for all $i\in\{1,\dots,p\}$,

$$\mu\otimes\mu(\partial_iP)=0. \tag{5.4.31}$$

Note that, by linearity, it is enough to prove this equality when $\mu(B_j) = 0$ for all j. Now, by definition, we have

$$\begin{aligned}\partial_i P &= \sum_{k:i_k=i,n_k>0}\sum_{l=1}^{n_k} U_{i_1}^{n_1}B_1\cdots B_{k-1}U_i^l \otimes U_i^{n_k-l}B_k\cdots U_{i_p}^{n_p}B_p\\ &\quad - \sum_{k:i_k=i,n_k<0}\sum_{l=0}^{n_k-1} U_{i_1}^{n_1}B_1\cdots B_{k-1}U_i^{-l} \otimes U_i^{n_k+l}B_k\cdots U_{i_p}^{n_p}B_p\,.\end{aligned}$$

Taking the expectation on both sides, since $\mu(U_j^i) = 0$ and $\mu(B_j) = 0$ for all $i \neq 0$ and j, we see that freeness implies that the trace of the right side vanishes (recall here that, in the definition of freeness, two consecutive elements have to be in free algebras but the first and the last element can be in the same algebra). Thus, $\mu \otimes \mu(\partial_i P) = 0$, which proves the claim. □

Proof of Corollary 5.4.11 The only point to prove is the first. By Theorem 5.4.10, we know that the normalized trace of any polynomial P in $\{U_i^N D_i^N (U_i^N)^*\}_{1\le i\le p}$ converges to $\tau(P(\{U_iD_iU_i\}_{1\le i\le p}))$ with the subalgebras generated by $\{D_i\}_{1\le i\le p}$ and $\{U_i, U_i^*\}_{1\le i\le p}$ free. Thus, if

$$P(\{X_i\}_{1\le i\le p}) = Q_1(X_{i_1})\cdots Q_k(X_{i_k}), \quad \text{with } i_{\ell+1}\neq i_\ell,\ 1\le \ell\le k-1$$

and $\tau(Q_\ell(X_{i_\ell})) = \tau(Q_\ell(D_{i_\ell})) = 0$, then

$$\tau(P(\{U_iD_iU_i\}_{1\le i\le p})) = \tau(U_{i_1}Q_1(D_{i_1})U_{i_1}^*\cdots U_{i_k}Q_k(D_{i_k})U_{i_k}^*) = 0,$$

since $\tau(Q_\ell(D_{i_\ell})) = 0$ and $\tau(U_i) = \tau(U_i^*) = 0$. □

Exercise 5.4.12 Extend Theorem 5.4.2 to the self-dual random matrices constructed in Exercise 2.2.4.

Exercise 5.4.13 In the case where the D_i are diagonal matrices, generalize the arguments of Theorem 5.4.2 to prove Theorem 5.4.5.

Exercise 5.4.14 Take $D^N(ij) = 1_{i=j}1_{i\le[\alpha N]}$ the projection on the first $[\alpha N]$ indices and X^N be an $N\times N$ matrix satisfying the hypotheses of Theorem 5.4.5. With I_n the identity matrix, set

$$\begin{aligned}Z^N &= D^N X^N(I_N - D^N) + (I_N - D^N)X^N D^N\\ &= \begin{pmatrix} 0 & X^{N-[\alpha N],[\alpha N]}\\ (X^{N-[\alpha N],[\alpha N]})^* & 0\end{pmatrix}\end{aligned}$$

with $X^{N-[\alpha N],[\alpha N]}$ the corner $(X^N)_{1\le i\le[\alpha N],[\alpha N]+1\le j\le N}$ of the matrix X^N. Show that $(Z^N)^2$ has the same eigenvalues as those of the Wishart matrix $W^{N,\alpha} :=$

$X^{N-[\alpha N],[\alpha N]}(X^{N-[\alpha N],[\alpha N]})^*$ with multiplicity 2, plus $N-2[\alpha N]$ zero eigenvalues (if $\alpha \geq 1/2$ so that $N-[\alpha N] \leq [\alpha N]$). Prove the almost sure convergence of the spectral measure of the Wishart matrix $W^{N,\alpha}$ by using Theorem 5.4.5.

Exercise 5.4.15 Continuing in the setup of Exercise 5.4.14, take $T_N \in \mathrm{Mat}_{[\alpha N]}$ to be a self-adjoint matrix with converging spectral distribution. Prove the almost sure convergence of the spectral measure of the Wishart matrix

$$X^{N-[\alpha N],[\alpha N]}T_N T_N^*(X^{N-[\alpha N],[\alpha N]})^*.$$

Exercise 5.4.16 Take $(\sigma(p,q))_{0\leq p,q\leq k-1} \in M_k(\mathbf{C})$ and put

$$\sigma_{ij}(N) = \sigma(p,q)1_{\substack{[pN/k]\leq i<[(p+1)N/k] \\ [qN/k]\leq j<[(q+1)N/k]}} \text{ for } 0 \leq p,q \leq k-1\,.$$

Take X^N to be an $N \times N$ matrix satisfying the hypotheses of Theorem 5.4.5 and put $Y^N_{ij} = N^{-\frac{1}{2}}\sigma_{ij}(N)X^N_{ij}$. Let A^N be a deterministic matrix in the noncommutative probability space $M_N(\mathbf{C})$ and D^N be the diagonal matrix $\mathrm{diag}(1/N,2/N,\ldots,1)$. Assume that $(A^N,(A^N)^*,D^N)$ converge in law towards τ, while the spectral radius of A^N stays uniformly bounded. Prove that $(Y^N+A^N)(Y^N+A^N)^*$ converges in law almost surely and in expectation.
Hint: Show that $Y^N = \sum_{1\leq i\leq k^2} a_i\Sigma^N_i X^N\tilde{\Sigma}^N_i$, with $\{\Sigma^N_i,\tilde{\Sigma}^N_i\}_{1\leq i\leq k^2}$ appropriate projection matrices. Show the convergence in law of $\{(\Sigma^N_i,\tilde{\Sigma}^N_i)_{1\leq i\leq k^2}, A^N, (A^N)^*\}$ by approximating the projections Σ^N_i by functions of D^N. Conclude by using Theorem 5.4.5.

Exercise 5.4.17 Another proof of Theorem 5.4.10 can be based on Theorem 5.4.2 and the polar decomposition $U^N_j = G^N_j(G^N_j(G^N_j)^*)^{-\frac{1}{2}}$ with G^N_j a complex Gaussian matrix which can be written, in terms of independent self-adjoint Gaussian Wigner matrices, as $G^N_j = X^N_j + \mathrm{i}\tilde{X}^N_j$.
(i) Show that U^N_j follows the Haar measure.
(ii) Approximating $G^N_j(G^N_j(G^N_j)^*)^{-\frac{1}{2}}$ by a polynomial in $(X^N_j,\tilde{X}^N_j)_{1\leq j\leq p}$, prove Theorem 5.4.10 by using Theorem 5.4.5.

Exercise 5.4.18 State and prove the analog of Theorem 5.4.10 when the U^N_i follow the Haar measure on the orthogonal group $O(N)$ instead of the unitary group $U(N)$.

5.5 Convergence of the operator norm of polynomials of independent GUE matrices

The goal of this section is to show that not only do the traces of polynomials in Gaussian Wigner matrices converge to the traces of polynomials in free semicircular variables, as shown in Theorem 5.4.2, but that this convergence extends to the operator norm, thus generalizing Theorem 2.1.22 and Exercise 2.1.27 to any polynomial in independent Gaussian Wigner matrices.

The main result of this section is the following.

Theorem 5.5.1 *Let $(X_1^N,\ldots,X_m^N)$ be a collection of independent matrices from the GUE. Let $(S_1,\ldots,S_m)$ be a collection of free semicircular variables in a C^*-probability space $(\mathscr{S},\sigma)$ equipped with a faithful tracial state. For any noncommutative polynomial $P \in \mathbb{C}\langle X_1,\ldots,X_m\rangle$, we have*

$$\lim_{N\to\infty} \|P(\frac{X_1^N}{\sqrt{N}},\ldots,\frac{X_m^N}{\sqrt{N}})\| = \|P(S_1,\ldots,S_m)\| \quad a.s.$$

On the left, we consider the operator norm (largest singular value) of the $N\times N$ random matrix $P(\frac{X_1^N}{\sqrt{N}},\ldots,\frac{X_m^N}{\sqrt{N}})$, whereas, on the right, we consider the norm of $P(S_1,\ldots,S_m)$ in the C^*-algebra $\mathscr{S}$. The theorem asserts a correspondence between random matrices and free probability going considerably beyond moment computations.

Remark 5.5.2 If $(\mathscr{A},\tau)$ is a C^*-probability space equipped with a faithful tracial state, then the norm of a noncommutative random variable $a\in\mathscr{A}$ can be recovered by the limit formula

$$\|a\| = \lim_{k\to\infty} \tau((aa^*)^k)^{\frac{1}{2k}}\,. \tag{5.5.1}$$

However, (5.5.1) fails in general, because the spectrum of aa^* can be strictly larger than the support of the law of aa^*. We assume faithfulness and traciality in Theorem 5.5.1 precisely so that we can use (5.5.1).

We pause to introduce some notation. Let $\mathbf{X} = (X_1,\ldots,X_m)$. We often abbreviate using this notation. For example, we abbreviate the statement $Q(X_1,\ldots,X_m)\in\mathbb{C}\langle X_1,\ldots,X_m\rangle$ to $Q(\mathbf{X})\in\mathbb{C}\langle\mathbf{X}\rangle$. Analogous "boldface" notation will often be used below.

Theorem 5.5.1 will follow easily from the next proposition. The proof of the proposition will take up most of this section. Recall that $\mathbb{C}\langle\mathbf{X}\rangle$ is equipped with the unique involution such that $X_i^* = X_i$ for $i=1,\ldots,m$. Recall also that the *degree* of $Q = Q(\mathbf{X}) \in \mathbb{C}\langle \mathbf{X} \rangle$ is defined to be the maximum of the lengths of the words in the variables X_i appearing in Q.

Proposition 5.5.3 *Let $\mathbf{X}^N := (X_1^N, \ldots, X_m^N)$ be a collection of independent matrices from the GUE. Let $\mathbf{S} := (S_1, \ldots, S_m)$ be a collection of free semicircular variables in a C^*-probability space $(\mathscr{S}, \sigma)$. Fix an integer $d \geq 2$ and let $P = P(\mathbf{X}) \in \mathbb{C}\langle \mathbf{X} \rangle$ be a self-adjoint noncommutative polynomial of degree $\leq d$. Then, for any $\varepsilon > 0$, $P(\frac{\mathbf{X}^N}{\sqrt{N}})$, for all N large enough, has no eigenvalue at distance larger than ε from the spectrum of $P(\mathbf{S})$, almost surely.*

We mention the state σ and degree bound d in the statement of the proposition because, even though they do not appear in the conclusion, they figure prominently in many formulas and estimates below. We remark that since formula (5.5.1) is not needed to prove Proposition 5.5.3, we do not assume faithfulness and traciality of σ. Note the *scale invariance* of the proposition: for any constant $\gamma > 0$, the conclusion of the proposition holds for P if and only if it holds for γP.

Proof of Theorem 5.5.1 (Proposition 5.5.3 granted). We may assume that P is self-adjoint. By Proposition 5.5.3, using $P(\mathbf{S})^* = P(\mathbf{S})$,

$$\limsup_{N\to\infty} \|P(\frac{\mathbf{X}^N}{\sqrt{N}})\| \leq (\text{spectral radius of } P(\mathbf{S})) + \varepsilon = \|P(\mathbf{S})\| + \varepsilon, \quad a.s.,$$

for any positive ε. Using Theorem 5.4.2, we obtain the bound

$$\sigma(P(\mathbf{S})^\ell) = \lim_{N\to\infty} \frac{1}{N} \mathrm{tr}(P(\frac{\mathbf{X}^N}{\sqrt{N}})^\ell) \leq \liminf_{N\to\infty} \|P(\frac{\mathbf{X}^N}{\sqrt{N}})\|^\ell, \ a.s.$$

By (5.5.1), and our assumption that σ is faithful and tracial,

$$\liminf_{N\to\infty} \|P(\frac{\mathbf{X}^N}{\sqrt{N}})\| \geq \sup_{\ell\geq 0} \sigma(P(\mathbf{S})^{2\ell})^{\frac{1}{2\ell}} = \|P(\mathbf{S})\|, \quad a.s.,$$

which gives the complementary bound. □

We pause for more notation. Recall that, given a complex number z, $\Re z$ and $\Im z$ denote the real and imaginary parts of z, respectively. In general, we let $1_{\mathscr{A}}$ denote the unit of a unital complex algebra $\mathscr{A}$. (But we let I_n denote the unit of $\mathrm{Mat}_n(\mathbb{C})$.) Note that, for any self-adjoint element a of a C^*-algebra $\mathscr{A}$, and $\lambda \in \mathbb{C}$ such that $\Im\lambda > 0$, we have that $a - \lambda 1_{\mathscr{A}}$ is invertible and $\|(a - \lambda 1_{\mathscr{A}})^{-1}\| \leq 1/\Im\lambda$. The latter observation is used repeatedly below.

For $\lambda \in \mathbb{C}$ such that $\Im\lambda > 0$, with $P \in \mathbb{C}\langle \mathbf{X} \rangle$ self-adjoint, as in Proposition 5.5.3, let

$$g(\lambda) = g^P(\lambda) = \sigma((P(\mathbf{S}) - \lambda 1_{\mathscr{S}})^{-1}), \tag{5.5.2}$$

$$g_N(\lambda) = g_N^P(\lambda) = E\frac{1}{N}\operatorname{tr}\left((P(\frac{\mathbf{X}^N}{\sqrt{N}}) - \lambda I_N)^{-1}\right). \tag{5.5.3}$$

Both $g(\lambda)$ and $g_N(\lambda)$ are analytic in the upper half-plane $\{\Im\lambda > 0\}$. Further, $g(\lambda)$ is the Stieltjes transform of the law of the noncommutative random variable $P(\mathbf{S})$ under σ, and $g_N(\lambda)$ is the expected value of the Stieltjes transform of the empirical distribution of the eigenvalues of the random matrix $P(\frac{\mathbf{X}^N}{\sqrt{N}})$. The uniform bounds

$$|g(\lambda)| \le \frac{1}{\Im\lambda}, \quad |g_N(\lambda)| \le \frac{1}{\Im\lambda} \tag{5.5.4}$$

are clear.

We now break the proof of Proposition 5.5.3 into three lemmas.

Lemma 5.5.4 *For any choice of constants $c_0, c_0' > 0$, there exist constants $N_0, c_1, c_2, c_3 > 0$ (depending only on P, c_0 and c_0') such that the following holds.*

For all integers N and complex numbers λ, if

$$N \ge \max(N_0, (c_0')^{-1/c_1}), \ |\Re\lambda| \le c_0, \ \text{and}\ N^{-c_1} \le \Im\lambda \le c_0', \tag{5.5.5}$$

then

$$|g^P(\lambda) - g_N^P(\lambda)| \le \frac{c_2}{N^2(\Im\lambda)^{c_3}}. \tag{5.5.6}$$

Now for any $\gamma > 0$ we have $\gamma g^{\gamma P}(\gamma\lambda) = g^P(\lambda)$ and $\gamma g_N^{\gamma P}(\gamma\lambda) = g_N^P(\lambda)$. Thus, crucially, this lemma, just like Proposition 5.5.3, is scale invariant: for any $\gamma > 0$, the lemma holds for P if and only if it holds for γP.

Lemma 5.5.5 *For each smooth compactly supported function $\phi : \mathbb{R} \to \mathbb{R}$ vanishing on the spectrum of $P(\mathbf{S})$, there exists a constant c depending only on ϕ and P such that $|E\frac{1}{N}\operatorname{tr}\phi(P(\mathbf{X}^N))| \le \frac{c}{N^2}$ for all N.*

Lemma 5.5.6 *With ϕ and P as above, $\lim_{N\to\infty} N^{\frac{4}{3}} \cdot \frac{1}{N}\operatorname{tr}\phi(P(\frac{\mathbf{X}^N}{\sqrt{N}})) = 0$, almost surely.*

The heart of the matter, and the hardest to prove, is Lemma 5.5.4. The main idea of its proof is the *linearization trick*, which has a strong algebraic flavor. But before commencing the proof of that lemma, we will present (in reverse order) the chain of implications leading from Lemma 5.5.4 to Proposition 5.5.3.

Proof of Proposition 5.5.3 (Lemma 5.5.6 granted) Let $D = sp(P(\mathbf{S}))$, and write $D^\varepsilon = \{y \in \mathbb{R} : d(y,D) < \varepsilon\}$. Denote by $\hat{\mu}_N$ the empirical measure of the eigenvalues of the matrix $P(\frac{\mathbf{X}^N}{\sqrt{N}})$. By Exercise 2.1.27, the spectral radii of the matrices $\frac{X_i^N}{\sqrt{N}}$ for $i = 1,\ldots,m$ converge almost surely towards 2 and therefore there exists a finite constant M such that $\limsup_{N\to\infty} \hat{\mu}_N([-M,M]^c) = 0$ almost surely. Consider a smooth compactly supported function $\phi : \mathbb{R} \to \mathbb{R}$ equal to one on $(D^\varepsilon)^c \cap [-M,M]$ and vanishing on $D^{\varepsilon/2} \cup [-2M,2M]^c$. We now see that almost surely for large N, no eigenvalue can belong to $(D^\varepsilon)^c$, since otherwise

$$\frac{1}{N}\mathrm{tr}\,\phi(P(\frac{\mathbf{X}^N}{\sqrt{N}})) = \int \phi(x)d\hat{\mu}_N(x) \geq N^{-1} \gg N^{-\frac{4}{3}},$$

in contradiction to Lemma 5.5.6. □

Proof of Lemma 5.5.6 (Lemma 5.5.5 granted) As before, let $\hat{\mu}_N$ denote the empirical distribution of the eigenvalues of $P(\frac{\mathbf{X}^N}{\sqrt{N}})$. Let ∂_i be the noncommutative derivative defined in (5.4.10). Let $\partial_{X_i^N(\ell,k)}$ be the derivative as it appears in (5.4.13) and (5.4.15). The quantity $\int \phi(x)d\hat{\mu}_N(x)$ is a bounded smooth function of $\mathbf{X}^N$ satisfying

$$\partial_{X_i^N(\ell,k)} \int \phi(x)d\hat{\mu}_N(x) = \frac{1}{N^{\frac{3}{2}}}((\partial_i P)(\frac{\mathbf{X}^N}{\sqrt{N}})\tilde{\sharp}\phi'(P(\frac{\mathbf{X}^N}{\sqrt{N}})))_{k,\ell} \tag{5.5.7}$$

where we let $A \otimes B\tilde{\sharp}C = BCA$. Formula (5.5.7) can be checked for polynomial ϕ, and then extended to general smooth ϕ by approximations. As a consequence, with d bounding the degree of P as in the statement of Proposition 5.5.3, we find that

$$\|\nabla \int \phi(x)d\hat{\mu}_N(x)\|_2^2 \;\leq\; \frac{C}{N^2}\sum_{i=1}^m(\|\frac{X_i^N}{\sqrt{N}}\|^{2d-2}+1)\frac{1}{N}\mathrm{tr}\left(|\phi'(P(\frac{\mathbf{X}^N}{\sqrt{N}}))|^2\right)$$

for some finite constant $C = C(P)$. Now the Gaussian Poincaré inequality

$$\mathrm{Var}(f(\mathbf{X}^N)) \leq cE\sum_{i,\ell,r}|\partial_{X_i^N(\ell,r)}f(\mathbf{X}^N)|^2 \tag{5.5.8}$$

must hold with a constant c independent of N and f since all matrix entries $X_i^N(\ell,r)$ are standard Gaussian, see Exercise 4.4.6. Consequently, for every sufficiently small $\varepsilon > 0$, we have

$$\begin{aligned}\mathrm{Var}(\int \phi(x)d\hat{\mu}_N(x)) &\leq cE(\|\nabla \int \phi(x)d\hat{\mu}_N(x)\|_2^2)\\ &\leq \frac{2cCmN^{\varepsilon}}{N^2}E(\int \phi'(x)^2 d\hat{\mu}_N(x))\\ &\quad +c\|\phi'\|^2_\infty E(\frac{C}{N^2}\sum_{i=1}^m \|\frac{X_i^N}{\sqrt{N}}\|^{2d-2}\mathbf{1}_{\|\frac{X_i^N}{\sqrt{N}}\|^{2d-2}\geq N^{\varepsilon}})\\ &\leq \frac{2cCm}{N^{2-\varepsilon}}E(\int \phi'(x)^2 d\hat{\mu}_N(x)) + \|\phi'\|^2_\infty \frac{C'}{N^4} \qquad (5.5.9)\end{aligned}$$

for a constant $C' = C'(\varepsilon)$, where we use the fact that

$$\forall 1 \leq p < \infty, \sup_N E\left\|\frac{X_i^N}{\sqrt{N}}\right\|^p < \infty \qquad (5.5.10)$$

by Lemma 2.6.7. But Lemma 5.5.5 implies that $E[\int \phi'(x)^2 d\hat{\mu}_N(x)]$ is at most of order N^{-2} since ϕ' vanishes on the spectrum of $P(\mathbf{S})$. Thus the right side of (5.5.9) is of order $N^{-4+\varepsilon}$ at most when ϕ vanishes on the spectrum of $P(\mathbf{S})$. Applying Chebyshev's inequality, we deduce that

$$P(|\int \phi(x)d\hat{\mu}_N(x) - E(\int \phi(x)d\hat{\mu}_N(x))| \geq \frac{1}{N^{\frac{4}{3}}}) \leq C''N^{\frac{8}{3}-4+\varepsilon}$$

for a finite constant $C'' = C''(P,\varepsilon,\phi)$. Thus, by the Borel–Cantelli Lemma and Lemma 5.5.5, $\int \phi(x)d\hat{\mu}_N(x)$ is almost surely of order $N^{-\frac{4}{3}}$ at most. □

Proof of Lemma 5.5.5 (Lemma 5.5.4 granted) We first briefly review a method for reconstructing a measure from its Stieltjes transform. Let $\Psi : \mathbb{R}^2 \to \mathbb{C}$ be a smooth compactly supported function. Put $\bar{\partial}\Psi(x,y) = \pi^{-1}(\partial_x + \mathrm{i}\partial_y)\Psi(x,y)$. Assume that $\Im\Psi(x,0) \equiv 0$ and $\bar{\partial}\Psi(x,0) \equiv 0$. Note that by Taylor's Theorem $\bar{\partial}\Psi(x,y)/|y|$ is bounded for $|y| \neq 0$. Let μ be a probability measure on the real line. Then we have the following formula for reconstructing μ from its Stieltjes transform:

$$\Re\int_0^{\infty} dy \int_{-\infty}^{+\infty} dx \left(\int \frac{\bar{\partial}\Psi(x,y)}{t-x-\mathrm{i}y}\mu(dt)\right) = \int \Psi(t,0)\mu(dt). \qquad (5.5.11)$$

This can be verified in two steps. One first reduces to the case $\mu = \delta_0$, using Fubini's Theorem, compact support of $\Psi(x,y)$ and the hypothesis that

$$|\bar{\partial}\Psi(x,y)|/|t-x-\mathrm{i}y| \leq |\bar{\partial}\Psi(x,y)|/|y|$$

is bounded for $y > 0$. Then, letting $|(x,y)| = \sqrt{x^2+y^2}$, one uses Green's Theorem on the domain $\{0 < \varepsilon \leq |(x,y)| \leq R,\ y \geq 0\}$ with R so large that Ψ is supported in the disc $\{|(x,y)| \leq R/2\}$, and with $\varepsilon \downarrow 0$.

Now let ϕ be as specified in Lemma 5.5.5. Let M be a large positive integer, later to be chosen appropriately. Choose the arbitrary constant c_0 in Lemma 5.5.4 so that ϕ is supported in the interval $[-c_0, c_0]$. Choose $c_0' > 0$ arbitrarily. We claim that there exists a smooth function $\Psi : \mathbb{R}^2 \to \mathbb{C}$ supported in the rectangle $[-c_0, c_0] \times [-c_0', c_0']$ such that $\Psi(t,0) = \phi(t)$ and $\bar{\partial}\Psi(x,y)/|y|^M$ is bounded for $|y| \neq 0$. To prove the claim, pick a smooth function $\psi : \mathbb{R} \to [0,1]$ identically equal to 1 near the origin, and supported in the interval $[-c_0', c_0']$. One verifies immediately that $\Psi(x,y) = \sum_{\ell=0}^{M} \frac{i^\ell}{\ell!}\phi^{(\ell)}(x)\psi(y)y^\ell$ has the desired properties. The claim is proved.

As before, let $\hat{\mu}_N$ be the empirical distribution of the eigenvalues of $P(\frac{\mathbf{X}^N}{\sqrt{N}})$. Let μ be the law of the noncommutative random variable $P(\mathbf{S})$. By hypothesis ϕ vanishes on the spectrum of $P(\mathbf{S})$ and hence also vanishes on the support of μ. By (5.5.11) and using the uniform bound

$$\left\|(P(\frac{\mathbf{X}^N}{\sqrt{N}}) - \lambda I_N)^{-1}\right\| \leq 1/\Im\lambda\,,$$

we have

$$\begin{aligned} E\int \phi d\hat{\mu}_N &= E\int \phi d\hat{\mu}_N - \int \phi(t)\mu(dt) \\ &= \Re \int_0^\infty \int_{-\infty}^{+\infty} (\bar{\partial}\Psi(x,y))(g_N(x+\mathrm{i}y) - g(x+\mathrm{i}y))dz\,. \end{aligned}$$

Let $c_4 = c_4(M) > 0$ be a constant such that

$$\sup_{(x,y)\in[-c_0,c_0]\times(0,c_0']} |\bar{\partial}\Psi(x,y)|/|y|^M < c_4\,.$$

Then, with constants N_0, c_1, c_2 and c_3 coming from the conclusion of Lemma 5.5.4, for all $N \geq N_0$,

$$|E\int \phi d\hat{\mu}_N| \leq 2c_4 \int_{-c_0}^{c_0}\int_0^{N^{-c_1}} y^{M-1}dxdy + \frac{c_4 c_2}{N^2}\int_{-c_0}^{c_0}\int_0^{c_0'} y^{M-c_3}dxdy\,,$$

where the first error term is justified by the uniform bound (5.5.4). With M large enough, the right side is of order N^{-2} at most. □

We turn finally to the task of proving Lemma 5.5.4. We need first to introduce suitable notation and conventions for handling block-decomposed matrices with entries in unital algebras.

Let $\mathscr{A}$ be any unital algebra over the complex numbers. Let $\mathrm{Mat}_{k,k'}(\mathscr{A})$ denote the space of k-by-k' matrices with entries in $\mathscr{A}$, and write $\mathrm{Mat}_k(\mathscr{A}) = \mathrm{Mat}_{k,k}(\mathscr{A})$. Elements of $\mathrm{Mat}_{k,k'}(\mathscr{A})$ can and will be identified with elements of the tensor product $\mathrm{Mat}_{k,k'}(\mathbb{C}) \otimes \mathscr{A}$. In the case that $\mathscr{A}$ itself is a matrix algebra, say $\mathrm{Mat}_n(\mathscr{B})$, we identify $\mathrm{Mat}_{k,k'}(\mathrm{Mat}_n(\mathscr{B}))$ with $\mathrm{Mat}_{kn,k'n}(\mathscr{B})$ by viewing each element of the latter space as a k-by-k' array of blocks each of which is an n-by-n matrix. Recall that the unit of $\mathscr{A}$ is denoted by $1_{\mathscr{A}}$, but that the unit of $\mathrm{Mat}_n(\mathbb{C})$ is usually denoted by I_n. Thus, the unit in $\mathrm{Mat}_n(\mathscr{A})$ is denoted by $I_n \otimes 1_{\mathscr{A}}$.

Suppose that $\mathscr{A}$ is an algebra equipped with an involution. Then, given a matrix $a \in \mathrm{Mat}_{k\times\ell}(\mathscr{A})$, we define $a^* \in \mathrm{Mat}_{\ell\times k}(\mathscr{A})$ to be the matrix with entries $(a^*)_{i,j} = a^*_{j,i}$. Suppose further that $\mathscr{A}$ is a C^*-algebra. Then we use the GNS construction to equip $\mathrm{Mat}_{k\times\ell}(\mathscr{A})$ with a norm by first identifying $\mathscr{A}$ with a C^*-subalgebra of $B(H)$ for some Hilbert space H, and then identifying $\mathrm{Mat}_{k\times\ell}(\mathscr{A})$ in compatible fashion with a subspace of $B(H^\ell, H^k)$. In particular, the rules enunciated above equip $\mathrm{Mat}_n(\mathscr{A})$ with the structure of a C^*-algebra. That structure is unique because a C^*-algebra cannot be renormed without destroying the property $\|aa^*\| = \|a\|^2$.

We define the *degree* of $Q \in \mathrm{Mat}_{k\times\ell}(\mathbb{C}\langle\mathbf{X}\rangle)$ to be the maximum of the lengths of the words in the variables X_i appearing in the entries of Q. Also, given a collection $\mathbf{x} = (x_1, \dots, x_m)$ of elements in a unital complex algebra $\mathscr{A}$, we define $Q(\mathbf{x}) \in \mathrm{Mat}_{k\times\ell}(\mathscr{A})$ to be the result of making the substitution $\mathbf{X} = \mathbf{x}$ in every entry of Q.

Given for $i = 1, 2$ a linear map $T_i : V_i \to W_i$, the tensor product $T_1 \otimes T_2 : V_1 \otimes V_2 \to W_1 \otimes W_2$ of the maps is defined by the formula

$$(T_1 \otimes T_2)(A_1 \otimes A_2) = T_1(A_1) \otimes T_2(A_2),\ A_i \in V_i\,.$$

For example, given $A \in \mathrm{Mat}_k(\mathscr{A}) = \mathrm{Mat}_k(\mathbb{C}) \otimes \mathrm{Mat}_N(\mathbb{C})$, one evaluates $(\mathrm{id}_k \otimes \frac{1}{N}\mathrm{tr})(A) \in \mathrm{Mat}_k(\mathbb{C})$ by viewing A as a k-by-k array of N-by-N blocks and then replacing each block by its normalized trace.

We now present the linearization trick. It consists of two parts summarized in Lemmas 5.5.7 and 5.5.8. The first part is the core idea: it describes the spectral properties of a certain sort of patterned matrix with entries in a C^*-algebra. The second part is a relatively simple statement concerning factorization of a noncommutative polynomial into matrices of degree ≤ 1.

To set up for Lemma 5.5.7, fix an integer $d \ge 2$ and let $k_1, \dots, k_{d+1}$ be positive integers such that $k_1 = k_{d+1} = 1$. Put $k = k_1 + \cdots + k_d$. For $i = 1, \dots, d$, let

$$K_i = \left\{ 1 + \sum_{\alpha<i} k_\alpha, \dots, \sum_{\alpha\le i} k_\alpha \right\} \subset \{1, \dots, k\} \tag{5.5.12}$$

and put $K_{d+1} = K_1$. Note that $\{1, \dots, k\}$ is the disjoint union of $K_1, \dots, K_d$. Let $\mathscr{A}$ be a C^*-algebra and for $i = 1, \dots, d$, let $t_i \in \mathrm{Mat}_{k_i\times k_{i+1}}(\mathscr{A})$ be given. Consider the block-decomposed matrix

$$T = \begin{bmatrix} & t_1 & & \\ & & \ddots & \\ & & & t_{d-1} \\ t_d & & & \end{bmatrix} \in \mathrm{Mat}_k(\mathscr{A}), \tag{5.5.13}$$

where for $i = 1, \ldots, d$, the matrix t_i is placed in the block with rows (resp., columns) indexed by K_i (resp., K_{i+1}), and all other entries of T equal $0 \in \mathscr{A}$. We remark that the GNS-based procedure we used to equip each matrix space $\mathrm{Mat}_{p,q}(\mathscr{A})$ with a norm implies that

$$\|T\| \geq \max_{i=1}^{d} \|t_i\|. \tag{5.5.14}$$

Let $\lambda \in \mathbb{C}$ be given and put $\Lambda = \begin{bmatrix} \lambda & 0 \\ 0 & I_{k-1} \end{bmatrix} \in \mathrm{Mat}_k(\mathbb{C})$. Below, we write $\Lambda = \Lambda \otimes 1_{\mathscr{A}}$, $\lambda = \lambda 1_{\mathscr{A}}$ and more generally $\zeta = \zeta \otimes 1_{\mathscr{A}}$ for any $\zeta \in \mathrm{Mat}_k(\mathbb{C})$. This will not cause confusion, and is needed to compress notation.

Lemma 5.5.7 *Assume that $t_1 \cdots t_d - \lambda \in \mathscr{A}$ is invertible and let c be a constant such that*

$$c \geq (1 + d\|T\|)^{2d-2} (1 + \|(t_1 \cdots t_d - \lambda)^{-1}\|).$$

Then the following hold.

(i) $T - \Lambda$ is invertible, the entry of $(T-\Lambda)^{-1}$ in the upper left equals $(t_1 \cdots t_d - \lambda)^{-1}$, and $\|(T-\Lambda)^{-1}\| \leq c$.

(ii) For all $\zeta \in \mathrm{Mat}_k(\mathbb{C})$, if $2c\|\zeta\| < 1$, then $T - \Lambda - \zeta$ is invertible and $\|(T-\Lambda-\zeta)^{-1} - (T-\Lambda)^{-1}\| \leq 2c^2\|\zeta\| < c$.

Proof Put $t_{\geq i} = t_i \cdots t_d$. The following matrix identity is easy to verify.

$$\begin{bmatrix} \lambda & -t_1 & & & \\ & 1 & -t_2 & & \\ & & \ddots & \ddots & \\ & & & 1 & -t_{d-1} \\ -t_d & & & & 1 \end{bmatrix} \begin{bmatrix} 1 & & & & \\ t_{\geq 2} & 1 & & & \\ \vdots & & \ddots & & \\ t_{\geq d-1} & & & 1 & \\ t_{\geq d} & & & & 1 \end{bmatrix}$$

$$= \begin{bmatrix} 1 & -t_1 & & & \\ & 1 & -t_2 & & \\ & & \ddots & \ddots & \\ & & & 1 & -t_{d-1} \\ & & & & 1 \end{bmatrix} \begin{bmatrix} \lambda - t_1 \cdots t_d & & & \\ & 1 & & \\ & & \ddots & \\ & & & 1 \end{bmatrix}$$

Here we have abbreviated notation even further by writing $1 = I_{k_i} \otimes 1_{\mathscr{A}}$. The first matrix above is $\Lambda - T$. Call the next two matrices A and B, respectively, and the last D. The matrices A and B are invertible since $A - I_k$ is strictly lower triangular and $B - I_k$ is strictly upper triangular. The diagonal matrix D is invertible by the hypothesis that $t_1 \cdots t_d - \lambda$ is invertible. Thus $\Lambda - T$ is invertible with inverse $(\Lambda - T)^{-1} = AD^{-1}B^{-1}$. This proves the first of the three claims made in point (i). For $i, j = 1, \dots, d$ let $B^{-1}(i, j)$ denote the $K_i \times K_j$ block of B^{-1}. It is not difficult to check that $B^{-1}(i, j) = 0$ for $i > j$, $B^{-1}(i,i) = I_{k_i}$, and $B^{-1}(i, j) = t_i \cdots t_{j-1}$ for $i < j$. The second claim of point (i) can now be verified by direct calculation, and the third by using (5.5.14) to bound $\|A\|$ and $\|B^{-1}\|$. Point (ii) follows by consideration of the Neumann series expansion for $(I_k - (T - \Lambda)^{-1}\zeta)^{-1}$. □

The second part of the linearization trick is the following.

Lemma 5.5.8 *Let $P \in \mathbb{C}\langle \mathbf{X} \rangle$ be given, and let $d \geq 2$ be an integer bounding the degree of P. Then there exists an integer $n \geq 1$ and matrices*

$$V_1 \in \mathrm{Mat}_{1\times n}(\mathbb{C}\langle \mathbf{X} \rangle), \;\; V_2, \dots, V_{d-1} \in \mathrm{Mat}_n(\mathbb{C}\langle \mathbf{X} \rangle), \;\; V_d \in \mathrm{Mat}_{n\times 1}(\mathbb{C}\langle \mathbf{X} \rangle)$$

of degree ≤ 1 such that $P = V_1 \cdots V_d$.

Proof We have

$$P = \sum_{r=0}^{d} \sum_{i_1=1}^{m} \cdots \sum_{i_r=1}^{m} c^r_{i_1,\dots,i_r} X_{i_1} \cdots X_{i_r}$$

for some complex constants $c^r_{i_1,\dots,i_r}$. Let $\{P^\nu\}_{\nu=1}^n$ be an enumeration of the terms on the right. Let $e^{(k,\ell)}_{i,j} \in \mathrm{Mat}_{k\times\ell}(\mathbb{C})$ denote the elementary matrix with entry 1 in position (i, j) and 0 elsewhere. Then we have a factorization

$$P^\nu = (e^{(1,n)}_{1,\nu} \otimes V^\nu_1)(e^{(n,n)}_{\nu,\nu} \otimes V^\nu_2) \cdots (e^{(n,n)}_{\nu,\nu} \otimes V^\nu_{d-1})(e^{(n,1)}_{\nu,1} \otimes V^\nu_d)$$

for suitably chosen $V^\nu_i \in \mathbb{C}\langle \mathbf{X} \rangle$ of degree ≤ 1. Take $V_1 = \sum_\nu e^{(1,n)}_{1,\nu} \otimes V^\nu_1$, $V_\ell = \sum_\nu e^{(n,n)}_{\nu,\nu} \otimes V^\nu_\ell$ for $\ell = 2, \dots, d-1$, and $V_d = \sum_\nu e^{(n,1)}_{\nu,1} \otimes V^\nu_d$. Then $V_1, \dots, V_d$ have all the desired properties. □

We continue to prepare for the proof of Lemma 5.5.4. For the rest of this section we fix a self-adjoint noncommutative polynomial $P \in \mathbb{C}\langle \mathbf{X} \rangle$ and also, as in the statement of Proposition 5.5.3, an integer $d \geq 2$ bounding the degree of P. For $i = 1, \dots, d$, fix $V_i \in \mathrm{Mat}_{k_i \times k_{i+1}}(\mathbb{C}\langle \mathbf{X} \rangle)$ of degree ≤ 1, for suitably chosen positive integers $k_1, \dots, k_{d+1}$, such that $P = V_1 \cdots V_d$. This is possible by Lemma 5.5.8. Any such factorization serves our purposes. Put $k = k_1 + \cdots + k_d$ and let K_i be as defined in (5.5.12). Consider the matrix

$$L = \begin{bmatrix} & V_1 & & \\ & & \ddots & \\ & & & V_{d-1} \\ V_d & & & \end{bmatrix} \in \mathrm{Mat}_k(\mathbb{C}\langle \mathbf{X} \rangle), \tag{5.5.15}$$

where, for $i = 1, \ldots, d$, the matrix V_i occupies the block with rows (resp., columns) indexed by the set K_i (resp., K_{i+1}), and all other entries of L equal $0 \in \mathbb{C}\langle \mathbf{X} \rangle$. It is convenient to write

$$L = a_0 \otimes 1_{\mathbb{C}\langle \mathbf{X} \rangle} + \sum_{i=1}^{m} a_i \otimes X_i, \tag{5.5.16}$$

for uniquely determined matrices $a_i \in \mathrm{Mat}_k(\mathbb{C})$. As we will see, Lemma 5.5.7 allows us to use the matrices $L(\frac{\mathbf{X}^N}{\sqrt{N}})$ and $L(\mathbf{S})$ to "code" the spectral properties of $P(\frac{\mathbf{X}^N}{\sqrt{N}})$ and $P(\mathbf{S})$, respectively. We will exploit this coding to prove Lemma 5.5.4.

We will say that any matrix of the form L arising from P by the factorization procedure above is a *d-linearization* of P. Of course P has many d-linearizations. However, the linearization construction is scale invariant in the sense that, for any constant $\gamma > 0$, if L is a d-linearization of P, then $\gamma^{1/d} L$ is a d-linearization of γP.

Put

$$\alpha_1 = \sup_{N=1}^{\infty} E(1 + d \left\| L(\frac{\mathbf{X}^N}{\sqrt{N}}) \right\|)^{8d-8}, \tag{5.5.17}$$

$$\alpha_2 = \|a_0\| + \sum_{i=1}^{m} \|a_i\|^2, \tag{5.5.18}$$

$$\alpha_3 = (1 + d\|L(\mathbf{S})\|)^{2d-2}. \tag{5.5.19}$$

Note that $\alpha_1 < \infty$ by (5.5.10). We will take care to make all our estimates below explicit in terms of the constants α_i (and the constant c appearing in (5.5.8)), in anticipation of exploiting the scale invariance of Lemma 5.5.4 and the d-linearization construction.

We next present the "linearized" versions of the definitions (5.5.2) and (5.5.3). For $\lambda \in \mathbb{C}$ such that $\Im\lambda > 0$, let $\Lambda = \begin{bmatrix} \lambda & 0 \\ 0 & I_{k-1} \end{bmatrix} \in \mathrm{Mat}_k(\mathbb{C})$. We define

$$G(\lambda) = (\mathrm{id}_k \otimes \sigma)((L(\mathbf{S}) - \Lambda \otimes 1_{\mathscr{S}})^{-1}), \tag{5.5.20}$$

$$G_N(\lambda) = E(\mathrm{id}_k \otimes \frac{1}{N}\mathrm{tr})((L(\frac{\mathbf{X}^N}{\sqrt{N}}) - \Lambda \otimes I_N)^{-1}), \tag{5.5.21}$$

which are matrices in $\mathrm{Mat}_k(\mathbb{C})$.

The next two lemmas, which are roughly parallel in form, give the basic properties of $G_N(\lambda)$ and $G(\lambda)$, respectively, and in particular show that these matrices are well defined.

Lemma 5.5.9 *(i) For $\lambda \in \mathbb{C}$ such that $\Im\lambda > 0$, $G_N(\lambda)$ is well defined, depends analytically on λ, and satisfies the bound*

$$\|G_N(\lambda)\| \leq \alpha_1(1+\frac{1}{\Im\lambda}). \tag{5.5.22}$$

(ii) The upper left entry of $G_N(\lambda)$ equals $g_N(\lambda)$.
(iii) We have

$$\left\| I_k + (\Lambda - a_0)G_N(\lambda) + \sum_{i=1}^{m} a_i G_N(\lambda) a_i G_N(\lambda) \right\| \leq \frac{c\alpha_1\alpha_2^2}{N^2}(1+\frac{1}{\Im\lambda})^4, \tag{5.5.23}$$

where c is the constant appearing in (5.5.8).

We call (5.5.23) the *Schwinger–Dyson approximation*. Indeed, as N goes to infinity, the left hand side of (5.5.23) must go to zero, yielding a system of equations which is closely related to (5.4.12). We remark also that the proof of (5.5.23) follows roughly the same plan as was used in Section 2.4.1 to give Proof #2 of the semicircle law.

Proof As before, let $e_{\ell,r} = e_{\ell,r}^{N,N} \in \mathrm{Mat}_N(\mathbb{C})$ denote the elementary matrix with entry 1 in position (ℓ, r), and 0 elsewhere. Given $A \in \mathrm{Mat}_{kn}(\mathbb{C})$, let

$$A[\ell,r] = (\mathrm{id}_k \otimes \mathrm{tr}_N)((I_k \otimes e_{r,\ell})A) \in \mathrm{Mat}_k(\mathbb{C}),$$

so that $A = \sum_{\ell,r} A[\ell,r] \otimes e_{\ell,r}$. (Thus, within this proof, we view A as an N-by-N array of k-by-k blocks $A[\ell,r]$.)

Since λ is fixed throughout the proof, we drop it from the notation to the extent possible. To abbreviate, we write

$$R_N = (L(\frac{\mathbf{X}^N}{\sqrt{N}}) - \Lambda \otimes I_N)^{-1}, \quad H_N = (\mathrm{id}_k \otimes \frac{1}{N}\mathrm{tr})R_N = \frac{1}{N}\sum_{i=1}^{N} R_N[i,i].$$

From Lemma 5.5.7(i) we get an estimate

$$\|R_N\| \leq (1+d\left\|L(\frac{\mathbf{X}^N}{\sqrt{N}})\right\|)^{2d-2}(1+\frac{1}{\Im\lambda}) \tag{5.5.24}$$

which, combined with (5.5.17), yields assertion (i). From Lemma 5.5.7(i) we also get assertion (ii).

Assertion (iii) will follow from an integration by parts as in (5.4.15). Recall that $\partial_{X_i^N(\ell,r)} X_{i'}^N(\ell',r') = \delta_{i,i'}\delta_{\ell,\ell'}\delta_{r,r'}$. We have, for $i \in \{1,\ldots,m\}$ and $\ell,r,\ell',r' \in \{1,\ldots,N\}$,

$$\partial_{X_i^N(r,\ell)} R_N[r',\ell'] = -\frac{1}{\sqrt{N}} R_N[r',r]a_i R_N[\ell,\ell']. \tag{5.5.25}$$

Recall that $E\partial_{X_i^N(r,\ell)} f(\mathbf{X}^N) = EX_i^N(\ell,r)f(\mathbf{X}^N)$. We obtain

$$-\frac{1}{\sqrt{N}} ER_N(\lambda)[r',r]a_i R_N(\lambda)[\ell,\ell'] = EX_i^N(\ell,r)R_N(\lambda)[r',\ell']. \tag{5.5.26}$$

Now left-multiply both sides of (5.5.26) by $\frac{a_i}{N^{3/2}}$, and sum on i, $\ell = \ell'$, and $r = r'$, thus obtaining the first equality below.

$$\begin{aligned}
-\sum_{i=1}^m E(a_i H_N a_i H_N) &= E(\mathrm{id}_k \otimes \frac{1}{N}\mathrm{tr})((L(\frac{\mathbf{X}^N}{\sqrt{N}}) - a_0 \otimes I_N)R_N) \\
&= E(\mathrm{id}_k \otimes \frac{1}{N}\mathrm{tr})\,(I_k \otimes I_N + ((\Lambda - a_0)\otimes I_N)R_N) \\
&= I_k + (\Lambda - a_0)G_N(\lambda).
\end{aligned}$$

The last two steps are simple algebra. Thus the left side of (5.5.23) is bounded by the quantity

$$\begin{aligned}
\Delta_N &= \left\| E[\sum_{i=1}^m a_i(H_N - EH_N)a_i(H_N - EH_N)] \right\| \\
&\le (\sum_i \|a_i\|^2)E\|H_N - EH_N\|_2^2 \le c(\sum_i \|a_i\|^2)E\sum_{i,\ell,r}\left\|\partial_{X_i^N(r,\ell)} H_N\right\|_2^2,
\end{aligned}$$

where at the last step we use once again the Gaussian Poincaré inequality in the form (5.5.8). For the quantity at the extreme right under the expectation, we have by (5.5.25) an estimate

$$\frac{1}{N^3}\sum_{i,r,\ell,r',\ell'} \mathrm{tr}\left(R_N[\ell',r]a_i R_N[\ell,\ell'] R_N[\ell,r']^* a_i^* R_N[r',r]^*\right) \le \frac{1}{N^2}(\sum_i \|a_i\|^2)\|R_N\|^4.$$

The latter, combined with (5.5.17), (5.5.18) and (5.5.24), finishes the proof of (5.5.23). □

We will need a generalization of $G(\lambda)$. For any $\Lambda \in \mathrm{Mat}_k(\mathbb{C})$ such that $L(\mathbf{S}) - \Lambda \otimes 1_{\mathcal{S}}$ is invertible, we define

$$\tilde{G}(\Lambda) = (\mathrm{id}_k \otimes \sigma)((L(\mathbf{S}) - \Lambda \otimes 1_{\mathcal{S}})^{-1}).$$

Now for $\lambda \in \mathbb{C}$ such that $G(\lambda)$ is defined, $\tilde{G}(\Lambda)$ is also defined and

$$\tilde{G}\left(\begin{bmatrix} \lambda & 0 \\ 0 & I_{k-1} \end{bmatrix}\right) = G(\lambda). \tag{5.5.27}$$

Thus, the function $\tilde{G}(\Lambda)$ should be regarded as an extension of $G(\lambda)$. Let $\mathscr{O}$ be the connected open subset of $\mathrm{Mat}_k(\mathbb{C})$ consisting of all sums of the form

$$\begin{bmatrix} \lambda & 0 \\ 0 & I_{k-1} \end{bmatrix} + \zeta,$$

where

$$\lambda \in \mathbb{C},\ \zeta \in \mathrm{Mat}_k(\mathbb{C}),\ \Im\lambda > 0,\ 2\alpha_3\|\zeta\|(1+\frac{1}{\Im\lambda}) < 1. \tag{5.5.28}$$

Recall that the constant α_3 is specified in (5.5.19).

Lemma 5.5.10 *(i) For $\lambda \in \mathbb{C}$ such that $\Im\lambda > 0$, $G(\lambda)$ is well defined, depends analytically on λ, and satisfies the bound*

$$\|G(\lambda)\| \le k^2\alpha_3(1+\frac{1}{\Im\lambda}). \tag{5.5.29}$$

(ii) The upper left entry of $G(\lambda)$ equals $g(\lambda)$.
(iii) More generally, $\tilde{G}(\Lambda)$ is well defined and analytic for $\Lambda \in \mathscr{O}$, and satisfies the bound

$$\left\|\tilde{G}\left(\begin{bmatrix} \lambda & 0 \\ 0 & I_{k-1} \end{bmatrix} + \zeta\right) - G(\lambda)\right\| \le 2k^2\alpha_3^2(1+\frac{1}{\Im\lambda})^2\|\zeta\| < k^2\alpha_3(1+\frac{1}{\Im\lambda}) \tag{5.5.30}$$

for λ and ζ as in (5.5.28).
(iv) If there exists $\Lambda \in \mathscr{O}$ such that $\Lambda - a_0$ is invertible and the operator

$$(L(\mathbf{S}) - a_0 \otimes 1_{\mathscr{S}})((\Lambda - a_0)^{-1} \otimes 1_{\mathscr{S}}) \in \mathrm{Mat}_k(\mathscr{S}) \tag{5.5.31}$$

has norm < 1, then

$$I_k + (\Lambda - a_0)\tilde{G}(\Lambda) + \sum_{i=1}^{m} a_i\tilde{G}(\Lambda)a_i\tilde{G}(\Lambda) = 0 \tag{5.5.32}$$

for all $\Lambda \in \mathscr{O}$.

In particular, $\tilde{G}(\Lambda)$ is by (5.5.32) invertible for all $\Lambda \in \mathscr{O}$. As we will see in the course of the proof, equation (5.5.32) is essentially a reformulation of the Schwinger–Dyson equation (5.4.12).

Proof Let us specialize Lemma 5.5.7 by taking $t_i = V_i(\mathbf{S})$ for $i = 1, \ldots, d$ and hence $T = L(\mathbf{S})$. Then we may take $\alpha_3(1+1/\Im\lambda)^{-1}$ as the constant in Lemma 5.5.7. We note also the crude bound $\|(\mathrm{id}_k \otimes \sigma)(M)\| \le k^2\|M\|$ for $M \in \mathrm{Mat}_k(\mathscr{S})$. By Lemma 5.5.7(i) the operator $L(\mathbf{S}) - \Lambda \otimes 1_{\mathscr{S}}$ is invertible, with inverse bounded in norm by $\alpha_3(1+1/\Im\lambda)^{-1}$ and possessing $(P(\mathbf{S}) - \lambda 1_{\mathscr{S}})^{-1}$ as its upper left entry.

Points (i) and (ii) of Lemma 5.5.10 follow. In view of the relationship (5.5.27) between $\tilde{G}(\Lambda)$ and $G(\lambda)$, point (iii) of Lemma 5.5.10 follows from Lemma 5.5.7(ii).

It remains only to prove assertion (iv). Since the open set $\mathcal{O}$ is connected, and $\tilde{G}(\Lambda)$ is analytic on $\mathcal{O}$, it is necessary only to show that (5.5.32) holds for all Λ in the nonempty open subset of $\mathcal{O}$ consisting of Λ for which the operator (5.5.31) is defined and has norm < 1. Fix such Λ now, and let M denote the corresponding operator (5.5.31). Put

$$b_i = a_i(\Lambda - a_0)^{-1} \in \mathrm{Mat}_k(\mathbb{C})$$

for $i = 1, \ldots, m$. By developing

$$(L(\mathbf{S}) - \Lambda \otimes 1_{\mathcal{S}})^{-1} = -((\Lambda - a_0)^{-1} \otimes 1_{\mathcal{S}})(I_k \otimes 1_{\mathcal{S}} - M)^{-1},$$

as a power series in M, we arrive at the identity

$$I_k + (\Lambda - a_0)\tilde{G}(\Lambda) = -\sum_{\ell=0}^{\infty} (\mathrm{id}_k \otimes \sigma)(M^{\ell+1}).$$

According to the Schwinger–Dyson equation (5.4.12),

$$b_i(\mathrm{id}_k \otimes \sigma)(S_i M^{\ell}) = b_i \sum_{p=1}^{\ell} (\mathrm{id}_k \otimes \sigma)(M^{p-1}) b_i (\mathrm{id}_k \otimes \sigma)(M^{\ell-p}),$$

whence, after summation, we get (5.5.32). □

Remark 5.5.11 In Exercise 5.5.15 we indicate a purely operator-theoretic way to prove (5.5.32), using a special choice of C^*-probability space.

Lemma 5.5.12 *Fix $\lambda \in \mathbb{C}$ and a positive integer N such that $\Im\lambda > 0$ and the right side of* (5.5.23) *is $< 1/2$. Put $\Lambda = \begin{bmatrix} \lambda & 0 \\ 0 & I_{k-1} \end{bmatrix} \in \mathrm{Mat}_k(\mathbb{C})$. Then $G_N(\lambda)$ is invertible and the matrix*

$$\Lambda_N(\lambda) = -G_N(\lambda)^{-1} + a_0 - \sum_{i=1}^{m} a_i G_N(\lambda) a_i \qquad (5.5.33)$$

satisfies

$$\|\Lambda_N(\lambda) - \Lambda\| \le \frac{2c\alpha_1\alpha_2^2}{N^2}(1 + \frac{1}{\Im\lambda})^4(|\lambda| + 1 + \alpha_2 + \alpha_1\alpha_2 + \frac{\alpha_1\alpha_2}{\Im\lambda}), \qquad (5.5.34)$$

where c is the constant appearing in (5.5.8).

Proof Let us write

$$I_k + (\Lambda - a_0)G_N(\lambda) + \sum_{i=1}^{m} a_i G_N(\lambda) a_i G_N(\lambda) = \varepsilon_N(\lambda).$$

By hypothesis $\|\varepsilon_N(\lambda)\| < 1/2$, hence $I_k - \varepsilon_N(\lambda)$ is invertible, hence $G_N(\lambda)$ is invertible, and we have an algebraic identity

$$\Lambda_N(\lambda) - \Lambda = (I_k - \varepsilon_N(\lambda))^{-1}\varepsilon_N(\lambda)(\Lambda - a_0 + \sum_{i=1}^m a_i G_N(\lambda) a_i).$$

We now arrive at estimate (5.5.34) by our hypothesis $\|\varepsilon_N(\lambda)\| < 1/2$, along with (5.5.23) to bound $\|\varepsilon_N(\lambda)\|$ more strictly, and finally (5.5.18) and (5.5.22). □

We record the last trick.

Lemma 5.5.13 *Let $z, w \in \mathrm{Mat}_k(\mathbb{C})$ be invertible. If*

$$z^{-1} + \sum_{i=1}^m a_i z a_i = w^{-1} + \sum_{i=1}^m a_i w a_i, \text{ and } \|z\|\|w\| \sum_{i=1}^m \|a_i\|^2 < 1,$$

then $z = w$.

Proof Suppose that $z \neq w$. We have $w - z = \sum_{i=1}^m z a_i (w-z) a_i w$ after some algebraic manipulation, whence a contradiction. □

Completion of the proof of Lemma 5.5.4 By the scale invariance of Lemma 5.5.4 and of the d-linearization construction, for any constant $\gamma > 0$, we are free to replace P by γP, and hence to replace the linearization L by $\gamma^{1/d} L$. Thus, without loss of generality, we may assume that

$$\alpha_1 < 2, \quad \alpha_2 < \frac{1}{18}, \quad \alpha_3 < 2. \tag{5.5.35}$$

The hypothesis of Lemma 5.5.10(iv) is then fulfilled. More precisely, with $\Lambda = \begin{bmatrix} i & 0 \\ 0 & I_{k-1} \end{bmatrix}$, the matrix $\Lambda - a_0$ is invertible, and the operator (5.5.31) has norm < 1. Consequently, we may take the Schwinger–Dyson equation (5.5.32) for granted.

Now fix $c_0, c_0' > 0$ arbitrarily. We are free to increase c_0', so we may assume that

$$c_0' > 3. \tag{5.5.36}$$

We then pick N_0 and c_1 so that:

> If (5.5.5) holds, then the right side of (5.5.23) is $< 1/2$ and the right side of (5.5.34) is $< \frac{1}{2\alpha_3}(1 + \frac{1}{3\lambda})^{-1}$.

Suppose now that N and λ satisfy (5.5.5). Then $\Lambda_N(\lambda)$ is well defined by formula (5.5.33) because $G_N(\lambda)$ is invertible, and moreover belongs to $\mathcal{O}$. We claim that

$$\tilde{G}(\Lambda_N(\lambda)) = G_N(\lambda). \tag{5.5.37}$$

To prove (5.5.37), which is an equality of analytic functions of λ, we may assume in view of (5.5.36) that

$$\Im\lambda > 2\,. \tag{5.5.38}$$

Put $z = G_N(\lambda)$ and $w = \tilde{G}(\Lambda_N(\lambda))$. Now

$$\|z\| < 3$$

by (5.5.22), (5.5.35) and (5.5.38), whereas

$$\|w\| < 6$$

by (5.5.29), (5.5.30), (5.5.35) and (5.5.38). Applying the Schwinger–Dyson equation (5.5.32) along with (5.5.35), we see that the hypotheses of Lemma 5.5.13 are fulfilled. Thus $z = w$, which completes the proof of the claim (5.5.37). The claim granted, for suitably chosen c_2 and c_3, the bound (5.5.6) in Lemma 5.5.4 holds by (5.5.30) and (5.5.34), along with Lemma 5.5.9(ii) and Lemma 5.5.10(ii). In turn, the proofs of Proposition 5.5.3 and Theorem 5.5.1 are complete. □

In the next two exercises we sketch an operator-theoretic approach to the Schwinger–Dyson equation (5.5.32) based on the study of Boltzmann–Fock space (see Example 5.3.3).

Exercise 5.5.14 Let T, π and S be bounded linear operators on a Hilbert space. Assume that T is invertible. Assume that π is a projector and let $\pi^\perp = 1 - \pi$ be the complementary projector. Assume that

$$\pi^\perp S \pi^\perp = S \text{ and } \pi^\perp T \pi^\perp = \pi^\perp T \pi^\perp S = \pi^\perp\,.$$

Then we have

$$\pi = \pi T^{-1}\pi(T - TST)\pi = \pi(T - TST)\pi T^{-1}\pi\,. \tag{5.5.39}$$

Hint: Use the block matrix factorization

$$\begin{bmatrix} a & b \\ c & d \end{bmatrix} = \begin{bmatrix} 1 & bd^{-1} \\ 0 & 1 \end{bmatrix} \begin{bmatrix} a - bd^{-1}c & 0 \\ 0 & d \end{bmatrix} \begin{bmatrix} 1 & 0 \\ d^{-1}c & 1 \end{bmatrix}$$

in the Hilbert space setting.

Exercise 5.5.15 Let V be a finite-dimensional Hilbert space with orthonormal basis $\{e_i\}_{i=1}^m$. Let $H = \bigoplus_{i=0}^\infty V^{\otimes i}$ be the corresponding Boltzmann–Fock space, as in Example 5.3.3. Let $v \in V^{\otimes 0} \subset H$ be the vacuum state. Equip $B(H)$ with the state $\phi = (a \mapsto \langle av, v\rangle)$. For $i = 1, \ldots, m$, let $\ell_i = e_i \otimes \cdot \in B(H)$ be the *left* creation operator previously considered. We will also consider the *right* creation operator $r_i = \cdot \otimes e_i \in B(H)$. For $i = 1, \ldots, m$ put $s_i = \ell_i + \ell_i^*$ and recall that $s_1, \ldots, s_m$ are

free semicircular elements in $B(H)$. Put $\mathbf{s} = (s_1, \ldots, s_m)$.
(i) For $\alpha = 1, \ldots, m$, show that $r_\alpha^* r_\alpha = 1_{B(H)}$ and $\pi_\alpha = r_\alpha r_\alpha^*$ is the orthogonal projection of H onto the closed linear span of all words $e_{i_1} \otimes \cdots \otimes e_{i_r}$ with terminal letter e_{i_r} equal to e_α.
(ii) Let $\pi_0 \in B(H)$ be the orthogonal projection of H onto $V^{\otimes 0}$. Show that we have an orthogonal direct sum decomposition $H = \bigoplus_{\alpha=0}^m \pi_\alpha H$.
(iii) Verify the relations

$$\pi_\alpha s_i \pi_\beta = \delta_{\alpha\beta} r_\alpha s_i r_\beta^*, \quad \pi_0 s_i r_\alpha = \delta_{i\alpha} \pi_0 = r_\alpha^* s_i \pi_0 \tag{5.5.40}$$

holding for $i, \alpha, \beta = 1, \ldots, m$.
(iv) Identify $\mathrm{Mat}_k(B(H))$ with $B(H^k)$. Let $L = a_0 + \sum_{i=1}^m a_i \otimes X_i \in \mathrm{Mat}_k(\mathbb{C}\langle \mathbf{X} \rangle)$ be of degree 1. Fix $\Lambda \in \mathrm{Mat}_k(\mathbb{C})$ such that $T = L(\mathbf{s}) - \Lambda \otimes 1_{B(H)} \in B(H^k)$ is invertible. Put $\pi = I_k \otimes \pi_0 \in B(H^k)$ and $S = \sum_{i=1}^m (I_k \otimes r_i) T^{-1} (I_k \otimes r_i^*) \in B(H^k)$. Put $\tilde{G}(\Lambda) = (\mathrm{id}_k \otimes \phi)(T^{-1})$. Use (5.5.39) and (5.5.40) to verify (5.5.32).

5.6 Bibliographical notes

For basics in free probability and operator algebras, we relied on Voiculescu's St. Flour course [Voi00b] and on [VoDN92]. A more combinatorial approach is presented in [Spe98]. For notions of operator algebras which are summarized in Appendix G, we used [Rud91], [DuS58], [Mur90], [Li92], [Ped79] and [Dix69]. For affiliated operators, we relied on [BeV93] and [DuS58], and on the paper [Nel74]. (In particular, the remark following Definition 5.2.28 clarifies that the notion of affiliated operators in these references coincide.) Section 5.3.2 follows closely [Spe03]. Many refinements of the relation between free cumulants and freeness can be found in the work of Speicher, Nica and co-workers, see the memoir [Spe98] and the recent book [NiS06] with its bibliography. A theory of cumulants for finite dimensional random matrices was initiated in [CaC06]. Subjects related to free probability are also discussed in the collection of papers [Voi97].

Free additive convolutions were first studied in [Voi86] and [BeV92] for bounded operators, then generalized to operators with finite variance in [Maa92] and finally to the general setting presented here in [BeV93]. A detailed study of free convolution by the semicircle law was done by Biane [Bia97b]. Freeness for rectangular matrices and related free convolution were studied in [BeG09]. The Markovian structure of free convolution (see [Voi00a] for a basic derivation) was shown in [Voi93] and [Bia98a, Theorem 3.1] to imply the existence of a unique *subordination function* $F : \mathbb{C} \to \mathbb{C}$ such that

- for all $z \in \mathbb{C} \backslash \mathbb{R}$, $G_{a+b}(z) = G_a(F(z))$,

- $F(\mathbb{C}^+) \subset \mathbb{C}^+$, $\bar{F}(z) = F(\bar{z})$, $\Im(F(z)) \geq \Im(z)$ for $z \in \mathbb{C}^+$ and $F(iy)/iy \to 1$ as y goes to infinity while staying in $\mathbb{R}$.

Note that, according to [BeV93, Proposition 5.2], the second set of conditions on F is equivalent to the existence of a probability measure ν on $\mathbb{R}$ so that $F = F_\nu$ is the reciprocal of a Stieltjes transform. Such a point of view can actually serve as a definition of free convolution, see [ChG08] or [BeB07].

Lemma 5.3.40 is a particularly simple example of infinite divisibility. The assumption of finite variance in the lemma can be removed by observing that the solution of (5.3.26) is infinitely divisible, and then using [BeV93, Theorem 7.5]. The theory of free infinite divisibility parallels the classical one, and in particular, a Lévy–Khitchine formula does exist to characterize infinitely divisible laws, see [BeP00] and [BaNT04]. The former paper introduces the Bercovici–Pata bijection between the classical and free infinitely divisible laws (see also the Boolean Bercovici–Pata bijection in [BN08]). Matrix approximations to free infinitely divisible laws are constructed in [BeG05].

The generalization of multiplicative free convolution to affiliated operators is done in [BeV93], see also [NiS97].

The relation between random matrices and asymptotic freeness was first established in the seminal article of Voiculescu [Voi91]. In [Voi91, Theorem 2.2], he proved Theorem 5.4.5 in the case of Wigner Gaussian (Hermitian) random matrices and diagonal matrices $\{D_i^N\}_{1\leq i\leq p}$, whereas in [Voi91, Theorem 3.8], he generalized this result to independent unitary matrices. In [Voi98b], he removed the former hypothesis on the matrices $\{D_i^N\}_{1\leq i\leq p}$ to obtain Theorem 5.4.5 for Gaussian matrices and Theorem 5.4.10 in full generality (following the same ideas as in Exercise 5.4.17). An elegant proof of Theorem 5.4.2 for Gaussian matrices which avoid combinatorial arguments appears in [CaC04]. Theorem 5.4.2 was extended to non-Gaussian entries in [Dyk93b]. The proof of Theorem 5.4.10 we presented follows the characterization of the law of free unitary variables by a Schwinger–Dyson equation given in [Voi99, Proposition 5.17] and the ideas of [CoMG06]. Other proofs were given in terms of Weingarten functions in [Col03] and with a more combinatorial approach in [Xu97]. For uses of master loop (or Schwinger–Dyson) equations in the physics literature, see e.g. [EyB99] and [Eyn03].

Asymptotic freeness can be extended to other models such as joint distribution of random matrices with correlated entries [ScS05] or to deterministic models such as permutation matrices [Bia95]. Biane [Bia98b] (see also [Śni06] and [Bia01]) showed that the asymptotic behavior of rescaled Young diagrams and associated representations and characters of the symmetric groups can be expressed in terms of free cumulants.

The study of the correction (central limit theorem) to Theorem 5.4.2 for Gaussian entries was performed in [Cab01] and [MiS06]. The generalization to non-Gaussian entries, as done in [AnZ05], is still open in the general noncommutative framework. A systematic study and analysis of the limiting covariance was undertaken in [MiN04]. The failure of the central limit theorem for a matrix model whose potential has two deep wells was shown in [Pas06].

We have not mentioned the notion of freeness with amalgamation, which is a freeness property where the scalar-valued state is replaced by an operator-valued conditional expectation with properties analogous to conditional expectation from classical probability theory. This notion is particularly natural when considering the algebra generated by two subalgebras. For instance, the free algebras $\{X_i\}_{1\le i\le p}$ as in Theorem 5.4.5 are free with amalgamation with respect to the algebra generated by the $\{D_i\}_{1\le i\le p}$. We refer to [Voi00b] for definitions and to [Shl98] for a nice application to the study the asymptotics of the spectral measure of band matrices. The central limit theorem for the trace of mixed moments of band matrices and deterministic matrices was done in [Gui02].

The convergence of the operator norm of polynomials in independent GUE matrices discussed in Section 5.5 was first proved in [HaT05]. (The norms of the limiting object, namely free operators with matrix coefficients, were already studied in [Leh99].) This result was generalized to independent matrices from the GOE and the GSE in [Sch05], see also [HaST06], and to Wigner or Wishart matrices with entries satisfying the Poincaré inequality in [CaD07]. It was also shown in [GuS08] to hold with matrices whose laws are absolutely continuous with respect to the Lebesgue measure and possess a strictly log-concave density. The norm of long words in free noncommutative variables is discussed in [Kar07a]. We note that a by-product of the proof of Theorem 5.5.1 is that the Stieltjes transform of the law of any self-adjoint polynomial in free semicircular random variables is an algebraic function, as one sees by applying the algebraicity criterion [AnZ08b, Theorem 6.1], to the Schwinger–Dyson equation as expressed in the form (5.5.32). Proposition 5.5.3 is analogous to a result for sample covariance matrices proved earlier in [BaS98a].

Many topics related to free probability have been left out in our discussion. In particular, we have not mentioned free Brownian motion as defined in [Spe90], which appears as the limit of the Hermitian Brownian motion with size going to infinity [Bia97a]. We refer to [BiS98b] for a study of the related stochastic calculus, to [Bia98a] for the introduction of a wide class of processes with free increments and for the study of their Markov properties, to [Ans02] for the introduction of stochastic integrals with respect to processes with free increments, and to [BaNT02] for a thorough discussion of Lévy processes and Lévy laws. Such a stochastic calculus was used to prove a central limit theorem in [Cab01], large deviation principles, see the survey [Gui04], and the convergence of the empirical distribution of interacting matrices [GuS08]. In such a noncommutative stochastic calculus framework, inequalities such as the Burkholder–Davis–Gundy inequality [BiS98b] or the Burkholder–Rosenthal inequalities [JuX03] hold.

Another important topic we did not discuss is the notion of free entropy. We refer the interested readers to the reviews [Voi02] and [HiP00b]. Voiculescu defined several concepts for an entropy in the noncommutative setup. First, the so-called microstates entropy was defined in [Voi94], analogously to the Boltzmann–Shannon entropy, as the volume of the collection of random matrices whose empirical distribution approximates a given tracial state. Second, in [Voi98a], the microstates-free free entropy was defined by following an infinitesimal approach based on the free Fisher information. Voiculescu showed in [Voi93] that, in the case of one variable, both entropies are equal. Following a large deviations and stochastic processes approach, bounds between these two entropies could be given in the general setting, see [CaG01] and [BiCG03], providing strong evidence toward the conjecture that they are equal in full generality. Besides its connections with large deviations questions, free entropies were used to define in [Voi94] another important concept, namely the free entropy dimension. This dimension is related with L^2-Betti numbers [CoS05], [MiS05] and is analogous to a fractal dimension in the classical setting [GuS07]. A long standing conjecture is that the entropy dimension is an invariant of the von Neumann algebra, which would settle the well known problem of the isomorphism between free group factors [Voi02, section 2.6]. Free entropy theory has already been used to settle some important questions in von Neumann algebras, see [Voi96], [Ge97], [Ge98] or [Voi02, section 2.5]. In another direction, random matrices can be an efficient way to tackle questions concerning C^*-algebras or von Neumman algebras, see e.g. [Voi90], [Dyk93a], [Răd94], [HaT99], [Haa02], [PoS03], [HaT05], [HaST06], [GuJS07] and [HaS09].

The free probability concepts developed in this chapter, and in particular free cumulants, can also be used in more applied subjects such as telecommunications, see [LiTV01] and [TuV04].

Appendices

A Linear algebra preliminaries

This appendix recalls some basic results from linear algebra. We refer the reader to [HoJ85] for further details and proofs.

A.1 Identities and bounds

The following identities are repeatedly used. Throughout, A,B,C,D denote arbitrary matrices of appropriate dimensions. We then have

$$\begin{aligned} \mathbf{1}_{\det A\neq 0}\det\begin{bmatrix} A & B \\ C & D \end{bmatrix} &= \det\left(\begin{bmatrix} A & 0 \\ C & D-CA^{-1}B \end{bmatrix}\begin{bmatrix} 1 & A^{-1}B \\ 0 & 1 \end{bmatrix}\right) \\ &= \det A\cdot\det[D-CA^{-1}B]\,, \end{aligned} \tag{A.1}$$

where the right side of (A.1) is set to 0 if A is not invertible.

The following lemma, proved by multiplying on the right by $(X-zI)$ and on the left by $(X-A-zI)$, is very useful.

Lemma A.1 (Matrix inversion) *For matrices X,A and scalar z, the following identity holds if all matrices involved are invertible:*

$$(X-A-zI)^{-1}-(X-zI)^{-1}=(X-A-zI)^{-1}A(X-zI)^{-1}\,.$$

Many manipulations of matrices involve their minors. Thus, let $I=\{i_1,\ldots,i_{|I|}\}\subset\{1,\ldots,m\}$, $J=\{j_1,\ldots,j_{|J|}\}\subset\{1,\ldots,n\}$, and for an m-by-n matrix A, let $A_{I,J}$ be the $|I|$-by-$|J|$ matrix obtained by erasing all entries that do not belong to a row with index from I and a column with index from J. That is,

$$A_{I,J}(l,k)=A(i_l,j_k)\,,\quad l=1,\ldots,|I|,\ k=1,\ldots,|J|\,.$$

The I,J minor of A is then defined as $\det A_{I,J}$. We have the following.

Theorem A.2 (Cauchy–Binet Theorem) *Suppose A is an m-by-k matrix, B a k-by-n matrix, $C = AB$, and, with $r \le \min\{m,k,n\}$, set $I = \{i_1,\dots,i_r\} \subset \{1,\dots,m\}$, $J = \{j_1,\dots,j_r\} \subset \{1,\dots,n\}$. Then, letting $\mathscr{K}_{r,k}$ denote all subsets of $\{1,\dots,k\}$ of cardinality r,*

$$\det C_{I,J} = \sum_{K \in \mathscr{K}_{r,k}} \det A_{I,K} \det B_{K,J}\,. \tag{A.2}$$

We next provide a fundamental bound on determinants.

Theorem A.3 (Hadamard's inequality) *For any column vectors $v_1,\dots,v_n$ of length n with complex entries, it holds that*

$$\det[v_1 \dots v_n] \le \prod_{i=1}^{n} \sqrt{\bar{v}_i^T v_i} \le n^{n/2} \prod_{i=1}^{n} |v_i|_\infty\,.$$

A.2 Perturbations for normal and Hermitian matrices

We recall that a normal matrix A satisfies the relation $AA^* = A^*A$. In particular, all matrices in $\mathscr{H}_N^{(\beta)}$, $\beta = 1,2$, are normal.

In what follows, we let $\|A\|_2 := \sqrt{\sum_{i,j} |A(i,j)|^2}$ denote the *Frobenius* norm of the matrix A. The following lemma is a corollary of Gersgorin's circle theorem.

Lemma A.4 (Perturbations of normal matrices) *Let A be an N by N normal matrix with eigenvalues λ_i, $i = 1,\dots,N$, and let E be an arbitrary N by N matrix. Let $\hat{\lambda}$ be any eigenvalues of $A+E$. Then there is an $i \in \{1,\dots,N\}$ such that $|\hat{\lambda} - \lambda_i| \le \|E\|_2$.*

For Hermitian matrices, more can be said. Recall that, for a Hermitian matrix A, we let $\lambda_1(A) \le \lambda_2(A) \le \cdots \le \lambda_N(A)$ denote the ordered eigenvalues of A. We first recall the

Theorem A.5 (Weyl's inequalities) *Let $A, B \in \mathscr{H}_N^{(2)}$. Then, for each $k \in \{1,\dots,N\}$, we have*

$$\lambda_k(A) + \lambda_1(B) \le \lambda_k(A+B) \le \lambda_k(A) + \lambda_N(B)\,. \tag{A.3}$$

The following is a useful corollary of Weyl's inequalities.

Corollary A.6 (Lipschitz continuity) *Let $A, E \in \mathscr{H}_N^{(2)}$. Then*

$$|\lambda_k(A+E) - \lambda_k(A)| \le \|E\|_2 \,. \tag{A.4}$$

Corollary A.6 is weaker than Lemma 2.1.19, which in its Hermitian formulation, see Remark 2.1.20, actually implies that, under the same assumptions,

$$\sum_k |\lambda_k(A+E) - \lambda_k(A)|^2 \le \|E\|_2^2 \,. \tag{A.5}$$

We finally note the following comparison, whose proof is based on the Courant–Fischer representation of the eigenvalues of Hermitian matrices.

Theorem A.7 *Let $A \in \mathscr{H}_N^{(2)}$ and $z \in \mathbb{C}^N$. Then, for $1 \le k \le N-2$,*

$$\lambda_k(A \pm zz^*) \le \lambda_{k+1}(A) \le \lambda_{k+2}(A \pm zz^*) \,. \tag{A.6}$$

A.3 Noncommutative matrix L^p-norms

Given $X \in \mathrm{Mat}_{k\times\ell}(\mathbb{C})$ with singular values $\mu_1 \ge \cdots \ge \mu_r \ge 0$, where $r = \min(k,\ell)$, and a constant $1 \le p \le \infty$, one defines the *noncommutative L^p-norm* of X by $\|X\|_p = \left(\sum_{i=1}^r \mu_i^p\right)^{1/p}$ if $p < \infty$ and $\|X\|_\infty = \lim_{p\to\infty} \|X\|_p = \mu_1$.

Theorem A.8 *The noncommutative L^p norms satisfy the following.*

$$\|X\|_p = \|X^*\|_p = \left\|X^T\right\|_p \,. \tag{A.7}$$

$$\|UX\|_p = \|X\|_p \text{ for unitary matrices } U \in \mathrm{Mat}_k(\mathbb{C}) \,. \tag{A.8}$$

$$\mathrm{tr}(XX^*) = \|X\|_2^2 \,. \tag{A.9}$$

$$\|X\|_p \ge \left(\sum_{i=1}^r |X_{i,i}|^p\right)^{1/p} \quad \text{for } 1 \le p \le \infty \,. \tag{A.10}$$

$$\|\cdot\|_p \text{ is a norm on the complex vector space } \mathrm{Mat}_{k\times\ell}(\mathbb{C}) \,. \tag{A.11}$$

Properties (A.7), (A.8) and (A.9) are immediate consequences of the definition. A proof of (A.10) and (A.11) can be found in [Sim05b, Prop. 2.6 & Thm. 2.7]. It follows from (A.10) that if X is a square matrix then

$$\|X\|_1 \ge |\,\mathrm{tr}(X)| \,. \tag{A.12}$$

For matrices X and Y with complex entries which can be multiplied, and exponents $1 \le p,q,r \le \infty$ satisfying $\frac{1}{p} + \frac{1}{q} = \frac{1}{r}$, we have the *noncommutative Hölder inequality*

$$\|XY\|_r \le \|X\|_p \|Y\|_q \,. \tag{A.13}$$

(See [Sim05b, Thm. 2.8].)

A.4 Brief review of resultants and discriminants

Definition A.9 Let

$$P = P(t) = \sum_{i=0}^{m} a_i t^i = a_m \prod_{i=1}^{m} (t - \alpha_i), \quad Q = Q(t) = \sum_{j=0}^{n} b_j t^j = b_n \prod_{j=1}^{n} (t - \beta_j),$$

be two polynomials where the *a*s, *b*s, αs and βs are complex numbers, the lead coefficients a_m and b_n are nonzero, and t is a variable. The *resultant* of P and Q is defined as

$$R(P,Q) = a_m^n b_n^m \prod_{i=1}^{m} \prod_{j=1}^{n} (\alpha_i - \beta_j) = a_m^n \prod_{i=1}^{m} Q(\alpha_i) = (-1)^{mn} b_n^m \prod_{j=1}^{n} P(\beta_j).$$

The resultant $R(P,Q)$ can be expressed as the determinant of the $(m+n)$-by-$(m+n)$ *Sylvester matrix*

$$\begin{bmatrix} a_m & \cdots & a_0 & & & \\ & \ddots & & \ddots & & \\ & & \ddots & & \ddots & \\ & & a_m & \cdots & & a_0 \\ b_n & \cdots & \cdots & \cdots & b_0 & \\ & \ddots & & & \ddots & \\ & & b_n & \cdots & \cdots & \cdots & b_0 \end{bmatrix}.$$

Here there are n rows of *a*s and m rows of *b*s. In particular, the resultant $R(P,Q)$ is a polynomial (with integer coefficients) in the *a*s and *b*s. Hence $R(P,Q)$ depends only on the *a*s and *b*s and does so continuously.

Definition A.10 Given a polynomial P as in Definition A.9, the *discriminant* of P is defined as

$$\begin{aligned} D(P) &= (-1)^{m(m-1)/2} R(P,P') = (-1)^{m(m-1)/2} \prod_{i=1}^{m} P'(\alpha_i) \\ &= a_m^{2m-1} \prod_{1 \le i < j \le n} (\alpha_i - \alpha_j)^2 . \end{aligned} \tag{A.14}$$

We emphasize that $D(P)$ depends only on the *a*s and does so continuously.

B Topological preliminaries

The material in Appendices B and C is classical. These appendices are adapted from [DeZ98].

B.1 Generalities

A family τ of subsets of a set $\mathscr{X}$ is a *topology* if $\emptyset \in \tau$, if $\mathscr{X} \in \tau$, if any union of sets of τ belongs to τ, and if any finite intersection of elements of τ belongs to τ. A topological space is denoted $(\mathscr{X}, \tau)$, and this notation is abbreviated to $\mathscr{X}$ if the topology is obvious from the context. Sets that belong to τ are called *open sets*. Complements of open sets are *closed sets*. An open set containing a point $x \in \mathscr{X}$ is a *neighborhood* of x. Likewise, an open set containing a subset $A \subset \mathscr{X}$ is a neighborhood of A. The *interior* of a subset $A \subset \mathscr{X}$, denoted A^o, is the union of the open subsets of A. The *closure* of A, denoted $\bar{A}$, is the intersection of all closed sets containing A. A point p is called an *accumulation point* of a set $A \subset \mathscr{X}$ if every neighborhood of p contains at least one point in A. The closure of A is the union of its accumulation points.

A *base* for the topology τ is a collection of sets $\mathscr{A} \subset \tau$ such that any set from τ is the union of sets in $\mathscr{A}$. If τ_1 and τ_2 are two topologies on $\mathscr{X}$, τ_1 is called stronger (or finer) than τ_2, and τ_2 is called weaker (or coarser) than τ_1 if $\tau_2 \subset \tau_1$.

A topological space is *Hausdorff* if single points are closed and every two distinct points $x, y \in \mathscr{X}$ have disjoint neighborhoods. It is *regular* if, in addition, any closed set $F \subset \mathscr{X}$ and any point $x \notin F$ possess disjoint neighborhoods. It is *normal* if, in addition, any two disjoint closed sets F_1, F_2 possess disjoint neighborhoods.

If $(\mathscr{X}, \tau_1)$ and $(\mathscr{Y}, \tau_2)$ are topological spaces, a function $f : \mathscr{X} \to \mathscr{Y}$ is a *bijection* if it is one-to-one and onto. It is *continuous* if $f^{-1}(A) \in \tau_1$ for any $A \in \tau_2$. This implies also that the inverse image of a closed set is closed. Continuity is preserved under compositions, i.e., if $f : \mathscr{X} \to \mathscr{Y}$ and $g : \mathscr{Y} \to \mathscr{Z}$ are continuous, then $g \circ f : \mathscr{X} \to \mathscr{Z}$ is continuous. If both f and f^{-1} are continuous, then f is a *homeomorphism*, and spaces $\mathscr{X}, \mathscr{Y}$ are called homeomorphic if there exists a homeomorphism $f : \mathscr{X} \to \mathscr{Y}$.

A function $f : \mathscr{X} \to \mathbb{R}$ is *lower semicontinuous* (*upper semicontinuous*) if its level sets $\{x \in \mathscr{X} : f(x) \leq \alpha\}$ (respectively, $\{x \in \mathscr{X} : f(x) \geq \alpha\}$) are closed sets. Clearly, every continuous function is lower (upper) semicontinuous and the pointwise supremum of a family of lower semicontinuous functions is lower semicontinuous.

A Hausdorff topological space is *completely regular* if for any closed set $F \subset \mathcal{X}$ and any point $x \notin F$, there exists a continuous function $f : \mathcal{X} \to [0,1]$ such that $f(x) = 1$ and $f(y) = 0$ for all $y \in F$.

A *cover* of a set $A \subset \mathcal{X}$ is a collection of open sets whose union contains A. A set is *compact* if every cover of it has a finite subset that is also a cover. A continuous image of a compact set is compact. A continuous bijection between compact spaces is a homeomorphism. Every compact subset of a Hausdorff topological space is closed. A set is *pre-compact* if its closure is compact. A topological space is *locally compact* if every point possesses a neighborhood that is compact.

Theorem B.1 *A lower (upper) semicontinuous function f achieves its minimum (respectively, maximum) over any compact set K.*

Let $(\mathcal{X}, \tau)$ be a topological space, and let $A \subset \mathcal{X}$. The *relative* (or induced) topology on A is the collection of sets $A \bigcap \tau$. The Hausdorff, normality and regularity properties are preserved under the relative topology. Furthermore, the compactness is preserved, i.e., $B \subset A$ is compact in the relative topology iff it is compact in the original topology τ. Note, however, that the "closedness" property is *not* preserved.

A nonnegative real function $d : \mathcal{X} \times \mathcal{X} \to \mathbb{R}$ is called a *metric* if $d(x,y) = 0 \Leftrightarrow x = y$, $d(x,y) = d(y,x)$, and $d(x,y) \leq d(x,z) + d(z,y)$. The last property is referred to as the *triangle* inequality. The set $B_{x,\delta} = \{y : d(x,y) < \delta\}$ is called the *ball* of center x and radius δ. The metric topology of $\mathcal{X}$ is the weakest topology which contains all balls. The set $\mathcal{X}$ equipped with the metric topology is a *metric* space $(\mathcal{X}, d)$. A topological space whose topology is the same as some metric topology is called *metrizable*. Every metrizable space is normal. Every regular space that possesses a countable base is metrizable.

A sequence $x_n \in \mathcal{X}$ *converges* to $x \in \mathcal{X}$ (denoted $x_n \to x$) if every neighborhood of x contains all but a finite number of elements of the sequence $\{x_n\}$. If $\mathcal{X}, \mathcal{Y}$ are metric spaces, then $f : \mathcal{X} \to \mathcal{Y}$ is continuous iff $f(x_n) \to f(x)$ for any convergent sequence $x_n \to x$. A subset $A \subset \mathcal{X}$ of a topological space is *sequentially compact* if every sequence of points in A has a subsequence converging to a point in $\mathcal{X}$.

Theorem B.2 *A subset of a metric space is compact iff it is closed and sequentially compact.*

A set $A \subset \mathcal{X}$ is *dense* if its closure is $\mathcal{X}$. A topological space is *separable* if it contains a countable dense set. Any topological space that possesses a countable base is separable, whereas any separable metric space possesses a countable base.

Even if a space is not metric, the notion of convergence on a sequence may be extended to convergence on *filters*, or *nets*, such that compactness, "closedness", etc. may be checked by convergence. The interested reader is referred to [DuS58] or [Bou87] for details.

Let J be an arbitrary set. Let $\mathcal{X}$ be the Cartesian product of topological spaces $\mathcal{X}_j$, i.e., $\mathcal{X} = \prod_j \mathcal{X}_j$. The *product topology* on $\mathcal{X}$ is the topology generated by the base $\prod_j U_j$, where U_j are open and equal to $\mathcal{X}_j$ except for a finite number of values of j. This topology is the weakest one which makes all projections $p_j : \mathcal{X} \to \mathcal{X}_j$ continuous. The Hausdorff property is preserved under products, and any countable product of metric spaces (with metric $d_n(\cdot,\cdot)$) is metrizable, with the metric on $\mathcal{X}$ given by

$$d(x,y) = \sum_{n=1}^{\infty} \frac{1}{2^n} \frac{d_n(p_n x, p_n y)}{1 + d_n(p_n x, p_n y)} .$$

Theorem B.3 (Tychonoff) *A product of compact spaces is compact.*

B.2 Topological vector spaces and weak topologies

A *vector space* over the reals is a set $\mathcal{X}$ that is closed under the operations of addition and multiplication by scalars, i.e., if $x, y \in \mathcal{X}$, then $x + y \in \mathcal{X}$ and $\alpha x \in \mathcal{X}$ for all $\alpha \in \mathbb{R}$. All vector spaces in this book are over the reals. A *topological vector space* is a vector space equipped with a Hausdorff topology that makes the vector space operations continuous. The *convex hull* of a set A, denoted $\mathrm{co}(A)$, is the intersection of all convex sets containing A. The closure of $\mathrm{co}(A)$ is denoted $\overline{\mathrm{co}}(A)$. $\mathrm{co}(\{x_1, \ldots, x_N\})$ is compact, and, if K_i are compact, convex sets, then the set $\mathrm{co}(\bigcup_{i=1}^N K_i)$ is closed. A *locally convex* topological vector space is a vector space that possesses a convex base for its topology.

Theorem B.4 *Every (Hausdorff) topological vector space is regular.*

A *linear functional* on the vector space $\mathcal{X}$ is a function $f : \mathcal{X} \to \mathbb{R}$ that satisfies $f(\alpha x + \beta y) = \alpha f(x) + \beta f(y)$ for any scalars $\alpha, \beta \in \mathbb{R}$ and any $x, y \in \mathcal{X}$. The *algebraic dual* of $\mathcal{X}$, denoted $\mathcal{X}'$, is the collection of all linear functionals on $\mathcal{X}$. The *topological dual* of $\mathcal{X}$, denoted $\mathcal{X}^*$, is the collection of all continuous linear functionals on the *topological* vector space $\mathcal{X}$. Both the algebraic dual and the topological dual are vector spaces. Note that, whereas the algebraic dual may be defined for any vector space, the topological dual may be defined only for a topological vector space. The product of two topological vector spaces is a topological vector space, and is locally convex if each of the coordinate spaces is locally convex. The topological dual of the product space is the product of the topological duals of the coordinate spaces. A set $\mathscr{H} \subset \mathscr{X}'$ is called *separating* if for any point $x \in \mathscr{X}$, $x \neq 0$, one may find an $h \in \mathscr{H}$ such that $h(x) \neq 0$. It follows from its definition that $\mathscr{X}'$ is separating.

Theorem B.5 (Hahn–Banach) *Suppose A and B are two disjoint, nonempty, closed, convex sets in the locally convex topological vector space $\mathscr{X}$. If A is compact, then there exists an $f \in \mathscr{X}^*$ and scalars $\alpha, \beta \in \mathbb{R}$ such that, for all $x \in A$, $y \in B$,*

$$f(x) < \alpha < \beta < f(y) . \tag{B.1}$$

It follows in particular that if $\mathscr{X}$ is locally convex, then $\mathscr{X}^*$ is separating. Now let $\mathscr{H}$ be a separating family of linear functionals on $\mathscr{X}$. The *$\mathscr{H}$-topology* of $\mathscr{X}$ is the weakest (coarsest) one that makes all elements of $\mathscr{H}$ continuous. Two particular cases are of interest.

(a) If $\mathscr{H} = \mathscr{X}^*$, then the $\mathscr{X}^*$-topology on $\mathscr{X}$ obtained in this way is called the *weak topology* of $\mathscr{X}$. It is weaker (coarser) than the original topology on $\mathscr{X}$.

(b) Let $\mathscr{X}$ be a topological vector space (not necessarily locally convex). Every $x \in \mathscr{X}$ defines a linear functionals f_x on $\mathscr{X}^*$ by the formula $f_x(x^*) = x^*(x)$. The set of all such functionals is separating in $\mathscr{X}^*$. The $\mathscr{X}$-topology of $\mathscr{X}^*$ obtained in this way is referred to as the *weak* topology* of $\mathscr{X}^*$.

Theorem B.6 *Suppose $\mathscr{X}$ is a vector space and $\mathscr{Y} \subset \mathscr{X}'$ is a separating vector space. Then the $\mathscr{Y}$-topology makes $\mathscr{X}$ into a locally convex topological vector space with $\mathscr{X}^* = \mathscr{Y}$.*

It follows in particular that there may be different topological vector spaces with the same topological dual. Such examples arise when the original topology on $\mathscr{X}$ is strictly finer than the weak topology.

Theorem B.7 *Let $\mathscr{X}$ be a locally convex topological vector space. A convex subset of $\mathscr{X}$ is weakly closed iff it is originally closed.*

Theorem B.8 (Banach–Alaoglu) *Let V be a neighborhood of 0 in the topological vector space $\mathscr{X}$. Let $K = \{x^* \in \mathscr{X}^* : |x^*(x)| \leq 1 , \forall x \in V\}$. Then K is weak* compact.*

B.3 Banach and Polish spaces

A *norm* $||\cdot||$ on a vector space $\mathscr{X}$ is a metric $d(x,y) = ||x-y||$ that satisfies the scaling property $||\alpha(x-y)|| = \alpha||x-y||$ for all $\alpha > 0$. The metric topology then yields a topological vector space structure on $\mathscr{X}$, which is referred to as a *normed* space. The standard norm on the topological dual of a normed space $\mathscr{X}$ is $||x^*||_{\mathscr{X}^*} = \sup_{||x||\leq 1}|x^*(x)|$, and then $||x|| = \sup_{||x^*||_{\mathscr{X}^*}\leq 1} x^*(x)$, for all $x \in \mathscr{X}$.

A *Cauchy sequence* in a metric space $\mathscr{X}$ is a sequence $x_n \in \mathscr{X}$ such that, for every $\varepsilon > 0$, there exists an $N(\varepsilon)$ such that $d(x_n, x_m) < \varepsilon$ for any $n > N(\varepsilon)$ and $m > N(\varepsilon)$. If every Cauchy sequence in $\mathscr{X}$ converges to a point in $\mathscr{X}$, the metric in $\mathscr{X}$ is called *complete*. Note that completeness is not preserved under homeomorphism. A complete separable metric space is called a *Polish* space. In particular, a compact metric space is Polish, and an open subset of a Polish space (equipped with the induced topology) is homeomorphic to a Polish space.

A complete normed space is called a *Banach* space. The natural topology on a Banach space is the topology defined by its norm.

A set B in a topological vector space $\mathscr{X}$ is *bounded* if, given any neighborhood V of the origin in $\mathscr{X}$, there exists an $\varepsilon > 0$ such that $\{\alpha x : x \in B, |\alpha| \leq \varepsilon\} \subset V$. In particular, a set B in a normed space is bounded iff $\sup_{x\in B}||x|| < \infty$. A set B in a metric space $\mathscr{X}$ is *totally bounded* if, for every $\delta > 0$, it is possible to cover B by a finite number of balls of radius δ centered in B. A totally bounded subset of a complete metric space is pre-compact.

Unlike in the Euclidean setup, balls need not be convex in a metric space. However, in normed spaces, all balls are convex. Actually, the following partial converse holds.

Theorem B.9 *A topological vector space is normable, i.e., a norm may be defined on it that is compatible with its topology, iff its origin has a convex bounded neighborhood.*

Weak topologies may be defined on Banach spaces and their topological duals. A striking property of the weak topology of Banach spaces is the fact that compactness, apart from closure, may be checked using sequences.

Theorem B.10 (Eberlein–Šmulian) *Let $\mathscr{X}$ be a Banach space. In the weak topology of $\mathscr{X}$, a set is sequentially compact iff it is pre-compact.*

B.4 Some elements of analysis

We collect below some basic results tying measures and functions on locally compact Hausdorff spaces. In most of our applications, the underlying space will be $\mathbb{R}$. A good reference that contains this material is [Rud87].

Theorem B.11 (Riesz representation theorem) *Let X be a locally compact Hausdorff space, and let Λ be a positive linear functional on $C_c(X)$. Then there exists a σ-algebra $\mathcal{M}$ in X which contains all Borel sets in X, and there exists a unique positive measure μ on $\mathcal{M}$ which represents Λ in the sense that*

$$\Lambda f = \int_X f d\mu \quad \text{for every } f \in C_c(X).$$

We next discuss the approximation of measurable functions by "nice" functions. Recall that a function s is said to be simple if there are measurable sets A_i and real constants $(\alpha_i)_{1\le i\le n}$ such that $s = \sum_{i=1}^n \alpha_i 1_{A_i}$.

Theorem B.12 *Let X be a measure space, and let $f : X \to [0,\infty]$ be measurable. Then there exist simple functions $(s_p)_{p\ge 0}$ on X such that $0 \le s_1 \le s_2 \cdots \le s_k \le f$ and $s_k(x)$ converges to $f(x)$ for all $x \in X$.*

The approximation of measurable functions by continuous ones is often achieved using the following.

Theorem B.13 (Lusin) *Suppose X is a locally compact Hausdorff space and μ is a positive Borel measure on X. Let $A \subset X$ be measurable with $\mu(A) < \infty$, and suppose f is a complex measurable function on X, with $f(x) = 0$ if $x \notin A$. Then, for any $\varepsilon > 0$ there exists a $g \in C_c(X)$ such that*

$$\mu(\{x : f(x) \neq g(x)\}) < \varepsilon.$$

Furthermore, g can be taken such that $\sup_{x\in X} |g(x)| \le \sup_{x\in X} |f(x)|$.

C Probability measures on Polish spaces

C.1 Generalities

The following indicates why Polish spaces are convenient when handling measurability issues. Throughout, unless explicitly stated otherwise, Polish spaces are equipped with their Borel σ-fields.

Theorem C.1 (Kuratowski) *Let Σ_1, Σ_2 be Polish spaces, and let $f : \Sigma_1 \to \Sigma_2$ be a measurable, one-to-one map. Let $E_1 \subset \Sigma_1$ be a Borel set. Then $f(E_1)$ is a Borel set in Σ_2.*

A *probability measure* on the Borel σ-field $\mathscr{B}_\Sigma$ of a Hausdorff topological space Σ is a countably additive, positive set function μ with $\mu(\Sigma) = 1$. The space of (Borel) probability measures on Σ is denoted $M_1(\Sigma)$. When Σ is separable, the structure of $M_1(\Sigma)$ becomes simpler, and conditioning becomes easier to handle; namely, let Σ, Σ_1 be two separable Hausdorff spaces, and let μ be a probability measure on $(\Sigma, \mathscr{B}_\Sigma)$. Let $\pi : \Sigma \to \Sigma_1$ be measurable, and let $\nu = \mu \circ \pi^{-1}$ be the measure on $\mathscr{B}_{\Sigma_1}$ defined by $\nu(E_1) = \mu(\pi^{-1}(E_1))$.

Definition C.2 A *regular conditional probability distribution given* π (referred to as r.c.p.d.) is a mapping $\sigma_1 \in \Sigma_1 \mapsto \mu^{\sigma_1} \in M_1(\Sigma)$ such that:
(a) there exists a set $N \in \mathscr{B}_{\Sigma_1}$ with $\nu(N) = 0$ and, for each $\sigma_1 \in \Sigma_1 \backslash N$,

$$\mu^{\sigma_1}(\{\sigma : \pi(\sigma) \neq \sigma_1\}) = 0;$$

(b) for any set $E \in \mathscr{B}_\Sigma$, the map $\sigma_1 \mapsto \mu^{\sigma_1}(E)$ is $\mathscr{B}_{\Sigma_1}$ measurable and

$$\mu(E) = \int_{\Sigma_1} \mu^{\sigma_1}(E)\nu(d\sigma_1).$$

It is property (b) that allows for the decomposition of measures. In Polish spaces, the existence of an r.c.p.d. follows from:

Theorem C.3 *Let Σ, Σ_1 be Polish spaces, $\mu \in M_1(\Sigma)$, and $\pi : \Sigma \to \Sigma_1$ a measurable map. Then there exists an r.c.p.d. μ^{σ_1}. Moreover, it is unique in the sense that any other r.c.p.d. $\overline{\mu}^{\sigma_1}$ satisfies*

$$\nu(\{\sigma_1 : \overline{\mu}^{\sigma_1} \neq \mu^{\sigma_1}\}) = 0.$$

Another useful property of separable spaces is their behavior under products.

Theorem C.4 *Let N be either finite or $N = \infty$.*
(a) $\prod_{i=1}^N \mathscr{B}_\Sigma \subset \mathscr{B}_{\prod_{i=1}^N \Sigma}$.
(b) *If Σ is separable, then $\prod_{i=1}^N \mathscr{B}_\Sigma = \mathscr{B}_{\prod_{i=1}^N \Sigma}$.*

We now turn our attention to the particular case where Σ is metric (and, whenever needed, Polish).

Theorem C.5 *Let Σ be a metric space. Then any $\mu \in M_1(\Sigma)$ is regular.*

Theorem C.6 *Let Σ be Polish, and let $\mu \in M_1(\Sigma)$. Then there exists a unique closed set C_μ such that $\mu(C_\mu) = 1$ and, if D is any other closed set with $\mu(D) = 1$, then $C_\mu \subseteq D$. Finally,*

$$C_\mu = \{\sigma \in \Sigma : \sigma \in U^o \ \Rightarrow \ \mu(U^o) > 0\}.$$

The set C_μ of Theorem C.6 is called the *support* of μ.

A probability measure μ on the metric space Σ is *tight* if, for each $\eta > 0$, there exists a compact set $K_\eta \subset \Sigma$ such that $\mu(K_\eta^c) < \eta$. A family of probability measures $\{\mu_\alpha\}$ on the metric space Σ is called a *tight family* if the set K_η may be chosen independently of α.

Theorem C.7 *Each probability measure on a Polish space Σ is tight.*

C.2 Weak topology

Whenever Σ is Polish, a topology may be defined on $M_1(\Sigma)$ that possesses nice properties; namely, define the *weak topology* on $M_1(\Sigma)$ as the topology generated by the sets

$$U_{\phi,x,\delta} = \{\nu \in M_1(\Sigma) : |\int_\Sigma \phi d\nu - x| < \delta\},$$

where $\phi \in C_b(\Sigma)$, $\delta > 0$ and $x \in \mathbb{R}$. If one takes only functions $\phi \in C_b(\Sigma)$ that are of compact support, the resulting topology is the *vague topology*.

Hereafter, $M_1(\Sigma)$ always denotes $M_1(\Sigma)$ equipped with the weak topology. The following are some basic properties of this topological space.

Theorem C.8 *Let Σ be Polish.*

(i) *$M_1(\Sigma)$ is Polish.*

(ii) *A metric compatible with the weak topology is the Lévy metric:*

$$d(\mu,\nu) = \inf\{\delta : \mu(F) \le \nu(F^\delta) + \delta \ \ \forall F \subset \Sigma \text{ closed}\}.$$

(iii) *$M_1(\Sigma)$ is compact iff Σ is compact.*

(iv) *Let $E \subset \Sigma$ be a dense countable subset of Σ. The set of all probability measures whose supports are finite subsets of E is dense in $M_1(\Sigma)$.*

(v) *Another metric compatible with the weak topology is the Lipschitz bounded metric:*

$$d_{\mathrm{LU}}(\mu,\nu) = \sup_{f \in \mathscr{F}_{\mathrm{LU}}} |\int_\Sigma f d\nu - \int_\Sigma f d\mu|, \qquad \text{(C.1)}$$

where $\mathscr{F}_{\mathrm{LU}}$ is the class of Lipschitz continuous functions $f : \Sigma \to \mathbb{R}$, with Lipschitz constant at most 1 *and uniform bound* 1.

The space $M_1(\Sigma)$ possesses a useful criterion for compactness.

Theorem C.9 (Prohorov) *Let Σ be Polish, and let $\Gamma \subset M_1(\Sigma)$. Then $\overline{\Gamma}$ is compact iff Γ is tight.*

Since $M_1(\Sigma)$ is Polish, convergence may be decided by sequences. The following lists some useful properties of converging sequences in $M_1(\Sigma)$.

Theorem C.10 (Portmanteau theorem) *Let Σ be Polish. The following statements are equivalent.*

(i) $\mu_n \to \mu$ *as* $n \to \infty$.

(ii) $\forall g$ *bounded and uniformly continuous,* $\lim_{n\to\infty} \int_\Sigma g\, d\mu_n = \int_\Sigma g\, d\mu$.

(iii) $\forall F \subset \Sigma$ *closed,* $\limsup_{n\to\infty} \mu_n(F) \le \mu(F)$.

(iv) $\forall G \subset \Sigma$ *open,* $\liminf_{n\to\infty} \mu_n(G) \ge \mu(G)$.

(v) $\forall A \in \mathscr{B}_\Sigma$, *which is a continuity set, i.e., such that* $\mu(\overline{A}\backslash A^o) = 0$, $\lim_{n\to\infty} \mu_n(A) = \mu(A)$.

A collection of functions $\mathscr{G} \subset B(\Sigma)$ is called *convergence determining* for $M_1(\Sigma)$ if

$$\lim_{n\to\infty} \int_\Sigma g d\mu_n = \int_\Sigma g d\mu\,, \quad \forall g \in \mathscr{G} \Rightarrow \mu_n \to_{n\to\infty} \mu\,.$$

For Σ Polish, there exists a countable convergence determining collection of functions for $M_1(\Sigma)$ and the collection $\{f(x)g(y)\}_{f,g\in C_b(\Sigma)}$ is convergence determining for $M_1(\Sigma^2)$.

Theorem C.11 *Let Σ be Polish. If K is a set of continuous, uniformly bounded functions on Σ that are equicontinuous on compact subsets of Σ, then $\mu_n \to \mu$ implies that*

$$\limsup_{n\to\infty} \sup_{\phi\in K} \Big\{ \Big| \int_\Sigma \phi d\mu_n - \int_\Sigma \phi d\mu \Big| \Big\} = 0\,.$$

The following theorem is the analog of Fatou's Lemma for measures. It is proved from Fatou's Lemma either directly or by using the Skorohod representation theorem.

Theorem C.12 *Let Σ be Polish. Let $f : \Sigma \to [0,\infty]$ be a lower semicontinuous function, and assume $\mu_n \to \mu$. Then*

$$\liminf_{n\to\infty} \int_\Sigma f d\mu_n \geq \int_\Sigma f d\mu \,.$$

D Basic notions of large deviations

This appendix recalls basic definitions and main results of large deviation theory. We refer the reader to [DeS89] and [DeZ98] for a full treatment.

In what follows, X will be assumed to be a Polish space (that is a complete separable metric space). We recall that a function $f : X \to \mathbb{R}$ is *lower semicontinuous* if the level sets $\{x : f(x) \leq C\}$ are closed for any constant C.

Definition D.1 A sequence $(\mu_N)_{N\in\mathbb{N}}$ of probability measures on X satisfies a *large deviation principle* with speed a_N (going to infinity with N) and rate function I iff

$$I : X \to [0,\infty] \text{ is lower semicontinuous.} \tag{D.1}$$

$$\text{For any open set } O \subset X,\ \liminf_{N\to\infty} \frac{1}{a_N} \log \mu_N(O) \geq -\inf_O I. \tag{D.2}$$

$$\text{For any closed set } F \subset X,\ \limsup_{N\to\infty} \frac{1}{a_N} \log \mu_N(F) \leq -\inf_F I. \tag{D.3}$$

When it is clear from the context, we omit the reference to the speed or rate function and simply say that the sequence $\{\mu_N\}$ satisfies the LDP. Also, if x_N are X-valued random variables distributed according to μ_N, we say that the sequence $\{x_N\}$ satisfies the LDP if the sequence $\{\mu_N\}$ satisfies the LDP.

Definition D.2 A sequence $(\mu_N)_{N\in\mathbb{N}}$ of probability measures on X satisfies a *weak large deviation principle* if (D.1) and (D.2) hold, and in addition (D.3) holds for all compact sets $F \subset X$.

The proof of a large deviation principle often proceeds first by the proof of a weak large deviation principle, in conjuction with the so-called exponential tightness property.

Definition D.3 (a) A sequence $(\mu_N)_{N\in\mathbb{N}}$ of probability measures on X is *exponentially tight* iff there exists a sequence $(K_L)_{L\in\mathbb{N}}$ of compact sets such that

$$\limsup_{L\to\infty} \limsup_{N\to\infty} \frac{1}{a_N} \log \mu_N(K_L^c) = -\infty.$$

(b) A rate function I is *good* if the level sets $\{x \in X : I(x) \leq M\}$ are compact for all $M \geq 0$.

The interest in these concepts lies in the following.

Theorem D.4 (a) ([DeZ98, Lemma 1.2.18]) *If $\{\mu_N\}$ satisfies the weak LDP and it is exponentially tight, then it satisfies the full LDP, and the rate function I is good.*
(b) ([DeZ98, Exercise 4.1.10]) *If $\{\mu_N\}$ satisfies the upper bound (D.3) with a good rate function I, then it is exponentially tight.*

A weak large deviation principle is itself equivalent to the estimation of the probability of deviations towards small balls.

Theorem D.5 *Let $\mathscr{A}$ be a base of the topology of X. For every $A \in \mathscr{A}$, define*

$$\Lambda_A = -\liminf_{N\to\infty} \frac{1}{a_N} \log \mu_N(A)$$

and

$$I(x) = \sup_{A\in\mathscr{A}:x\in A} \Lambda_A .$$

Suppose that, for all $x \in X$,

$$I(x) = \sup_{A\in\mathscr{A}:x\in A} \left\{ -\limsup_{N\to\infty} \frac{1}{a_N} \log \mu_N(A) \right\} .$$

Then μ_N satisfies a weak large deviation principle with rate function I.

Let d be the metric in X, and set $B(x,\delta) = \{y \in X : d(y,x) < \delta\}$.

Corollary D.6 *Assume that, for all $x \in X$,*

$$-I(x) = \limsup_{\delta\to 0} \limsup_{N\to\infty} \frac{1}{a_N} \log \mu_N(B(x,\delta)) = \liminf_{\delta\to 0} \liminf_{N\to\infty} \frac{1}{a_N} \log \mu_N(B(x,\delta)) .$$

Then μ_N satisfies a weak large deviation principle with rate function I.

From a given large deviation principle one can deduce a large deviation principle for other sequences of probability measures by using either the so-called contraction principle or Laplace's method.

Theorem D.7 (Contraction principle) *Assume that the sequence of probability measures $(\mu_N)_{N\in\mathbb{N}}$ on X satisfies a large deviation principle with good rate function I. Then, for any function $F : X \to Y$ with values in a Polish space Y which is* *continuous, the image* $(F\sharp\mu_N)_{N\in\mathbb{N}} \in M_1(Y)^{\mathbb{N}}$ *defined as* $F\sharp\mu_N(A) = \mu \circ F^{-1}(A)$ *also satisfies a large deviation principle with the same speed and rate function given for any* $y \in Y$ *by*

$$J(y) = \inf\{I(x) : F(x) = y\}.$$

Theorem D.8 (Varadhan's Lemma) *Assume that* $(\mu_N)_{N\in\mathbb{N}}$ *satisfies a large deviation principle with good rate function I. Let* $F : X \to \mathbb{R}$ *be a bounded continuous function. Then*

$$\lim_{N\to\infty} \frac{1}{a_N} \log \int e^{a_N F(x)} d\mu_N(x) = \sup_{x\in X}\{F(x) - I(x)\}.$$

Moreover, the sequence

$$\nu_N(dx) = \frac{1}{\int e^{a_N F(y)} d\mu_N(y)} e^{a_N F(x)} d\mu_N(x) \in M_1(X)$$

satisfies a large deviation principle with good rate function

$$J(x) = I(x) - F(x) - \sup_{y\in X}\{F(y) - I(y)\}.$$

Laplace's method for the asymptotic evaluation of integrals, which is discussed in Section 3.5.1, can be viewed as a (refined) precursor to Theorem D.8 in a narrower context. In developing it, we make use of the following elementary result.

Lemma D.9 (Asymptotics for Laplace transforms) *Let* $f : \mathbb{R}_+ \to \mathbb{C}$ *posses polynomial growth at infinity. Suppose that for some exponent* $\alpha > -1$ *and complex constant B,*

$$f(t) = At^\alpha + O(t^{\alpha+1}) \text{ as } t \downarrow 0.$$

Consider the Laplace transform

$$F(x) = \int_0^\infty f(t)e^{-tx}dt$$

which is defined (at least) for all real $x > 0$. *Then,*

$$F(x) = \frac{B\Gamma(\alpha+1)}{x^{\alpha+1}} + O\left(\frac{1}{x^{\alpha+2}}\right) \text{ as } x \uparrow \infty.$$

Proof In the special case $f(t) = Bt^\alpha$ we have $F(x) = \frac{B\Gamma(\alpha+1)}{x^{\alpha+1}}$, and hence the claim holds. To handle the general case we may assume that $B = 0$. Then we have $\int_0^1 e^{-tx} f(t)dt = O(\int_0^\infty t^{\alpha+1} e^{-tx}dt)$ and $\int_1^\infty e^{-tx} f(t)dt$ decays exponentially fast, which proves the lemma. □

Note that if $f(t)$ has an expansion in powers t^α, $t^{\alpha+1}$, $t^{\alpha+2}$ and so on, then iterated application of the claim yields an asymptotic expansion of the Laplace transform $F(x)$ at infinity in powers $x^{-\alpha-1}, x^{-\alpha-2}, x^{-\alpha-3}$ and so on.

E The skew field $\mathbb{H}$ of quaternions and matrix theory over $\mathbb{F}$

Whereas the reader is undoubtedly familiar with the fields $\mathbb{R}$ and $\mathbb{C}$, the skew field $\mathbb{H}$ of quaternions invented by Hamilton may be less familiar. We give a brief account of its most important features here. Then, with $\mathbb{F}$ denoting any of the (skew) fields $\mathbb{R}$, $\mathbb{C}$ or $\mathbb{H}$, we recount (without proof) the elements of matrix theory over $\mathbb{F}$, culminating in the spectral theorem (Theorem E.11) and its corollaries. We also prove a couple of specialized results (one concerning projectors and another concerning Lie algebras of unitary groups) which are well known in principle but for which references "uniform in $\mathbb{F}$" are not known to us.

Definition E.1 The field $\mathbb{H}$ is the associative (but not commutative) $\mathbb{R}$-algebra with unit for which 1, **i**, **j**, **k** form a basis over $\mathbb{R}$, and in which multiplication is dictated by the rules

$$\mathbf{i}^2 = \mathbf{j}^2 = \mathbf{k}^2 = \mathbf{ijk} = -1. \tag{E.1}$$

Elements of $\mathbb{H}$ are called *quaternions*. Multiplication in $\mathbb{H}$ is not commutative. However, every nonzero element of $\mathbb{H}$ is invertible. Indeed, we have $(a + b\mathbf{i} + c\mathbf{j} + d\mathbf{k})^{-1} = (a - b\mathbf{i} - c\mathbf{j} - d\mathbf{k})/(a^2 + b^2 + c^2 + d^2)$ for all $a, b, c, d \in \mathbb{R}$ not all vanishing. Thus $\mathbb{H}$ is a *skew field*: that is, an algebraic system satisfying all the axioms of a field except for commutativity of multiplication.

Remark E.2 Here is a concrete model for the quaternions in terms of matrices. Note that the matrices

$$\begin{bmatrix} \mathrm{i} & 0 \\ 0 & -\mathrm{i} \end{bmatrix}, \begin{bmatrix} 0 & 1 \\ -1 & 0 \end{bmatrix}, \begin{bmatrix} 0 & \mathrm{i} \\ \mathrm{i} & 0 \end{bmatrix}$$

with complex number entries satisfy the rules (E.1). It follows that the map

$$a + b\mathbf{i} + c\mathbf{j} + d\mathbf{k} \mapsto \begin{bmatrix} a + b\mathrm{i} & c + d\mathrm{i} \\ -c + d\mathrm{i} & a - b\mathrm{i} \end{bmatrix} \quad (a, b, c, d \in \mathbb{R})$$

is an isomorphism of $\mathbb{H}$ onto a subring of the ring of 2-by-2 matrices with entries in $\mathbb{C}$. The quaternions often appear in the literature identified with 2-by-2 matrices in this way. We do not use this identification in this book.

For every

$$x = a + b\mathbf{i} + c\mathbf{j} + d\mathbf{k} \in \mathbb{H} \quad (a, b, c, d \in \mathbb{R})$$

we define

$$\|x\| = \sqrt{a^2+b^2+c^2+d^2},\quad x^* = a - b\mathbf{i} - c\mathbf{j} - d\mathbf{k},\quad \Re x = a.$$

We then have

$$\|x\|^2 = xx^*,\ \|xy\| = \|x\|\,\|y\|,\ (xy)^* = y^*x^*,\ \Re x = \frac{x+x^*}{2},\ \Re xy = \Re yx$$

for all $x, y \in \mathbb{H}$. In particular, we have $x^{-1} = x^*/\|x\|^2$ for nonzero $x \in \mathbb{H}$.

The space of all real multiples of $1 \in \mathbb{H}$ is a copy of $\mathbb{R}$ and the space of all real linear combinations of 1 and $\mathbf{i}$ is a copy of $\mathbb{C}$. Thus $\mathbb{R}$ and $\mathbb{C}$ can be and will be identified with subfields of $\mathbb{H}$, and in particular both i and $\mathbf{i}$ will be used to denote the imaginary unit of the complex numbers. In short, we think of $\mathbb{R}$, $\mathbb{C}$ and $\mathbb{H}$ as forming a "tower"

$$\mathbb{R} \subset \mathbb{C} \subset \mathbb{H}.$$

If $x \in \mathbb{C}$, then $\|x\|$ (resp., x^*, $\Re x$) is the absolute value (resp., complex conjugate, real part) of x in the usual sense. Further, $\mathbf{j}x = x^*\mathbf{j}$ for all $x \in \mathbb{C}$. Finally, for all nonreal $x \in \mathbb{C}$, we have $\{y \in \mathbb{H} \mid xy = yx\} = \mathbb{C}$.

E.1 Matrix terminology over $\mathbb{F}$ and factorization theorems

Let $\mathrm{Mat}_{p\times q}(\mathbb{F})$ denote the space of p-by-q matrices with entries in $\mathbb{F}$. Given $X \in \mathrm{Mat}_{p\times q}(\mathbb{F})$, let $X_{ij} \in \mathbb{F}$ denote the entry of X in row i and column j. Let $\mathrm{Mat}_{p\times q} = \mathrm{Mat}_{p\times q}(\mathbb{R})$ and $\mathrm{Mat}_n(\mathbb{F}) = \mathrm{Mat}_{n\times n}(\mathbb{F})$. Let $0_{p\times q}$ denote the p-by-q zero matrix, and let $0_p = 0_{p\times p}$. Let I_n denote the n-by-n identity matrix. Given $X \in \mathrm{Mat}_{p\times q}(\mathbb{F})$, let $X^* \in \mathrm{Mat}_{q\times p}(\mathbb{F})$ be the matrix obtained by transposing X and then applying "asterisk" to every entry. The operation $X \mapsto X^*$ is $\mathbb{R}$-linear and, furthermore, $(XY)^* = Y^*X^*$ for all $X \in \mathrm{Mat}_{p\times q}(\mathbb{F})$ and $Y \in \mathrm{Mat}_{q\times r}(\mathbb{F})$. Similarly, we have $(xX)^* = X^*x^*$ for any matrix $X \in \mathrm{Mat}_{p\times q}(\mathbb{F})$ and scalar $x \in \mathbb{F}$. Given $X \in \mathrm{Mat}_n(\mathbb{F})$, we define $\mathrm{tr}\,X \in \mathbb{F}$ to be the sum of the diagonal entries of X. Given $X, Y \in \mathrm{Mat}_{p\times q}(\mathbb{F})$, we set $X \cdot Y = \Re\,\mathrm{tr}\,X^*Y$, thus equipping $\mathrm{Mat}_{p\times q}(\mathbb{F})$ with the structure of finite-dimensional real Hilbert space (Euclidean space). Given matrices $X_i \in \mathrm{Mat}_{n_i}(\mathbb{F})$ for $i = 1,\dots,\ell$, let $\mathrm{diag}(X_1,\dots,X_\ell) \in \mathrm{Mat}_{n_1+\cdots+n_\ell}(\mathbb{F})$ be the block-diagonal matrix obtained by stringing the given matrices X_i along the diagonal.

Definition E.3 The matrix $e_{ij} = e_{ij}^{(p,q)} \in \mathrm{Mat}_{p\times q}$ with entry 1 in row i and column j and 0s elsewhere is called an *elementary matrix*.

The set

$$\{ue_{ij} \mid u \in \mathbb{F} \cap \{1, \mathbf{i}, \mathbf{j}, \mathbf{k}\},\ e_{ij} \in \mathrm{Mat}_{p\times q}\}$$

is an orthonormal basis for $\mathrm{Mat}_{p\times q}(\mathbb{F})$.

Definition E.4 (i) Let $X \in \mathrm{Mat}_n(\mathbb{F})$ be a matrix. It is *invertible* if there exists $Y \in \mathrm{Mat}_n(\mathbb{F})$ such that $YX = I_n = XY$. It is *normal* if $X^*X = XX^*$. It is *unitary* if $X^*X = I_n = XX^*$. It is *self-adjoint* (resp., *anti-self-adjoint*) if $X^* = X$ (resp., $X^* = -X$). It is *upper triangular* (resp., *lower triangular*) if $X_{ij} = 0$ unless $i \le j$ (resp., $i \ge j$).
(ii) A matrix $X \in \mathrm{Mat}_n(\mathbb{F})$ is *monomial* if there is exactly one nonzero entry in every row and in every column; if, moreover, every entry of X is either 0 or 1, we call X a *permutation matrix*.
(iii) A self-adjoint $X \in \mathrm{Mat}_n(\mathbb{F})$ is *positive definite* if $v^*Xv > 0$ for all nonzero $v \in \mathrm{Mat}_{n\times 1}(\mathbb{F})$.
(iv) A matrix $X \in \mathrm{Mat}_n(\mathbb{F})$ is a *projector* if it is both self-adjoint and idempotent, that is, if $X^* = X = X^2$.
(v) A matrix $X \in \mathrm{Mat}_{p\times q}(\mathbb{F})$ is *diagonal* if $X_{ij} = 0$ unless $i = j$. The set of positions (i,i) for $i = 1, \ldots, \min(p,q)$ is called the *(main) diagonal* of X.

The group of invertible elements of $\mathrm{Mat}_n(\mathbb{F})$ is denoted $\mathrm{GL}_n(\mathbb{F})$, while the subgroup of $\mathrm{GL}_n(\mathbb{F})$ consisting of unitary matrices is denoted $\mathrm{U}_n(\mathbb{F})$. Permutation matrices in Mat_n belong to $\mathrm{U}_n(\mathbb{F})$.

We next present several factorization theorems. The first is obtained by the Gaussian elimination method.

Theorem E.5 (Gaussian elimination) *Let $X \in \mathrm{Mat}_{p\times q}(\mathbb{F})$ have the property that for all $v \in \mathrm{Mat}_{q\times 1}(\mathbb{F})$, if $Xv = 0$, then $v = 0$. Then $p \ge q$. Furthermore, there exists a permutation matrix $P \in \mathrm{Mat}_p(\mathbb{F})$ and an upper triangular matrix $T \in \mathrm{Mat}_q(\mathbb{F})$ with every diagonal entry equal to 1 such that PXT vanishes above the main diagonal but vanishes nowhere on the main diagonal.*

In particular, for square $A, B \in \mathrm{Mat}_p(\mathbb{F})$, if $AB = I_p$, then $BA = I_p$. It follows also that $\mathrm{GL}_n(\mathbb{F})$ is an open subset of $\mathrm{Mat}_n(\mathbb{F})$.

The Gram–Schmidt process gives more information when $p = q$.

Theorem E.6 (Triangular factorization) *Let $Q \in \mathrm{Mat}_n(\mathbb{F})$ be self-adjoint and positive definite. Then there exists a unique upper triangular matrix $T \in \mathrm{Mat}_n(\mathbb{F})$ with every diagonal entry equal to 1 such that T^*QT is diagonal. Further, T depends smoothly (that is, infinitely differentiably) on the entries of Q.*

Corollary E.7 (**UT factorization**) *Every $X \in \mathrm{GL}_n(\mathbb{F})$ has a unique factorization $X = UT$ where $T \in \mathrm{GL}_n(\mathbb{F})$ is upper triangular with every diagonal entry positive and $U \in \mathrm{U}_n(\mathbb{F})$.*

Corollary E.8 (Unitary extension) *If $V \in \mathrm{Mat}_{n\times k}(\mathbb{F})$ satisfies $V^*V = I_k$, then $n \geq k$ and there exists $U \in \mathrm{U}_n(\mathbb{F})$ agreeing with V in the first k columns.*

Corollary E.9 (Construction of projectors) *Let p and q be positive integers. Fix $Y \in \mathrm{Mat}_{p\times q}(\mathbb{F})$. Put $n = p+q$. Write $T^*(I_p + YY^*)T = I_p$ for some (unique) upper triangular matrix $T \in \mathrm{Mat}_p(\mathbb{F})$ with positive diagonal entries. Then $\Pi = \Pi(Y) = \begin{bmatrix} TT^* & TT^*Y \\ Y^*TT^* & Y^*TT^*Y \end{bmatrix} \in \mathrm{Mat}_n(\mathbb{F})$ is a projector. Further, every projector $\Pi \in \mathrm{Mat}_n(\mathbb{F})$ such that $\mathrm{tr}\,\Pi = p$ and the $p \times p$ block in upper left is invertible is of the form $\Pi = \Pi(Y)$ for unique $Y \in \mathrm{Mat}_{p\times q}(\mathbb{F})$.*

E.2 The spectral theorem and key corollaries

A reference for the proof of the spectral theorem in the unfamiliar case $\mathbb{F} = \mathbb{H}$ is [FaP03].

Definition E.10 (Standard blocks) A $\mathbb{C}$-*standard block* is any element of $\mathrm{Mat}_1(\mathbb{C}) = \mathbb{C}$. An $\mathbb{H}$-*standard block* is any element of $\mathrm{Mat}_1(\mathbb{C}) = \mathbb{C}$ with nonnegative imaginary part. An $\mathbb{R}$-*standard block* is either an element of $\mathrm{Mat}_1 = \mathbb{R}$, or a matrix $\begin{bmatrix} a & b \\ -b & a \end{bmatrix} \in \mathrm{Mat}_2$ with $b > 0$. Finally, $X \in \mathrm{Mat}_n(\mathbb{F})$ is $\mathbb{F}$-*reduced* if $X = \mathrm{diag}(B_1,\dots,B_\ell)$ for some $\mathbb{F}$-standard blocks B_i.

Theorem E.11 (Spectral theorem) *Let $X \in \mathrm{Mat}_n(\mathbb{F})$ be normal.*
(i) *There exists $U \in \mathrm{U}_n(\mathbb{F})$ such that U^*XU is $\mathbb{F}$-reduced.*
(ii) *Fix $U \in \mathrm{U}_n(\mathbb{F})$ and $\mathbb{F}$-standard blocks $B_1,\dots,B_\ell$ such that $\mathrm{diag}(B_1,\dots,B_\ell) = U^*XU$. Up to order, the B_i depend only on X, not on U.*

Corollary E.12 (Eigenvalues) *Fix a self-adjoint $X \in \mathrm{Mat}_n(\mathbb{F})$.*
(i) *There exist $U \in \mathrm{U}_n(\mathbb{F})$ and a diagonal matrix $D \in \mathrm{Mat}_n$ such that $D = U^*XU$.*
(ii) *For any such D and U, the sequence of diagonal entries of D arranged in nondecreasing order is the same.*

We call the entries of D the *eigenvalues* of the self-adjoint matrix X. (When $\mathbb{F} = \mathbb{R},\mathbb{C}$ this is the standard notion of eigenvalue.)

Corollary E.13 (Singular values) *Fix* $X \in \mathrm{Mat}_{p\times q}(\mathbb{F})$.
(i) *There exist* $U \in \mathrm{U}_p(\mathbb{F})$, $V \in \mathrm{U}_q(\mathbb{F})$ *and diagonal* $D \in \mathrm{Mat}_{p\times q}$ *such that* $D = UXV$.
(ii) *For any such* U, V *and* D, *the sequence of absolute values of diagonal entries of* D *arranged in nondecreasing order is the same.*
(iii) *Now assume that* $p \le q$, *and that* X *is diagonal with nonzero diagonal entries the absolute values of which are distinct. Then, for any* U, V *and* D *as in (i),* U *is monomial and* $V = \mathrm{diag}(V', V'')$, *where* $V' \in \mathrm{U}_p(\mathbb{F})$ *and* $V'' \in \mathrm{U}_{q-p}(\mathbb{F})$. *(We simply put* $V = V'$ *if* $p = q$.*) Furthermore, the product* UV' *is diagonal and squares to the identity.*

We call the absolute values of the entries of D the *singular values* of the rectangular matrix X. (When $\mathbb{F} = \mathbb{R}, \mathbb{C}$ this is the standard notion of singular value.) The squares of the singular values of X are the eigenvalues of X^*X or XX^*, whichever has $\min(p,q)$ rows and columns.

E.3 A specialized result on projectors

We present a factorization result for projectors which is used in the discussion of the Jacobi ensemble in Section 4.1. The case $\mathbb{F} = \mathbb{C}$ of the result is well known. But for lack of a suitable reference treating the factorization uniformly in $\mathbb{F}$, we give a proof here.

Proposition E.14 *Let* $0 < p \le q$ *be integers and put* $n = p + q$. *Let* $\Pi \in \mathrm{Mat}_n(\mathbb{F})$ *be a projector. Then there exists* $U \in \mathrm{U}_n(\mathbb{F})$ *commuting with* $\mathrm{diag}(I_p, 0_q)$ *such that* $U^*\Pi U = \begin{bmatrix} a & b \\ b^{\mathrm{T}} & d \end{bmatrix}$, *where* $a \in \mathrm{Mat}_p$, $2b \in \mathrm{Mat}_{p\times q}$ *and* $d \in \mathrm{Mat}_q$ *are diagonal with entries in the closed unit interval* $[0,1]$.

Proof Write $\Pi = \begin{bmatrix} a & \beta \\ \beta^* & d \end{bmatrix}$ with $a \in \mathrm{Mat}_p(\mathbb{F})$, $\beta \in \mathrm{Mat}_{p\times q}(\mathbb{F})$ and $d \in \mathrm{Mat}_q(\mathbb{F})$. Since every element of $\mathrm{U}_n(\mathbb{F})$ commuting with $\mathrm{diag}(I_p, 0_q)$ is of the form $\mathrm{diag}(v, w)$ for $v \in \mathrm{U}_p(\mathbb{F})$ and $w \in \mathrm{U}_q(\mathbb{F})$, we may by Corollary E.13 assume that a and d are diagonal and real. Necessarily the diagonal entries of a and d belong to the closed unit interval $[0,1]$. For brevity, write $a_i = a_{ii}$ and $d_j = d_{jj}$. We may assume that the diagonal entries of a are ordered so that $a_i(1-a_i)$ is nonincreasing as a function of i, and similarly $d_j(1-d_j)$ is nonincreasing as a function of j. We may further assume that whenever $a_i(1-a_i) = a_{i+1}(1-a_{i+1})$ we have $a_i \le a_{i+1}$, but that whenever $d_j(1-d_j) = d_{j+1}(1-d_{j+1})$ we have $d_j \ge d_{j+1}$.

From the equation $\Pi^2 = \Pi$ we deduce that $a(I_p - a) = \beta\beta^*$ and $d(I_q - d) = \beta^*\beta$. Let $b \in \mathrm{Mat}_p$ be the unique diagonal matrix with nonnegative entries such that $b^2 = \beta\beta^*$. Note that the diagonal entries of b appear in nonincreasing order, and in particular all nonvanishing diagonal entries are grouped together in the upper left. Furthermore, all entries of b belong to the closed interval $[0,1/2]$.

By Corollary E.13 there exist $v \in \mathrm{U}_p(\mathbb{F})$ and $w \in \mathrm{U}_q(\mathbb{F})$ such that $v[b\, 0_{p\times(q-p)}]w = \beta$. From the equation $b^2 = \beta\beta^*$ we deduce that v commutes with b^2 and hence also with b. After replacing w by $\mathrm{diag}(v, I_{q-p})w$, we may assume without loss of generality that $\beta = [b\, 0_{p\times(q-p)}]w$. From the equation

$$w^*\mathrm{diag}(b^2, 0_{q-p})w = \beta^*\beta = d(I_q - d)\,,$$

we deduce that w commutes with $\mathrm{diag}(b, 0_{q-p})$.

Let $0 \le r \le p$ be the number of nonzero diagonal entries of b. Write $b = \mathrm{diag}(\tilde{b}, 0_{p-r})$, where $\tilde{b} \in \mathrm{GL}_r(\mathbb{R})$. Since w commutes with $\mathrm{diag}(\tilde{b}, 0_{q-r})$, we can write $w = \mathrm{diag}(\tilde{w}, w')$, where $\tilde{w} \in \mathrm{U}_r(\mathbb{F})$ and $w' \in \mathrm{U}_{q-r}(\mathbb{F})$. Then we have $\beta = [\mathrm{diag}(\tilde{b}\tilde{w}, 0_{p-r})\; 0_{p\times(q-p)}]$ and, further, $\tilde{w}$ commutes with $\tilde{b}$.

Now write $a = \mathrm{diag}(\tilde{a}, a')$ with $\tilde{a} \in \mathrm{Mat}_r$ and $a' \in \mathrm{Mat}_{p-r}$. Similarly, write $d = \mathrm{diag}(\tilde{d}, d')$ with $\tilde{d} \in \mathrm{Mat}_r$ and $d' \in \mathrm{Mat}_{q-r}$. Both $\tilde{a}$ and $\tilde{d}$ are diagonal with diagonal entries in $(0,1)$. Both a' and d' are diagonal with diagonal entries in $\{0,1\}$. We have a block decomposition

$$\Pi = \left[\begin{array}{cccc} \tilde{a} & 0 & \tilde{b}\tilde{w} & 0 \\ 0 & a' & 0 & 0 \\ \tilde{w}^*\tilde{b} & 0 & \tilde{d} & 0 \\ 0 & 0 & 0 & d' \end{array}\right].$$

From the equation $\Pi^2 = \Pi$ we deduce that $\tilde{b}\tilde{a}\tilde{w} = \tilde{a}\tilde{b}\tilde{w} = \tilde{b}\tilde{w}(I_r - \tilde{d})$, hence $\tilde{a}\tilde{w} = \tilde{w}(I_r - \tilde{d})$, hence $\tilde{a}$ and $I_r - \tilde{d}$ have the same eigenvalues, and hence (on account of the care we took in ordering the diagonal entries of a and d), we have $\tilde{a} = I_r - \tilde{d}$. Finally, since $\tilde{d}$ and $\tilde{w}$ commute, with $U = \mathrm{diag}(I_p, \tilde{w}, I_{q-r})$, we have $U^*\Pi U = \left[\begin{array}{cc} a & b \\ b^{\mathrm{T}} & d \end{array}\right]$. □

E.4 Algebra for curvature computations

We present an identity needed to compute the Ricci curvature of the special orthogonal and special unitary groups, see Lemma F.27 and the discussion immediately following. The identity is well known in Lie algebra theory, but the effort needed to decode a typical statement in the literature is about equal to the effort needed to prove it from scratch. So we give a proof here.

Let $\mathfrak{su}_n(\mathbb{F})$ be the set of anti-self-adjoint matrices $X \in \mathrm{Mat}_n(\mathbb{F})$ such that, if $\mathbb{F} = \mathbb{C}$, then $\operatorname{tr} X = 0$. We equip the real vector space $\mathfrak{su}_n(\mathbb{F})$ with the inner product inherited from $\mathrm{Mat}_n(\mathbb{F})$, namely $X \cdot Y = \Re \operatorname{tr} XY^*$. Let $[X,Y] = XY - YX$ for $X,Y \in \mathrm{Mat}_n(\mathbb{F})$, noting that $\mathfrak{su}_n(\mathbb{F})$ is closed under the bracket operation. Let $\beta = 1,2,4$ according as $\mathbb{F} = \mathbb{R}, \mathbb{C}, \mathbb{H}$.

Proposition E.15 *For all $X \in \mathfrak{su}_n(\mathbb{F})$ and orthonormal bases $\{L_\alpha\}$ for $\mathfrak{su}_n(\mathbb{F})$, we have*

$$-\frac{1}{4}\sum_\alpha [[X,L_\alpha],L_\alpha] = \left(\frac{\beta(n+2)}{4} - 1\right)X. \tag{E.2}$$

Proof We have $\mathfrak{su}_1(\mathbb{R}) = \mathfrak{su}_1(\mathbb{C}) = 0$, and the case $\mathfrak{su}_1(\mathbb{H})$ can be checked by direct calculation with **i**, **j**, **k**. Therefore we assume that $n \geq 2$ for the rest of the proof.

Now for fixed $X \in \mathfrak{su}_n(\mathbb{F})$, the expression $[[X,L],M]$ for $L,M \in \mathfrak{su}_n(\mathbb{F})$ is an $\mathbb{R}$-bilinear form on $\mathfrak{su}_n(\mathbb{F})$. It follows that the left side of (E.2) is independent of the choice of orthonormal basis $\{L_\alpha\}$. We are therefore free to choose $\{L_\alpha\}$ at our convenience, and we do so as follows. Let $e_{ij} \in \mathrm{Mat}_n$ for $i,j = 1,\dots,n$ be the elementary matrices. For $1 \leq k < n$ and $u \in \{\mathbf{i},\mathbf{j},\mathbf{k}\}$, let

$$D_k^u = \frac{u}{\sqrt{k+k^2}}\left(-ke_{k+1,k+1} + \sum_{i=1}^k e_{ii}\right),\quad D_k = D_k^{\mathbf{i}},\quad D_n^u = \frac{u}{\sqrt{n}}\sum_{i=1}^n e_{ii}.$$

For $1 \leq i < j \leq n$ and $u \in \{1,\mathbf{i},\mathbf{j},\mathbf{k}\}$, let

$$F_{ij}^u = \frac{ue_{ij} - u^* e_{ji}}{\sqrt{2}},\quad E_{ij} = F_{ij}^1,\quad F_{ij} = F_{ij}^{\mathbf{i}}.$$

Then

$$\{E_{ij} : 1 \leq i < j \leq n\},$$
$$\{D_k : 1 \leq k < n\} \cup \{E_{ij}, F_{ij} : 1 \leq i < j \leq n\},$$
$$\{D_k^u : 1 \leq k \leq n,\ u \in \{\mathbf{i},\mathbf{j},\mathbf{k}\}\} \cup \{F_{ij}^u : 1 \leq i < j \leq n,\ u \in \{1,\mathbf{i},\mathbf{j},\mathbf{k}\}\}$$

are orthonormal bases for $\mathfrak{su}_n(\mathbb{R})$, $\mathfrak{su}_n(\mathbb{C})$ and $\mathfrak{su}_n(\mathbb{H})$, respectively.

We next want to show that, in proving (E.2), it is enough to consider just one X, namely $X = E_{12}$. We achieve that goal by proving the following two claims.

(I) Given $\{L_\alpha\}$ and X for which (E.2) holds and any $U \in \mathrm{U}_n(\mathbb{F})$, again (E.2) holds for $\{UL_\alpha U^*\}$ and UXU^*.

(II) The set $\{UE_{12}U^* \mid U \in \mathrm{U}_n(\mathbb{F})\}$ spans $\mathfrak{su}_n(\mathbb{F})$ over $\mathbb{R}$.

Claim (I) holds because the operation $X \mapsto UXU^*$ stabilizes $\mathfrak{su}_n(\mathbb{F})$, preserves the bracket $[X,Y]$, and preserves the inner product $X \cdot Y$. We turn to the proof of claim (II). By considering conjugations that involve appropriate 2-by-2 blocks, one can generate any element of the collection $\{F^u_{12}, D^u_1\}$ from E_{12}. Further, using conjugation by permutation matrices and taking linear combinations, one can generate $\{F^u_{ij}, D^u_k\}$. Finally, to obtain D^u_n, it is enough to show that $\mathrm{diag}(\mathbf{i},\mathbf{i},0,\dots,0)$ can be generated, and this follows from the identity

$$\mathrm{diag}(1,\mathbf{j})\mathrm{diag}(\mathbf{i},-\mathbf{i})\mathrm{diag}(1,\mathbf{j})^{-1} = \mathrm{diag}(\mathbf{i},\mathbf{i}).$$

Thus claim (II) is proved.

We are ready to conclude. The following facts may be verified by straightforward calculations:

- E_{12} commutes with D^u_k for $k > 1$ and $u \in \{\mathbf{i},\mathbf{j},\mathbf{k}\}$;
- E_{12} commutes with F^u_{ij} for $2 < i < j \le n$ and $u \in \{1,\mathbf{i},\mathbf{j},\mathbf{k}\}$;
- $[[E_{12},F^u_{ij}],F^u_{ij}] = -\frac{1}{2}E_{12}$ for $1 \le i < j < n$ such that $\#\{i,j\} \cap \{1,2\} = 1$ and $u \in \{1,\mathbf{i},\mathbf{j},\mathbf{k}\}$; and
- $[[E_{12},F^u_{12}],F^u_{12}] = [[E_{12},D^u_1],D^u_1] = -2E_{12}$ for $u \in \{\mathbf{i},\mathbf{j},\mathbf{k}\}$.

It follows that the left side of (E.2) with $X = E_{12}$ and $\{L_\alpha\}$ specially chosen as above equals cE_{12}, where the constant c is equal to

$$\frac{1}{4}\left(\frac{1}{2}\cdot 2\beta(n-2) + 2\cdot 2(\beta-1)\right) = \frac{\beta(n+2)}{4} - 1\,.$$

Since (E.2) holds with $X = E_{12}$ and specially chosen $\{L_\alpha\}$, by the previous steps it holds in general. The proof of the lemma is finished. □

F Manifolds

We have adopted in Section 4.1 a framework in which all groups of matrices we used were embedded as submanifolds of Euclidean space. This had the advantage that the structure of the tangent space was easy to identify. For completeness, we present in this appendix all notions employed, and provide in Subsection F.2 the proof of the coarea formula, Theorem 4.1.8. An inspiration for our treatment is [Mil97]. At the end of the appendix, in Subsection F.3, we introduce the language of connections, Laplace–Beltrami operators, and Hessians, used in Section 4.4. For the latter we follow [Hel01] and [Mil63].

F.1 Manifolds embedded in Euclidean space

Given a differentiable function f defined on an open subset of $\mathbb{R}^n$ with values in a finite-dimensional real vector space and an index $i = 1, \ldots, n$, we let $\partial_i f$ denote the partial derivative of f with respect to the ith coordinate. If $n = 1$, then we write $f' = \partial_1 f$.

Definition F.1 A *Euclidean space* is a finite-dimensional real Hilbert space E, with inner product denoted by $(\cdot,\cdot)_E$. A Euclidean set M is a nonempty locally closed subset of E, which we equip with the induced topology.

(A locally closed set is the intersection of a closed set with an open set.) We refer to E as the *ambient space* of M.

We consider $\mathbb{R}^n$ as Euclidean space by adopting the standard inner product $(x,y)_{\mathbb{R}^n} = x \cdot y = \sum_{i=1}^n x_i y_i$. Given Euclidean spaces E and F, and a map $f : U \to V$ from an open subset of E to an open subset of F, we say that f is *smooth* if (after identifying E with $\mathbb{R}^n$ and F with $\mathbb{R}^k$ as vector spaces over $\mathbb{R}$ in some way) f is infinitely differentiable.

Given for $i = 1, 2$ a Euclidean set M_i with ambient space E_i, we define the *product* $M_1 \times M_2$ to be the subset $\{m_1 \oplus m_2 \mid m_1 \in M_1,\, m_2 \in M_2\}$ of the orthogonal direct sum $E_1 \oplus E_2$.

Let $f : M \to N$ be a map from one Euclidean set to another. We say that f is *smooth* if for every point $p \in M$ there exists an open neighborhood U of p in the ambient space of M such that $f|_{U \cap M}$ can be extended to a smooth map from U to the ambient space of N. If f is smooth, then f is continuous. We say that f is a *diffeomorphism* if f is smooth and has a smooth inverse, in which case we also say that M and N are *diffeomorphic*. Note that the definition implies that every n-dimensional linear subspace of a Euclidean space is diffeomorphic to $\mathbb{R}^n$.

Definition F.2 (Manifolds) A *manifold* M of dimension n (for short: *n-manifold*) is a Euclidean set such that every point of M has an open neighborhood diffeomorphic to an open subset of $\mathbb{R}^n$.

We call n the *dimension* of M and write $n = \dim M$. A diffeomorphism $\Phi : T \to U$ where $T \subset \mathbb{R}^n$ is a nonempty open set and U is an open subset of M is called a *chart* of M. By definition M is covered by the images of charts. The product of manifolds is again a manifold. A subset $N \subset M$ is called a *submanifold* if N is a manifold in its own right when viewed as a subset of the ambient space of M.

Definition F.3 Let M be an n-manifold with ambient space E. Let $p \in M$ be a point. A *curve γ through $p \in M$* is by definition a smooth map $\gamma : I \to M$, where $I \subset \mathbb{R}$ is a nonempty open interval, $0 \in I$, and $\gamma(0) = p$. We define the tangent space $\mathbb{T}_p(M)$ of M at p to be the subset of E consisting of all vectors of the form $\gamma'(0)$ for some curve γ through $p \in M$.

The set $\mathbb{T}_p(M)$ is a vector subspace of E of dimension n over $\mathbb{R}$. More precisely, for any chart $\Phi : T \to U$ and point $t_0 \in T$ such that $\Phi(t_0) = p$, the vectors $(\partial_i \Phi)(t_0)$ for $i = 1, \dots, n$ form a basis over $\mathbb{R}$ for $\mathbb{T}_p(M)$. We endow $\mathbb{T}_p(M)$ with the structure of Euclidean space it inherits from E.

Let $f : M \to N$ be a smooth map of manifolds, and let $p \in M$. There exists a unique $\mathbb{R}$-linear transformation $\mathbb{T}_p(f) : \mathbb{T}_p(M) \to \mathbb{T}_{f(p)}(N)$ with the following property: for every curve γ with $\gamma(0) = p$ and $\gamma'(0) = X \in T_p(M)$, we have $(\mathbb{T}_p(f))(X) = (f \circ \gamma)'(0)$. We call $\mathbb{T}_p(f)$ the *derivative* of f at p. The map $\mathbb{T}_p(f)$ is an isomorphism if and only if f maps some open neighborhood of $p \in M$ diffeomorphically to some open neighborhood of $f(p) \in N$. If f is a diffeomorphism and $\mathbb{T}_p(f)$ is an isometry of real Hilbert spaces for every $p \in M$, we call f an *isometry*.

Remark F.4 Isometries need not preserve distances in ambient Euclidean spaces. For example, $\{(x,y) \in \mathbb{R}^2 \setminus \{(0,0)\} : x^2 + y^2 = 1\} \subset \mathbb{R}^2$ and $\{0\} \times (0, 2\pi) \subset \mathbb{R}^2$ are isometric.

Definition F.5 Let M be an n-manifold, with $A \subset M$. We say that A is *negligible* if for every chart $\Phi : T \to U$ of M the subset $\Phi^{-1}(A) \subset \mathbb{R}^n$ is of Lebesgue measure zero.

By the change of variable formula of Lebesgue integration, a subset $A \subset M$ is negligible if and only if for every $p \in M$ there exists a chart $\Phi : T \to U$ such that $p \in U$ and $\Phi^{-1}(A) \subset \mathbb{R}^n$ is of Lebesgue measure zero.

We exploit the change of variables formula to define a volume measure on the Borel subsets of M. We begin with the following.

Definition F.6 Let $\Phi : T \to U$ be a chart of an n-manifold M. Let E be the ambient space of M.
(i) The *correction factor* σ_Φ is the smooth positive function on T defined by the following formula, valid for all $t \in T$:

$$\sigma_\Phi(t) = \sqrt{\det_{i,j=1}^{n} ((\partial_i \Phi)(t), (\partial_j \Phi)(t))_E} \, .$$

(ii) The *chart measure* $\ell_{T,\Phi}$ on the Borel sets of T is the measure absolutely continuous with respect to Lebesgue measure restricted to T, ℓ_T, defined by

$$\frac{d\ell_{T,\Phi}}{d\ell_T} = \sigma_\Phi .$$

Lemma F.7 *Let A be a Borel subset of an n-manifold M, and let $\Phi : T \to U$ be a chart such that $A \subset U$. Then $\ell_{T,\Phi}(\Phi^{-1}(A))$ is independent of the chart Φ.*

Since a measure on a Polish space is defined by its (compatible) restrictions to open subsets of the space, one may employ charts and Lemma F.7 and define in a unique way a measure on a manifold M, which we call the *volume measure* on M.

Proposition F.8 (Volume measure) *Let M be a manifold.*
(i) *There exists a unique measure ρ_M on the Borel subsets of M such that for all Borel subsets $A \subset M$ and charts $\Phi : T \to U$ of M we have $\rho_M(A \cap U) = \ell_{T,\Phi}(\Phi^{-1}(A))$. The measure ρ_M is finite on compacts.*
(ii) *A Borel set $A \subset M$ is negligible if and only if $\rho_M(A) = 0$.*
(iii) *For every nonempty open subset $U \subset M$ and Borel set $A \subset M$ we have $\rho_U(A \cap U) = \rho_M(A \cap U)$.*
(iv) *For every isometry $f : M_1 \to M_2$ of manifolds we have $\rho_{M_1} \circ f^{-1} = \rho_{M_2}$.*
(v) *For all manifolds M_1 and M_2 we have $\rho_{M_1 \times M_2} = \rho_{M_1} \times \rho_{M_2}$.*

Clearly, $\rho_{\mathbb{R}^n}$ is Lebesgue measure on the Borel subsets of $\mathbb{R}^n$.

We write $\rho[M] = \rho_M(M)$ for every manifold M. We have frequently to consider such normalizing constants in the sequel. We always have $\rho[M] \in (0, \infty]$. (It is possible to have $\rho[M] = \infty$, for example $\rho[\mathbb{R}] = \infty$; but it is impossible to have $\rho[M] = 0$ because we do not allow the empty set to be a manifold.) If M is compact, then $\rho[M] < \infty$.

"Critical" vocabulary

Definition F.9 Critical and regular points Let $f : M \to N$ be a smooth map of manifolds. A $p \in M$ is a *critical point* for f if the derivative $\mathbb{T}_p(f)$ fails to be onto; otherwise p is a *regular point* for f. We say that $q \in N$ is a *critical value* of f if there exists a critical point $p \in M$ for f such that $f(p) = q$. Given $q \in N$, the *fiber* $f^{-1}(q)$ is by definition the set $\{p \in M \mid f(p) = q\}$. Finally, $q \in N$ is a *regular value* for f if q is not a critical value and the fiber $f^{-1}(q)$ is nonempty.

Our usage of the term "regular value" therefore *does not conform* to the traditions of differential topology. In the latter context, a regular value is simply a point which is not a critical value.

The following facts, which we use repeatedly, are straightforwardly deduced from the definitions.

Proposition F.10 *Let $f : M \to N$ be a smooth map of manifolds. Let M_{reg} (resp., M_{crit}) be the set of regular (resp., critical) points for f. Let N_{crit} (resp., N_{reg}) be the set of critical (resp., regular) values of f.*
(i) *The set M_{reg} (resp., M_{crit}) is open (resp., closed) in M.*
(ii) *The sets N_{crit} and N_{reg}, being σ-compact, are Borel subsets of N.*

Regular values are easier to handle than critical ones. Sard's Theorem allows one to restrict attention, when integrating, to such values.

Theorem F.11 (Sard) [Mil97, Chapter 3] *The set of critical values of a smooth map of manifolds is negligible.*

Lie groups and Haar measure

Definition F.12 A *Lie group* G is a manifold with ambient space $\mathrm{Mat}_n(\mathbb{F})$ for some n and $\mathbb{F}$ such that G is a closed subgroup of $\mathrm{GL}_n(\mathbb{F})$.

This *ad hoc* definition is of course not as general as possible but it is simple and suits our purposes well. For example, $\mathrm{GL}_n(\mathbb{F})$ is a Lie group. By Lemma 4.1.15, $\mathrm{U}_n(\mathbb{F})$ is a Lie group.

Let G be a locally compact topological group, e.g., a Lie group. Let μ be a measure on the Borel sets of G. We say that μ is *left-invariant* if $\mu A = \mu\{ga \mid a \in A\}$ for all Borel $A \subset G$ and $g \in G$. Right-invariance is defined analogously.

Theorem F.13 *Let G be a locally compact topological group.*
(i) *There exists a left-invariant measure on G (neither $\equiv 0$ nor infinite on compacts), called* Haar measure, *which is unique up to a positive constant multiple.*
(ii) *If G is compact, then every Haar measure is right-invariant, and has finite total mass. In particular, there exists a unique Haar probability measure.*

We note that Lebesgue measure in $\mathbb{R}^n$ is a Haar measure. Further, for any Lie group G contained in $\mathrm{U}_n(\mathbb{F})$, the volume measure ρ_G is by Proposition F.8(vi) and Lemma 4.1.13(iii) a Haar measure.

F.2 Proof of the coarea formula

In this subsection, we prove the coarea formula, Theorem 4.1.8. We begin by introducing the notion of f-adapted pairs of charts, prove a few preliminary lemmas, and then provide the proof of the theorem. Lemmas F.18 and F.19 can be skipped in the course of the proof of the coarea formula, but are included since they are useful in Section 4.1.3.

Let $f : M \to N$ be a smooth map from an n-manifold to a k-manifold and assume that $n \geq k$. Let $\pi : \mathbb{R}^n \to \mathbb{R}^k$ be the projection to the first k coordinates. Recall that a chart on M is a an open nonempty subset $S \subset \mathbb{R}^n$ together with a diffeomorphism Ψ from S to an open subset of M.

Definition F.14 A pair $(\Psi : S \to U, \Phi : T \to V)$ consisting of a chart of M and a chart of N is *f-adapted* if

$$S \subset \pi^{-1}(T) \subset \mathbb{R}^n, \ U \subset f^{-1}(V), \ f \circ \Psi = \Phi \circ \pi|_S,$$

in which case we also say that the open set $U \subset M$ is *good* for f.

The commuting diagram

$$\begin{array}{ccccccccccc} & \mathbb{R}^n & \supset & & S & \stackrel{\Psi}{\to} & U & & \subset & M & \\ \pi & \downarrow & & \pi|_S & \downarrow & & \downarrow & f|_U & & \downarrow & f \\ & \mathbb{R}^k & \supset & & T & \stackrel{\Phi}{\to} & V & & \subset & N & \end{array}$$

summarizes the relationships among the maps in question here.

Lemma F.15 *Let $f : M \to N$ be a smooth map from an n-manifold to a k-manifold. Let $p \in M$ be a regular point. (Since a regular point exists, necessarily $n \geq k$.) Then there exists an open neighborhood of p good for f.*

Proof Without loss we may assume that $M \subset \mathbb{R}^n$ and $N \subset \mathbb{R}^k$ are open sets. We may also assume that $p = 0 \in \mathbb{R}^n$ and $q = f(p) = 0 \in \mathbb{R}^k$. Write $f = (f_1, \ldots, f_k)$. Let $t_1, \ldots, t_n$ be the standard coordinates in $\mathbb{R}^n$. By hypothesis, for some permutation σ of $\{1, \ldots, n\}$, putting $g_i = f_i$ for $i = 1, \ldots, k$ and $g_i = t_{\sigma(i)}$ for $i = k+1, \ldots, n$, the determinant $\det_{i,j=1}^n \partial_j g_i$ does not vanish at the origin. By the inverse function theorem there exist open neighborhoods $U, S \subset \mathbb{R}^n$ of the origin such that $(\star) = (f_1|_U, \ldots, f_k|_U, t_{\sigma(k+1)}|_U, \ldots, t_{\sigma(n)}|_U)$ maps U diffeomorphically to S. Take Ψ to be the inverse of $(\star)$. Take Φ to be the identity map of N to itself. Then (Ψ, Φ) is an f-adapted pair of charts and the origin belongs to the image of Ψ.

□

Proposition F.16 *Let $f : M \to N$ be a smooth map from an n-manifold to a k-manifold. Let $M_{\mathrm{reg}} \subset M$ be the set of regular points of f. Fix $q \in N$ such that $f^{-1}(q) \cap M_{\mathrm{reg}}$ is nonempty. Then:*
(i) $M_{\mathrm{reg}} \cap f^{-1}(q)$ is a manifold of dimension $n-k$;
(ii) for every $p \in M_{\mathrm{reg}} \cap f^{-1}(q)$ we have $\mathbb{T}_p(M_{\mathrm{reg}} \cap f^{-1}(q)) = \ker(\mathbb{T}_p(f))$.

Proof We may assume that $M_{\mathrm{reg}} \neq \emptyset$ and hence $n \geq k$, for otherwise there is nothing to prove. By Lemma F.15 we may assume that $M \subset \mathbb{R}^n$ and $N \subset \mathbb{R}^k$ are open sets and that f is projection to the first k coordinates, in which case all assertions here are obvious. □

We pause to introduce some apparatus from linear algebra.

Definition F.17 Let $f : E \to F$ be a linear map between Euclidean spaces and let $f^* : F \to E$ be the adjoint of f. The *generalized determinant* $J(f)$ is defined as the square root of the determinant of $ff^* : F \to F$.

We emphasize that $J(f)$ is always nonnegative. If a linear map $f : \mathbb{R}^n \to \mathbb{R}^k$ is represented by a k-by-n matrix A with real entries, and the Euclidean structures of source and target f are the usual ones, then $J(f)^2 = \det(AA^{\mathrm{T}})$. In general, we have $J(f) \neq 0$ if and only if f is onto. Note also that, if f is an isometry, then $J(f) = 1$.

Lemma F.18 *For $i = 1,2$ let $f_i : E_i \to F_i$ be a linear map between Euclidean spaces. Let $f_1 \oplus f_2 : E_1 \oplus E_2 \to F_1 \oplus F_2$ be the orthogonal direct sum of f_1 and f_2. Then we have $J(f \oplus f') = J(f)J(f')$.*

Proof This follows directly from the definitions.

Lemma F.19 *Let $f : E \to F$ be a linear map between Euclidean spaces. Let $D \subset \ker(f)$ be a subspace such that $D^\perp$ and F have the same dimension. Let $x_1, \dots, x_n \in D^\perp$ be an orthonormal basis. Let $\Pi : E \to D^\perp$ be the orthogonal projection. Then:*
(i) $J(f)^2 = \det_{i,j=1}^n (fx_i, fx_j)_F$;
(ii) $J(f)^2$ is the determinant of the $\mathbb{R}$-linear operator $\Pi \circ f^* \circ f : D^\perp \to D^\perp$.

Proof Since $(fx_i, fx_j)_F = (x_i, \Pi f^* f x_j)_F$, statements (i) and (ii) are equivalent. We have only to prove statement (i). Extend $x_1, \dots, x_n$ to an orthonormal basis of $x_1, \dots, x_{n+k}$ of E. Let $y_1, \dots, y_n$ be an orthonormal basis of F. Let A be the n-by-n matrix with entries $(y_i, fx_j)_F$, in which case $A^{\mathrm{T}}A$ is the n-by-n matrix with entries $(fx_i, fx_j)_E$. Now make the identifications $E = \mathbb{R}^{n+k}$ and $F = \mathbb{R}^n$ such a way that $x_1,\ldots,x_{n+k}$ (resp., $y_1,\ldots,y_n$) becomes the standard basis in $\mathbb{R}^{n+k}$ (resp., $\mathbb{R}^n$). Then f is represented by the matrix $[A\ 0]$, where $0 \in \mathrm{Mat}_{n\times k}$. Finally, by definition, $J(f)^2 = \det[A\ 0][A\ 0]^{\mathrm{T}} = \det A^{\mathrm{T}}A$, which proves the result. □

Lemma F.20 *Let $f : E \to F$ be an onto linear map from an n-dimensional Euclidean space to a k-dimensional Euclidean space. Let $\{x_i\}_{i=1}^n$ and $\{y_i\}_{i=1}^k$ be bases (not necessarily orthonormal) for E and F, respectively, such that $f(x_i) = y_i$ for $i = 1,\ldots,k$ and $f(x_i) = 0$ for $i = k+1,\ldots,n$. Then we have*

$$J(f)^2 \det_{i,j=1}^{n}(x_i,x_j)_E = \det_{i,j=k+1}^{n}(x_i,x_j)_E \det_{i,j=1}^{k}(y_i,y_j)_F.$$

Proof Let A (resp., B) be the n-by-n (resp., k-by-k) real symmetric positive definite matrix with entries $A_{ij} = (x_i,x_j)_E$ (resp., $B_{ij} = (y_i,y_j)_F$). Let C be the $(n-k)$-by-$(n-k)$ block of A in the lower right corner. We have to prove that $J(f)^2 \det A = \det C \det B$. Make $\mathbb{R}$-linear (but in general not isometric) identifications $E = \mathbb{R}^n$ and $F = \mathbb{R}^k$ in such a way that $\{x_i\}_{i=1}^n$ (respectively, $\{y_i\}_{i=1}^k$) is the standard basis in $\mathbb{R}^n$ (respectively, $\mathbb{R}^k$), and (hence) f is projection to the first k coordinates. Let P be the k-by-n matrix with 1s along the main diagonal and 0s elsewhere. Then we have $fx = Px$ for all $x \in E$. Let Q be the unique n-by-k matrix such that $f^*y = Qy$ for all $y \in F = \mathbb{R}^k$. Now the inner product on E is given in terms of A by the formula $(x,y)_E = x^{\mathrm{T}}Ay$ and similarly $(x,y)_F = x^{\mathrm{T}}By$. By definition of Q we have $(Px)^{\mathrm{T}}By = x^{\mathrm{T}}A(Qy)$ for all $x \in \mathbb{R}^n$ and $y \in \mathbb{R}^k$, hence $P^{\mathrm{T}}B = AQ$, and hence $Q = A^{-1}P^{\mathrm{T}}B$. By definition of $J(f)$ we have $J(f)^2 = \det(PA^{-1}P^{\mathrm{T}}B) = \det(PA^{-1}P^{\mathrm{T}})\det B$. Now decompose A into blocks thus:

$$A = \begin{bmatrix} a & b \\ c & d \end{bmatrix}, \quad a = PAP^{\mathrm{T}}, \ d = C.$$

From the matrix inversion lemma, Lemma A.1, it follows that $\det(PA^{-1}P^{\mathrm{T}}) = \det A/\det C$. The result follows. □

We need one more technical lemma. We continue in the setting of Theorem 4.1.8. For the statement of the lemma we also fix an f-adapted pair $(\Psi : S \to U, \Phi : T \to V)$ of charts. (Existence of such implies that $n \geq k$.) Let $\pi : \mathbb{R}^n \to \mathbb{R}^k$ be projection to the first k coordinates. Let $\bar{\pi} : \mathbb{R}^n \to \mathbb{R}^{n-k}$ be projection to the last $n-k$ coordinates. Given $t \in T$ such that the set

$$S_t = \{x \in \mathbb{R}^{n-k} | (t,x) \in U\}$$

is nonempty, the map

$$\Psi_t = (x \mapsto \Psi(t,x)) : S_t \to U \cap f^{-1}(\Phi(t))$$

is chart of $M_{\mathrm{reg}} \cap f^{-1}(\Phi(t))$, and hence the correction factor σ_{Ψ_t}, see Definition F.6, is defined.

Lemma F.21 *With notation as above, for all $s \in S$ we have*

$$J(\mathbb{T}_{\Psi(s)}(f))\sigma_\Psi(s) = \sigma_{\Psi_{\pi(s)}}(\bar{\pi}(s))\sigma_\Phi(\pi(s)).$$

Proof Use Lemma F.20 to calculate $J(\mathbb{T}_{\Psi(s)}(f))$, taking $\{(\partial_i\Psi)(s)\}_{i=1}^n$ as the basis for the domain of $\mathbb{T}_{\Psi(s)}(f)$ and $\{(\partial_i\Phi)(\pi(s))\}_{i=1}^k$ as the basis for the range. □

Proof of Theorem 4.1.8 We may assume that $M_{\mathrm{reg}} \neq \emptyset$ and hence $n \geq k$, for otherwise there is nothing to prove. Lemma F.21 expresses the function $p \mapsto J(\mathbb{T}_p(f))$ locally in a fashion which makes continuity on M_{reg} clear. Moreover, $M_{\mathrm{crit}} = \{p \in M \mid J(\mathbb{T}_p(f)) = 0\}$. Thus the function in question is indeed Borel-measurable. (In fact it is continuous, but to prove that fact requires uglier formulas.) Thus part (i) of the theorem is proved. We turn to the proof of parts (ii) and (iii) of the theorem. Since on the set M_{crit} no contribution is made to any of the integrals under consideration, we may assume that $M = M_{\mathrm{reg}}$. We may assume that φ is the indicator of a Borel subset $A \subset M$. By Lemma F.15 the manifold M is covered by open sets good for f. Accordingly M can be expressed as a countable disjoint union of Borel sets each of which is contained in an open set good for f, say $M = \bigcup M_\alpha$. By monotone convergence we may replace A by $A \cap M_\alpha$ for some index α, and thus we may assume that for some f-adapted pair $(\Psi : S \to U, \Phi : T \to V)$ of charts we have $A \subset U$. We adopt again the notation introduced in Lemma F.21. We have

$$\begin{aligned}
\int_A J(\mathbb{T}_p(f))d\rho_M(p) &= \int_{\Psi^{-1}(A)} J(\mathbb{T}_{\Psi(s)}(f))d\ell_{S,\Psi}(s) \\
&= \int \left(\int_{\Psi_t^{-1}(A)} d\ell_{S_t,\Psi_t}(x)\right) d\ell_{T,\Phi}(t) \\
&= \int (\int_{A\cap f^{-1}(q)} d\rho_{f^{-1}(q)}(p))d\rho_N(q).
\end{aligned}$$

At the first and last steps we appeal to Proposition F.8(i) which characterizes the measures $\rho_{(\cdot)}$. At the crucial second step we apply Lemma F.21 and Fubini's Theorem. The last calculation proves both the measurability assertion (ii) and the integral formula (iii). □

F.3 Metrics, connections, curvature, Hessians, and the Laplace–Beltrami operator

We briefly review some notions of Riemannian geometry. Although in this book we work exclusively with manifolds embedded in Euclidean space, all formulas in this subsection can be understood in the general setting of Riemannian geometry.

Let M be a manifold of dimension m, equipped with a Riemannian metric g, and let μ be the measure naturally associated with g. By definition, g is the specification for every $p \in M$ of a scalar product g_p on $\mathbb{T}_p(M)$. In the setup of manifolds embedded in some Euclidean space that we have adopted, $\mathbb{T}_p(M)$ is a subspace of the ambient Euclidean space, the Riemannian metric g_p is given by the restriction of the Euclidean inner product to that subspace, and the volume measure μ coincides with the measure ρ_M given in Proposition F.8.

Let $C^\infty(M)$ denote the space of real-valued smooth functions on M.

Definition F.22 (i) A *vector field* (on M) is a smooth map X from M to its ambient space such that, for all $p \in M$, $X(p) \in \mathbb{T}_p(M)$. Given a vector field X and a smooth function $f \in C^\infty(M)$, we define the function $Xf \in C^\infty(M)$ by the requirement that $Xf(p) = \frac{d}{dt}f(\gamma(t))|_{t=0}$ for any curve γ through p with $\gamma'(0) = X(p)$.
(ii) If X,Y are vector fields, we define $g(X,Y) \in C^\infty(M)$ by

$$g(X,Y)(p) = g_p(X(p),Y(p)).$$

The *Lie bracket* $[X,Y]$ is the unique vector field satisfying, for all $f \in C^\infty(M)$,

$$[X,Y]f = X(Yf) - Y(Xf).$$

(iii) A collection of vector fields $L_1,\ldots,L_m$ defined on an open set $U \subset M$ is a *local frame* if $L_1(p),\ldots,L_m(p)$ are a basis of $\mathbb{T}_p(M)$ for all $p \in U$. The local frame $\{L_i\}$ is *orthonormal* if $g(L_i,L_j) = \delta_{ij}$.

Definition F.23 (i) For $f \in C^\infty(M)$, the *gradient* grad f is the unique vector field satisfying $g(X, \mathrm{grad}\, f) = Xf$ for all vector fields X. If $\{L_i\}$ is any local orthonormal frame, then $\mathrm{grad}\, f = \sum_i (L_i f) L_i$.
(ii) A *connection* ∇ is a bilinear operation associating with vector fields X and Y a vector field $\nabla_X Y$ such that, for any $f \in C^\infty(M)$,

$$\nabla_{fX} Y = f\nabla_X Y, \quad \nabla_X(fY) = f\nabla_X Y + X(f)Y.$$

The connection ∇ is *torsion-free* if $\nabla_X Y - \nabla_Y X = [X,Y]$.
(iii) The *Levi–Civita* connection is the unique torsion-free connection satisfying that, for all vector fields X,Y,Z,

$$Xg(Y,Z) = g(\nabla_X Y, Z) + g(Y, \nabla_X Z).$$

(iv) Given a vector field X, the *divergence* $\mathrm{div}X \in C^\infty(M)$ is the unique function satisfying, for any orthonormal local frame $\{L_i\}$,

$$\mathrm{div}X = \sum_i g(L_i, [L_i, X]).$$

Alternatively, for any compactly supported $f \in C^\infty(M)$,

$$\int g(\text{grad}\, f, X) d\mu = -\int f \text{div} X d\mu\,.$$

(v) The *Laplace–Beltrami* operator Δ on $C^\infty(M)$ is defined by $\Delta f = \text{div}\,\text{grad}\, f$. With respect to any orthonormal local frame $\{L_i\}$ we have

$$\Delta f = \sum L_i^2 f + \sum_{i,j} g(L_i, [L_i, L_j]) L_j f\,.$$

From part (iv) of Definition F.23, we have the classical integration by parts formula: for all functions $\varphi, \psi \in C^\infty(M)$ at least one of which is compactly supported,

$$\int g(\text{grad}\,\varphi, \text{grad}\,\psi) d\mu = -\int \varphi(\Delta\psi) d\mu\,. \tag{F.1}$$

In our setup of manifolds embedded in a Euclidean space, the gradient grad f introduced in Definition F.23 can be evaluated at a point $p \in M$ by extending f, in a neighborhood of p, to a smooth function $\tilde{f}$ in the ambient space, taking the standard gradient of $\tilde{f}$ in the ambient space at p, and finally projecting it orthogonally to $T_p(M)$. We also note (but do not use) that a connection gives rise to the notion of parallel transport of a vector field along a curve, and in this language the Levi–Civita connection is characterized by being torsion-free and preserving the metric g under parallel transport.

We use in the sequel the symbol ∇ to denote exclusively the Levi–Civita connection. It follows from part (iv) of Definition F.23 that, for a vector field X and an orthonormal local frame $\{L_i\}$, $\text{div} X = \sum_i g(\nabla_{L_i} X, L_i)$. Further, for all vector fields X, Y and Z,

$$\begin{aligned} 2g(\nabla_X Y, Z) &= Xg(Y,Z) + Yg(Z,X) - Zg(X,Y) \\ &\quad + g([X,Y],Z) + g([Z,X],Y) + g(X,[Z,Y])\,. \end{aligned} \tag{F.2}$$

Definition F.24 Given $f \in C^\infty(M)$, we define the *Hessian* Hess$f$ to be the operation associating with two vector fields X and Y the function

$$\text{Hess}(f)(X,Y) = (XY - \nabla_X Y) f = g(\nabla_X \text{grad}\, f, Y) = \text{Hess}(f)(Y,X)\,.$$

(The second and third equalities can be verified from the definition of the Levi–Civita connection.)

We have $\text{Hess}(f)(hX,Y) = \text{Hess}(f)(X,hY) = h\text{Hess}(f)(X,Y)$ for all $h \in C^\infty(M)$ and hence $(\text{Hess}(f)(X,Y))(p)$ depends only $X(p)$ and $Y(p)$.

With respect to any orthonormal local frame $\{L_i\}$, we have the relations

$$\begin{aligned}\operatorname{Hess}(f)(L_i,L_j) &= (L_iL_j - \nabla_{L_i}L_j)f\,,\\ \Delta f &= \sum_i (L_i^2 - \nabla_{L_i}L_i)f = \sum_i \operatorname{Hess}(f)(L_i,L_i)\,.\end{aligned} \tag{F.3}$$

In this respect, the Laplace–Beltrami operator is a *contraction* of the Hessian. The divergence, the Hessian and the Laplace–Beltrami operator coincide with the usual notions of gradient, Hessian and Laplacian when $M = \mathbb{R}^m$ and the tangent spaces (all of which can be identified with $\mathbb{R}^m$ in that case) are equipped with the standard Euclidean metric.

We are ready to introduce the *Riemannian curvature tensor* and its contraction, the *Ricci curvature tensor*.

Definition F.25 (i) The *Riemann curvature tensor* $R(\cdot,\cdot)$ associates with vector fields X,Y an operator $R(X,Y)$ on vector fields defined by the formula

$$R(X,Y)Z = \nabla_X(\nabla_Y Z) - \nabla_Y(\nabla_X Z) - \nabla_{[X,Y]}Z\,.$$

(ii) The *Ricci curvature tensor* associates with vector fields X and Y the function $\operatorname{Ric}(X,Y) \in C^\infty(M)$, which, with respect to any orthonormal local frame $\{L_i\}$, satisfies $\operatorname{Ric}(X,Y) = \sum_i g(R(X,L_i)L_i,Y)$.

We have $R(fX,Y)Z = R(X,fY)Z = R(X,Y)(fZ) = fR(X,Y)Z$ for all $f \in C^\infty(M)$ and hence $(R(X,Y)Z)(p) \in \mathbb{T}_p(M)$ depends only on $X(p)$, $Y(p)$ and $Z(p)$. The analogous remark holds for $\operatorname{Ric}(X,Y)$ since it is a contraction of $R(X,Y)Z$.

Many computations are simplified by the introduction of a special type of orthonormal frame.

Definition F.26 Let $p \in M$. An orthonormal local frame $\{L_i\}$ in a neighborhood of p is said to be *geodesic* at p if $(\nabla_{L_i}L_j)(p) = 0$.

A geodesic local frame $\{L_i\}$ in a neighborhood U of $p \in M$ can always be built from a given orthonormal local frame $\{K_i\}$ by setting $L_i = \sum_j A_{ij}K_j$ with $A : U \to \mathrm{Mat}_n$ a smooth map satisfying $A(p) = I_m$, $A^{\mathrm{T}}A = I_m$, and $(K_iA_{jk})(p) = -g(\nabla_{K_i}K_j,K_k)(p)$. With respect to geodesic frames $\{L_i\}$, we have the simple expressions

$$\operatorname{Hess}(f)(L_i,L_j)(p) = (L_iL_jf)(p), \quad \operatorname{Ric}(L_i,L_j)(p) = (\sum_k L_iC^j_{kk} - L_kC^j_{ik})(p)\,, \tag{F.4}$$

where $C^k_{ij} = g(\nabla_{L_i}L_j,L_k)$.

Curvature of classical compact Lie groups

Let G be a closed subgroup and submanifold of $\mathrm{U}_n(\mathbb{F})$, where the latter is as defined in Appendix E. In this situation both left- and right-translation in G are isometries. We specialize now to the case $M = G$. We are going to compute the Ricci curvature of G and then apply the result to concrete examples. In particular, we will provide the differential geometric interpretation of Proposition E.15.

The crucial observation is that, in this situation, "all computations can be done at the identity", as we now explain. For each $X \in \mathbb{T}_{I_n}(G)$, choose any curve γ through I_n such that $\gamma'(0) = X$ and let $\widetilde{X}$ be the vector field whose associated first order differential operator is given by $(\widetilde{X}f)(x) = \frac{d}{dt}f(x\gamma(t))|_{t=0}$ for all $f \in C^\infty(G)$ and $x \in G$. The vector field $\widetilde{X}$ does not depend on the choice of γ. Recall that $[X,Y] = XY - YX$ and $X \cdot Y = \Re \operatorname{tr} XY^*$ for $X,Y \in \mathrm{Mat}_n(\mathbb{F})$. For all $X,Y \in \mathbb{T}_{I_n}(G)$ one verifies by straightforward calculation that

$$[X,Y] \in \mathbb{T}_{I_n}(G),\ \ \widetilde{[X,Y]} = [\widetilde{X},\widetilde{Y}],\ \ g(\widetilde{X},\widetilde{Y}) = X \cdot Y.$$

It follows in particular from dimension considerations that every orthonormal basis $\{L_\alpha\}$ for $\mathbb{T}_{I_n}(G)$ gives rise to a global orthonormal frame $\{\widetilde{L}_\alpha\}$ on G.

Lemma F.27 *For all $X,Y,Z,W \in \mathbb{T}_{I_n}(G)$ we have*

$$\nabla_{\widetilde{X}}\widetilde{Y} = \frac{1}{2}\widetilde{[X,Y]},\ \ g(R(\widetilde{X},\widetilde{Y})\widetilde{Z},\widetilde{W}) = -\frac{1}{4}[[X,Y],Z]\cdot W,$$

and hence

$$\mathrm{Ric}(\widetilde{X},\widetilde{X}) \quad = \quad -\sum_\alpha \frac{1}{4}[[X,L_\alpha],L_\alpha]\cdot X, \tag{F.5}$$

where the sum runs over any orthonormal basis $\{L_\alpha\}$ of $\mathbb{T}_{I_n}(G)$.

Proof By formula (F.2) we have $g(\nabla_{\widetilde{X}}\widetilde{Y},\widetilde{Z}) = \frac{1}{2}[X,Y]\cdot Z$, whence the result after a straightforward calculation. □

We now consider the special cases $G = \{U \in \mathrm{U}_N(\mathbb{F}) \mid \det U = 1\}$ for $\mathbb{F} = \mathbb{R},\mathbb{C}$. If $\mathbb{F} = \mathbb{R}$, then G is the *special orthogonal group* $SO(N)$ whereas, if $\mathbb{F} = \mathbb{C}$, then G is the *special unitary group* $SU(N)$. Using now the notation of Proposition E.15, one can show that $\mathbb{T}_{I_N}(G) = \mathfrak{su}_N(\mathbb{F})$. Thus, from (E.2) and (F.5) one gets for $G = SO(N)$ or $G = SU(N)$ that

$$\mathrm{Ric}(X,X) = \left(\frac{\beta(N+2)}{4} - 1\right)g(X,X), \tag{F.6}$$

for every vector field X on G, where $\beta = 1$ for $SO(N)$ and $\beta = 2$ for $SU(N)$. We note in passing that if $G = \mathrm{U}_N(\mathbb{C})$ then $\mathrm{Ric}(\widetilde{X},\widetilde{X}) = 0$ for $X = \mathrm{i}I_N \in \mathbb{T}_{I_N}(\mathrm{U}_N(\mathbb{C}))$, and thus no uniform strictly positive lower bound on the Ricci tensor exists for $G = \mathrm{U}_N(\mathbb{C})$. We also note that (F.6) remains valid for $G = U_N(\mathbb{H})$ and $\beta = 4$.

G Appendix on operator algebras

G.1 Basic definitions

An *algebra* is a vector space $\mathscr{A}$ over a field F equipped with a multiplication which is associative, distributive and F-bilinear, that is, for $x, y, z \in \mathscr{A}$ and $\alpha \in F$:

- $x(yz) = (xy)z$,
- $(x+y)z = xz + yz, \;\; x(y+z) = xy + xz$,
- $\alpha(xy) = (\alpha x)y = x(\alpha y)$.

We will say that $\mathscr{A}$ is *unital* if there exists a unit element $e \in \mathscr{A}$ such that $xe = ex = x$ (e is necessarily unique because if e' is also a unit then $ee' = e' = e'e = e$).

A *group algebra* $F(G)$ of a group $(G, *)$ over a field F is the set $\{\sum_{g \in G} a_g g : a_g \in F\}$ of linear combinations of finitely many elements of G with coefficients in F (above, $a_g = 0$ except for finitely many g). $F(G)$ is the algebra over F with addition and multiplication

$$\sum_{g\in G} a_g g + \sum_{g\in G} b_g g = \sum_{g\in G} (a_g + b_g) g, \quad \left(\sum_{g\in G} a_g g\right)\left(\sum_{g\in G} b_g g\right) = \sum_{g,h\in G} a_g b_h g * h,$$

respectively, and with product by a scalar $b\sum_{g\in G} a_g g = \sum_{g\in G}(b a_g) g$. The unit of $F(G)$ is identified with the unit of G.

A *complex algebra* is an algebra over the complex field $\mathbb{C}$. A *seminorm* on a complex algebra $\mathscr{A}$ is a map from $\mathscr{A}$ into $\mathbb{R}^+$ such that for all $x, y \in \mathscr{A}$ and $\alpha \in \mathbb{C}$,

$$||\alpha x|| = |\alpha| ||x||, \quad ||x+y|| \le ||x|| + ||y||, \quad ||xy|| \le ||x|| \cdot ||y||,$$

and, if $\mathscr{A}$ is unital with unit e, also $||e|| = 1$. A *norm* on a complex algebra $\mathscr{A}$ is a seminorm satisfying that $||x|| = 0$ implies $x = 0$ in $\mathscr{A}$. A *normed complex algebra* is a complex algebra $\mathscr{A}$ equipped with a norm $||.||$.

Definition G.1 A complex normed algebra $(\mathscr{A}, ||.||)$ is a *Banach algebra* if the norm $||\cdot||$ induces a complete distance.

Definition G.2 Let $\mathscr{A}$ be a Banach algebra.

- An *involution* on $\mathscr{A}$ is a map * from $\mathscr{A}$ to itself that satisfies $(a+b)^* = a^* + b^*$, $(ab)^* = b^* a^*$, $(\lambda a)^* = \bar{\lambda} a^*$ (for $\lambda \in \mathbb{C}$), $(a^*)^* = a$ and $||a^*|| = ||a||$.

- $\mathscr{A}$ is a C^**-algebra* if it possesses an involution $a \mapsto a^*$ that satisfies $||a^*a|| = ||a||^2$.
- $\mathscr{B}$ is a *(unital)* C^**-subalgebra* of a (unital) C^*-algebra if it is a subalgebra and, in addition, is closed with respect to the norm and the involution (and contains the unit).

Here $\bar{\lambda}$ denotes the complex conjugate of λ. Note that the assumption $||a|| = ||a^*||$ ensures the continuity of the involution.

The following collects some of the fundamental properties of Banach algebras (see [Rud91, pp. 234–235]).

Theorem G.3 *Let $\mathscr{A}$ be a unital Banach algebra and let $G(\mathscr{A})$ denote the invertible elements of $\mathscr{A}$. Then $G(\mathscr{A})$ is open, and it is a group under multiplication. Furthermore, for every $a \in \mathscr{A}$, the* spectrum *of a, defined as*

$$\mathrm{sp}(a) = \{\lambda \in \mathbb{C} : \lambda e - a \notin G(\mathscr{A})\},$$

is nonempty, compact and, defining the spectral radius

$$\rho(a) = \sup\{|\lambda| : \lambda \in \mathrm{sp}(a)\},$$

we have that

$$\rho(a) = \lim_{n\to\infty} ||a^n||^{1/n} = \inf_{n\geq 1} ||a^n||^{1/n}.$$

(The last equality is valid due to sub-additivity.)

An element a of $\mathscr{A}$ is said to be *self-adjoint* (resp., *normal*, *unitary*) if $a^* = a$ (resp., $a^*a = aa^*$, $a^*a = e = aa^*$). Note that, if $\mathscr{A}$ is unital, its unit e is self-adjoint. Indeed, for all $x \in \mathscr{A}$, we have $e^*x = (x^*e)^* = x$, similarly $xe^* = x$, and hence $e^* = e$ by uniqueness of the unit.

A *Hilbert space H* is a vector space equipped with an inner product $\langle \cdot, \cdot \rangle$ that is complete for the topology induced by the norm $\|\cdot\| := \sqrt{\langle \cdot, \cdot \rangle}$.

Let H_1, H_2 be two Hilbert spaces with inner products $\langle \cdot, \cdot \rangle_{H_1}$ and $\langle \cdot, \cdot \rangle_{H_2}$ respectively. The *direct sum* $H_1 \oplus H_2$ is a Hilbert space equipped with the inner product

$$\langle (x_1, y_1), (x_2, y_2) \rangle_{H_1 \oplus H_2} = \langle x_1, x_2 \rangle_{H_1} + \langle y_1, y_2 \rangle_{H_2}. \tag{G.1}$$

The *tensor product* $H_1 \otimes H_2$ is a Hilbert space with inner product

$$\langle x_1 \otimes y_1, x_2 \otimes y_2 \rangle_{H_1 \otimes H_2} = \langle x_1, x_2 \rangle_{H_1} \langle y_1, y_2 \rangle_{H_2}. \tag{G.2}$$

Let $B(H)$ denote the space of bounded linear operators on the Hilbert space H. We define the adjoint T^* of any $T \in B(H)$ as the unique element of $B(H)$

satisfying

$$\langle Tx, y\rangle = \langle x, T^*y\rangle \quad \forall x, y \in H. \tag{G.3}$$

The space $B(H)$, equipped with the involution $*$ and the norm

$$\|T\|_{B(H)} = \sup\{|\langle Tx, y\rangle|, \|x\| = \|y\| = 1\},$$

has a structure of C^*-algebra, see Definition G.2, and *a fortiori* that of Banach algebra. Therefore, Theorem G.3 applies, and we denote by $\mathrm{sp}(T)$ the spectrum of the operator $T \in B(H)$.

We have (see [Rud91, Theorem 12.26]) the following.

Theorem G.4 *Let H be a Hilbert space. A normal $T \in B(H)$ is*

(i) *self-adjoint iff* $\mathrm{sp}(T)$ *lies in the real axis,*
(ii) *unitary iff* $\mathrm{sp}(T)$ *lies on the unit circle.*

The GNS construction (Theorem 5.2.24) discussed in the main text can be used to prove the following fundamental fact (see [Rud91, Theorem 12.41]).

Theorem G.5 *For every C^*-algebra $\mathscr{A}$ there exists a Hilbert space $H_{\mathscr{A}}$ and a norm-preserving $*$-homomorphism $\pi_{\mathscr{A}} : \mathscr{A} \to B(H_{\mathscr{A}})$.*

G.2 Spectral properties

We next state the spectral theorem. Let $\mathscr{M}$ be a σ-algebra in a set Ω. A *resolution of the identity* (on $\mathscr{M}$) is a mapping

$$\chi : \mathscr{M} \to B(H)$$

with the following properties.

(i) $\chi(\emptyset) = 0, \chi(\Omega) = I$.
(ii) Each $\chi(\omega)$ is a self-adjoint projection.
(iii) $\chi(\omega' \cap \omega'') = \chi(\omega')\chi(\omega'')$.
(iv) If $\omega' \cap \omega'' = \emptyset$, $\chi(\omega' \cup \omega'') = \chi(\omega') + \chi(\omega'')$.
(v) For every $x \in H$ and $y \in H$, the set function $\chi_{x,y}(\omega) = \langle \chi(\omega)x, y\rangle$ is a complex measure on $\mathscr{M}$.

When $\mathscr{M}$ is the σ-algebra of all Borel sets on a locally compact Hausdorff space, it is customary to add the requirement that each $\chi_{x,y}$ is a regular Borel measure (this is automatically satisfied on compact metric spaces). Then we have the following theorem. (For bounded operators, see [Rud91, Theorem 12.23], and for unbounded operators, see [Ber66] or references therein.)

Theorem G.6 *If T is a normal linear operator on a Hilbert space H with domain dense in H, there exists a unique resolution of the identity χ on the Borel subsets of* $\mathrm{sp}(T)$ *which satisfies*

$$T = \int_{\mathrm{sp}(T)} \lambda d\chi(\lambda).$$

We call χ the spectral resolution *of T.*

Note that $\mathrm{sp}(T)$ is a bounded set if $T \in B(H)$, ensuring that $\chi_{x,y}$ is a compactly supported measure for all $x, y \in H$. For any bounded measurable function f on $\mathrm{sp}(T)$, we can use the spectral theorem to define $f(T)$ by

$$f(T) = \int_{\mathrm{sp}(T)} f(\lambda) d\chi(\lambda).$$

We then have (see [Rud91, Section 12.24]) the following.

Theorem G.7

(i) *$f \to f(T)$ is a homomorphism of the algebra of all bounded Borel functions on* $\mathrm{sp}(T)$ *into $B(H)$ which carries the function* 1 *to I, the identity into T and which satisfies $\bar{f}(T) = f(T)^*$.*

(ii) $\|f(T)\| \leq \sup\{|f(\lambda)| : \lambda \in \mathrm{sp}(T)\}$, *with equality for continuous f.*

(iii) *If f_n converges to f uniformly on* $\mathrm{sp}(T)$, $\|f_n(T) - f(T)\|$ *goes to zero as n goes to infinity.*

The theory can be extended to unbounded operators as follows. An operator T on H is a linear map from H into H with domain of definition $\mathscr{D}(T)$. Two operators T, S are equal if $\mathscr{D}(T) = \mathscr{D}(S)$ and $Tx = Sx$ for $x \in \mathscr{D}(T)$. T is said to be *closed* if, for every sequence $\{x_n\}_{n\in\mathbb{N}} \in \mathscr{D}(T)$ converging to some $x \in H$ such that Tx_n converges as n goes to infinity to y, one has $x \in \mathscr{D}(A)$ and $y = Tx$. Equivalently, the graph $(h, Th)_{h\in\mathscr{D}(A)}$ in the direct sum $H \oplus H$ is closed. T is *closable* if the closure of its graph in $H \oplus H$ is the graph of a (closed) operator. The spectrum $\mathrm{sp}(T)$ of T is the complement of the set of all complex numbers λ such that $(\lambda I - T)^{-1}$ exists as an everywhere defined bounded operator. We next define the *adjoint* of a densely defined operator T; if the domain $\mathscr{D}(T)$ of the operator T is dense in H, then the domain $\mathscr{D}(T^*)$ consists, by definition, of all $y \in H$ such that $\langle Tx, y\rangle$ is continuous for $x \in \mathscr{D}(T)$. Then, by density of $\mathscr{D}(T)$, there exists a unique $y^* \in H$ such that $\langle Tx, y\rangle = \langle x, y^*\rangle$ and we then set $T^*y := y^*$.

A densely defined operator T is *self-adjoint* iff $\mathscr{D}(T^*) = \mathscr{D}(T)$ and $T^* = T$. We can now state the generalization of Theorem G.6 to unbounded operators.

Theorem G.8 [DuS58, p. 1192] *Let T be a densely defined self-adjoint operator. Then its spectrum is real and there is a uniquely determined regular countably additive self-adjoint spectral measure χ_T defined on the Borel sets of the real line, vanishing on the complement of the spectrum, and related to T by the equations*

$$\text{(a)} \qquad \mathscr{D}(T) = \{x \in H | \int_{\mathrm{sp}(T)} \lambda^2 d\langle \chi_T(\lambda)x, x\rangle < \infty\},$$

$$\text{(b)} \qquad Tx = \lim_{n \to \infty} \int_{-n}^{n} \lambda d\chi_T(\lambda)x.$$

Another good property of closed and densely defined operators (not necessarily self-adjoint) is the existence of a *polar decomposition*.

Theorem G.9 [DuS58, p. 1249] *Let T be a closed, densely defined operator. Then T can be written uniquely as a product $T = PA$, where P is a partial isometry, that is, P^*P is a projection, A is a nonnegative self-adjoint operator, the closures of the ranges of A and T^* coincide, and both are contained in the domain of P.*

Let $\mathscr{A}$ be a sub-algebra of $B(H)$. A self-adjoint operator T on H is *affiliated* with $\mathscr{A}$ iff it is a densely defined self-adjoint operator such that for any bounded Borel function $f$on the spectrum of A, $f(A) \in \mathscr{A}$. This is equivalent, by the spectral theorem, to requiring that all the spectral projections $\{\chi_T([n,m]), n \leq m\}$ belong to $\mathscr{A}$ (see [Ped79, p. 164]).

G.3 States and positivity

Lemma G.10 [Ped79, p. 6] *An element x of a C^*-algebra A is* nonnegative, $x \geq 0$, *iff one of the following equivalent conditions is true:*

(i) *x is normal and with nonnegative spectrum;*
(ii) *$x = y^2$ for some self-adjoint operator y in A;*
(iii) *x is self-adjoint and $||t1 - x|| \leq t$ for any $t \geq ||x||$;*
(iv) *x is self-adjoint and $||t1 - x|| \leq t$ for some $t \geq ||x||$.*

Lemma G.11 [Ped79, Section 3.1] *Let α be a linear functional on a C^*-algebra $(\mathscr{A}, *, ||.||)$. Then the two following conditions are equivalent:*

(i) *$\alpha(x^*x) \geq 0$ for all $x \in \mathscr{A}$;*
(ii) *$\alpha(x) \geq 0$ for all $x \geq 0$ in $\mathscr{A}$.*

When one of these conditions is satisfied, we say that α is nonnegative. *Then α is self-adjoint, that is, $\alpha(x^*) = \overline{\alpha(x)}$ and if $\mathscr{A}$ has a unit I, $|\alpha(x)| \leq \alpha(I)||x||$.*

Some authors use the term *positive* functional where we use nonnegative functional.

Lemma G.12 [Ped79, Theorem 3.1.3] *If α is a nonnegative functional on a C^*-algebra $\mathscr{A}$, then for all $x, y \in \mathscr{A}$,*

$$|\alpha(y^*x)|^2 \leq \alpha(x^*x)\alpha(y^*y).$$

G.4 von Neumann algebras

By Theorem G.5, any C^*-algebra can be represented as a C^*-subalgebra of $B(H)$, for H a Hilbert space. So, let us fix a Hilbert space H. $B(H)$ can be endowed with different topologies. In particular, the *strong* (resp., *weak*) topology on $B(H)$ is the locally convex vector space topology associated with the family of seminorms $\{x \to \|x\xi\| : \xi \in H\}$ (resp., the family of linear functionals $\{x \to \langle x\eta, \xi\rangle : \xi, \eta \in H\}$).

Theorem G.13 (von Neumann's double commutant theorem) *For a subset $\mathscr{S} \subset B(H)$ that is closed under the involution *, define the* commutant *of $\mathscr{S}$ as*

$$\mathscr{S}' := \{b \in B(H) : ba = ab, \forall a \in \mathscr{S}\}.$$

Then a C^*-subalgebra $\mathscr{A}$ of $B(H)$ is a W^*-algebra if and only if $\mathscr{A}'' = \mathscr{A}$.

We have also the following.

Theorem G.14 [Ped79, Theorem 2.2.2] *Let $\mathscr{A} \subset B(H)$ be a subalgebra that is closed under the involution * and contains the identity operator. Then the following are equivalent:*

(i) $\mathscr{A}'' = \mathscr{A}$;
(ii) *$\mathscr{A}$ is strongly closed;*
(iii) *$\mathscr{A}$ is weakly closed.*

In particular, $\mathscr{A}''$ is the weak closure of $\mathscr{A}$. The advantage of a von Neumann algebra is that it allows one to construct functions of operators which are not continuous.

A useful property of self-adjoint operators is their behavior under closures. More precisely, we have the following. (See [Mur90, Theorem 4.3.3] for a proof.)

Theorem G.15 (Kaplansky density theorem) *Let H be a Hilbert space and let $\mathscr{A} \subset B(H)$ be a C^*-algebra with strong closure $\mathscr{B}$. Let $\mathscr{A}_{sa}$ and $\mathscr{B}_{sa}$ denote the self-adjoint elements of $\mathscr{A}$ and $\mathscr{B}$. Then:*

(i) *$\mathscr{A}_{sa}$ is strongly dense in $\mathscr{B}_{sa}$;*
(ii) *the closed unit ball of $\mathscr{A}_{sa}$ is strongly dense in the closed unit ball of $\mathscr{B}_{sa}$;*
(iii) *the closed unit ball of $\mathscr{A}$ is strongly dense in the closed unit ball of $\mathscr{B}$.*

Von Neumann algebras are classified into three types: I, II and III [Li92, Chapter 6]. The class of *finite* von Neumann algebras will be of special interest to us. Since its definition is related to properties of projections, we first describe the latter (see [Li92, Definition 6.1.1] and [Li92, Proposition 1.3.5]).

Definition G.16 Let $\mathscr{A}$ be a von Neumann algebra.

(i) A *projection* is an element $p \in \mathscr{A}$ such that $p = p^* = p^2$.
(ii) We say that $p \leq q$ if $q - p$ is a *nonnegative* element of $\mathscr{A}$. We say that $p \sim q$ if there exists a $v \in \mathscr{A}$ so that $p = vv^*$ and $q = v^*v$.
(iii) A projection $p \in \mathscr{A}$ is said to be *finite* if any projection q of $\mathscr{A}$ such that $q \leq p$ and $q \sim p$ must be equal to p.

We remark that the relation $\sim$ in point (ii) of Definition G.16 is an equivalence relation.

Recall that, for projections $p, q \in B(H)$, the *minimum* of p and q, denoted $p \wedge q$, is the projection from H onto $pH \cap qH$, while the *maximum* $p \vee q$ is the projection from H onto $\overline{pH + qH}$. The minimum $p \wedge q$ can be checked to be the largest operator dominated by both p and q, with respect to the order $\leq$. The maximum $p \vee q$ has the analogous least upper bound property.

The following elementary proposition clarifies the analogy between the role the operations of taking minimum and maximum of projections play in noncommutative probability, and the role intersection and unions play in classical probability. This, and other related facts concerning projections, can be found in [Nel74, Section 1], see in particular (3) there. (For similar statements, see [Li92].) Recall the notions of tracial, faithful and normal states, see Definitions 5.2.9 and 5.2.26.

Proposition G.17 *Let $(\mathscr{A}, \tau)$ be a W^*-probability space, with τ tracial. Let $p, q \in \mathscr{A}$ be projections. Then $p \wedge q, p \vee q \in \mathscr{A}$ and $\tau(p) + \tau(q) = \tau(p \wedge q) + \tau(p \vee q)$.*

As a consequence of Proposition G.17, we have the following.

Property G.18 *Let $(\mathscr{A}, \tau)$ be a W^*- probability space, subset of $B(H)$ for some Hilbert space H. Assume that τ is a a normal faithful tracial state.*

(i) *Let $\varepsilon > 0$ and p, q be two projections in $\mathscr{A}$ so that $\tau(p) \geq 1-\varepsilon$ and $\tau(q) \geq 1-\varepsilon$. Then, with $r = p \wedge q$, $\tau(r) \geq 1-2\varepsilon$.*

(ii) *If p_i is an increasing sequence of projections converging weakly to the identity, then $\tau(p_i)$ goes to one.*

(iii) *Conversely, if p_i is an increasing sequence of projections such that $\tau(p_i)$ goes to one, then p_i converges weakly to the identity in $\mathscr{A}$.*

Proof of Property G.18 The first point is an immediate consequence of Proposition G.17. The second point is a direct consequence of normality of τ while the third is a consequence of the faithfulness of τ. □

Definition G.19 A von Neumann algebra $\mathscr{A}$ is *finite* if its identity is finite.

Von Neumann algebras equipped with nice tracial states are finite von Neumann algebras, as stated below.

Proposition G.20 [Li92, Proposition 6.3.15] *Let $\mathscr{A}$ be a von Neumann algebra. If there is a faithful normal tracial state τ on $\mathscr{A}$, $\mathscr{A}$ is a finite von Neumann algebra.*

We also have the following equivalent characterization of normal states on a von Neumann algebra, see [Ped79, Theorem 3.6.4].

Proposition G.21 *Let ϕ be a state on a von Neumann algebra $\mathscr{A}$ in $B(H)$. Let $\{\zeta_i\}_{i\geq 0}$ be an orthonormal basis for H and put, for $x \in B(H)$, $\mathrm{Tr}(x) = \sum_i \langle x\zeta_i, \zeta_i\rangle$. Then the following are equivalent:*

- *ϕ is normal;*
- *there exists an operator x of trace class on H such that $\phi(y) = \mathrm{Tr}(xy)$;*
- *ϕ is weakly continuous on the unit ball of $\mathscr{A}$.*

G.5 Noncommutative functional calculus

We take τ to be a linear form on a unital complex algebra $\mathscr{A}$ equipped with an involution $*$ such that, for all $a \in \mathscr{A}$,

$$\tau(aa^*) \geq 0. \tag{G.4}$$

Then, for all $a, b \in \mathscr{A}$, we have $\tau(a^*b) = \tau(b^*a)^*$ and the noncommutative version of the Cauchy–Schwarz inequality, namely

$$|\tau(a^*b)| \leq \tau(a^*a)^{\frac{1}{2}}\tau(b^*b)^{\frac{1}{2}}. \tag{G.5}$$

(See, e.g., [Ped79, Theorem 3.1.3].) Moreover, by an application of Minkowski's inequality,

$$\tau((a+b)^*(a+b))^{\frac{1}{2}} \le \tau(aa^*)^{\frac{1}{2}} + \tau(bb^*)^{\frac{1}{2}}. \tag{G.6}$$

Lemma G.22 *If τ is as above and, in addition, for some norm $\|\cdot\|$ on $\mathscr{A}$, $|\tau(a)| \le \|a\|$ for all $a \in \mathscr{A}$, then*

$$|\tau(b^*a^*ab)| \le \|a^*a\|\tau(b^*b). \tag{G.7}$$

Proof By the Cauchy–Schwarz inequality (G.5), the claim is trivial if $\tau(b^*b) = 0$. Thus, fix $b \in \mathscr{A}$ with $\tau(b^*b) > 0$. Define

$$\tau_b(a) = \frac{\tau(b^*ab)}{\tau(b^*b)}.$$

Note that τ_b is still a linear form on $\mathscr{A}$ satisfying (G.4). Thus, for all $a_1, a_2 \in \mathscr{A}$, by the Cauchy–Schwarz inequality (G.5) applied to $\tau_b(a_1^*a_2)$,

$$|\tau(b^*a_1^*a_2b)|^2 \le \tau(b^*a_1^*a_1b)\tau(b^*a_2^*a_2b).$$

Taking $a_1 = (a^*a)^{2^n}$ and a_2 the unit in $\mathscr{A}$ yields

$$\tau(b^*(a^*a)^{2^n}b)^2 \le \tau(b^*(a^*a)^{2^{n+1}}b)\tau(b^*b).$$

Chaining these inequalities gives

$$\tau(b^*(a^*a)b) \le \tau(b^*(a^*a)^{2^n}b)^{2^{-n}}\tau(b^*b)^{1-2^{-n}} \le \|b^*(a^*a)^{2^n}b\|^{2^{-n}}\tau(b^*b)^{1-2^{-n}}.$$

Using the sub-multiplicativity of the norm and taking the limit as $n \to \infty$ yields (G.7). □

We next assume that $(\mathscr{A}, *, \|\cdot\|)$ is a von Neumann algebra and τ a tracial state on $(\mathscr{A}, *)$. The following noncommutative versions of Hölder inequalities can be found in [Nel74].

For $a \in \mathscr{A}$, we denote $|a| = (aa^*)^{\frac{1}{2}}$. We have, for $a, b \in \mathscr{A}$, b a self-adjoint bounded operator,

$$|\tau(ab)| \le \|b\|\tau(|a|). \tag{G.8}$$

We have the noncommutative Hölder inequality saying that for all $p, q \ge 1$ such that $\frac{1}{p} + \frac{1}{q} = 1$, we have

$$|\tau(ab)| \le \tau(|a|^q)^{\frac{1}{q}}\tau(|b|^p)^{\frac{1}{p}}. \tag{G.9}$$

More generally, see [FaK86, Theorem 4.9(i)], for all $r \ge 0$ and $p^{-1} + q^{-1} = r^{-1}$,

$$|\tau(|ab|^r)|^{\frac{1}{r}} \le \tau(|a|^q)^{\frac{1}{q}}\tau(|b|^p)^{\frac{1}{p}}. \tag{G.10}$$

This generalizes and extends the matricial case of (A.13).

H Stochastic calculus notions

A good background on stochastic analysis, at a level suitable to our needs, is provided in [KaS91] and [ReY99].

Definition H.1 Let $(\Omega, \mathscr{F})$ be a measurable space.

(i) A *filtration* $\mathscr{F}_t, t \geq 0$, is a nondecreasing family of sub-σ-fields of $\mathscr{F}$.

(ii) A random time T is a *stopping time* of the filtration $\mathscr{F}_t$, $t \geq 0$, if the event $\{T \leq t\}$ belongs to the σ-field $\mathscr{F}_t$ for all $t \geq 0$.

(iii) A *process* X_t, $t \geq 0$, is *adapted* to the filtration $\mathscr{F}_t$ if, for all $t \geq 0$, X_t is an $\mathscr{F}_t$-measurable random variable. In this case, we say $\{X_t, \mathscr{F}_t, t \geq 0\}$ is an *adapted process*.

(iv) Let $\{X_t, \mathscr{F}_t, t \geq 0\}$ be an adapted process, so that $E[|X_t|] < \infty$ for all $t \geq 0$. The process $X_t, t \geq 0$ is said to be an $\mathscr{F}_t$ *martingale* (respectively, *submartingale*) if, for every $0 \leq s < t < \infty$,

$$E[X_t | \mathscr{F}_s] = X_s\,, \quad \text{resp.}, \; E[X_t | \mathscr{F}_s] \geq X_s\,.$$

(v) Let X_t, $t \geq 0$, be an $\mathscr{F}_t$ martingale, so that $E[X_t^2] < \infty$ for all $t \geq 0$. The *martingale bracket* $\langle X \rangle_t$, $t \geq 0$ of X_t is the unique adapted increasing process so that $X_t^2 - \langle X \rangle_t$ is a martingale for the filtration $\mathscr{F}_t$.

(vi) If X_t, $t \geq 0$, and Y_t, $t \geq 0$, are $\mathscr{F}_t$ martingales, their *cross-bracket* is defined as $\langle X, Y \rangle_t = [\langle X + Y \rangle_t - \langle X - Y \rangle_t]/4$.

In the case when the martingale X_t possesses continuous paths, $\langle X \rangle_t$ equals its quadratic variation. The usefulness of the notion of bracket of a continuous martingale is apparent in the following.

Theorem H.2 (Lévy) *Let $\{X_t, \mathscr{F}_t, t \geq 0\}$ with $X_0 = 0$ be a continuous, adapted, n-dimensional process such that each component is a continuous $\mathscr{F}_t$-martingale and the martingale cross bracket $\langle X^i, X^j \rangle_t = \delta_{i,j} t$. Then the components X_t^i are independent Brownian motions.*

Let $X_t, t \geq 0$ be a real-valued $\mathscr{F}_t$ adapted process, and let B be a Brownian motion. Assume that $E[\int_0^T X_t^2 dt] < \infty$. Then

$$\int_0^T X_t dB_t := \lim_{n \to \infty} \sum_{k=0}^{n-1} X_{\frac{Tk}{n}} (B_{\frac{T(k+1)}{n}} - B_{\frac{Tk}{n}})$$

exists, the convergence holds in L^2 and the limit does not depend on the choice of the discretization of $[0, T]$ (see [KaS91, Chapter 3]).

One can therefore consider the problem of finding solutions to the integral equation

$$X_t = X_0 + \int_0^t \sigma(X_s)dB_s + \int_0^t b(X_s)ds \tag{H.1}$$

with a given X_0, σ and b some functions on $\mathbb{R}^n$, and B a n-dimensional Brownian motion. This can be written under the differential form

$$dX_s = \sigma(X_s)dB_s + b(X_s)ds. \tag{H.2}$$

There are at least two notions of solutions: strong solutions and weak solutions.

Definition H.3 [KaS91, Definition 5.2.1] A *strong solution* of the stochastic differential equation (H.2) on the given probability space $(\Omega, \mathscr{F})$ and with respect to the fixed Brownian motion B and initial condition ξ is a process $\{X_t, t \geq 0\}$ with continuous sample paths so that the following hold.

(i) X_t is adapted to the filtration $\mathscr{F}_t$ given by $\mathscr{F}_t = \sigma(\mathscr{G}_t \cup \mathscr{N})$, with

$$\mathscr{G}_t = \sigma(B_s, s \leq t; X_0), \mathscr{N} = \{N \subset \Omega, \exists G \in \mathscr{G}_\infty \text{ with } N \subset G, P(G) = 0\}.$$

(ii) $P(X_0 = \xi) = 1$.

(iii) $P(\forall t, \quad \int_0^t (|b_i(X_s)| + |\sigma_{ij}(X_s)|^2)ds < \infty) = 1$ for all $i, j \leq n$.

(iv) (H.1) holds almost surely.

Definition H.4 [KaS91, Definition 5.3.1] A *weak solution* of the stochastic differential equation (H.2) is a pair (X, B) and a triple $(\Omega, \mathscr{F}, P)$ so that $(\Omega, \mathscr{F}, P)$ is a probability space equipped with a filtration $\mathscr{F}_t$, B is an n-dimensional Brownian motion, and X is a continuous adapted process, satisfying (iii) and (iv) in Definition H.3.

There are also two notions of uniqueness.

Definition H.5 [KaS91, Definition 5.3.4]

- We say that *strong uniqueness* holds if two solutions with common probability space, common Brownian motion B and common initial condition are almost surely equal at all times.
- We say that *weak uniqueness*, or *uniqueness in the sense of probability law*, holds if any two weak solutions have the same law.

Theorem H.6 *Suppose that b and σ satisfy*

$$\begin{aligned} \|b(t,x) - b(t,y)\| + \|\sigma(t,x) - \sigma(t,y)\| &\leq K\|x - y\|, \\ \|b(t,x)\|^2 + \|\sigma(t,x)\|^2 &\leq K^2(1 + \|x\|^2), \end{aligned}$$

for some finite constant K independent of t. Then there exists a unique solution to (H.2), and it is strong. Moreover, it satisfies

$$E[\int_0^T \|b(t,X_t)\|^2 dt] < \infty,$$

for all $T \geq 0$.

Theorem H.7 *Any weak solutions* $(X^i, B^i, \Omega^i, \mathscr{F}^i, P^i)_{i=1,2}$ *of (H.2) with* $\sigma = \mathbf{I}_n$, *so that*

$$E[\int_0^T \|b(t,X_t^i)\|^2 dt] < \infty,$$

for all $T < \infty$ *and* $i = 1,2$, *have the same law.*

Theorem H.8 (Burkholder–Davis–Gundy inequality) *There exist universal constants* λ_m, Λ_m *so that, for all* $m \in \mathbb{N}$, *and any continuous local martingale* $(M_t, t \geq 0)$ *with bracket* $(A_t, t \geq 0)$,

$$\lambda_m E(A_T^m) \leq E(\sup_{t \leq T} M_t^{2m}) \leq \Lambda_m E(A_T^m).$$

Theorem H.9 (Itô, Kunita–Watanabe) *Let* $f : \mathbb{R} \to \mathbb{R}$ *be a function of class* $\mathscr{C}^2$ *and let* $X = \{X_t, \mathscr{F}_t; 0 \leq t < \infty\}$ *be a continuous semi-martingale with decomposition*

$$X_t = X_0 + M_t + A_t,$$

where M is a local martingale and A the difference of continuous, adapted, nondecreasing processes. Then, almost surely,

$$f(X_t) = f(X_0) + \int_0^t f'(X_s) dM_s + \int_0^t f'(X_s) dA_s + \frac{1}{2} \int_0^2 f''(X_s) d\langle M \rangle_s, \quad 0 \leq t < \infty.$$

Theorem H.10 (Novikov) *Let* $\{X_t, \mathscr{F}_t, t \geq 0\}$ *be an adapted process with values in* $\mathbb{R}^d$ *such that*

$$E[e^{\frac{1}{2}\int_0^T \sum_{i=1}^d (X_t^i)^2 dt}] < \infty$$

for all $T \in \mathbb{R}^+$. *Then, if* $\{W_t, \mathscr{F}_t, t \geq 0\}$ *is a d-dimensional Brownian motion, then*

$$M_t = \exp\{\int_0^t X_u . dW_u - \frac{1}{2} \int_0^t \sum_{i=1}^d (X_u^i)^2 du\}$$

is an $\mathscr{F}_t$*-martingale.*

Theorem H.11 (Girsanov) *Let $\{X_t, \mathscr{F}_t, t \geq 0\}$ be an adapted process with values in $\mathbb{R}^d$ such that*

$$E[e^{\frac{1}{2}\int_0^T \sum_{i=1}^d (X_t^i)^2 dt}] < \infty.$$

Then, if $\{W_t, \mathscr{F}_t, P, 0 \leq t \leq T\}$ is a d-dimensional Brownian motion,

$$\bar{W}_t^i = W_t^i - \int_0^t X_s^i ds, \quad 0 \leq t \leq T,$$

is a d-dimensional Brownian motion under the probability measure

$$\bar{P} = \exp\{\int_0^T X_u dW_u - \frac{1}{2}\int_0^T \sum_{i=1}^d (X_u^i)^2 du\} P.$$

Theorem H.12 *Let $\{X_t, \mathscr{F}_t, 0 \leq t < \infty\}$ be a submartingale whose every path is right-continuous. Then, for any $\tau > 0$ and $\lambda > 0$,*

$$\lambda P(\sup_{0 \leq t \leq \tau} X_t \geq \lambda) \leq E[X_\tau^+].$$

We shall use the following consequence.

Corollary H.13 *Let $\{X_t, \mathscr{F}_t, t \geq 0\}$ be an adapted process with values in $\mathbb{R}^d$, such that*

$$\int_0^T \|X_t\|^2 dt = \int_0^T \sum_{i=1}^d (X_t^i)^2 dt$$

is uniformly bounded by the constant A_T. Let $\{W_t, \mathscr{F}_t, t \geq 0\}$ be a d-dimensional Brownian motion. Then, for any $L > 0$,

$$P(\sup_{0 \leq t \leq T} |\int_0^t X_u dW_u| \geq L) \leq 2e^{-\frac{L^2}{2A_T}}.$$

Proof We denote in short $Y_t = \int_0^t X_u . dW_u$ and write, for $\lambda > 0$,

$$\begin{aligned}
P(\sup_{0 \leq t \leq T} |Y_t| \geq A) &\leq P(\sup_{0 \leq t \leq T} e^{\lambda Y_t} \geq e^{\lambda A}) + P(\sup_{0 \leq t \leq T} e^{-\lambda Y_t} \geq e^{\lambda A}) \\
&\leq P\left(\sup_{0 \leq t \leq T} e^{\lambda Y_t - \frac{\lambda^2}{2}\int_0^t \|X_u\|^2 du} \geq e^{\lambda A - \frac{\lambda^2 A_T}{2}}\right) \\
&\quad + P\left(\sup_{0 \leq t \leq T} e^{-\lambda Y_t - \frac{\lambda^2}{2}\int_0^t \|X_u\|^2 du} \geq e^{\lambda A - \frac{\lambda^2 A_T}{2}}\right).
\end{aligned}$$

By Theorem H.10, $M_t = e^{-\lambda Y_t - \frac{\lambda^2}{2}\int_0^t \|X_u\|^2 du}$ is a nonnegative martingale. Thus, by Chebyshev's inequality and Doob's inequality,

$$P\left(\sup_{0\le t\le T} M_t \ge e^{\lambda A - \frac{\lambda^2 A_T}{2}}\right) \le e^{-\lambda A + \frac{\lambda^2 A_T}{2}} E[M_T] = e^{-\lambda A + \frac{\lambda^2 A_T}{2}}.$$

Optimizing with respect to λ completes the proof. □

The next statement, an easy consequence of the Dubins–Schwartz time change identities (see [KaS91, Thm. 3.4.6]), was extended in [Reb80] to a much more general setup than we need to consider.

Theorem H.14 (Rebolledo's Theorem) *Let $n \in \mathbb{N}$, and let M_N be a sequence of continuous centered martingales with values in $\mathbb{R}^n$ with bracket $\langle M_N \rangle$ converging pointwise (that is, for all $t \ge 0$) in L^1 towards a continuous deterministic function $\phi(t)$. Then, for any $T > 0$, $(M_N(t), t \in [0,T])$ converges in law as a continuous process from $[0,T]$ into $\mathbb{R}^n$ towards a Gaussian process G with covariance*

$$E[G(s)G^{\mathrm{T}}(t)] = \phi(t \wedge s).$$

References

[Ada69] F. Adams. *Lectures on Lie groups.* New York, NY, W. A. Benjamin, 1969.

[Adl05] M. Adler. PDE's for the Dyson, Airy and sine processes. *Ann. Inst. Fourier (Grenoble)*, **55**:1835–1846, 2005.

[AdvM01] M. Adler and P. van Moerbeke. Hermitian, symmetric and symplectic random ensembles: PDEs for the distribution of the spectrum. *Annals Math.*, **153**:149–189, 2001.

[AdvM05] M. Adler and P. van Moerbeke. PDEs for the joint distributions of the Dyson, Airy and sine processes. *Annals Probab.*, **33**:1326–1361, 2005.

[Aig79] M. Aigner. *Combinatorial Theory.* New York, NY, Springer, 1979.

[AlD99] D. Aldous and P. Diaconis. Longest increasing subsequences: from patience sorting to the Baik–Deift–Johansson theorem. *Bull. Amer. Math. Soc. (N.S.)*, **36**:413–432, 1999.

[AlKV02] N. Alon, M. Krivelevich and V. H. Vu. On the concentration of eigenvalues of random symmetric matrices. *Israel J. Math.*, **131**:259–267, 2002.

[AnAR99] G. E. Andrews, R. Askey and R. Roy. *Special Functions*, volume 71 of *Encyclopedia of Mathematics and its Applications.* Cambridge University Press, 1999.

[And90] G. W. Anderson. The evaluation of Selberg sums. *C.R. Acad. Sci. I.-Math.*, **311**:469–472, 1990.

[And91] G. W. Anderson. A short proof of Selberg's generalized beta formula. *Forum Math.*, **3**:415–417, 1991.

[AnZ05] G. W. Anderson and O. Zeitouni. A CLT for a band matrix model. *Probab. Theory Rel. Fields*, **134**:283–338, 2005.

[AnZ08a] G. W. Anderson and O. Zeitouni. A CLT regularized sample covariance matrices. *Ann. Statistics*, **36**:2553–2576, 2008.

[AnZ08b] G. W. Anderson and O. Zeitouni. A LLN for finite-range dependent random matrices. *Comm. Pure Appl. Math.*, **61**:1118–1154, 2008.

[AnBC$^+$00] C. Ané, S. Blachère, D. Chafï, P. Fougères, I. Gentil, F. Malrieu, C. Roberto and G. Scheffer. *Sur les inégalités de Sobolev logarithmique*, volume 11 of *Panoramas et Synthèse.* Paris, Societe Mathematique de France, 2000.

[Ans02] M. Anshelevich. Itô formula for free stochastic integrals. *J. Funct. Anal.*, **188**:292–315, 2002.

[Arh71] L. V. Arharov. Limit theorems for the characteristic roots of a sample covariance matrix. *Dokl. Akad. Nauk SSSR*, **199**:994–997, 1971.

[Arn67] L. Arnold. On the asymptotic distribution of the eigenvalues of random matrices. *J. Math. Anal. Appl.*, **20**:262–268, 1967.

[AuBP07] A. Auffinger, G. Ben Arous and S. Péché. Poisson convergence for the largest eigenvalues of heavy tailed random matrices. arXiv:0710.3132v3 [math.PR], 2007.

[Bai93a] Z. D. Bai. Convergence rate of expected spectral distributions of large random matrices. I. Wigner matrices. *Annals Probab.*, **21**:625–648, 1993.

[Bai93b] Z. D. Bai. Convergence rate of expected spectral distributions of large random matrices. II. Sample covariance matrices. *Annals Probab.*, **21**:649–672, 1993.

[Bai97] Z. D. Bai. Circular law. *Annals Probab.*, **25**:494–529, 1997.

[Bai99] Z. D. Bai. Methodologies in spectral analysis of large-dimensional random matrices, a review. *Stat. Sinica*, **9**:611–677, 1999.

[BaS98a] Z. D. Bai and J. W. Silverstein. No eigenvalues outside the support of the limiting spectral distribution of large-dimensional sample covariance matrices. *Annals Probab.*, **26**:316–345, 1998.

[BaS04] Z. D. Bai and J. W. Silverstein. CLT for linear spectral statistics of large-dimensional sample covariance matrices. *Annals Probab.*, **32**:553–605, 2004.

[BaY88] Z. D. Bai and Y. Q. Yin. Necessary and sufficient conditions for almost sure convergence of the largest eigenvalue of a Wigner matrix. *Annals Probab.*, **16**:1729–1741, 1988.

[BaY05] Z. D. Bai and J.-F. Yao. On the convergence of the spectral empirical process of Wigner matrices. *Bernoulli*, **6**:1059–1092, 2005.

[BaBP05] J. Baik, G. Ben Arous and S. Péché. Phase transition of the largest eigenvalue for nonnull complex sample covariance matrices. *Annals Probab.*, **33**:1643–1697, 2005.

[BaBD08] J. Baik, R. Buckingham and J. DiFranco. Asymptotics of Tracy–Widom distributions and the total integral of a Painlevé II function. *Comm. Math. Phys.*, **280**:463–497, 2008.

[BaDJ99] J. Baik, P. Deift and K. Johansson. On the distribution of the length of the longest increasing subsequence of random permutations. *J. Amer. Math. Soc.*, **12**:1119–1178, 1999.

[BaDS09] J. Baik, P. Deift and T. Suidan. *Some Combinatorial Problems and Random Matrix Theory.* To appear, 2009.

[Bak94] D. Bakry. *L'hypercontractivité et son utilisation en théorie des semigroupes*, volume 1581 of *Lecture Notes in Mathematics*, pages 1–114. Berlin, Springer, 1994.

[BaE85] D. Bakry and M. Emery. Diffusions hypercontractives. In *Séminaire de probabilités, XIX, 1983/84*, volume 1123 of *Lecture Notes in Mathematics*, pages 177–206. Berlin, Springer, 1985.

[BaNT02] O. E. Barndorff-Nielsen and S. Thorbjørnsen. Lévy processes in free probability. *Proc. Natl. Acad. Sci. USA*, **99**:16576–16580, 2002.

[BaNT04] O. E. Barndorff-Nielsen and S. Thorbjørnsen. A connection between free and classical infinite divisibility. *Infin. Dimens. Anal. Qu.*, **7**:573–590, 2004.

[BeB07] S. T. Belinschi and H. Bercovici. A new approach to subordination results in free probability. *J. Anal. Math.*, **101**:357–365, 2007.

[BN08] S. T. Belinschi and A. Nica. η-series and a Boolean Bercovici-Pata bijection for bounded k-tuples. *Adv. Math.*, **217**:1–41, 2008.

[BeDG01] G. Ben Arous, A. Dembo and A. Guionnet. Aging of spherical spin glasses. *Probab. Theory Rel. Fields*, **120**:1–67, 2001.

[BeG97] G. Ben Arous and A. Guionnet. Large deviations for Wigner's law and Voiculescu's non-commutative entropy. *Probab. Theory Rel. Fields*, **108**:517–542, 1997.

[BeG08] G. Ben Arous and A. Guionnet. The spectrum of heavy-tailed random matrices. *Comm. Math. Phys.*, **278**:715–751, 2008.

[BeP05] G. Ben Arous and S. Péché. Universality of local eigenvalue statistics for some sample covariance matrices. *Comm. Pure Appl. Math.*, **58**:1316–1357, 2005.

[BeZ98] G. Ben Arous and O. Zeitouni. Large deviations from the circular law. *ESAIM Probab. Statist.*, **2**:123–134, 1998.

[BeG05] F. Benaych-Georges. Classical and free infinitely divisible distributions and random matrices. *Annals Probab.*, **33**:1134–1170, 2005.

[BeG09] F. Benaych-Georges. Rectangular random matrices, related convolution. *Probab. Theory Rel. Fields*, **144**:471–515, 2009.

[BeP00] H. Bercovici and V. Pata. A free analogue of Hinčin's characterization of infinite divisibility. *P. Am. Math. Soc.*, **128**:1011–1015, 2000.

[BeV92] H. Bercovici and D. Voiculescu. Lévy-Hinčin type theorems for multiplicative and additive free convolution. *Pacific J. Math.*, **153**:217–248, 1992.

[BeV93] H. Bercovici and D. Voiculescu. Free convolution of measures with unbounded support. *Indiana U. Math. J.*, **42**:733–773, 1993.

[Ber66] S.J. Bernau. The spectral theorem for unbounded normal operators. *Pacific J. Math.*, **19**:391–406, 1966.

[Bia95] P. Biane. Permutation model for semi-circular systems and quantum random walks. *Pacific J. Math.*, **171**:373–387, 1995.

[Bia97a] P. Biane. Free Brownian motion, free stochastic calculus and random matrices. In *Free Probability Theory* (Waterloo, ON 1995), volume 12 of *Fields Inst. Commun.*, pages 1–19. Providence, RI, American Mathematical Society, 1997.

[Bia97b] P. Biane. On the free convolution with a semi-circular distribution. *Indiana U. Math. J.*, **46**:705–718, 1997.

[Bia98a] P. Biane. Processes with free increments. *Math. Z.*, **227**:143–174, 1998.

[Bia98b] P. Biane. Representations of symmetric groups and free probability. *Adv. Math.*, **138**:126–181, 1998.

[Bia01] P. Biane. Approximate factorization and concentration for characters of symmetric groups. *Int. Math. Res. Not.*, pages 179–192, 2001.

[BiBO05] P. Bianc, P. Bougerol and N. O'Connell. Littelmann paths and Brownian paths. *Duke Math. J.*, **130**:127–167, 2005.

[BiCG03] P. Biane, M. Capitaine and A. Guionnet. Large deviation bounds for matrix Brownian motion. *Invent. Math.*, **152**:433–459, 2003.

[BiS98b] P. Biane and R. Speicher. Stochastic calculus with respect to free Brownian motion and analysis on Wigner space. *Probab. Theory Rel. Fields*, **112**:373–409, 1998.

[BlI99] P. Bleher and A. Its. Semiclassical asymptotics of orthogonal polynomials, Riemann-Hilbert problem, and universality in the matrix model. *Annals Math.*, **150**:185–266, 1999.

[BoG99] S. G. Bobkov and F. G. Götze. Exponential integrability and transportation cost related to log-Sobolev inequalities. *J. Funct. Anal.*, **163**:1–28, 1999.

[BoL00] S. G. Bobkov and M. Ledoux. From Brunn–Minkowski to Brascamp–Lieb and to logarithmic Sobolev inequalities. *Geom. Funct. Anal.*, **10**:1028–1052, 2000.

[BoMP91] L. V. Bogachev, S. A. Molchanov and L. A. Pastur. On the density of states of random band matrices. *Mat. Zametki*, **50**:31–42, 157, 1991.

[BoNR08] P. Bougarde, A. Nikeghbali and A. Rouault. Circular jacobi ensembles and deformed verblunsky coefficients. arXiv:0804.4512v2 [math.PR], 2008.

[BodMKV96] A. Boutet de Monvel, A. Khorunzhy and V. Vasilchuk. Limiting eigenvalue distribution of random matrices with correlated entries. *Markov Process. Rel. Fields*, **2**:607–636, 1996.

[Bor99] A. Borodin. Biorthogonal ensembles. *Nuclear Phys. B*, **536**:704–732, 1999.

[BoOO00] A. Borodin, A. Okounkov and G. Olshanski. Asymptotics of Plancherel measures for symmetric groups. *J. Amer. Math. Soc.*, **13**:481–515, 2000.

[BoS03] A. Borodin and A. Soshnikov. Janossy densities. I. Determinantal ensembles. *J. Statist. Phys.*, **113**:595–610, 2003.

[Bou87] N. Bourbaki. *Elements of Mathematics – General Topology*. Berlin, Springer, 1987.

[Bou05] N. Bourbaki. *Lie Groups and Lie Algebras*. Berlin, Springer, 2005.

[BrIPZ78] E. Brézin, C. Itzykson, G. Parisi and J. B. Zuber. Planar diagrams. *Comm. Math. Phys.*, **59**:35–51, 1978.

[Bru91] M. F. Bru. Wishart processes. *J. Theoret. Probab.*, **4**:725–751, 1991.

[BrDJ06] W. Bryc, A. Dembo and T. Jiang. Spectral measure of large random Hankel, Markov and Toeplitz matrices. *Annals Probab.*, **34**:1–38, 2006.

[BuP93] R. Burton and R. Pemantle. Local characteristics, entropy and limit theorems for spanning trees and domino tilings via transfer-impedances. *Annals Probab.*, **21**:1329–1371, 1993.

[Cab01] T. Cabanal-Duvillard. Fluctuations de la loi empirique de grande matrices aléatoires. *Ann. Inst. H. Poincaré – Probab. Statist.*, **37**:373–402, 2001.

[CaG01] T. Cabanal-Duvillard and A. Guionnet. Large deviations upper bounds for the laws of matrix-valued processes and non-communicative entropies. *Annals Probab.*, **29**:1205–1261, 2001.

[CaMV03] M. J. Cantero, L. Moral and L. Velázquez. Five-diagonal matrices and zeros of orthogonal polynomials on the unit circle. *Linear Algebra Appl.*, **362**:29–56, 2003.

[CaC04] M. Capitaine and M. Casalis. Asymptotic freeness by generalized moments for Gaussian and Wishart matrices. Application to beta random matrices. *Indiana Univ. Math. J.*, **53**:397–431, 2004.

[CaC06] M. Capitaine and M. Casalis. Cumulants for random matrices as convolutions on the symmetric group. *Probab. Theory Rel. Fields*, **136**:19–36, 2006.

[CaD07] M. Capitaine and C. Donati-Martin. Strong asymptotic freeness for Wigner and Wishart matrices. *Indiana Univ. Math. J.*, **56**:767–803, 2007.

[ChG08] G. P. Chistyakov and F. Götze. Limit theorems in free probability theory I. *Annals Probab.*, **36**:54–90, 2008.

[Chv83] V. Chvàtal. *Linear Programming*. New York, NY, W. H. Freeman, 1983.

[Col03] B. Collins. Moments and cumulants of polynomial random variables on unitary groups, the Itzykson-Zuber integral, and free probability. *Int. Math. Res. Not.*, pages 953–982, 2003.

[CoMG06] B. Collins, E. Maurel-Segala and A. Guionnet. Asymptotics of unitary and orthogonal matrix integrals. arxiv:math/0608193 [math.PR], 2006.

[CoS05] A. Connes and D. Shlyakhtenko. L^2-homology for von Neumann algebras. *J. Reine Angew. Math.*, **586**:125–168, 2005.

[CoL95] O. Costin and J. Lebowitz. Gaussian fluctuations in random matrices. *Phys. Rev. Lett.*, **75**:69–72, 1995.

[DaVJ88] D. J. Daley and D. Vere-Jones. *An Introduction to the Theory of Point Processes*. Springer Series in Statistics. New York, NY, Springer, 1988.

[DaS01] K. R. Davidson and S. J. Szarek. Local operator theory, random matrices and Banach spaces. In *Handbook of the Geometry of Banach Spaces, Vol. I*, pages 317–366. Amsterdam, North-Holland, 2001.

[Dei99] P. A. Deift. *Orthogonal Polynomials and Random Matrices: a Riemann-Hilbert Approach*, volume 3 of *Courant Lecture Notes in Mathematics*. New York, NY, New York University Courant Institute of Mathematical Sciences, 1999.

[Dei07] P. Deift. Universality for mathematical and physical systems. In *International Congress of Mathematicians 2006. Vol. I*, pages 125–152. Zürich, Eur. Math. Soc., 2007.

[DeG09] P. A. Deift and D. Gioev. *Invariant Random Matrix Ensembles: General Theory and Universality*. Courant Lecture Notes in Mathematics. New York, NY, New York University Courant Institute of Mathematical Sciences, 2009. To appear.

[DeIK08] P. Deift, A. Its and I. Krasovsky. Asymptotics of the Airy–kernel determinant. *Comm. Math. Phys.*, **278**:643–678, 2008.

[DeIZ97] P. A. Deift, A. R. Its and X. Zhou. A Riemann–Hilbert approach to asymptotic problems arising in the theory of random matrix models, and also in the theory of integrable statistical mechanics. *Annals Math.*, **146**:149–235, 1997.

[DeKM+98] P. Deift, T. Kriecherbauer, K.T.-R. McLaughlin, S. Venakides and X. Zhou. Uniform asymptotics for orthogonal polynomials. *Doc. Math.*, **III**:491–501, 1998. Extra volume ICM 1998.

[DeKM+99] P. Deift, T. Kriecherbauer, K. T-R. McLaughlin, S. Venakides and X. Zhou. Uniform asymptotics for polynomials orthogonal with respect to varying exponential weights and applications to universality questions in random matrix theory. *Comm. Pure Appl. Math.*, **52**:1335–1425, 1999.

[DeVZ97] P. Deift, S. Venakides and X. Zhou. New results in small dispersion KdV by an extension of the steepest descent method for Riemann–Hilbert problems. *Int. Math. Res. Not.*, pages 286–299, 1997.

[DeZ93] P. A. Deift and X. Zhou. A steepest descent method for oscillatory Riemann–Hilbert problems. Asymptotics for the MKdV equation. *Annals Math.*, **137**:295–368, 1993.

[DeZ95] P. A. Deift and X. Zhou. Asymptotics for the Painlevé II equation. *Comm. Pure Appl. Math.*, **48**:277–337, 1995.

[DeZ98] A. Dembo and O. Zeitouni. *Large Deviation Techniques and Applications*. New York, NY, Springer, second edition, 1998.

[Dem07] N. Demni. The Laguerre process and generalized Hartman–Watson law. *Bernoulli*, **13**:556–580, 2007.

[DeS89] J. D. Deuschel and D. W. Stroock. *Large Deviations*. Boston, MA, Academic Press, 1989.

[DiE01] P. Diaconis and S. N. Evans. Linear functionals of eigenvalues of random matrices. *Trans. Amer. Math. Soc.*, **353**:2615–2633, 2001.

[DiS94] P. Diaconis and M. Shahshahani. On the eigenvalues of random matrices. *J. Appl. Probab.*, **31A**:49–62, 1994.

[Dix69] J. Dixmier. *Les C^*-algèbres et leurs Représentations*. Les Grands Classiques Gauthier-Villars. Paris, Jacques Gabay, 1969.

[Dix05] A. L. Dixon. Generalizations of Legendre's formula $ke' - (k-e)k' = \pi/2$. *Proc. London Math. Society*, **3**:206–224, 1905.

[DoO05] Y. Doumerc and N. O'Connell. Exit problems associated with finite reflection groups. *Probab. Theory Rel. Fields*, **132**:501–538, 2005.

[Dud89] R. M. Dudley. *Real Analysis and Probability*. Pacific Grove, CA, Wadsworth & Brooks/Cole, 1989.

[Due04] E. Dueñez. Random matrix ensembles associated to compact symmetric spaces. *Comm. Math. Phys.*, **244**:29–61, 2004.

[DuE02] I. Dumitriu and A. Edelman. Matrix models for beta ensembles. *J. Math. Phys.*, **43**:5830–5847, 2002.

[DuE06] I. Dumitriu and A. Edelman. Global spectrum fluctuations for the β-Hermite and β-Laguerre ensembles via matrix models. *J. Math. Phys.*, **47**:063302, 36, 2006.

[DuS58] N. Dunford and J. T. Schwartz. *Linear Operators, Part I*. New York, NY, Interscience Publishers, 1958.

[Dur96] R. Durrett. *Probability: Theory and Examples*. Belmont, MA, Duxbury Press, second edition, 1996.

[Dyk93a] K. Dykema. Free products of hyperfinite von Neumann algebras and free dimension. *Duke Math. J.*, **69**:97–119, 1993.

[Dyk93b] K. Dykema. On certain free product factors via an extended matrix model. *J.*

Funct. Anal., **112**:31–60, 1993.

[Dys62a] F. J. Dyson. A Brownian-motion model for the eigenvalues of a random matrix. *J. Math. Phys.*, **3**:1191–1198, 1962.

[Dys62b] F. J. Dyson. Statistical theory of the energy levels of complex systems. I. *J. Math. Phys.*, **3**:140–156, 1962.

[Dys62c] F. J. Dyson. Statistical theory of the energy levels of complex systems. II. *J. Math. Phys.*, **3**:157–165, 1962.

[Dys62d] F. J. Dyson. Statistical theory of the energy levels of complex systems. III. *J. Math. Phys.*, **3**:166–175, 1962.

[Dys62e] F. J. Dyson. The threefold way. Algebraic structure of symmetry groups and ensembles in quantum mechanics. *J. Math. Phys.*, **3**:1199–1215, 1962.

[Dys70] F. J. Dyson. Correlations between eigenvalues of a random matrix. *Comm. Math. Phys.*, **19**:235–250, 1970.

[DyM63] F. J. Dyson and M. L. Mehta. Statistical theory of the energy levels of complex systems. IV. *J. Math. Phys.*, **4**:701–712, 1963.

[Ede97] A. Edelman. The probability that a random real Gaussian matrix has k real eigenvalues, related distributions, and the circular law. *J. Multivariate Anal.*, **60**:203–232, 1997.

[EdS07] A. Edelman and B. D. Sutton. From random matrices to stochastic operators. *J. Stat. Phys.*, **127**:1121–1165, 2007.

[ERSY09] L. Erdös, J. Ramírez, B. Schlein and H.-T. Yau. Bulk universality for Wigner matrices. arXiv:0905.4176v1 [math-ph], 2009.

[ERS$^+$09] L. Erdös, J. Ramírez, B. Schlein, T. Tao, V. Vu and H.-T. Yau. Bulk universality for Wigner Hermitian matrices with subexponential decay. arXiv:0906.4400v1 [math.PR], 2009

[EvG92] L. C. Evans and R. F. Gariepy. *Measure Theory and Fine Properties of Functions*. Boca Raton, CRC Press, 1992.

[Eyn03] B. Eynard. Master loop equations, free energy and correlations for the chain of matrices. *J. High Energy Phys.*, **11**:018, 45 pp., 2003.

[EyB99] B. Eynard and G. Bonnet. The Potts-q random matrix model: loop equations, critical exponents, and rational case. *Phys. Lett. B*, **463**:273–279, 1999.

[FaK86] T. Fack and H. Kosaki. Generalized s-numbers of τ-measurable operators. *Pacific J. Math.*, **123**:269–300, 1986.

[FaP03] D. G. Farenick and B. F. Pidkowich. The spectral theorem in quaternions. *Linear Algebra and its Applications*, **371**:75–102, 2003.

[Fed69] H. Federer. *Geometric Measure Theory*. New York, NY, Springer, 1969.

[Fel57] W. Feller. *An Introduction to Probability Theory and its Applications, Part I*. New York, NY, Wiley, second edition, 1957.

[FeP07] D. Féral and S. Péché. The largest eigenvalue of rank one deformation of large Wigner matrices. *Comm. Math. Phys.*, **272**:185–228, 2007.

[FoIK92] A. S. Fokas, A. R. Its and A. V. Kitaev. The Isomonodromy approach to matrix models in 2D quantum gravity. *Comm. Math. Phys.*, **147**:395–430, 1992.

[FoIKN06] A. S. Fokas, A. R. Its, A. A. Kapaev and V. Yu. Novokshenov. *Painlevé Transcendents. The Riemann–Hilbert Approach*, volume 128 of *Mathematical Surveys and Monographs*. Providence, RI, American Mathematical Society, 2006.

[For93] P. J. Forrester. The spectrum edge of random matrix ensembles. *Nuclear Phys. B*, **402**:709–728, 1993.

[For94] P. J. Forrester. Exact results and universal asymptotics in the Laguerre random matrix ensemble. *J. Math. Phys.*, **35**:2539–2551, 1994.

[For05] P. J. Forrester. *Log-gases and Random Matrices*. 2005. http://www.ms.unimelb.edu.au/∼ matpjf/matpjf.html.

[For06] P. J. Forrester. Hard and soft edge spacing distributions for random matrix ensembles with orthogonal and symplectic symmetry. *Nonlinearity*, **19**:2989–3002, 2006.

[FoO08] P. J. Forrester and S. Ole Warnaar. The importance of the Selberg integral. *Bulletin AMS*, **45**:489–534, 2008.

[FoR01] P. J. Forrester and E. M. Rains. Interrelationships between orthogonal, unitary and symplectic matrix ensembles. In *Random Matrix Models and their Applications*, volume 40 of *Math. Sci. Res. Inst. Publ.*, pages 171–207. Cambridge, Cambridge University Press, 2001.

[FoR06] P. J. Forrester and E. M. Rains. Jacobians and rank 1 perturbations relating to unitary Hessenberg matrices. *Int. Math. Res. Not.*, page 48306, 2006.

[FrGZJ95] P. Di Francesco, P. Ginsparg and J. Zinn-Justin. 2D gravity and random matrices. *Phys. Rep.*, **254**:133, 1995.

[FuK81] Z. Füredi and J. Komlós. The eigenvalues of random symmetric matrices. *Combinatorica*, **1**:233–241, 1981.

[Ge97] L. Ge. Applications of free entropy to finite von Neumann algebras. *Amer. J. Math.*, **119**:467–485, 1997.

[Ge98] L. Ge. Applications of free entropy to finite von Neumann algebras. II. *Annals Math.*, **147**:143–157, 1998.

[Gem80] S. Geman. A limit theorem for the norm of random matrices. *Annals Probab.*, **8**:252–261, 1980.

[GeV85] I. Gessel and G. Viennot. Binomial determinants, paths, and hook length formulae. *Adv. Math.*, **58**:300–321, 1985.

[GiT98] D. Gilbarg and N. S. Trudinger. *Elliptic Partial Equations of Second Order*. New York, NY, Springer, 1998.

[Gin65] J. Ginibre. Statistical ensembles of complex, quaternion, and real matrices. *J. Math. Phys.*, **6**:440–449, 1965.

[Gir84] V. L. Girko. The circular law. *Theory Probab. Appl.*, **29**:694–706, 1984.

[Gir90] V. L. Girko. *Theory of Random Determinants*. Dordrecht, Kluwer, 1990.

[GoT03] F. Götze and A. Tikhomirov. Rate of convergence to the semi-circular law. *Probab. Theory Rel. Fields*, **127**:228–276, 2003.

[GoT07] F. Götze and A. Tikhomirov. The circular law for random matrices. arXiv:0709.3995v3 [math.PR], 2007.

[GrKP94] R. Graham, D. Knuth and O. Patashnik. *Concrete Mathematics: a Foundation for Computer Science*. Reading, MA, Addison-Wesley, second edition, 1994.

[GrS77] U. Grenander and J. W. Silverstein. Spectral analysis of networks with random topologies. *SIAM J. Appl. Math.*, **32**:499–519, 1977.

[GrMS86] M. Gromov, V. Milman and G. Schechtman. *Asymptotic Theory of Finite Dimensional Normed Spaces*, volume 1200 of *Lectures Notes in Mathematics*. Berlin, Springer, 1986.

[GrPW91] D. Gross, T. Piran and S. Weinberg. Two dimensional quantum gravity and random surfaces. In *Jerusalem Winter School*. Singapore, World Scientific, 1991.

[Gui02] A. Guionnet. Large deviation upper bounds and central limit theorems for band matrices. *Ann. Inst. H. Poincaré – Probab. Statist.*, **38**:341–384, 2002.

[Gui04] A. Guionnet. Large deviations and stochastic calculus for large random matrices. *Probab. Surv.*, **1**:72–172, 2004.

[GuJS07] A. Guionnet, V. F. R Jones and D. Shlyakhtenko. Random matrices, free probability, planar algebras and subfactors. arXiv:math/0712.2904 [math.OA], 2007.

[GuM05] A. Guionnet and M. Maïda. Character expansion method for a matrix integral. *Probab. Theory Rel. Fields*, **132**:539–578, 2005.

[GuM06] A. Guionnet and E. Maurel-Segala. Combinatorial aspects of matrix models. *Alea*, **1**:241–279, 2006.

[GuM07] A. Guionnet and E. Maurel-Segala. Second order asymptotics for matrix models. *Ann. Probab.*, **35**:2160–2212, 2007.

[GuS07] A. Guionnet and D. Shlyakhtenko. On classical analogues of free entropy dimension. *J. Funct. Anal.*, **251**:738–771, 2007.

[GuS08] A. Guionnet and D. Shlyakhtenko. Free diffusion and matrix models with strictly convex interaction. *GAFA*, **18**:1875–1916, 2008.

[GuZ03] A. Guionnet and B. Zegarlinski. Lectures on logarithmic Sobolev inequalities. In *Seminaire de Probabilités XXXVI*, volume 1801 of *Lecture Notes in Mathematics*. Paris, Springer, 2003.

[GuZ00] A. Guionnet and O. Zeitouni. Concentration of the spectral measure for large matrices. *Electron. Commun. Prob.*, **5**:119–136, 2000.

[GuZ02] A. Guionnet and O. Zeitouni. Large deviations asymptotics for spherical integrals. *J. Funct. Anal.*, **188**:461–515, 2002.

[GZ04] A. Guionnet and O. Zeitouni. Addendum to: "Large deviations asymptotics for spherical integrals". *J. Funct. Anal.*, **216**:230–241, 2004.

[Gus90] R. A. Gustafson. A generalization of Selberg's beta integral. *B. Am. Math. Soc.*, **22**:97–105, 1990.

[Haa02] U. Haagerup. Random matrices, free probability and the invariant subspace problem relative to a von Neumann algebra. In *Proceedings of the International Congress of Mathematicians, Vol. I (Beijing, 2002)*, pages 273–290, Beijing, Higher Education Press, 2002.

[HaS09] U. Haagerup and H. Schultz. Invariant subspaces for operators in a general II_1-factor. To appear in *Publ. Math. Inst. Hautes Etudes Sci.*, 2009.

[HaST06] U. Haagerup, H. Schultz and S. Thorbjørnsen. A random matrix approach to the lack of projections in $C^*(\mathbb{F}_2)$. *Adv. Math.*, **204**:1–83, 2006.

[HaT99] U. Haagerup and S. Thorbjørnsen. Random matrices and k-theory for exact C^*-algebras. *Doc. Math.*, **4**:341–450, 1999.

[HaT03] U. Haagerup and S. Thorbjørnsen. Random matrices with complex Gaussian entries. *Expo. Math.*, **21**:293–337, 2003.

[HaT05] U. Haagerup and S. Thorbjørnsen. A new application of random matrices: $\mathrm{Ext}(C^*(F_2))$ is not a group. *Annals Math.*, **162**:711–775, 2005.

[HaLN06] W. Hachem, P. Loubaton and J. Najim. The empirical distribution of the eigenvalues of a Gram matrix with a given variance profile. *Ann. Inst. H. Poincaré – Probab. Statist.*, **42**:649–670, 2006.

[Ham72] J. M. Hammersley. A few seedlings of research. In *Proceedings of the Sixth Berkeley Symposium on Mathematical Statistics and Probability (University of California, Berkeley, CA, 1970/1971), Vol. I: Theory of Statistics*, pages 345–394, Berkeley, CA, University of California Press, 1972.

[HaM05] C. Hammond and S. J. Miller. Distribution of eigenvalues for the ensemble of real symmetric Toeplitz matrices. *J. Theoret. Probab.*, **18**:537–566, 2005.

[HaZ86] J. Harer and D. Zagier. The Euler characteristic of the moduli space of curves. *Invent. Math.*, **85**:457–485, 1986.

[Har56] Harish-Chandra. Invariant differential operators on a semisimple Lie algebra. *Proc. Nat. Acad. Sci. U.S.A.*, **42**:252–253, 1956.

[HaTW93] J. Harnad, C. A. Tracy and H. Widom. Hamiltonian structure of equations appearing in random matrices. In *Low-dimensional Topology and Quantum Field Theory (Cambridge, 1992)*, volume 315 of *Adv. Sci. Inst. Ser. B Phys.*, pages 231–245, New York, NY, NATO, Plenum, 1993.

[HaM80] S. P. Hastings and J. B. McLeod. A boundary value problem associated with the second Painlevé transcendent and the Korteweg–de Vries equation. *Arch. Rational Mech. Anal.*, **73**:31–51, 1980.

[Hel01] S. Helgason. *Differential Geometry, Lie Groups, and Symmetric Spaces*, volume 34 of *Graduate Studies in Mathematics*. Providence, RI, American Mathematical Society, 2001. Corrected reprint of the 1978 original.

[HiP00a] F. Hiai and D. Petz. A large deviation theorem for the empirical eigenvalue distribution of random unitary matrices. *Ann. Inst. H. Poincaré – Probab. Statist.*, **36**:71–85, 2000.

[HiP00b] F. Hiai and D. Petz. *The Semicircle Law, Free Random Variables and Entropy*, volume 77 of *Mathematical Surveys and Monographs*. Providence, RI, American Mathematical Society, 2000.

[HoW53] A. J. Hoffman and H. W. Wielandt. The variation of the spectrum of a normal matrix. *Duke Math. J.*, **20**:37–39, 1953.

[HoJ85] R. A. Horn and C. R. Johnson. *Matrix Analysis*. Cambridge, Cambridge University Press, 1985.

[HoX08] C. Houdré and H. Xu. Concentration of the spectral measure for large random matrices with stable entries. *Electron. J. Probab.*, **13**:107–134, 2008.

[HoKPV06] J. B. Hough, M. Krishnapur, Y. Peres and B. Virág. Determinantal processes and independence. *Probab. Surv.*, **3**:206–229, 2006.

[HoKPV09] J. B. Hough, M. Krishnapur, Y. Peres and B. Virág. *Zeros of Gaussian Analytic Functions and Determinantal Point Processes*. Providence, RI, American Mathematical Society, 2009.

[Ism05] M. E. H. Ismail. *Classical and Quantum Orthogonal Polynomials in One Variable*, volume 98 of *Encyclopedia of Mathematics and its Applications*. Cambridge, Cambridge University Press, 2005.

[ItZ80] C. Itzykson and J. B. Zuber. The planar approximation. II. *J. Math. Phys.*, **21**:411–421, 1980.

[Jac85] N. Jacobson. *Basic Algebra. I.* New York, NY, W. H. Freeman and Company, second edition, 1985.

[JiMMS80] M. Jimbo, T. Miwa, Y. Môri and M. Sato. Density matrix of an impenetrable Bose gas and the fifth Painlevé transcendent. *Physica*, **1D**:80–158, 1980.

[Joh98] K. Johansson. On fluctuations of eigenvalues of random Hermitian matrices. *Duke Math. J.*, **91**:151–204, 1998.

[Joh00] K. Johansson. Shape fluctuations and random matrices. *Comm. Math. Phys.*, **209**:437–476, 2000.

[Joh01a] K. Johansson. Discrete orthogonal polynomial ensembles and the Plancherel measure. *Annals Math.*, **153**:259–296, 2001.

[Joh01b] K. Johansson. Universality of the local spacing distribution in certain ensembles of Hermitian Wigner matrices. *Comm. Math. Phys.*, **215**:683–705, 2001.

[Joh02] K. Johansson. Non-intersecting paths, random tilings and random matrices. *Probab. Theory Rel. Fields*, **123**:225–280, 2002.

[Joh05] K. Johansson. The arctic circle boundary and the Airy process. *Annals Probab.*, **33**:1–30, 2005.

[John01] I. M. Johnstone. On the distribution of the largest eigenvalue in principal components analysis. *Ann. Statist.*, **29**:295–327, 2001.

[Jon82] D. Jonsson. Some limit theorems for the eigenvalues of a sample covariance matrix. *J. Multivariate Anal.*, **12**:1–38, 1982.

[Juh81] F. Juhász. On the spectrum of a random graph. In *Algebraic Methods in Graph Theory, Coll. Math. Soc. J. Bolyai*, volume 25, pages 313–316. Amsterdam, North-Holland, 1981.

[JuX03] M. Junge and Q. Xu. Noncommutative Burkholder/Rosenthal inequalities. *Annals Probab.*, **31**:948–995, 2003.

[Kal02] O. Kallenberg. *Foundations of Modern Probability*. Probability and its Applications. New York, NY, Springer, second edition, 2002.

[KaS91] I. Karatzas and S. Shreve. *Brownian Motion and Stochastic Calculus*, volume 113 of *Graduate Texts in Mathematics*. New York, NY, Springer, second edition, 1991.

[Kar07a] V. Kargin. The norm of products of free random variables. *Probab. Theory Rel. Fields*, **139**:397–413, 2007.

[KaM59] S. Karlin and J. McGregor. Coincidence properties of birth and death processes. *Pacific J. Math.*, **9**:1109–1140, 1959.

[Kar07b] N. El Karoui. Tracy–Widom limit for the largest eigenvalue of a large class of complex sample covariance matrices. *Annals Probab.*, **35**:663–714, 2007.

[KaS99] N. M. Katz and P. Sarnak. *Random Matrices, Frobenius Eigenvalues, and Monodromy*, volume 45 of *American Mathematical Society Colloquium Publications*. Providence, RI, American Mathematical Society, 1999.

[Kea06] J. P. Keating. Random matrices and number theory. In *Applications of Random Matrices in Physics*, volume 221 of *NATO Sci. Ser. II Math. Phys. Chem.*, pages 1–32. Dordrecht, Springer, 2006.

[Kho01] A. M. Khorunzhy. Sparse random matrices: spectral edge and statistics of rooted trees. *Adv. Appl. Probab.*, **33**:124–140, 2001.

[KhKP96] A. M. Khorunzhy, B. A. Khoruzhenko and L. A. Pastur. Asymptotic properties of large random matrices with independent entries,. *J. Math. Phys.*, **37**:5033–5060, 1996.

[KiN04] R. Killip and I. Nenciu. Matrix models for circular ensembles. *Int. Math. Res. Not.*, pages 2665–2701, 2004.

[KiN07] R. Killip and I. Nenciu. CMV: the unitary analogue of Jacobi matrices. *Comm. Pure Appl. Math.*, **60**:1148–1188, 2007.

[KiS09] R. Killip and M. Stoiciu. Eigenvalue statistics for cmv matrices: from poisson to clock via circular beta ensembles. *Duke Math. J.*, **146**:361–399, 2009.

[KoO01] W. König and N. O'Connell. Eigenvalues of the Laguerre process as non-colliding squared Bessel processes. *Electron. Comm. Probab.*, **6**:107–114, 2001.

[KoOR02] W. König, N. O'Connell and S. Roch. Non-colliding random walks, tandem queues, and discrete orthogonal polynomial ensembles. *Electron. J. Probab.*, **7**, 24 pp., 2002.

[Kra90] C. Krattenthaler. Generating functions for plane partitions of a given shape. *Manuscripta Math.*, **69**:173–201, 1990.

[Led01] M. Ledoux. *The Concentration of Measure Phenomenon*. Providence, RI, American Mathematical Society, 2001.

[Led03] M. Ledoux. A remark on hypercontractivity and tail inequalities for the largest eigenvalues of random matrices. In *Séminaire de Probabilités XXXVII*, volume 1832 of *Lecture Notes in Mathematics*. Paris, Springer, 2003.

[Leh99] F. Lehner. Computing norms of free operators with matrix coefficients. *Amer. J. Math.*, **121**:453–486, 1999.

[Lév22] P. Lévy. *Lecons d'analyse Fonctionnelle*. Paris, Gauthiers-Villars, 1922.

[Li92] B. R. Li. *Introduction to Operator Algebras*. River Edge, NJ, World Scientific Publishing Co., 1992.

[LiTV01] L. Li, A. M. Tulino and S. Verdú. Asymptotic eigenvalue moments for linear multiuser detection. *Commun. Inf. Syst.*, **1**:273–304, 2001.

[LiPRTJ05] A. E. Litvak, A. Pajor, M. Rudelson and N. Tomczak-Jaegermann. Smallest singular value of random matrices and geometry of random polytopes. *Adv. Math.*, **195**:491–523, 2005.

[LoS77] B. F. Logan and L. A. Shepp. A variational problem for random Young tableaux. *Adv. Math.*, **26**:206–222, 1977.

[Maa92] H. Maassen. Addition of freely independent random variables. *J. Funct. Anal.*,

106:409–438, 1992.

[Mac75] O. Macchi. The coincidence approach to stochastic point processes. *Adv. Appl. Probability*, **7**:83–122, 1975.

[MaP67] V. A. Marčenko and L. A. Pastur. Distribution of eigenvalues in certain sets of random matrices. *Math. USSR Sb.*, **1**:457–483, 1967.

[Mat97] T. Matsuki. Double coset decompositions of reductive Lie groups arising from two involutions. *J. Algebra*, **197**:49–91, 1997.

[Mat94] A. Matytsin. On the large-*N* limit of the Itzykson–Zuber integral. *Nuclear Phys. B*, **411**:805–820, 1994.

[Mau06] E. Maurel-Segala. High order asymptotics of matrix models and enumeration of maps. arXiv:math/0608192v1 [math.PR], 2006.

[McTW77] B. McCoy, C. A. Tracy and T. T. Wu. Painlevé functions of the third kind. *J. Math. Physics*, **18**:1058–1092, 1977.

[McK05] H. P. McKean. *Stochastic integrals*. Providence, RI, AMS Chelsea Publishing, 2005. Reprint of the 1969 edition, with errata.

[Meh60] M. L. Mehta. On the statistical properties of the level-spacings in nuclear spectra. *Nuclear Phys. B*, **18**:395–419, 1960.

[Meh91] M.L. Mehta. *Random Matrices*. San Diego, Academic Press, second edition, 1991.

[MeD63] M. L. Mehta and F. J. Dyson. Statistical theory of the energy levels of complex systems. V. *J. Math. Phys.*, **4**:713–719, 1963.

[MeG60] M. L. Mehta and M. Gaudin. On the density of eigenvalues of a random matrix. *Nuclear Phys. B*, **18**:420–427, 1960.

[Mil63] J. W. Milnor. *Morse Theory*. Princeton, NJ, Princeton University Press, 1963.

[Mil97] J. W. Milnor. *Topology from the Differentiable Viewpoint*. Princeton, NJ, Princeton University Press, 1997. Revised printing of the 1965 edition.

[MiS05] I. Mineyev and D. Shlyakhtenko. Non-microstates free entropy dimension for groups. *Geom. Funct. Anal.*, **15**:476–490, 2005.

[MiN04] J. A. Mingo and A. Nica. Annular noncrossing permutations and partitions, and second-order asymptotics for random matrices. *Int. Math. Res. Not.*, pages 1413–1460, 2004.

[MiS06] J. A. Mingo and R. Speicher. Second order freeness and fluctuations of random matrices. I. Gaussian and Wishart matrices and cyclic Fock spaces. *J. Funct. Anal.*, **235**:226–270, 2006.

[Mos80] J. Moser. Geometry of quadrics and spectral theory. In *The Chern Symposium 1979 (Proc. Int. Sympos., Berkeley, CA., 1979)*, pages 147–188, New York, NY, Springer, 1980.

[Mui81] R. J. Muirhead. *Aspects of Multivariate Statistical Theory*. New York, NY, John Wiley & Sons, 1981.

[Mur90] G. J. Murphy. *C^*-algebras and Operator Theory*. Boston, MA, Academic Press, 1990.

[Nel74] E. Nelson. Notes on non-commutative integration. *J. Funct. Anal.*, **15**:103–116, 1974.

[NiS97] A. Nica and R. Speicher. A "Fourier transform" for multiplicative functions on non-crossing partitions. *J. Algebraic Combin.*, **6**:141–160, 1997.

[NiS06] A. Nica and R. Speicher. *Lectures on the Combinatorics of Free Probability*, volume 335 of *London Mathematical Society Lecture Note Series*. Cambridge, Cambridge University Press, 2006.

[NoRW86] J.R. Norris, L.C.G. Rogers and D. Williams. Brownian motions of ellipsoids. *Trans. Am. Math. Soc.*, **294**:757–765, 1986.

[Oco03] N. O'Connell. Random matrices, non-colliding processes and queues. In

Séminaire de Probabilités, XXXVI, volume 1801 of *Lecture Notes in Math.*, pages 165–182. Berlin, Springer, 2003.

[OcY01] N. O'Connell and M. Yor. Brownian analogues of Burke's theorem. *Stochastic Process. Appl.*, **96**:285–304, 2001.

[OcY02] N. O'Connell and M. Yor. A representation for non-colliding random walks. *Electron. Comm. Probab.*, **7**, 12 pp., 2002.

[Oko00] A. Okounkov. Random matrices and random permutations. *Int. Math. Res. Not.*, pages 1043–1095, 2000.

[Ona08] A. Onatski. The Tracy–Widom limit for the largest eigenvalues of singular complex Wishart matrices. *Ann. Appl. Probab.*, **18**:470–490, 2008.

[Pal07] J. Palmer. *Planar Ising Correlations*, volume 49 of *Progress in Mathematical Physics*. Boston, MA, Birkhäuser, 2007.

[Par80] B. N. Parlett. *The Symmetric Eigenvalue Problem*. Englewood Cliffs, N.J., Prentice-Hall, 1980.

[Pas73] L. A. Pastur. Spectra of random selfadjoint operators. *Uspehi Mat. Nauk*, **28**:3–64, 1973.

[Pas06] L. Pastur. Limiting laws of linear eigenvalue statistics for Hermitian matrix models. *J. Math. Phys.*, **47**:103303, 2006.

[PaL08] L. Pastur and A. Lytova. Central limit theorem for linear eigenvalue statistics of random matrices with independent entries. arXiv:0809.4698v1 [math.PR], 2008.

[PaS08a] L. Pastur and M. Shcherbina. Bulk universality and related properties of Hermitian matrix models. *J. Stat. Phys.*, **130**:205–250, 2008.

[Péc06] S. Péché. The largest eigenvalue of small rank perturbations of Hermitian random matrices. *Probab. Theory Rel. Fields*, **134**:127–173, 2006.

[Péc09] S. Péché. Universality results for largest eigenvalues of some sample covariance matrix ensembles. *Probab. Theory Rel. Fields*, **143**:481–516, 2009.

[PeS07] S. Péché and A. Soshnikov. Wigner random matrices with non-symmetrically distributed entries. *J. Stat. Phys.*, **129**:857–884, 2007.

[PeS08b] S. Péché and A. Soshnikov. On the lower bound of the spectral norm of symmetric random matrices with independent entries. *Electron. Commun. Probab.*, **13**:280–290, 2008.

[Ped79] G. Pedersen. *C^*-algebras and their Automorphism Groups*, volume 14 of *London Mathematical Society Monographs*. London, Academic Press, 1979.

[PeV05] Y. Peres and B. Virág. Zeros of the i.i.d. Gaussian power series: a conformally invariant determinantal process. *Acta Math.*, **194**:1–35, 2005.

[PlR29] M. Plancherel and W. Rotach. Sur les valeurs asymptotiques des polynomes d'hermite $H_n(x) = (-1)^n e^{x^2/2}(d/dx)^n e^{-x^2/2}$. *Comment. Math. Helv.*, **1**:227–254, 1929.

[PoS03] S. Popa and D. Shlyakhtenko. Universal properties of $L(\mathbf{F}_\infty)$ in subfactor theory. *Acta Math.*, **191**:225–257, 2003.

[PrS02] M. Prähofer and H. Spohn. Scale invariance of the PNG droplet and the Airy process. *J. Stat. Phys.*, **108**:1071–1106, 2002.

[Răd94] F. Rădulescu. Random matrices, amalgamated free products and subfactors of the von Neumann algebra of a free group, of noninteger index. *Invent. Math.*, **115**:347–389, 1994.

[Rai00] E. Rains. Correlation functions for symmetrized increasing subsequences. arXiv:math/0006097v1 [math.CO], 2000.

[RaR08] J. A. Ramírez and B. Rider. Diffusion at the random matrix hard edge. arXiv:0803.2043v3 [math.PR], 2008.

[RaRV06] J. A. Ramírez, B. Rider and B. Virág. Beta ensembles, stochastic airy spectrum, and a diffusion. arXiv:math/0607331v3 [math.PR], 2006.

[Reb80] R. Rebolledo. Central limit theorems for local martingales. *Z. Wahrs. verw. Geb.*, **51**:269–286, 1980.

[ReY99] D. Revuz and M. Yor. *Continuous Martingales and Brownian motion*, volume 293 of *Grundlehren der Mathematischen Wissenschaften*. Berlin, Springer, third edition, 1999.

[RoS93] L. C. G. Rogers and Z. Shi. Interacting Brownian particles and the Wigner law. *Probab. Theory Rel. Fields*, **95**:555–570, 1993.

[Roy07] G. Royer. *An Initiation to Logarithmic Sobolev Inequalities*, volume 14 of *SMF/AMS Texts and Monographs*. Providence, RI, American Mathematical Society, 2007. Translated from the 1999 French original.

[Rud87] W. Rudin. *Real and Complex Analysis*. New York, NY, McGraw-Hill Book Co., third edition, 1987.

[Rud91] W. Rudin. *Functional Analysis*. New York, NY, McGraw-Hill Book Co, second edition, 1991.

[Rud08] M. Rudelson. Invertibility of random matrices: norm of the inverse. *Annals Math.*, **168**:575–600, 2008.

[RuV08] M. Rudelson and R. Vershynin. The Littlewood–Offord problem and invertibility of random matrices. *Adv. Math.*, **218**:600–633, 2008.

[Rue69] D. Ruelle. *Statistical Mechanics: Rigorous Results*. Amsterdam, Benjamin, 1969.

[SaMJ80] M. Sato, T. Miwa and M. Jimbo. Holonomic quantum fields I-V. *Publ. RIMS Kyoto Univ.*, **14**:223–267, **15**:201–278, **15**:577–629, **15**:871–972, **16**:531–584, 1978-1980.

[ScS05] J. H. Schenker and H. Schulz-Baldes. Semicircle law and freeness for random matrices with symmetries or correlations. *Math. Res. Lett.*, **12**:531–542, 2005.

[Sch05] H. Schultz. Non-commutative polynomials of independent Gaussian random matrices. The real and symplectic cases. *Probab. Theory Rel. Fields*, **131**:261–309, 2005.

[Sel44] A. Selberg. Bermerkninger om et multipelt integral. *Norsk Mat. Tidsskr.*, **26**:71–78, 1944.

[Shl96] D. Shlyakhtenko. Random Gaussian band matrices and freeness with amalgamation. *Int. Math. Res. Not.*, pages 1013–1025, 1996.

[Shl98] D. Shlyakhtenko. Gaussian random band matrices and operator-valued free probability theory. In *Quantum Probability (Gdańsk, 1997)*, volume 43 of *Banach Center Publ.*, pages 359–368. Warsaw, Polish Acad. Sci., 1998.

[SiB95] J. Silverstein and Z. D. Bai. On the empirical distribution of eigenvalues of large dimensional random matrices. *J. Multivariate Anal.*, **54**:175–192, 1995.

[Sim83] L. Simon. *Lectures on Geometric Measure Theory*, volume 3 of *Proceedings of the Centre for Mathematical Analysis, Australian National University*. Canberra, Australian National University Centre for Mathematical Analysis, 1983.

[Sim05a] B. Simon. *Orthogonal Polynomials on the Unit Circle, I, II*. American Mathematical Society Colloquium Publications. Providence, RI, American Mathematical Society, 2005.

[Sim05b] B. Simon. *Trace Ideals and their Applications*, volume 120 of *Mathematical Surveys and Monographs*. Providence, RI, American Mathematical Society, second edition, 2005.

[Sim07] B. Simon. CMV matrices: five years after. *J. Comput. Appl. Math.*, **208**:120–154, 2007.

[SiS98a] Ya. Sinai and A. Soshnikov. Central limit theorem for traces of large random symmetric matrices with independent matrix elements. *Bol. Soc. Bras. Mat.*, **29**:1–24, 1998.

[SiS98b] Ya. Sinai and A. Soshnikov. A refinement of Wigner's semicircle law in a neighborhood of the spectrum edge for random symmetric matrices. *Funct. Anal. Appl.*,

32:114–131, 1998.

[Śni02] P. Śniady. Random regularization of Brown spectral measure. *J. Funct. Anal.*, **193**:291–313, 2002.

[Śni06] P. Śniady. Asymptotics of characters of symmetric groups, genus expansion and free probability. *Discrete Math.*, **306**:624–665, 2006.

[Sod07] S. Sodin. Random matrices, nonbacktracking walks, and orthogonal polynomials. *J. Math. Phys.*, **48**:123503, 21, 2007.

[Sos99] A. Soshnikov. Universality at the edge of the spectrum in Wigner random matrices. *Commun. Math. Phys.*, **207**:697–733, 1999.

[Sos00] A. Soshnikov. Determinantal random point fields. *Uspekhi Mat. Nauk*, **55**:107–160, 2000.

[Sos02a] A. Soshnikov. Gaussian limit for determinantal random point fields. *Annals Probab.*, **30**:171–187, 2002.

[Sos02b] A. Soshnikov. A note on universality of the distribution of the largest eigenvalues in certain sample covariance matrices. *J. Statist. Phys.*, **108**:1033–1056, 2002.

[Sos03] A. Soshnikov. Janossy densities. II. Pfaffian ensembles. *J. Statist. Phys.*, **113**:611–622, 2003.

[Sos04] A. Soshnikov. Poisson statistics for the largest eigenvalues of Wigner random matrices with heavy tails. *Electron. Comm. Probab.*, **9**:82–91, 2004.

[Spe90] R. Speicher. A new example of "independence" and "white noise". *Probab. Theory Rel. Fields*, **84**:141–159, 1990.

[Spe98] R. Speicher. Combinatorial theory of the free product with amalgamation and operator-valued free probability theory. *Mem. Amer. Math. Soc.*, **132**(627), 1998.

[Spe03] R. Speicher. Free calculus. In *Quantum Probability Communications, Vol. XII (Grenoble, 1998)*, pages 209–235, River Edge, NJ, World Scientific Publishing, 2003.

[SpT02] D. A. Spielman and S. H. Teng. Smooth analysis of algorithms. In *Proceedings of the International Congress of Mathematicians (Beijing 2002)*, volume I, pages 597–606. Beijing, Higher Education Press, 2002.

[Sta97] R. P. Stanley. *Enumerative Combinatorics*, volume 2. Cambridge University Press, 1997.

[Sze75] G. Szegő. *Orthogonal Polynomials*. Providence, R.I., American Mathematical Society, fourth edition, 1975. Colloquium Publications, Vol. XXIII.

[Tal96] M. Talagrand. A new look at independence. *Annals Probab.*, **24**:1–34, 1996.

[TaV08a] T. Tao and V. H. Vu. Random matrices: the circular law. *Commun. Contemp. Math.*, **10**:261–307, 2008.

[TaV08b] T. Tao and V. H. Vu. Random matrices: universality of esds and the circular law. arXiv:0807.4898v2 [math.PR], 2008.

[TaV09a] T. Tao and V. H. Vu. Inverse Littlewood–Offord theorems and the condition number of random discrete matrices. *Annals Math.*, **169**:595–632, 2009.

[TaV09b] T. Tao and V. H. Vu. Random matrices: Universality of local eigenvalue statistics. arXiv:0906.0510v4 [math.PR], 2009.

[t'H74] G. t'Hooft. Magnetic monopoles in unified gauge theories. *Nuclear Phys. B*, **79**:276–284, 1974.

[Tri85] F. G. Tricomi. *Integral Equations*. New York, NY, Dover Publications, 1985. Reprint of the 1957 original.

[TuV04] A. M. Tulino and S. Verdú. Random matrix theory and wireless communications. In *Foundations and Trends in Communications and Information Theory*, volume 1, Hanover, MA, Now Publishers, 2004.

[TrW93] C. A. Tracy and H. Widom. *Introduction to Random Matrices*, volume 424 of *Lecture Notes in Physics*, pages 103–130. New York, NY, Springer, 1993.

[TrW94a] C. A. Tracy and H. Widom. Level spacing distributions and the Airy kernel.

Commun. Math. Phys., **159**:151–174, 1994.

[TrW94b] C. A. Tracy and H. Widom. Level spacing distributions and the Bessel kernel. *Comm. Math. Phys.*, **161**:289–309, 1994.

[TrW96] C. A. Tracy and H. Widom. On orthogonal and symplectic matrix ensembles. *Commun. Math. Phys.*, **177**:727–754, 1996.

[TrW00] C. A. Tracy and H. Widom. Universality of the distribution functions of random matrix theory. In *Integrable Systems: from Classical to Quantum (Montréal, QC, 1999)*, volume 26 of *CRM Proc. Lecture Notes*, pages 251–264. Providence, RI, American Mathematical Society, 2000.

[TrW02] C. A. Tracy and H. Widom. Airy kernel and Painleve II. In A. Its and J. Harnad, editors, *Isomonodromic Deformations and Applications in Physics*, volume 31 of *CRM Proceedings and Lecture Notes*, pages 85–98. Providence, RI, American Mathematical Society, 2002.

[TrW03] C. A. Tracy and H. Widom. A system of differential equations for the Airy process. *Electron. Comm. Probab.*, **8**:93–98, 2003.

[TrW05] C. A. Tracy and H. Widom. Matrix kernels for the Gaussian orthogonal and symplectic ensembles. *Ann. Inst. Fourier (Grenoble)*, **55**:2197–2207, 2005.

[VaV07] B. Valko and B. Virag. Continuum limits of random matrices and the Brownian carousel. arXiv:0712.2000v3 [math.PR], 2007.

[VeK77] A. M. Vershik and S. V. Kerov. Asymptotics of the Plancherel measure of the symmetric group and the limiting form of Young tables. *Soviet Math. Dokl.*, **18**:527–531, 1977.

[VoDN92] D. V. Voiculescu, K. J. Dykema and A. Nica. *Free Random Variables*, volume 1 of *CRM Monograph Series*. Providence, RI, American Mathematical Society, 1992.

[Voi86] D. Voiculescu. Addition of certain non-commuting random variables. *J. Funct. Anal.*, **66**:323–346, 1986.

[Voi90] D. Voiculescu. Circular and semicircular systems and free product factors. In *Operator Algebras, Unitary Representations, Enveloping Algebras, and Invariant Theory (Paris, 1989)*, volume 92 of *Progr. Math.*, pages 45–60. Boston, MA, Birkhäuser, 1990.

[Voi91] D. Voiculescu. Limit laws for random matrices and free products. *Invent. Math.*, **104**:201–220, 1991.

[Voi93] D. Voiculescu. The analogues of entropy and of Fisher's information measure in free probability theory. I. *Comm. Math. Phys.*, **155**:71–92, 1993.

[Voi94] D. Voiculescu. The analogues of entropy and of Fisher's information measure in free probability theory. II. *Invent. Math.*, **118**:411–440, 1994.

[Voi96] D. Voiculescu. The analogues of entropy and of Fisher's information measure in free probability theory. III. The absence of Cartan subalgebras. *Geom. Funct. Anal.*, **6**:172–199, 1996.

[Voi97] D. Voiculescu, editor. *Free Probability Theory*, volume 12 of *Fields Institute Communications*. Providence, RI, American Mathematical Society, 1997. Papers from the Workshop on Random Matrices and Operator Algebra Free Products held during the Special Year on Operator Algebra at the Fields Institute for Research in Mathematical Sciences, Waterloo, ON, March 1995.

[Voi98a] D. Voiculescu. The analogues of entropy and of Fisher's information measure in free probability theory. V. Noncommutative Hilbert transforms. *Invent. Math.*, **132**:189–227, 1998.

[Voi98b] D. Voiculescu. A strengthened asymptotic freeness result for random matrices with applications to free entropy. *Int. Math. Res. Not.*, pages 41–63, 1998.

[Voi99] D. Voiculescu. The analogues of entropy and of Fisher's information measure in free probability theory. VI. Liberation and mutual free information. *Adv. Math.*,

146:101–166, 1999.

[Voi00a] D. Voiculescu. The coalgebra of the free difference quotient and free probability. *Int. Math. Res. Not.*, pages 79–106, 2000.

[Voi00b] D. Voiculescu. *Lectures on Probability Theory and Statistics: Ecole D'Été de Probabilités de Saint-Flour XXVIII - 1998*, volume 1738 of *Lecture Notes in Mathematics*, pages 283–349. New York, NY, Springer, 2000.

[Voi02] D. Voiculescu. Free entropy. *Bull. London Math. Soc.*, **34**:257–278, 2002.

[Vu07] V. H. Vu. Spectral norm of random matrices. *Combinatorica*, **27**:721–736, 2007.

[Wac78] K. W. Wachter. The strong limits of random matrix spectra for sample matrices of independent elements. *Annals Probab.*, **6**:1–18, 1978.

[Wey39] H. Weyl. *The Classical Groups: their Invariants and Representations*. Princeton, NJ, Princeton University Press, 1939.

[Wid94] H. Widom. The asymptotics of a continuous analogue of orthogonal polynomials. *J. Approx. Theory*, **77**:51–64, 1994.

[Wig55] E. P. Wigner. Characteristic vectors of bordered matrices with infinite dimensions. *Annals Math.*, **62**:548–564, 1955.

[Wig58] E. P. Wigner. On the distribution of the roots of certain symmetric matrices. *Annals Math.*, **67**:325–327, 1958.

[Wil78] H. S. Wilf. *Mathematics for the Physical Sciences*. New York, NY, Dover Publications, 1978.

[Wis28] J. Wishart. The generalized product moment distribution in samples from a normal multivariate population. *Biometrika*, **20A**:32–52, 1928.

[WuMTB76] T. T. Wu, B. M McCoy, C. A. Tracy and E. Barouch. Spin–spin correlation functions for the two-dimensional ising model: exact theory in the scaling region. *Phys. Rev. B.*, **13**, 1976.

[Xu97] F. Xu. A random matrix model from two-dimensional Yang–Mills theory. *Comm. Math. Phys.*, **190**:287–307, 1997.

[Zir96] M. Zirnbauer. Riemannian symmetric superspaces and their origin in random matrix theory. *J. Math. Phys.*, **37**:4986–5018, 1996.

[Zvo97] A. Zvonkin. Matrix integrals and maps enumeration: an accessible introduction. *Math. Comput. Modeling*, **26**:281–304, 1997.

General conventions and notation

Unless stated otherwise, for S a Polish space, $M_1(S)$ is given the topology of weak convergence, that makes it into a Polish space.

When we write $a(s) \sim b(s)$, we assert that there exists $c(s)$ defined for $s \gg 0$ such that $\lim_{s\to\infty} c(s) = 1$ and $c(s)a(s) = b(s)$ for $s \gg 0$. We use the notation $a_n \sim b_n$ for sequences in the analogous sense. We write $a(s) = O(b(s))$ if $\limsup_{s\to\infty} |a(s)/b(s)| < \infty$. We write $a(s) = o(b(s))$ if $\limsup_{s\to\infty} |a(s)/b(s)| = 0$. $a_n = O(b_n)$ and $a_n = o(b_n)$ are defined analogously.

The following is a list of frequently used notation. In case the notation is not routine, we provide a pointer to the definition.

$\forall$	for all
a.s., a.e.	almost sure, almost everywhere
$\mathrm{Ai}(x)$	Airy function
$(\mathscr{A}, \|\cdot\|, *, \phi)$	C^*-algebra (see Definition 5.2.11)
$\overline{A}, A^{\mathrm{o}}, A^{\mathrm{c}}$	closure, interior and complement of A
$A \backslash B$	set difference
$B(H)$	space of bounhed operators on a Hilbert space H
$C^k(S)$, $C^k_{\mathrm{b}}(S)$	functions on S with continuous (resp., bounded continuous) derivatives up to order k
$C^\infty(S)$	infinitely differentiable functions on S
$C^\infty_{\mathrm{b}}(S)$	bounded functions on S possessing bounded derivatives of all order
$C^\infty_{\mathrm{c}}(S)$	infinitely differentiable functions on S of compact support
$C(S, S')$	Continuous functions from S to S'
$C^\infty_{\mathrm{poly}}(\mathbb{R}^m)$	infinitely differentiable functions on $\mathbb{R}^m$ all of whose derivatives have polynomial growth at infinity.
CLT	central limit theorem
$\stackrel{\mathrm{Prob}}{\to}$	convergence in probability
$d(\cdot,\cdot), d(x,A)$	metric and distance from point x to a set A
$\det(M)$	determinant of M
$\Delta(x)$	Vandermonde determinant, see (2.5.2)
$\Delta(K)$	Fredholm determinant of a kernel K, see Definition 3.4.3
Δ_N	open $(N-1)$-dimensional simplex
$\mathscr{D}(\mathscr{L})$	domain of $\mathscr{L}$
$\emptyset$	the empty set
$\varepsilon(\sigma)$	the signature of a permutation σ
$\exists, \exists!$	there exists, there exists a unique

$f(A)$	image of A under f
f^{-1}	inverse image of f
$f \circ g$	composition of functions
$\mathrm{Flag}_n(\lambda, \mathbb{F})$	Flag manifold, see (4.1.4)
$\mathrm{GL}_n(\mathbb{F})$	invertible elements of $\mathrm{Mat}_n(\mathbb{F})$
$\mathbb{H}$	skew-field of quaternions
$\mathscr{H}_n(\mathbb{F})$	elements of $\mathrm{Mat}_n(\mathbb{F})$ with $X^* = X$
i	basis elements of $\mathbb{C}$ (together with 1)
$\mathbf{i}, \mathbf{j}, \mathbf{k}$	basis elements of $\mathbb{H}$ (together with 1)
i.i.d.	independent, identically distributed (random variables)
$\mathbf{1}_A(\cdot)$, $\mathbf{1}_a(\cdot)$	indicator on A and on $\{a\}$
I_n	identity matrix in $\mathrm{GL}_n(\mathbb{F})$
$\lfloor t \rfloor$, $\lceil t \rceil$	largest integer smaller than or equal to t, smallest integer greater than or equal to t
LDP	large deviation principle (see Definition D.1)
Lip($\mathbb{R}$)	Lipschitz functions on $\mathbb{R}$
LLN	law of large numbers
$\log(\cdot)$	logarithm, natural base
LSI	logarithmic Sobolev inequality (see Subsection 2.3.2 and (4.4.13))
$\mathrm{Mat}_{p \times q}(\mathbb{F})$	p-by-q matrices with entries belonging to $\mathbb{F}$ (where $\mathbb{F}$=$\mathbb{R}$, $\mathbb{C}$ or $\mathbb{H}$)
$\mathrm{Mat}_p(\mathbb{F})$	same as $\mathrm{Mat}_{p \times p}(\mathbb{F})$
$M_1(S)$	probability measures on S
μ, ν, ν'	probability measures
$\mu \circ f^{-1}$	composition of a (probability) measure and a measurable map
N(0,I)	zero mean, identity covariance standard multivariate normal
$\wedge, \vee$	(pointwise) minimum, maximum
PI	Poincaré inequality (see Definition 4.4.2)
$P(\cdot), E(\cdot)$	probability and expectation, respectively
$\mathbb{R}, \mathbb{C}$	reals and complex fields
$\mathbb{R}^d$	d-dimensional Euclidean space (where d is a positive integer)
$R_\mu(z)$	R-transform of a measure μ (see Definition 5.3.37)
ρ_M	volume on Riemannian manifold M
$\mathrm{sp}(T)$	spectrum of an operator T
$S_a(z)$	S-transform of a (see Definition 5.3.29)
$S_\mu(z)$	Stieltjes transform of a measure μ (see Definition 2.4.1).
S^{n-1}	unit sphere in $\mathbb{R}^n$
$SO(N)$, $SU(N)$	special orthogonal group (resp., special unitary group)
$\mathfrak{su}_n(\mathbb{F})$	anti-self-adjoint elements of $\mathrm{Mat}_n(\mathbb{F})$, with vanishing trace if $\mathbb{F} = \mathbb{C}$
$\Sigma(\mu)$	noncommutative entropy of the measure μ, see (2.6.4)
$\mathrm{tr}(M), \mathrm{tr}(K)$	trace of a matrix M or of a kernel K
v'	transpose of the vector (matrix) v
$v*$	transpose and complex conjugate of the vector (matrix) v
$U_n(\mathbb{F})$	unitary matrices in $\mathrm{GL}_n(\mathbb{F})$
$\{x\}$	set consisting of the point x
$\mathbb{Z}_+$	positive integers
$\subset$	contained in (not necessarily properly)
$\langle \cdot, \cdot \rangle$	scalar product in $\mathbb{R}^d$
$\langle f, \mu \rangle$	integral of f with respect to μ
$\oplus$	direct sum
$\otimes$	tensor product
$\boxplus$	free additive convolution (see Definition 5.3.20)
$\boxtimes$	free multiplicative convolution (see Definition 5.3.28

Index